Disordered Materials

Science and Technology

Institute for Amorphous Studies Series

Series editors

David Adler†
Massachusetts Institute of Technology
Cambridge, Massachusetts

and

Brian B. Schwartz
Institute for Amorphous Studies
Bloomfield Hills, Michigan
and Brooklyn College of the City University of New York
Brooklyn, New York

†Deceased

Disordered Materials

Science and Technology

Selected papers by
Stanford R. Ovshinsky

Edited by

David Adler
Late of Massachusetts Institute of Technology
Cambridge, Massachusetts

Brian B. Schwartz
Institute for Amorphous Studies
Bloomfield Hills, Michigan
and Brooklyn College of the City University of New York
Brooklyn, New York

and

Marvin Silver
University of North Carolina
Chapel Hill, North Carolina

Plenum Press • New York and London

Library of Congress Cataloging-in-Publication Data

Ovshinsky, Stanford R.
    Disordered materials : science and technology : selected papers /
by Stanford R. Ovshinsky ; edited by David Adler, Brian B. Schwartz,
and Marvin Silver.
        p.   cm. -- (Institute for Amorphous Studies series)
    Papers written or co-authored by Ovshinsky between 1968 and 1988.
    Includes bibliographical references.
    ISBN 0-306-43385-0
    1. Order-disorder models--Congresses.   2. Amorphous
semiconductors--Congresses.   3. Amorphous substances--Congresses.
I. Adler, David, 1935-1987.   II. Schwartz, Brian B., 1938-
III. Silver, M. (Marvin), 1924-   IV. Title.   V. Series.
QC173.4.073097   1991
530.4'2--dc20   89-28735
                                                             CIP

© 1991 Plenum Press, New York
A Division of Plenum Publishing Corporation
233 Spring Street, New York, N.Y. 10013

Printed in the United States of America

To
Stanford R. Ovshinsky

in honor of his pioneering contributions
to the science and technology of Disordered Materials

and his continued interest in serving humanity
and his many friendships

EDITORIAL COMMENT

The production of this book is the result of an extraordinary and close relationship between two exceptional scientists, Stanford R. Ovshinsky and David Adler.

As colleagues who know both of them, we were always in awe of the symbiotic relationship of the two scientists. They cared about each other and worked with each other in a loving and interdependent style. Their enthusiasm for the subject matter and their willingness to present bold and sometimes controversial subjects added to the strength of this relationship. Unfortunately, David Adler died suddenly in late March 1987, just before his 53rd birthday. The scientific community was shocked by the loss, and we, his friends and colleagues, will always miss his wisdom, humor, and grace.

At the time of Dave's death, we were working on this volume of selected papers by Stan Ovshinsky. We decided to leave much of the introduction as written by Dave, only adding some comments to update it and to include the most recent papers of Stan Ovshinsky on the occasion of his 65th birthday. This book is dedicated to the memory of David Adler and represents a celebration of the exceptional relationship between Stan and Dave.

We'd very much like to thank Ms. Ghazaleh Koefod for her continuous administrative assistance and loyalty to both Stan Ovshinsky and Dave Adler. Ghazaleh's commitment and skills have been key to the completion of this book in a timely and accurate fashion.

-- Brian B. Schwartz, New York
—Marvin Silver, North Carolina

CONTENTS

PART II: DEVICE APPLICATIONS

EPILOGUE

INTRODUCTION

Landmark contributions to science and technology are rarely recognized at the time of publication. Few people, even in technical areas, recognized the importance of developments such as the transistor, the laser, or electrophotography until we'l after their successful demonstration. So-called experts, in fact, tend to resist new inventions, a natural instinct based on a combination of fear of obsolescent expertise and jealousy arising from lack of active participation in the discovery.

Denigration of new ideas is a relatively safe modus operandi, since the vast majority eventually are abandoned well short of commerciality. However, a successful device can be identified by its completion of a three-phase history, dependent on the current reaction of the cognoscenti: (1) "I never heard of it!"; (2) "It is unstable, it is uneconomical, and it doesn't work!"; and (3) "I invented it first!"

The science and technology of disordered materials has not proceeded down the ordinary path. Stanford R. Ovshinsky, a self-taught genius who was previously known in scientific circles primarily for his contributions to automation and neurophysiology, began working in the field in 1955, when almost all physicists believed that amorphous semiconductors could not even exist. Ovshinsky studied the high-field properties of a large number of these materials and between 1958 and 1961 discovered and developed the two types of reversible switching phenomena which now bear his name. He not only investigated the switching behavior in detail and proposed a wide number of potential applications, but he also developed the chemical and metallurgical basis for choosing the composition of the materials in order to optimize the desired characteristics, proposed mechanisms for the origin of the phenomena, and reached important conclusions about the physical nature of the materials at equilibrium and their electronic nonequilibrium properties. Many of these ideas were condensed into a publication for Physical Review Letters, paper 1 in this collection. This paper immediately attracted attention to the field, and directly lead to the initiation of large research efforts at both industrial laboratories and universities throughout the world. Inevitably, there was the usual amount of controversy, with many experts simultaneously taking positions (2) and (3) above.

It has now been well over 20 years since the original publication date, and an objective view can be taken in hindsight. The most eloquent testimony to the importance of Ovshinsky's work is the fact that his paper has become one of the five most cited publications in the history of Physical Review Letters, that most prestigious of all physics journals. Prior to the appearance of this paper, amorphous semiconductors were considered to be just an inferior alternative to crystalline semiconductors, with no commercial potential and only of interest as scientific curiosities. Conventional solid-state physics was intimately based on crystalline periodicity, and theories relating to amorphous solids were taken to be extensions of this approach in which a glass was just a crystal with non-periodic perturbations. Ovshinsky's paper not only completely turned around the commercial prognosis for amorphous-based devices but also focused on the electronic behavior of chalcogenide glasses, materials which exhibit unique properties not possessed by other crystalline or amorphous solids. One of these properties, first discovered by Ovshinsky and described in paper 1, is the existence of a high-conductivity state which can be initiated by the application of a critical electric field. The rapid, reversible electronic transition from the low-conductivity (OFF) state to the high-conductivity (ON) state has come to be known as Ovonic threshold switching. Ovshinsky also discovered a bistable switching phenomenon in amorphous chalcogenide with less cross-linking. The latter materials exhibit reversible, electronically induced transitions between the amorphous and crystalline phases, both of which exist indefinitely in the absence of any applied field. This type of transition has come to be known as Ovonic memory switching and was developed by ECD into a host of applications which make use of either reversible amorphous-to-crystalline or reversible crystalline-to-amorphous transitions. Paper 1 summarized the properties of both types of switching and suggested mechanisms for the phenomena which later proved to be correct.

The effects of this work on solid-state physics were enormous, particularly because many theoreticians did not even believe in the existence of amorphous semiconductors. There was a hard core of physicists who had been considering the effects of disorder on the electronic density of states, mainly with an eye towards crystalline-alloy problems, but the behavior observed in chalcogenide glasses proved difficult to understand with the current theories. In these materials, electrical conduction occurred with a well-defined activation energy essentially equal to half the optical gap, just like in an intrinsic crystalline semiconductor. However, the Fermi energy, E_F, appeared to be completely pinned. In conventional terms, the absence of extrinsic conduction down to very low temperatures could be understood only if the material had a defect concentration in the parts per trillion range. Yet the electrical conductivity did not change significantly even if several percent of impurities were added to the glass. The first explanation of how this could come about was given by Cohen, Fritzsche, and Ovshinsky in paper 2, which proposed what is now called the CFO model. CFO showed that the overlap between extensive valence and conduction band tails, which arises from positional and compositional disorder, could strongly pin E_F. However, if well-defined mobility edges, as first postulated by Mott, exist, the materials would remain semiconducting and would exhibit what appears to be intrinsic behavior. Paper 2 intro-

duced the now universally used terms, mobility edge and mobility gap, and also many new concepts such as the existence of charged traps resulting from the overlap of localized gap states and a finite density of localized states at the Fermi energy, $g(E_F)$. The paper also discussed hopping conduction in the vicinity of E_F, thermally stimulated currents, photoluminescence, the field effect, the effects of contacts, and even noted that the existence of well-defined defect configurations would necessitate a generalization of the proposed model. The CFO model remained the backbone of amorphous semiconductor analysis for more than seven years, and the paper is also one of the most cited publications in the history of Physical Review Letters.

Ovshinsky's work stimulated an international symposium, Semiconductor Effects in Amorphous Solids, held in New York City in May 1969. The work of Ovshinsky and his coworkers dominated the entire conference, whose proceedings filled all of Volume 2 of the then new Journal of Non-Crystalline Solids. In paper 3 of the present collection, Evans, Helbers, and Ovshinsky presented a detailed experimental study of switching in a memory material in which a chalcogen plays an important role. They found the important result that the transition is induced after application of a critical energy, analyzed the nature of the conducting filament in the ON state, and presented some of the first data on light-induced effects in amorphous semiconductors. In paper 4, Bienenstock, Betts, and Ovshinsky demonstrated that the local environments of atoms in a Ge-Te glass were different from those in the corresponding crystal, a very surprising result at the time. They proposed the random covalent model, in which the covalent valence requirements of each atom were locally satisfied even in a multicomponent alloy. This represented the first hint of the great flexibility that can be provided by amorphous networks, once local structure is freed from crystalline constraints. This flexibility, which has always been one of the cornerstones of Ovshinsky's views, was not generally recognized until much later. Paper 3 also reported the results of an analysis of the structure of the ON state of a memory switch, in which it was determined that crystallites of Te and sometimes also GeTe were responsible for the high conduction. This confirmed Ovshinsky's original suggestion that memory switching is simply an amorphous-to-crystalline transition. Other work by Fritzsche and Ovshinsky (not reprinted here) at the same symposium detailed the thermal properties of this transition.

As early as 1962, Ovshinsky was experimenting with the effect of radiation on amorphous materials to generate an amorphous-to-crystalline transformation (see ref. 7 of paper 37), and a preliminary discussion was given in paper 3. In paper 5 of this collection, Feinleib and Ovshinsky published the results of optical studies of the crystalline and amorphous phases of a memory alloy, showing a sizeable increase in reflectivity upon crystallization. These were steps in the development of an optical-memory device capable of very high bit densities. Another link in this chain was provided by paper 7, in which Feinlieb, deNeufville, Moss and Ovshinsky demonstrated that the amorphous-to-crystalline transition and its inverse could both be accomplished through the use of a laser in the extremely short time of approximately 10^{-6} seconds. This paper also presented a detailed model for both transitions, invoking the important concept of photocrystallization.

Work in the field of amorphous semiconductors continued to grow exponentially. The first two biennial International Conferences on Amorphous and Liquid Semiconductors, held in Prague and Bucharest, were accommodated in single small rooms and had no published proceedings. But the Third Conference, which took place in Cambridge, England in September 1969, had simultaneous sessions and its proceedings filled all of Volume 4 of the Journal of Non-Crystalline Solids.

Paper 6 by Stanford and Iris Ovshinsky, written in honor of Sir Nevill Mott's 65th birthday, discussed the possible connection between both analog and digital techniques for information storage and nerve-cell operation in living organisms. This analogy between switching in inorganic and biological systems has been another cornerstone of Ovshinsky's viewpoint since his days at Wayne Medical School.

In 1973, amorphous semiconductor theory was essentially dominated by the CFO model. However, the origin of the uniqueness of chalcogenides was not at all evident from this model. At the Fifth International Conference on Amorphous and Liquid Semiconductors held in Garmisch-Partenkirchen in September 1973, Ovshinsky and Sapru (paper 8) presented a model in which the outer lone-pair electrons in chalcogenide glasses played a crucial role. They pointed out that in chalcogenides, as opposed to alloys involving only atoms from columns I-V in the Periodic Table, the excitation of free carriers does not require the breaking of bonds, and this provided a simple explanation of the reliability of threshold switching in chalcogenide glasses. This paper also noted the possibility of threefold and singly coordinated chalcogen atoms in the glassy phase, an idea which represented the seeds of the Kastner-Adler-Fritzsche (KAF) model for the electronic structure of chalcogenides developed three years later. Ovshinsky elaborated on this idea in paper 9, in which he was able to explain the then recently discovered photo-induced electron spin-resonance effect as a metastable excitation involving charged defect states which originate from lone-pair interactions between neighboring chalcogen atoms. A general approach, in which chalcogenides were distinguished from tetrahedrally bonded materials by their lower average coordination as well as by the presence of the lone pair, was presented by Ovshinsky in paper 10. This concept was the progenitor of Phillips' topological view of overconstrained and underconstrained amorphous network, an important element in our current understanding of the electronic structure of amorphous semiconductors.

The unique properties of chalcogenides are of vital importance to their switching behavior. However, these same properties preclude the possibility of doping these semiconductors and thereby using them in the conventional manner, e.g., as transistors, lasers, or solar cells. Ovshinsky and his coworkers developed a novel technique, known as chemical modification, which successfully overcomes this problem. Even in the 1950's, Ovshinsky diffused elements from columns I and II in the Periodic Table into amorphous thin films of transition metal compounds such as tantalum oxide to achieve reversible conductivity increases of factors of the order of 10^{14}; this process was the basis of the Ovitron switch, developed in 1958 (see paper 6). Modification by means of photodiffusion had been accomplished by 1968. This was the technique now referred to as photodoping, described in detail in talks by Ovshinsky at the Electrotechnical

Laboratory in Tokyo in 1971 and 1972. The first published report of chemical modification of amorphous materials, using lithium as the modifier, appeared in paper 10. A general description of this concept, including both a theoretical analysis and a wide range of experimental results utilizing many different elements as modifiers, was presented at the Seventh International Conference on Amorphous and Liquid Semiconductors, held in Edinburgh in June 1977, and is reprinted here as papers 11 and 12. The idea is to replace the doping effect, in which parts per million of a chemically active impurity are introduced in order to vary the position of E_F, by a different approach in which up to several percent of a carefully chosen impurity are added in a nonequilibrium manner. Only in this latter way can E_F be unpinned in chalcogenides. Ovshinsky and his colleagues were able to use chemical modification to change the electrical conductivity by many orders of magnitude, not only of chalcogenides but also of alloys containing elements from columns III-V of the Periodic Table. Ovshinsky also was able to decrease the electrical conductivity, in certain cases, when small amounts of a particular modifier were used, a useful result when high dark resistance is important, e.g., in devices based on photoconductivity.

Returning to the idea that less common forms of chemical bonding can be obtained in solids once atoms are freed from the tyranny of periodic constraints, Ovshinsky and David Adler discussed the effects on the electronic structure of amorphous semiconductors of dative bonding, multi-centered bonding, multiple bonding, transition-metal complexing, and hybridization with empty orbitals, among other examples. This general approach to the amorphous state is reprinted here as paper 13. At the time, a-Si:H alloys were attracting a great deal of attention because of their doping characteristics. The introduction of H into a-Si lowers the average coordination, thereby sharply reducing the concentration of strain-induced defects. In paper 14, Ovshinsky and Madan announced the development of a new alloy, a-Si:F:H, which exhibits still lower defect concentrations and better structural stability. There are several chemical reasons for this— not only is the overall bonding modified but also the additional introduction of strong ionic Si-F bonds plays an important role. Further properties

of this interesting new alloy were detailed by Madan and Ovshinsky in paper 16, an invited paper at the Eight International Conference on Amorphous and Liquid Semiconductors.

Meanwhile, the experimental data on threshold switching demonstrated clearly that the ON state is electronically induced, exhibits only minimal temperature increases, and represents a constant carrier concentration, all of which had been predicted by Ovshinsky many years before. In paper 15, Shur, Silver, Ovshinsky, and Adler presented a detailed model of switching which emphasized the unique properties of chalcogenides and showed how both the OFF-to-ON and the ON-to-OFF transition evolved in a natural way from these properties.

One of the difficulties with a-Si:H alloys is the overconstrained nature of the system. The high average coordination precludes the possibility of attaining both optimal bond lengths and optimal bond angles without crystalline periodicity. As discussed previously, introduction of ionic Si-F bonds ameliorates this problem by freeing some of the bond-angle constraints. This increases the possibility of optimizing some of the dihedral angles, extending the order out to third neighbors. Indeed, a-Si:F:H alloys exhibit intermediate range order, as evidenced by the result of Tsu, Izu, Ovshinsky, and their coworkers, reprinted here as papers 17 and 18; the latter was presented at the Ninth International Conference on Amorphous and Liquid Semiconductors, held in Grenoble in July 1981.

We have come a long way since 1968. It is now clear that the electronic properties of amorphous solids depend primarily on the chemical nature of the constituent atoms. This chemical nature includes the average coordination, which, together with the preparation conditions, determines the concentration of strain-related defects, the possible existence of outer lone pairs, which lead to thermodynamically induced defects, and the details of the bonding and local environment, which can result in localized states not related to defects per se. In the papers in Part I, it is clear that Ovshinsky was aware of all of these effects long before they were generally recognized by others in the field.

One of Ovshinsky's first excursions into the effect of alloying on superconducting properties was presented in 1982 (paper 19). This was followed by a series of papers dealing with superconductivity (papers 20 and 21). It was natural therefore that, with his interest in superconductivity and the effects of fluorine, work would have proceeded leading to high T_c material. A 155K superconducting material was reported in Physical Review Letters in June 1987 (papers 30 and 31). The new high temperature superconductors present a considerable challenge for the theoretical community. In a very physical approach, Ovshinsky presents a structural chemical model for the high T_c superconductors which takes into account the interplay of the 2^+ and 3^+ valence states of the chains and planes (paper 32).

The relationship between structure, chemistry, and properties continued to interest Ovshinsky in the 1980's and a series of papers followed regarding growth of amorphous films by plasma deposition (papers 22, 23 and 28). Study of order parameters was discussed in a paper authored with R. Tsu, J. Gonzalez-Hernandez and J. Doehler. Dangling bonds, always a problem, were shown to be passivated by gas adsorption (paper 25). Chemical bonding and structure in chalcogenide switching materials was discussed (paper 26). Critical material and bonding configuration were discussed in papers 27 and 29.

Part II deals with device applications. Paper 33, by Ovshinsky and Fritzsche, summarized Ovshinsky's original ideas for the use of memory switches in computer memory and logic arrays. The first commercial device, a 256-bit electronically erasable programmable read-only memory (EEPROM, formerly known as EAROM), developed in 1970, is discussed in detail, as is the optical mass memory. Mechanisms for both the electronic and optical transitions are presented.

One of the major advantages of amorphous devices is their resistance to radiation. In paper 34, Ovshinsky, Evans, Nelson, and Fritzsche reported on the radiation hardness of threshold switches subjected to heavy doses of X-rays and high neutron fluences. The devices continued to operate during the exposure and did not change their characteristics. Independent testing has continually confirmed these results in a wide array of other amorphous devices. Chalcogenide switches are particularly resistant to radiation because of the pinned E_F. In fact, it has been found that the connecting cables are destroyed by the radiation before the switches fail!

Following the development of the electronic memory, Ovshinsky turned his attention to the area of optical memory and imaging. He found that reversible amorphous-to-crystalline and crystalline-to-amorphous transitions could be induced by a variety of techniques, e.g., by application of light (including irradiation from a laser, an electronic flash, etc.), an electron beam, stress, or even catalytically. These techniques were first described by ovshinsky at the Gordon Conference on the Chemistry and Metallurgy of Semiconductors in July 1969, and were subsequently discussed in detail in a series of papers between 1969 and 1972. In one of these, paper 35, he and Klose showed how memory switching can be used in high-speed, high-resolution imaging and display devices, using the concept of photonucleation. Together with Feinleib and colleagues, he also developed various materials which exhibit not only archival memory but also other reversible optical effects such as laser-created vapor bubbles in a glassy material. These effects are discussed in paper 36 and represent the underlying principles of an optical mass memory, which can also be used as part of a video disk system. An array of different films, including high-contrast and continuous-tone films, was developed, as discussed in paper 37, which also analyzed the photographic mechanisms. This led to the development of other interesting optical media, for example, one in which the application of light leads to modification of either the local order or the overall morphology, making possible the continuous addition of information. This last concept forms the basis of the updatable microfiche system known as the MicrOvonic File.

Both threshold and memory switches are two-terminal devices in their simplest form, although three- and four-terminal devices can also be fabricated from them. An ingenious tunnel triode, developed by Shaw, Fritzsche, Silver, Smejtek, Holmberg, and Ovshinsky, was described in paper 38. This device not only serves as an amplifier with low power consumption, but it can also be used to measure carrier transit times and thermalization times in semiconductors without the usual complications.

The development of holography opened up the possibility of enormous bit capacities in three-dimensional memory blocks. In paper 39, deNeufville, Seguin, Moss, and Ovshinsky showed how photostructural effects in a-As_2S_3, and other amorphous chalcogenides could be used in such an application. Experimental results detailing the underlying mechanisms were also reported in this paper, which was presented at the Fifth International Conference on Amorphous and Liquid Semiconductors.

The most significant application of amorphous semiconductors will most likely turn out to be in low-cost solar cells. The a-Si:F:H alloy and its derivatives developed by Ovshinsky and his coworkers have emerged as the leading candidates in this field. Solar cells with greater than 13% conversion efficiency have been announced by Energy Conversion Devices, and foot-wide panels in a continuous array are being produced. Ovshinsky discussed the general approach to solar-energy conversion using the new alloy in papers 40 and 41; Madan, McGill, Czubatyj, Yang, and Ovshinsky reported on the first amorphous solar cell with an efficiency greater than 6% in paper 42, while deNeufville, Izu, and Ovshinsky discussed the development of batch-fabricated one-foot-square solar panels in paper 43. The tremendous progress that has occurred since the publication of these last two papers strikingly makes evident the speed of the current solar-cell development program. Indeed, thousand-foot long, one-foot wide rolls of Ovonic solar cells are now being produced routinely on a continuous-web machine.

Stan Ovshinsky and David Adler presented a picture of the application of amorphous materials in photovoltaic devices in paper 44. Jeff Yang and Ovshinsky showed that amorphous silicon could be a far better photovoltaic material than anyone previously expected (paper 45). Microelectronic and optical storage devices did not escape his attention as shown in papers 46 and 48.

Part III consists of sixteen review articles published by Ovshinsky between 1969 and 1989. In paper 48, which is a transcription of a talk given at the Fifth Annual National Conference on Industrial Research held in Chicago, Illinois in September 1969, he summarized the new science of amorphous semiconductors and gave his views on the computer of the future, utilizing amorphous technology. Now many of his predictions are becoming a reality. For example, with the recent rise of interest in high-bit-density storage using video-disk technology, much attention has been focused on laser-induced amorphous-to-crystalline and crystalline-to-amorphous transitions, described and analyzed in detail by Ovshinsky in this and other papers so many years ago. Paper 49, an invited paper at the Symposium on Semiconductor Effects in Amorphous Solids, represents an early discussion of the development of switching devices. The concept of imaging was analyzed in great detail by Ovshinsky and Fritzsche in paper 50, which also introduced a classification scheme for noncrystalline semiconductors that remains useful today. Ovshinsky was also invited to address the Sixth International Conference on Amorphous and Liquid Semiconductors, held in Leningrad in November 1975, and his contribution to the proceedings is reprinted here as paper 51. The theme of this review was the possibility of using the interactions between various forms of energy and amorphous materials to accomplish the processing and storage of information. In paper 52, the imaging applications were highlighted and an example of a continuous-tone photograph was given. The material used as the film in this case was a new organotelluride tailored for such photographic applications and also showed amplification not found in other nonsilver photography.

Paper 53 is a review written in honor of the award of the 1977 Nobel Prize in Physics to Sir Nevill Mott. It deals with amorphous materials in general and contains many insights which reward a careful reading. Ovshinsky was an invited speaker at both the International Conference on the Frontiers of Glass Science, held in Los Angeles in July 1980, and the Ninth International Conference on Amorphous and Liquid Semiconductors in Grenoble, and his contributions to the proceedings are reprinted here as papers 54 and 55, respectively. The former concentrates primarily on energy conversion, while the latter focuses on information processing. Eloquent testimony to the growth of the field of amorphous materials is the transition from the First International Conference on Amorphous and Liquid Semiconductors held in Prague in 1965 with about 20 attendees, primarily from Eastern Europe, to the present conferences with well over 500 scientists from more than 50 countries around the world and with published proceedings which fill two large volumes. In paper 56, published in honor of Professor Radu Grigorovici's 70th birthday,

Ovshinsky discussed the relationships between the structure and applications of amorphous materials, concentrating on the chemical basis of both. The new concept of a total interactive environment was first introduced in this paper, which also contains many insights into Ovshinsky's scientific philosophy.

An overall review of amorphous materials is presented in papers 57 and 58 with focus on the emergence from the dark ages of amorphous materials into a renaissance period past the classical era into modern times where amorphous materials are being used in devices routinely.

In years past, Ovshinsky had written for Sir Nevill Mott's 65th and 75th birthday festschrifts. In paper 59, Ovshinsky continues his earlier discussions of "disorder" and structure in a contribution to Sir Nevill's 80th birthday festschrift published as a three-volume set by Plenum Press. Paper 60, Ovshinsky's paper at the Nobel Laureate Symposium on Applied Quantum Chemistry, focuses on the role of intuition in creating new models and concepts in quantum chemistry.

To honor Hellmut Fritzsche, another hero in this struggle to unravel the mysteries of amorphous materials, Stan wrote on the quantum nature of amorphous solids (paper 61). (Note by editors: it is fitting that one of the last papers, number 62, referenced here is one Stan Ovshinsky wrote with Dave Adler on the progress made in understanding and using amorphous materials.) Paper 63 gives a personal view of the role of local order and defects in order to present a model for the behavior of the new high temperature superconductors, and the last paper in this section (paper 64) shows that high quality "epitaxial" superconductivity films can be grown with the aid of fluorine on a sapphire base.

Finally, as an epilogue we include a recent paper on high-temperature superconductivity by Stan Ovshinsky which contains a very personal tribute to David Adler and a book review of Hideki Yukawa's Creativity and Intuition: A Physicist Looks at East and West, in which Ovshinsky presents his views on the inventive process.

These papers represent only a fraction of Ovshinsky's complete works, which fortunately are continuing at an unabated pace. I have tried to choose those that emphasize his original insights into both science and technology and his comprehensive view of materials as the servant, rather than the master, of mankind. His unique ability to stimulate colleague while collaborating as a peer should be clear from these papers.

PART I

CHEMISTRY AND PHYSICS OF DISORDERED MATERIALS

REVERSIBLE ELECTRICAL SWITCHING PHENOMENA IN DISORDERED STRUCTURES

Stanford R. Ovshinsky

Energy Conversion Devices, Inc., Troy, Michigan 48084

(Received 23 August 1968)

We describe here a rapid and reversible transition between a highly resistive and a conductive state effected by an electric field which we have observed in various types of disordered materials, particularly amorphous semiconductors[1,2] covering a wide range of compositions. These include oxide- and boron-based glasses and materials which contain the elements tellurium and/or arsenic combined with other elements such as those of groups III, IV, and VI.

Such amorphous materials can be described as intrinsic semiconductors[3,4] with an optical energy gap E_g typically between 0.6 and 1.4 eV and an activation energy[5] for electrical conduction ΔE between 0.7 and 1.6 eV depending on composition.

Field-effect measurements[6] indicate that the "band gap" contains a distribution of trapping and recombination centers with densities in excess of 10^{19} cm^{-3} eV^{-1}.

The measurements shown below have been obtained with an amorphous semiconductor containing (in atomic percent) 48 at.% tellurium, 30 at.% arsenic, 12 at.% silicon, and 10 at.% germanium. The specimen was an evaporated film, 5×10^{-5} cm thick, between two carbon electrodes with a contact area of about 10^{-4} cm^2. This material has a resistivity at 300°K of $\rho = 2 \times 10^7$ Ω cm, $\Delta E = 1.0$ eV, and a positive thermopower.

Figure 1 shows oscilloscope pictures of (a) the I-V characteristic, (b) the voltage V across the unit, and (c) the current I passing through the above unit as a function of time. In this case, a 60-Hz ac voltage was applied across the unit and a 10^4-Ω load resistor was used. The I-V curve is independent of frequenc to at least 10^6 Hz.

The major features of the switching phenomena shown are the following: (1) The I-V characteristic is symmetrical with respect to the reversal of the applied voltage and current. (2) The same switching characteristic is observed when the semiconducting materials is sputtered or hot-pressed between the electrodes. It remains symmetrical even when the electrodes are of different contacting areas or of different materials. (3) In the highly resistive state, the material is Ohmic at fields below about 10^4 V/cm. At higher fields, the dynamic resistance decreases monotonically with increasing voltage. (4) When the applied voltage exceeds a threshold voltage V_t, the unit switches along the load line to the conducting state.[7] (5) In the conducting state, the current can be increased or decreased without significantly affecting the voltage drop, termed the holding voltage V_h, across the unit. Here the dynamic resistance is close to zero. (6) As the current is reduced below a characteristic value termed the holding current I_h, the unit switches back to the original highly resistive state along the load line. (7) The switching process is repeatable. Alternating or full-wave unfiltered rectified voltages have been applied continuously to units over periods of many months without noticeable change in their characteristics.

In the highly resistive state, the threshold voltage V_t of the unit increases nearly linearly with the thickness of the film.[8] This observation and also the fact that the symmetry of the I-V characteristic is preserved even when different electrode materials are used indicate that the current in the highly resistive state is not electrode limited but that the field is essentially uniform throughout the bulk of the material. In contrast to this, the holding voltage is weakly dependent on thickness, which suggests that in the conductive state the voltage drop occurs predominantly

---

Reprinted by permission from *Physical Review Letters*, Vol. 21, No. 20, pp. 1450–1453 (11 November 1968).

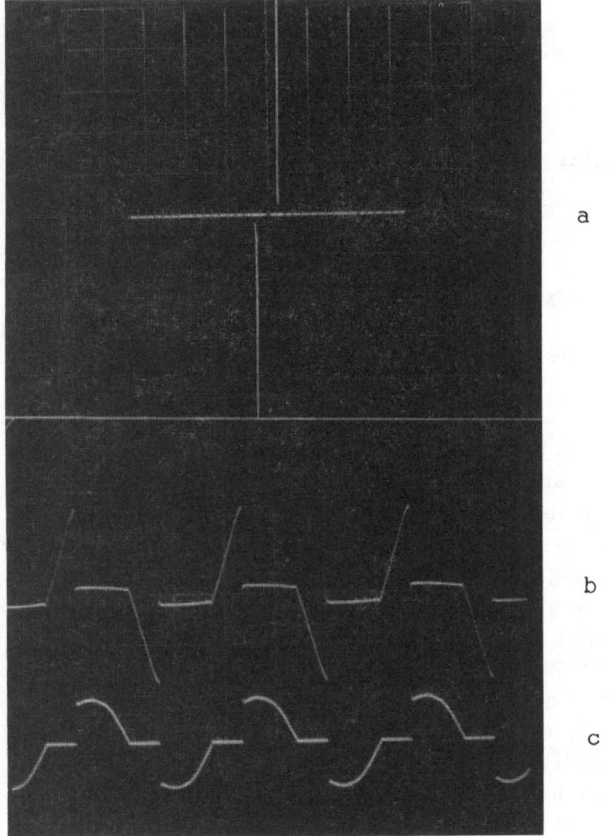

FIG. 1. Response of switching unit to 60-Hz voltage.
(a) I-V characteristic: vertical, 2 mA/div; horizontal,
5 V/div. (b) Voltage: vertical, 5 V/div; horizontal,
5 msec/div. (c) Current: vertical, 20 mA/div; horizontal, 5 msec/div.

at one or both electrodes. A dependence of V_h on
electrode material has also been observed.[9] Upon switching into the conductive state, a current
filament appears to form, growing in diameter
with increasing current flow.[10] Diameters of
about 5×10^{-3} cm were observed by thermal probing for $I \approx 100$ mA. Similar current filaments
have been observed and discussed by Böer and
co-workers[11] and are quite generally associated
with S-shaped instabilities.[12]

Figure 2 shows the switching phenomenon which
follows the application of a rectangular voltage
pulse [Fig. 2(a)]. Because the load resistance is
much smaller than that of the unit before switching, the applied voltage appears across the unit
until switching occurs, as seen in Fig. 2(b). Note
that switching occurs after a delay time t_d. The
duration of the switching process is less than 1.5
$\times 10^{-10}$ sec. After switching, the holding voltage
V_h is observed across the unit until the end of
the applied pulse. Figure 2(c) shows the abrupt
increase of current after t_d and the sudden drop

at the end of the pulse. Figure 2(d) shows the decrease of t_d with increasing voltage V. The functional form is approximately $t_d = \text{const} \times \exp(-V/V_2)$.

The switching processes can be analyzed in
terms of a nucleation theory wherein the nucleation rate is dependent on the applied voltage. In
this model,[13] nucleation precedes the process
leading to the delay time t_d. A statistical analysis of the distribution of threshold voltages for
30 000 successive switching events recorded on
one unit operated with 60-Hz ac voltage showed
the nucleation probability per unit time to be a
constant for a given applied voltage and to depend
on voltage as $\exp(V/V_1)$, where V_1 was about 0.1
V for the unit examined. The peak of the probability distribution occurred at 13.9 V; we take
this as the definition of V_t.[14,15] The nucleation
rate at the peak is 7×10^4 sec^{-1}.

The conduction and switching processes appear
closely tied to the structures of disordered and
amorphous materials which in turn are associated with the local bonding characteristics of the
constituent chemical elements. A high density of
localized states in the forbidden energy gap is
split from the conduction and valence bands by

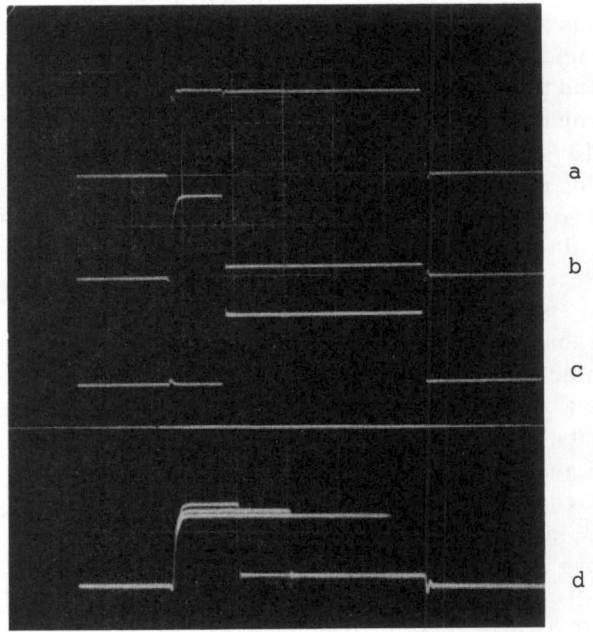

FIG. 2. Response of switching unit to square-wave
pulse. Horizontal time scale 500 nsec/div. (a) Voltage
applied across unit and 1000-Ω series resistor: vertical, 5 V/div. (b) Voltage across unit: vertical, 5 V/
div. (c) Current through unit: vertical, 10 mA/div.
(d) Same as (b) for different amplitudes of applied voltage pulse.

the translational and compositional disorder. The disorder allows most atoms to complete their valence state and thereby establish a high degree of compensation, yielding the observed intrinsic behavior of these materials. A large overlap in energy of those levels stemming from the valence band with those from the conduction band is expected with self-compensation of charges producing a large density of traps. These traps when occupied can easily be ionized at moderate fields by lowering the trap depth and reducing the capture cross section.[16] A rapid increase of carrier concentration results, and can explain the observed non-Ohmic behavior in the low-conductivity state.

The relatively close spacing of these localized states in both space and energy makes it likely that internal-field ionization and emission from these localized states play an important role in the carrier generation necessary to initiate the conducting state at threshold values of the order of 10^5 V/cm. We feel the exponential dependence of the current on voltage in the highly resistive state before switching supports this contention. The low mobility and high number of localized states offer, in principle, the possibility of the dramatic conductance changes observed.

We believe that after switching, a redistribution of carriers as a result of oppositely charged carriers having very different mobilities and transition rates through the electrode interfaces gives rise to a space charge and field enhancement near one electrode and to the small value of V_h.

An unusual memory effect is observed[2,17] in materials in which structural changes are facilitated by the removal of cross-linking elements from the above formula—for example, the reduction of arsenic to 5 %. After switching from a highly resistive state, structural changes result in the preservation of a conductive state even when the current is totally removed. The material can be reversibly switched back to the highly resistive state by application of a current pulse of either polarity exceeding a threshold value.

I particularly wish to acknowledge the early advice, help, and encouragement given by H. Fritzsche, the valuable assistance and discussions of M. H. Cohen and K. W. Böer and also of D. Turnbull and A. Bienenstock, and the devoted help of my wife and colleague, I. M. Ovshinsky, and the staff of Energy Conversion Devices, Inc.

[1]Parts of this work were presented earlier: S. R. Ovshinsky, at the Fourth Symposium on Vitreous Chalcogenide Semiconductors, sponsored by the Academy of Sciences of the USSR, Leningrad, 23-27 May 1967 (unpublished), and at the International Colloquium on Amorphous and Liquid Semiconductors, sponsored by the Rumanian Academy of Science, Bucharest, 28 September-3 October 1967 (unpublished), and in Proceedings of the Electronic Components Conference, Washington, D. C., May 1968 (McGregor and Werner, Inc., Washington, D. C., 1968), p. 313 ff.

[2]S. R. Ovshinsky, U. S. Patent No. 3 271 591.

[3]A discussion of the field of amorphous semiconductors and earlier references can be found in the review article by N. F. Mott, Advan. Phys. 16, 49 (1967).

[4]The intrinsic behavior of amorphous semiconductors was first reported by A. F. Ioffé and B. T. Kolomiets; cf. Ref. 3.

[5]The conductivity is expressed as $\sigma = \sigma_0 \exp(-\Delta E / kT)$.

[6]H. Fritzsche, E. A. Fagen, and S. R. Ovshinsky, to be published.

[7]Even in a circuit that is current stabilized by a 10^8-Ω load resistor, the unit cannot be held at an operating point between the highly resistive and the conducting state. In some cases, relaxation oscillations governed by the load resistor and the unit's capacitance ($C \approx 3$ pF) have been observed.

[8]By changing the film thickness, values of V_t between 2.5 and 300 V have been obtained.

[9]Nichrome electrodes yield a particularly low value, $V_h = 0.5$ V. A more detailed study is in progress.

[10]H. Fritzsche, private communication.

[11]K. W. Böer, E. Jahne, and E. Neubauer, Phys. Status Solidi 1, 231 (1961).

[12]B. K. Ridley, Proc. Phys. Soc. (London) 81, 996 (1963).

[13]M. H. Cohen, to be published.

[14]This unit is not the one used for Figs. 1 and 2.

[15]M. H. Cohen, R. G. Neale, and S. R. Ovshinsky, to be published.

[16]G. A. Dussel and R. H. Bubé, J. Appl. Phys. 37, 2797 (1966).

[17]S. R. Ovshinsky, to be published.

SIMPLE BAND MODEL FOR AMORPHOUS SEMICONDUCTOR ALLOYS

Morrel H. Cohen,[1] H. Fritzsche,[1] and S.R. Ovshinsky[2]

[1]James Franck Institute and Department of Physics, University of Chicago, Chicago, Illinois 60637
[2]Energy Conversion Devices, Inc., Troy, Michigan 48084

(Received 21 March 1969)

Amorphous covalent alloys particularly of group-IV, -V, and -VI elements are readily formed over broad ranges of composition.[1-6] They have been described as low-mobility electronic intrinsic semiconductors with a temperature-activated electrical conductivity $\sigma = \sigma_0 \times \exp(-\Delta E/kT)$ which sometimes extends well into the molten state.[2,3,7] They remain intrinsic with changed ΔE when their composition is changed.[1,5,7] These alloys transmit infrared light up to an exponential absorption edge from which an energy gap E_g is estimated.[1,2] The value of E_g usually is smaller than $2\Delta E$, often by as much as 10-20%.[7,8] Photoconductivity[9] and recombination-radiation[10] measurements have been interpreted as giving evidence for the presence of localized states in the gap.

This Letter describes a simple band model based on the common features of the covalent amorphous alloys which is able to explain many of their properties. Among the novel features of the model are overlapping conduction and valence-band tails of localized states and sharp mobility edges.

We suppose that in these alloys most atoms are in sites satisfying their valence requirements.[11] This leads to the notion of a valence band of extended states despite the randomly differing valences of the constituent atoms and to an energy separation between valence- and conduction-band states corresponding roughly to an energy for breaking valence bonds. However, there is compositional disorder, and the translational disorder is enhanced over that in an elemental or compound glass because the coordination number varies from site to site to accommodate the varying valences of the atoms. One

therefore expects a high density of localized states tailing in from the conduction and valence bands, as shown in Fig. 1(a).[12] By a localized state, we mean one with probability amplitude decreasing exponentially with distance from the center of localization for sufficiently large distance. The fluctuations in potential on the atomic scale caused by the disorder give rise to these localized states. There may be more than one localized state associated with a given large potential fluctuation. In general, only one such state can be occupied at a time; double occupancy may occasionally be possible.

In our model of amorphous alloys, the tails of the valence and conduction bands overlap, which means that an electron in a valence band in some region of the material may have a higher energy than an extra electron in a nonbonding state in another part of the material. Such electrons from the top of the valence-band tail fall into spatially distinct states in the lower conduction-band tail. The Fermi level E_F thus falls near the center of the "gap" where the total density of states is near its minimum.

Conduction-band states are locally neutral when unoccupied, and valence-band states are locally neutral when occupied by an electron. The empty valence-band tail states therefore give rise to a random distribution of localized positive charges neutralized on the average by a corresponding distribution of an equal number of localized negative charges which are associated with the occupied conduction-band tail states. The resulting Coulomb potential fluctuations of course alter the energies of all states so that the occupancy of the tail states has to be considered self-consistently. These charged states above and below

Reprinted by permission from *Physical Review Letters*, Vol. 22, No. 20, pp. 1065–1068 (19 May 1969).

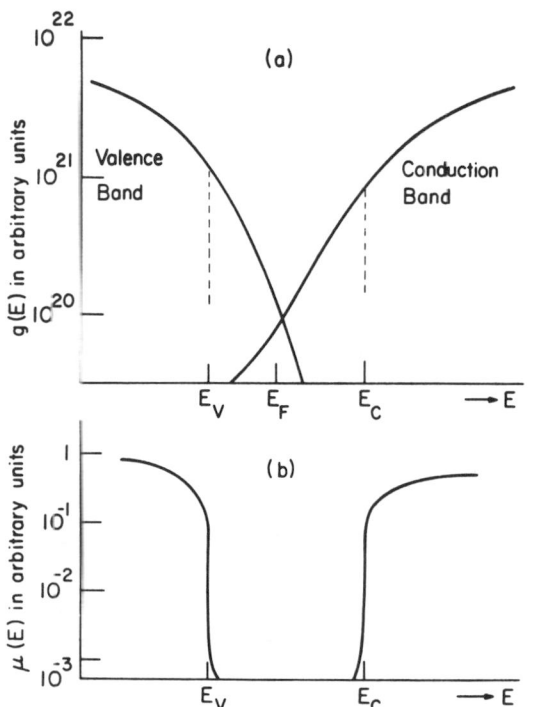

FIG. 1. Sketch of (a) the partial densities of states of the valence and conduction bands and (b) the electron and hole mobility, respectively. The units of the ordinates are indicated as arbitrary because no quantitative calculation has been made. States which are neutral when occupied are associated with the valence band, the others with the conduction band; they overlap in the mobility gap.

E_F act as efficient trapping centers for electrons and for holes, respectively.

We suppose that the transition from extended band states to localized gap states would occur at an energy E_v for the valence band and E_c for the conduction band if it were not for the spatially varying electrostatic potential V caused by the density N of randomly distributed charged gap states. Ignoring V, we would expect the mobility to drop sharply at E_c and E_v as shown in Fig. 1(b). At these mobility edges, the conduction process changes from a low-mobility band transport with finite mobility at $T=0$ to a thermally activated hopping[13] between localized gap states which disappears at $T=0$.

A possible interpretation of the low, temperature-independent Hall mobility[14] is that the mean free paths in those portions of the bands of extended states relevant to transport is of the order of the interatomic separation in these materials. The mean free path is the distance over which the phase coherence of the wave functions extends, and we take it to be small compared to

the distance over which V varies, $N^{-1/3}$ or 45 Å for $N=10^{19}$ cm^{-3}. V is therefore locally constant and the band model depicted in Figs. 1(a) and 1(b) applies locally except that E_v and E_c fluctuate together relative to E_F by $\delta E \approx 2e^2 N^{1/3}/\epsilon$, or 0.1 eV for $N=10^{19}$ cm^{-3} and a dielectric constant ϵ of 10.

This band model correlates and is supported by a variety of observations.

The presence of the "mobility edges" near E_c and E_v explains why a well-defined activation energy ΔE is observed for σ despite the lack of sharp band edges. A further elaboration of the model permits us to estimate $2\Delta E$ to be about $(E_c-E_v)+2\delta E$ and somewhat larger than $E_g < E_c -E_v$. A conduction showing an activation energy significantly lower than E_g has been observed to set in at lower temperatures.[15] This conduction is believed to be phonon-assisted hopping near E_F rather than band conduction by carriers excited from shallow donor or acceptor levels because its magnitude is very sensitive to minor heat treatments. Furthermore, measurements of electron tunneling from a metal through an oxide barrier layer into an amorphous semiconductor show[16] that E_F remains near the gap center as T is lowered from 300 to 78°K. From the magnitude of the low-temperature hopping conduction and from the change in conductivity which results when a thin film of amorphous semiconductor is charged as in an insulated-gate field-effect transistor, one can estimate the density of gap states near the Fermi level. A value $g(E_F) \sim 6 \times 10^{19}$ cm^{-3} eV^{-1} was obtained for a chalcogenide glass of composition Te$_{0.5}$As$_{0.3}$Si$_{0.1}$Ge$_{0.1}$ for which $\Delta E = 0.45$ eV. This magnitude for $g(E_F)$ is consistent with a magnitude for N of 10^{19} cm^{-3}. Deviations from the local satisfaction of valence requirements are unimportant until of order N.

A gap state density of this magnitude agrees with the observation that blocking or rectifying Schottky barriers do not appear to exist at contacts of any metal with an amorphous covalent alloy.[17] This model predicts space charge layers sufficiently thin for carriers to tunnel through easily, their thickness being roughly $(4\pi g e^2/\epsilon)^{-1/2}$, or ~30 Å with a dielectric constant ϵ of 10.

The recombination radiation[10] observed at 78°K in vitreous As$_2$Te$_3 \cdot$As$_2$Se$_3$ at 78°K shows a weak peak at 1.16 eV and a strong peak at 0.67 eV. The first is clearly due to interband recombination. We attribute the stronger low energy peak to the efficient capture of carriers by the charged electron and hole traps near E_F.

The decay of photoconductivity after electron-hole pair-producing illumination shows a fast and slow component.[9,18] The long relaxation time can be hours at low temperatures, leading to a photo-stimulated excess conductivity which can be orders of magnitude larger than the conductivity prior to illumination. A similar excess conductivity can be stimulated at low temperatures by a transient increase of the electric field into the non-Ohmic range.[18] Such a stored and presumably trap-activated conductivity appears to contradict a model with a large number of recombination centers distributed throughout the gap. However, as pointed out above, the electron traps are predominantly above E_F and the corresponding hole traps below E_F. The initial rapid decay of the photoconductivity we attribute to the capture of carriers by the high density of traps below (or above for holes) the mobility edges. Nonequilibrium distributions of electrons and holes are thereby established leading to the observed excess conductivity. The long relaxation time is governed by their decay towards equilibrium via the slow processes of hopping and of capture of carriers by deep traps.

This band model is probably inadequate for elemental and compound amorphous semiconductors on the one hand and for molecular amorphous solids with large band gaps on the other. In the former case, there are likely to be well-defined structural defects leading to localized states of well-defined energy, contradicting the monotonic variation of g with E we have supposed to occur in the tails, and moving the Fermi energy out of the middle of the gap. Examples for the latter case are silicon oxide and solutions of metal in liquid ammonia[19] in which dissolved atoms neither are incorporated in nor greatly change the principal molecular units which make up the amorphous material and hence may give rise to donor or acceptor states within narrow energy regions of the gap.

This band model does not resolve the dilemma of the positive thermopower versus the negative Hall coefficient observed in covalent alloy glasses,[3,14,20,21] nor does it explain the magnitude of the density of band states near the mobility edges which is needed to explain the magnitude of the factor σ_0 in the conductivity.[12,22,23]

---

*Work supported in part by the National Aeronautics and Space Administration Grant No. NGL 14-001-009, Air Force Office of Scientific Research, U. S. Air Force Contract No. AF 49(638)-1653, and support of materials science by the Advanced Research Projects Agency.

[1]R. Frerichs, J. Opt. Soc. Am. 43, 1153 (1953).

[2]B. T. Kolomiets, Phys. Status Solidi 7, 359, 713 (1964).

[3]A. D. Pearson, in Modern Aspects of the Vitreous State, edited by J. D. Mackenzie (Butterworths Scientific Publications, Ltd., London, 1964), Vol. 3.

[4]J. A. Savage and S. Nielsen, Phys. Chem. Glasses 5, 82 (1964).

[5]J. T. Edmond, Brit. J. Appl. Phys. 17, 979 (1966).

[6]A. R. Hilton et al., Phys. Chem. Glasses 7, 105, 112, 116 (1966).

[7]T. N. Vengel and B. T. Kolomiets, Zh. Tekh. Fiz. 27, 2484 (1957) [translation: Soviet Phys.–Tech. Phys. 2, 2314 (1957)].

[8]B. V. Pavlov and B. T. Kolomiets, Vitreous State (Izdatel'stvo Akademii Nauk Ukrainskoi SSR, Kiev, 1960), p. 201.

[9]A. M. Andriesh and B. T. Kolomiets, Fiz. Tverd. Tela 5, 1461 (1963) [translation: Soviet Phys.–Solid State 5, 1063 (1963)].

[10]B. T. Kolomiets, T. N. Manontova, and V. V. Negreskul, Phys. Status Solidi 27, K15 (1968).

[11]N. F. Mott, Advan. Phys. 16, 49 (1967).

[12]The fact that as a result of disorder the sharpness of the band edges disappears and localized states tail into the gap has been pointed out by many authors. A review of the early work is given by Mott in Ref. 11. See also N. F. Mott, to be published.

[13]N. F. Mott and W. D. Twose, Phil. Mag. 10, 107 (1961); N. F. Mott, Phil. Mag. 17, 1259 (1968).

[14]J. C. Male, Brit. J. Appl. Phys. 18, 1543 (1967).

[15]E. A. Fagen, S. R. Ovshinsky, and H. Fritzsche, Bull. Am. Phys. Soc. 14, 311 (1969), and to be published.

[16]A. Nwachuku and M. Kuhn, Appl. Phys. Letters 12, 163 (1968); J. W. Osmun and H. Fritzsche, Bull. Am. Phys. Soc. 14, 311 (1969).

[17]S. R. Ovshinsky, E. J. Evans, D. L. Nelson, and H. Fritzsche, IEEE Trans. Nucl. Sci. NS-15, 311 (1968).

[18]H. Fritzsche, E. A. Fagen, and S. R. Ovshinsky, Bull. Am. Phys. Soc. 14, 311 (1969), and to be published.

[19]M. H. Cohen and J. C. Thompson, Advan. Phys. 17, 857 (1968).

[20]B. T. Kolomiets and T. F. Nazarova, Fiz. Tverd. Tela 2, 395 (1960) [translation: Soviet Phys.–Solid State 2, 369 (1960)].

[21]H. L. Uphoff and J. H. Healy, J. Appl. Phys. 32, 950 (1961), and 33, 2770 (1962); J. F. Dewald and W. F. Peck, Jr., J. Electrochem. Soc. 111, 561 (1964).

[22]R. L. Myuller, J. Appl. Chem. USSR 3, 519 (1962).

[23]J. Stuke, in Proceedings of the Conference on Low Mobility Semiconductors, Sheffield, England, 1966 (unpublished).

REVERSIBLE CONDUCTIVITY TRANSFORMATIONS IN CHALCOGENIDE ALLOY FILMS

E.J. Evans, J.H. Helbers, and S.R. Ovshinsky

Energy Conversion Devices, Inc., 1675 W. Maple, Troy, Michigan 48084

1. INTRODUCTION

Certain amorphous chalcogenide alloys, such as $Ge_{15}Te_{85}$, exhibit metastable conductivity. These reversible structural changes in amorphous alloys entailing variation in order and conductivity have been made the basis of memory devices.[1] In this paper, we present our observations on induced conductivity transformation from one state to another. Because the conductivity involves structural changes in the bulk we relate our results to those obtained from calorimetric [2] and structural investigation [3] on the same materials. We also present measurements on the electrical characteristics of memory switches.

Metastable conductivity transformation in chalcogenide alloy films were produced by energy deposition via Joule heating in the case of electrical experiments and by temperature cycling of the film. In the latter case the temperature time profiles were generated by programming a heat source and by illuminating a film sample with a xenon flashlamp. The main results found are:

(1) The flashlamp technique for energy deposition provides a reversible method for varying the degree or order and conductivity within a film sample.

(2) In these materials the conductivity changes are related to the energy absorbed and the rate of absorption.

(3) The conductivity activation energy changes from a value of ~ 0.5 eV in the resistive state to less than 0.05 eV in the conductive state, indicating a bulk memory phenomenon in the memory switching device.

(4) Electrical pulse measurements on memory devices indicate that the transformation from low to high conductance states (SET) occurs at constant energy. Saturation resistance of the SET state is current controlled. Similarly, the reverse transition (RESET) is shown to occur at constant energy.

2. EXPERIMENTAL RESULTS

Thin film samples 1 μm to 5 μm thick were sputtered from pressed cathodes of composition $Ge_{15}Te_{81}X_4$, where X gives the remaining compositional variable. By introducing cross-linking elements, designated by X above, the rate of crystallization can be decreased and the subsequent disordering which occurs in memory action can be enhanced. Controlling the concentration of cross-linking atoms in the germanium-tellurium matrix permits control of the memory characteristics via gradual or more abrupt structural transformations. Among the effective cross-linking materials are Group V elements such as arsenic and phosphorus. The films were deposited on 1 inch x 1 inch glass substrates with edge contacts of evaporated NiCr, providing a sample of one square geometry. Electrical leads were attached to the NiCr electrodes to permit sheet resistivity to be measured directly.

Typical of the materials used were 706, which contained phosphorus, and 729 in which arsenic was substituted for the phosphorus. Samples were placed in an inert atmosphere oven, and log resistance was plotted versus temperature. The scan rate was approximately 5°C/min. A bias voltage of 1.0 V was applied to the sample and current measured with a Keithley 413A log ammeter. A sketch of these data is shown in Fig. 1. The samples were heated from 25°C to an elevated temperature and then cooled to ambient. In this manner, the peak temperature was increased on successive runs. Material 729 showed a conductivity transition at (Fig. 1A) about 180°C. If this temperature is not exceeded, the material remains in the OFF state without substantial change. If the transition temperature is exceeded, the material undergoes a change from an intrinsic semiconductor to a highly degenerate semiconductor. In 729, thermal cycling with a peak temperature less than T_4 (of Fig. 1A) did not appreciably change the OFF state activation energy for this scan rate.

A similar experiment was performed on a film sample of 706 memory material which is known to have the adaptive resistance property, i.e., there is a continuous variation of resistance between the well-defined OFF state and a totally ON state. The data sketched in Fig. 1B indicated this adaptive memory behavior to correspond to a shifting of the bulk resistivity and activation energy resulting from increasing the peak cycling temperature. This material does not show a "sharp" transition as does the 729.

Reprinted by permission from *Journal of Non-Crystalline Solids*, Vol. 2, pp. 334–346 (1970).

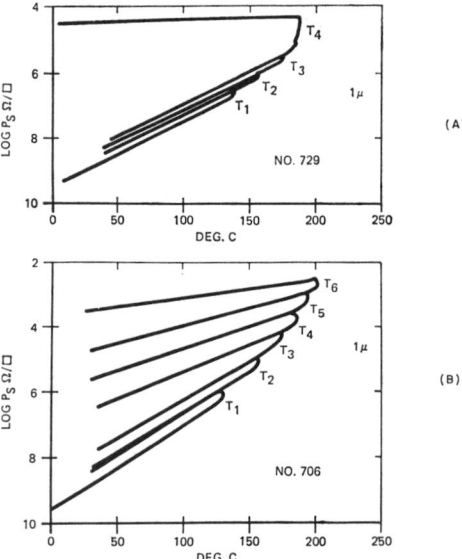

(A)

(B)

Fig. 1   Log resistance versus temperature for thin film samples of $Ge_{15}Te_{81}X_4$. X represents a composition variable. In (A) No. 729 contains arsenic and in (B) No. 706 contains phosphorus. Peak temperatures T_1 were increased each run until the conductive state was observed. The sketched data is representative of several samples of each material.

Dilatometer runs were made on bulk samples of both 706 and 729. The 729 data are sketched in Fig. 2A along with a sketch of the 706 data in Fig. 2B. Several comments can be made.(4) Scan rate was 3°C/sec.

(1) In a film of 729 material in Fig. 2A, the resistance transition to a degenerate semiconductor occurs at about 180°C, which is about 40°C higher than the classical softening point of the bulk sample. In addition, the temperature difference between the initial softening and the point where the dilatometer run terminated due to excessive softening (or slump) was 11°C.

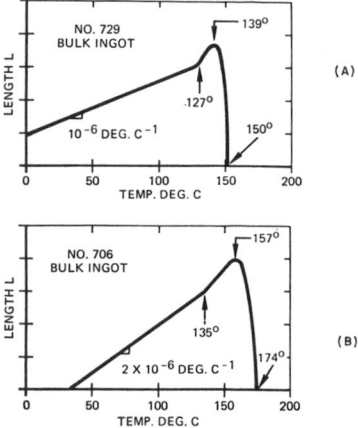

(A)

(B)

Fig. 2   Dilatometric   measurements   on   bulk samples of (A) No. 729 and (B) No. 706.

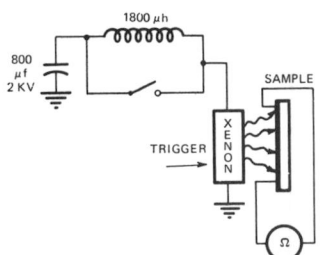

Fig. 3   Circuit   diagram   for   the   xenon flashlamp apparatus. The shorting switch is used to decrease the maximum pulse width of 3 msec to 0.6 msec.

(2) In the 706 material, the change in resistivity and activation energy appears to be most pronounced in the region of 155°C to 185°C. This temperature range is near the softening range of the dilatometer run for 706. In addition, it appears that the temperature region from classical softening at 157°C to the slump region at 174°C is 50% broader than that of 729. This would suggest that the adaptive nature of the material is associated with the temperature range of softening and viscosity of the material.

Additional   experimentation   consisted   of measuring   resistivity   as   a   function   of   energy absorbed   from   a   xenon   flashlamp   source.   The experimental   arrangement   is   shown   in   Fig. 3.   The capacitor bank is capable of providing 300J input to the lamp through an inductor and the network is designed to allow the light output pulse shape to be adjusted for a width of 600 µsec to 3 msec.

Several 729 samples were prepared on test substrates to a thickness of 3 µm. The film sheet resistivity after sputtering was about 3×10^7 $\Omega/\square$ which corresponds to a bulk resistivity of about 10^4 Ω-cm.

One sample consisting of a 3 µm film of 729 on a 1 inch x 1 inch glass slide was exposed to 40 successive broad pulses of maximum energy and the resistance recorded as a function of number of pulses. These data are shown as the solid curve in Fig. 4.   It can be seen that the resistance decreased by four orders of magnitude. When the SET was completed, the inductive bank was shorted to provide RESET pulses. Resistance versus number of pulses was again recorded and is plotted as the dashed curve (5) in Fig. 4. The resistance in the final state was approximately the original value. The form of the curve is only approximate; its only value is to note the decrease and increase in resistance. Continued experiments with the sample indicated an increasing degradation of sample OFF resistance and film appearance apparently caused by oxidation in the atmospheric environment. No flashlamp data were taken under inert conditions. It is indicated that the narrow pulses cause heating with a quench to RESET the memory film to a disordered state, and that the wide pulse produced by inductor broadening enables the film to SET to a more ordered state.

Several   samples   of   memory   material   on   glass slides have been flashlamp treated to SET, RESET and

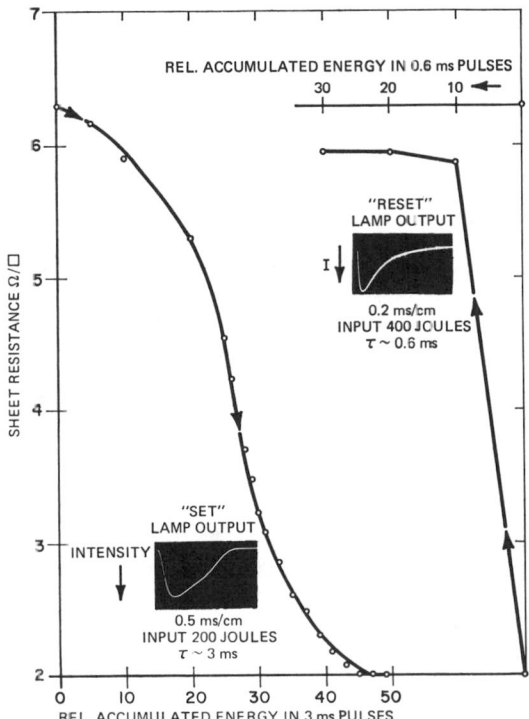

Fig. 4   Resistance versus relative accumulated energy in a thin-film of memory material. The 3 msec pulses decrease film resistance while the 0.6 msec pulses increase the film resistance.

intermediate states, and analyzed by X-ray diffraction.(3) It was determined that the conductive state is characterized by the presence of crystalline tellurium. The minimum crystallite size was estimated to be 400Å. The RESET flash was found to vitrify substantially all the crystallized tellurium, and produce an X-ray pattern characteristic of a disordered material. Samples which were left in intermediate states of resistivity showed weak tellurium peaks. In substantially all materials SET by the flashlamp technique the presence of crystalline GeTe was not observed.

Differential thermal analysis was performed (2) on bulk samples of both memory materials. It was determined that an exothermic transformation from the amorphous to a more ordered state occurs near 210°C. Cooling from this temperature either slowly or rapidly produces a conducting state in the bulk sample. These materials also showed an endothermic transformation near 350°C. The material is more ordered below this temperature and disordered above it. As indicated in ref. 2, slow cooling from above 350°C produces a conductive state by permitting a sufficient time for the transformation at 210°C to generate the more ordered state. Quenching the sample from above 350°C retains the disordered state and produces the amorphous highly resistive state.

3. ELECTRICAL CHARACTERISTICS OF THE OVONIC MEMORY SWITCH (OMS)

The OMS is a two-terminal device with an electrically alterable conductance. In an early

form, the devices were produced by depositing a one to ten micron film of memory material on two graphite hemispheres having a radius of curvature of approximately 10mm. The two hemispheres were spring loaded against each other with about 10 g force to form a conduction area of approximately 10^{-6}cm^2. Later structures consist entirely of thin films.(6) The device data presented here are representative for all devices although the thermal time constant depends on the particular structure. The device has two fairly distinct conductivity regions; a high resistance OFF or RESET state and low resistance ON or SET state. Either state can be maintained without the application of power.

The I-V characteristic of a typical (OMS) is shown in Fig. 5A. The high resistance state becomes nonlinear as the threshold voltage is approached. When the threshold voltage (V_T) and current (I_T) are exceeded, a sudden decrease in resistance occurs. The transition time is sub-nanosecond. Electronic breakdown occurs (7,8) in the nonconducting amorphous state of the memory device as it does in the threshold switch. The energy provided by this phenomenon is not sufficient, however, to cause the conductivity transformation to the permanently conductive state. This is due to bond energies of the material constituents and the use of steric hindrances. If the required transformation energy is applied for a minimum ordering time, the device will SET in the low resistance state. If power is removed, the device will remain stably in that condition. The switching effect at the threshold voltage and the transformation to the permanently conductive state are indicated to be entirely different phenomena. Switching is found only in disordered materials, and is not accompanied by a structural change. The SET, RESET transformation is a structural change.

The ON state characteristic can vary from a linear to a slightly s-shaped characteristic as seen in Fig. 5B. Increasing SET current decreases the SET resistance and produces a more ohmic state.

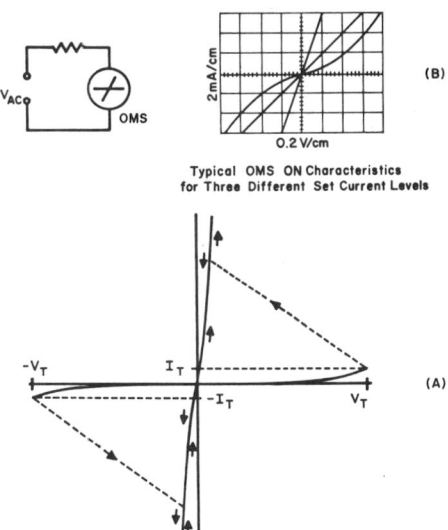

Fig. 5   Current-voltage characteristic (A) of an OMS. The variation of the ON state with SET current is shown in (B).

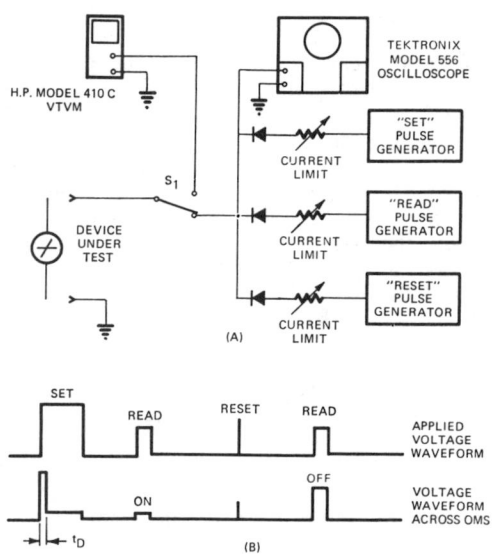

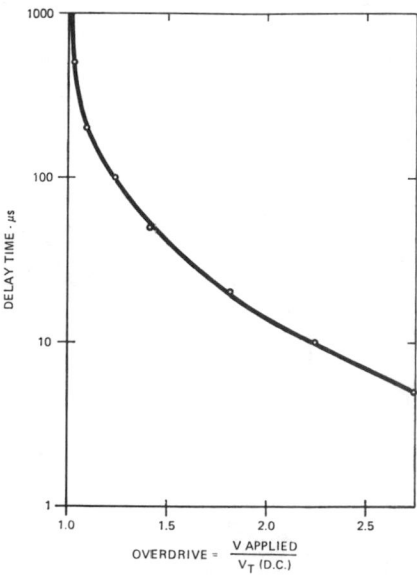

Fig. 7 Switching delay time versus voltage overdrive for a 10 μm-thick OMS.

Fig. 6 Electrical parameter test apparatus for OMS devices. The open circuit waveform to the device is shown in (B) along with a typical waveform during measurement.

Switching the OMS from the SET to the RESET state is achieved by applying a narrow current pulse of either polarity with sufficient energy to disorder the material and short enough in time to permit a thermal quench. The magnitude of this current must be significantly greater than the SET current in order to provide comparable energy.

Rectangular pulses applied by the arrangement in Fig. 6A were used to evaluate the following device parameters:

(1) SET time – τ_S – time of conduction during the SET operation,

(2) SET current – I_s,

(3) ON resistance – R_{on},

(4) RESET time – τ_r – RESET pulse width,

(5) RESET current – I_R,

(6) OFF resistance – R_{off}.

The pulse generators allowed either single or multiple pulses of varying width, amplitude, and source impedance to be applied to the device. Resistance was measured with the H.P. 410C. The output voltage and current of this instrument on all resistance ranges are well below the levels which would alter the state of the device. When a rectangular pulse with an amplitude greater than the dc threshold voltage is applied to an OMS, the rapid switching transition occurs after a delay time t_D, as shown in Fig. 6B. The delay time decreases rapidly as the amplitude of the applied pulse is increased. Figure 7 shows t_D versus voltage overdrive for devices with a 10 μm thick memory film. For the same percentage of overdrive the delay time also decreases rapidly as the thickness of the active film is reduced.

The recovery of threshold voltage after the application of a RESET pulse is measured with a

waveform similar to that shown in Fig. 6B. Interrogation (READ) pulses of different amplitudes are brought closer to the trailing edge of the RESET pulse until they cause the device to switch. A plot of the amplitude of the interrogation pulses and the minimum time between the trailing edge of the RESET pulse and the leading edge of the READ pulse which does not cause switching is shown for a 10 μm thick device in Fig. 8. The device recovery time t_r is defined as the time required for the threshold to reach one-half of its dc value and decreases rapidly as film thickness is reduced. The recovery time is also strongly affected by the thermal time constant of the structure.

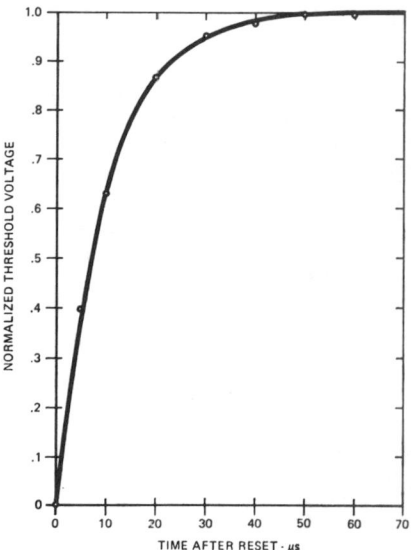

Fig. 8 Normalized threshold voltage recovery versus time for a 10 μm-thick OMS. Time is measured from the cessation of a RESET pulse.

Additional experiments to observe the SET and RESET characteristics of the OMS consisted of dissipating pulse energy in the device and measuring the change in resistance. Since material 706 has a broad transformation range, data is presented on these samples.

A completely RESET OMS was pulsed with SET pulses of varying current and pulse width. After each SET pulse the device was fully RESET. SET pulse widths from 1 msec to 10 sec and SET currents from 1 mA to 50 mA were investigated. These data are plotted in Fig. 9 as SET energy ($V_H I_S \tau_S$) versus device resistance. V_H is the voltage across the sample during conduction. The curves for various set currents coincide over a large range of set energies. This indicates that the device is energy controlled in this region. The lower portion of the curves indicates a saturation region where, for a given current, no further decreases in resistance is caused by an increase in SET energy. This saturation phenomena is related to the thermal characteristics of the package since power dissipation will prevent heating additional film volume to the transformation temperature previously indicated. RESET experiments were performed by saturating an OMS to the SET state, applying RESET pulses of varying energy and measuring the final resistance. After each pulse the SET state was again saturated. RESET pulse widths from 50 nsec to 10 μsec and RESET currents from 20 mA to 1 A were investigated. When these data were plotted as RESET energy ($V_H I_R \tau_r$) versus device resistance the curves illustrated in Fig. 10 resulted. All points of constant energy coincided within a small experimental error (± 5%). Each of the curves represents a different SET current as indicated in the figure. It was observed that when a device is SET to a lower saturation level more energy is required to RESET the device. One further point of interest is that RESET pulse energies are cumulative over long periods of time. That is, five pulses of

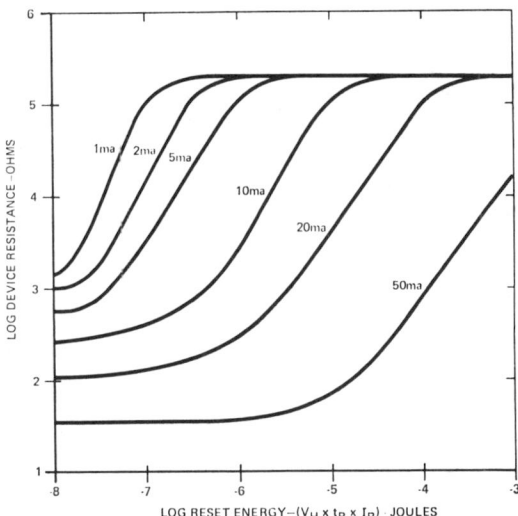

Fig. 10. Log resistance versus log RESET energy for a 10 μm-thick OMS. Material is No. 706. The family of curves are generated by varying the SET current prior to application of the RESET energy.

one microjoule each produce the same change in resistance as one pulse of five microjoules. The time between RESET pulses did not affect the resistance change. This indicates the RESET state to be determined by the total accumulated RESET energy as long as each RESET pulse is no longer than 10 μsec. Devices made from material 729 show similar results except that all curves are much steeper; the transition from no effect to fully ON or fully OFF typically occurs with less than one order of magnitude change in energy compared to 3 or 4 orders of magnitude for material 706.

5. OMS MODEL

The OMS utilizes two independent phenomena. The first is an electronic breakdown region characteristic of covalent amorphous alloys and the second requires a structural transformation between states of metastable conductivity. The transformation in the $Ge_{15}Te_{81}X_4$ system have been related to order-disorder of Te. The transformation temperature of the ordered state is indicated to be approximately 200°C whereas above approximately 375°C the materials are again disordered. These compounds have been balanced between the ordered or conductive state and the amorphous or highly resistive state by introduction of cross-linking elements, thereby maximizing stability of both states.

The device structure and associated thermal time constant have significant effects on the OMS electrical behavior. In operation, the SET voltage pulse induces the breakdown and permits a current channel or filament to form. During this pulse, and after breakdown, sufficient energy is delivered to the material in the channel to cause the ordering transformation at about 200°C. The time required for the conductivity transformation is probably controlled by atomic mobility or diffusion, and nucleation probability. The required diffusion

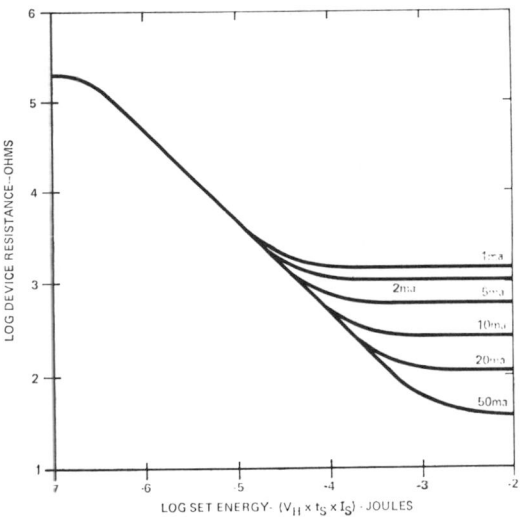

Fig. 9 Log resistance versus log SET energy for a 10 μm-thick OMS. Material is No. 706. The various saturation resistances in the ON state are controlled by the indicated SET current.

length need not be greater than a few angstroms. After the SET pulse is over the conductive state is retained. Application of the RESET pulse in a time shorter than the thermal time constant of the package confines the RESET energy to the conductive channel, initiates the disordering transformation at the elevated temperature and the electrodes provide dissipation of the energy to quench the film. This restores the amorphous and high resistance state.

Acknowledgment

The authors would like to express gratitude to E.A. Fagen, H. Fritzsche and A. Bienenstock for providing certain of the reported data and physical interpretation, and to the Technical Staff of Energy Conversion Devices, Inc. for the many discussions and reviews.

References

1. S.R. Ovshinsky, Phys. Rev. Lett. 21 (1968) 1450; also U.S. Patent No. 3,271,591.
2. H. Fritzsche and S.R. Ovshinsky, J. Non-Crystalline Solids 2 (1970) 148.
3. A. Bienenstock, F. Betts and S.R. Ovshinsky, J. Non-Crystalline Solids 2 (1970) 347.

4. The dilatometric and calorimetric measurements were made on bulk samples whereas the resistivity measurements were taken on sputtered films. There is no certainty that the composition of the samples was identical. In addition, bulk samples used for the dilatometric studies were cooled from the melt and hence are expected to have different structure from films deposited on cold substrates.
5. It is not implied that considerably more energy is required to SET than RESET. Since the thermal time constant of the thin-film sample on glass is shorter than the minimum flashlamp pulse width, the energy lost from the broad pulse will be significantly higher than the narrow-pulse due to heat conduction. The indicated energy is flashlamp input and has not been quantitatively related to absorbed energy in the sample. Electrical pulse measurements indicate that RESET and SET energies are comparable.

6. R.G. Neale, J. Non-Crystalline Solids 2 (1970) 558.
7. H. Fritzsche, Physics of Instabilities in Amorphous Semiconductors, presented at Symp. on Instabilities in Semiconductors, IBM, Watson Res. Lab., Yorktown Heights, N.Y. (1969).
8. S.R. Ovshinsky, Ovonic Switching Devices, presented at Intern. Colloq. on Amorphous and Liquid Semiconductors, Bucharest, Rumania (1967).

STRUCTURAL STUDIES OF AMORPHOUS SEMICONDUCTORS

A. Bienenstock,[1] F. Betts,[1] and S.R. Ovshinsky[2]

[1]Department of Materials Science, Stanford University, Stanford, California 94305
[2]Energy Conversion Devices, Inc., Troy, Michigan 48084

1. Introduction

This paper describes structural studies of the reversible memory type electrical transition[1]) in amorphous films of Ge–Te[2]) based alloys. The films, as evaporated, are high resistance intrinsic-like semiconductors. After subjection to a sufficiently high electrical field to bring about a transition to a low resistance state, and a further delay during which there is current flow, a portion of the material is transformed to a semipermanent low resistance state. That is, the material remains in a low resistance state without further application of an electric field. The material may be switched back to a high resistance state through application of a current pulse with a rapid turn off. This paper reports initial results of studies designed to elucidate structural aspects of the phase transition in these materials.

Initial efforts were based on the hypothesis that the memory effect involved the vitrification and devitrification of GeTe. A possible mechanism for the transformation was suggested by the structural model proposed by Hilton et al.[3]). On the basis of infrared and radial distribution studies, they deduced a chain structure in which coordination is two-fold, the interatomic distance is 2.57 Å and the bond angle is approximately 110°. This should be contrasted with the only slightly distorted rocksalt structure of crystalline GeTe, where the coordination is six-fold, the interatomic separation is approximately 3 Å and the bond angle is about 90°. This difference in local coordinations suggests that an appreciable portion of the difference in conductivities is due to a change in bonding with a corresponding change in the effective band gap.

Unfortunately, the chain model of Hilton et al.[3]) can only be viewed as a conjecture. The infrared studies, in themselves, do not yield an unambiguous structure. The supporting radial distribution work was performed on a sample of the composition $Ge_{15}As_{45}Te_{40}$. It shows a nearest neighbor peak at 2.50 Å and a second neighbor peak at 4.02 Å. The peak at 2.50 Å would be appropriate for both As–Te and Ge–Te nearest neighbor pairs. Since As and Ge have approximately the same atomic scattering factors, it would be expected that the As–Te peaks would dominate the radial distribution at 2.50 Å. On the other hand, the radial distribution shows a minimum at 3.1 Å, indicating that there is very little bonding of the type associated with crystalline GeTe. This might, however, be accounted for by the small quantity of Ge contained in the sample.

Doubt about a significant difference in the coordinations of amorphous and crystalline GeTe was also raised by the work of Chopra and Bahl[4]). They state: "The optical absorption edge is the same for both structures, suggesting the validity of the same energy band diagram for the 2 cases." It should be noted, however, that Tsu et al.[5]) suggest that the band gap in crystalline GeTe is approximately 0.1–0.3 eV and that "... the observed optical absorption edges for our films at 0.7–1.0 eV are strongly Burstein-shifted". It is a crystalline absorption edge at approximately 0.8 eV to which Chopra and Bahl have compared their amorphous absorption edge measurements. The possibility of a significant Burstein shift in the crystalline absorption edge led us to believe that Chopra and Bahl's measurements do not rule out structural differences. The radial distribution work described in this paper is an attempt at clarifying this point.

While the radial distribution work was underway, techniques for the reversible thermal transformation of appreciable volumes of samples were developed at Energy Conversion Devices, Inc. The second portion of this paper deals with X-ray diffraction identifications of the heat treatment products.

2. Radial distribution studies of amorphous Ge_xTe_{1-x} alloys

2.1. EXPERIMENTAL PROCEDURES AND DATA ANALYSIS

Three films of amorphous Ge_xTe_{1-x} alloys were prepared at Energy Conversion Devices, Inc. by R. Nowicki. Sample 1 was in the form of a bulk powder, while samples 2 and 3 were thin films, of thicknesses about 34 μm and 50 μm, respectively, obtained by vapor deposition. Approximate values of x were obtained for each sample through X-ray emission measurements, based on comparisons with bulk samples. The values were 0.11, 0.66 and 0.72 respectively for samples 1, 2 and 3. The possible error in these numbers due to the finite thicknesses of samples 2 and 3 has not been determined. This error is not expected to be large, however, because both films are sufficiently thick to cause a factor of 10 attenuation of the appropriate wave length X-rays.

X-ray diffraction patterns from these samples were obtained with molybdenum K_α radiation on a Picker diffractometer using a LiF diffracted beam monochromator and a pulse-height analysis window sufficiently narrow to eliminate $\frac{1}{2}\lambda$ components of the X-ray beam. The intensity data were obtained using the step scan technique with a step size of 0.267° $(2-\theta)$ and 4000 counts per step. Measurements were made over the angular range 10° to 100° $(2-\theta)$, corresponding to a range in s $(4\pi \sin \theta/\lambda)$ of 1.54 to 13.68. Measurements of the passband of the monochromator indicated that, to a good approximation, the Compton scattering could be neglected in the data analysis. Fig. 1 shows the diffraction pattern from sample 3, after scaling and polarization corrections.

Atomic radial distributions were obtained from the observed intensity data, $I(s)$, by Fourier analysis using the relation[6])

$$4\pi r^2 \rho(r) = 4\pi r^2 \rho_0 + \frac{2r}{\pi} \int_{s_{min}}^{s_{max}} s\, i(s) \exp(-as^2) \sin(rs)\, ds. \quad (1)$$

Here, $\rho(r)$ is the atomic radial distribution function in units of electrons[2]) per Å[3]. ρ_0, its average value, is normally obtained from density measure-

* Work supported in part by the Advanced Research Projects Agency through the Center for Materials Research, Stanford University, by the Department of Materials Science, Stanford University, and by Energy Conversion Devices, Inc.

Reprinted by permission from *Journal of Non-Crystalline Solids*, Vol. 2, pp. 347–357 (1970).

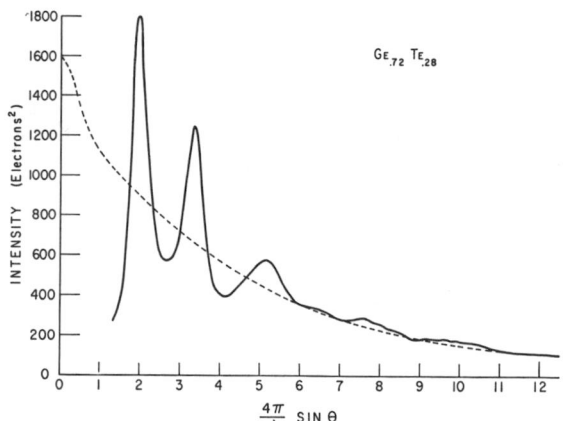

Fig. 1. Scaled, polarization corrected diffracted intensity as a function of s ($= 4\pi \sin\theta/\lambda$). The dashed curve represents the scattering which is independent of atomic configurations.

ments. $i(s)$ is given by the equation

$$i(s) = \left[I_{\text{corr}} - \sum_j x_j f_j^2(s) \right] \left[\frac{\sum_j x_j f_j(0)}{\sum_j x_j f_j(s)} \right]^2, \qquad (2)$$

where x_j is the mole fraction of component j, while f_j is its atomic or ionic scattering factor and I_{corr} is the scaled, polarization and background corrected intensity. The $\exp(-as^2)$ in eq. (1) is an arbitrary temperature factor inserted to aid the convergence of the incomplete Fourier transform of eq. (1). a has been chosen in this work so that $as^2_{\max} = 1.85$. In evaluating eq. (2), the atomic scattering factors of Cromer and Waber[7] were used for Ge and Te. Dispersion corrections listed in the International Tables for X-ray Crystallography[8] were also included in the calculation.

Sample densities necessary for the determination of ρ_0 were not determined experimentally. Instead, a density of 5.6 for sample 1 and 5.1 for samples 2 and 3 were used as a result of some considerations which will not be presented here. The effect of the approximate nature of the densities will be discussed below.

Eq. (1) was evaluated numerically using the IBM 360 computer of the Stanford Computation Center.

2.2. RADIAL DISTRIBUTION CURVES AND THEIR ANALYSIS

The resulting radial distribution curves for samples 1, 2 and 3 are shown as figs. 2, 3 and 4 respectively. Before commenting on their analysis, it is worthwhile to indicate their shortcomings. All three curves show structure at r values less than 2 Å. In principle, $\rho(r)$ should be zero in this region. This structure indicates that there are errors in the data or their analysis. These errors are likely to influence the area and shapes of the peaks in the distribution, but should not have a major effect on the peak positions. The dotted line shows the contribution of the estimated ρ_0 term. It should be noted that the fluctuations about ρ_0 in the small r region are extremely small, indicating that it is the computed density which is in error. We have found, however, that densities approximately one-half those used would be needed to bring $\rho(r)$ to more reasonable values. It is highly unlikely that the densities of the films are that small. Thus, we assume that there are other errors involved. Systematic attempts to determine the nature and the means of elimination of these errors are underway. It should also be noted that the effect of the temperature factor is to introduce a slight broadening of the radial distribution peaks. That is, we expect that the peaks are slightly broader than they would have been had we been able to extend the upper limit of the integral closer to infinity.

In spite of the distributions' limitations, some of their features are well worth noting. Over the broad composition range studied, the first and second neighbor peaks always appear at 2.7 Å and 4.2 Å, respectively. These peak positions are not expected to be appreciably affected by the errors. This indicates that the bulk of the nearest neighbor and next nearest neighbor distances are at approximately 2.6 and 4.1 Å respectively. Thus, in no case do we see any indication of the bonding in crystalline GeTe, with nearest

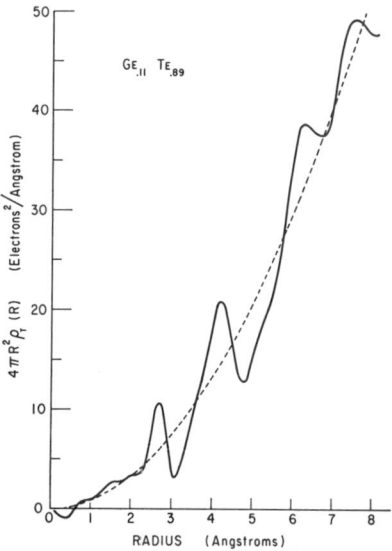

Fig. 2. Calculated radial distribution for sample 1, of composition Ge₁₁Te₈₉. The dashed curve represents the contribution of the average electron density.

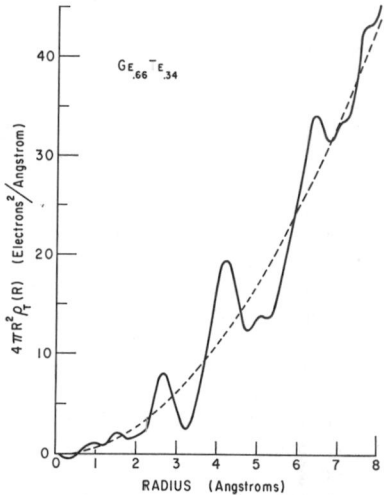

Fig. 3. Calculated radial distribution for sample 2, of composition Ge₆₆Te₃₄. The dashed curve represents the contribution of the average electron density.

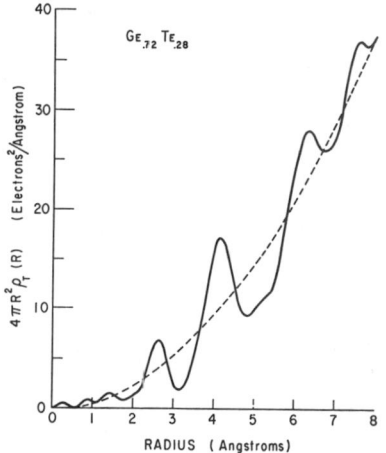

Fig. 4. Calculated radial distribution for sample 3, of composition Ge₇₂Te₂₈. The dashed curve represents the contribution of the average electron density.

neighbor and next nearest neighbor separations of 3.0 and 4.2 to 4.3 Å respectively. On the other hand, there is a noticeable broadening of the first neighbor peak in going from $x = 0.11$ to $x = 0.66$. This may be due to contributions of Ge–Te separations which are appreciably larger than 2.6 Å. Alternatively, the broadening may be due to the above mentioned errors in the data or their analysis.

In all three distributions, the area under the first peak suggests that the coordination is two or three fold, implying chain-like structures. As discussed above, however, these areas are uncertain due to the errors. In addition, there is an ambiguity in their analysis which is impossible to remove with X-ray measurements alone. That is, the nearest neighbor peak could be the superposition of Te–Te, Ge–Ge and Ge–Te nearest neighbor pairs. The contribution to the area from each pair is proportional to the product of the number of electrons in each atom. Without a knowledge of the number of each type of pair, the coordination cannot be determined uniquely. When it is possible that a single peak in the distribution has contributions from different types of pairs, it is impossible to determine the coordination uniquely without a knowledge of the relative number of each type of pair. This, of course, is the information to be determined.

In summary, then, the most important result of this study is the conclusion that there is not an appreciable number of Ge–Te pairs in the amorphous materials at the interatomic separation associated with crystalline GeTe. This result is inconsistent with Chopra and Bahl's conclusion that the same energy band diagram is appropriate for the crystalline and amorphous materials. Our efforts are now aimed in two directions. First, we are concentrating on improving the quality of the radial distributions. Attempts will be made to measure the densities of the films. In addition, possible sources of systematic errors in the measured diffracted intensities are being studied. In addition, however, we are searching out more specific tools for the analysis of coordination changes. While quadrupole broadening of nuclear magnetic resonance lines would appear to be ideal, it is not well suited for Ge and Te. Hence, we are turning to an investigation of X-ray absorption spectroscopy techniques. With this tool, it may be possible to study the changes in coordination of Ge and Te separately when the samples are transformed from the amorphous to crystalline state.

3. X-ray diffraction identification of bulk crystallization products

Fritzsche and Ovshinsky[2]) in their DTA studies of Ovonic memory materials, have noted two types of behaviors in finely powdered bulk materials. In one series of materials, two exothermic peaks are evident in the first heating cycle. The low temperature peak is observed between 200 and 250 °C, while a higher temperature peak is found between 250 and 300 °C. In other materials, only one peak between 200 and 250 °C is observed.

We have studied a number of these materials, based on the Ge–Te system, by X-ray diffractometry. In all cases, the materials were, for the most part, amorphous prior to heating. Occasionally, extremely weak crystalline peaks were observed, indicating that a small number of crystallites were formed as the melt cooled.

The heating products were determined in the following manner. Fritzsche supplied three types of samples: 1) single peak samples in which DTA heating cycle was terminated immediately after the peaks, 2) double peak samples in which the cycle was terminated immediately after the first peak, 3) double peak samples in which the cycle was terminated after the second peak.

Samples of types (1) and (2) always showed sharp Bragg peaks which were readily identifiable as crystalline Te, as well as an amorphous background.

Samples of type (3) showed sharp Bragg peaks from crystalline Te and GeTe, as well as an amorphous background.

Thus, the DTA peaks between 200 and 250 °C can be associated with the crystallization of Te, while the higher temperature peaks, when they appear, are due to the crystallization of GeTe.

4. X-ray diffraction studies of thin films

While studies of bulk samples indicated that the switching process is associated with crystallization, it is not justifiable to extrapolate bulk sample properties to explain thin film devices. Two objections to such an extrapolation are: 1) The recrystallization process requires separation of the Te and GeTe phases. Such phase separation might occur as the bulk melt cooled, or in the relatively slow heating associated with the DTA process. The films, as deposited, however, are most probably homogeneous. It is not at all apparent that phase separation can take place in the short switching times of the devices. 2) Given that crystallization and the corresponding phase separation can take place in the conversion of high resistivity to low resistivity material, it is not obvious that it would be possible to obtain the reverse transformation to the high resistance, amorphous state without a remixing of the material. In order to obtain more information about these questions, a partially completed program of measurements built around thin film samples was undertaken. In this section, the completed experiments will be described and experiments to be performed will be discussed.

A series of films of the one DTA peak type materials were vapor deposited onto glass microscope slides by J. Evans. Metallic electrodes were also deposited on the slide, underneath the film. The composition of these films has not been determined, but they can be characterized by their crystallization behavior. Film thicknesses were all of the order of 2 to 4 μm. It was verified that all films were amorphous as prepared.

The initial experiments were performed on samples which were converted to the conducting state thermally. This was achieved by placing the slide on a hotplate, whose surface temperature was 400 °C, for approximately 2 min in air. This process reduced the resistance of the film from $10^7 \Omega$ to 250 Ω. The resulting X-ray diffraction pattern showed a number of intense sharp peaks, all of which could be associated with Te. In addition, extremely weak peaks associated with the electrodes were observed. The angular breadth of the diffraction peaks indicated an average crystallite size of at least 400 Å perpendicular to the microscope slide. No determination of the crystallite size parallel to the slide could be made.

A similarly treated sample, whose resistance was reduced from $10^7 \Omega$ to 500 Ω by heating was then reset to a final resistance of $10^6 \Omega$ through subjection to one 500 μsec light pulse by Evans, Helbers and Ovshinsky[9]). Their procedure is described in an accompanying paper. The resulting X-ray diffraction pattern showed a broad maximum typical of amorphous films.

These results should be contrasted with those obtained from two samples transformed to the conducting state by J. Evans with the flash lamp technique. The first sample showed a film sheet resistance change from $10^7 \Omega$ to 200 Ω as a result of the flashing. The diffraction pattern was that of polycrystalline Te, but the ratios of intensities of different peaks were far from that expected for randomly oriented crystallites. What is more, the intensities

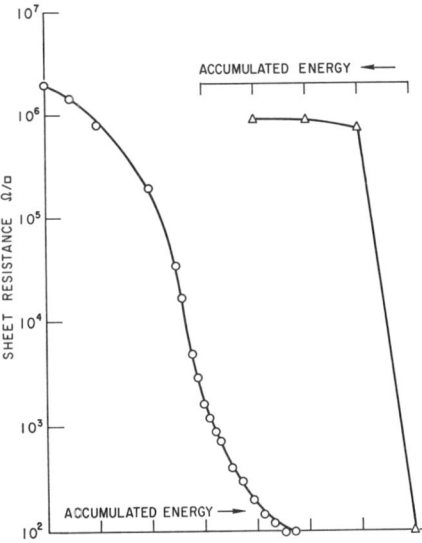

Fig. 5. Record of measured sheet resistance of a thin film memory material as a function of flash lamp pulsing. The circles show the decrease of resistance through application of long duration pulses. The triangles show the increase of resistance through application of short duration, sharp cut-off pulses.

of the peaks varied with rotation of the microscope slide about its normal. In one orientation of the slide, the (110) reflection is considerably stronger than the (101) reflection, although the latter peak is expected to be approximately three times as intense for a sample consisting of randomly oriented crystallites. As the sample was rotated through an angle of 90° about the normal, both peaks disappeared. We believe that these effects are due to needle crystallites oriented in the plane of the microscope slide. Reinvestigation showed that similar, but much less marked, effects can be observed in the thermally transformed samples.

A second sample, which was first transformed from a film sheet resistance of $2 \times 10^6 \Omega$ to $10^2 \Omega$ by the flash lamp technique, was transformed back to a resistance of $9 \times 10^5 \Omega$ through a succession of short duration pulses. (See fig. 5 for a description of the change of resistance with pulsing.) The diffraction pattern from this sample shows an extremely weak (101) reflection while the (110) peak is nearly as strong as it was for the conducting sample described in the preceding paragraph. This (110) peak could also be made to disappear by rotation of the sample. A model for these results would be one in which the smallest of the crystallites in the conducting samples are relatively randomly oriented, yielding the (101) reflection. These small crystallites are revitrified with the short duration pulses. Larger, relatively well oriented crystallites which account for the strong (110) reflection remain crystalline.

These results indicate, first of all, that it is Te which crystallizes and vitrifies in the reversible memory transformation. They also show that it is possible to convert the system from one consisting of small crystallites of Te to an amorphous state in an extremely short time. This is shown quite clearly by the reconversion of the thermally treated samples to the amorphous state with the application of one short duration pulse. It should be noted that after the application of the first short duration pulse to the flash lamp converted sample, there is no further appreciable increase of resistance with pulsing. A model for this would be one in which the small crystallites are vitrified by the first short duration pulse, but the large crystallites are never revitrified. Finally, it should be noted that many more pulses are required to convert an evaporated high resistance film to the conducting state than are needed to return it to the amorphous state. This effect may be explained by the necessity to phase separate the material in the process of crystallization. One remaining problem is to determine whether the material remains phase separated in the revitrification process. Attempts are now being made to determine this through transmission electron microscopy. We also hope to have, in the near future, a more complete picture of crystallite orientation and sizes associated with these processes.

References

1) S. R. Ovshinsky, Control Engineering **11** (1964) 69; Talk at Intern. Colloq. on Amorphous and Liquid Semiconductors, Bucharest, 1967; U.S. Patent No. 3,271,591.
2) H. Fritzsche and S. R. Ovshinsky, J. Non-Crystalline Solids **2** (1970) 148, 393.
3) A. R. Hilton, C. E. Jones, R. D. Dobrott, H. M. Klein, A. M. Bryant and T. D. George, Phys. Chem. Glasses **4** (1966) 116.
4) K. L. Chopra and S. K. Bahl, Bull. Am. Phys. Soc. **14** (1968) 98.
5) R. Tsu, W. E. Howard and L. Esaki, Phys. Rev. **172** (1968) 779.
6) R. F. Kruh, in: *Handbook of X-Rays*, Ed. E. F. Kaelble (McGraw-Hill, New York, 1967).
7) D. T. Cromer and J. T. Waber, Acta Cryst. **18** (1965) 104.
8) *International Tables for X-Ray Crystallography*, Vol. 3 (Kynoch Press, Birmingham, 1962).
9) E. J. Evans, J. H. Helbers and S. R. Ovshinsky, J. Non-Crystalline Solids **2** (1970) 334.

REFLECTIVITY STUDIES OF THE Te(Ge,As)-BASED AMORPHOUS SEMICONDUCTOR IN THE CONDUCTING AND INSULATING STATES

J. Feinleib and S.R. Ovshinsky

Energy Conversion Devices, Inc., Troy, Michigan 48084

1. Reflectivity studies of simpler amorphous semiconductors

In the first paper of this conference, Prof. Stuke presented a detailed review of the optical and electrical properties of the basic materials, Se, Ge and Te, which can be formed in both the amorphous and crystalline form[1]). In the final day of this meeting we would like to tie the basic physics learned from these discussions to more material ends – that is to the design of materials which can be utilized for the basic differences between crystal and amorphous forms and, in particular, the differences in optical properties.

Let us recall some of the optical data which have now become well established as intrinsic to the crystalline and amorphous states of the simpler materials. In fig. 1 there is presented the now familiar data of Tauc et al.[2]) of the reflectivity spectra of single crystal, polycrystal and amorphous Ge. As has been well demonstrated, the sharp structure caused by the distinct K-space critical points in the crystal are broadened and greatly reduced in magnitude for many interband transitions when going from crystal to amorphous material. But we shall be concerned with a subtler aspect of these spectra, namely the change in absolute magnitude of the overall reflectivity and the corresponding changes in the optical constants:

$$\varepsilon_2 = 2nk \quad \text{and} \quad \varepsilon_1 = n^2 - k^2, \tag{1}$$

which are related to the magnitude of reflectivity, R, by

$$\sqrt{R}\, e^{i\theta} = \frac{n - 1 - ik}{n + 1 - ik},$$

where θ is a phase change on reflection. The more familiar form is:

$$R = \frac{(n-1)^2 - k^2}{(n+1)^2 - k^2}. \tag{2}$$

In amorphous Ge the overall reflectivity in the range shown is somewhat reduced in magnitude from that of the crystal over much of the spectrum. But note that in this instance the reflectivity of both amorphous and crystalline Ge appear to be nearly the same at low energies, below the absorption edge, where the absorption k is very low and the reflectivity is dominated by the static dielectric constant or, alternatively, by the refractive index $n(0)$. The values of ε_2 determined by Kramers–Kronig analysis of these data are shown in fig. 1. The imaginary part of the dielectric constant, ε_2, is plotted since this is most closely related to the oscillator strengths of the transition[3]). For Ge, the shape of the ε_2 curve is changed from the crystal to the amorphous phase but has no overall displacement or large magnitude change. The effect of this on reflectivity is obtained from a Kramers–Kronig calculation of $\varepsilon_1(\omega)$ using the relation

$$\varepsilon_1(\omega) - 1 = \frac{2}{\pi} \int_0^\infty \frac{\omega' \varepsilon_2(\omega')\, d\omega'}{\omega'^2 - \omega^2}, \tag{3}$$

for $\omega = 0$

$$\varepsilon_1(0) - 1 = \frac{2}{\pi} \int_0^\infty \frac{\varepsilon_2(\omega')\, d\omega'}{\omega'}. \tag{4}$$

The analysis for amorphous and crystal Ge shows that $\varepsilon_1(0)$ does not change and therefore $n(0)$ and R have approximately the same magnitudes in crystal and amorphous phases at low photon energies. In contrast, the compound amorphous semiconductor, Cd Ge As$_2$, studied by Tauc[4]) shows a large change in ε_2 as seen in fig. 2. We chose this compound as an example because

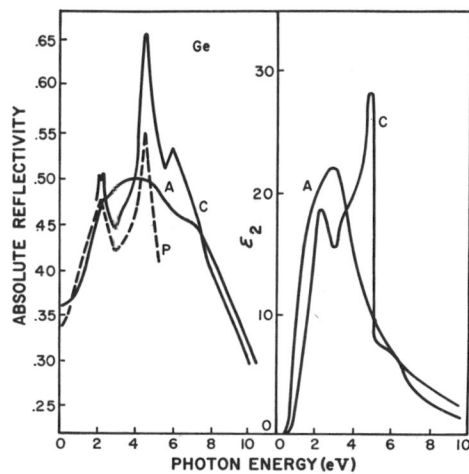

Fig. 1. Reflectivity and ε_2 of crystalline (C), amorphous (A) and polycrystalline (P) germanium[2]).

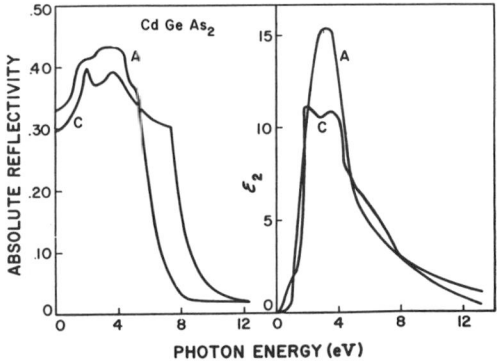

Fig. 2. Optical properties of CdGeAs$_2$[4]).

Reprinted by permission from *Journal of Non-Crystalline Solids*, Vol. 4, pp. 564–572 (1970).

this material is somewhat unusual in that over an appreciable energy range, ε_2 is larger for the amorphous phase than for the crystal. The result is that $\varepsilon_1(0)$ as calculated by (4) is larger for amorphous material than crystal and, therefore, the reflectivity is correspondingly larger, by about 14%, at low energies. This particular example seems to answer the question of whether or not this effect is caused by surface contamination which might affect the data: The amorphous material has reproducibly higher reflectivity than the crystalline material. Usually the reverse is found and it was believed that the susceptibility of disordered material to surface contamination was the reason for low amorphous state reflectivities. In this connection, Tauc[4,5]) has recently shown that the sum rule for the oscillator strengths is upheld for the amorphous and crystalline phases of both Ge and $\overline{\text{Cd}}$GeAs$_2$ within experimental error. This rule states that the integral of ε_2 is conserved:

$$\int_0^\infty \omega \varepsilon_2(\omega)\, d\omega = \frac{4\pi N e^2}{m}, \qquad (5)$$

where N equals the total density of valence electrons, which must be the same for both states of the material, when material density is taken into account. We see then that although ε_2 may change strongly from one phase to the other in some spectral regions, a decrease at low energies must be made up by an increase at higher energies. In the calculation of ε_2 from reflectivity it is not fully appreciated that not only does ε_2 depend on the relative reflectivity, as derived by the usual Kramers–Kronig procedure using (1) and the relation:

$$\theta(\omega) = \frac{\omega}{\pi} \int_0^\infty \frac{\ln\left[R(\omega')/R(\omega)\right] d\omega'}{\omega'^2 - \omega^2}, \qquad (6)$$

but also that the *absolute* magnitude of R enters the expression via relation (1). Thus, in order for the sum rule to be satisfied, the *absolute* value of reflectivity is determined, and is intimately related to the change in band structure from the crystal to the amorphous state.

2. Reflectivity of Te and Te-based materials

This lengthy background is required to understand the motivation for the materials research program that has been conducted to explore the possibility of utilizing amorphous semiconductors for a high bit density memory element which can be read optically when information is stored in the two states of the material. Selection of such a material from the many possible amorphous–crystal systems is guided by two main requirements: (1) stable glass and crystal states at the operating temperature, usually room temperature, and (2) a maximum change in the optical properties between the two phases. A most promising system has been found from an examination of the optical constants of Te, which are displayed in fig. 3. Te has a trigonal structure when crystalline and therefore has a polarization dependent ε_2, as Stuke has shown[6]. As seen here, the amorphous phase has a striking change in ε_2 in the low energy region where it appears much smaller than that in the crystal. There is also a relatively large shift of ε_2 toward higher energies. In fact, the energy gap for the absorption edge nearly doubles, going from 0.4 to 0.8 eV between crystal and amorphous states. This causes an unusually large 40% change in $n(0)$, from 5.3 in the crystal to 3.3 in the amorphous state. The corresponding change in reflectivity is thus also about 40% lower in the amorphous phase, and the large difference persists over the spectral range to at least 3.5 eV. There has not been sufficient work on the amorphous phase to verify that the ε_2 sum rule is satisfied equally in the crystal and amorphous phases. However, it is expected that the large shift of ε_2 to higher energies will account for its large magnitude change on going to the amorphous phase. Because of this strikingly large change in optical properties in the low energy range, device applications utilizing the optical changes between the two states of Te are attractive. A severe drawback, however, is the fact that the amorphous phase of Te is unstable above 10 °C. We have found, however, that the Te based chalcogenide glasses with more than 50% Te, have stable amorphous and crystalline phases at room temperature, and they still retain the sharp changes in optical properties exhibited by pure Te[7].

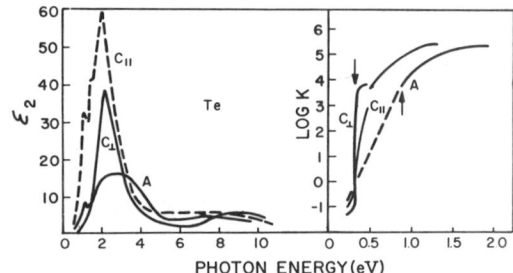

Fig. 3. Changes in the absorption edge and ε_2 from the crystalline and amorphous states of Te[1]).

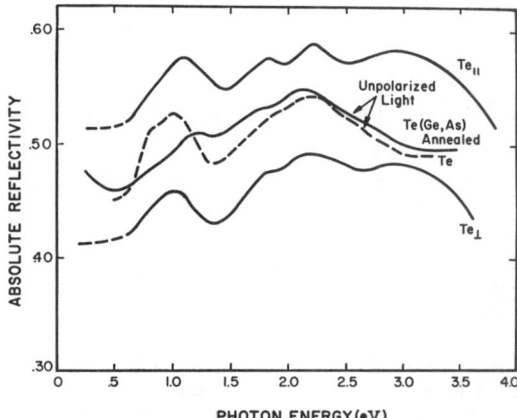

Fig. 4. Reflectivity of pure single crystal Te and of the crystalline state of the Te (Ge, As) amorphous semiconductor. Polarized light data are from ref. 6.

Fig. 4 shows the reflectivity data of Stuke and Keller[6]) for single crystal Te with the light polarized parallel and perpendicular to the trigonal axis. We have measured the absolute reflectivity of a cleaved single crystal of Te in unpolarized light, and the resulting data are seen to fall between the polarized light results, as would be expected. Indeed, this gives us confidence to within a few percent in the absolute values measured. The essential structure in the Te reflectivity is clearly observed and consists of a relatively sharp peak (~ 0.2 eV wide) centered at 1.0 eV and then a broad maximum at 2.1 eV with superimposed weaker structure.

The fourth curve plotted in this figure is a Te-based amorphous semiconductor which has been annealed to produce a crystalline state. This particular material was chosen because it has a stable glass structure with repeated thermal cycling and is one of many materials composed of more than 50% Te which shows memory effects[8]). The resistivity changes from crystal to glass phases by a factor of 10^4. The composition of this material was 81% Te and the remainder made up with additions of Ge and As. The samples were prepared in bulk form from the melt, and the slowly cooled samples show microcrystalline structure[9]). The samples were polished with diamond paste, and the final surface was prepared by an etching polish. The data shown in fig. 4 clearly show structure similar to pure Te, except that the sharp peak at 1.0 eV is considerably reduced in magnitude and appears somewhat shifted to higher energy. Also to be noted is that the absolute magnitude of reflectivity of this material has nearly the same value as pure Te over this spectral range, except that for this material there is an increase in reflectivity below 0.5 eV, whereas the reflectivity of pure Te remains essentially constant at low energies.

The reflectivity of this material in the amorphous state is shown in fig. 5. The original crystalline bulk material was divided into parts and an amorphous sample was produced from one piece by heating it in inert gas filled quartz ampule to ~ 475 °C and rapidly quenching it in water. The resulting bulk glassy samples were then polished and etched in the same manner as the annealed samples. As seen in the figure, the major change in reflectivity is the striking decrease in absolute value over the spectral range studied as compared to the annealed material. There is no sharp structure but two

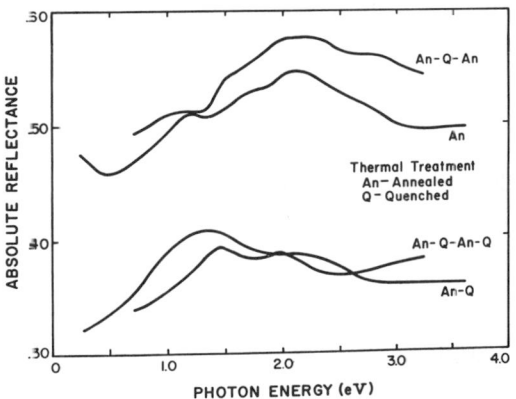

Fig. 5. Comparison of reflectivity in amorphous and crystalline states of Te (Ge, As).

TABLE 1

Comparison of optical refractive index and reflectivity of amorphous and crystalline solids

	n_c	n_A	$(R_c - R_A)/R_c$ (0.5 eV)
Ge	4.0	4.0	0 %
CdGeAs$_2$	3.4	3.85	− 14 %
Se	2.81	2.46	+ 22 %
Te	5.3	3.3	+ 38 %
Te 80 % (Ge, As)			+ 30 %

broad peaks are seen near 1.2 eV and near 2.0 eV, somewhat displaced from the peaks in the crystalline phase, if indeed they are directly related. Another notable feature is that the reflectivity below 0.5 eV continues to decrease in the quenched material whereas it has an upturn in the crystalline phase.

In order to determine the reproducibility of this effect, samples of the bulk material were thermally cycled a second time and new samples prepared by polishing. As indicated in fig. 5, there are some noticeable changes in measured reflectivity, but the differences between the two annealed samples are only about 10%, and the changes in the quenched samples are less than 5%. These differences can well be attributed to variations in surface treatment, and to a lesser extent, inhomogeneities produced by this severe method of materials preparation. The major change in the reflectivity between quenched and annealed samples is the real, intrinsic difference between the two states of the material.

3. Discussion

From an examination of the optical constants of the basic glass forming semiconductors, we can expect that between the amorphous and crystalline states, there is a wide variation in the change in the intrinsic low energy reflectivity. The data are summarized in table 1. The Te system has been found to have the largest changes, and by suitable alloying, a stable room temperature glass is obtained which maintains the large optical changes.

Stuke[10] has described in a general way the apparent reason for the large change in the optical constants of Te. Basically it arises from an increase in the electronic interaction within the Te chain structure as the interaction between chains are weakened on going to the amorphous phase. Apparently the inter-chain interaction gives the lowest band gap in the crystal, and these transitions are weakened in the amorphous phase so that the first energy gap arises from different transitions at higher energy. Thus the change in coordination number between the two phases results in the shift in oscillator strength. Going back to eq. (4), we see immediately that if the initial rise in $\varepsilon_2(\omega')$ moves to higher energies, the effect is to greatly reduce $\varepsilon_1(0)$ because of the energy denominator. In the case of Te, although the absorption edge changes by 0.4 eV on going from crystal to amorphous phase, about the same as for Ge, the energy denominator increases by a factor of two because the crystal gap is only 0.4 eV. Thus we get the large effect observed in the reflectivity of Te as compared to the higher gap materials.

The Te (Ge, As) glass preserves these characteristics of Te. The NMR studies reported by Adler[11] here and the radial distribution studies of Bienenstock[12] have given direct evidence that for similar material studied here, the Te-constituent changes coordination with phase change as it does in the pure material. In fact, in the crystalline state we have evidence that one effect of the additional elements to the Te base of the material is to act primarily as a large concentration of donors or acceptors in the crystalline phase. This is seen in fig. 5, where the low energy rise in reflectivity is very likely due to free carrier effects that are not seen at these energies in pure Te. This is consistent with a large concentration ($\sim 10^{19}$) of ionized impurities. It thus appears that this system can provide the large optical changes observed in Te and, in addition, form a stable room temperature glass.

Acknowledgments

We would like to thank K. Iding, R. Nowicki and R. Seguin for preparing the samples and we are grateful to J. Thompson and R. Shaw for helpful discussions.

References

1) J. Stuke, J. Non-Crystalline Solids 4 (1970) 1.
2) J. Tauc, A. Abraham, L. Pajasova, R. Grigorovici and A. Vancu, in: Proc. Intern. Conf. on Physics of Non-Crystalline Solids, Delft, 1964, Ed. J. A. Prins (North-Holland, Amsterdam, 1965) p. 606.
3) H. Ehrenreich and M. H. Cohen, Phys. Rev. 115 (1959) 786.
4) T. Tauc, L. Stourac, V. Vorlicek and M. Zavetova, in: Proc. Intern. Conf. on Phys. of Semicond., Moscow, 1968, p. 1251.
5) T. Tauc and A. Abraham, Czech. J. Phys. (to be published).
6) J. Stuke and H. Keller, Phys. Status Solidi 7 (1964) 189.
7) E. A. Fagen and H. Fritzsche, J. Non-Crystalline Solids 2 (1970) 180.
8) S. R. Ovshinsky, Phys. Rev. Letters 21 (1968) 1450.
    H. Fritzsche and S. R. Ovshinsky, J. Non-Crystalline Solids 2 (1970) 148, 393.
9) A. Bienenstock, F. Betts and S. R. Ovshinsky, J. Non-Crystalline Solids 2 (1970) 347.
10) J. Stuke, Proc. German Phys. Soc., Munich (1969).
11) D. Adler, J. M. Franz, C. R. Hewes, B. P. Kraemer, D. J. Sellmyer and S. D. Senturia, J. Non-Crystalline Solids 4 (1970) 330.
12) F. Betts, A. Bienenstock and S. R. Ovshinsky, J. Non-Crystalline Solids 4 (1970) 554.

ANALOG MODELS FOR INFORMATION STORAGE AND TRANSMISSION IN PHYSIOLOGICAL SYSTEMS

Stanford R. Ovshinsky and Iris M. Ovshinsky

Energy Conversion Devices, Inc., Troy, Michigan 48084

(Received 4 June 1970)

In this paper we wish to examine in a qualitative way the problem of encoding information in atomic structures with, effectively, a high signal-to-noise ratio. How can such information be retained as memory and in what forms can it be detected and used? These are great unsolved problems of biology and their solution would also be of real value in electronic technology.

We propose certain hypotheses based upon our work and these suggest at least some plausible answers to the above questions. It is our starting point that there are physical processes which are common to inorganic materials and organisms. This does not imply the crude and misleading analogy to computers that has plagued the neurophysiological field almost since the computer's inception. On the contrary, it is our hope to provide from our knowledge of model systems a physical base for the storage and exchange of information which is directly relevant to real biological structures.

Schroedinger (1) has postulated that "order comes from order." This means, for instance, that a specific item of genetic information must be retained by a structure of considerable rigidity and stability; the original function is lost as the structure is altered. Solid state physicists can appreciate this point by reference to an illustration from their own field: impurities or structural defects in periodic systems alter their properties and thereby, in a sense, the "information content" of such systems. The kind of change considered by Schroedinger was somewhat different, namely that associated with the transition from one relatively stable molecular configuration to another. Information of the type contained in reflex action will not be considered here. It can be compared to the wired-in instructions of a computer. In biology, it has its corresponding counterpart in the function of DNA.

It is our belief that in the field of memory which is not genetically controlled but relies upon information-imprinting events, one is dealing with a form of disorder to order phenomenon, the two phases having the ability to coexist within a prescribed temperature range. The ordered regions extend probably to only tens of angstroms. The structural changes have electrical, optical and mechanical consequences which can serve for read-out purposes. Of course, our analog models are inorganic while the nervous system is organic, but similar kinds of mechanisms should be present and the working areas should have similar dimensions. This means that we limit ourselves here to models which can have the same density of information storage as the central nervous system (CNS). Accordingly, realistic neuronal models should not have areas larger than a few square microns, nor thicknesses greater than hundreds of angstroms. Our models (see below) have such dimensions. This is of great importance when one starts treating the traffic problems of an informational system; quantity is all too quickly reflected in quality.

We know from comparative neurophysiology that nerve cells in a dog's brain number 3 billion, in a chimpanzee 5.5 billion and in a man 10 billion. The multitude of interconnections associated with each cell creates complexities and choices for various types of switching events. Subthreshold activity, we believe, determines the point in time of switch activation and whether it is excitatory or inhibitory. For example, a cell can be altered by chemical and electrical events taking place in the synaptic region. The traffic flow problems of the CNS are essentially problems of the timing of switching events (2) and the introduction into the circuit of switches with specific threshold values offers a further facility, namely that of keying the selection of useful information. However, we will here concern ourselves primarily with the aspect of intelligence handling which involves memory. In physical terms we think of this as the ability of a matrix material to have impressed upon it (by some energy source) a chemical or structural order which reflects the input information and which, under proper conditions, can give up that information without substantial distortion.

We have attempted since 1957 to elucidate, by experimental model building,(3-7) disorder to order mechanisms which would switch from a high impedance to a low impedance state. Several types of switching action are necessary for informational control. The first is a monostable switch wherein a change of resistance is effected through a control signal. At the termination of such a sustaining energy, the device will switch off. The only type

Reprinted by permission from *Materials Research Bulletin*, Vol. 5, pp. 681–690 (1970).

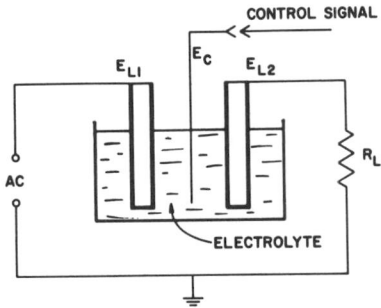

Fig. 1 Schematic diagram of amorphous dielectric film switch and modulator. The load resistor R_L and the amorphous film on the anodized tantalum electrodes E_{L1} and E_{L2} submersed in electrolyte form the load circuit. Current flows through the load circuit if a positive signal is applied to the control electrode E_C. Gain and memory are observed when metallic ions influence the blocking properties of the amorphous dielectric films which are shown for AC operation back to back.

of memory action in which this switch can participate is a volatile one.

The second and third types of switches, both based on nonvolatility and bistability are required for true memory purposes. One can be reversibly altered between the nonconducting and the conducting state. The other changes its resistance in a varying manner in response to energy events from the outside. The latter is an analog memory.

Little is known about the memory mechanism in organisms. As regards nerve impulses, we know that the nerve fiber is surrounded by a semipermeable membrane charged positively on the outside and negatively on the inside. Nerve cell action is initiated at the surface of the cell which is a region for the reception and integration of various physical stimuli. When the stimulus reaches the surface of the fiber, its permeability to certain ions increases, which can be considered as a reduction in electrical resistance.

Crystalline models do not fit into this picture because cell surfaces are by definition disordered and ionic movements are not compatible with PN junctions. We felt that various disordered materials could be utilized for models. We picked tantalum with an anodically grown amorphous oxide film, some hundreds of angstroms thick. It had to be a stable film that would not chemically react with a surrounding electrolyte that was to serve as a reservoir for selected ions. The intention was to direct these ions into at least portions of the amorphous film. The electrolyte was considered to be the equivalent of the synaptic gap, a chemically inactive electrode (Au or Pt) controlled the ion transport to and from the oxide film by the application of a voltage between it and the tantalum substrate (Fig. 1). Of course, neither the electrolyte nor the film were chosen for their resemblance to physiological material as such, but only as a means of studying possible mechanisms. The electrolyte was hydrochloric acid (20%) which was then saturated with zinc chloride to provide the metal ions. Upon the application of a small signal, typically 2 Volts (inert electrode positive), and

with a small amount of current flowing there would be a drastic change in resistance. The tantalum oxide whose resistivity was ordinarily 10^{14} ohm-cm was changed to a conductor. This change persisted as long as the signal remained applied. Without the metallic ions the control process was lost. Therefore, the interaction of the metallic ion with the amorphous film was responsible for the functioning of an active switching device.

Apart from being a source of ions, the electrolyte served merely as a conductor and can be considered simply as an extension of the inert electrode. This was proved by potential probing. Oxidation and reduction processes were present as well as double-layer effects, but were shown not to be primarily responsible for the amplification and switching action. The device could also be used as a field effect modulator since the flow of current could be controlled in response to variable changes in the control circuit.

The above described device, the "Ovitron," met the criteria of nerve cell and synaptic action in that an amorphous film displayed a change of resistance concomitant with an increase of permeability to ions in response to a stimulus. It also indicated how some memory effects could be achieved. They were obtained when the electrolyte was made more neutral and essentially became a metal electrode such as zinc. A transient pulse of one polarity turned the device on and a pulse of opposite polarity turned it off. Figure 2 shows the voltage-current characteristics on AC of such a device. [Simmons and Verderber (8) later reported on their observation of some similar effects.] We welcomed repeated reversibility of the switching effect as a demonstration of the fact that dielectric breakdown was not a contributing factor for in our earlier bipolar memory experiments with amorphous materials reversibility was difficult to control since memory action and dielectric breakdown seemed often associated.(9) Ovitron switching and modulation of electrical resistance by a metal-dielectric amorphous film-metal structure indicated that storage of information was polar in nature even though the conduction process was electronic. It was felt that the information was being stored by ionic distortion of local chemical bonds.

Pursuing our study of both unistable and bistable switches, we began in January of 1960 to investigate other types of amorphous materials. Semiconductor aspects were emphasized in order to

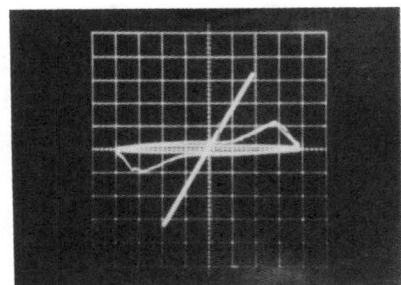

Horizontal 5V/div.
Vertical 50mA/div.
3 Stages of Memory Action
On—Transient—Off
(Superimposed)

Fig. 2 Memory action of amorphous oxide film. (Two devices back to back.)

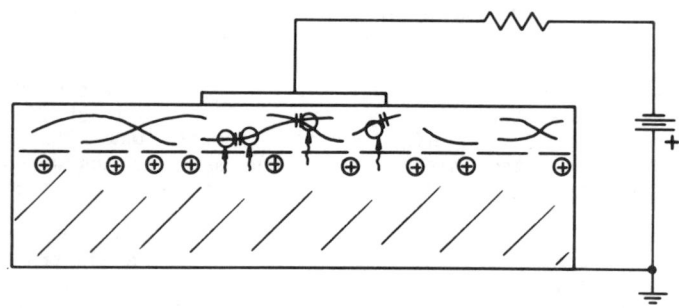

Fig. 3 Ion initiated semiconductor switch and modulator. Schematic view of structural changes caused by field induced ion migration.

assure electronic breakdown possibilities. It was also felt that memory action could be achieved by inducing phase changes, locking in the information in structure. This has since been proved to be true for a large range of materials. The most appropriate materials for our purposes turned out to be chalcogenides. Reversible changes of resistance which can be indefinitely maintained without external energy consumption have been amply demonstrated in devices which have come to be known as Ovonic Memory Switches.(10-16) Changes of resistance are associated with the disordered or high resistance state being transformed into a more ordered or crystalline conductive state. We originally chose tellurium and selenium because they were the most similar (by virtue of their chain structure) to the basic informational configuration which was just then changing the outlook of the biological world: the helix. This is illustrative of the fact that a synthesis of seemingly unrelated fields is often the basis for new scientific advances.

We have previously described (17-20) how electronic and structural changes are responsive to temperature, electric fields, and light in these materials. Here we will discuss a memory which utilizes some of our earlier concepts of ion interaction, with chalcogenide glasses being the materials altered. For instance, a long chain polymeric material such as selenium is deposited on a conducting substrate (Fig. 3). Another contact is deposited on top of the selenium film and completes the structure in its simplest form. One of the contacts is a source of alkali ions. When an electric field is applied ions move into the selenium chains, thereby causing crystallization as the final result, possibly through the intermediate process of chain shortening. Before crystallization, changes can occur which alter the resistivity of the material in a less drastic manner. To disorder the material again a pulse of opposite polarity is used. This also helps to restore the amorphous phase by transient heat dissipation.

Where structural changes are desired for memories one has various processes available, either singly or together: 1. Thermal--When amorphous materials which have not sufficiently cross-linked to prevent atomic diffusion are heated above a critical temperature (the glass transition temperature), crystallization takes place. 2. Chemical--Various elements which may be internally present in small

amounts or made available by diffusion from outside can quasi-catalytically cause the crystallization process to take place with less input of thermal energy. 3. Electric--It is known (10,21,22) that at least some materials crystallize in the presence of an electric field at much lower temperatures than ordinarily expected. Thus, tantalum oxide crystallizes at approximately 100°C in the presence of a strong electric field, as compared with its normal crystallization temperature of 650°C. Moreover, any other process that weakens chemical bonds favors crystallization effects, for example, light, shock waves, etc. Structural changes in response to all of the above can create images, opening up new fields of application.

Structural changes often have other clear-cut effects besides changes of conductivity which can be readily detected by external means ranging from differential thermal analysis to electron microscopy. When a material goes from the amorphous to the crystalline condition, there are changes of band gap, optical absorption and reflectivity,(19) volume, permeability to ions, surface charge and adhesive properties.(23) Some similar changes can indeed be seen in connection with nerve action.(24,25) To account for the electronic behavior of amorphous covalent alloys, an unconventional band theory has been proposed (26) which may also have implications for biological systems. Band gaps of the type associated with crystalline structures have no obvious applicability, and attempts to utilize conventional semiconductor approaches to these problems are not, therefore, likely to succeed. We have seen switching in layered structures indicating that anisotropic effects may be a link between crystalline and amorphous semiconductor theory.

Structure-dependent conducting mechanisms associated with disordered systems will teach us much about amorphous semiconductors, and we feel will be a bridge to the molecular biochemist in the search for unifying concepts of information control.

Acknowledgment

We admire the scientific insight of Professor Sir Nevill Mott who still is the youngest of all of us.

References

1. E. Schroedinger, What is Life?, p. 69. Macmillan, New York (1947).
2. F. Morin, G. LaMarche and S.R. Ovshinsky, Laval Medical 26, 3 (1958).
3. J.D. Conney, Control Engineering 5, 82 (1958).
4. J.D. Conney, Control Engineering 6, 121 (1959).
5. L. Young, Anodic Oxide Films, p. 147. Academic Press, New York (1961).
6. S.R. Ovshinsky, The Physical Base of Intelligence-Model Studies, presented at Detroit Physiological Society, Dec. 17, 1959.
7. A. Hoffer and H. Osmond, The Hallucinogens, p. 520. Academic Press, New York (1967).
8. J.G. Simmons and R.R. Verderber, Proc. Roy. Soc. A 301, 77 (1967).
9. S.R. Ovshinsky, unpublished data (1958).
10. S.R. Ovshinsky, Symmetrical Current Controlling Device, U.S. Patent No. 3,271,591.

11. Automation 10, 45 (1963).
12. M.P. Southworth, Control Engineering 11, 69 (1964).
13. S.R. Ovshinsky, Phys. Rev. Lett. 21, 1450 (1968).
14. S.R. Ovshinsky, J. Non-Cryst. Solids 2, 99 (1970).
15. H. Fritzsche, IBM J. Res. Develop. 13, 515 (1969).
16. H. Fritzsche and S.R. Ovshinsky, J. Non-Cryst. Solids 4, 464 (1970).
17. S.R. Ovshinsky, in Bull. Acad. Sci. USSR 9, 91 (1967).
18. H. Fritzsche and S.R. Ovshinsky, J. Non-Cryst. Solids 2, 148 (1970).
19. J. Feinleib and S.R. Ovshinsky, J. Non-Cryst. Solids 4, 564 (1970).
20. E.J. Evans, J.H. Helbers and S.R. Ovshinsky, J. Non-Cryst. Solids 2, 334 (1970).
21. D.A. Vermilyea, J. Electrochem. Soc. 102, 207 (1955).
22. C.H. Sie, J. Non-Cryst. Solids 4, 548 (1970).
23. H.K. Henisch, Scientific American 221, 30 (1969).
24. A.V. Hill and J.V. Howarth, Proc. Roy. Soc. B/149, 167 (1958).
25. I. Singer and I. Tasaki, in Biological Membranes (D. Chapman, ed.) p. 347. Academic Press, New York (1968).
26. M.H. Cohen, H. Fritzsche and S.R. Ovshinsky, Phys. Rev. Lett. 22, 1065 (1969).

RAPID REVERSIBLE LIGHT-INDUCED CRYSTALLIZATION OF AMORPHOUS SEMICONDUCTORS

J. Feinleib, J. deNeufville, S.C. Moss, and S.R. Ovshinsky

Energy Conversion Devices, Inc., Troy, Michigan 48084

(Received 11 November 1970)

We have observed a high-speed crystallization of amorphous semiconductor films and the reversal of this crystallization back to the amorphous state using short pulses of laser light and evidenced by a sharp change in optical transmission and reflection. This optical switching behavior is analogous to the memory-type electrical switching effect in these materials which has received wide attention[1] since the observation by S. R. Ovshinsky[2] of both threshold and memory switching in amorphous semiconductors. In this letter, we propose a model which closely relates the optical and electrical switching behavior, and shows that the phase change from amorphous to crystalline state is not only a thermal phenomenon but is directly influenced by the creation of excess electron-hole carriers by either the light, or, for the electrical device, by the electric field. The reversibility of the phenomenon in this model is obtained through the large difference in crystallization rates with the light on or off.

The high-speed light-induced transition and its reversibility has been observed in several amorphous chalcogenide semiconducting films, and we report here results for the system $Te_{81}Ge_{15}Sb_2S_2$, since this is a typical memory switching material for which extensive electrical,[3] optical,[4] and thermodynamic data[5] have been accumulated. The optical changes between the crystalline and amorphous phases have been measured on bulk material[4] and indicate that there is a substantial change in reflectivity caused by a shift in the absorption edge to lower energy accompanying the phase change from the amorphous to crystalline state. This change is similar to the changes observed for pure Te.[6] It is possible to observe such light-induced changes in reflection and transmission in sputtered thin films by optical microscopy. In our experimental arrangement, the 5145-Å line of an electrically pulsed argon-ion laser was focused by the 10 power objective of a microscope on to very thin films (~1000 Å) of material sputtered on several transparent substrates. Optical changes were observed for various pulse lengths of from 1 to 16 μsec and of corresponding peak intensities from 100 to 10 mW in a spot of about 3 μm in diameter.

In order to further explore the nature of the laser-induced optical changes in these glasses, selected laser pulsing of thin films mounted in a configuration convenient for transmission electron microscopy was attempted. This entailed sputtering a $Te_{81}Ge_{15}Sb_2S_2$ composition onto carbon-coated 200 mesh nickel Athene microscope grids of the standard 3.0-mm diameter. The sputtering was at a rate of ~300 Å/min onto the grids which were partially shadowed by a microscope slide to allow for a variation in final thickness of from 500 to 1000 Å.

After observing the effects of laser pulsing under an optical microscope, the grids were then placed, chalcogenide side up, on the cooling stage holder of a JEM-7 electron microscope and were examined in transmission. A liquid-nitrogen stage helped keep the sample cool, and care was taken to avoid crystallizing the films in the electron beam. The results can be illustrated by the sequence of photos in Figs. 1 and 2. Figure 1 shows a typical isolated laser dot in which the center was somewhat overheated and was rather severely disturbed or possibly evaporated. The major result is the crystallization of the laser-irradiated area. The crystallite size at the perimeter of the dot can be estimated to be (at most) a few hundred angstroms. One can also see a corner of the dot bitten off, i.e., reset to the amorphous phase, by the succeeding pulse. The obvious test for crystal-

Reprinted by permission from *Applied Physics Letters*, Vol. 18, No. 6, pp. 254–257 (15 March 1971).

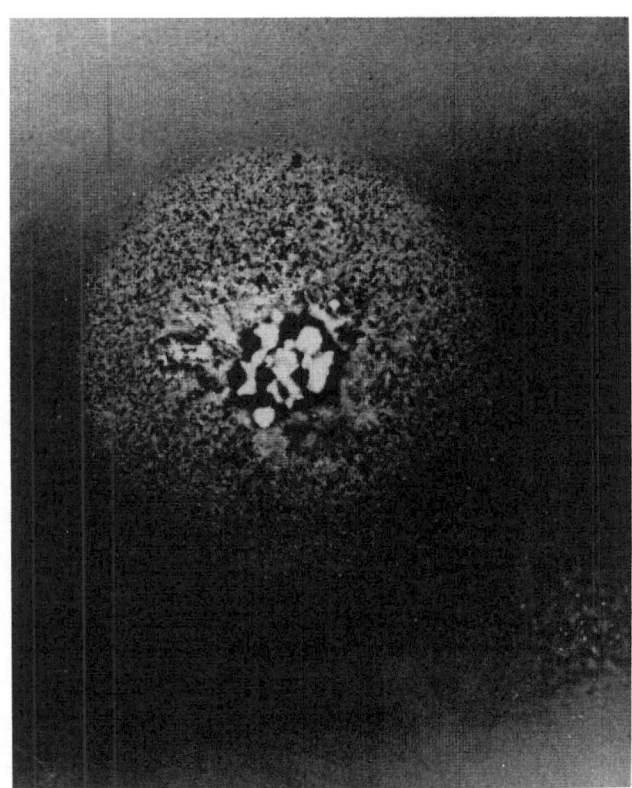

FIG. 1. Electron micrograph of laser pulse-induced crystallization of a 3-μm spot of an amorphous chalcogenide film.

lization is selected area diffraction and Figs. 2(a) and 2(b) show those results. In Fig. 2(a), the selected area aperture has been placed to include only the crystallized dot. The corresponding diffraction pattern, while spotty, is similar to Debye ring patterns which we have obtained on both electrically and thermally crystallized material. The selected area pattern in Fig. 2(b) is from an area away from the dot and is characteristic of the amorphous chalcogenide in any unirradiated region.

We have shown rather striking evidence of the ability of a pulse of a few microseconds to produce a crystallized dot in a chalcogenide film which can ordinarily be quenched to the amorphous state rather slowly (degrees per minute) without initiating any crystallization. The purely thermal behavior of this material has been observed by differential thermal analysis which shows that crystallization of the glass on heating occurs between the glass transition at 125 °C and the crystalline melting temperature of 380 °C. The amorphous state is obtained by a quench from the liquid at a minimum rate of about 50°/min. Bulk samples of similar materials have been annealed (by JdeN) in the supercooled liquid region between the glass transition temperature T_g and the melting point

T_m on a time scale of minutes without any indication of crystallization. Thus, in some of these materials, the time involved for purely thermal crystallization while passing through the crystallizing temperature range is of the order of minutes. In contrast, the material can be switched from the amorphous to crystalline state by an electrical pulse of the order of milliseconds,[3] and our experiments showed that light initiated crystallization in the order of microseconds or less.

Evans *et al.*[3] and Bienenstock *et al.*[7] have previously shown that light pulses of the order of milliseconds can produce crystallization of this memory material and that subsequent pulses take it back to the amorphous state, but the optical and thermal effects were not separable. Dresner and Stringfellow[8] and others[9] have measured light-enhanced crystallization rates in amorphous Se. Dresner and Stringfellow observed enhancement by a factor of 20 at 84 °C which is well below the op-

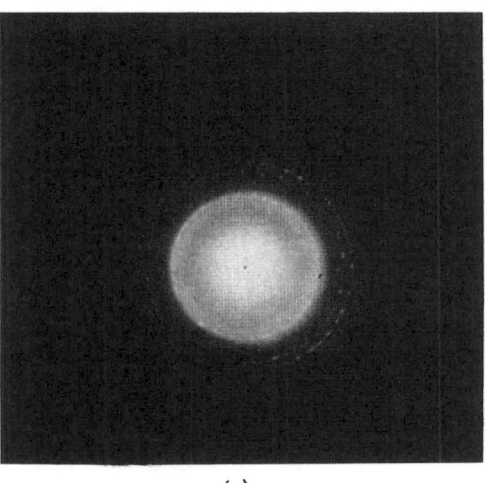

(a)

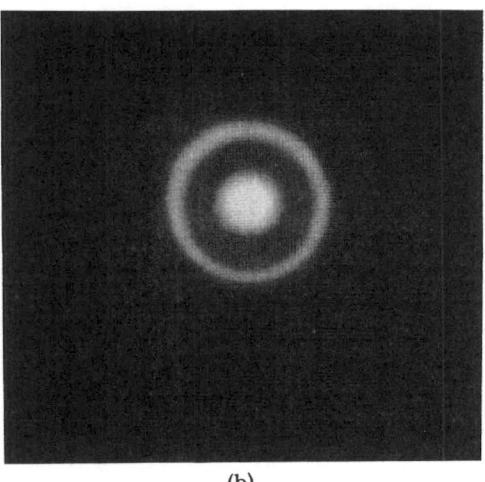

(b)

FIG. 2. Diffraction pattern of (a) crystallized region and (b) as-deposited amorphous region.

timum crystallization temperature so that growth times were still of the order of minutes. They showed by the wavelength dependence that the enhancement depends on the number of free carriers generated by the photons absorbed rather than on the total energy absorbed. The effect of the light on crystal nucleation rates has not, however, been studied so that the total effect of the light (including both nucleation and growth) on crystallization rates of amorphous materials may greatly exceed the enhancement that has in the past been measured.

The experimental observations strongly suggest that the light-induced (or current-induced) phase change is caused by a direct effect of the electron-hole pairs formed by the photon absorption at wavelengths shorter than the absorption edge or, for the electrical mechanism, by the electric field produced excess electron-hole carriers above the thermal equilibrium value. This would imply that in an optically, or electrically[10] (as opposed to purely thermally) reversible configuration, the crystallization is initiated while the light or current pulse is applied, rather than at some later time.

We illustrate this behavior with the model heating and cooling curves of Fig. 3. If a short light pulse is applied with a photon energy above the absorption edges of either state of the material, the material temperature will change according to the energy absorbed and at a rate determined by a characteristic thermal time constant. In curve C, the material is initially in the crystalline state. If the light pulse is to reverse the state to the amorphous

phase, then the material must be heated above the melting point T_m. After the material reaches this temperature, the light pulse is turned off and it cools at the rate determined by the characteristic time constant of the material in the configuration of the experiment. If this rate is rapid enough, the disordered material will pass through the crystallization temperature range between T_m and T_g so that no crystallization occurs within the time t_3. In this manner the crystalline phase starting from ambient temperature is reversed to the amorphous phase and retained there.

In the reverse process, the material is initially amorphous and the light pulse heats the material by absorption. The material reaches the crystallization temperature range above T_g while the light is still on. The rate of temperature rise changes because of the evolution of heat on crystallizing and by changes in the specific heat of the amorphous material above T_g. In order to achieve crystallization, the illuminated area of the material must not exceed the temperature T_m while irradiated by the light pulse and must undergo nucleation and crystal growth under the influence of light while being heated into the crystallization temperature range. Crystallization can take place for a total time t_1 which is only about twice as long as t_3. If only the time at temperature were important in producing crystallization, then we would expect to observe some crystallization under the condition of curve C, since t_3 and t_1 are not greatly different.

We must conclude from this simple argument that the crystallization process is appreciably enhanced while the material is passing through the crystallization temperature range in time t_2. The asymmetry in the crystallization rates between t_2 and t_3 must be very large to allow for copious crystallization after cycling through curve A while permitting no crystallization after cycling through curve C. This necessarily large asymmetry and the evidence that the time scale of microseconds is much too short to allow crystallization by purely thermal means suggests that the asymmetry is directly produced by the light. Our explanation is also in accord with the observation that haloes of crystallized material do not appear around a spot that has been crystallized and reset in the amorphous state by successive laser pulses. This is direct evidence that the time required for the heat to dissipate after the end of the erasing pulse is too short to allow thermal transformation of the surrounding amorphous material. The same argument may be applied to the electrically induced crystallization, although the fact that filament formation occurs in this configuration makes the t

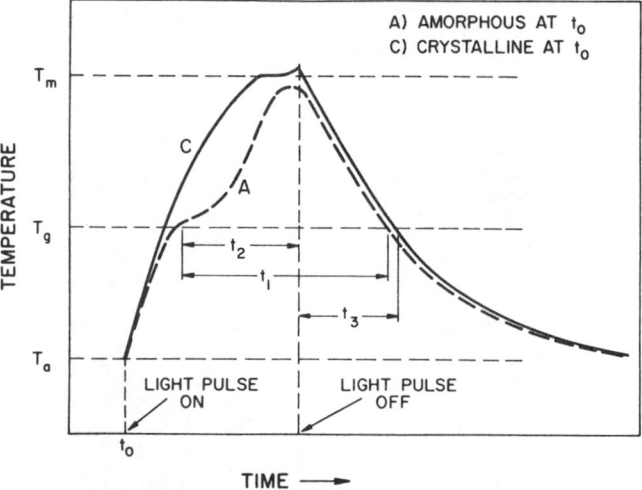

FIG. 3. Model heating and cooling curves of an amorphous material irradiated by a short, square light pulse. T_a is the ambient temperature, T_g is the glass transition temperature, and T_m is the conventional melting point for the crystallized phase.

thermal effects more difficult to analyze. The apparent similarity of the mechanisms suggest a common origin which is the creation of high density of excess electron-hole pairs either by the electric field or by photon absorption.

The suggestion that the presence of excess carriers can enhance crystallization has been made by Ovshinsky.[2,11] Dresner and Stringfellow,[8] Bienenstock,[12] and others have attributed the enhancement to the flow of excess holes. The amorphous state in these covalent materials exists either because of the crosslinking of the molecular chain structure that appears in the crystallized state or because of the polymeric tangling of these chains. The creation of an electron-hole pair implies the weakening of a bond, i.e., a valence bond electron is excited to an antibonding state (conduction band) so that in the case of the crosslinked structure it can more readily be broken and reconnected in an ordered state. The creation of these electron-hole pairs is more efficiently effected by the electric field or by photons above the band edge than by the low-energy phonons. The light thereby has a twofold effect in enhancing the crystallization (1) the recombination of light-produced carriers raises the temperature of the glass well above T_g where atomic mobility is high and (2) the carrier creation contributes an efficient bond-weakening mechanism. Our experiments show that the appropriate light intensity can satisfy these two requirements and the nucleation and crystal growth may take place in less than 1 μsec. With the proper thermal time constants, these crystallized states can be recycled into the amorphous state by the photon or electrical sources in about the same times. We do not inteend, however, to imply that the above mechanism is universal. Indeed, we have observed laser-induced optical changes in other configurations and materials which appear not to be related to devitrification but rather to other types of structural modification on which we shall report at some later time.

We wish to thank D. Adler, A. Bienenstock, M.H. Cohen, and H. Fritzsche for several interesting discussions.

[1]H.K. Henisch, Sci. Am. <u>221</u>, 30 (1969); see also Vols. 2 and 4, J. Non-Cryst. Solids (1970) for extensive review.

[2]S.R. Ovshinsky, Phys. Rev. Letters <u>21</u>, 1450 (1968); Proc. of 1968 Elec. Comp. Conf., Washington, D.C., p. 313 (unpublished).

[3]E.J. Evans, J.H. Helbers, and S.R. Ovshinsky, J. Non-Cryst. Solids <u>2</u>, 334 (1970).

[4]J. Feinleib and S.R. Ovshinsky, J. Non-Cryst. Solids <u>4</u>, 564 (1970).

[5]H. Fritzsche and S.R. Ovshinsky, J. Non-Cryst. Solids <u>2</u>, 148 (1970).

[6]J. Stuke, J. Non-Cryst. Solids <u>4</u>, 1 (1970).

[7]A. Bienenstock, F. Betts, and S.R. Ovshinsky, J. Non-Cryst. Solids <u>2</u>, 347 (1970).

[8]J. Dresner and G.B. Stringfellow, J. Phys. Chem. Solids <u>29</u>, 303 (1968).

[9]I.A. Paribok-Aleksandrovich, Soviet Phys. Solid State <u>11</u>, 1631 (1970).

[10]H. Fritzsche, IBM J. Res. Develop. <u>13</u>, 515 (1969).

[11]S.R. Ovshinsky, Patent No. 3,530,441.

[12]A. Bienenstock, Bull. Am. Phys. Soc. <u>15</u> (1970).

THREE DIMENSIONAL MODEL OF STRUCTURE AND ELECTRONIC PROPERTIES OF CHALCOGENIDE GLASSES

S.R. Ovshinsky and K. Sapru

Energy Conversion Devices, Inc., Troy, Michigan 48084

An important difference between crystalline and amorphous structures is that while in crystals the local environment is the same everywhere, in amorphous materials one finds a large spectrum of 3-dimensionally varying spatial and bonding relationships. This gives rise to unique energy interactions, orbital overlaps and bonding configurations of the lone pair electrons associated with the group VI chalcogen atoms which do not occur in crystals. We present the 'ball and spoke' models we have built to test out the concepts of lone pair relationships discussed earlier (Ovshinsky 1972, Ovshinsky 1973, Ovshinsky and Fritzsche 1973).

Amorphous chalcogenide materials can be divided into two categories, one in which structural change is prohibited (threshold material), and the other into which structural change is designed (memory material). In the former, a chalcogen such as tellurium can be chemically incorporated into an amorphous matrix through its covalent bonding properties and that of its alloying elements.

Silicon, germanium and arsenic act to cross-link the tellurium. Such a matrix has as its primary functions the prevention of atomic movement or crystallization through the steric hindrances and energy barriers of the strong matrix bonds. Another effect is that the nonbonded lone pairs are positioned spatially and energetically by their covalently bonded surroundings as well as the strongly repulsive lone pair-lone pair interactions which distort some of the nonbonded lone pair electrons in tellurium into unusual orbital configurations. These would affect the identity and distribution of localized states described in the CFO model (Cohen et al. 1969).

Figure 1 shows a three-dimensional model illustrating our threshold material, $Te_{40}As_{35}Si_{18}Ge_7$. Note the heavy cross-linking and lack of long tellurium chains. Figure 2 indicates schematically the positioning of the lone pairs by the local environment and the lone pair-lone pair interaction. Hybridization can be expected.

Kastner has stated that in the chalcogenide semiconductors the unshared p electrons of the group VI elements form the valence band, and the antibonding states the conduction band (Kastner 1972). We emphasize that in our materials the spatial and energetic relationships of some nonbonding p orbitals are distorted, shifting energy positions both lower and higher. A consequence is

Fig. 1   Photograph of the 3-dimensional model of the threshold material. Light balls represent Te atoms and dark sticks represent their lone-pairs. The dark balls are Ge, Si and As. The coordination can vary from site to site; however, in the model we have shown Ge and Si as 4-coordinated and As 3-coordinated. Some dangling bonds can be seen at the surface.

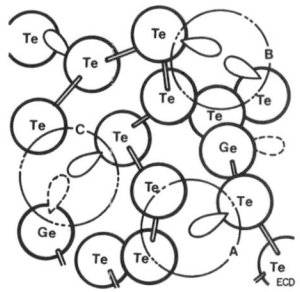

Fig. 2   Schematic representation of (A) lone pair interaction with the surrounding bonds, (B) lone pair-lone pair interaction, and (C) formation of the coordinate bond.

Reprinted by permission from the *Proceedings of the Fifth International Amorphous and Liquid Semiconductors Conference,* Garmisch-Partenkirchen, Germany (1974), pp. 447–452.

Electronic Compensation In The Threshold Switch

(A) Recombination (B) Injection of the
of the excited   compensating carrier
carrier utilizing   utilizing the hole.
the hole.

Fig. 3 Schematic representation of the threshold excitation and recombination emphasizing the hole. The upper half shows lone pair excitation with one electron removed, leaving a hole. Lower part shows that recombination can take place by (A) relaxation, (B) injection or addition of carriers.

that lower ionization potential can be utilized for removing them from the molecular systems. The introduction of selenium and sulphur systems or mixtures with tellurium is a designed parameter helpful in accenting certain types of transitions in response to differing types of energy input, e.g., the resulting larger band gap is used more often for optical interactive materials.

The lone pair carriers are paired even while they are unshared so that they are chemically compensated, not reactive; all bonds including lone pair orbitals are locally satisfied. There is in some regions charge imbalance and polarization from the lone pairs' interaction with nearest neighbors.

Structural integrity during reversible electronic breakdown becomes physically understandable since the electrons do not come from breaking bonds between atoms but from the ionizable nonbonded lone pairs which fill the space between the rigid bonds. If a motivating potential such as an electric field is placed across such a material, there will be at least electronic n to π^* transitions involving particularly the higher energy tail of the band of lone pair states, providing free carriers for current flow.

So far we are describing a bulk phenomenon in which the major electronic effect is the excitation from the lone pair configuration and recombination of the excited carriers as they return statistically to their equilibrium sites. These excitations could be the fast oscillations observed in our switches under low load conditions (Ovshinsky 1968a). However, there is an important concept that flows from n to π^* transition. If a lone pair is excited, the remaining lone pair electron is also affected in energy and the point of origin of the excited electron creates an unsatisfied orbital (a hole). Unlike the placement of donors and acceptors in crystalline material, this hole is predominant only transiently under excited conditions (see Fig. 3). In the more heavily cross-linked materials such as we describe for an Ovonic threshold switch, there are no permitted structural relaxations or atomic diffusions which can utilize the hole. This vacancy is an electronically deficient active site which creates a charge imbalance that can be compensated electronically.

While the mechanical model does not teach us about transport properties, it suggests that there can be coexisting electronic mechanisms not permitted in crystalline materials. For when lone pairs are viewed in a varying three-dimensional amorphous matrix, there is a spread of separations between them. Some are within tunnelling distance of each other, others are far enough apart for avalanche and impact ionization to take place. The full gamut of field emission processes is possible (Ovshinsky 1968a), including injection which could be one method of electronically compensating the hole as shown in Fig. 3. When injection is reduced, bulk recombination completes the return to the high resistance state.

The Ovonic memory materials, e.g., $Te_{81}Ge_{15}Sb_2S_2$, depend upon similar principles. Because of their structural differences due to fewer cross-links, longer chain structures (more tellurium and therefore more lone pairs) and weaker and varied bonds (including Van der Waals) (see Fig. 4), structural rearrangements can occur as a result of excitation. The n to π^* transition is a structurally reactive one in such a material. Its high energy content and the hole's attraction to nearby electrons are of great structural importance. The free radical nature of the excited electron itself causes structural accommodations through molecular vibrational forces of the more flexible chain segments and weaker bond arrangements of the matrix structure around the excited configuration. The unsatisfied orbital can be utilized by whatever bond scissions and interactions have been generated by the excitation process including any concomitant thermal excitation which may occur (Ovshinsky 1968b). Since any increase of thermal energy such as Joule's heat raises the overall energy of the environment, additional nearby bonds are broken. Increased temperature also changes viscosity, increases vibrational, translational and rotational movement and permits diffusion. The reactive hole can be utilized by these processes to form new localized structures,

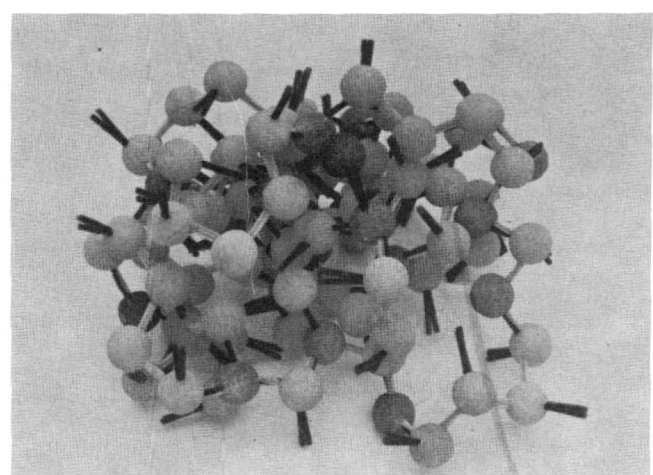

Fig. 4 Photograph of the 3-dimensional model of the memory material. The lower part is shown in a relaxed position in order to clearly see how with subsequent chain folding and cross-linking lone pair interaction takes place. Light balls represent Te atoms and dark sticks represent their lone pairs. Dark balls are Ge atoms. The 4 darkest balls are Sb and S.

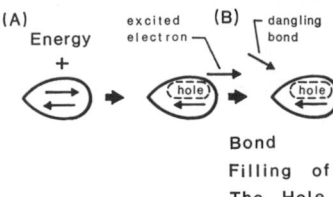

Chemical Compensation In The Memory Switch

Bond
Filling of
The Hole

Fig. 5 Schematic representation of the memory switching action leading to structural change. One electron excited leaving behind a hole. Compensation can occur either electronically or chemically by movement of surrounding atoms which fill the hole by chemical bonding.

i.e., nucleating points. Additional crystal growth can be thermally controlled around the site.

Figure 2(C) also shows the possibility of coordinate bonds being formed by the lone pairs in regions where germanium can be 2-fold coordinated. The germanium can then act as an acceptor for the lone pair donor electrons. The coordinate bond is another weak cross-link in the spectrum of bonds available in a memory material (Ovshinsky and Fritzsche 1973). Clustering of active tellurium atoms could be accelerated through the formation of 1- and 3-electron bonds, if present even transiently.

The placement and interactions of the lone pair configuration differ in space and energy as a function of material design. Excitonic type states should also result from some of these considerations. We suggest that they are possibly involved with the mechanism resulting in the critical electric field effect as well as the optical interactions observed. (Adler and Mathur 1973, Buckley 1973, deNeufville et al. 1973, Ovshinsky 1968a, Ovshinsky 1973, Shaw et al. 1973, Smith and Henisch 1973.)

Lone pairs play many roles in amorphous materials. They are a link to our work with amorphous organochalcogenides (Ovshinsky 1973; Chang and Ovshinsky 1973; Ovshinsky et al. 1973). They act as structural 'buffers' in their interactions with surrounding bonds and, through their nonbonding and bonding characteristics, play unique parts in both excitation and structural change.

ACKNOWLEDGMENTS

We owe special thanks to Iris M. Ovshinsky, L. Pellier and Hellmut Fritzsche for helpful discussions.

REFERENCES

D. Adler and M. Mathur, 1973, private communications.

D. Buckley, 1973, private communication.

Y. Chang and S.R. Ovshinsky, 1973, unpublished.

M.H. Cohen, H. Fritzsche and S.R. Ovshinsky, 1969, Phys. Rev. Lett. 22, 1065.

J. deNeufville, S.C. Moss and S.R. Ovshinsky, 1973, Proceedings of the Fifth International Conference on Amorphous and Liquid Semiconductors, Garmisch-Partenkirchen.

M. Kastner, 1972, Phys. Rev. Lett. 28, 355.

S.R. Ovshinsky, 1968a, Phys. Rev. Lett. 21, 1450; 1968b, Proceedings of the Electronic Components Conference, Washington, D.C., p. 313; 1972, unpublished manuscript; 1973, invited presentation at the Topical Meeting on Optical Storage of Digital Data, Aspen, Colorado.

S.R. Ovshinsky, Y. Chang and A. Eisenberg, 1973, to be published.

S.R. Ovshinsky and H. Fritzsche, 1973, I.E.E.E. Trans. Electron. Devices 20, 91.

M.P. Shaw, S.H. Holmberg and S.A. Kostylev, 1973, Phys. Rev. Lett., to be published.

W. Smith and H.K. Henisch, 1973, Phys. Stat. Solidi A K81.

For a general discussion of the role of lone pairs in chemistry, the following references may be useful:

C.A. Coulson, 1961, Valence (Oxford University Press) p.186.

H. Krebs and P. Fischer, 1970, Discussion of the Faraday Society, No. 50: The Vitreous State (London: The Faraday Society) p.35.

Sir John Lennard-Jones and J.A. Pople, 1950, Proceedings of the Royal Society of London, 202, 166.

Linus Pauling, 1960, The Nature of the Chemical Bond (Cornell University Press).

J.A. Pople, 1950, Proceedings of the Royal Society of London, 202, 323.

R.T. Sanderson, 1967, Inorganic Chemistry (New York: Reinhold Publishing Corporation).

LOCALIZED STATES IN THE GAP OF AMORPHOUS SEMICONDUCTORS

Stanford R. Ovshinsky

Energy Conversion Devices, Inc., Troy, Michigan 48084

(Received 30 July 1975; revised manuscript received 5 May 1976)

Now that the initiation and maintenance of bias-induced switching in threshold-type,[1] chalogenide-based, amorphous semiconductor materials have been definitively established as electronic processes,[2-8], there is one major remaining problem—the nature of the localized states in the gap of these materials. In this Comment I address this subject, with emphasis on their role in the switching transition.

The discovery of optically induced, localized, paramagnetic states in chalcogenide glasses by Bishop, Strom, and Taylor[9,10] can be explained by utilizing a recently proposed model for the localized states in these glasses.[11] The model, which extends that of Mott[12] and Cohen, Fritzsche, and Ovshinsky[13] by identifying the nature and origin of localized states, is briefly the following. In lone-pair, amorphous, chalcogenide semiconductors,[14] of which the chalcogenide glasses are the chief example,[15] a spectrum of localized states tailing into the gap from the valence and conduction bands is introduced by the interaction of lone pairs with each other and with their local environment. Because the lone-pair orbitals are filled at equilibrium, these tail states are paired and thus diamagnetic. Excitation creates empty orbitals (holes) which are compensated by electronic or structural means.[11,16]

In noncompensated, tetrahedral, amorphous materials such as a germanium and silicon, an equilibrium density of dangling bonds should exist, and these should be reflected in paramagnetically detectable, singly occupied states. The experiments of Agarwal[17] confirm their existence in amorphous germanium.

However, while no paramagnetism is observed in the chalcogenide samples whose electron distribution is at equilibrium, Bishop, Strom, and Taylor[9,10] found ESR signals at low temperatures after excitation with sub-band-gap illumination. They interpret the spins as associated with holes localized on chalcogen atoms and electrons localized on arsenic atoms. The nonequilibrium state brought about by illumination produces, in addition, a broad optical absorption band below gap energies. This nonequilibrium state is stable for several hours at 6°K.

I believe that the explanation for this finding is related to the interaction of the lone-pair orbitals with the structural bonding configuration (matrix or "lattice"), so that when a lone-pair excitation takes place, a metastable, localized hole is created through the breaking up of the lone pair on the chalcogen atom. In order to understand the origin of the metastable hole, the question of the possibility of dangling bonds in these materials should be considered.

Street and Mott[13] recently proposed the existence of defect states arising from dangling bonds on the chalcogen atoms. As I have discussed,[19] such states are energetically unfavorable. Crosslinked chalcogenides are "ideal" glasses since they have the structural flexibility to intrinsically compensate dangling bonds. Such compensation is uniquely associated with the nature of the lone-pair atom in most amorphous chalcogenide materials since it has a primary divalency based on any two of the p orbitals, and the remaining p orbitals have a spectrum of energetic interactions which are spin compensated as will be detailed later in this paper. Choices of these interactions are allowed by the amorphous state since they are dependent upon the *varying local environment* which provides three-dimensional orbital relationships found in no crystalline or nonchalcogenide amorphous material.

Reprinted by permission from *Physical Review Letters*, Vol. 36, No. 24, pp. 1469–1472 (14 June 1976).

The primary divalency provides for a maximum of these orbital interactions since the bends and twists of the resulting helical configurations place the remaining lone pairs in various bonding and nonbonding electronic interactions. While stabilizing the material, these provide for structural flexibility and the placement of a range of localized states in the gap. Such a polymeric structure leads to a strong interaction between the lone-pair localized states and their matrix environment. A change of charge and occupancy of the localized states acts upon the matrix just as the matrix helps position the localized states originally. This is unlike tetrahedral materials in which there is a great deal more structural rigidity since all four bonding positions must find mates or have some remain as dangling bonds or voids. It is this ability to have conformational changes which sets up the conditions for metastability since the presence or movement of lone-pair configurations has a profound effect in determining molecular shape.[20]

I think that at room temperature thermal fluctuations reduce the effect of stabilizing bonding distortions around the localized hole, and rapid recombination occurs. Thus, paramagnetism cannot be easily detected except under strong nonequilibrium conditions. However, at the temperature used by Bishop *et al.*, thermal bonding fluctuations are small and, therefore, local charge distributions associated with lone-pair interactions can be metastable. For example, excitation in the form of n to π^* or to σ^* transitions in such an environment can lead to a metastable, localized hole since the new electronic configuration interacts with the nearest-neighbor environment so that the nonequilibrium holes, electrons, or both are stabilized and prevented from recombination by atomic distortions created around the carriers. Structural relaxation processes ultimately re-establish the equilibrium electronic configurations, although these can take hours or more at very low temperature. Either thermal or optical excitation (infrared) in the new absorption band restores equilibrium by exciting either electrons from the valence band into the metastable hole states or metastable electrons into the conduction band. This explains the photobleaching effect observed by Bishop *et al.* with low-energy photons.

The transient appearance of localized states associated with a disequilibration of the carrier distribution should also occur at *room temperature* under proper excitation conditions. This has been the leitmotif of my work on switching in chalcogenide glasses.[21]

I now detail the model which is based upon an interaction between two different kinds of electronic configurations specific to amorphous chalcogenide materials.[11,16] The first is the arrangement of bonding electrons (ordinarily lying much lower in energy than the nonbonded lone pairs) which are responsible for the overall structural integrity of the material even when a very large density of lone-pair electrons is excited. This matrix is associated with the thermal fluctuations described. The second is the distribution of lone-pair electrons spread over a large energy range, but not primarily responsible for the cohesive energy of the material. It is among these nonbonded, lone-pair configurations that low-energy excitation processes play a role. In some cases to be described, some lone pairs have secondary structural effects. The various configurations, due to the variety of local environments afforded by the disordered state, have interesting and unique properties such as repulsive interactions of lone pairs with nearby filled orbitals, including other lone pairs, which spread the density of valence-band states into the gap; attractive interactions including the donation of the complete lone pair to an acceptor configuration—the dative bond and the additional coordination of the lone pair to nearest neighbors forming one- and three-electron bonds. These nearest neighbors can include other lone-pair atoms as well as the alloying elements and are attractive due to the charge configurations present, e.g., of either a dangling bond of the chalcogen or of the crosslinking atom. As a variation of these electronic configurations, I suggest that in some areas lone pairs can lose their electrons through their nearest-neighbor interactions and still be compensated through the replenishment of electrons coming from other nearby clusters of lone pairs so that there is a continuous mixing in which the parentage of the electrons may be lost. The p orbitals, then, in both primary and secondary valency, pick off (compensate) any available, unsatisfied bond and the nonbonded lone pairs themselves, by their charge interactions, "buffer" the structural bonds. That is, the molecular bond shapes reflect their presence.

Therefore, in a lone-pair amorphous material, there are varying types of localized states, all dependent upon nearest-neighbor relationships which in some cases have an attractive effect, in others repulsive. Such a spectrum of states ranges from strong bonds to weak bonds, to varying charge configurations including inert lone-pair positions.[22,23] Even in the inert pair state, the lone pairs have steric effects since they help set up bond angles by their Coulombic interaction with nearest neighbors. In any case, whether strongly or weakly localized, the lone-pair configurations described are spin compensated.

The understanding of the nature of the localized states provides a basis to explain in principle the nondestructive, reversible breakdown, and the high resistance, as well as the highly conductive state, in amorphous switches. The density of these localized states is so great ($10^{17} - 10^{19}$ eV^{-1} cm^{-3}) that their field-induced excitation can lead to a sufficient concentration of free electrons to provide a metalliclike conducting state in what previously was a high-resistance, nonconducting material.[1] It is clear that, while switching is a bulk phenomenon, the redistribution of charge resulting from the availability and stability of a large density of localized holes and electrons, together with the excitation of a large concentration of free electrons, introduces the possibility of high-level injection processes.[11]

In memory-type materials, which are designed for bistability by reducing the composition of crosslinking group-IV and -V atoms, the excited lone-pair carriers interact with the structural bonds and the stabilizing distortions caused by excitation around the localized holes become nucleating centers for structural transformations.[11,16,24] Although the excitation processes in such materials are similar to those in threshold (unstable) switches in which structural change is prohibited by material design due to strong and heavy covalent crosslinking, structural rearrangements can occur because the memory class of materials contains much less crosslinking, weaker bonds, and more lone-pair interactions. They are, thus, much less structurally stable in the amorphous state, especially to electronic excitations.[25-27]

Unlike an impurity in the lattice of a crystalline material or dangling bonds in some types of amorphous materials, the orbital configurations described in this model are only transiently exposed under nonequilibrium conditions. However, as discussed earlier, they are observable under conditions where thermal fluctuations are minimized and the distorted orbitals created by nearest-neighbor and lone-pair interactions are frozen in position sufficiently long enough to be detected. This explains the metastable conditions that can be observed. I suggest this is just what has been accomplished by the work of Bishop, Strom, and Taylor.[9,10]

[1]S. R. Ovshinsky, Phys. Rev. Lett. 21, 1450 (1968).

[2]H. K. Henisch and W. R. Smith, Phys. Status Solidi (a) 17, K81 (1973).

[3]M. P. Shaw, S. H. Holmberg, and S. A. Kostylev, Phys. Rev. Lett. 31, 542 (1973).

[4]S. H. Holmberg and M. P. Shaw, in *Proceedings of the Fifth International Conference on Amorphous and Liquid Semiconductors, Garmisch-Partenkirchen, West Germany, 1974*, edited by J. Stuke and W. Brenig, (Taylor and Francis, London, 1974), p. 687.

[5]W. D. Buckley and S. H. Holmberg, Solid State Electron. 18, 127 (1975), and Phys. Rev. Lett. 32, 1429 (1974).

[6]K. E. Petersen, D. Adler, and M. P. Shaw, Appl. Phys. Lett. 25, 1585 (1974).

[7]D. M. Kroll, Phys. Rev. B 11, 3814 (1975).

[8]K. E. Petersen and D. Adler, J. Appl. Phys. 47, 256 (1976).

[9]S. G. Bishop, U. Strom, and P. C. Taylor, Phys. Rev. Lett. 34, 1346 (1975).

[10]S. G. Bishop, U. Strom, and P. C. Taylor, Phys. Rev. Lett. 36, 543 (1976).

[11]S. R. Ovshinsky and K. Sapru, in *Proceedings of the Fifth International Conference on Amorphous and Liquid Semiconductors, Garmisch-Partenkirchen, West Germany, 1974*, edited by J. Stuke and W. Brenig (Taylor and Francis, London, 1974), p. 447.

[12]N. F. Mott, Adv. Phys. 16, 49 (1967).

[13]M. H. Cohen, H. Fritzsche, and S. R. Ovshinsky, Phys. Rev. Lett. 22, 1065 (1969).

[14]M. Kastner, Phys. Rev. Lett. 28, 355 (1972).

[15]There are amorphous semiconductors other than chalcogenide glasses which contain, for example, large concentrations of arsenic, which have s-electron lone pairs and which, while lower in energy than the chalcogenide p-electrons, can be affected by the disordered state and contribute to the switching process (see Ref. 11).

[16]S. R. Ovshinsky, in *Proceedings of the Fourth International Congress on Reprography and Information, Hanover, Germany, 1975*, (Helwich Verlag, Darmstadt, Germany, 1975), p. 109.

[17]S. C. Agarwal, Phys. Rev. B 7, 685 (1973).

[18]R. A. Street and N. F. Mott, Phys. Rev. Lett. 35, 1293 (1975).

[19]S. R. Ovshinsky, in Proceedings of the Sixth International Conference on Amorphous and Liquid Semiconductors, Leningrad, 1975 (to be published).

[20]W. J. Orville-Thomas, *Structure of Small Molecules* (Elsevier, Amsterdam, 1966), p. 50.

[21]For complete references see S. R. Ovshinsky and H. Fritzsche, IEEE Trans. Electron Devices 20, 91 (1973).

[22]J. E. Fergusson, *Stereochemistry and Bonding in Inorganic Chemistry* (Prentice-Hall, Englewood Cliffs, New Jersey, 1974), p. 166.

[23]K. Mislow, *Introduction to Stereochemistry* (Benjamin, Menlo Park, California, 1965), p. 38.

[24]S. R. Ovshinsky, in Topical Meeting on Optical Storage of Digital Data, Aspen, Colorado, 1973 (unpublished).

[25]S. R. Ovshinsky, in *Electronic Components Conference Proceedings, Washington, D. C., 1968* (Institute of Electrical and Electronics Engineers, New York, 1968), p. 313.

[26]S. R. Ovshinsky, J. Non-Cryst. Solids $\underline{2}$, 99 (1970).

[27]S. R. Ovshinsky and I. M. Ovshinsky, Mater. Res. Bull. $\underline{5}$, 681 (1970).

LONE-PAIR RELATIONSHIPS AND THE ORIGIN OF EXCITED STATES IN AMORPHOUS CHALCOGENIDES

S.R. Ovshinsky

Energy Conversion Devices, Inc., Troy, Michigan 48084

What are the chemical and structural bases of localized states in amorphous materials? In the case of tetrahedral amorphous materials, chemistry and structure can be closely related and the localized states can be clearly distinguished, for example, materials which have not reached their lowest free energy can have dangling bonds, structurally dependent voids and other kinds of vacancy defects. We classify these as extrinsic localized states which can be annealed out. When the materials are at their lowest free energy, as represented by the ideal model of Polk,(1) the localized states which remain and are the result of chemical bonding and positional disorder we consider intrinsic. In such materials, substitutional impurities can insert new extrinsic localized states and the possibility of ordinary doping exists, as shown by the work. of Spear and LeComber.(2) While this is an important subject, we believe that it will follow the concepts of classical crystalline tetrahedral materials.

We will limit our discussion to describing the unique properties of the amorphous state of lone-pair materials which provide intrinsic localized states that are available to electrical and optical excitation. We report on our ability to obtain extrinsic states in these materials. Understanding the origin of localized states leads to new insights into the excitations available.

Amorphous lone-pair semiconductors (3) are ideal model systems to understand the intrinsic nature of localized states.(4) We view the lone-pair materials as composed of two different systems interacting with each other.(5,6) One is the atomic network composed of covalently bonded atoms which makes up the matrix (lattice) which provides the structural configuration of the materials. The other is the lone-pair orbitals distributed in 3-dimensional space by their interaction with the matrix and with each other. These electronic configurations are compressed into whatever space is available to them by their nearest neighbor local environment. Since there are varying environments, there are varying types of electronic configurations that may give rise to localized states. This is unlike a crystalline lone-pair material in which lone-pair orbitals are periodically spaced, and obviously differs drastically from amorphous or crystalline tetrahedral materials. In this picture, there is a minimum of dangling bonds and an important new class of localized states may be introduced by the compositional and positional disorder which affect the position of the lone pairs in space and energy. The reason that there are no dangling bonds in a heavily crosslinked polymeric structure is also uniquely associated with the divalency of the lone-pair atom which acts to tie up any available bonding sites and positions the remaining p-orbitals in a spectrum of lone-pair configuration ranging from weak lone-pair bonding to those that are "inert." A primarily divalent polymeric structure with lone pairs leads to strong interaction between the localized states and their matrix environment. A change of charge and occupancy of the localized states acts upon the matrix just as the matrix helps position the localized states originally. This makes physically understandable the Stokes shift in these materials.

We have pointed out in several papers,(5,6,7) the importance of the hole in such lone-pair amorphous configurations. Briefly, the reasoning is that there is a spectrum of localized states tailing into the gap from the valence band which is introduced by the interaction of lone pairs with each other and with their environment. Most of these are in paired configuration and thus diamagnetic; however, there are some singly occupied paramagnetic centers in which a localized hole exists on the chalcogen atom. Excitation of the material by either an electric field or light leads to a greatly augmented density of such holes.(5,6) We interpreted the experiments of Bishop et al. (8) in this light and note that their latest paper (9) indeed agrees with our prediction of the identity of the hole. We point out that our previous explanations,(5,6,10) which are emphasized here, explain the unique nature of the amorphous state which produces the optically induced, localized, paramagnetic centers.

Such differing lone-pair configurations in varying 3-dimensional orbitals within an interactive matrix do not exist in crystalline chalcogenide materials, they do not exist in crystalline tetrahedral materials and they do not exist in amorphous tetrahedral materials. It is, therefore, clear that the picture of a rigid undisturbed

Reprinted by permission from the *Proceedings of the International Topical Conference on Structure and Excitation of Amorphous Solids*, Williamsburg, Virginia, 24–27 March 1976, pp. 31–36.

distribution of impurity and defect states with which we are familiar in crystalline and amorphous tetrahedral semiconductors is not applicable.

Identifying the various kinds of inherent localized states in lone-pair amorphous materials requires not only an understanding by inference of states connected with the covalent bonding of the amorphous matrix and those connected with the lone pairs, but a means of probing their position in space and energy. In such amorphous materials, by going from elemental to compound to alloy, differing local environments are provided for the lone pairs and they, therefore, are differently positioned in space and energy in the matrix. Affecting these localized states in a specific manner so as to alter their electronic environment would be most valuable. We have accomplished this and created an extrinsic conductivity level in these materials. By the addition of a small percentage of lithium ions to our lone-pair semiconductors, we are able to affect the charge distribution of some lone-pair state energy levels at low temperature. The resultant conductivity varies with temperature (Fig. 1) in a manner very similar to doped silicon, exhibiting a distinct intrinsic and a distinct extrinsic region.(11) The addition of other elements can have a similar effect on the lone-pair state interactions since such extrinsic behavior has also been observed in the thallium-tellurium system.(12)

The localization of states in amorphous materials which contain lone pairs has interesting implications, for lone pairs obviously exist in oxides as well as sulfides, selenides and tellurides. It would, therefore, be useful to find means of characterizing lone pair materials by parameters which have universal significance. We, therefore, characterize the tetrahedral and lone-pair materials by three parameters. One describes the compositional or positional disorder P. This increases from elemental to compound to alloy glasses. The second is the connectedness C (13,14) which describes the average covalent coordination of the material. It increases from C=2 to C=4 as one moves from the Group VI elements through increasingly stronger, crosslinked alloy structures to the tetrahedral materials. A high C results in a rigid matrix which is unlikely to deform or change its bonding appreciably around a defect. The third is the degree of localization L which increases as one moves from the tellurides to the oxides. With increasing L the effect of randomness of the potential on the band states, and thus the number and relative spread of the tail

states, decreases. Large L materials have small polarizabilities, small dielectric constants and the interband optical constants are hardly affected by disorder. This classification scheme appears helpful to explain why the nature of extrinsic and intrinsic localized states changes as one moves from tellurides to oxide glasses or from chalcogenide to tetrahedral materials.

Chalcogenide glasses having low C offer, in addition, the structural flexibility for energy saving deformations and new bonding arrangements around the localized states. These become particularly active when a lone-pair state is excited, loses one electron and acts like a free radical.(5,6) Large localization L seems to favor these distortions. The electron excited into a localized state causes similar distortions and bond angle changes. When the excited electron remains in the neighborhood of the remaining hole, luminescence is thus expected with an appreciable Stokes shift.

In this picture, it is plausible that one finds the materials which show irreversible and reversible photostructural changes (15,16) among those which have large L and small C. The large L provides a large recombination energy and localizes this energy in a volume of atomic dimension.(16) The small C provides the flexibility and mobility for bond switching and rearrangement.

Our thesis is that in the field of amorphous materials one can identify and distinguish the origin and parentage of localized states which are intrinsic to the material from those which are extrinsic. We believe that in lone-pair amorphous semiconductors there are various types of intrinsic states some of which are the tail states originating from the randomness of the potential or Emin's polaron states.(17) In alloy glasses further intrinsic states are created by the compositional disorder which in particular pushes locally some high lying, lone-pair states into the gap which are affected by the proximity of electro-positive molecular configurations. The availability of p-orbitals for bonding, lone-pair bonding, and other lone-pair interactions minimizes dangling bonds in crosslinked amorphous materials differing from the models of Mott et al.(18) We have shown that extrinsic states can be introduced in amorphous lone-pair materials. We have explained the unique nature of some of the localized states in the gap and have emphasized the importance of holes in such lone-pair configurations, especially in the excited state.(5) We have illustrated the connection between compositional disorder, connectedness and localization. The antibonding orbitals of these materials by their complexity introduce a new area of investigation which cannot be adequately treated here.(19)

I wish to thank Hellmut Fritzsche for his helpful comments.

REFERENCES

1. D.E. Polk, J. Non-Cryst. Solids 5, 365 (1971).

2. W.E. Spear and P.G. LeComber, Solid State Comm. 17, 1193 (1975).

3. M. Kastner, Phys. Rev. Lett. 28, 355 (1972).

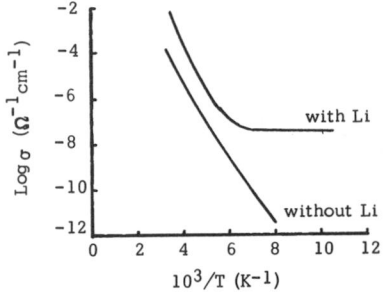

Fig. 1 Doping of lone-pair amorphous semiconductors.

4. M.H. Cohen, H. Fritzsche and S.R. Ovshinsky, Phys. Rev. Lett. <u>22</u>, 1065 (1969).

5. S.R. Ovshinsky and K. Sapru, Proc. of 5th Intl. Conf. on Amorphous & Liquid Semiconductors, edited by J. Stuke and W. Brenig, Taylor and Francis, p. 447, 1974.

6. S.R. Ovshinsky, Proc. of 6th Intl. Conf. on Amorphous & Liquid Semiconductors, November 18-24, 1975, Leningrad, USSR.

7. S.R. Ovshinsky, 4th Intl. Congress for Reprography and Information 1975, Hannover, Germany, April 1975.

8. S.G. Bishop, U. Strom and P.C. Taylor, Phys. Rev. Lett. <u>34</u>, 1346 (1975).

9. S.G. Bishop, U. Strom and P.C. Taylor, Phys. Rev. Lett. <u>36</u>, 543 (1976).

10. S.R. Ovshinsky, Phys. Rev. Lett., submitted July 1975.

11. S.R. Ovshinsky, R. Flasck and H. Fritzsche, to be published.

12. H. Fritzsche and M.A. Paesler, (at this conference).

13. M. Kastner, Phys. Rev. B <u>7</u>, 5237 (1973).

14. J.P. deNeufville and H.K. Rockstad, Proc. of 5th Intl. Conf. on Amorphous & Liquid Semiconductors, edited by J. Stuke and W. Brenig, Taylor and Francis, p. 419, 1974.

15. J.P. deNeufville, R. Seguin, S.C. Moss and S.R. Ovshinsky, Proc. of 5th Intl. Conf. on Amorphous & Liquid Semiconductors, edited by J. Stuke and W. Brenig, Taylor and Francis, p. 737, 1974.

16. H. Fritzsche, J. Jap. Soc. Appl. Phys. <u>43</u> (suppl.) 32 (1974).

17. D. Emin, Electronic and Structural Properties of Amorphous Semiconductors, edited by P.G. LeComber and J. Mort, Academic Press, p. 261, 1973.

18. N.F. Mott, E.A. Davis and R.A. Street, Phil. Mag. <u>31</u>, 961 (1975).

19. S.R. Ovshinsky, to be published.

DISCUSSION AND COMMENTS

J. Stuke: What material was it for which you found the influence of Li on the temperature dependence of the conductivity?

S.R. Ovshinsky: The chalcogenide was evaporated $Ge_{50} Te_{50}$.

CHEMICAL MODIFICATION OF AMORPHOUS CHALCOGENIDES

Stanford R. Ovshinsky

Energy Conversion Devices, Inc., Troy, Michigan 48084

It is well known in crystalline tetrahedral semiconductors that conventional incorporation of atoms by substitutional doping increases the electrical conductivity by many orders of magnitude. Such doping is the basis of the crystalline semiconducting industry and has recently been extended to amorphous silicon and germanium, with hydrogen appearing to play an important compensating role.(1-3)

This paper describes how a new approach can be used to achieve desired modification of the localized gap states and, thereby, control the electrical properties of a large variety of amorphous semiconductors. Because these concepts differ greatly from substitutional doping, we use the term modification.

We have achieved extrinsic conduction by ovonically modifying amorphous semiconductors consisting of one or more elements of Groups III, IV V or VI. The materials can be put together in innumerable combinations, the optical band gap can be tailor-made to design specifications and the electrical activation energy can be independently altered so that conductivity changes over many orders of magnitude can be controlled. Not only chalcogenides but materials based upon elements such as boron, carbon, silicon, etc. have been modified. Materials which are stable over 500°C have been made.

Our premise is that amorphicity is most interesting when it does not mimic crystalline analogs. Once destroying the constraints of crystalline symmetry, we open up the possibilities of nonstoichiometric mixtures, unique orbital configurations and variant bondings. We can synthesize new materials which emphasize desired physical, chemical and electrical properties.

Normal structural bonding (NSB) characterizes the primary bonding configuration of amorphous materials, e.g., the bonding of amorphous silicon is primarily tetrahedral, while that of lone-pair materials is mostly divalent. However, due to their intrinsic disorder, amorphous materials have a spectrum of bonding, nonbonding and antibonding states.(4-9) Rather than trying to overcome this by making a material more ordered, as in the compensated and doping approach,(1,2) we introduce new bonding and antibonding states as well as alter the ones that are already there, i.e., make it less ordered. The NSB arrangements result in background gap states (tail states), whereas alternative bonding arrangements involve localized charges which can be detected against the numerous neutral bonds by their deviations from the expected atomic coordinations.(5,7-9) These latter configurations, and again there is a spectrum of them,(10) including those in the KAF model for chalcogenides,(11) we call deviant or defect electronic configurations (DECs). They can also be chemically "donors" or "acceptors." NSB is responsible for the optical gap, and changes to it can be made smoothly by following our approach of substituting elements whose structural bonds can increase or decrease the band gap. In dealing with localized gap states which control the electronic conduction, we are interested in the interaction between DECs and NSB. Since DECs depend upon the creation of alternative bonding and charge configurations which create new active centers in the gap, they vary with the type of amorphous material. There are two major types of DECs—those that pin the Fermi level, and those that are created by inserting the modifier in unusual configurations, thus tending to move the Fermi level.

If the modifiers are added to the melt, they can most easily form NSBs. Also, compound formation, phase separation and crystallization are often observed upon cooling. Entirely different electronic configurations and nearest neighbor interactions are achieved when the modifiers are incorporated below the glass transition and/or melting point. However, some alloys, especially of the light elements, can still form the proper DEC states at high temperatures by virtue of special multiorbital bonding. When the matrix (host) material is deposited separately from the modifier on a cold substrate, they will intermingle in such a manner that the resulting DECs will represent new bonds and charge configurations. [We have also, in some cases, been able to diffuse-in the modifier.(12)]

Modifiers are chosen so that they can bond to the structure in a strongly covalent manner and yet create a variety of new electronic configurations through the interactions of their other orbitals with their local environment. They, therefore, are

---

Reprinted by permission from the *Proceedings of the Seventh International Conference on Amorphous and Liquid Semiconductors*, Edinburgh, Scotland, 27 June–1 July 1977, pp. 519–523.

preferably atoms that can provide or form multiple orbitals. Transition metals and rare earths, because of their d- and f-bands, electron-deficient elements, such as those of Group III, e.g., indium, and especially light elements, such as carbon with its multiple choice bonding, and boron, because of its unique attribute of forming multi-center bonds, can be used. Boron's three-center bond has various configurations, some of which have electronic structures similar to chalcogens except that the nonbonding states are empty. We have previously pointed out (4,5) that coordinate or dative bonding plays an important role in many of these materials. Furthermore, we have shown that DEC forming can occur when alkali metals, such as lithium, are introduced into an amorphous solid, and can interact with charged states in the gap through ionization.(6)

Since the placement of the modifying atoms can result in metastable bonding configurations, we choose those atoms that preferably have directional interactions which are responsive to and can affect the local environment. Exemplary materials are the transition metals. They have the ability to amplify DEC creation by virtue of their many multidirectional orbitals. Amorphicity through its lack of crystalline constraints permits the varied atomic positions needed so that sufficient defect states are created. The lobed orbitals have a "pin cushion" effect which results in partially filled and unfilled orbitals which can act as chemical acceptors to nearby charge configurations, and can also act as donors by inserting electrons into oppositely charged orbitals as well as placing other states into the gap. This occurs without interfering with the functional ability to take part in NSB. The result is that interactive transition-metal orbital configurations yield electron states in the gap in different density and energy than could occur in an intrinsic material. Our principle is not limited to lone-pair configurations, but is applicable to DECs created by other means, i.e., charge configurations of various types, including dangling bonds. We describe the experimental results of the interaction of transition metals with chalcogens in a separate paper at this conference.(13)

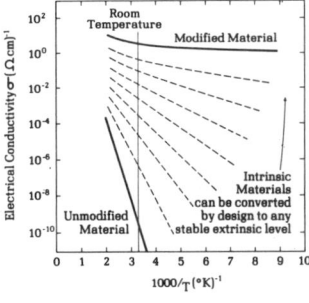

Fig. 1 Comparison of electrical conductivity for typical modified and unmodified Ovonic materials.

Intrinsic materials, such as the quarternary GeTeSeAs system, As, SiC, B and Si can be converted to extrinsic form by the addition of a transition metal, boron, carbon, etc. In all cases, the percentage of the multiorbital materials added can range from approximately .1% to over 30%.(12)

Fig. 1 is a generalized curve illustrative of the changes that occur in amorphous materials which are made extrinsic by modification whether they are from Groups III, IV, V or VI.

The chalcogenide materials have a spectrum of DECs (9-11,14) involving lone-pair centers. These include valence alternation pair centers involving Groups V and VI elements. The DECs originate from new bonding configurations of the lone-pair electrons and their concentration equilibrates in a self-compensating manner at high temperatures. They must be changed to achieve enhanced conductivity. Their self-compensating mechanism can be overcome by utilizing special defect chemistry, e.g., transition-metal orbitals interacting with lone pairs, creating new DECs which result in new electrically active centers; d-orbitals are available for increased directional interaction with lone pairs and are more numerous than in s and p electron systems, and more complicated spatial interactions can occur. This relationship is beyond that of a simple chemical acceptor and there can be an additional pairing of orbitals where the d-bands interact with the charged states so that the charged states that ordinarily exist in the solid can be altered, e.g., a d-electron or hole can compensate a lone-pair charged state and still place d-related states in the gap. The resulting new states are likely to compensate and dominate the active centers if the modifiers are incorporated at low temperatures as described above. As a result, the Fermi level is shifted and extrinsic conduction is observed. In addition, these new centers allow hopping conduction, showing that the negative effective correlation energy is a property only of the specific kind of DECs described by Street et al. (14) and Kastner et al.(11)

The same general principles have been successfully applied to materials of Groups III through VI since the d-bands can interact with other charge configurations whatever their origin. We have modified primarily tetrahedral materials as well as semiconductors containing a majority of electron-deficient elements. Examples are amorphous semiconductors containing boron, carbon, silicon, arsenic, boron-silicon alloys, silicon-carbon alloys, etc. The modifying species are capable of multiorbital interactions such as hybridization, complexing and π bonding.

Our results indicate that hydrogen is not necessary for the development of defect states in boron, silicon or other materials. (While boron hydride can provide multiorbital states,(15) multiorbital states of boron can be formed without hydrogen.) In fact, hydrogen may limit the conductivity in silicon-based materials; for example, amorphous silicon can be modified by a number of materials. When transition metals were utilized for modification, over a thousand times the conductivity of that of doped amorphous silicon was achieved.

In summary, novel interactions created by independently introducing modifiers into amorphous materials produce new electronic structures, in which the localized gap states that control the conductivity differ from those of both the matrix and the modifier. Different preparation methods enable one to place in an amorphous material different types of interactive modifiers.

Transition metals, boron, carbon, and other elements which can expand their coordination can be used, depending on the matrix material. The ability to expand coordination number in response to nearest neighbor relationships is important to DEC formation. Since amorphous solids tend to differ locally in coordination sites, the introduction of the modifying atom, ion or molecule during the formation of the film offers the maximum possibilities for the interactions.

We have described for the first time Ovonic modification of amorphous materials that contain as their chief elements those ranging from Group III to Group VI. These modifications have allowed us to manipulate the optical band gap and to control the electrical activation energy resulting in conductivity changes of up to 9 orders of magnitude. New devices which can operate from room temperature to over $500^{o}C$, ranging from energy conversion to magnetic devices, are made possible by applying the principles of stereochemistry and coordination chemistry to solid state physics in a unique manner.

I owe special thanks to Iris Ovshinsky. Richard Flasck, Krishna Sapru and Thomas Anderson have been most helpful. My thanks to Masa Izu for his fruitful suggestions, and to Hellmut Fritzsche, David Adler and Melvin Shaw for useful discussions and clarifications.

REFERENCES

1. W.E. Spear and P.G. LeComber, Solid State Comm. 17, 1193 (1975).

2. W. Paul, A.J. Lewis, G.A.N. Connell and T.D. Moustakas, Solid State Comm. 20, 969 (1976).

3. A. Triska, D. Dennison and H. Fritzsche, Bull APS 20, 392 (1975).

4. S.R. Ovshinsky and H. Fritzsche, IEEE Trans. on Electron Devices ED-20, 91 (1973).

5. S.R. Ovshinsky and K. Sapru, in Amorphous and Liquid Semiconductors, ed. by J. Stuke and W. Brenig (Taylor and Francis, London, 1974), p. 447.

6. S.R. Ovshinsky, in Structure and Excitations of Amorphous Solids, ed. by G. Lucovsky and F.L. Galeener (AIP, New York, 1976), p.31.

7. S.R. Ovshinsky, in proceedings of the 4th International Congress on Reprography 1975, Hannover, Germany, p. 109.

8. S.R. Ovshinsky, in proceedings of the VI International Conference on Amorphous and Liquid Semiconductors, ed. by B.T. Kolomiets, Leningrad, USSR, 1976, p. 426.

9. S.R. Ovshinsky, Phys. Rev. Lett. 36, 1469 (1976).

10. It should be kept in mind that there is a spectrum of localized states created by lone-pair interaction with each other and with their nearest neighbors. These play important roles in electronic relationships in these materials and they are far more numerous than the KAF states.

11. M. Kastner, D. Adler and H. Fritzsche, Phys, Rev. Lett. 37, 1504 (1976)

12. S.R. Ovshinsky et al., to be published.

13. R. Flasck, M. Izu, K. Sapru, T. Anderson, S.R. Ovshinsky and H. Fritzsche, paper at this conference, p. 524.

14. R.A. Street and N.F Mott, Phys. Rev. Lett. 35, 1293 (1975).

15. W.N. Lipscomb, Boron Hydrides (W.A. Benjamin, Inc. 1963).

OPTICAL AND ELECTRONIC PROPERTIES OF MODIFIED AMORPHOUS MATERIALS

R. Flasck,[1] M. Izu,[1] K. Sapru,[1] T. Anderson,[1] S.R. Ovshinsky,[1] and H. Fritzsche[2]

[1]Energy Conversion Devices, Inc., Troy, Michigan 48084
[2]Department of Physics, University of Chicago, Chicago, Illinois 60637

1. INTRODUCTION

The thermal activation energy, ΔE, of the conductivity in amorphous chalcogenide semiconductors is usually about half of the optical gap energy E_o. This is commonly interpreted as (intrinsic) conduction at the mobility edge by carriers whose concentration is governed by the Fermi energy E_F which is pinned near the gap center. Small changes in glass composition and the presence of impurities usually have little effect on the properties except for small changes in E_o and ΔE with composition. The insensitivity to impurities is ascribed to the fact that in a material lacking long-range order each foreign atom can satisfy its valency requirement,(1) and, thus, will not act as an electrically active center. The origin of the various states in the gap (DECs) (2,3) and the pinning of E_F is attributed to those DECs which are inherent valence alternation defect centers.(4,5) The effect of certain additives has been noted and explained in terms of a small shift of E_F toward one of the bands, and, thus, decreasing ΔE by at most one third.(6) As an exception to this commonly observed behavior, we reported earlier the occurrence of an extrinsic conductivity branch below 150K when lithium had been diffused into a chalcogenide glass film.(7)

In another paper at this conference,(2) the principles for Ovonic modification of the film reported on here as well as various other amorphous films are discussed, some of which yield even more drastic decreases of ΔE and, hence, conductivity increases by many orders of magnitude. Since these modifications change the optical gap E_o only in a minor way, we believe that the experimental results reported here indicate that the Fermi energy has been unpinned.

2. EXPERIMENTAL DETAILS

Thin (0.1-2 μm) films of $Ge_{32}Te_{32}Se_{32}As_4$ plus desired amounts of nickel additive were deposited by r.f. sputtering onto sodium-free glass substrates in a background pressure of 2×10^{-6}torr. The nickel concentration was varied by changing the number of small nickel disks attached to the face of a 3-1/2" diameter sputtering cathode of the host material. The r.f. power of about 55 watts resulted in deposition rates of ~ 300Å/minute and a substrate temperature of 80°C.

The conductivity and Seebeck coefficient were measured in a planar sample geometry with carbon electrodes. No evidence of a contact resistance was noted. Samples on transmission electron microscopy grids were produced in the same deposition run as the optical and electrical samples. Selected area electron diffraction and high magnification electron microscopy were employed to determine the homogeneity and lack of crystallinity within the samples. X-ray diffraction studies have also verified the lack of crystallinity.

The samples were subject to Auger and Esca compositional depth profiling. The compositions of the four samples discussed here were $(Ge_{0.32}Te_{0.32}Se_{0.32}As_{0.04})_{1-x}Ni_x$, with x = 0, 0.07, 0.091 and 0.114, respectively.

3. RESULTS

Figure 1 shows that the optical absorption curves are only slightly changed. One notices a larger absorption tail as the Ni concentration is increased.

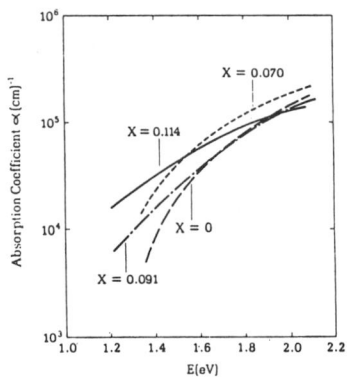

Fig. 1 Optical absorption of modified materials.

Reprinted by permission from the *Proceedings of the Seventh International Conference on Amorphous and Liquid Semiconductors*, Edinburgh, Scotland, 27 June–1 July 1977, pp. 524–528.

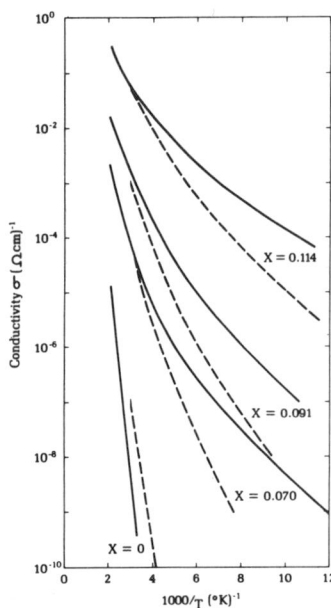

Fig. 2 Electrical conductivity of modified materials.

The conductivity curves of the samples are shown in Fig. 2 as dashed lines before annealing and as full lines after annealing at 200°C for 15 minutes. It is interesting to note that σ of the unmodified films decreases with annealing, which is the normal behavior for amorphous semiconductors, whereas σ of the other samples increases by a substantial amount. The data listed in Table 1 refer to annealed samples. ΔE is the slope of the conductivity curves near T = 500K. In view of the fact that the Arrhenius plots of the modified samples do not show a straight line portion at high T, we are not certain whether ΔE is a measure of $E_F - E_V$, the separation of the Fermi energy from the mobility edge. It is possible that upon further heating, the σ curves would become steeper and finally extrapolate to the value $\sigma_0 \approx 10^3 - 10^4 \, \text{ohm}^{-1} \text{cm}^{-1}$ commonly observed for conduction in extended states. If that were so, our interpretation would be that we presently observe phonon-assisted hopping in a tail of localized states and variable range hopping at low T.

On the other hand, these conductivity curves closely resemble those of doped a-Si (8,9) where one finds, as we do, $\sigma_0 \sim 10 \, \text{ohm}^{-1} \text{cm}^{-1}$. Such low values of σ_0 are possible even for band conduction if E_F is close to the tail states whose density increases rapidly toward the mobility edge. Because of the energy dependence of the density of states function, E_F increases with T. The linear term of this increase gives rise to a prefactor of order 10^{-2} which may account for the small σ values.

The low T portions of the σ curves yield straight lines if plotted according to Mott's law σ ~ exp – $(T_0/T)^{1/4}$ as shown in Fig. 3. Table I lists the density of states $N(E_F)$ obtained from $T_0 = 16 \, \alpha^3/k \, N(E_F)$, where k is the Boltzmann constant.

Table I Properties of annealed modified materials.

X	ΔE (eV)	N(E_F) (eV⁻¹cm⁻³)	Δσ (Ω⁻¹cm⁻¹)	S (μV/K)
0	0.74	—	$<10^{-10}$	+ 2000
0.070	0.33	10^{18}	3×10^{-9}	+ 54
0.091	0.25	2.6×10^{18}	8×10^{-7}	- 82
0.114	0.18	3.8×10^{18}	1×10^{-5}	- 84

For the radius of the localized state wave functions we chose $\alpha^{-1} = 10^{-7}$cm.

The photoconductivity Δσ listed in Table I was measured with a flux of 10^{17} photons/sec cm² of white light. The large increase of Δσ with Ni content is tentatively attributed to increased photo-excitation involving localized states. The Seebeck coefficients are listed without comment. Its complete temperature dependence will be measured before an interpretation is attempted.

4. SUMMARY AND DISCUSSION

The important result of this work is the experimental verification that extrinsic conduction in a chalcogenide glass can be achieved by the addition of a modifying element and that the magnitude of the change is dependent upon the relative amount of the element used. We use the term modifier instead of dopant because of the relatively large concentrations of Ni added. Of these, probably only a fraction produce electrically active centers. The conductivity changes far exceed those predicted (6,10) when charged centers are added to a lone-pair semiconductor at equilibrium. An essential feature of our preparation method appears to be the fact that the modifying atoms are

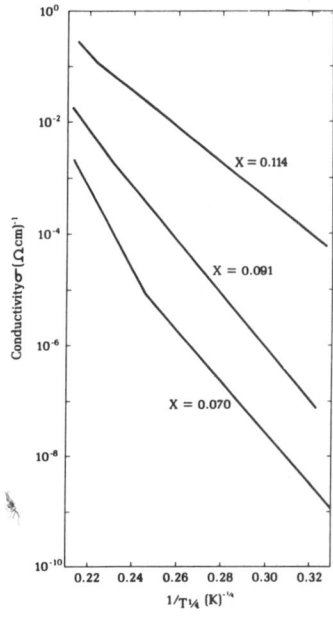

Fig. 3 Electrical conductivity of modified material as a function of $T^{-1/4}$.

incorporated at a sufficiently low temperature so that the material could not reach a structural or configurational equilibrium.

If one adds these Ni concentrations to the melt of the chalcogenide material, one obtains partial crystallization even after rapid quenching. The modified films do not crystallize more readily than unmodified films. From this we conclude that inserting the atoms into the material in a manner which permits new bonding configurations that cannot normally be obtained by cooling the substance from the melt is responsible for the modification observed.

Hopping conduction should not be observable in chalcogenide glasses because of the negative effective correlation energy of the dominant inherent defect species.(4) We do observe hopping conduction and large conductivity changes presumably because the low temperature modification method used here enables one to introduce active centers without the compensating increase of valence alternation centers.(5,6) Very similar effects have been observed with different transition-metal modifiers in a number of chalcogenide glasses.

The authors would like to thank Dr. Edward Benn and Dr. Wolodymyr Czubatyj for helpful discussion, Mr. Arthur Myatt and Ms. Krystyna Dec for sample preparation and Mr. John Tyler for structural work.

References

1. N.F. Mott, Adv. in Phys. $\underline{16}$, 49 (1967).
2. S.R. Ovshinsky, paper at this conference, p. 519.
3. S.R. Ovshinsky, Phys. Rev. Lett. $\underline{36}$, 1469 (1976).
4. R.A. Street and N.F. Mott, Phys. Rev. Lett. $\underline{35}$, 1293 (1975).
5. M. Kastner, D. Adler and H. Fritzsche, Phys. Rev. Lett. $\underline{37}$, 1504 (1976).
6. N.F. Mott, Phil. Mag. $\underline{34}$, 1101 (1976).
7. S.R. Ovshinsky, in Structure and Excitations of Amorphous Solids, ed. by G. Lucovsky and F. L. Galeener, (AIP, New York, 1976), p. 31.
8. W.E. Spear and P.G. Le Comber, Phil. Mag. $\underline{33}$, 935 (1976).
9. J. Stuke et al., to be published.
10. H. Fritzsche, paper at this conference, p.3.

LOCAL STRUCTURE, BONDING, AND ELECTRONIC PROPERTIES OF COVALENT AMORPHOUS SEMICONDUCTORS

Stanford R. Ovshinsky[1] and David Adler[2]

[1]Energy Conversion Devices, Inc., Troy, Michigan 48084
[2]Department of Electrical Engineering and Computer Science, Massachusetts Institute of Technology, Cambridge, Massachusetts 02139

1. INTRODUCTION

Amorphous solids, as opposed to crystalline solids, have an atomic structure with no long-range periodicity. Since the quantum theory of solids was developed in the period before the importance of amorphous materials was evident, the symmetries resulting from periodicity were exploited to simplify the calculation of physical properties. This led to a rapid understanding of the general behavior of crystals, but also obscured both the fundamental reasons for this behavior and the greater range of properties which could be achieved in amorphous solids. Even after amorphous semiconductors were discovered and characterized, an inordinately large effort was expended into understanding the properties of those amorphous materials with simple crystalline analogues. It is the purpose of this paper to emphasize the new modes of behavior when the constraints imposed by long-range periodicity are removed.

When an amorphous solid has a crystalline analogue, such as silicon, the electrical and optical properties of the two forms are generally similar (see, for example, Mott 1977, Adler 1977). The major reason for this is that the short-range environment around any particular atom in either type of solid is predominantly determined by the chemical nature of the relevant atoms, and the local relaxations which can occur at all temperatures tend to minimize the energy in the same way. Amorphous solids are prepared by freezing-in long-range disorder, which can be accomplished because there is an activation barrier preventing large-scale atomic relaxations at ordinary temperatures. Thus, small deviations of covalent bond angles from their optimal values ultimately result in the complete loss of long-range order.

However, it is now clear that amorphous solids can exhibit not only the same qualitative range of electronic behavior found in crystalline materials, but also a wide array of unique properties. These include reversible switching phenomena (Ovshinsky

1968), a strong diamagnetism (Di Salvo et al. 1972), photostructural effects (Ovshinsky and Klose 1972, deNeufville 1975), and large linear terms in the specific heat capacity (Zeller and Pohl 1971), among others. Recently, this point has been reemphasized by the achievement of extrinsic conduction in amorphous films by both crystalline-like substitutional doping (Spear and Le Comber 1976) and chemical modification (Ovshinsky 1977, Flasck et al. 1977). The relative difficulty of attaining substitutional doping in amorphous as compared to crystalline semiconductors can be attributed to the absence in the former of the steric constraints imposed by the necessity of long-range periodicity.

On the other hand, we wish to emphasize that the removal of constraints can, in most respects, be regarded _positively_, as the introduction of new degrees of freedom, i.e., the lack of periodicity in amorphous semiconductors allows for the presence of a multitude of unusual local structural effects that are extremely difficult to reproduce in crystals in any meaningful quantities. The demonstration of extrinsic conduction in a wide array of materials (Ovshinsky 1977) has focused on some of these new possibilities.

In this paper, we show that an approach based on the diverse chemical bonding that characterizes the component atoms of the amorphous solid leads to a unifying method for understanding a wide range of observable phenomena. We discuss in detail the types of local structure which can be expected in both crystalline and amorphous semiconductors, and show how some of the more unusual configurations can result in extrinsic conduction. These are used to explain the observed experimental results. In Section 2, we briefly discuss normal structural bonding in both tetrahedral and chalcogenide materials. This bonding represents the source of the major contributions to the cohesive energy of the solid and the electronic density of states. It is primarily responsible for both the optical and dielectric properties of the materials. On the other hand, the electrical behavior is ordinarily controlled by deviations from normal structural bonding. In Section 3, we discuss those deviant electronic configurations which are encouraged by the steric constraints of periodic crystals, i.e., substitutional impurities and defects. Some of the

---

† Research supported, in part, by the National Science Foundation.

Reprinted by permission from _Contemporary Physics_, Vol. 19, No. 2, pp. 109–126 (1978).

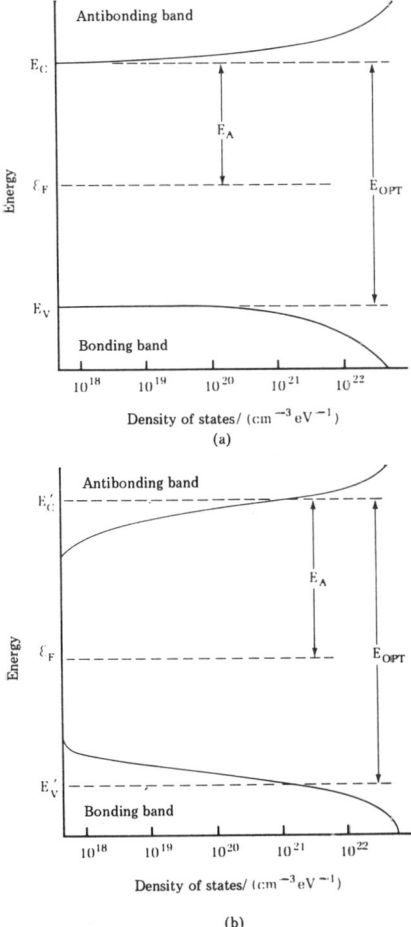

(a)

(b)

Fig. 1   (a) Sketch of the electronic density of states of a perfect crystalline tetrahedrally bonded solid. E_V and E_C are the valence and conduction band edges, respectively, and ϵ_F is the Fermi energy: E_{OPT} is the approximate value of the optical band gap, while E_A represents the activation energy for free-electron conduction. (b) Sketch of the electronic density of states of an ideal amorphous tetrahedrally bonded solid. E_V' and E_C' are the valence and conduction band mobility edges, respectively; the other notation is the same as in (a).

many novel bonding possibilities which can characterize amorphous solids with carefully chosen compositions and preparation techniques are enumerated in Section 4. The electronic structure of each of these is discussed in detail and then used to analyze the conductivity data of Ovshinsky (1977) and of Flasck et al. (1977). Some general conclusions are presented in Section 5.

2.   NORMAL STRUCTURAL BONDING

Normal structural bonding (NSB) characterizes the local environment surrounding nearly all the atoms in a crystalline material and the vast majority of atoms in an amorphous solid as well. For the non-transition-metal and non-rare-earth atoms, the lowest-energy covalent configuration is

one in which each atom in Groups I–IV in the Periodic Table bonds with N nearest neighbors (N = 1, 2, 3, or 4) and each atom in Groups IV–VIII is surrounded by 8–N nearest neighbors (N = 4, 5, 6, 7, or 8). [We are assuming that no significant charge transfer occurs in these covalent solids; otherwise, it is the electronic configuration of the ion which determines the NSB.] In each case, there are also optimal bond angles which depend on not only the particular valence but also the nature of both the atom itself and of its neighbors. In crystalline materials, the local configuration, the bond lengths, and the bond angles are usually all optimized, giving a solid of maximum cohesive

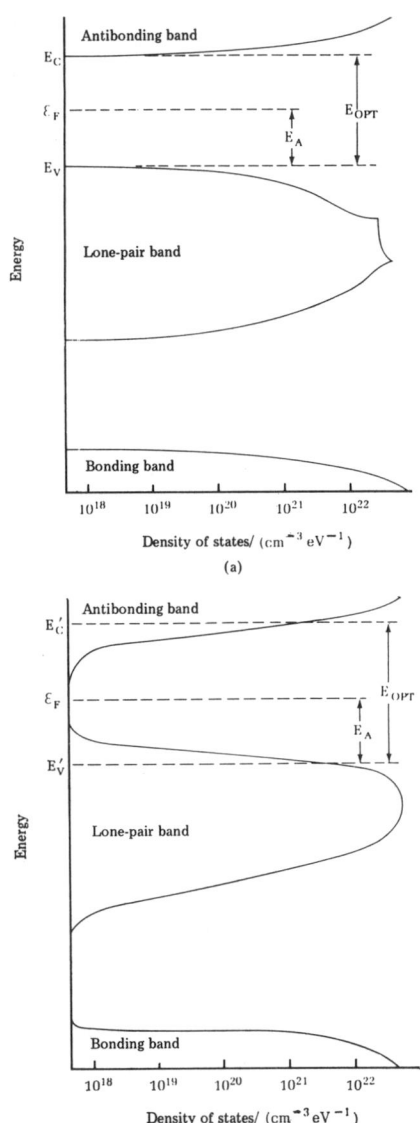

(a)

(b)

Fig. 2   (a) Sketch of the electronic density of states of a perfect crystalline chalcogen solid. The notation is the same as in Fig. 1 (a), except that it is assumed that free-hole rather than free-electron conduction predominates. (b) Sketch of the electronic density of states of an ideal amorphous chalcogen solid. The notation is the same as in Fig. 1 (b), except for the assumption that free-hole rather than free-electron conduction predominates.

energy. Amorphous solids ordinarily have a range of deviations of the covalent bond angles from their optimal values (Moss and Graczyk 1969, Shevchik and Paul 1972), but these do not ordinarily result in large decreases in the cohesive energy. There do not appear to be any significant deviations in bond lengths (Shevchik and Paul 1972), which would cost much more energy. Since normal structural bonding characterizes almost all of the atoms, it controls the optical properties of the material, such as the absorption and photoemission spectra. The detailed nature of the electronic structure of any particular solid depends on the composition, but the general features of the density of states are determined primarily by the predominant form of bonding. For example, tetrahedrally bonded solids, such as silicon, are characterized by bonding and antibonding bands near the Fermi energy; sketches of the resulting electronic density of states for both crystalline and amorphous tetrahedral solids are given in Fig. 1. The major difference between the two forms is the possible existence of tails of localized states in the bands of the latter (Mott 1967, Cohen et al. 1969).

On the other hand, if the material contains a large percentage of chalcogen (i.e., Group VI) atoms, the presence of high-energy lone-pair electrons (Kastner 1972) leads to qualitatively different electronic densities of states, as shown in Fig. 2. The valence band is now non-bonding, and does not contribute significantly to the cohesive energy of the solid (Ovshinsky and Sapru 1974).

Since sharp band edges are the result of long-range periodicity (van Hove 1953), they cannot exist in truly amorphous solids. However, the extent of the band tails in amorphous semiconductors and the energy dependence of the carrier mobility in these tails have not yet been determined.

In both tetrahedral and chalcogenide materials, if only NSB is present, we obtain a structure called either a perfect crystal or an ideal glass, depending on the nature of the solid. Such idealizations do not exist in reality, but they can be approached by careful preparation techniques. The electronic properties of a semiconductor are usually controlled by small concentrations of states resulting from deviations from NSB. However, if near-ideal material is produced, intrinsic conduction can be observed. The densities of states of Figs. 1 and 2 then show that the activation energy for conduction in extended states is about half of the optical gap in crystalline semiconductors and of the order of half of the mobility gap (i.e., the gap between the extended states in the valence and conduction bands) in amorphous semiconductors.

The structural stability of a solid also follows from its NSB. Only tetrahedrally bonded materials have rigid three-dimensional stability. Trigonal bonding, which characterizes solids such as phosphorus and arsenic, tends to produce two-dimensional covalent arrays (for example, puckered sheets) with only weak van der Waals forces between adjacent sheets. Chalcogenides such as selenium and tellurium, exhibit divalent bonding and only one-dimensional stability (that is, chain-like structures). However, the presence of atoms from Groups IV or V of the periodic table cross-link the chalcogen chains, so that three-dimensional

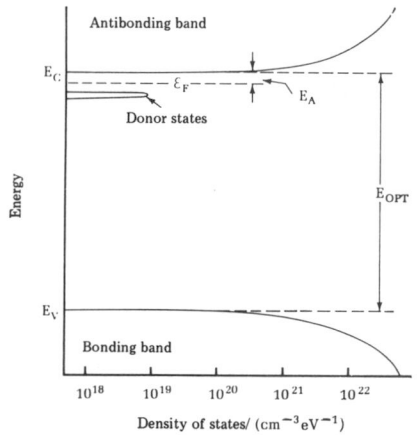

Fig. 3   Sketch of the electronic density of states of phosphorus-doped crystalline silicon. The notation is the same as that of Fig. 1 (a).

stability can be attained in an amorphous alloy (Ovshinsky 1968).

3.   CONVENTIONAL DEVIANT ELECTRONIC CONFIGURATIONS

All solids, amorphous or crystalline, contain atoms which are coordinated differently from their NSB. We call them deviant electronic configurations (DECs). Some of these DECs (e.g., substitutional impurities, vacancies, interstitials) are well known in crystalline semiconductors; it is these conventional DECs which we discuss briefly in this section. In the next section, we describe those DECs which characterize primarily amorphous solids.

3.1   Substitutional Impurities

If the impurity atoms is isoelectronic with the atom for which it is substituting, it is ordinarily not electrically active, although it could yield deep donor or acceptor levels in the gap (Adler 1975). However, if the impurity atom has a different valence, it can result in significant changes in the Fermi level. A familiar example is phosphorus-doped crystalline silicon. If a phosphorus atom substitutes for a silicon atom in the diamond lattice, it is forced into tetrahedral (sp^3) coordination; but since phosphorus contains five outer electrons, one must enter an energetically unfavorable antibonding orbital. All such antibonding electrons can reduce their energy somewhat by spreading on to those silicon atoms surrounding the phosphorus impurities, thus forming shallow donor levels; the resulting density-of-states diagram is shown in Fig. 3. The energy of a phosphorus atom which is tetrahedrally coordinated is much larger than that of one trigonally (p^3) coordinated, but the constraints of the crystal preclude the latter at low phosphorus concentrations.

In an analogous manner, if small concentrations of boron are introduced into crystalline silicon, the boron is forced to bond tetrahedrally instead of its energetically more favorable trigonal (sp^2) coordination. However, in this case, the missing electron on the boron introduces a localized hole.

The spreading of this hole on to neighboring silicon atoms leads to a shallow acceptor level.

For many years, it was believed that substitutional doping of electrically active impurities was impossible in amorphous solids because the resulting high-energy configurations would not be expected to occur without the steric constraints of the crystal. However, Spear and Le Comber (1975) successfully demonstrated extrinsic conduction in silane-decomposed amorphous silicon doped with small concentrations of either phosphorus or boron. It now appears likely that the material in which this extrinsic conduction has been obtained is actually not amorphous silicon but rather amorphous silicon-hydrogen alloys, and that the presence of hydrogen is essential to the doping (Knights 1977, Paul et al. 1976, Fritzsche 1977). The exact nature of the coordination of the impurities introduced into doped amorphous silicon-hydrogen alloys is not completely clear at the present time, but we shall discuss an alternative possibility to the conventionally accepted one in Section 4.

3.2 Defects

Defects, such as vacancies and interstitials, are unique to crystalline solids, since they are possible only because of the constraints of a periodic structure. They are not always stable at room temperature; e.g., the monovacancy in silicon anneals away at 140 K. Defects in covalent crystals generally result in strained bonds, i.e., bond lengths or bond angles at different values from the optimal ones. Local structural rearrangements produce a metastable configuration, but the defects ordinarily lead to donor or acceptor levels. These arise because non-optimal bond angles and dilated bond lengths always raise the energies of the bonding states and lower the energies of the antibonding states; the former states are filled and thus act as donors, while the latter are empty and act as acceptors. Since only the bonding states are ordinarily filled, the presence of these strained bonds increases the energy of the crystal; however, since the overall defect density is not generally very large, the resulting decrease of cohesive energy is small. The resulting band structure is shown in Fig. 4.

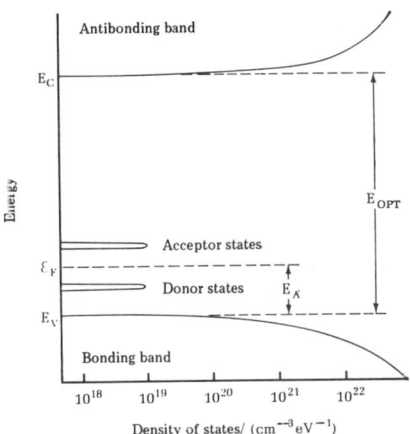

Fig. 4 Sketch of the electronic density of states of a tetrahedrally bonded crystalline solid containing about 10^{18} vacancies/cm^3. The notation is the same as that of Fig. 2 (a).

3.3 Surface States

Both crystalline and amorphous solids have non-ideal structural configurations near their surfaces, yielding states which are relatively localized in those regions. In crystals, the long-range periodic constraints tend to lead to the existence of strained bonds on the surface. For example, each atom on a silicon (111) surface does not have a fourth atom present to complete its sp^3 bonding requirements. A surface reconstruction takes place, the exact nature of which is still not completely clear. The apparent absence of free spins associated with the surface atoms suggests that either strained bonds exist or there is a charge transfer near the surface, producing Si$^+$ and Si$^-$ ions, each of which needs only three bonds to complete its valence requirements. This configuration is very similar to that of a valence-alternation pair (Kastner et al. 1976), which will be discussed in Section 4. In the case of strained bonds, the electronic density of states is qualitatively the same as that shown in Fig. 4. If charge transfer occurs, the effective negative correlation energy complicates the situation somewhat, although the resultant electrical properties (i.e., deep donor or acceptor levels) remain similar (Adler and Yoffa 1976).

For a solid such as GaAs, the neutral atoms by themselves optimally bond with three-fold coordination, sp^2 for Ga and p^3 for As. Thus, a surface reconstruction can avoid the presence of strained bonds, and no surface states need appear in the energy gap.

In amorphous solids, the absence of long-range periodicity often allows for a suppression of strained surface bonds beyond that obtained after crystalline surface reconstructions. For example, single-crystalline tellurium necessarily has two faces in which the parallel chains end, and thus strained bonds must exist. However, one can envisage an amorphous tellurium structure in which all surfaces are essentially parallel to the distorted chains, and, in fact, this would be the optimal structure--only van der Waals forces are broken at the surfaces. Thus, amorphous surfaces are generally less electrically active than those of corresponding crystals.

4. DEVIANT ELECTRONIC CONFIGURATIONS WHICH CHARACTERIZE AMORPHOUS SOLIDS

4.1 Undercoordination (Dangling Bonds)

When the coordination of any atom is less than optimal, the missing bonds are referred to as dangling. The energy of an atom with a single dangling bond is E_b greater than the atom in its ground-state configuration, where E_b is the bond strength. Since E_b is typically about 2-6 eV, a dangling bond is energetically very unfavorable. Nevertheless, they characterize tetrahedral amorphous semiconductors because of a combination of two effects: (1) since five-fold coordination is impossible within s-p bonding, the lowest-energy deviant configuration of a tetravalent atom contains a dangling bond; (2) the rigid three-dimensional structure required by tetrahedral bonding leads,

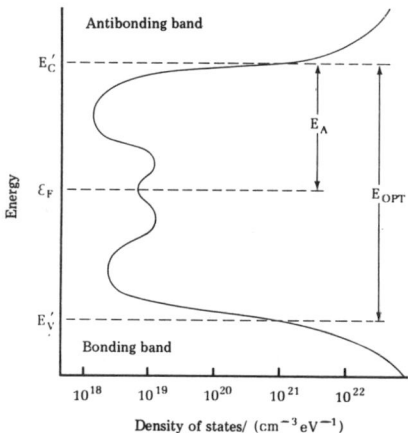

Fig. 5 Sketch of the electronic density of states of as-deposited evaporated amorphous silicon. The notation is the same as that of Fig. 1 (b). The Fermi energy is pinned by the large dangling-bond density ($\sim 10^{19}$ cm^{-3}).

particularly in amorphous solids deposited from the vapor phase, to structures in which it is difficult to avoid some three-fold coordination, since the latter tends to relieve the stresses inherent in a rigid tetrahedral network. In practice, evaporated films tend to contain relatively large voids, and these must contain some dangling bonds in addition to many strained bonds. The strained bonds are equivalent to the surface states discussed in Section 3, but dangling bonds are qualitatively different in several respects. To begin with, they each involve an unpaired spin, and thus should contribute to both an EPR signal and a Curie term in the magnetic susceptibility. But, in addition, they produce a very different electronic density of states, as is sketched in Fig. 5. The occupied states are non-bonding, and so are higher in the gap than those resulting from strained bonds. Since the dangling bond states are quite localized, the unoccupied states are still higher in energy, on the average by the correlation energy, U (Hubbard 1963). When large densities of dangling bonds are present, as in unannealed, evaporated amorphous silicon, the two bands overlap, yielding a finite density of states at the Fermi energy. This can result in the predominance of variable-range hopping conduction (Mott 1969). As the material is annealed, the dangling-bond density decreases, presumably due to a coalescence of voids. Since the localized-state density and the density-of-states at the Fermi energy decrease concomitantly, both the EPR signal and the variable-range hopping conduction also anneal away (Brodsky 1971). In amorphous silicon prepared from plasma-decomposition of silane, sufficient hydrogen remains to tie up almost all of the dangling bonds in the material. The silicon-hydrogen bond is sufficiently strong effectively to remove the localized states from the gap, thus allowing the Fermi energy to be moved. This appears to be the reason why ordinary doping is possible in this material (Paul et al. 1976).

Significant densities of neutral dangling bonds would not be expected in chalcogenide glasses for several reasons: (1) the bond energies, E_b, are relatively large; (2) much lower energy DECs are possible, as will be discussed in subsection 3; and (3) the lack of structural rigidity relative to that of tetrahedrally bonded networks allows for structural rearrangements which can eliminate potential dangling bonds, even at low temperatures (Ovshinsky and Sapru 1974).

4.2 Overcoordination

For all atoms in Groups V through VIII in the Periodic Table, overcoordination relative to the optimum 8-N is possible. Overcoordination effectively coverts a lone pair into a bonding-antibonding pair. This always results in an increase in energy (Kauzmann 1957), but generally by a smaller amount than a dangling bond. Thus, overcoordination is often the lowest-energy neutral DEC. If it occurs, the antibonding electron would then spread on to neighboring atoms and lower its energy somewhat, resulting in a density of states completely analogous to that sketched in Fig. 3. All such electrons, however, would then contribute to an EPR signal, in conflict with the experimental observations. Thus, such configurations, in fact, do not occur in significant concentrations. The reason for this is given in the next subsection.

4.3 Valence-Alternation Pairs

Street and Mott (1975) and Mott et al. (1975) proposed that a neutral defect, which they called a dangling bond, would be unstable towards a transformation into a negatively and positively charged pair. In this way, they accounted for the pinning of the Fermi energy without free spins, the photoluminescence properties, and the nature of some electron and hole traps. Kastner et al. (1976) considered the chemical nature of these defects quantitatively, and showed that the lowest-energy neutral defect in chalcogenide glasses is a three-fold coordinated chalcogen atom, which they called C_3^0. Using plausible simplifications, they were able to show that the reaction,

$$2C_3^0 \rightarrow C_3^+ + C_1^- \tag{1}$$

(where C_3^+ represents a positively charged, three-fold coordinated chalcogen atom and C_1^- represents a negatively charged, singly coordinated chalcogen atom), is indeed exothermic, as postulated by Street and Mott. Kastner et al. called such a C_3^+-C_1^- pair a valence-alternation pair (VAP). The reaction, (1), can take place because spontaneous bond-breaking can effectively convert a C_3^- center into a C_1^- center. Due to the ease of this or the reverse conversions the presence of VAPS ordinarily pins the Fermi level (Adler and Yoffa 1977). Since it takes a low energy to produce VAPs ($\sim$ 0.5 eV), as many as $\sim 10^{19}$ cm^{-3} can be present in a typical glass. The resulting band structure is shown in Fig. 6.

In order to obtain extrinsic conduction when large densities of VAPS are present, a sufficient concentration of an electrically active chemical modifier must be introduced to convert all of the C_3^+ centers to C_1^- or vice versa. Later in this

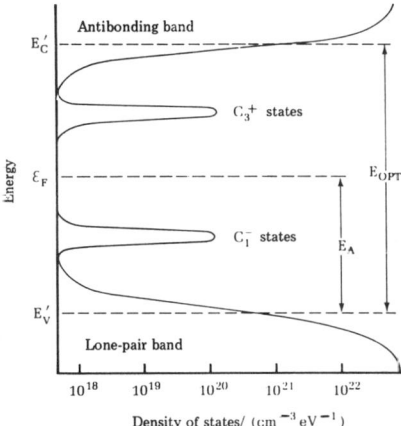

Fig. 6 Sketch of the effective one-electron density of states of a typical chalcogenide glass. The notation is the same as that of Fig. 2 (b). The Fermi energy is pinned by the valence alternation pairs, despite the low density of one-electron states, at ε_F.

section, we shall show how this has been accomplished by several different techniques.

When the dielectric constant of the glass is relatively low, the C_3^+-C_1^- pairs could be intimate ones (VAPs) due to their internal electrostatic attraction. For example, this is expected to be important in glassy selenium and As_2Se_3 (Kastner et al. 1975). In this case, the Fermi energy is unpinned to the extent of twice the electrostatic attraction energy (Adler and Yoffa 1977), and extrinsic conduction is easier to obtain.

4.4 Lone-Pair/Lone-Pair Interactions

Chalcogen atoms have high-energy lone pairs, which do not ordinarily participate in covalent bonding. In a crystal, they are oriented

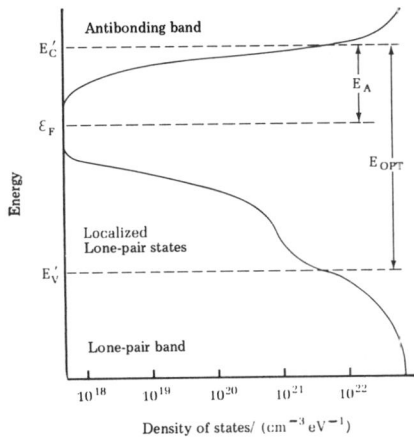

Fig. 7 Sketch of the electronic density of states of a chalcogenide glass with strong lone-pair/lone-pair interactions. The notation is the same as that of Fig. 1 (b). The localized states due to any VAPs which may be present are not shown.

identically. However, in an amorphous solid, they have a wide spectrum of orientations, resulting in diverse mutual lone-pair interactions (Ovshinsky 1976a). These can give a distribution of states in the gap which could yield n-type or even metallic conduction under extreme conditions of strong coupling. This possibility exists in chalcogenide glasses because it involves interactions among non-bonding electrons, and thus does not significantly increase the energy of the solid. The resulting localized states are very important because of the large density of lone pairs in the glass (one per chalcogen atom). The density of states of a chalcogenide glass in which such interactions are large is sketched in Fig. 7. In real materials, the additional existence of significant densities of valence-alternation pairs ordinarily compensates for any increase of the Fermi level, and conversion of some of the C_3^+ centers to C_1^- centers keeps ε_F pinned. However, even if they are electrically inactive, the importance of these states should not be underemphasized (Ovshinsky 1976b, Kastner 1977).

4.5 Dative Bonds

A dative bond is one in which lone pairs bond with empty orbitals. It is a two-center, two-electron bond, just like an ordinary covalent bond, but both electrons originate from the same atom. An atom from Group III, such as Ga, can form a dative bond with the lone pair on a Group V atom, such as As, at the cost of only the As s-p promotion energy. Thus, GaAs bonds tetrahedrally rather than trigonally. Similarly, a Group II atom can form two dative bonds with a Group VI atom (at the cost of only one promotion energy), and thus solids such as CdTe bond tetrahedrally. A Group III atom can form a dative bond with a Group VI atom without any cost in energy, due to the presence of the high-energy lone pair on the latter.

Dative bonds have corresponding (empty) antibonding levels which could give localized acceptor states in the gap. Ordinarily, they would appear above ε_F, but in certain cases could give shallower acceptors.

Although dative bonds are common in crystalline materials, they have a unique role in amorphous solids, particularly when large concentrations of chalcogens are present (Ovshinsky and Sapru 1974). When a glass is quenched from the melt, dative bonds result in structures analogous to those described above. However, when, for example, Group III atoms such as In or Ga are subsequently introduced into an amorphous matrix, the geometric restriction that only an even number of half-bonds can be broken leads to several new possibilities. The most energetically favorable of these is likely to be an In^+ or a Ga^+ ion forming three dative bonds with surrounding chalcogen atoms. The extra electrons donated by the In or Ga atoms upon ionization convert some of the C_3^+ centers to C_1^- via the reaction,

$$C_3^+ + 2e^- \rightarrow C_1^- \qquad (2)$$

Initially, the Fermi level remains pinned (Adler and Yoffa 1976). However, once a sufficient

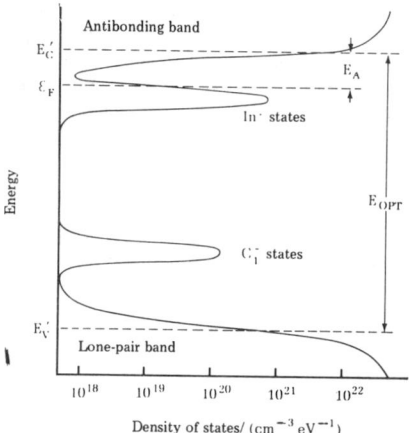

Fig. 8   Sketch of the electronic density of states of a typical chalcogenide glass which has been chemically modified by a sufficient concentration of In to obtain extrinsic conduction. The notation is the same as that of Fig. 1 (b).

impurity concentration is introduced to convert all of the C_3^+ centers to C_1^-, any additional impurities lead to extrinsic n-type conduction. The density of states corresponding to this situation is sketched in Fig. 8. Extrinsic conduction in chalcogenide glasses with both In and Ga used as chemical modifiers has been achieved (Ovshinsky 1977, Flasck et al. 1977). The resulting conduction is indeed n-type, in accordance with the above prediction. We should note that atoms from Groups I and II can also form dative bonds with chalcogen atoms.

4.6   Charge Compensation

If a strongly electropositive atom such as Li is introduced into a previously prepared amorphous solid by, for example, diffusion or implantation (Ovshinsky 1976c, 1977), it often occupies a position in which no additional covalent bonds are formed. In this case, the uppermost filled electronic energy level of the modifier is generally located well above the middle of the energy gap. The situation is analogous to the introduction of Li into crystalline Si--the Li enters the lattice interstitially, and produces shallow donor levels 0.03 eV below the conduction-band edge (Aggarwal et al. 1965). Since the short-range order in corresponding crystalline and amorphous tetrahedrally bonded solids is essentially identical, we might expect that introduction of Li into amorphous Si should also lead to donor levels. However, this assumes that the solid does not contain large densities of dangling bonds. If the latter are present, it would be more energetically favorable for the Li to form covalent bonds and thus decrease the conductivity, similar to the effect when hydrogen is introduced into amorphous Si. After the dangling bonds are saturated, however, any additional Li should indeed yield donor levels. If there are still large densities of localized states in the gap, the outer electrons on the Li donors merely fill up the previously unoccupied states just above the Fermi level, thus increasing the conductivity only slightly. However, when there is only a small density of localized states near the Fermi level, as

in amorphous Si decomposed from SiH_4 and deposited on high-temperature substrates, the introduction of Li should yield extrinsic n-type conduction.

When Li is introduced into a chalcogenide glass, the energy level of the outer electron is again located well above the middle of the gap. But in this case, the presence of valence alternation pairs initially suppresses an increase in conductivity, since the outer electrons can lower their energy by transferring in pairs to the C_3^+ centers, thereby converting them to C_1^- centers [see reaction (2)]. Thus, at low concentrations of Li, the Fermi level remains pinned. However, if sufficient Li is introduced, the C_3^+ centers eventually become saturated, and extrinsic n-type conduction results. [We should note that the Li^+ ions can stabilize their positions in the glass by forming four sp^3 dative bonds with lone pairs on neighboring chalcogen atoms.] The electronic band structure for this situation is shown in Fig. 9. Extrinsic conduction resulting from the diffusion of Li into chalcogenide glasses has been achieved by Ovshinsky (1976c).

If a strongly electronegative atom such as F is introduced into an amorphous matrix, its lowest unoccupied electronic energy level will generally lie well below the middle of the gap. If so, extrinsic p-type conduction should result. In chalcogenide glasses, initially pairs of electrons will leave C_1^- centers to occupy the lower empty orbitals, converting them to C_3^+ centers. After saturation of the C_1^- centers, extrinsic p-type conduction should be attainable. In general, introduction of halogen impurities is more difficult to accomplish experimentally than introduction of alkali impurities because of the larger size of the former, but it should be possible using implantation techniques.

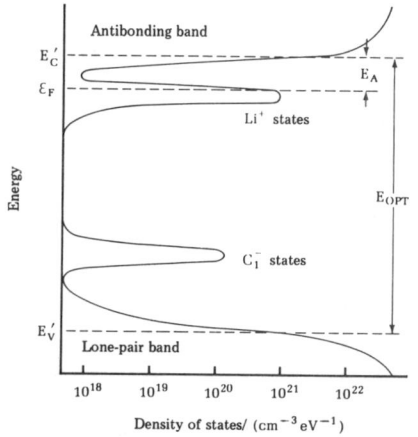

Fig. 9   Sketch of the electronic density of states of a typical chalcogenide glass which has been chemically modified by diffusion of a sufficient concentration of Li to obtain extrinsic conduction. The notation is the same as that of Fig. 1 (b).

4.7 Multi-Center Bonding

The possibility of exploiting multi-center bonding in amorphous solids opens up a whole spectrum of new methods for obtaining extrinsic conduction (Ovshinsky 1977). The most studied examples of three-center bonding are the boron hydrides (Lipscomb 1963). Boron often exhibits this type of bonding, in which two electrons are shared by three atoms in a single covalent bond. The key to understanding the resulting electronic effects is that three-center bonding also yields two non-bonding states which are ordinarily empty. If these states are sufficiently low in energy, they can then act as acceptor levels, as shown in Fig. 10.

Recently, boron has been used to obtain extrinsic p-type conduction in amorphous silicon prepared from decomposition of SiH_4 (Spear and Le Comber 1976). The boron itself is decomposed from diborane, B_2H_6, a molecule in which three-center bonding via two bridging hydrogens is predominant. It has been clear form the beginning that the residual hydrogen from the decomposition of silane efficiently removes almost all of the localized (dangling bond) states from the gap (see, for example, Adler 1971). However, it has been more difficult to understand why boron should enter the amorphous solid in tetrahedral coordination, with the resulting sharp increase in energy. It has been suggested (Ovshinsky 1977) that even the boron atoms which are electrically active do not bond tetrahedrally, but rather form three-center bonds with a bridging hydrogen. If so, the two empty non-bonding orbitals resulting from such a bond then act as acceptors, yielding the observed extrinsic p-type conduction. It was further shown that silicon-boron alloys without hydrogen yielded the same conductivity behavior as the films deposited from silane-diborane mixtures (Ovshinsky 1977), indicating the presence of three-center bonding with bridging boron or silicon atoms.

Although there is no equally simple analysis of the phosphorus doping of amorphous silicon-hydrogen

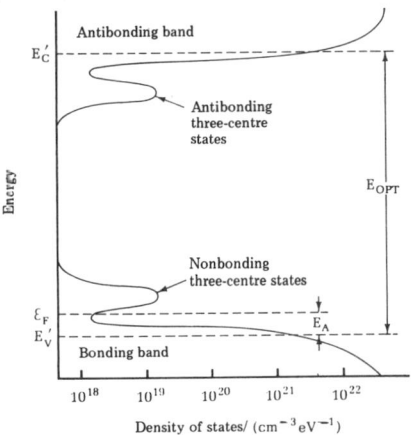

Fig. 10 Sketch of the electronic density of states of amorphous silicon-hydrogen alloys doped by the presence of small concentration of boron. The bonding three-center states are within the valence band. The notation is the same as that of Fig. 2 (b).

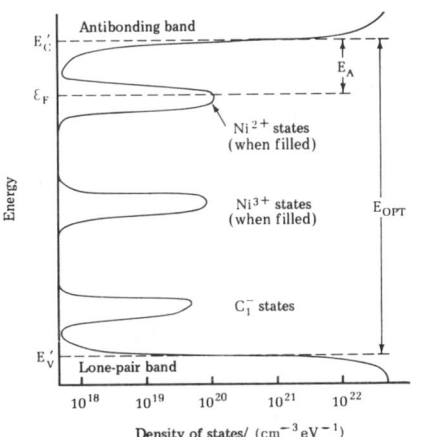

Fig. 11 Sketch of the effective one-electron density of states of a typical chalcogenide glass which has been chemically modified by a sufficient concentration of Ni to obtain extrinsic conduction. Phonon-assisted hopping of localized holes from Ni^{3+} ions to Ni^{2+} ions could exceed the free-electron conduction. The notation is the same as that of Fig. 1 (b).

alloys, it is to be noted that the existence of three-center phosphorus-hydrogen-phosphorus bonds would yield donor levels and thus extrinsic n-type conduction. Such a bond would be stable, having a structure isoelectronic to XeF_2. Although no hydride of phosphorus analogous to B_2H_6 is known, diphosphine decomposes into a polymer of stoichiometry approximating P_2H (Fehner 1968).

4.8 Multiple Bonding

Carbon, nitrogen, and oxygen can easily form double or even triple bonds, which can be exploited in amorphous solids to tie up potential dangling bonds. The electronic effect of multiple-bond formation is to shift localized states farther away from the Fermi level. In this sense, the use of carbon as a modifier is similar to the introduction of hydrogen into amorphous silicon.

4.9 Transition-Metal and Rare-Earth Modifiers

When a transition-metal or rare-earth atom is introduced into an amorphous solid, the presence of strongly correlated d or f electrons opens up many new possibilities. The first effects to consider result from the variable valencies of all of these atoms. Let us consider Ni as an example. The electronic states of Ni in different oxidation states (e.g., Ni^{2+}, Ni^{3+}, etc.) are usually rather close in energy in semiconducting compounds; thus, the charge state can depend on the exact position of the Fermi energy. In any event, we ordinarily expect that the transition-metal or rare-earth atom is positively charged in chalcogenide glasses, so that there will be an excess of C_1^- centers over C_3^+. Thus, if a sufficient concentration of the transition metal or rare earth is present, the C_3^+ centers can be saturated, and the Fermi level increases until it enters a band arising from states of the modifier. This would ordinarily result in extrinsic n-type conduction. The corresponding density-of-states diagram is shown in Fig. 11. However, if the activation energy for creating a free electron in the conduction band still remains large, the presence of a finite density of single-particle states at the Fermi energy could instead result in the predominance of variable-range-hopping conduction. This appears to be the case, for example, in $Ge_{32}Te_{32}Se_{32}As_4$ co-sputtered with Ni, where the electrical conductivity varies exponentially with $T^{-1/4}$ (Ovshinsky 1977). This type of conduction is generally difficult to obtain in unmodified chalcogenide glasses because of the negative effective correlation energy associated with a major defect (Street and Mott 1975, Adler and Yoffa 1976).

In a tetrahedral amorphous solid, there is no need to saturate the C_3^+ centers, since they are not present. Instead, after a sufficient density of electrons is donated by the modifier to fill the unoccupied localized states which are pinning the Fermi energy (see Fig. 5), extrinsic n-type conduction should be observable. This has been achieved by co-sputtering transition metals such as Ni with Si (Ovshinsky 1977), which yields a room-temperature conductivity of the order of 10 Sm $= 0.1$ $\Omega^{-1}cm^{-1}$, a factor of 10^3 greater than that resulting from conventional doping techniques (Spear and Le Comber 1976).

Other possibilities arise when transition-metal or rare-earth atoms are introduced into amorphous solids. Within each electronic configuration, there are ordinarily a large number of ligand-field-split terms. Although these do not affect the electrical conduction, they could be observed optically at moderate concentrations and they can strongly influence the magnetic properties. EPR and susceptibility measurements are needed to determine whether the transition element is in its high-spin or low-spin state; intraionic correlations favor the former, while ligand-field effects favor the latter.

An important additional consideration is the possibility of metal-metal bonding, particularly for transition elements in low oxidation states. Ordinary two-center bonds can exist, but even these can be complex due to the large orbital degeneracies and the possibility of π and δ bonding. Furthermore, multi-center bonding involving transition metals is common and a wealth of different chemical complexes have been identified (Cotton and Wilkinson 1972). In many compounds, chelation of the transition-metal atoms occurs, and this should be particularly common when significant concentrations of atoms such as sulphur or nitrogen are present; chelation always results in rings of greatly enhanced stability. Because of the many complexing possibilities, transition-metal impurities readily dissolve in amorphous solids in large concentrations. The resulting bonds are often quite delocalized, and can lead to enhanced screening effects. In general, metal-metal bonding tends sharply to reduce the unpaired-spin density, and thus the EPR signal. Preliminary measurements (Fritzsche, private communication) indicate that metal-metal bonding could be significant in modified amorphous semiconductors.

In addition to the above effects, for transition metals with less than five d electrons, dative bonding can also exist with chalcogen lone pairs. This has been observed when, for example, tungsten and molybdenum are introduced into chalcogenides. It should be emphasized that the introduction of transition-metal and rare-earth modifiers can decrease as well as increase the conductivity of the amorphous solid. For example, 5 per cent of Gd has been shown to reduce the conductivity of amorphous $Ge_{32}Se_{32}Te_{32}As_4$ by a factor of about 30 (Ovshinsky, to be published).

4.10 Compensation of Lone Pairs

There are several ways to compensate the lone pairs in a chalcogenide glass, thus reducing the density of VAPs and making it easier to achieve extrinsic conduction. One method would be to introduce large concentrations of atoms from Groups II or III, in which case the lone pairs of the chalcogen atoms form dative bonds with the empty orbitals of the modifier atoms. A second possibility is to introduce sufficient concentrations of a strongly electronegative atom (e.g., O, F) to ionize the chalcogens and thus reduce the lone-pair density. The resulting material should be capable of extrinsic conduction with much smaller concentrations of modifiers. This has been accomplished by co-sputtering Ni with TeO_2, resulting in a conductivity enhancement by a factor of 10^6 (Ovshinsky 1977).

4.11 Hybridization with Empty Orbitals

In many chemical compounds, atoms such as As, Te, and even Si find it energetically favorable to promote some s or p electrons to d orbitals and take advantage of the extra bonds resulting from five- and six-fold coordination. This opens up the additional possibility of forming dative bonds between the empty d orbitals and the remaining lone pairs in the solid. This could result in a lower VAP density and ultimately extrinsic conduction, in the same way as discussed in the previous subsection.

5. CONCLUSIONS

We have shown how the removal of crystalline constraints inherent in all amorphous solids can allow a wealth of unusual chemical-bonding possibilities, many of which can be exploited to yield extrinsic conduction. Several of the ideas discussed already have been used to enhance the electrical conductivity of amorphous semiconductors by factors of up to 10^9. The overall chemical approach used here provides a unified method for analyzing the electronic structure of all of these materials.

Acknowledgment

We should like to thank Sir Nevill Mott and William Paul for critical readings of this manuscript.

References

ADLER, D., 1971, Amorphous Semiconductors (London: Buttersworths).
ADLER, D., 1975, Treatises on Solid State Chemistry, Vol. 2, edited by N.B. Hannay (New York: Plenum Press), p. 237.

ADLER, D., 1977, Scientific American, 236, No. 5, 36.

ADLER, D., and MOSS, S.C., 1973, Comments on Solid State Phys., 5, 63.

ADLER, D., and YOFFA, E.J., 1976, Phys. Rev. Letters, 36, 1197.

ADLER, D., and YOFFA, E.J., 1977, Canad. J. Chem. 55, 1920.

AGGARWAL, R.L., FISHER, P., MOURZINE, V., and RAMDAS, A.K., 1965, Phys. Rev., 138, A882.

BRODSKY, M.H., 1971, J. Vac. Sci. Technol., 8, 125.

COHEN, M.H., FRITZSCHE, H., and OVSHINSKY, S.R., 1969, Phys. Rev. Letters, 22, 1065.

COTTON, F.A., and WILKINSON, G., 1972, Advanced Inorganic Chemistry (New York: Interscience).

DENEUFVILLE, J.P., 1975, Optical Properties--New Developments, edited by B.O. Seraphin (Amsterdam: North-Holland), p. 437.

DI SALVO, F.J., MENTH, A., WASZCZAK, J.V., and TAUC, J., 1972, Phys. Rev. B, 6, 4574.

FEHNER, T.P., 1968, J. Amer. Chem. Soc., 90, 6062.

FLASCK, R., IZU, M., SAPRU, K., ANDERSON, T., OVSHINSKY, S.R., and FRITZSCHE, H., 1977, Amorphous and Liquid Semiconductors, edited by W.E. Spear (Edinburgh: Centre for Industrial Consultancy and Liaison), p. 524.

FRITZSCHE, H., 1977, Amorphous and Liquid Semiconductors, edited by W.E. Spear (Edinburgh: C.I.C.L.), p. 3.

HUBBARD, J., 1963, Proc. Roy. Soc. (London) A, 276, 238.

KASTNER, M., 1972, Phys. Rev. Letters, 28, 355.

KASTNER, M., 1977, Amorphous and Liquid Semiconductors, edited by W.E. Spear (Edinburgh: C.I.C.L.), p. 504.

KASTNER, M., ADLER, D., and FRITZSCHE, H., 1976, Phys. Rev. Letters, 37, 1504.

KAUZMANN, W., 1957, Quantum Chemistry (New York: Academic Press), p. 392.

KNIGHTS, J.C., 1977, Amorphous and Liquid Semiconductors, edited by W.E. Spear (Edinburgh: C.I.C.L.), p. 433.

LIPSCOMB, W.N., 1963, Boron Hydrides (New York: W.A. Benjamin).

MOSS, S.C., and GRACZYK, J.F., 1969, Phys. Rev. Letters, 23, 1167.

MOTT, N.F., 1967, Adv. Phys., 16, 49.

MOTT, N.F., 1969, Phil. Mag. 19, 835.

MOTT, N.F., 1977, Contemp. Phys. 18, 225.

MOTT, N.F., DAVIS, E.A., and STREET, R.A., 1975, Phil. Mag. 32, 961.

OVSHINSKY, S.R., 1968, Phys. Rev. Letters, 21, 1450.

OVSHINSKY, S.R., 1976a, Phys. Rev. Letters, 36, 1469.

OVSHINSKY, S.R., 1976b, Structure and Properties of Non-Crystalline Semiconductors, edited by B.T. Kolomiets (Leningrad: Nauka), p. 426.

OVSHINSKY, S.R., 1976c, Structure and Excitation of Amorphous Solids, edited by G. Lucovsky and F.L. Galeener (New York: AIP), p. 31.

OVSHINSKY, S.R., 1977, Amorphous and Liquid Semiconductors, edited by W.E. Spear (Edinburgh: C.I.C.L.), p. 519.

OVSHINSKY, S.R., and KLOSE, P., 1972, J. Non-Cryst. Solids, 8-10, 892.

OVSHINSKY, S.R., and SAPRU, K., 1974, Amorphous and Liquid Semiconductors, edited by J. Stuke and W. Brenig (London: Taylor & Francis), p. 447.

PAUL, W., LEWIS, A.J., CONNELL, G.A.N., and MOUSTAKIS, T.D., 1976, Solid State Commun., 20, 969.

SHEVCHIK, N.J., and PAUL, W., 1972, J. Non-Cryst. Solids, 8-10, 381.

SPEAR, W.E., and LE COMBER, P.G., 1975, Solid State Commun., 17, 1193.

SPEAR, W.E., and LE COMBER, P.G., 1976, Phil. Mag., 33, 935.

STREET, R.A., and MOTT, N.F., 1975, Phys. Rev. Letters, 35, 1293.

VAN HOVE, L., 1953, Phys. Rev., 89, 1189.

ZELLER, R.C., and POHL, R.O., 1971, Phys. Rev. B, 4, 2029.

A NEW AMORPHOUS SILICON-BASED ALLOY FOR ELECTRONIC APPLICATIONS

Stanford R. Ovshinsky and Arun Madan

Energy Conversion Devices, Inc., 1675 West Maple Road, Troy, Michigan 48084

(Received 16 October 1978; accepted 6 November 1978)

THERE is a need for alternative energy sources. Photovoltaics are an attractive possibility, but their application has been limited by economic considerations in single-crystal materials, and for physical reasons such as grain boundaries in poly-crystalline materials. Amorphous semiconductors are especially attractive in this regard because they are basically much less expensive than their crystalline counterparts and because they possess a direct band gap with a high value for the optical absorption coefficient. We report here the development of a new alloy that eliminates the physical problems associated with the silicon-hydrogen alloys.

Amorphous semiconductors ordinarily have a large density of localised states which act as traps which lead to low values for the drift mobility and the recombination time of free carriers. For example, elemental amorphous silicon is of little technological importance because it contains a density of localised states in excess of $10^{20} cm^{-3} eV^{-1}$. Madan

Table 1 Comparison of the new alloy with the best silicone–hydrogen alloy fabricated

Type of amorphous alloy	Si–F–H	Si–H	Comments
Optical band gap E_0	1.65 eV	1.55 eV (refs. 11, 12)	Measured using photoconductive data
Minimum density of localised states $N(E)$	$10^{16} cm^{-3} eV^{-1}$	$10^{17} cm^{-3} eV^{-1}$	As measured by the field-effect experiments. The analysis[1] neglects surface states and hence these numbers represent an upper limit to $N(E)$
Intrinsic behaviour	$\Delta E(\sim 0.8 eV)$ and σ_{RT} $(10^{-10}$–$10^{-8}(\Omega cm)^{-1})$ are weakly dependent on deposition temperature between 400 and 700 K	$\Delta E(0.8$–$0.5 eV)$ and $\sigma_{RT}(10^{-11}$–$10^{-7}(\Omega cm)^{-1})$ are strongly dependent on deposition temperature between 350 and 600 K (ref. 13)	σ_{RT} and ΔE refer to the room temperature conductivity and conductivity activation energy
Doping efficiency	Values of $\sigma_{RT} \sim 5(\Omega cm)^{-1}$ and $\Delta E' \sim 0.05 eV$ attained using very small concentrations of PH_3 (500 v.p.p.m.) and AsH_3 (50 v.p.p.m.)	Values of $\sigma_{RT} \sim 10^{-2}$ $(\Omega cm)^{-1}$ and $\Delta E \sim 0.20 eV$ attained using 4×10^3 v.p.p.m. PH_3 in SiH_4	$\Delta E'$ refers to an estimate of the activation energy at room temperature by drawing a tangent to the σ–T curve
Mode of deposition	Radio frequency glow discharge in SiF_4/H_2	Radio frequency glow discharge in SiH_4 gas	
Photostructural changes	Absent	Major[6]	
Photoconductivity σ_L	$(1$–$3)10^{-4}(\Omega cm)^{-1}$	$(1$–$3)10^{-4}(\Omega cm)^{-1}$ (ref. 14)	Measured under $AM-1$ incident radiation of power 90 mW cm^{-2}
Infrared peaks			Thick deposits ($\sim 5 \mu m$) made on crystalline Si substrates
2,100 cm^{-1}	Very small	Present	Usually atributed to SiH_2 (ref. 15)
2,000 cm^{-1}	Absent	Present	Usually attributed to Si–H stretch[15]
1,000 cm^{-1}	Present	Present	Usually attributed to Si–O stretch[15]
900 cm^{-1}	Absent	Present	Usually attributed to H–Si–H scissors[15]
830 cm^{-1}	Present	Absent	Attributed to Si–F stretch[11]
640 cm^{-1}	Present	Present	Attributed to Si–H wag[15]
Hydrogen content	~0.5%	>5%	Estimated from the integrated absorption coefficient by using infrared techniques[16]
Fluorine content	~2 to 6% dependent on ratio of SiF_4/H_2	None	Estimated from ESCA measurements

Reprinted by permission from *Nature*, Vol. 276, No. 5687, pp. 482–484 (30 November 1978).

et al.[1,2] reported that amorphous-silicon films decomposed from silane gas by an r.f. glow discharge and deposited on a heated substrate have a much reduced density of states at the Fermi level, of the order of $10^{17} cm^{-3} eV^{-1}$. This material has subsequently been shown to be an amorphous silicon-hydrogen alloy[3]. In addition, Spear and Le Comber[4] showed that this alloy could be doped with phosphorous and boron; Carlson and Wronski[5] further demonstrated that it had a sufficiently high carrier lifetime to yield solar cells of up to 5.5% efficiency. Nevertheless, the still relatively high density of localised states throughout the gap leads to inefficient doping and limits the carrier lifetime. In addition, undesirable photostructural changes have been observed[6] and the material effuses hydrogen above 450°C (ref. 3). Despite many efforts, there has been no success in overcoming these basic problems.

The new alloy has silicon and fluorine as its main structural components. This alloy is multi-elemental and includes hydrogen, and can also include other elements such as oxygen without deleterious effects. The steric chemical and electrical properties associated with our new alloys will be described elsewhere[7,8]. Table 1, which condenses data taken from more than 600 sample depositions, compares our Si-F-H alloy (fabricated from the gas ratio of $SiF_4/H_2 = 10/1$) with the best Si-H alloy.

Table 1 shows that our material is in almost all properties, except for photoconductivity, distinctly different from the best Si-H alloy which makes it attractive for electronic device applications. Of special interest is the fact that the intrinsic properties of our material are only weakly dependent on the substrate temperature, much like multi-component chalcogenide glasses, and in sharp distinction from the strong dependence observed in silane-decomposed films.

With regard to the infrared spectrum, it is important that the peak at 2,000 cm^{-1} corresponding to the silicon-hydrogen stretch does not appear in our films. The peak at 2,100 cm^{-1} is ordinarily assigned to the SiH$_2$ group which has a wag at 640 cm^{-1} and a H-Si-H scissors mode at 900 cm^{-1}. The scissors mode is also absent in our film. Hydrogen through its various bonding configurations with other elements can be responsible for localised states in the gap[9,10]. A full interpretation of this point and the above data will be given elsewhere[11].

The new alloy reported here is superior to silane-based films for electronic applications. In particular, there is a much lower density of states at the Fermi level. This leads to much larger doping efficiencies. In fact, effective n-type doping has been achieved with only 50 v.p.m. AsH$_3$ in the premix. Furthermore, no photostructural effects have been observed. Because of its highly desirable characteristics, this material seems to have a potential for general semiconducting applications as well as for photovoltaics.

We thank David Adler, Melvin Shaw, Hellmut Fritzsche, Richard Flasck and Morrel Cohen for valuable discussions, and also Edward Benn, Larry Christian, and Thomas Anderson for technical assistance.

1. Madan, A., Le Comber, P.G. & Spear, W.E. J. Non. Cryst. Solids 11, 219 (1976).
2. Spear, W.E. in Amorphous and Liquid Semiconductors (eds Stuke, J. & Brenig, W.) (Taylor & Francis, London, 1974).
3. Fritzsche, H., Tsai, C.C. & Persans, P. Solid St. Tech. 55 (1978).
4. Spear, W.E. & Le Comber, P.G. Phil. Mag. 33, 935 (1976).
5. Carlson, D.E. & Wronski, C.R. J. electr. Mater. 6, 95 (1977).
6. Staebler, D. & Wronski, C.R. Appl. Phys. Lett. 31, 292 (1977).
7. Ovshinsky, S.R. J. Non. Cryst. Solids (in the press).
8. Ovshinsky, S.R. & Madan, A. (to be published).
9. Ovshinsky, S.R. in Amorphous and Liquid Semiconductors, 519 (ed. Spear, W.E.) (C.I.C.L. Edinburgh, 1977).
10. Ovshinsky, S.R. & Adler, D. Contemp. Phys. 19, 109 (1978).
11. Madan, A., Anderson, T. & Ovshinsky, S.R. (to be published).
12. Loveland, R.J., Spear, W.E. & Al-Sharbaty, A. J. Non. Cryst. Solids 13, 55 (1973/1974).
13. Le Comber, P.G., Madan, A. & Spear, W.E. J. Non. Cryst. Solids 11, 219 (1972).
14. Zanzucchi, P.J., Wronski, C.R. & Carlson, D.E. J. appl. Phys. 48, 5227 (1977).
15. Knights, J.C., Lucovsky, G., & Nemanich, R.J. Phil. Mag. (in the press.)
16. Levin, I.W. & King, W.T. J. chem. Phys. 37, 1375 (1962).

THRESHOLD SWITCHING IN CHALCOGENIDE-GLASS THIN FILMS

D. Adler,[1] M.S. Shur,[2] M. Silver,[3] and S.R. Ovshinsky[4]

[1]Department of Electrical Engineering and Computer Science, and Center for Materials Science and Engineering, Massachusetts Institute of Technology, Cambridge, Massachusetts 02139
[2]Department of Electrical Engineering, University of Minnesota, Minneapolis, Minnesota 55455
[3]Department of Physics and Astronomy, University of North Carolina, Chapel Hill, North Carolina 27514
[4]Energy Conversion Devices, Inc., Troy, Michigan 48084

(Received 10 April 1978; accepted for publication 28 December 1979)

I. INTRODUCTION
A. Electronic and thermal effects

The application of sufficiently high electric fields to any material eventually results in deviations from linearity in the observed current-voltage $I(V)$ characteristic. There are two general classes of explanations for such non-Ohmic effects—thermal and electronic. Thermal effects arise because the electrons accelerated by the field always emit phonons in an attempt to return to equilibrium. Electronic effects are due to changes in the response of the charged carriers to high applied fields. In general, both effects must be considered in any quantitative analysis, and the two can produce a coupled response ofter called "electrothermal." The use of the terminology electrothermal encompasses predominantly thermal and predominantly electronic processes as well as all intermediate cases, and therefore should not prejudice the casual observer into concluding that both effects are necessarily important. In a discussion of the physical mechanism in a particular sample, the major parameters controlling its operation must be identified and separated out from the less significant features. For example, although transistors often become hot in actual circuit conditions, there is no question that they are electronic devices and that, in fact, excess heat is deleterious to their operation. In this paper, we are concerned with electric-field-induced threshold and memory switching in thin films of chalcogenide glasses, the phenomenon reported by one of us (S.R. Ovshinsky) in 1968.[1] It is the purpose of the paper to (1) review the present understanding of these phenomena (Sec. I), (2) analyze mechanisms by which current filamentation can be induced via the switching transition (Sec. II), (3) present an isothermal analysis of the threshold process based on the unique electronic properties of the amorphous chalcogenide (Sec. III), and (4) discuss some interesting possibilities concerning the coupled electronic and structural aspects of these materials (Sec. IV).

There has been a great deal of discussion about the physical mechanism of amorphous-semiconductor switching over the past ten years.[2] In this paper, we do not feel it necessary to discuss the large body of work related to potential switching mechanisms,[2] generally thermal, which either (1) have not been observed in the samples under consideration here or (2) apply only when a special device geometry is used to mask the ordinary operation. As will become clear shortly, the increased understanding of both the electronic structure of amorphous semiconductors and their switching behavior that has been attained over the last few years has made a substantial segment of the previous literature unimportant for our present purposes. We will, therefore, limit our discussion to those experiments and models which elucidate the switching mechanism in actual thin-film chalcogenide-glass switches under ordinary operations.

B. Switching parameters

The current-voltage $I(V)$ characteristics of a typical threshold switch[1-3] are shown in Fig. 1. It is evident that when voltage is first applied to the sample, a high resistance is observed (approximately $10^7 \ \Omega$); this is called the *OFF-state* regime. When the threshold voltage V_t, typically 10–100 V, is exceeded, the sample switches to a different low-resistance operating point on the load line, with the dynamic resistance falling to about 1–10 Ω; this regime is known as the *ON* state. As long as a minimum current I_h, called the holding current, is maintained, the sample remains in the

a)Supported by the U.S. Army Research Office under Grant No. DAAG 29-78-G-0035.
b)Supported by the U.S. Army Research Office under Grant No. DAAG 29-77-G-0081.

Reprinted by permission from *Journal of Applied Physics,* Vol. 51, No. 6, pp. 3289–3309 (June 1980).

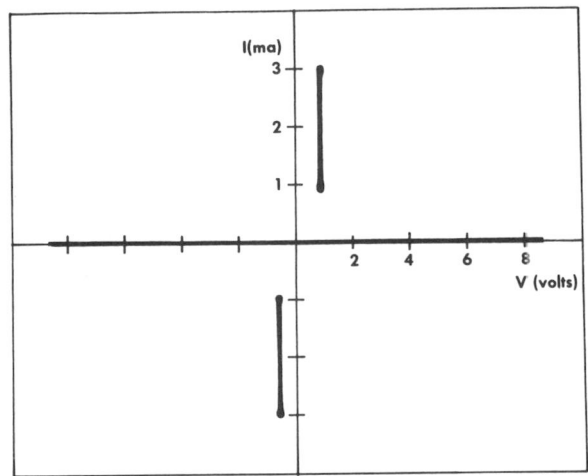

FIG. 1. Current as a function of voltage for a 1-μm-thick film of amorphous T$_{39}$As$_{36}$Si$_{17}$Ge$_7$P, sandwiched between Mo electrodes. This is a trace from a Tektronix curve-tracer oscilloscope, which implies a 60-Hz ac signal. (After Ref. 1.)

ON state. However, if the current falls below I_h, the sample either switches back to an OFF-state operating point on the load line or undergoes relaxation oscillations between the ON and OFF states, depending on the value of the load resistance.[4] Since I_h depends on the circuit conditions, it is better to treat an essentially circuit-independent parameter, the holding voltage V_h, as more fundamental; V_h is usually about 1–2 V.

The $I(V)$ characteristics shown in Fig. 1 are representative of a class of systems that possess so called "S-shaped" negative differential conductance (SNDC) characteristics,[4] and, as such, require the operation of a *positive feedback* mechanism during the transition from the OFF to ON states. Furthermore, SNDC samples are unstable against the formation of *high-current-density filaments*[4,5] during the OFF to ON transition. The *radius of the filament*[6] r_f is another parameter of importance in characterizing such samples (r_f is typically 2–25 μm).

Voltage-pulse experiments[6-9] provide a major source of information about threshold switching, and lead to the introduction of other parameters of interest. After a voltage pulse is applied, a delay time t_d typically less than 10 μsec, elapses before the onset of switching. The delay time just above threshold is subject to wide statistical fluctuations; however, these fluctuations disappear at voltages in excess of $1.2V_t$ and t_d then decreases exponentially with further overvoltage. Above a critical voltage, t_d appears to go to zero.[8,9] The switching time t_0 has proven to be faster than any means found of measuring it, but is known to be less than 1.5×10^{-10} sec.[1] It has also been convenient to define a pulse interruption time[6,7] t_s as the time between removal of V_h and application of an ensuing pulse with $V \geqslant V_h$. After V_h is removed for a time t_{sm}, the maximum benign interruption time, only V_h is required to restore the ON state (t_{sm} is typically about 250 nsec, but varies with the original ON-state operating point).

For longer values of t_s, the voltage required to reswitch the sample approaches the original threshold V_t the latter being completely restored in a recovery time t_r.

C. OFF-state characteristics

For conciseness, we will emphasize a typical switching material, Te$_{39}$As$_{36}$Si$_{17}$Ge$_7$P$_1$, which has been perpaps the most thoroughly studied sample.[3,6] At low fields, less than about 10^3 V/cm, the $I(V)$ characteristics are linear and the resistivity varies with temperature as $\rho(t) = 5 \times 10^{-3}$ exp(0.5 [eV]/kT) Ω cm, which yields a room-temperature resistivity of the order of 10^7 Ω cm. The optical energy gap is approximately 1.1 eV, or about twice the thermal activation energy, a result typical of amorphous chalcogenide materials. Holes ordinarily dominate the conduction process[10,11]; the hole drift mobility is trap controlled and is about 2×10^{-5} cm^2/V sec at room temperature. The band conductivity mobility is much higher, about 10 cm^2/V sec. Thus, although there are approximately 10^{16} holes/cm^3 below the Fermi level at room temperature, only about 10^{11} cm^{-3} of these are below the valence band mobility edge at any instant of time. The carrier lifetime[10] is of the order of 2×10^{-11} sec.

When, for example, Mo electrodes are put in contact with the chalcogenide, the bands in the latter bend upward by approximately 0.15 eV. Under very low applied bias voltage, the current is linear, but as the applied bias is increased into the field range 10^3–10^5 V/cm, a Schottky barrier forms and the current becomes contact limited[12]; it is controlled by various tunneling contributions from field and thermionic field emission.[13] However, in the 10^5 V/cm range, a high-field characteristic appears in which the conduction is bulk limited and of the form $\sigma = \sigma_0$ exp (ϵ/ϵ_a), where ϵ is the field, and ϵ_a is a constant. In fact, in the field region above about 10^4 V/cm, the OFF-state current-voltage characteristics can be fit rather well[14] by using an expression for the conductivity[15] given by $\sigma = \sigma_0$ exp $[- (\Delta E - \beta \epsilon/kT)$ $ - \epsilon_0/\epsilon]$, where ΔE is the thermal activation energy, β is a constant representing a field-dependent decrease in activation energy, and ϵ_0 is a constant associated with carrier multiplication. The enhanced conductivity at high fields may be associated with Poole-Frenkel emission from traps of sufficiently large density ($\simeq 10^{19}$ cm^{-3}) such that the electronic wave functions associated with nearest-neighbor traps overlap. (Photoconductivity results suggest that the traps which limit the mobility may not be the ones responsible for the field enhancement of the conductivity.[10])

D. The switching transition

Voltage-pulse measurements have been very useful in elucidating several important aspects of the switching transition, with major contributions coming from the use of nanosecond pulses.[8,9] First, it has been shown that for sufficiently short pulses, t_d reaches t_0, and the switching transition occurs at a critical value of applied bias.[8] Since this demon-

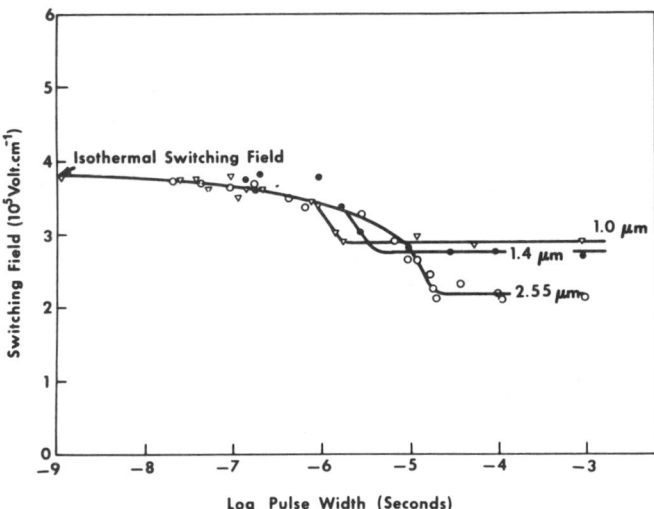

FIG. 2. Variation of the switching field with voltage pulse width for virgin samples of $Ge_{17}Te_{79}Sb_2S_2$ having a 20-μm pore diameter and three different thicknesses. Note that for the shortest pulse the switching field asymptotically approaches the same value independent of sample thickness. (After Ref. 9.)

strates that the ON state can be reached in a *total* elapsed time of a fraction of a nanosecond while the current maintains its OFF-state value, then the switching *initiation* event is clearly electronic rather thermal in origin. (In what follows, the word "initiation" is used to mean the sharp increase in conduction that occurs after t_d.) More significantly, it has also been demonstrated[8,9] that for short pulses the switching transition occurs at a critical value of *electric field* [16] ϵ_c independent of sample thickness (Fig. 2 shows this behavior); the existence of ϵ_c suggests that t_d itself may have an electronic component, resulting from the propagation of ϵ_c through the material. A model for such an isothermal component of t_d, briefly described in Ref. 14 and detailed in Sec. III, relies on these features and the experimentally determined implication that the switching process is initiated when the field in some region of the material (ordinarily near one electrode) exceeds ϵ_c. Since a Schottky barrier forms under bias and the material is p type in the OFF state, the highest field just prior to switching will be located near the anode contact. When this field reaches ϵ_c, carrier generation in the vicinity of the anode is induced, leading to a redistribution of the field and the attainment of ϵ_c in regions near the cathode and downstream from the anode. In the regions where ϵ_c has been reached, some of the generated carriers fill all the charged traps, thus leading to an enhanced mobility. A decreased-resistance region thus propagates through the sample; when it extends throughout, a sharp increase in current results. In the nonstatistical regime, t_d represents the time of propagation of the enhanced conductance regions through the sample. There is experimental evidence, both direct[17] and indirect,[18] that such propagation effects occur. In the statistical regime, the large observed fluctuations in t_d

could have many origins. For example, small variations in thickness, inhomogeneities, or thermal effects could lead to wide variations in t_d very near to threshold. Since switching in the statistical regime has not been investigated in sufficient detail to reach any definitive conclusions, we shall not consider it further. The Schottky barrier considered in this discussion does not affect the value of ϵ_c measured in short-pulse experiments since it forms only because of the inability of the electrode to supply the necessary holes at fields *below* ϵ_c.[2] Since in our model ϵ_c is the field sufficient for bulk generation of free carriers, no such barriers form once a voltage of $\epsilon_c l$ is exceeded, where l is the thickness of the film. We shall return to these points later.

E. ON-state characteristics

The principal features of the electronically sustained ON state have recently been described.[6] In particular, the current dependence of r_f has been experimentally exposed by several independent methods. The results are shown in Fig. 3. First, a study of velocity saturation in crystalline-Si/amorphous-chalcogenide heterojunctions provided a means of determining the current density in the chalcogenide in the 2–9-mA range (prior to an avalanche breakdown in the Si depletion region). It was found that the filament area A_f increases more or less proportionally to the increase in current, indicating that the current density remains constant in the filament over a wide range of current. Second, the

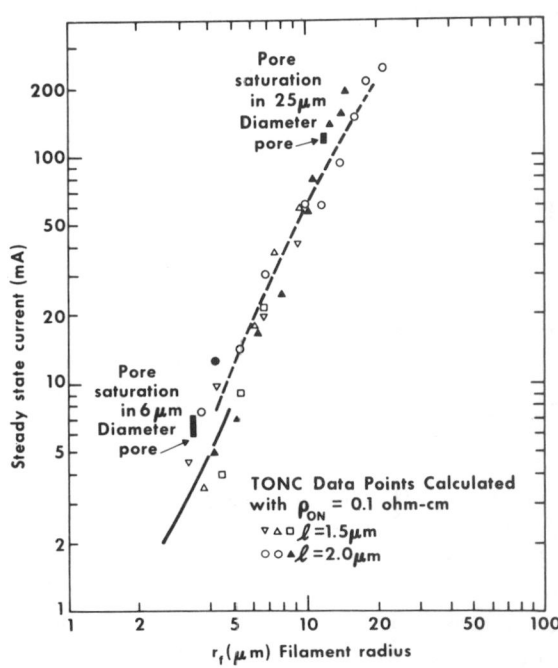

FIG. 3. Filament radius as a function of steady-state current determined by four methods. The solid line represents the results of the velocity-saturation analysis, the data points are the TONC results, and the dashed line is calculated from Shanks'[20] carbon/chalcogenide/carbon results. Two pore saturation points are also indicated. (After Ref. 6.)

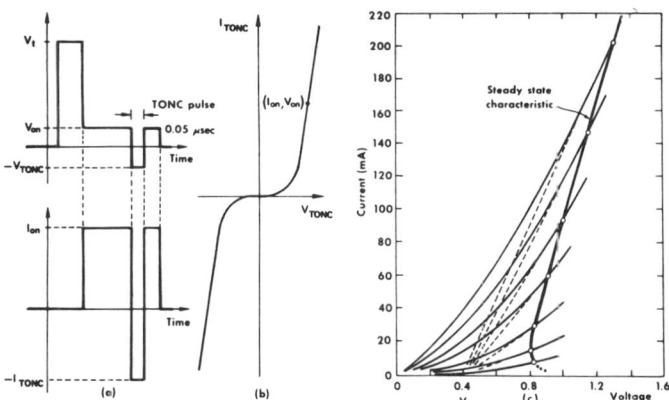

FIG. 4. (a) Transient ON-state characteristic (TONC) measurements; (b) a typical result; (c) different Mo/amorphous/Mo sample TONC curves taken from different I_{on}, V_{on} points. The sample is 50 μm in diameter. (After Ref. 6.)

transient-ON-$I(V)$-characteristic (TONC) technique[7] was used to analyze the ON-state behavior (see Fig. 4). It was found that the TONC was stable for only about 50 nsec, after which the response gradually relaxed to the *steady*-ON-state-$I(V)$ characteristics. Therefore, for TONC pulses less than 50 nsec, we expect that the area of the filament remains the same as in the steady state; the shape of the TONC should then depend upon the value of the operating steady-ON-state current for which a particular TONC is taken. This is in fact the case.

In general, we expect three contributions to the voltage drop across the sample in the ON state: the resistance of the ON-state material, the contact resistance R_c, and the interfacial barrier V_B. The TONC curves should obey

$$V_{\text{TONC}} = V_B + I \left[R_c + \rho_{on} l / A_f (I_{dc}) \right], \quad (1)$$

where ρ_{on} is the ON-state resistivity, l is the thickness of the material, and $A_f (I_{dc})$ is the area of the filament at the steady-state operating point. Extrapolation of the sub-50-nsec TONC curves should yield the same value for V_B, and this value should be the same as the metal/amorphous-chalcogenide barrier measured by other means. The agreement is good.[6] The TONC slopes then determine the variation in A_f with steady-state current.

Further, the steady-ON-state voltage is

$$V_{DC} = V_B + I R_c + \rho_{on} l J(I), \quad (2)$$

where J is the current density. If, as expected, J is independent of I, extrapolation of the steady-ON-state characteristics to $I = 0$ will yield an effective barrier voltage

$$V_B(\text{eff}) = V_B + \rho_{on} l J, \quad (3)$$

and this should vary linearly with the thickness of the chalcogenide film. Experiments show this correlation rather well.[6] Furthermore, extrapolation of V_B (eff) to $l = 0$ should yield V_B. Again, this is the experimentally observed situation.[6] These results also yield $\rho_{on} \simeq 0.07 \, \Omega$ cm.

As a further test of the above conclusions, samples with relatively small cross-sectional areas, 6 and 25-μm pore di-

ameters, were also studied. According to Fig. 3, the ON-state filament should fill (saturate) the entire sample for currents of 5 and 100 mA for 6 and 25-μm diameter samples, respectively. Further increases of current must then lead to an increased current density, which eventually leads to irreversible changes in the ON-state characteristics due to deleterious sample heating effects.[6] Points representing the onset of such changes are shown in Fig. 3. Not only do the ON-state characteristics change once pore saturation currents are exceeded, but for currents above about 30% greater than this value, a permanent drop in threshold voltage occurs, often by as much as a factor of 2. This suggests that such formation effects[18] are artifacts of operating the samples at too large a current load. This is rather a subtle point, because even when some samples are operated at steady-ON-state current levels well below pore saturation, formation effects are readily observed. The reason for this is that although the *steady-state* currents may be sufficiently low, the *transient* current "spike" that occurs during the switching transition, which is controlled by the local lumped-element circuit parameters,[4] may be many times greater than the steady-ON-state current. Such current spikes can produce intense local heating that can alter the morphology of the material. Therefore, to avoid formation effects, three factors must be considered: (1) the material must not have a tendency to crystallize; (2) the steady-ON-state currents must be kept below pore saturation; and (3) the transient current spike must be kept sufficiently low [by maximizing a circuit damping parameter $(I/R_0)(L/C)^{1/2}$, where R_0 is the OFF-state resistance, L is the intrinsic plus package inductance, and C is the package capacitance].[4,19]

Shanks' results[20] for the ON-state $I(V)$ characteristics of a chalcogenide film having pyrolytic graphite electrodes were also used to determine the filament area as a function of current. The results are shown as the dotted line in Fig. 3.

Analysis of the gain observed in an N (ON-state)/P/N amorphous-crystalline heterojunction transistor[21] as a function of the crystalline-Si base doping concentration showed that the free-carrier concentration in the ON-state is of the order of 10^{19} cm^{-3}. This implies that the ON-state free-carrier mobility is about 10 cm^2/V sec, the order the band conductivity mobility in the OFF state.

For long pulses, the average field at threshold is about $(2-3) \times 10^5$ V/cm. The critical fields observed in these materials are usually $(3-8) \times 10^5$ V/cm.[8,9] The total of the cathode and anode interface barriers for Mo/amorphous/Mo samples is 0.4 eV.[6] If this is distributed evenly between cathode and anode, and if in the ON state the critical field must be maintained near both electrodes, the band bending will then extend about 30-70 Å into the amorphous material. This is sufficiently narrow that it is possible to sustain the ON state by either strong-field-emission or thermionic-field-emission tunneling[13] through the electrode barriers. However, if the barriers are asymmetric, the depletion regions can be larger in extent, and tunneling processes become less likely. Alternatively, the ON state can be maintained from carrier generation in the high-field regions themselves. Since the potential drop in these regions is less than E_g, such generation would have to be from localized states rather than from across-the-gap excitation. In either event, it is likely that both electrons and holes contribute to the ON-state current, just as they do in the OFF state. However, just as holes predominate in the OFF state, there is evidence that electrons predominate in the ON state.[21,22]

Using the value 0.07 Ω cm for the ON-state resistivity of the amorphous film, we find that the field across the bulk of the film in the ON state is only about 1.4×10^3 V/cm. If the carrier mobility is about 10 cm^2/V sec, the transit time will be about 10 nsec. In the OFF state, the trapping time is only about 20 psec,[10] but if we model switching as an event that occurs via the filling of the charged traps after the capture of carriers generated when the critical field is exceeded, then the carrier lifetime would be expected to increase by a factor of 10^3 or more when the material is induced into the ON state. (In the ON state, we expect the recombination to proceed via capture by neutral traps.)

Implicit in the above discussion is the assumption that E_g is substantially preserved in the ON state. This is consistent with a wide range of heterojunction[22] and heterojunction transistor[21] data, which show that the interface barriers persist after switching occurs, and under these conditions the amorphous material can inject hot electrons into crystalline Si[6] and GaAs.[23] Furthermore, Walsh *et al.*[24] have evidence for the emission of ON-state nonblackbody radiation in the 0.5-eV range, just about the energy separation between the charged defect states when the sample is in the OFF state (see Secs. I G and IV). Thus we can conclude that E_g is not reduced substantially after switching. Numerical calculations[14] indicate that the OFF-state $I(V)$ characteristics can be fit with a field-dependent activation-energy term that reduces E_g by about 0.1–0.2 eV *prior* to switching. In any event, there is no hard evidence that E_g is significantly reduced because of the switching transition, as suggested in Ref. 25.

F. Recovery properties

When the current is reduced below I_h, the sample switches back to the OFF state. We expect that there might exist a minimum r_f for which radial diffusion would break the filament. (See Sec. III f) This would set an *absolute* minimum value for the current that can be maintained in the ON state I_{hm}. However, observation of I_{hm} is normally difficult to achieve because of the reactive components in the circuit.[4] If we define I_h as that current below which circuit-controlled relaxation oscillations occur,[4,20] then for most sample configurations there will always be a range of currents between I_h and I_{hm} that are unstable against relaxation oscillations. The package capacitance C and intrinsic plus package inductance L will always produce $I_h > I_{hm}$. (For large C, I_h can also be varied via the interaction between the transient response and the recovery behavior.[26]) On the other hand, if an intimate double-pulse technique is employed,[27]

where the sample is forced to remain in the ON state after switching by first rapidly reducing the applied bias, then by minimizing C and maximizing the load resistor R_L, values of I_h as low as 10 μA can be observed. (These are currents that would produce relaxation oscillation were the ON state not "held" by the second pulse.) For current densities in the filament in the range of 10^3 A/cm^2, such low values of current imply that filament radii in the 0.5-μm regime can be stabilized.[6] In this case, V_h is rather high (in Sec. III, we calculate that minimum filament radii of about 1 μm are expected) since the ON-state characteristic itself exhibits a long NDC region for these values of current.[27] The fields are therefore sufficiently high to keep the transit time sufficiently short that the ON state is maintained in an extremely narrow filament; as $V_h \rightarrow V_t$, r_f approaches its minimum value. There is evidence[28] that the minimum filament radius may, in fact, be in the fractional-micrometer regime; thus, the possibility that I_{hm} exists is real. However, for essentially all circumstances where the battery voltage is kept constant after switching occurs, I_h should be treated as a completely circuit-controlled parameter.[4,20,27]

Once the voltage across the sample is removed, the recovery curve can be studied. As shown in Fig. 5, the recovery process depends upon the steady-state operating point. The behavior can be explained as follows.[28] After the voltage is removed, the field at the anode adjusts almost instantaneously but the cathode field decays slowly (for reasons we shall explain in Sec. I G), maintaining carrier generation or tunneling near that contact. Since the applied voltage is now zero, a counter field will be built up near the anode within a dielectric relaxation time; this explains the symmetry of the TONC results shown in Fig. 4(b). The limiting feature of the recovery process is then the ambipolar diffusion of carriers radially out of the conducting filament. As the diffusion proceeds, the radius of the filament decreases. As long as any filament remains, only V_h is required to resuscitate the ON

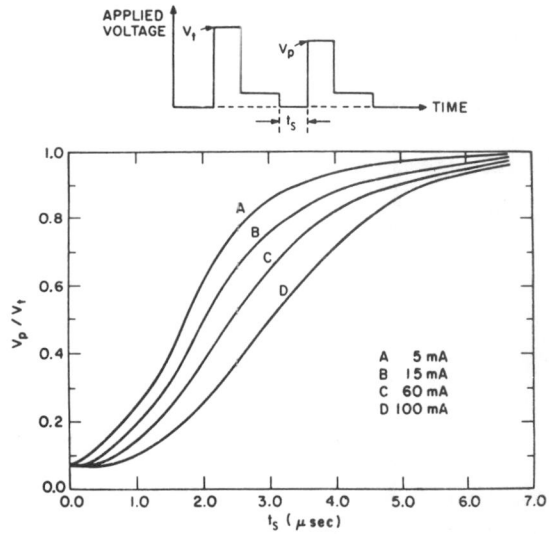

FIG. 5. Recovery of threshold voltage as a function of interruption time t_s for several values of ON-state current. (After Ref. 6.)

state. However, after a time which depends on the orignal r_f (and thus I_{on}), the filament shrinks to zero radius and the contact barriers begin to decay. (This is the origin of the parameter t_{sm} discussed in Sec. IB) Once the equilibrium contact barriers are re-established, the original V_t is completely restored.

The resistance of the sample as a function of time after the removal of the voltage has been measured.[28] A sharp increase in resistance is observed near t_{sm}, the point at which we expect the filament to have shrunk to zero radius. Furthermore, the mobility can be estimated from the value of the diffusion constant necessary to fit the recovery curves, and an Einstein relation. Although the mobility for ambipolar diffusion need not be related to the single-carrier mobility, it is suggestive that the obtained value is 10 cm^2/V sec, the band conductivity mobility in the OFF state.

G. Electronic properties of amorphous chalcogenides

Although threshold switching has been observed in a wide variety of crystalline as well as amorphous semiconductors,[2] it has particularly desirable properties, especially with regard to reproducibility and durability in chalcogenide-glass films. Homma et al.[29] have recently shown that many of the switching parameters of a pnictide glass are superficially very similar to those of chalcogenide glasses. However, the delay-time characteristics, maximum recovery times, TONC behavior, and filament radius were not investigated, and it was found that the nonchalcogenide alloys exhibited asymmetric behavior and had a much shorter lifetime. Particularly in view of the latter, we might ask if there are any unique characteristics of chalcogenide glasses which are important to threshold switching. In this subsection, we discuss the electronic properties of this unusual class of materials and show how these lead to the particularly desirable switching characteristics experimentally observed.

One of the major features of amorphous chalcogenide films prepared by sputtering or evaporation onto substrates kept at room temperature is their lack of a measurable density of unpaired spins.[30] Anderson[31] explained this by invoking a negative correlation energy for these materials. Street and Mott[32] then showed that a single defect center with a negative correlation energy would explain not only the absence of unpaired spins, but a host of other physical properties of the chalcogenides, such as the absence of variable-range hopping, the large Stokes shift in the photoluminescence, and the induced EPR signal at low temperatures.[33] These models, however, failed to distinguish between the chalcogenides and the tetrahedrally bonded amorphous solids because they did not focus on the chemical nature of the component atoms. In order to understand the precise nature of the defect states in the chalcogenides that produce such effects, we must realize that chalcogen atoms have nonbonding lone-pair p electrons that form the valence band.[34,35] Ovshinsky and Sapru[36] suggested that interactions between the lone-pair electrons on different atoms and interactions with

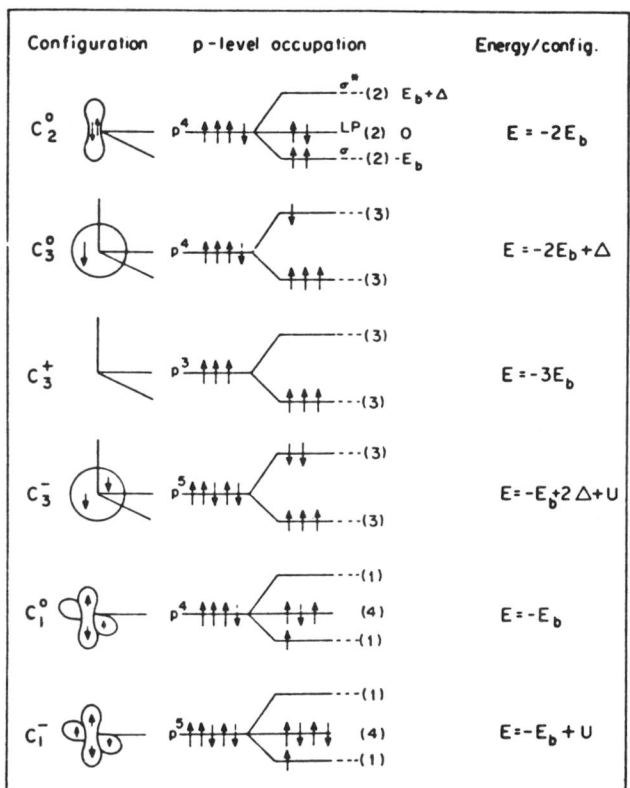

FIG. 6. Structure and energy of simple bonding configurations for chalcogen atoms in covalent amorphous semiconductors. In the configuration column, straight lines represent bond (σ) orbitals, lobes represent lone-pair (nonbonding) orbitals, and circles represent antibonding (σ^*) orbitals. Each bonding electron is paired with another from a neighboring atom. The energy of a lone-pair electron is taken as the zero of energy. U is the correlation energy and Δ the antibonding repulsive energy. (After Ref. 38.)

their local environment can create a spectrum of localized gap states, some of which could be charged defects. Ovshinsky[37] further showed how this could explain the paramagnetic behavior of these solids. Kastner et al.[38] elucidated the nature of the charged defects and supported the view of Street and Mott[32] that it is their presence that dominates the electronic properties of these materials.

Kastner et al.[38] showed that the lowest energy neutral defect in an amorphous chalcogenide is probably not a dangling bond (a singly coordinated chalcogen atom, which we call C_1^0), but rather is more likely a threefold coordinated chalcogen atom, C_3^0 (Fig. 6 shows the structure and energy of a variety of chalcogen bonding configurations). The former costs a bond energy E_b while the latter costs the antibonding repulsive energy Δ. However, the lowest-energy defect is *not* neutral. The C_3^0 centers are unstable against the reaction

$$2\,C_3^0 \rightarrow C_3^+ + C_1^-. \tag{4}$$

The reaction of Eq. (4) is possible because a threefold-coordinated chalcogen together with a neighboring twofold-coordinated chalcogen can spontaneously transform to neighboring singly and twofold-coordinated centers by breaking one of the bonds on C_3, leaving a C_1 neighbor. This is represented by

$$C_3 + C_2 \rightarrow C_2 + C_1. \qquad (5)$$

The presence of an additional electron on an atomic site, as is the case with C_1^-, requires an additional energy because of the Coulomb repulsion between two electrons on the same site, which is just the correlation energy U. The approximate energies of all C_1, C_2, and C_3 states are shown in Fig. 6. It is clear that the reacton of Eq. (4) is exothermic provided that

$$2\Delta > U, \qquad (6)$$

which appears to be the case.[38] Furthermore, since positively and negatively charged centers attract one another via the Coulomb interaction, the defect centers with two oppositely charged chalcogens as nearest neighbors have the lowest energy of all. These nearest neighboring pairs are called IVAP's (intimate valence alternation pairs). Nonintimate pairs (NVAP's) are also possible, but they will have energies greater than IVAP's.[39] It is expected from free-energy considerations that NVAP's predominate in threshold-type chalcogenides.[38,40]

Bulk dipole densities in the range proposed by Street and Mott[32] and Kastner et al.[38] have been detected by an ac technique in amorphous Si and several of its As alloys.[41] Observation of such dipolar activity is suggestive of the presence of VAP's and indicates that ferroelectricity is a possibility in some amorphous chalcogenides, as discussed in Sec. IV B.

Threshold-switching-type amorphous chalcogenide films are expected to have large densities of VAP's,[38] probably between 10^{18} and 10^{19} cm^{-3}. Furthermore, the C_1^- and C_3^+ charged centers should (1) be present in equal concentrations, (2) pin E_f near the gap center,[41] and (3) very likely be responsible for the observed trap-limited mobility in these materials. If these traps can be filled by either field-induced generation or tunneling of electrons and holes under high-field condition, then the ON-state carrier mobility will become equal to the OFF-state band conductivity mobility. As is evident from our discussion in the last two sections, such a model[14] can account for the ON-state behavior. As we shall show in Sec. III, it can also explain many of the major characteristics of switching in these materials. For example, an important point in understanding the recovery process is the reason why the barrier near the cathode decays very rapidly. We can explain this via the asymmetric nature of the C_3^+ and C_1^- traps. The cathode barrier must have excess positive charge, which we contend is obtained by the trapping of holes at the C_1^- sites, thus converting them to C_3^0 sites. The anode barrier has excess negative charge, obtained from the trapping of electrons at C_3^+ sites, thus converting them to C_3^0. A return to equilibrium at the anode after removal of the bias occurs via the trapping of holes, according to the reaction

$$C_3^0 + e^+ \rightarrow C_3^+. \qquad (7)$$

This process requires no bonding changes, and thus would be expected to occur very rapidly. At the cathode, restoration of equilibrium takes place via trapping of electrons,

$$C_3^0 + e^- \rightarrow C_1^-, \qquad (8)$$

which requires a local bonding re-arrangement and thus a change in coordination. (Note that this is only a very short-range displacement effect and does not require diffusion. Further, the maximum density of such local events involves at most 0.1% of the component atoms.) Consequently, the field near the cathode persists for a time after removal of the applied voltage. As we have already pointed out in Sec. I F, however, a counter field at the anode can build up rapidly [via the raction of Eq. (7), without bonding changes]. Carrier generation or tunneling then continues near both contacts until the continual radial diffusion of carriers eventually shrinks the filament to zero.

We should emphasize that the defect centers that characterize chalcogenide glasses are important only because of the presence of high-energy lone-pair electrons on chalcogen atoms. The utilization of these lone-pair electrons in forming additional bonds as in the C_3^+ center, or to avoid the presence of energetically unfavorable antibonding electrons as in the C_1^- center provides an additional flexibility unavailable, for example, in tetrahedrally bonded amorphous solids. The fact that the lone-pair electrons are ordinarily nonbonding results in no net weakening of the amorphous matrix via processes (7) or (8). Furthermore, since lone-pair electrons in their lowest-energy configuration form the valence band of chalcogenide glasses, as discussed previously, even the excitation of free carriers in these materials leads neither to any broken bonds nor any loss in structural stability;[36] again, this is in contrast to the situation in tetrahedrally bonded materials. (We should note that pnictide glasses are intermediate between chalcogenide and tetrahedrally bonded materials, and defects similar to C_3^+ and C_1^- probably exist.[40] Indeed, Homma et al.[29] find that a pnictide glass superficially behaves quite similar to chalcogenide glasses in many respects. However, in accordance with our model, the lower lifetime of switches based on pnictide glasses is very likely connceted with the lower density of VAP's.)

To summarize, it is now clear why threshold switching has particularly desirable characteristics in multicomponent amorphous chalcogenide films: (1) the high resistivity of the OFF state resulting from the trap-limited mobility inhibits joule-heating effects (discussed in Sec. II), (2) the nearly equal densities of positively and negatively charged traps inhibit large ON-state bulk space-charge effects (detailed in Sec. III), (3) the presence of large concentrations of cross-linking atoms inhibits crystallization (discussed in Sec. IV), and (4) the fact that the valence electrons are nonbonding allows for the excitation of large free-carrier concentrations without any decrease in structural stability (discussed further in Sec. III).

II. CURRENT FILAMENTATION

A. Introduction

It has been suspected for many years that SNDC elements are unstable against the formation of high-current-density filaments.[43] A dominant mechanism by which SNDC and filamentation occurs is through the carrier concentration dependence of the generation and/or recombination rates. Examples are impurity breakdown in compensated semiconductors,[44] impact ionization in Gunn diodes,[45] microplasmas in PNPN structures.[20,46,47] Here, the SNDC characteristics are due to the nonlinear dependence of the field-induced carrier generation or recombination on the carrier concentration; a positive feedback mechanism results and leads to filamentation. Ridley[5] postulated the occurrence of (1) high-electric-field domains in NNDC elements, and (2) high-current-density filaments in SNDC elements using both thermodynamic and electrical stability arguments. For NNDC elements, several subsequent detailed one-dimensional analyses[48,49] have successfully supported and augmented Ridley's principal conclusions. However, the electrical stability of SNDC elements under isothermal conditions has not as yet been treated in similar detail. Using Maxwell's equations, Shaw et al.[4] demonstrated that SNDC elements are unstable against the formation of inhomogeneous current-density distributions under high-frequency excitation. In this section, we demonstrate that filamentation is expected in general for SNDC elements under isothermal conditions.

We should emphasize that thermal effects can also lead to current filamentation.[50-53] Depending upon the composition and geometry, either electronic or thermal effects can predominate. In the ON state, the current density is sufficiently small that thermal effects are not important.[6] However, the transient current spike that occurs during the switching transition, discussed in Sec. I, can result in a thermal contribution to t_d. In this section, we show that current filamentation is independent of the relative importance of these effects. (The presence and position of a primary filament-nucleation site will depend upon the transverse and longitudinal thermal and electronic boundary conditions.)

B. Filamentation in a semiconductor having one type of charge carrier

In this section, we demonstrate isothermal filament formation by applying the Poisson and current-continuity equations to the case of a system consisting of a semiconductor having only one type of carrier, which we choose to be n type in this discussion. (A similar result holds for two types of carriers as we shall show in Sec. II C.) We consider a cylindrical sample under a uniform bias applied in the z direction (using cylindrical coordinates). Mobile electrons are supplied to the conduction band from shallow ionized donors and are trapped by deep states which are neutral when empty and negatively charged when filled by electrons. The population of the trapping centers depends on ϵ_z and n. The

basic equations that describe the radial component of electric field in steady state are the Poisson equation

$$\frac{1}{r}\frac{\partial r \epsilon_r}{\partial r} = \frac{q}{\epsilon}(N_D - N^- - n), \tag{9}$$

the current-continuity equation

$$qn\mu_n\epsilon_r + qD_n\frac{\partial n}{\partial r} = 0, \tag{10}$$

which ensures no radial current flow, and the steady-state kinetic equation for the filled traps

$$g(n,\epsilon_z)nN^- = \alpha(n,\epsilon_z)n(N_T - N^-). \tag{11}$$

Here N_D is the donor concentration, N_T is the trap concentration, N^- is the concentration of filled traps, D_n is the diffusion coefficient, and $g(n,\epsilon_z)$ and $\alpha(n,\epsilon_z)$ are the functions describing electron generation and trapping, respectively. Equations (9) and (11) yield

$$\frac{1}{r}\frac{\partial}{\partial r}(r\epsilon_z) = \frac{q}{\epsilon}[N_D - F(n,\epsilon_z) - n], \tag{12}$$

where

$$F(n,\epsilon_z) \equiv N^- = N_T \Big/ \left(1 + \frac{g(n,\epsilon_z)}{\alpha(n,\epsilon_z)}\right). \tag{13}$$

Equations (10) and (12) have a uniform solution ϵ_{r0}, n, given by

$$\epsilon_{r0} = 0 \tag{14}$$

and

$$n_0 + F(n_0,\epsilon_z) = N_D. \tag{15}$$

We can then solve Eq. (15) to find n_0 as a function of ϵ_z and thus the current-density-field characteristics of a uniform sample $j_z = qn_0(\epsilon_z)\mu_n\epsilon_z$.

We next investigate small radial perturbations from the uniform solution: $n = n_0 + \delta n(r)$; $\epsilon_r = \epsilon_{r0} + \delta\epsilon(r)$, where $\delta n \ll n_0$. Equations (10) and (12) yield

$$\frac{\partial^2}{\partial r^2}\delta n(r) + \frac{1}{r}\frac{\partial}{\partial r}\delta n(r) - \frac{q\gamma n_0\mu_n}{\epsilon D_n}\delta n(r) = 0, \tag{16}$$

where

$$\gamma = \frac{\partial F}{\partial n}\bigg|_{n=n_0} + 1. \tag{17}$$

B introducing the dimensionless variables, $N \equiv \delta n/n_0$ and $X \equiv r/R_D$, where $R_D = \epsilon D_n/qn_0\mu_n|\gamma|$ is an effective Debye radius, Eq. (16) becomes

$$\frac{\partial^2 N}{\partial X^2} + \frac{1}{X}\frac{\partial N}{\partial X} - N = 0 \tag{18}$$

for $\gamma > 0$, and

$$\frac{\partial^2 N}{\partial X^2} + \frac{1}{X}\frac{\partial N}{\partial X} + N = 0 \tag{19}$$

for $\gamma < 0$. The boundary conditions corresponding to filamentary solutions are

$$\frac{\partial N}{\partial X}\bigg|_{r=0} = 0 \tag{20}$$

73

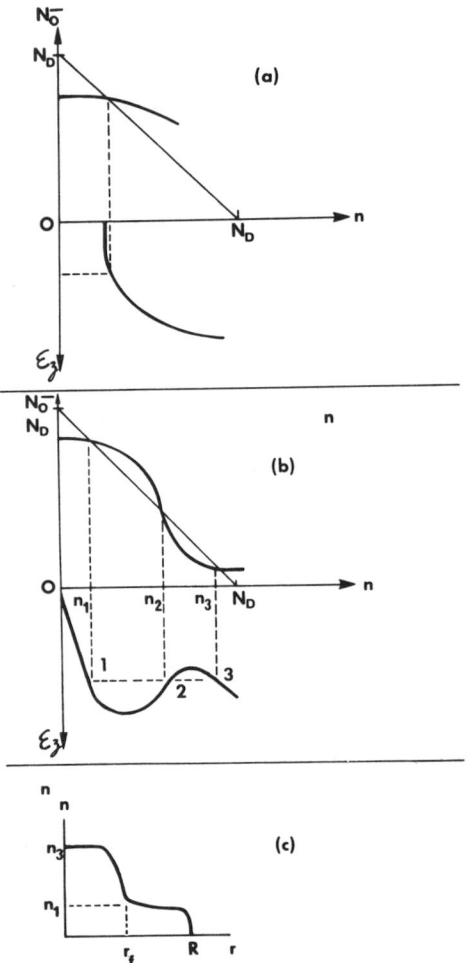

FIG. 7. N_0^- versus n and n versus ϵ_z (see text for definition): (a) normal case for positive differential conductance; (b) SNDC case (the number of filled traps decreases as n increases); (c) qualitative carrier density profiles for a filament in a semiconductor having one type of carrier.

and

$$N|_{x\to\infty} = 0. \qquad (21)$$

The solution of Eq. (18) is

$$N = Z_0(ix), \qquad (22)$$

where

$$Z_0(ix) = C_1 I_0(X) + C_2 K_0(X). \qquad (23)$$

I_0 is a modified Bessel function of the first kind and K_0 is a modified Bessel function of the second kind; the C's are constants. From Eq. (21) we have that $C_1 = 0$, but then we cannot fulfill the other boundary condition [Eq. 20]. We thus conclude that filamentation does not occur for $\gamma > 0$.

For $\gamma < 0$, Eq. (19) has a solution,

$$\delta n = C J_0(X), \qquad (24)$$

which satisfies the boundary conditions. Again, J_0 is a Bessel function of zeroth order. Equation (24) implies that a nonuniform radial carrier density distribution (current filamentation) can occur if for some value of n a value of $\gamma < 0$ results.

We can appreciate the physical significance of the above result by considering Eqs. (12) and (10) in the steady state and graphically solving the two equations

$$N_0^- = N_D - n_0 \qquad (25)$$

and

$$N_0^- = F(n_0, \epsilon_z). \qquad (26)$$

The solution also determines the dependence of n on ϵ_z; both N_0^- and ϵ_z as functions of n are shown in Fig. 7. Note that condition $\gamma < 0$ [point 2 in Figure 7(b)] implies that an SNDC curve determines the relation between n and ϵ_z. Hence SNDC directly leads to filamentation.

It follows from the small-signal solution presented above that the characteristic size of the filament walls is approximately the Debye radius R_D, which is characteristically quite small. We can, therefore, expect the following concentration profile to result (see Fig. 7): an extensive high-density core of radius $r_f \gg R_D$, a thin charged filament wall several R_D's thick, a carrier concentration within the core of n_3, and a carrier concentration far from the filament of n_1.

Although the above argument is based on small-signal considerations, we can understand large-signal solutions in the regime where the wall of the filament is small compared to the diameter of the filament. In this regime, we can approximate Eq. (11) as

$$\frac{\partial \epsilon r}{\partial r} \simeq \frac{q}{\epsilon}[N_D - F(n, \epsilon_z) - n], \qquad (27)$$

and make use of Eq. (9) to obtain

$$\frac{\partial}{\partial n}\epsilon_r^2 = -\frac{D_n}{\mu_n}\frac{N_D - F(n, \epsilon_z) - n}{n}, \qquad (28)$$

or

$$\epsilon_r^2 = -\frac{D_n}{\mu_n}\int_{n_1}^n \frac{N_D - F(n, \epsilon_z) - n}{n}dn. \qquad (29)$$

By considering Fig. 7(b), we see that such a solution is only possible if the $n(\epsilon_z)$ relation exhibits a region of SNDC. Furthermore, since from Eq. (9) and Fig. 7(c) we see that ϵ_r will vanish except in the walls of the filament, then

$$\int_{n_1}^{n_3} \frac{N_D - F(n, \epsilon_z) - n}{n}dn = 0, \qquad (30)$$

which is the analogue of the "equal-areas rule" developed for NNDC systems.[48] It must also be true that

$$n_1 + F(n_1, \epsilon_z) = N_D \qquad (31a)$$

and

$$n_3 + F(n_3, \epsilon_z) = N_D. \qquad (31b)$$

By solving Eqs. (30) and (31), we can find the values n_1, n_3, and ϵ_z that are uniquely specified for a given function $f(n_1, \epsilon_z)$. This means that an SNDC sample containing a filament is a voltage limiter. The current in the sample can increase at a fixed value of ϵ_z simply by having r_f expand, i.e.,

$$I = q\mu_n\epsilon_z\pi[n_3 r_f^2 + n_1(R - r_f)^2]. \qquad (32)$$

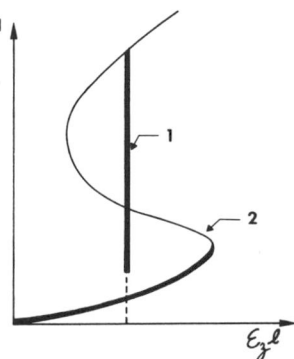

FIG. 8. Qualitative current-voltage characteristic for uniform applied fields for an SNDC sample containing a filament (curve 1) and in the absence of filamentation (curve 2, the homogeneous solution, which is unstable against the formation of a filament).

Figure 8 depicts the homogeneous and filamentary characteristics expected from the above considerations.

C. Filamentation in a semiconductor having two types of charge carriers

We can extend these results to the case in which two types of charge carriers are present. Consider the cylindrical geometry again and imagine, e.g., that electron-hole pairs are produced by impact ionization. Assume quasineutrality holds. The equations describing the radial motion in such a system are then

$$j_n = q\left(\mu_n n\epsilon_r + D_n \frac{\partial n}{\partial r}\right), \tag{33a}$$

$$j_p = q\left(\mu_p p\epsilon_r + D_p \frac{\partial p}{\partial r}\right), \tag{33b}$$

$$\frac{1}{q}\frac{1}{r}\frac{\partial}{\partial r}(rj_p) = g_n(\epsilon_z,n) + g_p(\epsilon_z,p)p - Rnp - p/\tau(n,p), \tag{34}$$

$$\frac{\partial}{\partial r} r(j_p + j_n) = 0, \tag{35}$$

$$n = n_0 + p, \tag{36}$$

where n and p are the electron and hole concentrations, n_0 is the electron concentration produced by the shallow donors, g_n and g_p are the generating terms describing impact ionization by electrons and holes, R is the coefficient of binary recombination, and $\tau(n, p)$ is a linear recombination time. Equations (33)–(36) yield

$$j_p = -qD_A(p)\frac{\partial p}{\partial r} \tag{37}$$

and

$$\frac{\partial j_p}{\partial r} + \frac{j_p}{r} = -qf(p), \tag{38}$$

where

$$D_A(p) \equiv \frac{D_n\mu_p p + D_p\mu_n(n_0 + p)}{\mu_n(n_0 + p) + \mu_p p} \tag{39}$$

is the ambipolar diffusion coefficient and $f(p)$ is given by

$$f(p) \equiv Rnp + (p/\tau) - g_n - g_p p \equiv R_e - G, \tag{40}$$

where R_e and G are the total recombination and generation rates, respectively.

As before, we first treat the small-signal case, $p = p_0 + \delta p(r)$, where $\delta p(r) \ll p_0$ and p_0 is given by the solution of Eq. (38) for a uniform sample, i.e., $f(p_0) = 0$. Equations (37) and (38) yield

$$\frac{\partial^2 \delta p(r)}{\partial r^2} + \frac{1}{r}\frac{\partial}{\partial r}\delta p(r) - \frac{f'_0}{D_A(p_0)}\delta p(r) = 0, \tag{41}$$

where $f'_0 = \partial f/\partial p \big|_{p=p_0}$. In dimensionless form, with $N = \delta p/p_0$ and $x = r/L_D$, where $L_D = [D_A(p_0)/|f'_0|]^{1/2}$, we again obtain Eqs. (18) and (19), depending on the sign of f'_0. Identical arguments as those following Eq. (19) apply in this case. A filamentary solution is possible if $f'_0 < 0$; an example of the variation of f and ϵ_z on p is shown in Fig. 9. A similar large-signal discussion also follows. For a filament with a wide flat core ($r_f \gg L_D$), Eq. (38) can be approximated as

$$\frac{\partial j_p}{\partial r} = -qf(p); \tag{42}$$

from Eqs. (42) and (37), we obtain

$$\frac{\partial}{\partial p}j_p^2 = 2q^2 D_A(p)f(p), \tag{43}$$

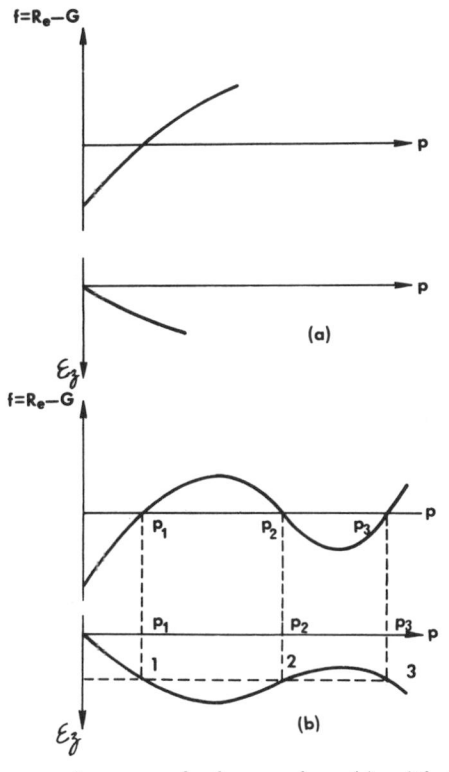

FIG. 9. f versus p and p versus ϵ_z for the normal case (a) and the SNDC case (b). (See text for definitions.)

which yields

$$j_p^2 = 2q^2 \int_{p_1}^{p} D_A f(p)dp, \tag{44}$$

where

$$\int_{p_1}^{p_\lambda} D_A f(p)dp = 0 \tag{45}$$

is the analogue of the equal-areas rule and determines the relationship between the carrier concentrations inside and outside the filament. The $I(\epsilon_z)$ characteristic that results is similar to the one shown in Fig. 8, i.e.,

$$I = q\epsilon_z \pi \{ [\mu_p p_1 + \mu_n (p_1 + n_0)](R - r_f)^2 + [\mu_p p_3 + \mu_n (p_3 + n_0)]r_f^2 \}. \tag{46}$$

The major difference in the properties of the current filaments in systems with one or two types of carriers lies in the concentration profile within the walls of the filament. In the latter case, the walls are generally wider, the width being determined by the effective ambipolar diffusion length.

III. ISOTHERMAL ANALYSIS

A. Trapping and switching in semiconductors

Our model of switching in amorphous chalcogenide semiconductors involves the field-induced filling of positively and negatively charged traps. Hence, we first demonstrate that the filling of traps in general can lead to an SNDC curve (Fig. 8, curve 2) by considering a simple idealized system where there are N_D ionized shallow donors and N_A deep traps which are negatively charged when filled by electrons.

In the absence of field-induced carrier generation the concentration of electrons in the conduction band is

$$n = N_D - N_A. \tag{47}$$

When electron-hole pairs are generated, the system can be described by the Shockley-Read equations

$$G = \alpha_n n N_A (1 - f), \tag{48}$$

$$N_D - n - N_A f + p = 0, \tag{49}$$

and

$$G = \alpha_p p N_A f, \tag{50}$$

where G is the field-induced generation rate and $f \equiv N_A^- / N_A$, where N_A^- is the concentration of filled acceptors. For simplicity, we assume that the generation coefficients for electrons and holes are equal, so that

$$G = A(n + p)\lambda_\epsilon, \tag{51}$$

where A is a constant and λ_ϵ is a dimensionless monotonically increasing function of the applied field. We can now rewrite Eqs. (48)–(50) in dimensionless form as

$$\lambda(N + P) = N(1 - f), \tag{52}$$

$$P - N + N_0 + 1 - f = 0, \tag{53}$$

and

$$\xi(N + P) = Pf, \tag{54}$$

where $N = n/N_A$, $N_0 = (N_D - N_A)/N_A$, $P = P/N_A$, $\lambda = A\lambda_\epsilon / \alpha_n N_A$, and $\xi = \alpha_n / \alpha_p$. From Eq. (52), we have

$$f = 1 - \lambda - \lambda P/N. \tag{55}$$

Substituting Eq. (55) into Eqs. (53) and (54), we find

$$PN - N^2 + N_0 N + \lambda N + \lambda P = 0 \tag{56}$$

and

$$(P^2/N) - (\lambda^{-1} - 1 - \xi)(P/N) + \xi = 0. \tag{57}$$

The solutions of Eqs. (56) and (57) yield

$$P/N|_{+,-} = \tfrac{1}{2}(\lambda^{-1} - 1 - \xi) \pm \{ [\tfrac{1}{2}(\lambda^{-1} - 1 - \xi)]^2 - \xi \}^{1/2} \tag{58}$$

and

$$N = \frac{N_0 + \lambda(1 + P/N)}{1 - P/N}. \tag{59}$$

The existence of two possible solutions for N, P for a given valuve of λ indicates that an SNDC characteristic is present. Analysis of Eq. (58) shows that two solutions exist for

$$\frac{1}{2(1 + \xi)} \leqslant \lambda \leqslant \frac{1}{[1 + (\xi)^{1/2}]^2}. \tag{60}$$

Equation (60) determines the range of voltages over which NDC exists (see Fig. 8, curve 2). A physical explanation of this phenomenon is the following. Under conditions of impact ionization, the generation rate is the proportional to the carrier concentration [Eq. (51)]. For small concentrations at a given voltage, the generation rate is small, but if a larger concentration of electron-hole pairs can also exist for the same voltage, then a larger generation rate will also occur. Furthermore, the recombination mechanism depends not only on the carrier concentration, but also on whether or not the traps are filled by holes. For example, we might have the following two states corresponding to one value of voltage: (1) low electron concentration, traps not filled by holes (OFF); (2) high electron concentration, traps substantially filled by holes (NDC). In essence, carrier generation produces pairs; the holes neutralize the traps that were compensated by electrons and the generated electrons enhance the conductivity of the material by increasing the denisty of free carriers. In our model, both the generated electrons and holes fill the charged traps, thereby increasing the mobility, and also provide excess carriers. The conductivity is enhanced by both a mobility and a carrier-concentration increase.

B. The delay time

Experimental evidence[8,9,20,47] supports the view that for times greater than about 1 μsec, observed t_d's in both memory and threshold samples of amorphous chalcogenide films can be explained by a critical field, electrothermal model. However, numerical calculations[14] and short-pulse experiments[8] strongly indicate that it is still reasonable to

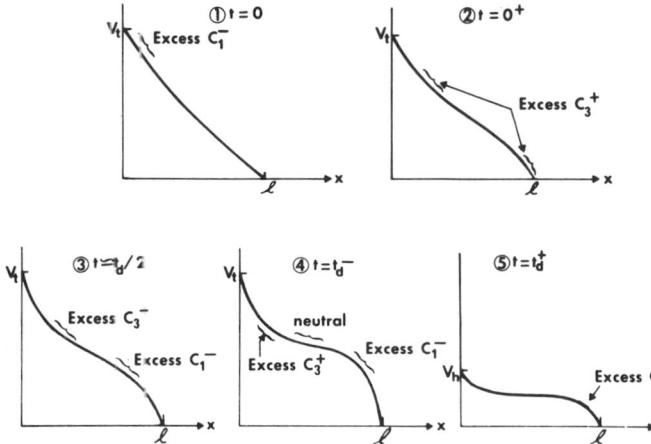

FIG. 10. Suggested voltage versus distance profiles before ($t < 0$), during ($0 \leqslant t \leqslant t_d$), and after ($t > t_d$) an isothermal switching event. At $t = 0$, excess C_1^-'s are fixed near the anode in order to maintain the high anode field. At $t = 0^+$, excess C_3^+'s appear on the anode side since holes produced in the high-field region neutralize the C_1^- centers. The excess C_3^+'s produced near the cathode maintain the high field there. At $t = \frac{1}{2}t_d$, excess C_1^-'s are produced as electrons neutralize C_3^+ centers.

assume that t_d's observed in the range 10^{-7}–10^{-9} sec[8,54] have an appreciable electronic component. A mechanism that we propose for the isothermal contribution to t_d involves carrier generation in a high-field region near one electrode and the subsequent filling of charged C_3^+ and C_1^- traps[37] in the bulk.[14,46] If we consider a sample 1 μm in thickness, a delay time of 10^{-7} sec, and a field of a10^6 V/cm, we obtain for the mobility of the carriers that fill the traps during t_d a value of 10^{-5} cm^2/V sec, which is consistent with observed mobilities.[55]

We suggest that an isothermal delay-time process can proceed as follows. For partially blocking metal contacts[48] that form narrow barriers[56] on p-type material,[35,39] somewhat higher fields exist near the anode than the cathode when bias is applied. The high anode field is maintained by fixed excess C_1^- sites[38] near the anode ($C_3^+ \rightarrow e^+ + C_3^0$; holes are depleted from the anode region). Figure 10 shows what can occur when sufficient bias is applied so that the critical field is reached near the anode. At $t = 0$, holes are generated by the high field near the anode. (Note that this generation need not involve the valence-alternation pairs, but rather can originate from localized lone-pair electrons in the valence band tail.[36]) The generated holes drift into the bulk where they are captured by and neutralize C_1^- sites ($t = 0^+$). Hole capture, therefore, produces excess C_3^+ sites in the bulk, causing a high field to develop near the cathode in order to provide electrons that will drift toward the anode in an attempt to restore space-charge neutrality. The high field at the cathode region becomes depleted of electrons. Generated electrons entering the bulk from the cathode neutralize C_3^+ centers ($t \simeq \frac{1}{2}t_d$), thereby producing excess C_1^- centers near the cathode. These trapping events occur faster than the production of excess C_3^+ centers near the anode, since here no

coordination change is required. This causes more holes to enter from the anode, again to restore space-charge neutrality. Feedback is established. At $t = t_d^-$, we are just prior to switching; the isothermal contribution to t_d is the time required to fill all the charged traps, as in a PIN diode.[48,57]

Once all the excess C_3^+ sites are neutralized, a electron can transit the sample without being trapped. Switching to a high-current state occurs; after the C_1^- sites are neutralized, electrons and holes traverse the sample with an enhanced mobility. Note that neutralization of the C_1^- sites involves local atomic displacement since C_1^0 is unstable with respect to a bond adjustment to C_3^0. Aside from the current "spike" that occurs during the switching event,[4] electronically induced "polymeric-matrix relaxation"[58] as well as ON-state heating can also lead to forming[18] or memeory-type[2,42] effects. We discuss the possibility of such effects further in Sec. IV.

In the ON state, narrow sheets of fixed charges (depletion layers) sustain the high electrode fields; the anode has excess C_1^- (as it did in the OFF state) and the cathode has excess C_3^+. Hot electrons and holes are produced at the cathode and anode, respectively, and if the barrirs are sufficiently thin, strong-field-emission and thermionic-field-emission tunneling[59] can also take place from the metallic electrodes. The carriers will drift through the bulk without much recombination. The configuration is similar to avalanche injection in crystalline semiconductors.[60,61] The ON state is maintained by double avalanche injection, but only one type of mobile carrier is generated at each electrode, as in the breakdown of compensated Ge at low temperatures.[49] The above and ensuing description of the ON state has several features that are different from previously proposed electronic models.[62-67]

Although trapping events in the ON state are few because of the absence of charged traps, the trapping rates for electrons and holes at the neutral sites are similar and will lead to recombination. If a hole is trapped first, the new center will rapidly trap an electron and neutralize, and vice versa. C_3^0 sites trap holes and electrons at the same rate. As was pointed out in Sec. I E, when the sample is rapidly turned OFF ($V \rightarrow 0$), the anode field can readjust rapidly by capturing holes ($C_3^0 \rightarrow C_3^+$) whereas the cathode response is slower because of the $C_3^0 \rightarrow C_1^-$ transition.[28]

C. Formulation of the phenomenological kinetic equations

We have shown that carrier generation followed by a change in concentration of the filled traps can produce an SNDC characteristic which in turn leads to current filamentation. For the case of one type of carrier, it is trap emptying that produces SNDC and filamentation. For two types of carriers, we can have SNDC with either trap emptying or trap filling, depending on the details of the generation and recombination processes. We now use these ideas to generate and solve a set of phenomenological kinetic equations for an amorphous chalcogenide film containing equal numbers of C_3^+ and C_1^- trapping centers.

As discussed in Sec. I G and III B, we assume that in the OFF state the following recombination processes dominate: (1) hole capture by C_1^- centers ($C_1^- + e^+ \rightarrow C_1^0 \rightarrow C_3^0$) and (2) electron capture by C_3^+ centers ($C_3^+ + e^- \rightarrow C_3^0$). The second process should be faster since it does not involve a local bond rearrangement and thus has no activation barrier. This effect further enhances the p-type nature of the OFF state[42] if we assume that emission from the charged centers is a rare process. In our model, the charged centers act as recombination centers.

We further assume that in the ON state the C_3^+ and C_1^- centers are filled, so that the following processes are important: (1) hole capture by a neutral center ($C_3^0 + e^+ \rightarrow C_3^+$) and (2) electron capture by a neutral center ($C_3^0 + e^- \rightarrow C_3^- \rightarrow C_1^-$). The latter process is slower since it requires a local bond rearrangement which involves an activation barrier.[35] However, immediately after either hole or electron capture by a neutral center, the charged centers so created will rapidly capture a mobile carrier of opposite charge from the dense background of mobile carriers present in the ON state: $C_3^0 + e^+ \rightarrow C_3^+ + e^- \rightarrow C_3^0$, a slow and then very fast process with no bond modification; $C_3^0 + e^- \rightarrow C_1^- + e^+ \rightarrow C_3^0$, a very slow and then very fast process with two-bond modifications. Both processes will be much slower than the OFF-state events because the C_3^0 centers are neutral. The ON state should therefore be intrinsic in the sense that the same number of oppositely charged mobile carriers are present. It could, however, appear as n or p type in nature depending upon the relative band mobilities and/or the nature of the contact conditions, which, e.g., could be more blocking for one of the carrier species.[21]

Taking into account the recombination processes described above, we suggest that the following set of phenomenological kinetic equations be used to describe a thin amorphous chalcogenide film. For holes,

$$\frac{1}{q}\operatorname{div}j_p + \frac{\partial p}{\partial t} = G_t - \gamma_p N_3^0 p - \beta_p N_1^- p, \tag{61}$$

where G_t is the total generation rate (thermal plus electronic) for electrons and holes, N_3^0 is the concentration of C_3^0 centers, N_1^- is the concentration of C_1^- centers, and γ_p and β_p are capture coefficients. The second term on the right-hand side describes the capture of holes by neutral centers and the third term describes the capture by C_1^- centers. As stated above, we expect that $\gamma_p < \beta_p$. For electrons,

$$-\frac{1}{q}\operatorname{div}j_n + \frac{\partial p}{\partial t} = G_t - \gamma_n N_3^0 p - \beta_n N_3^+ n. \tag{62}$$

According to the above arguments we expect that $\beta_n > \beta_p > \gamma_n$; electron capture by C_3^+ is faster than hole capture by C_1^-, which is faster than hole capture by C_3^0, which in turn is faster than electron capture by C_3^0. In general,

$$\frac{\partial N_3^+}{\partial t} = \gamma_p N_3^0 p - \beta_n N_3^+ n - A N_3^+ N_1^- + B(N_3^0)^2, \tag{63}$$

where the last two terms on the right-hand side take into account the reaction,

$$2C_3^0 \rightleftarrows C_3^+ + C_1^-, \tag{64}$$

where A and B are reaction coefficients. Similarly,

$$\frac{\partial N_1^-}{\partial t} = \gamma_n N_3^0 n - \beta_p N_1^- p - A N_3^+ N_1^- + B(N_3^0)^2. \tag{65}$$

Also, the total number of traps N_T is given by

$$N_T = N_3^0 + N_3^+ + N_1^-. \tag{66}$$

Adding Eqs. (61) and (63) and then substracting Eqs. (62) and (65) yields the continuity equation,

$$\operatorname{div}j + \frac{\partial p}{\partial t} = 0, \tag{67}$$

where

$$\rho = q(p - n + N_3^+ - N_1^-) \tag{68}$$

is the space charge, and

$$j = j_p + j_n \tag{69}$$

is the total current density.

In the absence of mobile charge carriers the steady-state equations are

$$A N_3^+ N_1^- = B(N_3^0)^2; \tag{70}$$

$$N_T = N_3^+ + N_3^0 + N_1^-; \tag{66}$$

$$N_3^+ = N_1^-. \tag{71}$$

Equation (71) shows that in the absence of mobile carriers the C_3^+ and C_1^- centers can be created or destroyed only pair-wise [see Eq. (64)]; it follows from the condition of charge neutrality in the absence of mobile carriers. According to the OFF-state model of chalcogenide materials[38] $B \gg A$, so that when p and n are small enough, the first two terms on the right-hand side of Eqs. (63) and (65) are negligible, and

$$N_3^+ = N_1^- \simeq \tfrac{1}{2} N_T. \tag{72}$$

D. Kinetic equations for the OFF state

In the OFF state, p, n, and N_3^0 are all very small compared to N_3^+ and N_1^-. Under these conditions, the kinetic equations, Eqs. (61)–(63) and Eq. (65), become

$$\frac{1}{q}\operatorname{div}j_p + \frac{\partial p}{\partial t} \simeq G - \frac{p - p_0}{\tau_p}, \tag{73}$$

$$-\frac{1}{q}\operatorname{div}j_n + \frac{\partial n}{\partial t} \simeq G - \frac{n - n_0}{\tau_n}, \tag{74}$$

$$\frac{\partial N_1^-}{\partial t} \simeq -\frac{p - p_0}{\tau_p} + G, \tag{75}$$

$$\frac{\partial N_3^+}{\partial t} \simeq -\frac{n - n_0}{\tau_n} + G, \tag{76}$$

where $G = G_t - G_{th}$ is the generation rate due to the electric field, which for the moment we assume is uniform across the sample, and

$$\tau_p \simeq 2/\beta_p N_T, \tag{77a}$$

$$\tau_n \simeq 2/\beta_n N_T; \tag{77b}$$

$$n_0 = G_{th}\tau_n, \tag{78a}$$

$$p_p = G_{th}\tau_p. \tag{78b}$$

Equations (73)–(78) represent the following model. Holes are trapped by C_1^- centers and electrons by C_3^+ centers; their concentrations diminish and the reaction of Eq. (64) proceeds to the right in order to replenish the C_3^+ and C_1^- centers that have been neutralized via trapping. Re-emission of carriers from the C_3^0 centers is assumed negligible with respect to the reacton of Eq. (64). Again, in this limit, the charged centers are recombination rather than trapping centers.

Since $\beta_n > \beta_p$, then $\tau_n < \tau_p$ and $n_0 < p_0$; again, there are more free holes than electrons and the OFF state is p type.

Equations (73)–(78) are valid as long as

$$\beta_n \tfrac{1}{2} n N_T \ll A (\tfrac{1}{2} N_T)^2. \tag{79}$$

The left-hand side of Eq. (79) represents the rate of production of C_3^0 centers by electron trapping and the right-hand side is the rate of production of C_3^0 centers via the reation of Eq. (64). If we make the reasonable assumption that Eq. (78) is valid up to the relatively high field-induced generation level $G \gg G_{th}$, then the condition of validity of Eqs. (73)–(78) becomes

$$G \overset{\sim}{<} \tfrac{1}{4} A N_T^2. \tag{80}$$

In the range of validity of Eq. (80), we may assume that $n \ll p$; the steady OFF-state conductivity is then

$$\sigma \simeq \sigma_p = q\mu_p p. \tag{81}$$

Furthermore, if we assume that G is related to an impact ionization process, then

$$G = pg(\epsilon) \tag{82}$$

in steady state. Equation (73) yields

$$p = p_0/[1 - g(\epsilon)\tau_p], \tag{83}$$

and thus

$$\sigma = q\mu_p p_0/[1 - g(\epsilon)\tau_p]. \tag{84}$$

Equation (84) describes avalanche breakdown in the OFF state. The critical field for breakdown ϵ_c is determined by the condition

$$g(\epsilon_c) = \tau_p^{-1}. \tag{85}$$

For voltages below V_t the conductivity will be field dependent and, for $g(\epsilon)\tau_p \ll 1$, Eq. (84) yields

$$\sigma \simeq q\mu_p p_0[1 + g(\epsilon)\tau_p]. \tag{86}$$

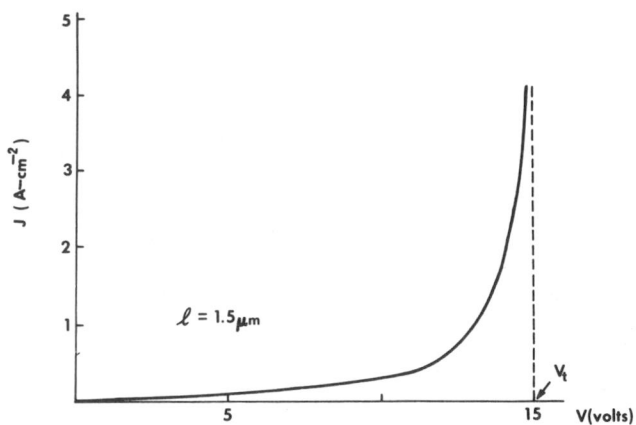

FIG. 11. Estimated $I(V)$ characteristic in the isothermal OFF state for a uniform applied electric field. $L = 1.5\ \mu$m.

There are several forms of the conductivity expression that can be used for modeling carrier generation effects. For example, in Eq. (9) we used a form $\sigma \sim \sigma \exp(-\epsilon_0/\epsilon)$. Another possibility is to write

$$g(\epsilon) = A' \exp(\epsilon/\epsilon_0'), \tag{87}$$

where $\epsilon_0' = CT$.[68] Taking $T = 300\ °$K, $C = 200$ V/cm $°$K($\epsilon_0' = 6 \times 10^4$ V/cm), the band mobility $\mu_p \simeq 10$ cm^2/V sec, $\tau_p \simeq 10^{-8}$ sec, $\epsilon_c \simeq 10^5$ V/cm, and $\sigma_0 \simeq 1.5 \times 10^{-6}\ \Omega$ cm^{-1} yields $p_0 \simeq 10^{12}$ cm^{-3} and $A' \simeq 2 \times 10^7$ sec^{-1}. (If we were to choose μ_p as the drift mobility, then p_0 would be the free plus trapped hole concentrations at states near the valence-band edge.) The resulting $I(V)$ characteristic for this *uniform field* OFF-state case is shown in Fig. 11. Note that the $I(V)$ characteristic is of the form obtained experimentally for short (2 nsec) pulses.[9] Agreement is expected here since (1) isothermal conditions prevail, (2) the fields across the sample will be relatively uniform, and (3) the exciting voltage pulse is turned off prior to the time it takes for the generated carriers to propagate and fill traps, the condition where *all* the traps are filled in the sample. A feedback loop is not yet established and the resulting characteristics are simply those representative of an avalanche diode. Were we to model an actual sample where thermal and contact effects are treated, delay-time-mode switching would then occur at points on the OFF state $I(V)$ characteristic below V_t (see Fig. 11), and the avalanche characteristic would be masked.

E. Kinetic equations for the ON state

In the ON state, the electron and hole concentrations are sufficiently large so that the C_3^+ and C_1^- centers are almost completely neutralized. When p and n are sufficiently large, we assume that the rates of production and decay of the charged centers due to the capture of mobile carriers are much larger than the corresponding rates due to the reaction of Eq. (64). This situation is opposite to that of the OFF state, where p and n are both sufficiently small so that the reaction of Eq. (64) is almost completely responsible for the small value of N_3^0 and the fact that $N^+ = N^- \simeq \frac{1}{2}N_T$. Now, since p, n, and N_3^0 are large compared to N_3^+ and N_1^-, and $N_3^0 \simeq N_T$, Eqs. (61)–(63) and Eq. (65) become

$$\frac{1}{q}\,\mathrm{div}j_p + \frac{\partial p}{\partial t} \simeq G_t - \gamma_p N_T p - \beta_p N_1^- p, \tag{88}$$

$$-\frac{1}{q}\,\mathrm{div}j_n + \frac{\partial n}{\partial t} \simeq G_t - \gamma_n N_T n - \beta_n N_3^+ n, \tag{89}$$

$$\frac{\partial N_3^+}{\partial t} = \gamma_p N_T p - \beta_n N_3^+ n, \tag{90}$$

$$\frac{\partial N_1^-}{\partial t} = \gamma_n N_T n - \beta_p N_1^- p. \tag{91}$$

In the steady state, these become:

$$N_3^+ = \gamma_p N_T p / \beta_n n, \tag{92}$$

$$N_1^- = \gamma_n N_T n / \beta_p p, \tag{93}$$

$$\frac{1}{q}\,\mathrm{div}j_p = G_t - \gamma_p N_T p - \gamma_n N_T n, \tag{94}$$

$$-\frac{1}{q}\,\mathrm{div}j_n = G_t - \gamma_n N_T n - \gamma_p N_T p. \tag{95}$$

Furthermore, when either (1) the current is uniformly distributed in the steady ON state (i.e., the filament fills the entire sample), or (2) we consider only the central part of the filament where the carrier density is uniform, we have

$$G_t - \gamma_p N_T p - \gamma_n N_T n = 0 \tag{96}$$

and

$$p - n + \frac{\gamma_p N_T p}{\beta_n n} - \frac{\gamma_n N_T n}{\beta_p p} = 0. \tag{97}$$

For large generation levels, Eqs. (96) and (97) have as a solution

$$p = n = G_t \tau, \tag{98}$$

where

$$\tau^{-1} = N_T(\gamma_p + \gamma_n) \simeq N_T \gamma_p, \tag{99}$$

provided $\gamma_p \gg \gamma_n$. Equation (98) is valid if

$$p = n \gg \frac{\gamma_p N_T}{B_n}, \quad \frac{\gamma_n N_T}{B_p}. \tag{100}$$

Equations (98)–(100) then yield

$$G \gg \frac{\gamma_p^2 N_T^2}{B_n}, \quad \frac{\gamma_p \gamma_n N_T^2}{\beta_p}, \tag{101}$$

which determines the lower limit of the generation rate required to maintain the ON state.

We now estimate order of magnitude values for γ_p and β_n. If we assume that the capture cross section S_p for hole trapping by a C_3^0 center is about 10^{-20} cm^2 (a reasonable value), and the thermal velocity v_t is about 10^7 cm/sec, then $\gamma_p \simeq S_p v_t \simeq 10^{-13}$ cm^{-3} sec^{-1}. Furthermore, for $N_T \simeq 10^{19}$ cm^{-3} and $\tau_n \simeq 10^{-11}$ sec, we find $\beta_n \simeq 2/N_T \tau_n \simeq 2 \times 10^{-8}$ cm^{-3} sec^{-1}. With these values and Eqs. (98) and (99), we can obtain a portrait of the ON state. First, the mobile-carrier density $p = n \gg 10^{14}$ cm^{-3}. Next, for band mobilities of the order of 10 cm^2/V sec and $p = n \simeq 10^{19}$ cm^{-3}, we obtain a minimum conductivity in the ON state of $\sigma_{\min} = q(\mu_p + \mu_n)p \simeq 20\,\Omega^{-1}$ cm^{-1}. Finally, for holding voltages of about 1 V and bulk holding fields in the ON state ϵ_h of about 10^3 V/cm, we obtain for the minimum holding current density $j_{h\,\min} = \sigma_{\min}\epsilon_h \simeq 2\times 10^4$ A/cm^2, which is a typical value observed experimentally.[6]

F. A model for the ON state

We may now write a set of equations for the ON state as follows:

$$p = n,$$
$$\frac{1}{q}\,\mathrm{div}j_p = G_t - p/\tau,$$
$$-\frac{1}{q}\,\mathrm{div}j_n = G_t - n/\tau, \tag{102}$$
$$j_p = q\mu_p p\epsilon - qD_p\,\mathrm{grad}\,p,$$
$$j_n = q\mu_n n\epsilon - qD_p\,\mathrm{grad}\,n.$$

Our thrust will be to solve this set of equations for a system where field-induced carrier generation occurs in a very narrow region near the electrodes and the current is nonuniform in a direction transverse to the direction of current flow. Prior to this, however, we will estimate numerical values for some important parameters that characterize the ON state and discuss a qualitative model.

We first estimate the trapping time $\tau \simeq (\gamma_p N_T)^{-1} \simeq 10^{-6}$ sec, and dielectric relaxation time $\tau_\mu \simeq \epsilon/\sigma \simeq 3 \times 10^{-13}$ sec, in the ON state. Since $\tau_\mu \ll \tau$, the ON state behaves like a "lifetime" rather than "relaxation" semiconductor.[69,70] The product of τ_μ and v_t yields the characteristic length of the charge operation, which in our case is about 10^{-6} cm. We have, therefore, a quasineutral situation since the charge separation is much less than the length of the sample.

Next, we estimate the ambipolar diffusion coefficient, $D_A \simeq \mu kT/q \simeq 0.3$ cm^2/sec, and diffusion length $L_D = (D_A \tau)^{1/2} \simeq 6 \times 10^{-4}$ cm. Note that *typical* sample lengths l and radii R are such that $L_D > l$ and $R > L_D$.

Since V_h is independent of l over a wide range, we expect that the field-induced generation of carriers in the ON state takes place in narrow regions near the electrodes, perhaps even less than 100 Å in thickness. The field, at least in part, of these regions is at ϵ_c; electrons and holes created by ϵ_c drift and diffuse out of these regions without recombining to any great extent *anywhere* in the sample (since $L_D > l$). To determine which of the two transport processes (drift or diffusion) dominate, we compare the drift current density $j_{\mathrm{drift}} \simeq \sigma_{\mathrm{on}} V_h/l \simeq 2q\mu_p \epsilon_h$, with the diffusion current density $j_{\mathrm{diff}} = qD_A \partial p/\partial x \simeq qD_A p/l$. Using the Einstein relation, their ratio is $2q\epsilon_h l/kT \simeq 100$; the drift current is clearly dominant.

We reinforce the following basic model for the ON state. Two very thin regions at the contacts generate electrons (at the cathode) and holes (at the anode) at equal rates; the holes drift toward the cathode and the electrons toward the anode (double avalanche injection[60,61]). Their concentrations are equal and the electron-hole plasma (droplet?[64]) is uniform in the direction of current flow, since $L_D > l$. Because we have a homogeneous plasma, we can introduce, instead of G, an effective generation rate G_{eff} which does not depend on position. The simplest case is to assume that G_{eff} does not depend upon the applied voltage; it is just a constant. The conditions for the steady ON state then yield the following estimate: $G_{eff} \simeq p/\tau \simeq 10^{24}$ sec^{-1}. (We shall use this simple model in what follows. If required, however, a dependence of G_{eff} on bias or carrier density can readily be incorporated into the model.)

It is useful now to compare the inverse of the characteristic generation time at *threshold*, $g(\epsilon_c) \simeq \tau_p^{-1} \simeq 10^8$ sec^{-1}, with the analogous ON-state parameters at *holding*, $g_{on} \simeq G_{eff}/p = \tau^{-1} \simeq 10^6$ sec^{-1}. We see that due to the much longer trapping time in the ON state, g_{on} is two orders of magnitude smaller than $g(\epsilon_c)$. Hence the holding voltage can be much smaller than the threshold voltage. [Furthermore, since $\tau_p \simeq 2/\beta_p N_T$, we expect that $g(\epsilon_c)$ should be relatively insensitive to the carrier concentration. The results of photoexcitation experiments[16] suggest that this is indeed the case over a moderate range of photoinduced conductance changes; V_t is independent of light intensity in this range. However, V_t can be lowered by sufficiently intense optical radiation.[71]] At threshold, we have seen that a current runaway (avalanche) will occur with *no* voltage switchback until the bulk traps fill. When the traps are filled (at high current levels, about the same order of magnitude as I_h) the voltage across the sample drops because a smaller inverse generation time is sufficient to compensate for recombination via the neutral traps. If ϵ_c is reached uniformly across the sample, then excess current will flow while the traps fill. If ϵ_c is reached in only a small part of the sample, the amount of excess current that will flow during t_d could depend upon whether one or both types of carriers are produced locally. If we are biased in the statistical regime[8,18,72] just at threshold, it is possible to observe no excess current flow at all during t_d.[72]

The foregoing arguments define two distinct states in which an amorphous chalcogenide film can carry current: an ON state with a current density $j_{on} \simeq 10^4$ A/cm^2 and an OFF state where the current density is several orders of magnitude smaller. The maximum ON state current will be given by $I_{max} = j_{on}A$, where A is the cross-sectional area of the sample.

Suppose that we supply current to the system such that it has a value below I_{max} but greater than the critical switching current I_t. The sample can accommodate such a current by forming a current filament, as discussed in Sec. II. A qualitative model for the filament ("pancake") is illustrated in Fig. 12. In the regions of the sample which are in the OFF state, the carrier concentration is very small; the center of the filament is in the ON state. Carriers diffuse from the core of the filament to the transition region between the ON and OFF states. The steady state radially dependent equations that describe the *transition* region (i.e., the walls of the filament) are

$$j_{pr} = q\mu p\epsilon - qD\partial p/\partial r, \tag{103}$$

$$j_{nr} = q\mu n\epsilon + qD\partial n/\partial r, \tag{104}$$

$$\frac{1}{r}\frac{\partial}{\partial r} rj_{nr} + \frac{1}{r}\frac{\partial}{\partial r} rj_{pr} = 0, \tag{105}$$

$$n = p, \tag{106}$$

$$\frac{1}{q}\frac{1}{r}\frac{\partial}{\partial r} rj_{pr} = -p/\tau_{eff}, \tag{107}$$

where τ_{eff} is an effective lifetime in the transition region comprising the walls of the filament (we estimate $\tau_{eff} \simeq \tau\tau_p \simeq 10^{-7}$ sec). Equations (103) and (104) describe the radial flux of holes and electrons, where we have assumed that $\mu_n = \mu_p$ and $D_n = D_p = D$. The sample is cylindrical and the ON-state plasma is homogeneous in the z direction. Equation (105) is the continuity equation, Eq. (106) represents quasineutral conditions, and Eq. (107) is the continuity equation for holes. Since $J_{r\,total} \equiv j_{nr} + j_{pr} = 0$, then $j_{nr} = -j_{pr}$. Using this fact, we obtain from Eqs. (103), (104), (106), and (107):

$$j_{pr} = -qD\partial p/\partial r, \tag{108}$$

$$\frac{1}{q}\frac{\partial}{\partial r}j_{pr} + \frac{1}{q}\frac{j_{pr}}{r} = \frac{p}{\tau_{eff}}. \tag{109}$$

These yield

$$D\frac{\partial^2 p}{\partial r^2} + \frac{D}{r}\frac{\partial p}{\partial r} - \frac{p}{\tau_{eff}} = 0, \tag{110}$$

the Bessel equation, which was analyzed in Sec. III. As before, we employ dimensionless variables: $y = p/p_0$, where $p_0(= G\tau)$ is the concentration inside the core of the filament; $x = r/D\tau_{eff})^{1/2}$. Equation (110) then becomes

$$\frac{\partial^2 y}{\partial x^2} + \frac{1}{x}\frac{\partial y}{\partial x} - y = 0, \tag{111}$$

whose solution is

$$y = z_0(ix), \tag{112}$$

where

$$z_0(ix) = C_1 I_0(x) + C_2 k_0(x). \tag{113}$$

I_0 is a modified Bessel function of the first kind and k_0 is a modified Bessel function of the second kind. C_1 and C_2 are constants determined from the boundary conditions

$$y(x_f) = 1; \tag{114a}$$

$$y(x \to \infty) = 0. \tag{114b}$$

Here, $x_f = r_f/D\tau_{eff})^{1/2}$ is the dimensionless well width. From the boundary conditions, we have that $C_1 = 0$ because

$I_0(x)$ increases exponentially for large r. The solution is

$$y(x) = k_0(x)/k_0(x_f).\qquad(115)$$

For $x \gg 1$, $k_0(x) \simeq (\pi/2x)^{1/2}e^{-x}$. Thus, if the core of the filament is much larger than the diffusion length ($x, x_1 \gg 1$), then

$$y(x) = (x_f/x)^{1/2}e^{-x}.\qquad(116)$$

This solution describes the diffusion of carriers radially outward from the filament and their subsequent recombination. The dimension of the filament walls should be of the order of a diffusion length; we therefore obtain for the minimum filament radius, $r_{f,\text{min}} \simeq (D\tau_{\text{eff}})^{1/2}/2 \simeq 10^{-4}$ cm, in reasonable agreement with experiment.[6] The model also predicts that the cross-sectional area of the filament increases linearly with current,[1] as experimentally observed[6] (Fig. 3). Furthermore, the maximum benign interruption time of the ON state, t_m, (see Sec. IA), which is about 100–500 nsec,[6,7,28] is determined by the lifetime in the transition region

$$t_{\text{sm}} \simeq \frac{\pi r_f^2}{4\pi D} = \tfrac{1}{4}\pi D j_{\text{on}}.$$

An approximate proportionality of t_{sm} and I has been experimentally observed.[7,28]

The ON state consists of about 10^{19}cm^{-3} electrons and holes moving through the conduction and valence bands, respectively.[6,21] If these are excited only from the C_1^- and C_3^+ centers, rather than from interband generation, then it restricts the maximum free-carrier concentration to the density of VAP's.[36] In this model, when additional current is applied after the material is in the ON state, the carrier concentration cannot increase; instead, the filament grows in area, maintaining a constant current density.[6] Such a result, however, is expected for systems that are effective voltage limiters, as the ON state is.

Petersen and Adler[28] have recently proposed another possible, quite similar model for the ON state in which the conduction takes place within the C_3^0 (centered about 0.05 eV below the conduction-band mobility edge) rather than in the valence and conduction bands. In this model, the high conductivity is initiated by a Mott transition in the C_3^0 band, which should occur at a critical carier concentration very close to that obvserved in the ON state. The most evident experimentally observable difference between that model and the one presented here is the dependence of the ON-state carrier concentration on the VAP density. In our model, the carrier concentration in the ON state should vary with the VAP density, while in the alternate model of Petersen and Adler it should be independent of the VAP density; however, a minimum VAP density exists in the latter, below which no switching can occur. At present, insufficient data exists on the VAP densities of different switching materials to even begin to critically test these models, but we should note that the two descriptions of the ON state are analogous to the two possibilities which arise when a semiconductor such as crystalline silicon is heavily doped—the material becomes de-

generate when either a Mott transition takes place in the impurity band or the Fermi energy enters the conduction band. There is no *a priori* way to determine which occurs first.

The Mott-transition model[28] has some further appealing features. First, it is consistent with having an ON-state barrier ($\simeq 0.4$–0.5 eV) less than the band gap in that this configuration allows only one type of ON-state carrier (electrons) to dominate, which appears to be the case.[6,21] Here, space-charge neutrality is maintained in the C_3^0 band by electrons donated from C_3^0 sites. (Holes will be less likely to move off C_3^0 sites because a change in local coordination would be required.) Second, it also provides for the second type of carrier required by the rather successful phenomenological equations of Walsh and Vezzoli.[67] (The first type is required to fill the C_3^+ and C_1^- sites; this generates the second type which moves in the C_3^0 band. Switching occurs when enough of the latter are present to cause a Mott transition.) Last, the ON-state luminescence observed by Walsh *et al.*[24] can be explained via the $2C_3^0 \rightarrow C_3^+ + C_1^-$ transition [Eq. (64)], which produces about 0.5 eV of energy.

The above model assumes that electrons dominate; hence the cathode barrier is assumed to drop the largest portion of the holding voltage. Another possibility for the ON state in this case would be simply the thermal activation from the C_3^0 impurity band to the conduction band. The activation energy for this case, about 0.05 eV, is consistent with the temperature dependence of the conductivity observed in the ON state. Finally, the ON-state luminescence[24] centered at about 0.5 eV could be associated with a collapsed band gap.[25] If this were the case, then the ON-state barrier would be comparable with the reduced band gap and either avalanche or double injection could sustain the ON-state current. (A partially collapsed band gap could result if the filling of traps decreased the electronic disorder in the system; localized states would transform to extended states.)

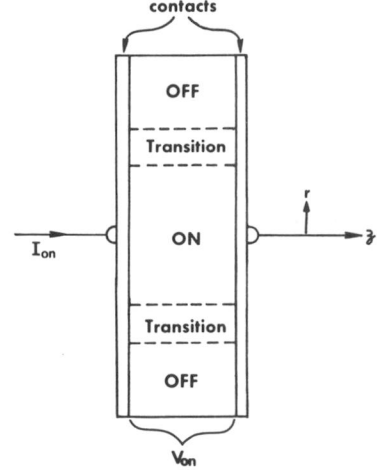

FIG. 12. Sketch of the ON-state current density configuration.

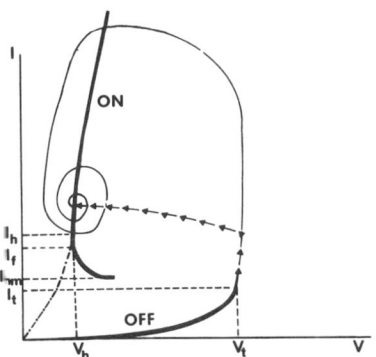

FIG. 13. Steady-state and time-dependent current-voltage characteristics of a threshold switch. The details are found in Sec. IV A of the text.

IV. SUMMARY, CONCLUSIONS, AND SPECULATIONS
A. Summary

The results of our analysis, coupled with experiments[6] and the circuit aspects,[4,19,27,73] are summarized in Figs. 12 and 13. Figure 12 depicts the filamentary current distribution in the ON state and Fig. 13 the complete $I(V)$ characteristics of an amorphous chalcogenide switch. The thick solid line represents the steady-state characteristics of a threshold switch under normal operating conditions,i.e., as observed on a curve-tracer oscilloscope. Here we may expect some thermal contributions to the OFF-state $I(V)$ curve, depending on the sample geometry, composition, and thermal boundary conditions. The thin solid line shows the actual current and voltage during the very rapid switching transition as determined by the local lumped-element circuit parameters that are always present,[4,19] but excluding the capacitances associated with the measuring circuit.[73] At any value of current the difference between the thin and thick lines represents the inductive voltage drop in the sample. Note the damped oscillatory development of the ON state. The near-vertical arrows (dashed) represent the critical field-induced avalanche and trap-filling processes, i.e., they represent the initial transport response of the carriers to the presence of a critical field, neglecting inductive and capacitive effects. In the usual delay-time mode, the avalanche process is masked by the trapping of carriers at the C_3^+ and C_1^- centers during the last part of the delay time. In this mode, Joule heating effects near the center of the sample first cause the conductivity there to rise and the field to rearrange. An enhanced concentration of thermally generated carriers fills some additional traps, so that when the critical field is reached near one of the electrodes, fewer carriers have to be field induced in order to fill the bulk traps. (In electrothermal calculations,[14] it is assumed that the ON state is achieved immediately once the field in any region reaches ϵ_c; no electronic delay time mechanism is ordinarily introduced. Hence the calculated t_d's should be, and are, shorter than those observed experimentally.)

For the case where relatively uniform fields above a critical value can be established essentially isothermally over the entire sample, the avalanche region can be observed in virgin samples.[9,74] (The departure of the arrows from the near-vertical line in Fig. 13 represents the response of the carriers to the field during the very rapid switching transition. At this point, all the traps are filled.) Since carriers are generated throughout the sample in this mode, some can enter the external circuit prior to trap filling and produce quite large prethreshold currents.[9,74] These, in turn, can produce filamentation prior to the switching transition, with filamentary current densities of the same order of magnitude as those in the ON state.[6] However, the temperature reached inside the preswitched filament ($<60\,°C$) is well below that necessary to induce the ON state thermally ($\approx 600\,°C$), and thus these results are consistent with the fundamentally electronic nature of the switching phenomenon.

For a given sample and reactive-component environment, when the load resistor is increased such that $I_h < I_{on} < I_{hm}$, circuit-controlled relaxation oscillations are observed.[4,73] The region between I_h and I_{hm} can be revealed by an intimate-double-pulse technique,[27] where situations such that $I_{hm} = I_t$ have been observed.

If a sample is switched to the steady ON state and then kept there for a sufficiently long time, either of two things may happen. For a relatively high-resistance material, where the temperature at the core of the filament is only moderately above ambient, the sample switches OFF once the bias is removed in an oscillatory manner that is essentially the same as shown in Fig. 13 during the OFF- to ON-state transition. However, for cases where the material and operating conditions are such that a crystalline phase is induced in the material, the ON state becomes fixed and the steady-state characteristics for $I_{on} < I_f$ are then as shown by the dashed-dot line in Fig. 13. A high-current reset pulse must then be established to restore the amorphous OFF state.[1,43]

The above represents our present view of the switching transition and conventional threshold and memory behavior. However, there appears to be a variety of further intriguing aspects to the problem. In particular, there is some evidence that a form of memory can be observed even under isothermal conditions, or, rather, when there is no evidence of crystallization within the filament. We speculate on these aspects in Sec. IV B.

B. On the possibility of isothermal "dc" memory, forming, and dipolar effects

In addition to ordinary threshold and memory switching, another type of behavior has been observed which suggests that in some cases the induced high-conductance state may remain effectively noncrystalline. What occurs in this "dc effect" is that, for dc-driven systems, some threshold switches develop metastable ON states that persist for long times (hours) after the dc bias is removed. After this time they *recover*, in the sense that the high-resistance OFF state is restored and threshold switching behavior is once again observed. Although it is certainly possible that field and

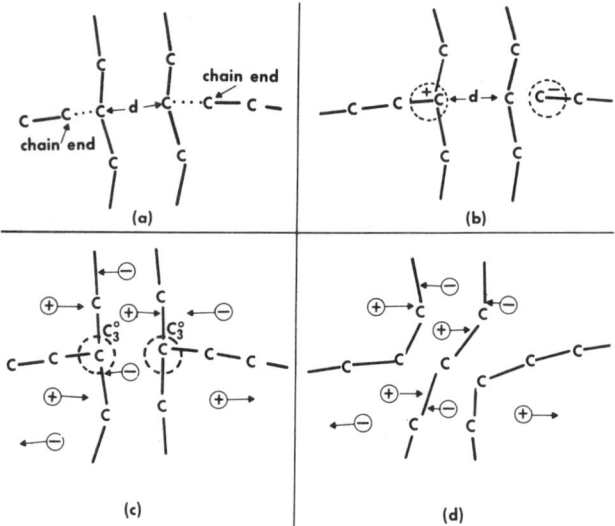

FIG. 14. Two-dimensional sketch of some possible atomic configurations in a chalcogenide film. C denotes chalcogen atoms: (a) during the process of film growth; (b) the OFF state; (c) the ON state; (d) a possible crystallization mode.

thermally induced electrode migration effects are involved in these occurrences,[75] our use of molybdenum and vitreous carbon electrodes suggest that such effects would be minimized in our samples. It has been suggested[36,38] that the trapping of electrons and holes produced by the switching event might be responsible for the dc effect by inducing bonding distortions within the material. Furthermore, it has also been shown that the chain lengths, bond strengths, and flexibility of chalcogen atoms in the amorphous matrix can determine whether a given material will be prone to exhibiting memory effects.[36,58,76] We now indicate how such processes might occur.

Consider the sequence of events shown in Fig. 14. Figure 14(a) shows a situation that might occur during the growth of a sputtered chalcogenide film.[77] The C's represent chalcogen atoms, which ordinarily form covalent chains. (We ignore for the moment the presence of nonchalcogen atoms.) The interchain separaton is represented by d in the figure; it can be as small as an average interatomic separation within a chain. Also shown in Fig. 14(a) are two chains where chain-end atoms (C_1^0) are near to C_2^0 atoms located within another chain. The film is not under bias; we are in the OFF state and p and n are very small. We know that C_1^0 centers are unstable with respect to a $C_1^0 \rightarrow C_3^0$ transition and that the $2C_3^0 \rightarrow C_3^+ + C_1^-$ transition is also energetically favored.[38] However, rather than actually having C_3^0 centers created during the film-growth process, we envision a charge-transfer reaction occurring directly so that $2C_1^0 \rightarrow C_3^+ + C_1^-$, as shown in Fig. 14(b). VAP's form, and d becomes the average separation between charges within each dipole. (If they are sufficiently close, the VAP's can be described as IVAP's.[38]) We induce the ON state next, which floods the sample with carriers and neutralizes the C_1^- and

C_3^+ centers. Here, $p, n \gg N_3^+, N_1^-$; the situation shown in Fig. 14(c) results after the $C_1^0 \rightarrow C_3^0$ bonding modification occurs.

Now suppose the material is flexible enough and/or the filament hot enough to allow the C_3^0's to convert to C_2^0's, which would mean a tendency toward crystallization. Note that when the C_3^+ and C_1^- centers are neutralized, the strong Coulomb attraction between them disappears; this also aids the potential flexibility of the structure. The chain end structures that formed the C_3^0's could now shift somewhat and a crystallization mode could develop, as shown in Fig. 14(d), where all the chalcogens are now C_2^0's. In this model, if the material is designed properly, such a transformation can in principle be induced isothemally; we have a charge-induced crystallization effect.

From the above discussion, we see that there are three separate and distinct dominant configurations possible: (1) C_3^+ and C_1^- (OFF), (2) C_3^0 (ON), and (3) C_2^0 (memory). But now suppose that the VAP's are reasonably well separated and the $C_3^0 \rightarrow C_2^0$ transition is difficult because of the relatively rigid structure of a particular material (a sufficient density of crosslinking atoms is present). Imaging that the ON state has just been removed and all the excess field-induced free carriers have recombined. The OFF state will be re-established by the $2C_3^0 \rightarrow C_3^+ + C_1^-$ transition. But suppose this process is slow in a material once C_3^0's have actually been formed (remember that we do not develop them when the film is being grown). Under these conditions, a high-conductance state will be maintained during the relatively long time when the C_3^+ and C_1^- states are being re-established. The reformation of the C_3^+ and C_1^- centers represent a room-temperature annealing process that returns the material to its higher resistance OFF state.[78]

In addition to the above consideration, when a C_1^- center captures a hole and becomes a C_1^0 center, it is unstable toward a $C_1^0 \rightarrow C_3^0$ transformation. Thus, hole capture alone can lead to tighter bonding configurations and more structural rigidity, if the material is flexible enough to permit such configurations to exist. We also suggest that materials containing VAP's should not readily show a dc effect, since when the ON-state bias is removed, a short-range charge-transfer reaction will rapidly reform the C_3^+ and C_1^- centers. Materials with VAP's that have their charge fairly well separated may be the best exhibitors of the dc effect. This would be favored in multicomponent chalcogenide glasses with large dielectric constants.

We would like to point out another interesting consequence of our model of the development of the C_3^+ and C_1^- centers. Suppose that during the film-growth process shown in Figs. 14(a) and 14(b) the development of the charged defects can in some manner be inhibited. Rather than have a large density of C_3^+ and C_1^- centers, we would have large densities of C_1^0 and/or C_3^0 centers, both of which are paramagnetic. Hauser et al.[30] have observed EPR signals in a variety of chalcogen materials when they are deposited on substrates held at liquid nitrogen temperatures. Upon annealing to room temperature, the EPR signal disappears. It is likely that the low-temperature deposition inhibits the necessary bond re-arrangements; however, the annealing process allows the charge-transfer reactions, $2C_1^0 \rightarrow C_3^+ + C_1^-$ and $2C_3^0 \rightarrow C_3^+ + C_1^-$, to occur. If this is the case, it would seem possible to effectively tie up the C_1^0 and/or C_3^0 centers in other ways; e.g., if hydrogen were introduced into the system during the growth process (in a manner similar to that used for $Si_x H_{1-x}$ alloys[79,80]). In this manner, chalcogenide films of lower dark conductance but better photoconductance could be produced.

Needless to say, aside from the large-current spike[4] that causes forming effects[18] upon switching, such effects could also result from the electronically induced events discussed above. However, at the present time sufficient data to make any such conclusions firm are not available.

It is clear from the previous discussion that observable dipolar effects may be present in chalcogenide films.[41] In fact, a dipole-dipole interaction between the charged defects could lead to a ferroelectric transition if the dipoles are sufficiently strong and their separation large compared to interatomic spacings. As a crude approximation, we note that the dipole-dipole interaction is of the same order as the thermal energy when

$$k^2 N \simeq k T_c,$$ (117)

where the dipole moment $d \simeq qa$, a being an average charge-separation distance and q the electron charge. T_c, the critical temperature at which the ferroelectric transition would occur, is below about 10 °K for $d \simeq 10^{-17}$ CGS and $N \simeq 10^{19}$ cm^{-3}. A more accurate estimate can be obtained by using a Langevin-type model.[81] In this model, the dielectric constant of the material is $\epsilon = 1 + 4\pi N d^2/3k(T - T_c)$, where $T_c = 4\pi N d^2/9k$. Here, values of T_c near 10 °K are also predicted. However, if we also include the electronic polarizability, T_c would be somewhat higher.

Considerations of the above kind are not appropriate for polar liquids, for example water; they would predict that water should become ferroelectric which is, evidently, not true. But the reason that other types of molecular interactions are more important in water and in other polar liquids is that the separation between dipoles is just equal to the interatomic distance. In this aspect, the unique feature of a chalcogenide glass is that the dipoles are comparatively strong and the average distance between them $(N^{-1/3})$ is much larger than interatomic distances. Consequently, a long-range dipole interaction could prevail and may cause a ferroelectric transition.

Our above estimates were made for a quite moderate values of N. In certain glasses, larger values of N are possible, leading to a higher transition temperature. At higher temperature $(T \gg T_c)$, the dipole defects should contribute to the dielectric constant. In order to estimate this contribution, we can use the theory of a dilute dipole solution, which predicts

that

$$\Delta\epsilon = \frac{4\pi d^2 N}{3kT} \left(\frac{\epsilon_0(\epsilon_\infty + 2)}{2\epsilon_0 + \epsilon_\infty} \right)^2,$$ (118)

where ϵ_0 and ϵ_∞ are the dc and high-frequency dielectric constants, respectively. Assuming $\epsilon_0 \simeq 10$ and $\epsilon_\infty \simeq 6$, Eq. (118) yields

$$\Delta\epsilon \simeq \frac{290}{T(°K)}.$$ (119)

Thus, even at temperatures well above T_c, the dipolar defect states should lead to a measurable temperature dependence of the dielectric constant.

C. Conclusions

From the results of this study we conclude that the initiation of switching in thin amorphous chalcogenide films is fundamentally an electronic process. Furthermore, the maintenance of the filamentary ON state in threshold switches is also electronic in nature. The switching transition occurs when a critical electric field is reached somewhere in the sample, usually near an electrode. Field-induced carrier generation then causes the charged traps to fill (neutralize). When all the taps are filled, carriers can transit the sample with an enhanced mobility and the generation rate required to keep the traps filled is reduced from its threshold value; switching then occurs.

It is obvious that switching, as all other phenomena involving power dissipation, exhibits thermal effects. However, in threshold-type samples, these effects can be minimized by a suitable choice of material and geometry. In such cases, the initiation and maintenance of the ON state are predominantly electronic processes. Only the delay time then can have a significant thermal component, and even this can be minimized under operation. In contrast, conventional memory switching is the result of a reversible amorphous-crystalline transition induced by a combination of electronic and thermal effects. In general, thermal effects can be controlled (as, for example, by varying the length of the pulse driving the device) to provide the desired behavior. More subtle types of electronic memory, formation, and dc effects may also occur, but their origin remains somewhat speculative at this time.

Finally, we emphasize that switching in chalcogenide glasses has particularly desirable properties because of the flexibility resulting from the presence of lone-pair valence electrons. These enable the excitation of large concentrations of free carriers without bond breaking, while at the same time lead to a large density of valence-alternation pairs. The latter retard the OFF-state electrical conductivity by both pinning the Fermi energy and introducing positively as well as negatively charged traps, thus minimizing Joule heating. The switching phenomena follow from electronic effects; threshold switching results from the field stripping of chalcogen lone-pair electrons and subsequent filling of lone-

85

pair induced traps; memory switching results from local structural rearrangements in less crosslinked chalcogenides under excited conditions.

ACKNOWLEDGMENTS

We gratefully acknowledge the many contributions of Melvin P. Shaw at all stages of this work. We should also like to thank Sir Nevill Mott and Hellmut Fritzsche for useful discussions.

[1] S.R. Ovshinsky, Phys. Rev. Lett. **21**, 1450 (1968).

[2] For comprehensive reviews of the extensive work done prior to 1971, see D. Adler, *Amorphous Semiconductors* (CRC Press, Cleveland, 1971), p. 96; H. Fritzsche, in *Amorphous and Liquid Semiconductors*, edited by J. Tauc (Plenum, New York, 1974), p. 313. For a more recent review, see D. Adler, H.K. Henisch, and N.F. Mott, Rev. Mod. Phys. **50**, 209 (1978).

[3] D. Adler, in *Amorphous and Liquid Semiconductors*, edited by W.E. Spear (C.I.C.L., Edinburgh, Scotland, 1977), p. 695; H. Fritzsche and S.R. Ovshinsky, J. Non-Cryst. Solids **2**, 393 (1970).

[4] M.P. Shaw, H.L. Grubin, and I.J. Gastman, IEEE Trans. Educ. **ED-20**, 169 (1973).

[5] B.K. Ridley, Proc. Phys. Soc. **82**, 954 (1963); H.K. Rockstad, and M.P. Shaw, IEEE Trans. Educ. **ED-20**, 593 (1973).

[6] K.E. Petersen and D. Adler, J. Appl. Phys. **47**, 256 (1976).

[7] R.W. Pryor and H.K. Henisch, J. Non-Cryst. Solids **7**, 181 (1972).

[8] M.P. Shaw, S.H. Holmberg, and S.A. Kostylev, Phys. Rev. Lett. **31**, 542 (1973).

[9] W.D. Buckley and S.H. Holmberg, Solid State Electron. **18**, 127 (1975).

[10] D.K. Reinhard, D. Adler, and F.O. Arntz, J. Appl. Phys. **47**, 1560 (1976).

[11] E.J. Yoffa and D. Adler, Phys. Rev. B **15**, 2311 (1977).

[12] B.P. Mathur, Ph.D. thesis, MIT, 1973.

[13] F.A. Padovani and R. Stratton, Solid State Electron. **9**, 695 (1966).

[14] K. Subhani, M.S. Shur, M.P. Shaw, and D. Adler, in *Amorphous and Liquid Semiconductors*, edited by W.E. Spear (C.I.C.L., Edinburgh, Scotland, 1977), p. 712.

[15] D.K. Reinhard, F.O. Arntz, and D. Adler, Appl. Phys. Lett. **23**, 521 (1973).

[16] H.K. Henisch, W.R. Smith, and W. Wihl, in *Amorphous and Liquid Semiconductors*, edited by J. Stuke and W. Brenig (Taylor & Francis, London, 1974), p. 567.

[17] K. Homma, Appl. Phys. Lett. **18**, 198 (1971).

[18] M.P. Shaw, S.C. Moss, S.A. Kostylev, and L.H. Slack, Appl. Phys. Lett. **22**, 114 (1973).

[19] M.P. Shaw and I.J. Gastman, Appl. Phys. Lett. **19**, 243 (1971); J. Non-Cryst. Solids **8-10**, 999 (1972).

[20] R.R. Shanks, J. Non-Cryst. Solids **2**, 505 (1970).

[21] K.E. Petersen, D. Adler, and M.P. Shaw, IEEE Trans. Educ. **ED-23**, 471 (1976).

[22] K.E. Petersen and D. Adler, Appl. Phys. Lett. **25**, 211 (1974).

[23] R.C. Frye, D. Adler, and M.P. Shaw, J. Appl. Phys. **50** (1979).

[24] P.J. Walsh, S. Ishioka, and D. Adler, Appl. Phys. Lett. **33**, 593 (1978).

[25] K.B. Ma, J. Non-Cryst. Solids **24**, 345 (1977).

[26] D. Adler and L.P. Flora, in *Amorphous & Liquid Semiconductors*, edited by J. Stuke and W. Brenig (Taylor & Francis, London, 1974), p. 1407.

[27] A.J. Hughes, P.A. Holland, and A.H. Lettington, J. Non-Cryst. Solids **17**, 89 (1975).

[28] K.E. Petersen and D. Adler in *Amorphous & Liquid Semiconductors*, edited by W.E. Spear (C.I.C.L., Edinburgh, Scotland, 1977), p. 707; K.E. Petersen and D. Adler, J. Appl. Phys. **50**, C.I.C.L. (1979).

[29] K. Homma, H.K. Henisch, and S.R. Ovshinsky, J. Non-Cryst. Solids, **35/36**, 1105 (1980).

[30] J.J. Hauser, F.J. DiSalvo, Jr., and R.S. Hutton, Philos. Mag. **35**, 1557 (1977).

[31] P.W. Anderson, Phys. Rev. Lett. **34**, 953 (1975).

[32] R.A. Street and N.F. Mott, Phys. Rev. Lett. **35**, 1293 (1975).

[33] S.G. Bishop, U. Strom, and P.C. Taylor, Phys. Rev. Lett. **34**, 1346 (1975).

[34] M. Kastner, Phys. Rev. Lett. **28**, 355 (1972).

[35] D. Adler, Sci. Am. **236**, 36 (1977); K.L. Ngai, T.L. Reinecke, and E.N. Economou, Phys. Rev. B **17**, 790 (1978); D. Adler and M. Kastner, (unpublished).

[36] S.R. Ovshinsky and K. Sapru, in *Amorphous and Liquid Semiconductors*, edited by J. Stuke and W. Brenig (Taylor & Francis, London, 1974), p. 447.

[37] S.R. Ovshinsky, Phys. Rev. Lett. **36**, 1469 (1976).

[38] M. Kastner, D. Adler, and H. Fritzsche, Phys. Rev. Lett. **37**, 1504 (1976).

[39] D. Adler and E.J. Yoffa, Can. J. Chem. **55**, 1920 (1977).

[40] D. Adler, J. Non-Cryst. Solids **35/36**, 819 (1980); M. Kastner and H. Fritzsche, Philos. Mag. **37B**, 199 (1978).

[41] M. Abkowitz and D.M. Pai, Phys. Rev. Lett. **38**, 1412 (1977).

[42] D. Adler and E.J. Yoffa, Phys. Rev. Lett. **36**, 1197 (1976).

[43] A.M. Barnett, IBM J. Res. Dev. **13**, 522 (1969).

[44] I. Melngailis and A.G. Milnes, J. Appl. Phys. **33**, 995 (1962).

[45] B.L. Gelmont and M.S. Shur, J. Phys. D **6**, 842 (1973).

[46] A. Blicher, *Thyristor Physics* (Springer-Verlag, New York, 1976).

[47] I.V. Varlamov and V.V. Osipov, Sov. Phys. -Semicond. **3**, 893 (1970); I.V. Varlamov, V.V. Osipov, and E.A. Poltoratskii, *ibid*, 978.

[48] See, for example, M.P. Shaw, H.L. Grubin, and P.R. Solomon, *The Gunn-Hilsum Effect* (Academic, New York, 1979).

[49] P.N. Butcher, Rep. Prog. Phys. **30**, 97 (1967); B.W. Knight and G.A. Peterson, Phys. Rev. **155**, 393 (1967).

[50] D.L. Thomas and J.C. Male, J. Non-Cryst. Solids **8-10**, 522 (1972).

[51] C. Popescu and N. Croitoru, J. Non-Cryst. Solids **8-10**, 531 (1972).

[52] T. Kaplan and D. Adler, J. Non-Cryst. Solids **8-10**, 538 (1972).

[53] D.N. Knoll and M.H. Cohen, J. Non-Cryst. Solids **8-10**, 544 (1972).

[54] S.H. Holmberg and M.P. Shaw in *Amorphous & Liquid Semiconductors*, edited by J. Stuke and W. Brenig (Taylor & Francis, London, 1974), p. 687.

[55] D.K. Reinhard, D. Adler, and F.O. Arntz in *Amorphous & Liquid Semiconductors*, edited by J. Stuke and W. Brenig (Taylor & Francis, London, 1974), p. 745.

[56] H. Wey, Phys. Rev. B **13**, 3495 (1976).

[57] W.H. Weber and G.W. Ford, Solid State Electron **13**, 1333 (1970).

[58] S.R. Ovshinsky, *Proc. 6th Int. conf. Amorphous & Liquid Semiconductors*, edited by B.T. Kolomeits (Nauka, Leningrad, U.S.S.R., 1976), p. 426.

[59] M.P. Shaw, *Handbook of Semiconductors* (North-Hollnad, Amsterdam, 1979), Vol. 4, Chap. 1.

[60] J.B. Gunn, Proc. Phys. Soc. **69,8-B**, 781 (1956).

[61] M.C. Steele, H. Ando, and M.A. Lampert, J. Phys. Soc. Jpn. **17**, 1729 (1962).

[62] H.K. Henisch, Sci. Am. **221**, 2 (1969).

[63] H.K. Henisch, E.A. Fagen, and S.R. Ovshinsky, J. Non-Cryst. Solids **4**, 538 (1979).

[64] N.F. Mott, Philos. Mag. **26**, 1015 (1972); **32**, 159 (1975).

[65] I. Lucas, J. Non-Cryst. Solids **6**, 136 (1971).

[66] H. Haberland, Solid State Electron. **13**, 207 (1970).

[67] P.J. Walsh and G.C. Vezzoli, in *Amorphous & Liquid Semiconductors*, edited by J. Stuke and W. Brenig (Taylor & Francis, London 1974), p. 1391.

[68] K.W. Boer and R. Haislip, Phys. Rev. Lett. **29**, 230 (1970).

[69] W. van Roosbroeck and H.C. Casey, Jr., Phys. Rev. B **5**, 2154 (1972).

[70] C. Popescu and H.K. Henisch, Phys. Rev. B **11**, 1563 (1975).

[71] B.D. Rodgers, C.B. Thomas, and H.S. Reehal, Philos. Mag. **31**, 1013 (1976).

[72] S.H. Lee, H.K. Henisch, and W.D. Burgess, J. Non-Cryst. Solids **8-10**, 422 (1972).

[73] R. Callarotti and P. Schmidt, in *Amorphous & Liquid Semiconductors*, edited by W.E. Spear (C.I.C.L., Edinburgh, Scotland, 1977), p. 717.

[74] D. Allsopp, M.J. Thompson, and J. Allison, in *Amorphous & Liquid Semiconductors*, edited by W.E. Spear (C.I.C.L., Edinburgh, Scotland, 1977), p. 732.

[75] J.M. Mackowski, J.P. Thomas, P. Kumurdjian, and J. Tousset, in *Amorphous & Liquid Semiconductors*, edited by W.E. Spear, (C.I.C.L., Edinburgh, Scotland, 1977), p. 570.

[76] S.R. Ovshinsky, Proc. 4th Int. Congress for Reprographie and Information, Hanover, Germany, 1975 (unpublished).

[77] This argument was developed with the assistance of R. Flasck.

[78]M.J. Thomson, J. Allison, and S.R. Jones, in *Amorphous & Liquid Semiconductors*, edited by W.E. Spear (C.I.C.L., Edinburgh, Scotland, 1977), p. 727.

[79]W.E. Spear and P.G. LeComber, Solid State Commun. **17**, 1193 (1975).

[80]W. Paul, A.J. Lewis, G.A.N. Connell, and T.D. Moustakas, Solid State Commun. **20**, 969 (1976).

[81]C. Kittel, *Introduction to Solid State Physics* (Wiley, New York, 1968).

PROPERTIES OF AMORPHOUS Si:F:H ALLOYS

A. Madan and S.R. Ovshinsky

Energy Conversion Devices, Inc., 1675 West Maple Road, Troy, Michigan 48084

INTRODUCTION

Previously we have shown that amorphous silicon-based alloys containing F and H (1-7) possess desirable properties for photovoltaic application. Amorphous tetrahedral semiconductors ordinarily possess a very large density of states which act as traps leading to low values for drift mobility and low recombination lifetimes of free carriers. However, Spear and his group (8-10) reported that a-Si decomposed from SiH_4 gas by r.f. glow discharge and deposited on a heated substrate produces a film which has a comparatively low density of states. However, materials prepared in this way possess a large concentration of H.(11) Because of the relatively low density of states, these types of films can be doped n type, albeit using relatively large concentration of the dopant. However, p-type doping using B_2H_6 is generally

accompanied with the decrease in the band gap, indicating alloying action rather than conventional doping.(9) Furthermore, the material has apparently a sufficiently large carrier lifetime that efficiencies of photovoltaic devices exceeding 5% have been reported.(12)

In this paper we give further details as regards the properties of a-Si:F:H alloys. We show that the density of states is lower than in amorphous Si:H alloy, and hence its doping characteristics represent a significant improvement over the a-Si:H alloy. The consequence of lowering the state density is that the depletion width is wider which is crucial for obtaining high efficiency devices.

EXPERIMENTAL DETAILS

A capacitive radio frequency glow discharge apparatus was used to fabricate amorphous thin films of Si:F:H from SiF_4 and H_2 mixtures. Full details as regards deposition conditions have been detailed elsewhere.(6) The experimental aspects involved in electrical and optical measurements have also been detailed in reference 5.

RESULTS AND DISCUSSION

Electrical Properties

Figure 1 shows the dark conductivity, σ_D $(\Omega \text{ cm})^{-1}$ plotted as a function of reciprocal temperature for several samples using gas ratios r ($=SiF_4/H_2$) ranging from 10 to 99. These samples were deposited using nominal r.f. power of 50 watts and substrate temperature, T_s, of 380°C. Evidently, changing the gas ratio, r, alters the transport behavior from an inactivated process to an activated process. As will be discussed later, these samples exhibited n-type behavior. For samples of type C, in Fig. 1, the pre-exponent $\sigma_0 > 10^3$ $(\Omega \text{ cm})^{-1}$, and hence we conclude that dominant conduction process is extended state conduction above the conduction band edge, E_c. Samples of the type A can be fitted to Mott's $T^{1/4}$ (13) law as shown in the inset of

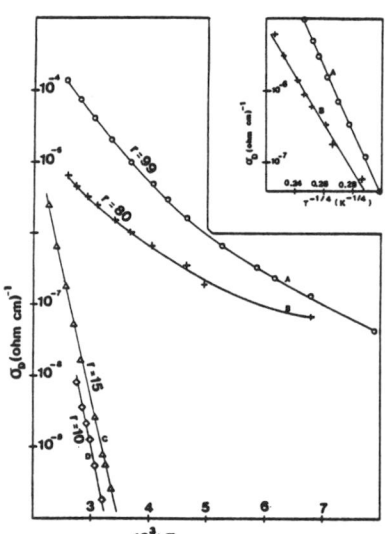

Fig. 1 The dark conductivity, σ_D vs. $10^3/T$, plotted for samples fabricated from different gas ratios, r = SiF_4/H_2. The inset shows σ_D plotted as a function of $T^{-1/4}$ for high values of r.

Reprinted by permission from *Journal of Non-Crystalline Solids*, Vols. 35/36, pp. 171–181 (1980).

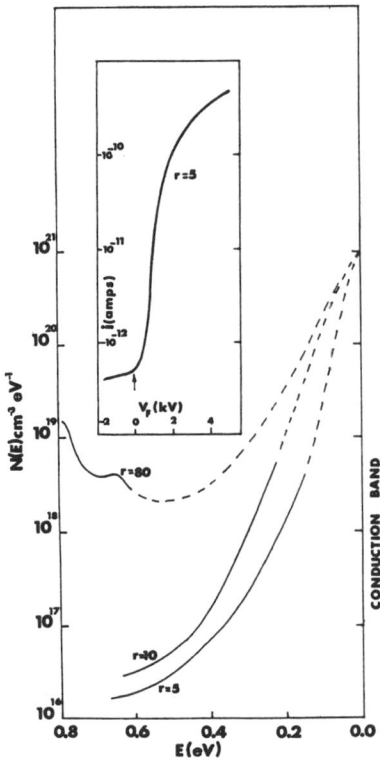

Fig. 2 The localized density of states N(E) $(cm^{-3}eV^{-1})$ plotted for different gas ratios r. The inset shows a typical $i(V_F)$ curve and the arrow indicates the assumed flat band position.

Fig. 1. From the slope, the density of states around the Fermi level, E_F, can be estimated to be about $10^{19}cm^{-3}eV^{-1}$ which is in close agreement with the estimates derived from field effect data, as considered below.

Localized State Density

The analysis in estimating the density of states, N(E), from field-effect experiments neglects any effect due to surface states, and hence the numbers quoted below represent upper limits. We should emphasize that the same procedure has been used for a-Si:F:H as was used for a-Si:H alloy.(10)

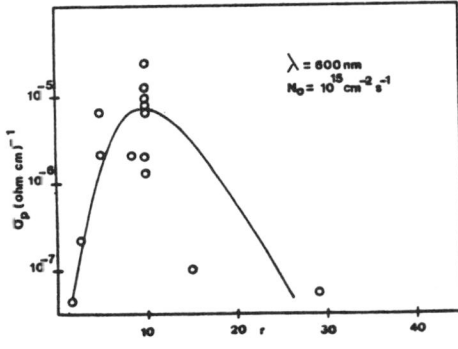

Fig. 3 The photoconductivity, σ_p $(\Omega\ cm)^{-1}$, plotted as a function of r.

A typical source to drain current (i) as a function of gate voltage (V_F) for a-Si:F:H alloy is shown in the inset of Fig. 2. The arrow indicates the assumed flat band position. For samples fabricated with 5 < r < 99, n-type response was observed and we should note that N(E) is reduced substantially by orders of magnitude from ≃ 80 to 5, when r changes from 80 to 5.

We previously noted that the value of r determines the type of conduction process. This becomes understandable since high r values, as shown in Fig. 2, possess a large density of states. Hence, the dominant conduction process is via localized states in the vicinity of E_F. As r is decreased, N(E) decreases sufficiently so that the conduction changes to an activated process.

We should also note that the density of states is considerably lower than reported for a-Si:H alloy, and, furthermore, the peak in N(E) for a-Si:H is absent in this new alloy.(10)

Photoconductivity

Amorphous films fabricated from r > 30 did not exhibit any significant photoconductivity. This is not surprising since N(E) is quite large for these types of films, as shown in Fig. 3. With decreasing r, the photoconductivity of the sample increases markedly as shown in Fig. 3. This figure shows the photoconductivity parameter $\sigma_p = \sigma_L - \sigma_D$, where σ_L is the conductivity under illumination of intensity $10^{15}cm^{-2}s^{-1}$ at λ = 600nm.

Optical Band Gap, E_0

From the photoconductivity data we can obtain an estimate of E_0 by plotting $[i_p h\nu/eN_0(1-R)]^{1/2}$ vs. $h\nu$ (14) where i_p = photocurrent, $h\nu$ = photon energy, N_0 = photon flux, R = reflectivity; this is shown in the inset of Fig. 4, where the intercept on the energy axis is E_0. The variation of E_0 with r is

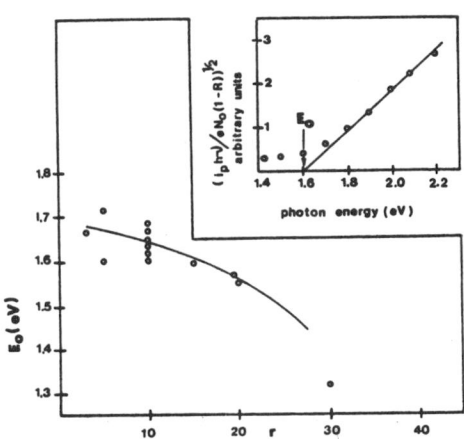

Fig. 4 The optical parameter, E_0, as defined in the text, plotted as a function of r. The inset shows the plot of $[i_p h\nu/eN_0(1-R)]^{1/2}$ as a function of the photon energy, $h\nu$.

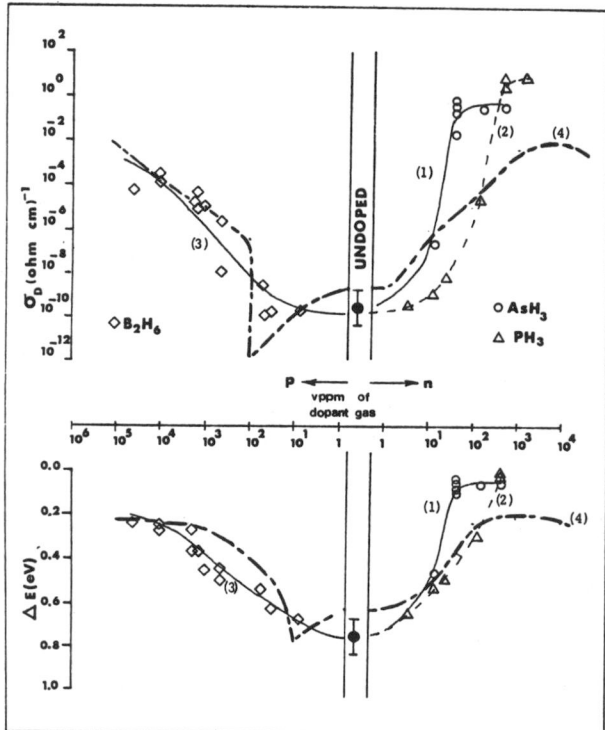

Fig. 5   The   room   temperature   dark conductivity, σ_D $(\Omega \text{ cm})^{-1}$ and the conductivity activation energy, ΔE (eV), plotted as a function of vppm of AsH_3 (curve 1), PH_3 (curve 2) and B_2H_6 (curve 3) introduced into the premix gas ratio of r = 10/1, $T_s \simeq 380^0$C. The p to n transition indicated refers to a-Si:F:H alloy. For comparison purposes (curve 4), the reported doping results (9) for the a-Si:H alloy are also included.

shown in Fig. 4. It is found in general that when alloys are formed, the optical band gap increases if the alloying action introduces a stronger bond. As considered in reference 5, the F content increases with decreasing r. In addition, we expect that upon decreasing r, more H would be incorporated into the film. With the greater inclusion of F and H, strong Si-F and Si-H bonds are formed, and since these are stronger than Si-Si bond, the optical gap is therefore expected to increase, as observed.

Returning to the question of band gap, a more conventional approach is to plot $(\alpha h\nu)^{1/2}$ vs. $h\nu$, where α is the absorption coefficient. This leads to an estimate of ~ 1.9 eV for films fabricated using gas ratio of r ~ 5/1. Using a similar technique for a-Si:H, the band gap is often quoted to be about 1.8 eV.

n- and p-Type Doping

The lower density of states, especially in the upper half of the band gap, achieved in a-Si:F:H alloy reflects in the ease of n-type doping as shown in Fig. 5. The figure shows the dark conductivity (σ_D) and its activation energy (ΔE) as a function of vppm of PH_3/AsH_3 which were added to the premix. We should note that very high dark $\sigma_D > 5$ $(\Omega \text{ cm})^{-1}$

with corresponding low activation energies (~ 0.05 eV) can be achieved with very small addition of the dopant gas. The apparent difference between the two n-type dopant curves for AsH_3 and PH_3 is merely a reflection of the different bond strengths of As-H and P-H. Since the former bond is weaker than the P-H bond, then for a particular power employed, As would be preferentially deposited in relation to P.

The above doping characteristics for a-Si:F:H represents a significant improvement over the reported doping characteristics of a-Si:H alloy as shown in Fig. 5 where we have included the doping characteristic of a-Si:H alloy;(9) we should note that by using a much greater concentration of dopant gas, the most heavily doped sample in the a-Si:H system has yielded dark conductivities which are ~ 10^3 less than attained in a-Si:F:H system.

As shown in Fig. 5, the p-type doping characteristics using B_2H_6 in the premix gave very similar results as reported for a-Si:H alloy which is also shown in this figure. However, it seems somewhat curious that the doping curves for the two alloys should show such similarities.

To reconcile this, we should perhaps consider whether the addition of B in any way affects the Si-based material. Figure 6 shows $(\alpha h\nu)^{1/2}$ vs. $h\nu$ for intrinsic n- and p-type doped a-Si:F:H alloys. With the addition of 500 vppm PH_3, corresponding to a n^+ layer, no apparent change between this and the intrinsic material is evident. However, with the addition of B_2H_6 in the premix, vast changes in optical absorption takes place, i.e., the optical band gap narrows from ~ 1.9 eV to ~ 1.3 eV with the addition of 1% B_2H_6. These results are similar to those obtained for a-Si:H alloy.(15) The implication of this is that a new alloy involving B has been synthesized which possesses a narrower band gap and exhibits p-type characteristics. It is

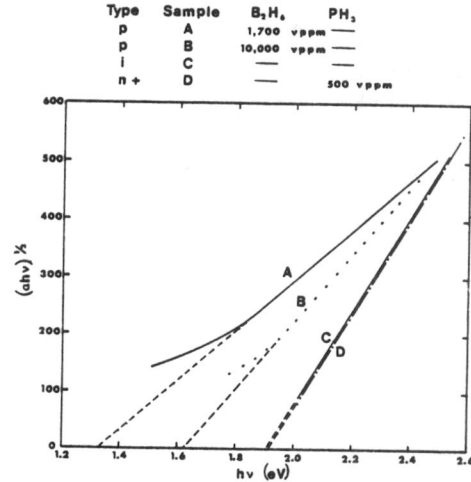

Fig. 6 $(\alpha h\nu)^{1/2}$ plotted   as   a   function of photon energy, $h\nu$, for n^+, intrinsic (i) and p-type layers.

possible that three-center bonds may be responsible in part for this behavior.(16,17) This is in contrast to the results obtained when P or As are added where conventional n type seems to take place. Results using other dopants for p-type doping will be presented elsewhere.(18)

Infrared

As considered in reference 5, the infrared spectra for sample prepared at $T_c \simeq 400^0C$ using gas ratio $r \simeq 10$ exhibited a small peak at 2100 cm^{-1} and 2000 cm^{-1}, together with a prominent peak at 830 cm^{-1}. The peak at 2100 cm^{-1} in the Si-H film has been usually attributed to the SiH$_2$ group which possesses a scissors mode at 900 cm^{-1};(19) however, we have not been able to observe this in a-Si:F:H alloy. It is possible that because of induction effects,(4,20) 2100 cm^{-1} peak could be associated with H-Si-F complex. For this alloy, F content is of the order of 4%, as deduced from Auger results. We, therefore, suggest that the peak at 830 cm^{-1} could be associated with the stretching of Si-F bond.

Some Device Aspects

Many device properties, including solar cell efficiency, depend on the recombination characteris-

tics of photogenerated carriers in the semiconducting material. These can be investigated by such techniques as photoconductivity, photoluminescence, drift mobility and photovoltaic experiments. All of these have been carried out on a-Si:H films without a general consensus as to the predominant recombination mechanisms.(21,22) However, Crandall et al. (23) have presented evidence in favor of geminate recombination from a photovoltaic study of a-Si:H alloy. If this process were to dominate, it would be a serious impediment to the development of an efficient a-Si:H solar cell since the quantum efficiency of photogenerated carriers in the peak region of the solar spectrum would then be considerably reduced.

Geminate recombination is important when the photoexcited electron and hole do not escape their initial Coulomb attraction. There are three important characteristics when geminate recombination is important:(24) (1) a rapidly increasing photoconductive yield as the frequency of the light increases; (2) a decreasing activation energy for photoconduction as the frequency increases; and (3) an increase in yield at applied fields in excess of $\sim 10^4$Vcm^{-1}. In Fig. 7 we show the photoconductivity as function of temperature, incident photon energy, and applied field. In these figures the curves have been normalized which takes into account the different absorption for blue and red excitation, ($\alpha \simeq 10^5$cm^{-1} at $\lambda = 4880$Å and $\alpha \simeq 10^4$cm^{-1} at $\lambda = 6328$Å). Since none of the three characteristics of geminate recombination mentioned above are present in a-Si:F:H alloy, we conclude that this mechanism is not important for photon energies in excess of 1.9 eV.

Depletion Width

The depletion width has been determined on Au Schottky barrier devices by a simple analytical technique which has involved considering the density of states in the form of hyperbolic functions, by C-V technique,(25) and by considering the wavelength dependence of short-circuit currents.

For the latter technique, we fabricated devices in the following configuration: a thin n$^+$ layer was deposited on a glass substrate, followed by $\simeq 0.5$ µm thick active layer, and finally contacted with a semitransparent Au Schottky barrier contact. Because of the highly conducting n$^+$ layer ($\sigma_D \geq 5$ (Ω cm)$^{-1}$), the omission of a back metallic contact did not represent too serious a problem. Figure 8 shows the device configuration and indicates the various parameters used. As indicated in the figure, the front illumination is defined as light incident on the Au side of the device and back illumination as the light incident on the n$^+$ side of the device.

For a typical device, Table I shows the short-circuit current density for front and back illuminations for different wavelengths. We should note that for blue illumination ($\lambda = 4880$Å), the short-circuit current was reduced by about an order of magnitude when illuminated from the back in comparison with the front illumination case. This is despite the fact that the incident flux was greater when illuminated from back since it had not

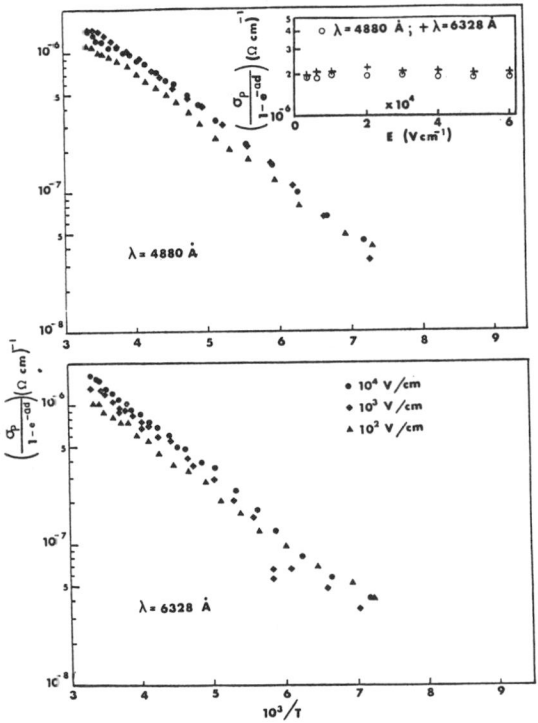

Fig. 7 Normalized photoconductivity $(\sigma_p/(1-e^{-\alpha d}))$ as a function of reciprocal temperature is plotted for fields in the range 10^2Vcm^{-1} to 10^4Vcm^{-1} for incident illumination of (a) $\lambda = 4880$Å and (b) $\lambda = 6328$Å. The inset shows the normalized photoconductivity at room temperature for fields up to 6×10^4Vcm^{-1}.

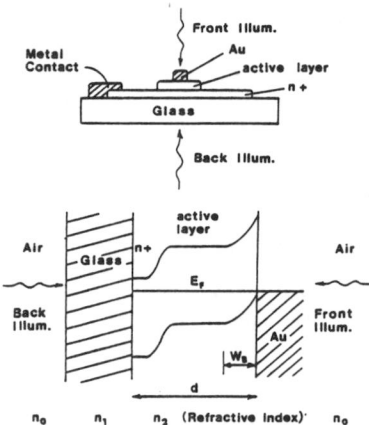

Fig. 8 The band diagram for the Schottky barrier device is shown. Also included is the device geometry of various parameters as defined in the text.

been attenuated by back metallic contacts. This result can be explained if we consider that the width of the depletion layer $W_B < d$, where d is the thickness of n^+ and active layer. Since $\alpha \simeq 10^5 cm^{-1}$ for $\lambda = 4880\text{Å}$, illuminating the device from the back will, therefore, lead to a small absorption within the depletion region since the field separation region mainly exists at Au/active layer interface. However, there may be a small contribution from n^+/active layer interface. Because of the small absorption, the short-circuit current will be low, which was observed. However, using front illumination, large absorption occurs in the depletion region. Therefore, the current should be much larger in comparison with back illumination case, as observed.

Similar reasoning can be extended for the case of red excitation, where $\alpha \simeq 10^4 cm^{-1}$ at $\lambda = 6328\text{Å}$. In this case, as shown in Table 1, the current generated for the back illumination case is larger in comparison with the front illumination. This is readily explained if we consider the device characteristics, i.e., $d \simeq 0.55$ μm, $T_{Au} = 0.15$.

Table I The change in short-circuit current for an Au Schottky barrier device at two different wavelengths for front and back illuminations. $E_c - E_F = 0.47$ eV for the active layer.

λ	T_{Au}	I_{sc}	
4880Å	26%	Front Illumination	300 nA
		Back Illumination	42 nA
6328Å	15%	Front Illumination	150 nA
		Back Illumination	600 nA

Illuminating from the back, the intensity was not attenuated by any metallic contacts; further, because of the low value of α and the highly reflecting nature of the Au contact, we should consider multiple passes through the depletion width. This leads to a large effective absorption, and hence large generation of short-circuit current. Illuminating from the front, apart from attenuation due to low transmissive contacts, there is effectively one pass of radiation through the depletion width because of the omission of a back contact, which leads to a small effective absorption, thus leading to a relatively low value for the short-circuit current.

We, therefore, propose that the short-circuit produced in these devices can be mostly accounted for by the effective absorption which takes place within the depletion layer.

Because of the low mobility of holes, to a first approximation we neglect any contribution due to diffusion currents. Hence, we can write down the short-circuit current density for the front and back illumination in the following way:

$$I_{sc}(F,\lambda) = eN_o(\lambda)T_{Au}(\lambda)\left\{1 - \exp \alpha(\lambda)W_B\right\}$$
$$\times \left\{1 + \frac{(n_2 - n_1)^2}{(n_2 + n_1)^2} \exp - \alpha(\lambda)W_B \exp - 2\alpha(\lambda)(d-W_B)\right.$$
$$\left. \times (1+R_{Au}(\lambda) \exp - \alpha(\lambda)W_B\right\} \qquad (1)$$

$$I_{sc}(B_1\lambda) = N_o\left\{1 - \frac{(n_1 - n_o)^2}{(n_1 + n_o)^2}\right\}\left\{1 - \frac{(n_2 - n_1)^2}{(n_2 + n_1)^2}\right\}$$
$$\times \left\{1 - \exp - \alpha(\lambda)W_B\right\}\left\{1 + R_{Au}(\lambda) \exp - \alpha(\lambda)W_B\right\} \qquad (2)$$
$$\times \exp - \alpha(\lambda)(d-W_B)$$

where B = back illumination; F = front illumination; λ = wavelength; N_o = photon flux intensity; α = absorption coefficient; W_B = width of depletion layer; d = thickness of intrinsic and active layer; T_{Au} = transmission of Au contact; R_{Au} = reflection of Au contact; n_o = refractive index of air; n_1 = refractive index of glass; and n_2 = refractive index of a-Si.

Using eqs. (1) and (2) at different wavelengths and independently measuring $N_o(\lambda)$, $T_{Au}(\lambda)$, $R_{Au}(\lambda)$, $\alpha(\lambda)$, d, n_o, n_1, n_2, and $I_{sc}(\lambda)$, we can then determine the width of the depletion layer, W_B.

In Fig. 9, the crosses show variation of barrier width, W_B, as a function of the position of the Fermi level. As is to be expected, W_B depends very much on the position of E_F since a-Si:F:H possesses a nonuniform density of localized states.

We have shown (25) that the measured density of states spectrum can be reasonably approximated by hyperbolic functions. This assumes that the density of states is composed of donor and acceptor type states within the gap. Using this, we have analytically derived the space charge density as a function of the position of E_F. The shape and width of the Schottky barrier can, therefore, be analyzed by solving Poisson's equation and the solution to these can, therefore, be described in a graphical form, as shown in Fig. 9; the continuous lines show the width of the depletion layer as a function of E_F

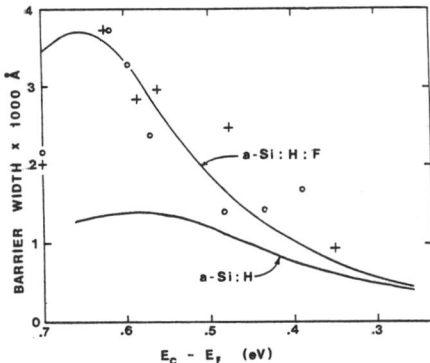

Fig. 9   The depletion width, W_B, is plotted as a function of the Fermi level position in the active layer. The full lines indicate the results derived from the analytical model for a-Si:H and a-Si:F:H alloys.(25) The circles are the points taken from C-V data,(25) and crosses from the wavelength dependence of short-circuit current.

position for a-Si:H and a-Si:F:H materials. Figure 9 also indicates by circles some experimental points which were deduced from C-V measurements for a-Si:F:H devices.(25)

As is evident from Fig. 9, for a-Si:F:H alloy, a reasonable agreement exists for the depletion width estimates between the analytical model, C-V data, and from the consideration of the wavelength dependence of the short-circuit current.

CONCLUSION

We have shown that a-Si:F:H material possesses a low N(E), high photoconductivity, can be easily doped, and that it has a wide depletion region when fabricated within a device configuration.

In conclusion, we should mention that the photovoltaic devices fabricated from a-Si:H is often quoted to possess a depletion width of the order of 0.2 μm which is in good agreement with the results shown in Fig. 9 for that material. The inherent large density of states near the band edges leads to low values for the diffusion lengths of holes and electrons. Consequently, the predominant component of short-circuit current density is due to the drift within the depletion layer rather than the diffusion of carriers. The drift current is related to the amount of effective absorption occurring within the depletion layer whose width in turn is limited by the density of states.

In the above, we have shown that in a-Si:F:H material, the depletion width is indeed widened. This aspect provides for a potential for this material in photovoltaic applications.

Acknowledgments

The authors would sincerely like to thank Professors D. Adler, H. Fritzsche, M. Shur, M. Silver and M. Shaw for stimulating discussions. We also thank W. Czubatyj, E. Benn, T. Anderson, L. Christian and R. Himmler for preparation of samples and their assistance in measurements.

References

1. S.R. Ovshinsky and A. Madan, Nature 276 (1978) 482.
2. S.R. Ovshinsky and A. Madan, in Proceedings of the 1978 meeting of the American Section of the International Solar Energy Society, ed. K.W. Böer and A.F. Jenkins (AS of ISES, University of Delaware, 1978) p. 69.
3. S.R. Ovshinsky, New Scientist 80 (1978) 647.
4. S.R. Ovshinsky, J. Non-Cryst. Solids 32 (1979) 17.
5. A. Madan, S.R. Ovshinsky and E. Benn, Phil. Mag., in press.
6. A. Madan and S.R. Ovshinsky, to be published in the Proceedings of the Symposium on Applied Technology to Solar Energy Systems, Jurica, Queretaro, Mexico, January 29-February 3, 1979.
7. A. Madan, S.R. Ovshinsky, W. Czubatyj and M. Shur, to be published in the Proceedings of the 21st Electronic Materials Conference, Boulder, Colorado, June 1979.
8. W.E. Spear and P.G. Le Comber, J. Non-Cryst. Solids 8-10 (1972) 727.
9. W.E. Spear and P.G. Le Comber, Phil. Mag. 33 (1976) 935.
10. A. Madan, P.G. Le Comber and W.E. Spear, J. Non-Cryst. Solids 11 (1976) 219.
11. H. Fritzsche, C.C. Tsai and P. Persans, Solid State Tech. (January 1978) 55.
12. D.E. Carlson and C.R. Wronski, J. Electr. Mater. 6 (1977) 95.
13. N.F. Mott, Phil. Mag. 19 (1969) 835.
14. R.J. Loveland, W.E. Spear and A. Al-Sharbaty, J. Non-Cryst. Solids 13 (1973/74) 55.
15. C.C. Tsai, Phys. Rev. B 19 (1979) 2041.
16. S.R. Ovshinsky, in Proceedings of the 7th International Conference on Amorphous and Liquid Semiconductors, ed. W.E. Spear (CICL, Edinburgh, 1977) p. 519.
17. S.R. Ovshinsky and D. Adler, Contemp. Phys. 19 (1978) 109.
18. S.R. Ovshinsky, M. Izu and V. Cannella, to be published.
19. J. Knights, G. Lucovsky and P. Nemanich, Phil. Mag., in press.
20. G. Lucovsky, Solid State Commun. 29 (1979) 571.
21. D. Engemann and R. Fischer, in Proceedings of the 7th International Conference on Amorphous and Liquid Semiconductors, ed. W.E. Spear (CICL, Edinburgh, 1977) p. 947.
22. D. Anderson and W.E. Spear, Phil. Mag. 36 (1977) 695.
23. R.S. Crandall, R. Williams and B.E. Tompkins, J. Appl. Phys., in press; Bull. Am. Phys. soc. 24 (1979) 273.
24. A. Madan, W. Czubatyj, D. Adler and M. Silver, to be published.
25. M. Shur, W. Czubatyj and A. Madan, this Proceedings.

ELECTROREFLECTANCE AND RAMAN SCATTERING INVESTIGATION OF GLOW-DISCHARGE AMORPHOUS Si:F:H

R. Tsu,[1] M. Izu,[1] S.R. Ovshinsky,[1] and F.H. Pollak[2]

[1]Energy Conversion Devices, Inc., Troy, Michigan 48084
[2]Physics Department, Brooklyn College of CUNY, Brooklyn, New York 11210

(Received 16 June 1980 by J. Tauc)

1. INTRODUCTION

A NEW TYPE of amorphous semiconductor alloy has recently been reported [1]. This new material has silicon and fluorine as its main structural components. It is multi-elemental and includes hydrogen and can also include other elements such as oxygen without deleterious effects. This Si : F : H alloy, prepared by the glow-discharge decomposition of SiF_4 mixed with hydrogen, has been reported to overcome a number of problems of a-Si and a-Si : H. The a-Si : F : H alloy is highly photoconductive, possesses a lower density of states in the upper half of the band gap than the a-Si : H alloy and is devoid of any photostructural changes. In order to gain further information about the electronic and vibrational states of this amorphous semiconductor, we have investigated the electrolyte electroreflectance (EER) and Raman spectra of several samples of this material.

During the past decade, electric field modulated reflectivity has been developed into a powerful tool for a detailed analysis of the electronic states of crystalline semiconductors [2, 3]. To date, little work has been done in the modulated spectroscopy of disordered systems. Although some elecroreflectance (ER) spectra have been reported near the absorption edge of materials in the amorphous state, no ER response has been observed for higher lying transitions [2, 4, 5].

Raman spectra gives an unambiguous distinction between the amorphous and crystalline state [6]. It has been reported that the observation of the *sharp* 522 cm^{-1} line of crystalline silicon provides a unique tool for detecting traces of crystallinity. Any presence of this line is conclusive evidence for the sample not being totally amorphous [6]. For a-Si prepared by sputtering or ion-bombardment, a broad peak is found near 465 cm^{-1} having a typical line-width of ~ 100 cm^{-1} [6–8]. Similar results are obtained for a-Si:H [7, 8]. Crystalline silicon exhibits a sharp peak at 522 cm^{-1} with a line width of ~ 4 cm^{-1}. Polycrystalline material shows a peak at the same position but somewhat broader (~ 5 cm^{-1}) [9]. To date, peak positions intermediate between the two (465 and 522 cm^{-1}) have not been reported for a disordered tetrahedrally-bonded semiconductor except in ion-damaged amorphous Si which has been partially annealed by pulsed laser irradiation [10–12]. The lowest observed Raman frequency in this laser-annealed material is at 512 cm^{-1}.

In this letter, we present the observation in several samples of a-Si : F : H of (a) structure in the EER spectra in the range 1–5 eV and (b) Raman peaks intermediate between 465 and 522 cm^{-1}. This is the first report of modulated optical structure above the fundamental gap [13] and intermediate Raman peaks in a disordered tetrahedral semiconductor. In all samples investigated, EER features were seen at 3.4 and 4.5 eV, energies which correspond to optical and modulated optical structure of crystalline silicon [2, 3]. In particular, the 4.5 eV feature can be related to the degree of local disorder in the material. In one sample, having a very high electrical conductivity obtained by As-doping, we have also observed prominent EER structure near 1.2 and 2.0 eV. These features may be related to the $\Gamma_{25'}-\Delta_1$ and $\Gamma_{25'}-L_1$ transitions, respectively, of the corresponding

Reprinted by permission from *Solid State Communications*, Vol. 36, pp. 817–822 (1980).

Table 1. Relevant parameters of our samples. All samples contain approximately 5% F and 5% H except sample a-Si:H which has about 10% H. The substrate 7059 is the Corning glass identification number

Sample	Thickness (A)	Substrate	Conductivity ($\Omega^{-1}\,cm^{-1}$)	Doping
a-Si:H	1000	Si	10^{-10}	—
088	3000	7059	10^{-10}	—
101	1000	7059 + Moly	4×10^{-3}	As
102	500	7059 + Moly	4	As
103	2000	7059	25	As

crystalline material [14]. A relatively sharp ($\sim 30\,cm^{-1}$ wide) Raman peak at $512\,cm^{-1}$ was obtained for the same sample. The observations of the 3.4 and 4.5 eV EER structures (particularly the latter) as well as Raman peaks intermediate between 465 and $522\,cm^{-1}$ strongly indicated that *a*-Si:F:H possesses "microcrystalline" or some other intermediate range order (IRO) (i.e. an improved connectivity between the elements of the random network).

2. EXPERIMENTAL RESULTS

Our samples were prepared by the glow-discharge process, using an RF power of $\sim 50\,W$ to decompose SiF_4 gas with a gas mixture of SiF_4:H of $\sim 5:1$. Typically, a flow of $0.1\,l\,min^{-1}$ at ~ 0.5 torr was used. Arsenic doping was accomplished by the introduction of AsH_3. The substrate material was 7059 Corning glass. The temperature of the substrate was typically $300°C$. Sometimes a molybdenum layer of thickness several hundred angstroms was evaporated onto the 7059 glass substrate. This layer serves as a bottom electrode and also eliminates possible luminescence from the glass in the Raman measurements. Table 1 gives some relevant information about our samples.

Shown in Fig. 1 are the EER spectra of samples 101 (bottom), 102 (center), and 103 (top) in the energy range 1.0–5.5 eV. Above about 3 eV, all the s spectra are quite similar with pronounced features at 3.4 and 4.5 eV. It should be noted that in this range of photon energies, the penetration depth of the light is equal to or less than 200 A and thus these features cannot be due to interference effects. Although modulates reflectivity or transmission results have been reported for disordered semiconductors in the region of the absorption edge [2, 4, 5], the results of Fig. 1 are the first observation of modulated spectra for energies above

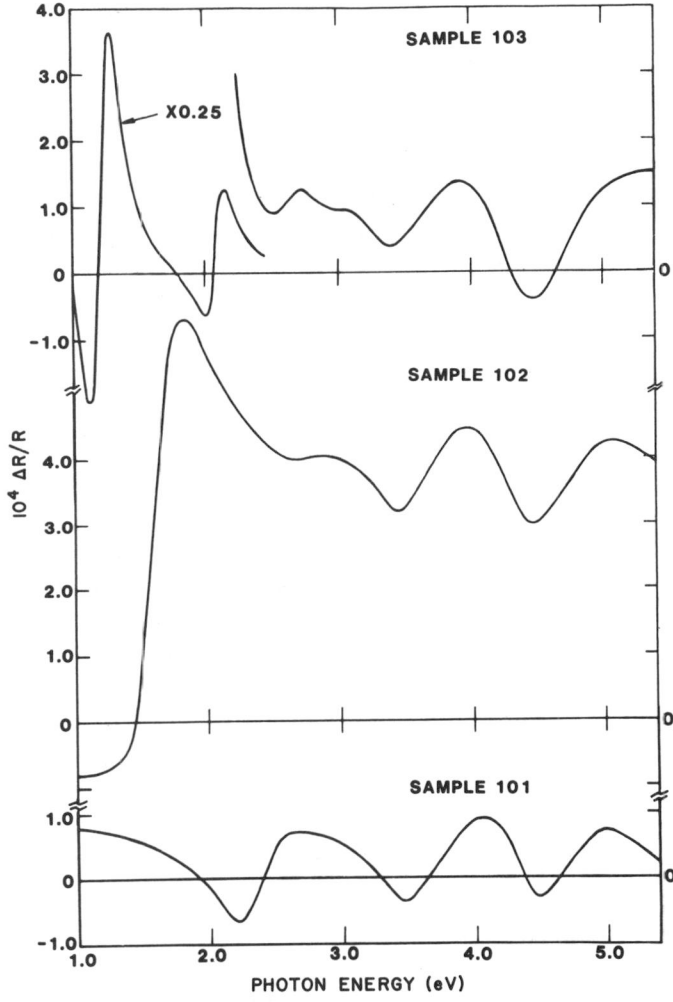

Fig. 1. Electrolyte electroreflectance spectra in the range 1–5.5 eV for samples 101, 102 and 103. For sample 101 the modulating voltage was 3 V peak-to-peak, zero bias; for sample 102, 4 V peak-to-peak with zero bias was used while for sample 103, 5 V peak-to-peak with a d.c. bias of -2.5 V (relative to the platinum counterelectrode) was employed.

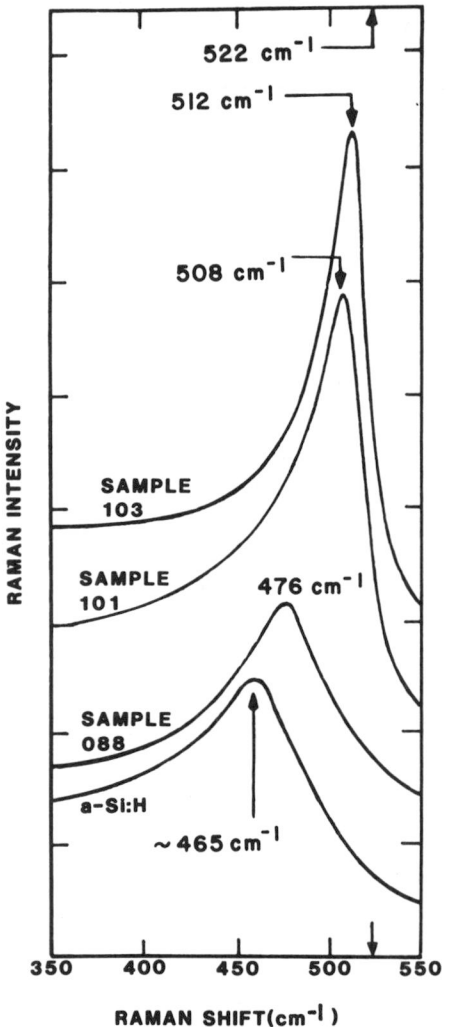

Fig. 2. Scattering intensity as a function of Raman shift for the samples used in this investigation. The curves are displayed for clarity. The zone-center phonon frequency for crystalline silicon (522 cm^{-1}) has been indicated by arrows.

Figure 2 gives the results of the Raman scattering experiment. A cylindrical lens was used to avoid heating effects, and to match the image to the entrance split of SPEX II double monochromator with standard cooled photomultiplier and photon counting system. Typically, several hundred milliwatts at 4880 A was used and spectra were taken in the near backscattering geometry. For the a-Si:H sample, the broad spectrum peaking at about 465 cm^{-1} is similar to that reported previously for a-Si:H [6−8]. The spectrum of sample 088 exhibits a peak at 476 cm^{-1}. Note that the peak positions of samples 101 and 103 shift dramatically higher to 508 cm^{-1} and 512 cm^{-1}, respectively. These intermediate peak positions between amorphous and crystalline material have not been reported for disordered silicon except for ion-damaged amorphous Si which has been partially annealed by pulsed laser irradiation [10−12]. The amount of shift is too large to be accounted for by strain arguments. Note that none of the sampes in Fig. 2 shows any traces of a feature at 522 cm^{-1}. Using this criterion, we conclude that our samples contain no traces of crystallinity and are totally disordered [6]. We shall discuss the significance of our Raman measurements in the next section.

3. DISCUSSION OF RESULTS

Although EER structure has been reported in the vicinity of the absorption edge for hydrogenated a-Si [4, 5], ER response has not been obtained involving transitions above the absorption edge. The structures we observe at 3.4 and 4.5 eV correspond to the interband transitions in crystalline Si [2, 3]. Regardless of the thickness of these samples, the energy positions were not changed. On this basis, together with the small depth of penetration of the light at these energies, we have ruled out any interference effects.

It has been established that the main contribution to the 3.4 eV peak in silicon is due to transitions along the ⟨111⟩ regions ($L'_3 - L_1$) of the Brillouin zone (BZ) [15]. Hence it is the same as the E_1 peak in other diamond- and zinc-blend type materials [2, 16]. The ⟨111⟩ directions are precisely the directions of the bonds in the unit tetrahedron. Therefore, any feature related to them is expected to be retained as long as the short-range order is preserved. For example, the large asymmetric single bump ϵ_2 (imaginary part of complex dielectric constant) of amorphous tetrahedrally bonded semiconductors coincides with the E_1 features of the crystalline material [17−19]. Several theoretical works have also demonstrated this relationship [20−22]. On

this region. Below 3 eV, the spectra are different and we shall return to this point later.

The electroreflectance measurements were made using the electrolyte technique which has been described extensively in the literature [2, 3]. The electrolyte used was a 0.001 N solution of KCl. The modulating voltage was in the form of a square wave of amplitude ~ 5 V peak-to-peak at 220 Hz. For sample 103, which was deposited onto the 7059 glass substrate, electrical contact was made directly to the amorphous film by means of Viking metal (an amalgam of mercury, thallium and indium). Whenever a Mo-metallic layer was between the film and glass substrate (sample 101 and 102) electrical contact was made directly to the metal using silver paste.

the other hand, the 4.5 eV feature (E_2) in crystalline silicon is due to transitions at the X point (X_4-X_1) and Σ regions ($\Sigma_2-\Sigma_3$) of the BZ [2]. The E_2 feature in ϵ_2 is washed out in going from the crystalline to amorphous material [16−19]. Based on this fact, as well as on theoretical calculations [17, 18, 21−23], the E_2 feature is to be associated with an ordering which is longer range than that related to E_1. For example, theoretical investigations of the polymorphs of Ge [21, 23] and Si [22] show that the E_2 peak is selectively lowered with respect to the E_1 structure when the local disorder is increased. Thus the observation of the E_2 structure in our EER experiment strongly suggests for IRO in the a-Si:F:H.

In all three EER spectra, we have observed some EER feature around 2 eV (see Fig. 1). For one sample having an unusually high electrical conductivity (sample 103) we have also seen sharp structure at 1.2 eV as well as at 2 eV. There is a possibility that the structures below 3 eV may be due to interference effects, nevertheless, close examination shows that the line shape is quite different from ER signals attributed to interference [5, 24]. It is interesting to note that in crystalline silicon the $\Gamma_{25'}-L_1$ gap is 1.7 eV while the $\Gamma_{25'}-\Delta_1$ gap is 1.1 eV [13]. Thus, it may be that the 2 eV feature is related to the former gap while the 1.2 eV peak is associated with the latter gap. It is possible that $\langle 111 \rangle$ nature of the $\Gamma_{25'}-L_1$ gap, i.e. the tetrahedral bonding symmetry, may require less ordering in order to be observed in comparison to features related to the $\Gamma_{25'}-\Delta_1$ gap.

Structure at 2 eV and around 3 eV has been reported in a resonance Raman scattering (RRS) experiment on a-Si [7]. Resonance Raman scattering can be considered as a form of "internal" modulation of the optical constants by the phonons [25]. The experimental results have been interpreted with a polarizability theory in terms of the measured derivatives of the the dielectric constant [7]. Since a-Si has only short-range order, this experimental result is additional evidence for the association of the 2 and 3.4 eV structures with this degree of ordering.

The observation of ER structure is generally more sensitive to the details of the sample microstructure in comparison to other modulation techniques (thermo-modulation, piezo-modulation, etc.) which provide first derivative-like features [2, 3]. For the case of ER in disordered semiconductors, Esser has pointed out the relation between the externally applied electric field and the spatial variations in the random internal fields [26]. This latter quantity can be considered as a

"broadening parameter". This may explain why the broad 2 and 3 eV features of the RRS of a-Si (which is interpreted in terms of a first derivative modulation technique) [7] have not been previously observed in ER of a-Si or a-Si:H. The previously mentioned "broadening parameter" may be too large in relation to the externally applied electric field to observe ER. Therefore, the fact that we have observed ER structure above the fundamental absorption edge in a disordered semiconductor is also an important aspect of our interpretation of IRO in a-Si:F:H.

We have also studied several samples of a-Si:H and found no evidence for EER structure at 3.4 or 4.5 eV.

In addition to the observation of the 4.5 eV EER peak the fact that we have seen Raman peaks intermediate between amorphous (465 cm^{-1}) and crystalline material (522 cm^{-1}) is the other major piece of evidence for our interpretation of IRO in a-Si:F:H. No such Raman phenomenon has been observed in a disordered tetrahedrally-bonded semiconductor except for ion-damaged laser-annealed silicon [10−12]. Note also that for sample 103, which shows a Raman peak closest to crystalline material, we observe EER structure at 1.2 eV. It is possible that these results are consistent with the increased degree of ordering of this sample in relation to 101 and 102.

It is difficult to account for the down shift of our observed Raman peaks from 522 cm^{-1} in terms of strain. A uniform tensile strain of several percent is needed to produce a down shift of 10 cm^{-1} [27]. Although we cannot rule out the possibility of a few percent increase in lattice constants, it is more likely due to the appearance of "pseudo-microcrystals" of the order of 10 Å. Calculations of the phonon structure of a 64 atoms Si cluster by Alben *et al.* [28] gives a frequency about 505 cm^{-1}. One may also use the rule of $kl \sim 1$ to estimate the particle size from the neutron scattering data [29] for single crystalline silicon, assuming all of the down shift (522−508 cm^{-1}) may be accounted for by a final size-effect, we obtained again ~ 10 Å. Therefore, it is certainly consistent with our EER and Raman findings that IRO is present in our amorphous Si:F:H.

In summary, our observations of EER structure above the fundamental gap (particularly the 4.5 eV feature) as well as Raman peaks intermediate between 465 and 522 cm^{-1} strongly indicate that a-Si:F:H possesses "microcrystalline" or some other intermediate range order. Since these effects are not seen in a-Si:H, either in our own experiments or as reported by other workers, it appears that IRO is introduced by the presence of the fluorine.

Acknowledgements — We are grateful to S. Solin for the use of the Raman apparatus used to measure two of our samples and to H. Fritzsche for his kind cooperation. Several very useful conversations have been conducted with D.E. Aspnes and A. Bienenstock. Fred H. Pollak wishes to acknowledge the support of PSC/BHE grant 13106.

REFERENCES

1. See, for example A. Madan, S.R. Ovshinksy & E. Benn, *Phil. Mag.* **B40**, 259 (1979) and references therein.

2. See, for example, Y. Hamakawa & T. Nishino, *Optical Properties of Solids: New Developments* (Edited by B.O. Seraphin) North Holland, New York (1979) and references therein.

3. See, for example, D.E. Aspnes, *Handbook on Semiconductors* (Edited by M. Balkanski), Vol. 2, p. 109. North Holland, New York (1978) and references therein.

4. H. Okamato, Y. Nitta, T. Adachi & Y. Hamakawa, *Surf. Sci.* **86** (1979) (to be published).

5. E.C. Freeman, D.A. Anderson & W. Paul, *Phys. Rev.* (to be published).

6. See, for example, M.H. Brodsky, *Light Scattering in Solids* (Edited by M. Cardona), p. 208. Springer-Verlag, New York (1975) and references therein.

7. D. Bermejo, M. Cardona & M.H. Brodsky, *Proc. 7th Int. Conf. Amorph. Liq. Semicond.* (Edited by W.E. Spear), p. 343. Edinburgh (1977).

8. M.H. Brodsky, M. Cardona & J.J. Cuomo, *Phys. Rev.* **B16**, 3556 (1977).

9. G.J. Jan, F.H. Pollak & R. Tsu, (private communication).

10. R. Tsu, J.E. Baglin, T.Y. Tan, M.Y. Tsai, K.C. Park & R. Hodgson, *AIP Conference Proc. No. 50* (Edited by S.D. Ferris, H.J. Leamy & J.M. Poate). *Am. Inst. Phys.*, New York (1979).

11. J.F. Morhange, G. Kanellis, M. Balkanski, F.F. Peray, J. Icole & M. Croset, *AIP Conference Proc. No. 50, Loc. cit.* [3].

12. R. Tsu, J.E. Baglin T.Y. Tan & R.J. von Gutfeld, *Proc. Elect. Chem. Soc. 1980*, (Edited by C.L. Anderson, G.K. Celler & G.A. Rozgonyi), Vol. 80-1, p. 382 (1980); R. Tsu & S.S. Jha, *J. de Phys. (Paris)*, Colloque, March (1980).

13. R. Tsu, M. Izu, S.R. Ovshinsky & F.H. Pollak, *Bull. Am. Phys. Soc.* **25**, 295 (1980).

14. R.A. Forman, W.R. Thurber & D.E. Aspnes, *Solid State Commun.* **14**, 1007 (1974).

15. F.H. Pollak & G.W. Rubboff, *Phys. Rev. Lett.* **29**, 789 (1972); K. Kondo & A. Moritani, *Phys. Rev.* **B14**, 1577 (1976).

16. See, for example, M. Cardona, *Semiconductors and Semimetals* (Edited by R.K. Willardson & A.C. Beer), Vol. 3, p. 125. Academic Press, New York (1976) and references therein.

17. See, for example, G.A.N. Connell, *Amorphous Semiconductors* (Edited by M.H. Brodsky), p. 73. Springer-Verlag, New York (1979) and references therein.

18. See, for example, M.L. Theye, *Optical Properties of Solids – New Developments* (Edited by O.B. Seraphin), p. 353. North Holland, Amsterdam, (1976) and references therein.

19. J. Stuke & G. Zimmerer, *Phys. Status Solidi (b)* **49**, 513 (1972).

20. D.E. Aspnes, G.K. Celler, J.M. Poate, G.A. Rozgonyi & T.T. Sheng, *Proc. Symposium on Laser and Electron Beam Processing of Electronic Materials.* J. Electrochem. Soc. Proc. vol. 80-1 **414** 1980.

21. B. Kramer, *Phys. Status Solidi (b)* **47**, 501 (1971).

22. J.D. Joannopoulous & M.L. Cohen, *Phys. Rev.* **88**, 2733 (1973).

23. I.B. Ortenberger & D. Henderson, *Proc. 11th Int. Conf. Phys. Semicond., Warsaw, 1972*, p. 465 Polish Scientific Publishers, Warsaw (1972).

24. F.H. Pollak (private communication).

25. M. Cardona, *Surf. Sci.* **37**, 100 (1973); W. Richter, *Solid State in Physics*, Vol. 78, p. 121. Springer-Verlag, New York (1976) and references therein.

26. B. Esser, *Phys. Status Solidi (b)* **51**, 735 (1972).

27. E. Anastassakis, A. Pinczuk, E. Burstein, F.H. Pollak & M. Cardona, *Solid State Commun.* **8**, 133 (1970).

28. R. Alben, S. Weire, J.E. Smith & M.H. Brodsky, *Phys. Rev.* **B11** 2271 (1975).

29. G. Dalling, *Inelastic Scattering of Neutrons in Solids and Liquids.* Symposium on Inel. Scattering, Chalk River, Canada, **2**, 37 (1962).

THE NATURE OF INTERMEDIATE RANGE ORDER IN Si:F:H:(P) ALLOY SYSTEMS

R. Tsu,[1] S.S. Chao,[1] S.R. Ovshinsky,[1] G.J. Jan,[2] and F.H. Pollak[2]

[1]Energy Conversion Device, Inc., Troy, Michigan 48084
[2]Brooklyn College of CUNY, Brooklyn, New York 11210

INTRODUCTION

Previously, we have reported, in heavily As- or P-doped Si:F:H alloy systems, the appearance of a Raman peak lying intermediate between 522 cm^{-1} for c-Si and ∠80 cm^{-1} for amorphous Si.(1,2) Whenever such Raman peak is observed, electrolyte-electro-reflectance (EER) peaks appear around 2 eV, together with those associated with c-Si at 3.4 eV and 4.5 eV. We have explained these observations in terms of an intermediate range order or a "microcrystalline phase." Now we have found similar observations in moderately P-doped samples. On glass substrates EER may be observed when the volume fraction of crystallinity has passed 0.16, the critical density, ρ_{cr}, in percolation processes.(3,4) However, on stainless steel substrates, EER has been observed for $\rho_{cr} < 0.16$, indicating that unlike conductivity, EER requires only the existence of relatively sharp electronic density of states. Extensive structural studies indicate that this intermediate order is distinct from amorphous state and is similar to the microcrystalline state discussed by Moss and Graczyk.(5) Transmission electron diffraction shows that the first, second and third rings coincide with those of (111), (220) and (311) of crystalline silicon. The average particle size determined from Raman peak position gives 20 to 60Å. Particle sizes in this range or even smaller have been reported by Goodman,(7) however, substantial amounts of oxygen were introduced in silicon. Therefore, our samples offer a unique opportunity for the investigation of small particle and grain boundary effects.

Subsequent to our report on the As-doped Si:F:H alloy system,(8) Tanaka et al. (9) reported the observation of a crystalline phase in highly P-doped Si:F:H. Applying higher RF-power, Hamasaki et al. (10) has produced crystallization in P-doped a-Si:F. It is important for us to point out that the occurrence of crystallinity in fairly heavily n-doped region is actually desirable for solar cell purposes because of higher electrical conductivity and lower optical absorption. The latter property allows maximum light transmission through the contact to the intrinsic Si:F:H which is without trace of crystallinity.

RESULTS AND DISCUSSIONS

Our Si:F:H(P) samples are prepared by a glow discharge decomposition of SiF$_4$ and H$_2$. Phosphorus doping is introduced by mixing PH$_3$ with the gas mixture having the ratio of PH$_3$ to SiF$_4$ between 20 to 500 ppm. Composition is determined by Auger, SIMS and resonant nuclear reaction.(11) Typically, our films contain 1-4% of F, 4-6% of H, and 1-10% of P. Figure 1 shows the measured Raman linewidth Δω versus the Raman frequency. Note that the crystalline Raman peak at 522 cm^{-1} for Si is shifted down to peak position as low as 508 cm^{-1} due to small particle size effects or possibly also by strain. On the other hand, the amorphous peak, marked a, has a linewidth of ~ 80 cm^{-1} centered at 480 cm^{-1}. The sputtered a-Si has noticeably broader linewidth and lower Raman frequency. Also plotted in Fig. 1 is the particle size ℓ versus frequency,

Reprinted by permission from *Journal de Physique*, Vol. 42, Suppl. to No. 10, C4, pp. 269–272 (October 1981).

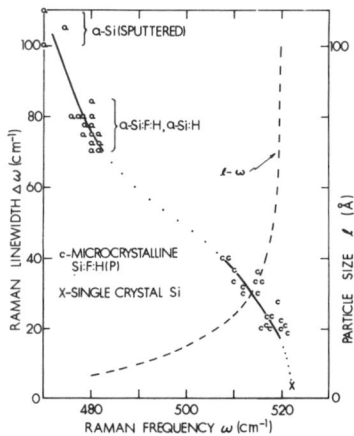

Fig. 1   Raman   linewidth   and   particle   size
(dashed) versus frequency.

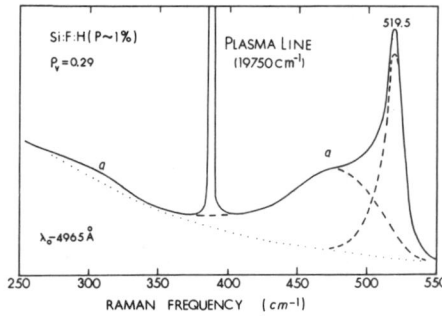

Fig. 3   Raman   intensity   versus   Raman   frequency
in cm^{-1}.   Portion marked a is due to the amorphous
part   of   the   film   while   the   crystalline   peak   is
marked 519.5. The volume fraction of crystallinity
at the critical density for percolation is found to
be 0.16.

obtained   from $\ell \sim 2\pi/k$ using   the   phonon
dispersion curve from neutron scattering.(12)   Thus,
we   found   that   the   particle   size   of   our
microcrystalline is in the range of 20Å to 60Å.   It
is   significant   that   no   Raman   peak   has   been   found
between   483   cm^{-1} and   508   cm^{-1},   indicating   the
absence of a continuous transition.

Figure 2 shows a typical TED (a) and TEM (b).
Note that the second amorphous ring lies between the
(220) and (311) of c-Si. The diffraction pattern is
a   superposition   of   c-Si   and   a-Si.   In   the   bright
field TEM micrograph (b), there appear clusters of a
few   hundred   angstroms.   However,   within   each

cluster, there are fine structures of the order of
50Å.   Therefore,   the   particle   size   of   Fig.   1
corresponds to the fine structures in TEM.

The Raman spectrum of a typical Si:F:H (P ~ 1%)
sample, shown in Fig. 3, may be resolved into an
amorphous   component,   marked   a,   and   a   crystalline
component having a peak at 519.5 cm^{-1}.   The plasma
line of the Ar-laser at 19750 cm^{-1} is left in during
the   measurements   as   an   accurate   calibration   for   the
Raman   frequency.   The   volume   fraction   of
crystallinity for this sample is found to be, $\rho_v \sim$
0.25.   Since   the   critical   density   in   percolation
processes for these samples was determined to be
0.16 (3), this sample is highly conductive having σ
~ 4 $(\Omega cm)^{-1}$, and shows EER response in spite of the
use of a glass substrate.   Note that this two-phase
behavior   was   not   seen   on   the   samples   reported   in
ref. (1) and ref. (2) because in our previous
investigations,   we   dealt   with   essentially
single-phase   "microcrystals,"   having ρ_v much
higher.   From   the   preparation   point   of   view,   in
addition   to   lower   phosphorus   contents,   these
two-phase   samples   are   generally   thinner,   having
thicknesses ~ 600Å.

Part of our present success in identifying the
nature   of   these   films   is   due   to   the   extreme
sensitivity of Raman scattering.   We are able to
produce a good Raman spectrum of a film no thicker
than 200Å.   These thinner films also yield better
transmission   electron   micrographs   and   diffraction
rings.

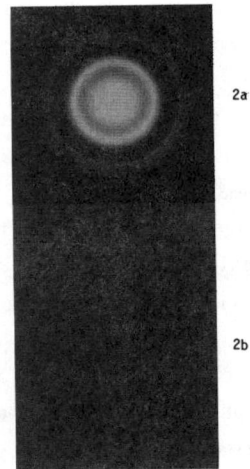

Fig. 2   TED (a), showing the (111), (220) and
(311) rings of c-Si. Note that the amorphous second
ring lies between (220) and (311). TEM (b) showing
clusters   and   grains   inside   the   clusters.
Magnification: 80,000.

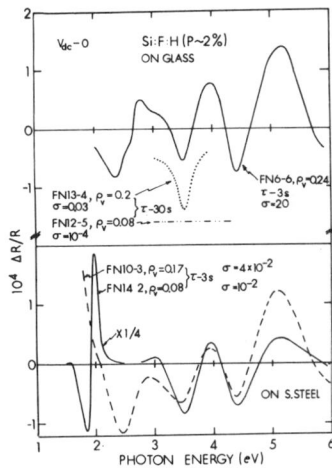

Fig. 4   EER spectra of two types of samples:
Top, on glass; and bottom, on stainless steel
substrates. The electrolyte was 0.001N solution of
KCl. Modulating frequency is between 17 and 36 Hz.
The time-constant used was 3 s; however, 30 s was
used for weak and noisy responses.

Figure 4 shows the EER response of two groups of
samples, one on glass substrates, and the other on
stainless steel substrates. For the former, EER is
strong for ρ_v = 0.24 taken at 3 s time constant;
and weak for ρ_v = 0.2 which required a 30 s time
constant to show up the usual 3.4 eV structure. For
ρ_v = 0.08, which is below the percolation limit, no
EER response was observed. On the other hand, for
the latter samples, EER response is strong for both
samples, ρ_v = 0.17 and 0.08, above and below the
critical density for a percolation process. Note
that the usual structure near 2 eV, 3.4 eV and 4.5
eV are essentially similar to those reported by us
previously. The difference between the two groups
can be understood. In order to apply a high
electric field for these samples on glass
substrates, one of the two electrodes was on the
film itself, therefore requiring conductivity;
whereas, on stainless steel, the metal serves as one
of the two electrodes which does away the need for
higher conductivity. Therefore, our results
indicate that EER response, unlike conductivity,
does not depend on percolation processes. In fact,
as long as there exists a sufficient amount of
microcrystals, EER response should appear whenever a
high electric field can be applied.

CONCLUSION

To summarize, we have now specified in detail
the nature of the intermediate range order in our
Si:F:H(P) samples. The range is approximately 20 to
60Å, having TED similar to microcrystals. The
combination of Raman scattering and EER provides us
means to make detailed studies. The minimum
detectability of trace crystallinity appears to be
about 2% crystallinity. On the other hand, EER may
ultimately be a very sensitive tool to detect trace
crystallinity. Unlike Raman measurements, at
present EER does not give a quantitative
determination of crystallinity. However, the
possibility of using EER to study the electronic
structure of these extremely small particles should
be very valuable.

ACKNOWLEDGMENTS

We like to thank ARCO for their support. F.H.
Pollak and G.J. Jan also wish to acknowledge the
support of PSC/BHE grant 13106.

REFERENCES

1. R. Tsu, M. Izu, S.R. Ovshinsky and F.H. Pollak,
   Solid State Comm. 36, 817 (1980).

2. R. Tsu, M. Izu, V. Cannella, S.R. Ovshinsky,
   G.J. Jan and F.H. Pollak, Proc. 15th Int. Conf.
   Semicond., Kyoto, 1980; J. Phys. Soc. Japan 49,
   Suppl A, 1249 (1980).

3. "Critical Density in Percolation Process: Volume
   Fraction of Crystallinity," R. Tsu, J.
   Gonzalez-Hernandez, S.C. Lee, S.S. Chao and K.
   Tanaka, to be published.

4. H. Scher and R. Zallen, J. Chem. Phys. 53, 3759
   (1970).

5. S.C. Moss and J.F. Graczyk, Phys. Rev. Lett. 23,
   1167 (1969).

6. Z. Iqbal, A.P. Webb and S. Veprek, Appl. Phys.
   Lett. 36, 163 (1980).

7. A.M. Goodman, Inst. Phys. Conf. Ser. 43, 805
   (1979).

8. R. Tsu, M. Izu, S.R. Ovshinsky and F.H. Pollak,
   Bull Am. Phys. Soc. 25, 295 (1980).

9. K. Tanaka, K. Nakagawa, A. Matsuda, M.
   Matsumura, H. Yamamoto, S. Yamasaki, H. Okushi
   and S. Iizima, Proc. 12th Conf. Solid State
   Devices, 1989, Tokyo.

10. T. Hamasaki, H. Kurata, M. Hirose and Y. Osaka,
    Appl. Phys. Lett. 37, 1084 (1980).

11. M.H. Brodsky, M.A. Frisch, J.F. Ziegler and W.A.
    Lanford, Appl. Phys. Lett. 30, 561 (1977).

12. G. Dolling, Elastic Scattering of Neutron, Symp.
    Chalk River 2, 37 (1972).

CORRELATION BETWEEN THE SUPERCONDUCTING AND NORMAL STATE PROPERTIES OF AMORPHOUS MOLYBDENUM–SILICON ALLOYS

A.S. Edelstein, S.R. Ovshinsky, H. Sadate-Akhavi, and J. Wood

Energy Conversion Devices, Inc., 1675 West Maple Road, Troy, Michigan 48084

(Received 3 September 1981 by A.R. Miedema)

We have made a study of the resistivity, $\rho(T)$, and the structural and superconducting properties of amorphous $Mo_{1-x}Si_x$. Besides the usual interest in studying a metal to nonmetal transition, this system has the additional feature that Si changes from acting metallic for $x \ll 1$ to acting nonmetallic for $x \sim 1$. Murarka et al. (1) measured the resistivity of both amorphous and crystallized cosputtered films of MoSi and a comparison of their data with ours will be made below. We have observed a large increase in $d\rho(297)/dx$ with increasing x at $x = x_0 \equiv 0.63 \pm 0.05$ which is most likely associated with an electronic transition involving the Si valence electrons. A tentative model for this transition is presented below. Besides observing this increase in $d\rho(297)/dx$, the specific results pertinent to superconductivity are:

1. The initial linear decrease of the superconducting transition temperature T_c is 0.075 K/at.% Si. This is consistent with the e/a dependence found by Collver and Hammond (2) for near neighbor amorphous 4d metals if one assumes that each Si atom contributes 4 electrons to the conduction band.

2. Superconductivity persists up to at least 64 at.%, i.e., slightly into the concentration region where the resistivity begins to increase. As $x \rightarrow x_0$ from below, the superconducting transition width ΔT_c increases rapidly until at $x \simeq x_0$ the width ΔT_c becomes comparable to T_c.

3. For $x > x_0$, T_c, if it exists, is less than 1.5K.

The amorphous $Mo_{1-x}Si_x$ samples were prepared by RF sputtering. The background pressure, prior to filling the system with 6μ of Argon for sputtering, was 4×10^{-6} mm of Hg. The quartz substrates were sometimes cooled to liquid nitrogen temperatures without noticeable effect during the deposition of the one-micron thick film samples. The Mo and Si starting materials were 99.95 and 6-9's pure, respectively. The compositions of the films are estimated to be accurate to 5 at.% and were determined with 1 at.% precision by Auger analysis using theoretical sensitivity factors. These numbers were then corrected using data obtained on $MoSi_2$ powder samples. The correction factor (.903) is also consistent with EDS data taken on a splat quenched sample of $MoSi_2$. The carbon and oxygen concentrations were less than 1 at.%. All the films were determined to be amorphous using Cu $K\alpha$ radiation. Transmission electron diffraction measurements on 400-500Å thick samples sputtered onto carbon with $0.65 \leq x \leq 0.75$ showed evidence for a second phase only for samples with the highest Si concentrations. For such samples, the electron diffraction pattern showed an additional ring which corresponded approximately to the d spacing of a-Si, i.e., d = 3.19Å for one sample and to a smaller d spacing of approximately 2.8Å in a second sample. Thus, the samples apparently become two-phase for x somewhat greater than x_0 with one of the phases resembling a-Si. The dark field TEM images could not reveal the scale of a possible second phase at a resolution of ~ 50Å, though this result may simply be due to a lack of intensity. The X-ray spectrum, which is primarily due to scattering from Mo atoms, showed two broad peaks which are typical of amorphous metals. The angular position of these two peaks is plotted in Fig. 1a as a function of x. One sees that the position of the higher angle peak changes continuously with x while the position of the lower angle peak is approximately constant. A simple dilution model, which assumes that the higher angle peak is due solely to nearest neighbor contributions, fails since it predicts the wrong sign for the shift. The structural analysis suggests that: 1) A single amorphous phase exists for $x < x_0$. 2) This amorphous phase evolves continuously in structure as its composition is changed. 3) For $x = 0.75$ a second phase resembling a-Si is observed. One must add, however, that measurements (4) of the superconducting critical field H_{c2} show that RF sputtered Mo-Si films with $x < x_0$ may be inhomogeneous.

Reprinted by permission from *Solid State Communications*, Vol. 41, No. 2, pp. 139–142 (1982).

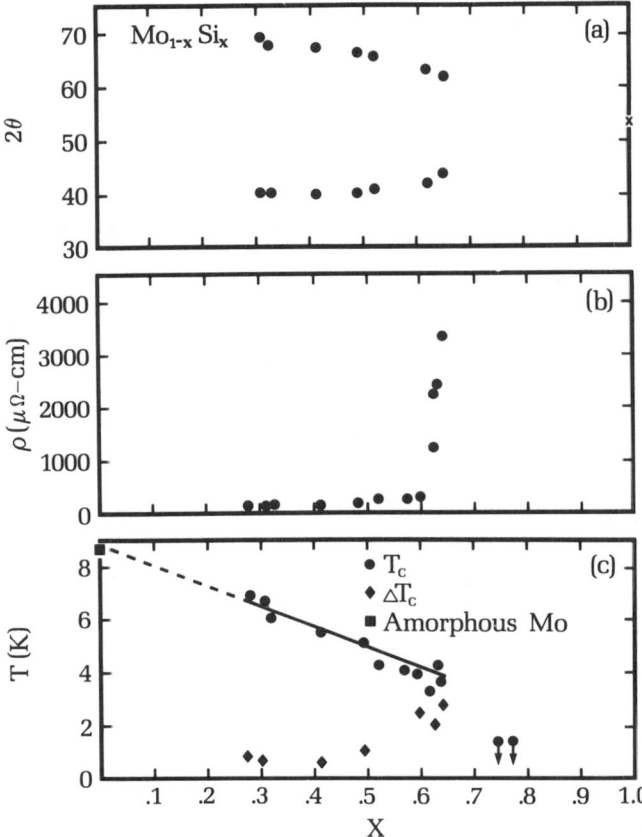

Fig. 1a Angular position of the two X-ray peaks as a function of composition. The symbol x denotes the position of one of the peaks of a-Si inferred from the electron diffraction data of Moss and Graczyk.(3)

Fig. 1b Room temperature resistivity as a function of composition.

Fig. 1c Superconducting transition temperature T_c and superconducting transition temperature width ΔT_c as a function of composition.

Measurements of the room temperature resistivity ρ of $Mo_{1-x}Si_x$ as a function of x, shown in Fig. 1b, show a large increase at x_0. Data of Murarka et al. (1) also show that the room temperature values of ρ increase with increasing x but their data do not show the abrupt increase at x_0. Figure 2 shows a comparison of the room temperature conductivity σ of our samples as a function of Mo concentration 1-x with the data of Romer et al. (5) taken on La-Ar mixtures. The dashed curve shows the prediction of percolation theory (6), $\sigma \propto (x_p - x)^{1.6}$ where x_p was chosen equal to 0.67. Other choices of the parameter x_p do not improve the quality of fit. One sees that the amorphous Mo-Si conductivity crosses Mott's threshold (7) for minimum metallic conductivity σ_{min} at a lower concentration of the metallic species than occurs for La-Ar mixtures, and the transition is more abrupt than that of the La-Ar mixtures. Of more significance, the conductivity of amorphous Mo-Si alloys changes more abruptly as a function of the Mo concentration than is predicted

by percolation theory. In contrast, the concentration dependence of the conductivity of La-Ar mixture is in qualitative agreement with percolation theory. In order to investigate the applicability of Mott's theory (8) of variable range hopping, log ρ versus $T^{-1/4}$ of several samples (9) is plotted in Fig. 3. None of the samples have the temperature dependence predicted by Mott over the entire temperature range investigated (4.2 $\leq$ T < 294K). The data on the 0.77 $\pm$ 0.05 Si concentration sample could be fitted to $\exp[-(T_0/T)^{1/4}]$ over a limited temperature range. These data could also be fitted to $\rho = \rho_0 T^{-3/2}$. Such a temperature dependence is consistent with a model for ionized impurity conduction in a semiconductor (10) in which the number of carriers is independent of T but the mobility varies as $T^{3/2}$. One can possibly explain the fact that $\rho(T)$ does not increase as rapidly as $\exp[-(T_0/T)^{1/4}]$ at low temperatures by assuming the samples are inhomogeneous and that the increase in ρ is limited by regions of higher conductivity. The values of ρ for samples with x $\simeq$ x_0 increase approximately as $-\log T$ with decreasing T, while samples with x < 0.59 have a small negative $d\rho/dT$, for temperatures above the superconducting region. The fact that different temperature dependences for ρ were observed for different composition samples indicates that different mechanisms are involved. In general, these data satisfied the correlation observed in disordered and amorphous metals (11) that $-\, d\rho/dT$ increases with increasing ρ.

The slope $d\rho/dT$ is negative for x < x_0, i.e., for films which display superconductivity. Since ρ then has a maximum ρ_m above the superconducting

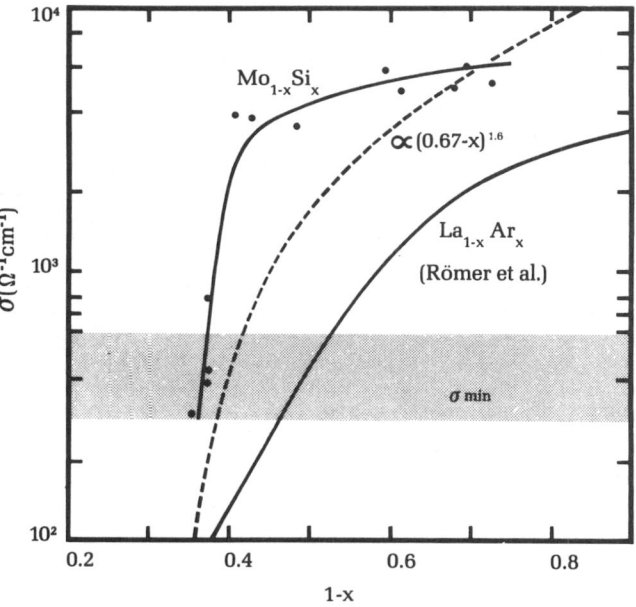

Fig. 2 Conductivity as a function of the concentration of the metallic species 1-x of $Mo_{1-x}Si_x$ and $La_{1-x}Ar_x$. The latter curve was taken from Fig. 1 of ref. 5. The dashed curve is the prediction of percolation theory.

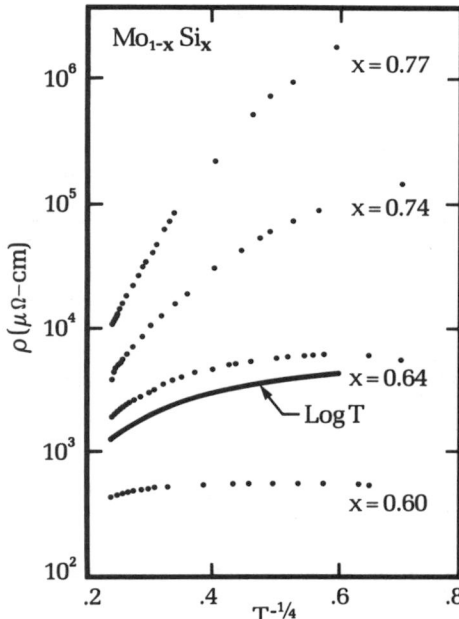

Fig. 3   Resistivity of $Mo_{1-x}Si_x$ as a function of $T^{-1/4}$.   The solid curve was calculated from $\rho(\mu\Omega-cm) = -788.12 \ln T + 5710$.

transition temperature T_c, we can define T_c by the relation $\rho(T_c) = \rho_m/2$. The transition width ΔT_c is defined as the difference between the temperatures where ρ takes on the values $0.9\rho_m$ and $0.1\rho_m$. For large x, ρ did not always decrease to $0.1\rho_m$ at our lowest available temperature, 1.5K. In these cases, we took ΔT_c as twice the temperature interval between the temperature where $\rho = 0.9\rho$. and T_c. The values of T_c and ΔT_c obtained according to these definitions are plotted in Fig. 1c as a function of x. The values of T_c decrease linearly as a function of x for $0.26 \leq x \leq 0.65$ and can be extrapolated to the x = 0 value of 8.8K reported (12) for amorphous Mo. The slope of the decrease, dT_c/dx, is 0.075 K/at.%. The values of T_c measured by Collver and Hammond (2) for amorphous near neighbor 4d elements depend upon the average group number or electron/atom ratio (e/a). If one assumes that Mo and Si contribute 6 and 4 electrons, respectively, our T_c's fall near those obtained by Collver and Hammond for the same e/a. Though one might not expect that Si should contribute so many electrons to the conduction band, and, further, it will not contribute d electrons, the high values of T_c certainly suggest that for $x < x_0$ each Si atom contributes several electrons.

As $x \rightarrow x_0$ from below there is an increase in the transition width. For x > 0.63 there is a rapid decrease in T_c. The points with x = 0.74 and 0.77 have T_c less than our lowest measuring temperature, 1.5K. Actually, it seems likely that T_c is zero for these values of x, since the resistivity of these samples was monotonically increasing with decreasing

T down to our lowest measuring temperature, which is half the value of ΔT_c for the high x superconducting samples. These effects on T_c and ΔT_c appear to be associated with the onset of the increase in the room temperature resistivity.

We have observed that several phenomena occur in the vicinity of the concentration at which there is a large increase in the room temperature resistivity of $Mo_{1-x}Si_x$. This increase cannot be fitted to standard percolation theory. There is a transition from an amorphous state to a two-phase amorphous state and a rapid decrease in the superconducting transition temperature. There is also a rapid increase in the superconducting transition width.

The following is a tentative model for interpreting these effects. For $x < x_0$, the Si atoms hybridize with Mo orbitals to form a metallic band. Even though one should be cautious in applying the e/a model to interpret the T_c's, since Si is not a transition element, the model suggests that Si may contribute approximately four electrons to this conduction band. For $x \sim 1$, the Si atoms are covalently bonded and the system is non-metallic. One could envision that the system would remain basically metallic as the average local environment evolved with increasing x until on an average it resembled that of the metal $MoSi_2$. This would occur at x = 0.67, which is approximately equal to x_0. As one increases x beyond this concentration, there would be more and more Si-rich local environments with covalent bonding. This could give rise to small local regions or inhomogeneities containing primarily covalently bonded Si atoms. These regions would decrease the conductivity in two ways. First, their formation would take electrons from the conduction band. Second, the conduction electrons would tend to avoid these regions since their electrical potential energy is higher within these regions. If the addition of a single Si atom could create one of these regions, then, as observed in Fig. 2, the conductivity would fall very rapidly for $x > x_0$. These regions may be the α-Si-like regions observed in the TED measurements at x = 0.75.

These covalently bonded Si regions could also affect the superconducting properties. If, after their formation, there remained superconducting regions of average size L, then there could be thermodynamic superconducting fluctuations. These fluctuations would result if the free energy difference between the normal and superconducting states per unit volume times L_3 in these small regions is comparable to kT. To estimate the size of L such that $T_c \simeq \Delta T_c$, we shall assume that L satisfies the condition for a zero-dimensional superconductor (13), namely, $L < \xi$ where ξ is the coherence length. With this assumption, $L \sim 20\text{Å}$. In addition, the a-Si-like phase, if it occurs in this compositional range, could by the proximity effect depress and broaden the superconducting transition.

The authors wish to acknowledge the assistance of R. Seguin, who prepared the samples, S. Flessa, the personnel of ECD's Structures Laboratory, and helpful conversations with D. Allred and J. deNeufville. This research benefited from the overall support of energy conversion at ECD by the Atlantic Richfield Company.

References

1. S.P. Murarka, D.B. Fraser, T.F. Retajczyk, Jr. and T.T. Sheng, J. Appl. Phys. 51, 5380 (1980).
2. M.M. Collver and R.H. Hammond, Phys. Rev. Lett. 30, 92 (1973).
3. S.C. Moss and J.F. Graczyk, Phys. Rev. Lett. 23, 1167 (1969).
4. A.S. Edelstein, F.P. Missell and P.M. Tedrow, unpublished data.
5. R. Romer, F. Siebers and H. Micklitz, Solid State Commun. 36, 881 (1980).
6. S. Kirkpatrick, Rev. Mod. Phys. 45, 574 (1973).
7. N.F. Mott, Phil. Mag. 26, 1015 (1972).
8. N.F. Mott, Phil. Mag. 19, 835 (1969).
9. The room temperature values of ρ shown in Fig. 2 do not agree exactly with those shown in Fig. 1b since these samples were prepared under slightly different conditions. This sensitivity to preparation conditions might explain the difference between our results and those of Murarka et al. (1) mentioned in the text.
10. S.M. Sze, "Physics of Semiconductor Devices," Wiley, New York, p. 39.
11. J.H. Mooij, Phys. Stat. Sol. (a) 17, 521 (1973).
12. D.B. Kimhi and T.H. Geballe, Phys. Rev. Lett. 45, 1039 (1980).
13. M. Tinkham, "Introduction to Superconductivity," McGraw Hill, New York, p. 237.

SUPERCONDUCTING PROPERTIES OF AMORPHOUS MULTILAYER METAL–SEMICONDUCTOR COMPOSITES

A.M. Kadin,[1] R.W. Burkhardt,[1] J.T. Chen,[2] J.E. Keem,[1] and S.R. Ovshinsky[1]

[1]Energy Conversion Devices, 1675 West Maple Road, Troy, Michigan 48084
[2]Department of Physics, Wayne State University, Detroit, Michigan 48202

INTRODUCTION

There has been considerable interest in the past decade in developing novel materials using precision microlayering techniques. Much of this has been oriented towards crystalline epitaxy, using pairs of materials with lattice spacings that almost match. However, for most other material pairs, layer interfaces tend to be dominated by dislocations, agglomeration and interdifussion, if allowed to thermally equilibrate.

Our approach (1) has been quite different. Recognizing the difficulties and limitations in obtaining perfect crystalline multilayer films, we have attempted to make multilayer films that are amorphous not only at the interfaces but within the layers themselves. Since we can then deposit the metastable layers onto substrates held at room temperature, we can make very smooth and repeatable layers of many diverse materials. Because of the short electronic mean free path, these multilayer structures (which are not properly superlattices because the crystalline lattice itself is not well defined) exhibit a different set of phenomena than those in the crystalline superlattices.

Although such an amorphous multilayer structure may show no sharp x-ray structure indicative of long-range atomic order within the component materials, it can exhibit very sharp x-ray diffraction peaks corresponding to the repeat spacing d of the amorphous layers. In fact, using this principle together with the alternation of light and heavy elements, ECD has developed, and is currently marketing, OVONYX^{TM} structures that show superior performance as high-reflectivity x-ray devices, with reflectances that exceed 80% at near-normal incidence.

We have fabricated and studied similar samples for their superconducting properties. Recently, several groups (2-7) have investigated artificially layered superconducting structures, including metal-metal systems,(2-4) with emphasis on effects of crystalline epitaxy, and metal-semiconductor systems.(5-7) We have chosen to study the latter type of system, consisting of a refractory transition element (e.g., Nb, Mo, W) alternating with a semiconducting element (Si, Ge, C), with repeat spacings ranging from 10Å to over 100Å. We will survey a number of such layered samples that exhibit a wide range of superconducting behavior.

SAMPLE FABRICATION AND STRUCTURE

To produce these multilayer structures, an unheated substrate, usually an oxidized Si wafer or glass slide, was transported periodically past two targets in either an rf magnetron sputtering system (in an Ar plasma) or an Ar-ion-beam sputtering system. Background pressures were in the low 10^{-7} torr range. All indications suggested that the deposition rate (of order 1Å/sec), once established, could remain constant to within a few percent over the several hours needed to achieve total thickness up to 1 μm.

Several analytical methods were at our disposal to determine the ordering on the scales of both the atoms and the layers. X-ray diffraction was used to determine the degree of crystalline order within the layers, as well as to look for peaks corresponding to coherent layering. X-ray fluorescence was also available, particularly with thicker samples, to determine the elemental composition. Finally, samples were analyzed by Auger depth profiling, in a semi-quantitative way, to estimate the composition and to confirm the presence of layering.

The primary tool to investigate the degree of layering was low-angle x-ray diffraction, usually using a Cu K-α source. Since many of our layered samples had interfaces that were sharp on the scale of the layer spacing, they diffract up to many orders, as the example in Fig. 1a indicates. This sample, consisting of approximately 70Å of Mo alternating with about 35Å of Si, exhibited peaks up to about 15th order in diffraction. This suggests that the interface width is sharper than about 7Å, and a simple model of absolute x-ray reflectance has suggested comparable interface mixing or roughness for similar samples.

An Auger depth profile is shown for the same sample in Fig. 1b. Note that the apparent

Reprinted by permission from J.M. Gibson and L.R. Dawson, eds., *Layered Structures Epitaxy and Interfaces* (Materials Research Society Symposia Proceedings, Vol. 37), pp. 503–508 (1985).

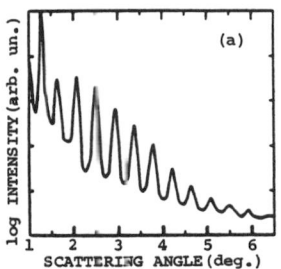 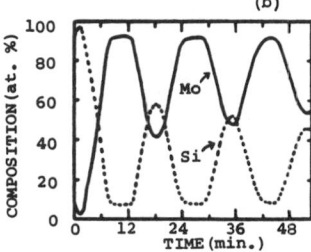

Fig. 1 Structural data for Mo-Si multilayer #302 with d_s=70Å, d_i=35Å and 160 bilayers. a) X-ray diffraction spectrum for λ=1.54Å. b) Auger depth profile for 1.5 keV Ar ion sputter beam.

compositional modulation here is incomplete, and gradually diminishes in each successive layer. We believe that this is an artifact of the Auger depth profile itself. First, the Auger electrons have an escape depth of 20 to 30Å, limiting the resolution. Second, the depth profile involves sputter etching by Ar ions (of energy 1.5 keV in Fig. 1b), which may produce uneven etching and ion mixing. Higher energy Ar ions can obscure the layering almost completely.

The semiconducting element, particularly Si, normally deposits in the amorphous state, even for rather thick films, when the substrate is held at room temperature. This is not always true, however, for the metallic layers; for layer thicknesses greater than about 40Å, even refractory metallic films come down as disordered, but rather oriented microcrystalline films. For thinner metallic layers, however, strains at interfaces may tend to make the amorphous metallic structure relatively more stable. Some evidence for this was taken on a set of Nb-Si and W-Si multilayers, using x-ray diffraction in the conventional reflection mode, which probes structure perpendicular to the layers. The evidence clearly indicates that there is no crystalline coherence from one metal layer to the next, and that for the thinner layers, the broadening of the crystalline peaks seems to be somewhat greater than that directly attributable to the size of a single layer. The amount of disorder, and perhaps the development of amorphous structure, does appear to increase substantially as the thickness of the metallic layer is decreased below about 30Å. Furthermore, for the thinnest layers, without epitaxy, it would certainly be surprising if crystalline order continued to exist along the planes. However, a more definitive probe of crystallinity within a layer could be made by directly probing structure within the plane, perhaps by transmission, and we are pursuing this approach. Finally, there seems to be very little evidence of metal-silicide formation, except perhaps for the very smallest layer spacing, where the atoms are largely mixed.

CRITICAL TEMPERATURE AND NORMAL-STATE PROPERTIES

Samples were generally measured in a four-probe configuration using Cu contacts pressed into the films, with sample geometry (defined by photolithographic lift-off) consisting of wide pads connected by a long narrow section 3.8mm x 0.3mm. The resistivity of these multilayers, measured

parallel to the film and attributed to the metallic layers, is of order 100μ ohm-cm or less, and either falls or is essentially constant (arising at most a few percent) down to low temperatures. Even single layers down to 10Å appear to be continuous, although we cannot rule out the presence of defects (e.g., holes) in such thin layers. Generally, resistivity increases as the thickness of the metal layer is decreased, suggesting an electronic mean free path that is limited in part by film thickness.

As we have noted above, the structure of the very thin metallic layers is highly disordered, perhaps even amorphous. This raises the critical temperature of the Mo and W composites substantially (from less than 1K to greater than 4K), while lowering that in the Nb multilayers. Both of these trends are consistent with other studies of T_c in bulk amorphous or disordered samples(8). It is not totally clear, however, whether these effects in the present case are a consequence more directly of structural modification or of compositional mixing that certainly exists to some degree at layer interfaces.

Taking a set of Nb-Si multilayer samples as a particular example, several generalizations can be made concerning the dependence of T_c on the three independent parameters d_s, d_i (the thickness of superconducting and insulating layers, neglecting to first order the interface smearing), and the number of bilayers N. First, holding d_s and N constant and varying d_i, larger d_i tends to be associated with lower values of T_c and with broader transitions. This can be understood by noting that large d_i yields essentially decoupled thin superconducting films, and either intrinsic 2-D fluctuations or film defects might be expected to broaden the transition. If d_i and N are held constant, then increasing d_s tends to raise T_c for Nb, a feature we believe is a consequence of the effect of order on T_c. Finally, for d_s and d_i constant, increasing N also tends to increase T_c slightly, together with a slightly decreasing normal state resistivity. This result may be due to a tendency toward improved layer smoothness and order as the number of layers increases.

CRITICAL FIELDS AND CURRENTS

Superconducting critical fields were measured in-house in a superconducting magnet dewar with fields up to 90 kG and temperatures down to 1.4K, and some samples were also taken to the Francis Bitter National Magnet Laboratory to allow fields up to 190 kG. In both cases, samples could be mounted with the film (and the layers) oriented either parallel or perpendicular to the direction of the field. We define the critical field as that field for which the resistance has reached one-half its normal-state value.

In general, the parallel upper critical magnetic field of these metal-semiconductor multilayer films is substantially larger than that for the field perpendicular to the layers. Essentially, this occurs because screening supercurrent act as pair-breaking perturbations, and the presence of the insulating layers tends to restrict the flow of

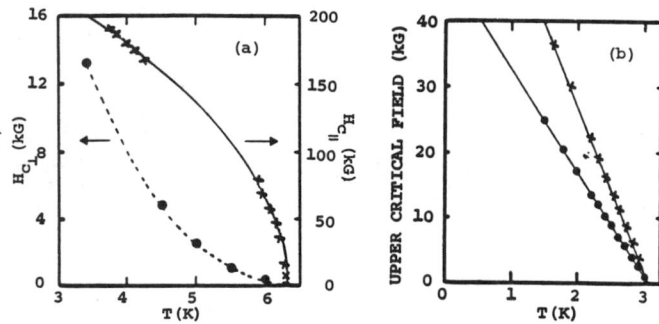

Fig. 2   Parallel (x) and perpendicular (dots) critical magnetic fields for a) Mo-Si #302 and b) Nb-Si #902-18.

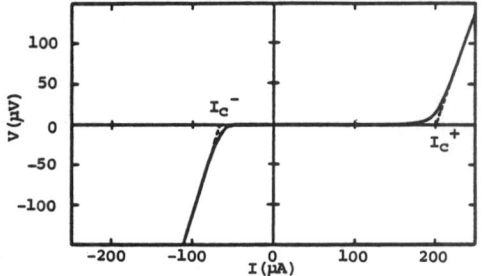

Fig. 3   Asymmetric I-V curve for Nb-Si #920-18 for a parallel field H=20 kG at T=1.5K.

supercurrent perpendicular to the layers in response to a parallel field. The perpendicular critical field is then the same as the usual upper critical field of a superconductor

$$H_{c\perp} = H_{c2} = \phi_0/2\pi\xi^2(T) \qquad (1)$$

where ϕ_0=hc/2e=2.07x10^{-7}gaus-cm^2 is the flux quantum, and $\xi(T)\sim(T_c-T)^{-1/2}$ is the Ginzburg-Landau coherence length, typically of order 100Å or greater in our films. This implies that $H_{c\perp}$ should go to zero linearly at T_c.

For the parallel critical field, there are several relevant regimes. In the "2-D" case, the coupling between adjacent metallic layers is so weak that the critical field is that which corresponds to a single metallic layer. For a layer of thickness $d_s < \xi$,

$$H_{c\parallel} = \sqrt{3}\ \phi_0/\pi\xi d_s \qquad (2)$$

Eq. (2) is an upper limit; for very small d_s, spin paramagnetism may produce a lower critical field.(9)

The critical fields for Mo-Si sample #302 are presented in Fig. 2a. The solid curve fit to the parallel field data is of the form $(T_c-T)^{1/2}$, in agreement with Eq. (2). Note the very high critical field for this orientation (greater than 200 kG at T = 0), which exceeds substantially the paramagnetic limiting field of 120 kG and indicates the presence of substantial spin-orbit scattering.(9) A parallel field of this dependence was the general rule for samples with insulator thickness d_i greater than about 30Å. However, this sample exhibits some discrepancies with the theory. The perpendicular critical field, instead of being linear near T_c, curves concave upward. If we apply Eq. (1) anyway at T=4k, one obtains ξ = 200Å, whereas, if we take a value d_s = 70Å, then from Eq. (2) and the parallel field, we obtain ξ(4K) = 90Å. Some similar deviations were seen in refs. (5) and (6), and they remain somewhat of a puzzle.

The other limit, the anisotropic 3D case, has been treated in terms of a model of Josephson coupling between the metallic layers, which is then mapped onto an anisotropic Ginzburg-Landau equation.(10) The result is an effective coherence length $\xi_\perp$ that is reduced from the value $\xi = \xi_\parallel$ within the plane by a factor proportional to the square root of the interfilm coupling, or roughly to the square root of the anisotropy in the normal state resistivity. This yields a parallel critical field

$$H_{c\parallel} = \phi_0/2\pi\xi_\parallel\xi_\perp \qquad (3)$$

which is larger than the perpendicular critical field, but exhibits the same linear temperature dependence.

Figure 2b shows the critical field behavior for a Nb-Si sample with d_i = d_s = 20Å. The critical fields are far less anisotropic than those in Fig. 2a, and follow straight lines. Using Eqs. (1) and (3), we infer values of $\xi_\parallel$ and $\xi_\perp$ of 115 and 80Å, respectively, both substantially larger than the repeat spacing d = 40Å and thus in the anisotropic 3D regime. A dependence like this was generally seen for insulating layers thinner than about 20Å, as long as d_s was not too great.

Qualitatively, one might expect Eq. (3) to remain valid as long as $\xi_\perp$ was much greater than the metal film thickness. According to ref. (10), the crossover between the two regimes of Eqs. (2) and (3) should occur when $\xi_\perp$ = d/√2. We have seen some evidence of such crossover behavior for insulating layers of intermediate thickness (or for thicker metal layers).

For subcritical magnetic fields applied parallel to the layers and perpendicular to the flow of current, some of these films exhibit behavior that appears very much like classic flux-flow characteristics. This corresponds to vortices moving perpendicular to the layers. An interesting feature of these characteristics that we have recently reported (7) is that they may be strongly asymmetric, and in particular that the critical current may be much larger (by up to a factor of 5) for one sign of the current than for the other sign (see Fig. 3b). Furthermore, this asymmetry reverses exactly if the sign of the magnetic field is reversed. This can be understood in terms of a preferred direction of vortex flow, either toward or away from the substrate. It is still unclear whether this effect is due to asymmetry of vortex entry at the film surfaces, or an asymmetric compositional profile at layer interfaces, or some alternative asymmetry. It has appeared in a number of layered samples, and only for this particular flux flow configuration of currents and fields. Investigation is continuing.

CONCLUSIONS

Finally, we believe that layering may be a useful tool to fabricate new and metastable materials, even without crystalline epitaxy, particularly in the limit of small, nearly atomic layer thicknesses. Since many superconducting compounds with high critical temperatures are metastable at best, this approach may be a fruitful one to investigate new higher-T_c superconductors.

ACKNOWLEDGEMENTS

We wish to thank J. Wood of ECD and the staff of the Francis Bitter National Magnet Laboratory for assistance with some of the high-field measurements. Some of the early work was funded in part by ARCO.

REFERENCES

1. S.R. Ovshinsky, U.S. Patent No. 4,343,044 (1982) and others pending.

2. I.K. Schuller and C.M. Falco, Thin Solid Films 90, 221 (1982); C.S.L. Chun, C.G. Zheng, J. Vincent and I.K. Schuller, Phys. Rev. B 29, 4915 (1984).

3. D.B. McWhan, M. Gurvitch, J.M. Powell and L.R. Walker, J. Appl. Phys. 54, 3886 (1983).

4. W.P. Lowe and T.H. Geballe, Phys. Rev. B 29, 4961 (1984).

5. T.W. Haywood and D.G. Ast, Phys. Rev. B 18, 2225 (1978).

6. S.T. Ruggiero, T.W. Barbee and M.R. Beasley, Phys. Rev. Lett. 45, 1299 (1980); Phys. Rev. B 26, 4894 (1982).

7. A.M. Kadin, R.W.Burkhardt, J.T. Chen, J.E. Keem and S.R. Ovshinsky, Proc. 17th Int. Conf. on Low Temp. Phys., Karlsruhe, West Germany, 1984, ed. by U. Eckern et al., p. 579 (Elsevier North-Holland, Amsterdam, 1984).

8. M.M. Collver and R.H. Hammond, Phys. Rev. Lett. 30, 92 (1973).

9. N.R. Werthamer, E. Helfand and P.C. Hohenberg, Phys. Rev. 147, 295 (1966).

10. R.A. Klemm, A. Luther and M.R. Beasley, Phys. Rev. B 12, 877 (1975).

SUPERCONDUCTING PROPERTIES OF SPUTTERED Mo–C FILMS WITH COLUMNAR MICROSTRUCTURE

J. Wood,[1] J.E. Keem,[1] J.T. Chen,[2] A.M. Kadin,[1] R.W. Burkhardt,[1] and S.R. Ovshinsky[1]

[1]Energy Conversion Devices, Inc., 1675 West Maple Road, Troy, Michigan 48084
[2]Department of Physics, Wayne State University, Detroit, Michigan 48202

(Received 10 September 1984)

INTRODUCTION

It has been evident for some time that a fine microstructure on the order of the superconducting coherence length (typically of order 100Å in relevant materials) is useful in optimizing critical currents and fields in technological superconductors. In the present paper, we discuss the superconducting properties of sputtered Mo-C films having nonequilibrium columnar microstructure on this scale. Some of the samples exhibit behavior in large critical fields that compares very favorably with commercial superconducting wire. We also present some preliminary data on multilayered films based on Mo-C, which also exhibit outstanding high-field behavior.

The cubic B1 (or NaCl) phase of transition metal carbides and nitrides tends to favor superconductivity, most notably for NbN and $NbN_{1-x}C_x$. MoC also forms in this phase, with quite respectable critical temperatures as high as 14K.(1,2) This phase has a rather limited range of equilibrium stability, however, and is formed most easily when it is sub-stoichiometric in C, so that it has also been characterized as the MoC_{1-x} phase.(3,4) We have found that we can fabricate this phase rather easily in our laboratory using rf sputtering onto room-temperature substrates, and the presence of a small percentage of impurities appears to enhance the formation of this metastable structure without substantially depressing superconductivity.

Oriented columnar microstructure tends to form in many systems during condensation from the vapor, particularly in cases where growth starts in small isolated clusters epitaxially on the substrate. Even where epitaxy does not occur, this type of microstructure tends to be stabilized by the presence of impurities at grain boundaries. It has proven possible in several systems to take advantage of this tendency toward columnar growth for improved superconducting properties. The significance of this for NbN-based materials has been emphasized in refs. 5-8, where critical fields have been measured that approach the highest known. We will present

evidence that a similar approach may explain the dramatically enhanced critical fields and critical currents in our MoC samples.

EXPERIMENTAL PROCEDURE

The samples were fabricated using either a single-target rf diode sputtering system with a composite target, or alternatively a multi-target rf magnetron sputtering system with substrate transport past the two elemental targets. For both systems, background pressures were normally in the low 10^{-6} torr range, and sputtering was carried out dynamically in 2-7 mtorr of argon. Deposition rates were typically on order 10-100Å/min for sample thicknesses up to a micron or more. Multilayer films, which will be discussed briefly toward the end of the paper, could be fabricated either by periodic injection of a reactive gas into the sputtering mixture, or by the use of additional targets. Substrates used, which were generally unheated during deposition, included silicon wafers, glass, Kapton (9) tape, and copper-coated Kapton tape. Estimated temperatures during unheated deposition were in the range of 100C or below. Some samples were cooled well below room temperature using liquid nitrogen. Structure and composition were analyzed using a number of techniques, including angle-resolved x-ray diffraction for crystal structure, x-ray fluorescence and Auger depth profiling for composition, and scanning electron microscopy for visual examination of grain structure.

Experimental samples were defined by photolithography and liftoff to form a set of narrow strips about 0.35cm long by 0.03cm wide, which were measured resistively using a standard four-probe configuration. The critical temperature T_c was defined as the midpoint of the resistive transition, and likewise the critical fields were defined by the midpoint in the resistance vs. field curves. Critical field data was taken largely using a 90kG superconducting solenoid and a Janis Supervaritemp dewar (with temperatures down to 1.5K in liquid He and above 4.2K in the vapor). Some samples were

Reprinted by permission from *IEEE Transactions on Magnetics,* Vol. MAG-21, No. 2, pp. 842–845 (March 1985).

also measured at the Francis Bitter National Magnet Lab using conventional Bitter magnets capable of reaching 180kG. For both systems, the sample could be oriented parallel as well as perpendicular to the field direction. Critical currents were measured immersed in liquid He using clamped copper current leads to prevent heating, and were defined by the criterion of one microvolt across the sample.

RESULTS AND DISCUSSION

We have prepared Mo-C in a wide range of compositions, largely using a sputtering target composed of carbon on one end and molybdenum on the other. In this way, a continuous gradient in composition was obtained on the substrate located relatively close to the target, where the direction of this gradient was perpendicular to the narrow strips used to determine superconducting properties. The values of T_c obtained for two such runs are shown in Fig. 1 as a function of composition. Although there is a variation between the two runs (due to differences in sputtering rate, impurity inclusion, etc.), there is clearly a maximum critical temperature for a composition of about 45-60% C (all compositions in this paper are in atomic percent). For such a composition, samples with T_c up to 13K have been prepared. There was also up to several percent oxygen "contamination" in many of the samples, which varied somewhat depending on the details of the preparation conditions, but did not seriously degrade the critical temperatures up to concentrations of 5% or more.

X-ray diffraction of the material with the highest T_c indicated that it was largely crystalline, essentially single-phase in the B1 structure, with a lattice spacing of 4.31Å. As the concentration of C increased beyond 60%, the X-ray peaks started to broaden substantially, and eventually the material became amorphous. Conversely, on the Mo-rich side, the evidence suggests the presence of amorphous Mo, together with some hexagonal Mo_2C. As the C concentration

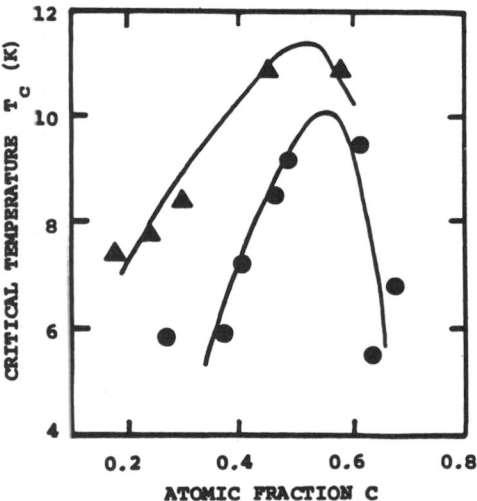

Fig. 1 Critical temperatures vs. composition for two Mo-C gradient runs; 214 (triangles) and 254 (circles). The lines are guides to the eye. T_c is defined by the middle of the resistive transition; typical transition widths are 0.5K.

Table I Data for representative Mo-C sputtered films.

SAMPLE	254 -8c	264 -4a	315* -5a	499* -7c
Composition (%C)	54	54	56	53
Thickness (µm)	0.94	0.67	0.74	0.94
T_c (K)	8.9	9.7	9.0	8.1
ρ_N (µΩ-cm)	420	390	460	980
$dH_{c2\parallel}/dT$ (kG/K)	31	31	31	37
$dH_{c2\perp}/dT$ (kG/K)	51	49	36	57

*Layered with N_2--see text.

exceeded about 40%, the B1 phase started to appear strongly. For the optimized material (with about 45% C), direct observation by scanning electron microscopy indicated the presence of columnar microstructure perpendicular to the substrate, with column diameters 1000Å or smaller, while X-ray diffraction exhibited preferential crystallite orientation (in the [111] direction) and line-broadening corresponding to 100Å minimum crystallite size. Taken together, this suggests a range of column diameters between about 100 and 1000Å.

Although the composition of our measured samples was macroscopically uniform, the superconducting transition was generally rather broad (of order 0.5K or more), pointing to a microscopic spatial variation of T_c on the scale of the column diameters. This is also suggested in our recent work indicating a large inverse ac Josephson effect in these samples over this temperature regime,(10) corresponding to a large number of weakly coupled superconducting grains. The resistivity of the samples is also consistent with this picture. The best crystalline MoC with the highest T_c still has a resistivity of several hundred µohm-cm, and this becomes much higher as the C concentration is increased. We speculate that for these extremely resistive samples, much of the resistivity can be attributed to boundary resistance between the grains, perhaps due to the presence of amorphous C. This might account for the C atoms in excess of stoichiometric MoC. Alternatively, we cannot rule out the possible presence of inter-column voids as observed in ref. 5.

Critical field data for several representative samples is given in Table I, and the more complete dependence for two of these is presented in Figs. 2 and 3. (Samples 315 and 499 were multilayered with nitrogen as discussed below.) A notable feature of these films is the fact that the perpendicular critical field was substantially greater than the parallel critical field (by as much as a factor of 1.7), in contrast to the more usual situation for thin films. This anisotropy can then be attributed semi-quantitatively to the fact that the perpendicular orientation effectively provides a measurement of the surface critical field H_{c3} of the columns, while the parallel orientation measures the more usual H_{c2}. Essentially, the same behavior has

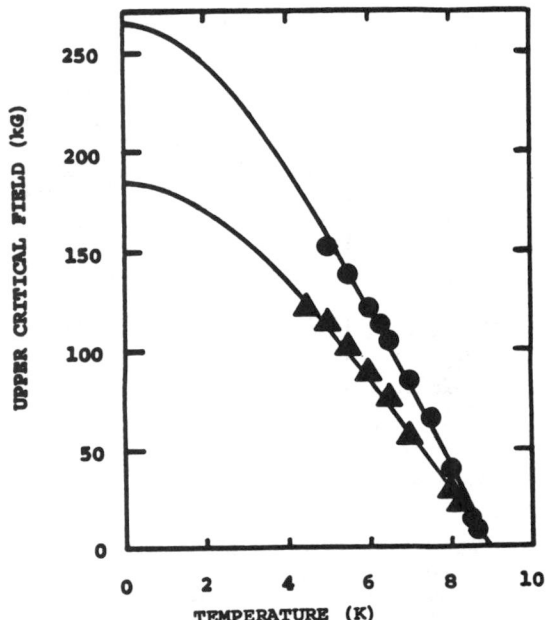

Fig. 2 Temperature dependence of upper critical magnetic fields for parallel (triangles) and perpendicular (circles) orientations for sample 254-8c. The lines indicate the Ginzburg-Landau dependence fit to the slope at T_c, assuming no paramagnetic limiting.(14)

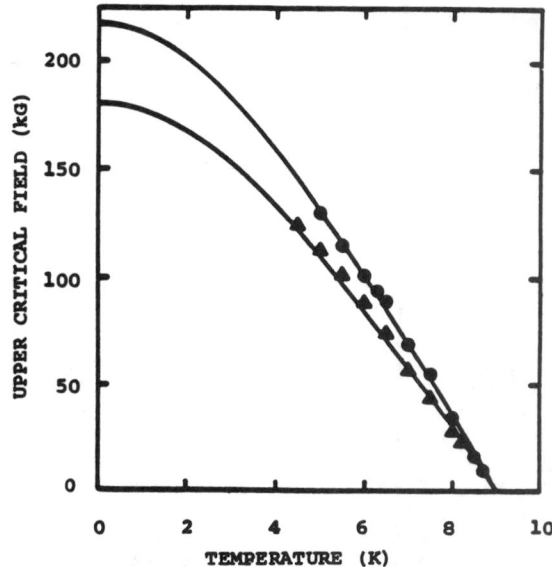

Fig. 3 Critical fields for parallel (triangles) and perpendicular (circles) orientations for sample 315-5a (Mo-C layered with N_2). Theoretical lines as in Fig. 2.

previously been seen with NbN films that contain similar microstructure.(6-8)

In addition to their remarkable anisotropy, the critical field curves exhibited very high slopes near T_c, with values that approach 60kG/K. Values this large are comparable or greater than those seen even in the most extreme Type-II superconductors, including amorphous alloys.(11-13) Although for the Mo-C system, these highest values occurred in samples which had somewhat less than the highest critical temperature (T_c was typically 8-10K), the extrapolated critical fields were nonetheless impressive. Using the Ginzburg-Landau theory, in the absence of paramagnetic limiting (14) $[H_{c2}(0)=0.69T_c(dH_{c2}/dT)|T_c]$ gives several values for $H_{c2}(0)$ in excess of 300kG. These values are far in excess of previously measured values of H_{c2} (100kG or less) listed in the literature on B1-phase MoC.(15) Some of the data at the highest fields fall somewhat below that given by the simple Ginzburg-Landau theory. It is unclear whether this is a manifestation of paramagnetic limiting, or alternatively may reflect some size effects as suggested for NbN films having a column-void structure.(7-8) Size effects may well be relevant, however, since the Ginzburg-Landau coherence length covers the range from 30 to 100Å (corresponding to H_{c2} from 30-300kG) for samples and temperatures of interest, comparable to the estimated minimum column dimension of order 100Å.

In attempting to understand the magnitudes of these critical field slopes, the assumption is made that the slopes for parallel magnetic field actually represent H_{c2}. We then use the formula appropriate for a dirty BCS weak-coupling superconductor:

$$dH_{c2}/dT=12\rho_N\gamma e/\pi^3 k_B=4.5\rho_N\gamma$$

where ρ_N is the normal state resistivity (in ohm-cm), γ the electronic specific heat coefficient (in ergs/cm^{3}K^2), and H_{c2} is in Teslas. This assumption of weak coupling is consistent with electron tunneling measurements on these materials.(16) Taking an electronic heat capacity coefficient (17) 3700 ergs/cm^{3}K^2 and the maximum slope of 35 kG/K gives a resistivity of about 210 μohm-cm. This is not unreasonable, even though the measured resistivity is several times higher, because it is likely, and consistent with our understanding of these systems, that much of this resistivity comes from between the grains, whereas the internal grain resistivity in the metallic grains is unlikely to be more than the maximum metallic resistivity (typically 200-300μohm-cm) in this material. A similar discrepancy was observed for columnar NbN.(6,7) The material within the columns is thus highly disordered, but without the drastic suppression of critical temperature that tends to occur in the presence of disorder in other classes of superconductors (e.g., A-15s).

We have also measured critical currents of some of these samples as a function of magnetic field, obtaining some rather large values for critical current density J_c. In Fig. 4, J_c is plotted for sample 254-8c (see Fig. 2 for critical fields for this sample) as a function of parallel magnetic field (J and H are also parallel) for T=4.2K. Preliminary evidence (not shown) suggests some anisotropy in the critical currents, with that for perpendicular fields being substantially greater than that for parallel fields. All this is

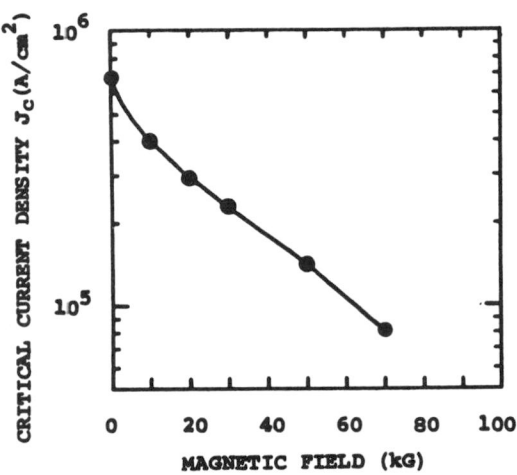

Fig. 4 Critical current density vs. magnetic field for sample 254-8c at T=4.2K. The field was oriented parallel to the film and the current. The line is a guide to the eye.

consistent with our picture that the magnetic and transport properties of these films are dominated by the presence of the columnar microstructure.

In order to further investigate questions of anisotropy in these materials, we have fabricated artificially layered structures based on Mo-C, either by pulsed gas injection of N_2 into the sputtering chamber or by alternately sputtering layers of Si. Samples 315 and 499, prepared by the former method, yielded an effective layer repeat spacing of about 800Å (and 16 bilayers), with a maximum nitrogen concentration over 30%, as indicated by Auger depth profiling. The composition figures for these samples in Table I actually represent estimated %C+N in the nitrogen-deficient layers. The structure of the nitrogen-rich layers has not been fully characterized, but we believe them to be superconducting with a lower T_c than that of the MoC. Fig. 3 shows critical field data for sample 315-5a, where the degree of anisotropy was somewhat less than in typical non-layered samples, although the perpendicular critical fields were still the larger ones. It is unclear whether this lessened anisotropy followed directly from the layering, or indirectly from a disturbance of the columnar growth process. With the alternative method of producing multilayered composites, preliminary samples obtained by sequential sputtering from Mo, C and Si targets exhibited very large parallel critical fields, due to the layering, together with rather depressed perpendicular critical fields due to incomplete formation of the desired Mo-C phase and associated columnar microstructure. Work in these areas is continuing.

CONCLUSIONS

We have demonstrated the feasibility of fabricating high-critical-field superconducting Mo-C films on room-temperature substrates, including preparation as a tape for possible large-scale technological application. The critical fields and currents of these films compare favorably with several currently commercial superconducting

materials. The high-performance properties are due to the presence of stabilized columnar growth on a scale comparable to the superconducting coherence length. Preliminary evidence suggests that the enhanced critical fields perpendicular to the substrate may be extended to other orientations through the use of controlled artificial layering on the same scale. Overall, we have shown that close control over microstructure is possible in sputtered films and can lead to substantial improvement in superconducting performance.

Acknowledgements

This work was supported in part by a grant from ARCO. We also wish to thank Dr. Lawrence Rubin and the staff of the Francis Bitter National Magnet Laboratory for their cooperation in making some of the critical field measurements possible.

References

1. L.E. Toth, E. Rudy, J. Johnston and E.R. Parker, J. Phys. Chem. Solids 26, 517 (1965).

2. R.H. Willens and E. Buehler, Appl. Phys. Lett. 7, 25 (1965).

3. W. Krauss and C. Politis, in Superconductivity in d- and f-band metals 1982, ed. by W. Buckel and w. Weber, Kernforschungszentrum Karlsruhe 1982, p. 439.

4. E.K. Storms, The Refractory Carbides, Chapter VIII, Academic Press (1967).

5. W. Wagner, D. Ast and J.R. Gavaler, J. Appl. Phys. 45, 465 (1974).

6. M. Ashkin and J.R. Gavaler, J. Appl. Phys. 49, 2449 (1978).

7. J.R. Gavaler, A.T. Santhanam, A.I. Braginski, M. Ashkin and M.A. Janocko, IEEE Trans. MAG-17, 573 (1981).

8. M. Ashkin, J.R. Gavaler, J. Greggi and M. Decroux, J. Appl. Phys. 55, 1044 (1984).

9. Trademark for Dupont Polyimide film.

10. H. Sadate-Akhavi, J.T. Chen, A.M. Kadin, J.E. Keem and S.R. Ovshinsky, Solid State Commun. 50, 975 (1984).

11. R. Koepke and G. Bergmann, Solid State Commun. 19, 435 (1976).

12. F.P. Missell, S. Frota-Pessoa, J. Wood, J. Tyler and J.E. Keem, Phys. Rev. B27, 1596 (1983).

13. M. Ikeba, Y. Muto, S. Ikeda, H. Fujimori and K. Suzuki, Physica 107B, 387 (1981).

14. N.R. Werthamer, E. Helfand and P.C. Hohenberg, Phys. Rev. 147, 295 (1966).

15. B.W. Roberts, J. Phys. Chem. Ref. Data 5, 743 (1976); NBS Technical Note 983 (1978).

16. H. Sadate-Akhavi, J.T. Chen, F.P. Missell and J.E. Keem, Bull. Am. Phys. Soc. 27, 381 (1982), and additional unpublished data.

17. L.E. Toth and J. Zbasnik, Acta Metall. 16, 1177 (1968).

THE ROLE OF FREE RADICALS IN THE FORMATION OF AMORPHOUS THIN FILMS

Stanford R. Ovshinsky

Energy Conversion Devices, Inc., 1675 West Maple Road, Troy, Michigan 48084

A subject of great scientific and technological interest, amorphous materials is intimately connected with the topic of this meeting. This is especially so since one of the preferred methods of making amorphous films for photovoltaic devices utilizes plasmas in which dissociation of gases is associated with dc electric fields or radio frequency excitation. Such creation of plasmas by nonthermal means generates excited states in the gas, ultimately producing different chemical species from that of the ground state molecules. The excited states have a high energy content and, in some cases, a unique electronic distribution, and are thus much more reactive. The excited states and the chemical reactions are far more selective and provide a different route of excitation and relaxation than simple thermal activation, since thermal mechanisms achieve an overall increase in the energy content of the plasma which affects all molecular structures equally whereas the preferred methods for making photovoltaic devices are involved with chemical processes directly attributable to specific electronically excited states.

It is the thesis of this paper that it is the free radicals, e.g., atomic hydrogen and/or fluorine, which play decisive roles both in silicon-containing plasmas and in the condensed state during and after film formation. Such films have the prerequisite chemical, structural, and electronic parameters necessary for a material with an inherently low density of states in the gap, the signature of photovoltaic grade amorphous materials.

A plasma is as of now the preferred method of manufacturing amorphous solar cells. While there is much plasma diagnostic work (1) attempting to identify the various species in the plasma and correlate them with actual device performance, there are few experimentally definitive identifications of the many species generated or of their lifetimes, or of recombination and quenching mechanisms.

As a contribution to the basic understanding of this important subject, I will further develop here my simplifying premise that it is the free radicals, or, more precisely, free atoms of hydrogen and/or fluorine, which are the key catalytic features of the plasma and the important bridge between the plasma and the condensed state which is the film. In the film-forming process, it is these free radicals which combine with the primary element silicon in such ways as to decrease the concentration of dangling bonds. Especially, it is atomic fluorine which generates new beneficial configurations with low densities of states in the gap (2) not only by its bonding, but by its influence on other elements and by the generation of additional hydrogen and fluorine free radicals which further aid in making better quality materials.

Amorphous silicon alloys are direct band gap materials and efficiently absorb light of energy greater than 1.6 to 1.8 eV. They thus absorb in a film 0.5 μm thick as much sunlight as does crystalline silicon (an indirect band gap material) in a 50 μm sample. This makes them ideal for the application of thin film technology to photovoltaics.

The key problem in making amorphous photovoltaic devices has been how to lower the concentration of defects in amorphous silicon alloys so as to remove the dangling bonds and other defects which act as traps and recombination centers, and thus sinks for electrons and holes. Ordinarily, sputtered amorphous silicon is useless as a photovoltaic material since dangling bonds are ubiquitous and result in an extraordinary number of states in the gap (10^{20}cm^{-3}). The reason for this is that tetrahedral structure is not completed due to energetic considerations, the lowest free energy condition favoring dangling bonds and voids.(3) The relationship between structure and function is clearly illustrated in such an amorphous material since the addition of hydrogen (at first accidentally) created a silicon-hydrogen alloy through the decomposition of silane gas in the form of a plasma with a density of states that is now as low as $\sim 10^{16}$cm^{-3}. These alloys have a low enough density of gap states that they can be substitutionally doped and therefore have various electronic applications such as thin-film transistors and solar cells. Together with a group of collaborators, I showed that the further addition of fluorine would create a new silicon alloy with an even smaller dangling bond and defect concentration ($\sim 10^{15}$cm^{-3}).(4-6) This material has superior

Reprinted by permission from the *Proceedings of the International Ion Engineering Congress,* ISIAT '83 & IPAT '83, Kyoto, 12–16 September 1983, pp. 817–828.

physical properties as well. The ability to develop such a material evolved from my understanding that free radicals are not just terminators of dangling bonds but play a dual role as structural elements, since they enter the structure as bridging atoms, thus providing flexibility to the amorphous silicon alloy by relieving stresses while concomitantly preserving the tetrahedral local order and generating a _new local order_ without dangling bonds. Fluorine, the element that I chose, is therefore not just a compensator but permits new local order configurations which eliminate residual dangling bonds.(2)

The problem that has puzzled many investigators new to the amorphous field has been how a film made of the same chemical elements as another film prepared by a different technique can have such different properties. The answer lies in the three-dimensional freedom of atoms in amorphous materials which permits and in fact encourages bonding options that do not exist in crystalline material by virtue of the restrictions of crystalline symmetry. This freedom generates local order environments which are directly affected by the preparation techniques involved. It is therefore the manner in which the atoms are able to attach to each other, that is, their local configuration, that is so important. It follows from this that the deposition parameters are directly involved with the reactivity of the elements. The deposition conditions control the bonding and therefore the configurations of amorphous films. Plasma techniques, the most reactive of the processes utilized, offer a new dimension in the design of synthetic amorphous materials. The constituents and reactivity of a plasma are of paramount importance. There are obviously a large variety, a veritable zoo, of molecular fragments and excited species in a plasma. Utilizing the concept of parsimony and basic chemical principles, one can predict the most important species that would be generated in such a plasma.

If one has a gas mixture of silicon tetrafluoride plus hydrogen, or silane plus fluorine, or silane plus silicon tetrafluoride, and generates a plasma in a typical manner, i.e., through a capacitively coupled 13.56 MHz RF glow discharge decomposition process with a pressure of 0.1-0.3 Torr and a substrate temperature of approximately 250°C, then one of the chief chemical reactions is the formation of free atoms of fluorine and hydrogen by the abstraction of hydrogen by fluorine, since it is known that atomic fluorine reacts with hydrogen so as to generate not only HF but free hydrogen and fluorine atoms as well. The low energy necessary for the dissociation of the fluorine molecule and the catalytic effect of hydrogen on the decomposition of silicon tetrafluoride indicate that the films deposited from such a highly reactive plasma contain silicon, fluorine, and hydrogen in unusual configurations.(2) Indeed, Lee and deNeufville of our laboratory have confirmed this prediction and we reported at the 10th International Conference on Amorphous and Liquid Semiconductors in Tokyo that we have made the first observation of HSiF produced by the reaction of silane with fluorine and detected by laser-induced fluorescence.(7) This is an important result which has many implications. It would appear that the fluorination of silane involves the abstraction of free radical hydrogen atoms by fluorine gas and that the probabilities are that the fluorine atoms aid as well in subsequently generating a wide variety of intermediaries.

The standard plasma decomposition of silane offers limited atomic species in the solid and the hydrogen combines in several different configurations which do not make for an optimally low density of states in the gap.(8) Fluorinated materials, on the other hand, not only specifically tie up dangling bonds by strong bonding, but also restrict the bonding choices of hydrogen so as to make for stronger hydrogen bonding to silicon, and generate new configurations of silicon, fluorine and hydrogen.(2,7)

There has been an oversimplification by those who consider that hydrogen and fluorine are merely bond terminators. The amount of hydrogen necessary to compensate the dangling bonds of silicon is less than 0.1%. The rest of the hydrogen is used to provide structural links similar to the crosslinks which I have previously described in chalcogenides.(2) One must have a built-in flexibility permitting the completion of chemically saturated local structures,(9) for if one had only elemental silicon and all tetrahedral bonds were completed, then a rigid tetrahedral crystal structure would ensue. Hydrogen and fluorine both act to permit the completion of a tetrahedral structure but distort bonds. This is made possible by the high reactivity, small size, and single bonding specificity of both elements as well as their ability to be structurally involved as bridges and links.(2) Fluorine has several other advantages, including the introduction of some ionicity, and therefore acts to take away the strain and compensate the tetrahedral structure while still utilizing its chemical bonding toward preserving the amorphous state. In addition, its extreme electronegativity is of paramount importance in organizing local order structures.(2)

Tsu and colleagues at our laboratory (10) have been able to support this thesis experimentally by showing that there is an increase of local order as one progresses from sputtered amorphous silicon to amorphous silicon alloys with hydrogen and finally to the most complete local order amorphous silicon alloys with fluorine and hydrogen. We used Raman scattering to characterize the change in structural order under heat treatment and correlated this with the optimal reduction of electronic gap states. We called this improved form of order without crystallinity intermediate range order.

This result explains experimentally how improved substitutional doping becomes possible when the background noise of density of gap states is low enough so that the chemical "information" of the substitutional elements can be detected. The greater the intermediate range order, the easier it is to accomplish substitutional doping with small amounts of dopant. In fact, the chemical reactions that cause fluorine to be instrumental in making intermediate range order amorphous materials can also lead to structures that result in microcrystallinity under certain conditions.(10,11) We therefore have evidence to support my original contention that fluorine produces the most order in amorphous materials.(2,4)

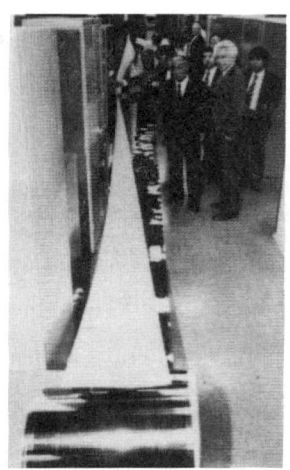

Fig. 1 A sixteen inch by one thousand foot (40cm x 300m) continuous roll of Ovonic photovoltaic cells

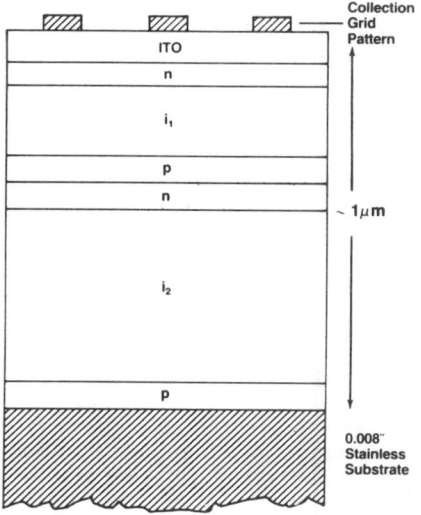

Fig. 2 Schematic of a Sharp-ECD Solar, Inc. tandem photovoltaic cell

To provide a more profound understanding of how the reactions in plasmas are related to the amorphous state, I wish to emphasize the role of order both in the plasma and in materials. Obviously, a plasma does not have periodicity, has transient nonequilibrium states, and is able to have unusual bonding configurations determined only by chemical reactivity in three-dimensional space. Gases, liquids and amorphous solids have much in common in that their bonding configurations are not dictated by the tyranny of the crystalline lattice. The creation and freezing in of nonequilibrium configurations is a property unique to amorphous solids.(8,12) Unlike in gases, transient configurations can be stabilized in an amorphous solid and therefore one has in such a solid a total interactive environment (TIE) which reflects local chemical bonding relationships.(13) It should therefore be clear that when one speaks of amorphicity one is speaking not only of the absence of long-range order but also of the presence of a large number of different types of short-range order, the richness of which provides the possibility for new phenomena. I have stated that "Since amorphous solids tend to differ locally in coordination sites, the introduction of the modifying atom, ion or molecule during the formation of the film offers the maximum possibilities for the interactions."(8)

With this understanding, one can now see why two other means of making photovoltaic materials, chemical vapor deposition (CVD) and ion implantation, have not proven successful. The CVD process of thermal decomposition of silane or disilane has always produced inferior electronic materials. The reason for this is that the bonding of silicon and hydrogen that can be generated at $600^{\circ}C$ is simply very limited.(14) Very little hydrogen can be incorporated and the density of gap states is high. Hydrogen diffusion into the material has been attempted to rectify the problem. However, once a solid has been formed having its own local configuration, there are thermodynamic difficulties in restructuring the local order which by this time exhibits rigidity. From the principles that I have outlined here, one could predict limited success since it is the bonding of the silicon and the

hydrogen that takes place within a relaxation time during which it freezes in the configuration needed. Hydrogen should not only be considered as compensator. It must be incorporated in its preferred intrinsic configuration with silicon in order to have the local order which has the minimum density of states. Ion implantation of hydrogen has similar problems. The local order generated by the implantation process is different from that of the plasma process due to the relaxation times of the atoms and excited species. Merely annealing after implantation does not permit the local order structural rearrangements necessary. It is therefore clear that there are preferred local order geometries and chemical bonding and these are dynamically formed.

I believe, as I originally predicted in Williamsburg (15), that as we make better and better tetrahedral amorphous materials, they will more and more resemble the classical crystalline tetrahedral materials and therefore, at some point, 'amorphous silicon,' i.e., amorphous silicon alloys, will resemble crystalline silicon in many ways even though there will still be some important differences, e.g., electron-phonon interactions. This holds promise for the future of electronics since, as we will now show, we are making high quality amorphous silicon alloys over 1 foot wide and over 1000 feet long. See Fig. 1.

Figure 2 shows a schematic of a tandem photovoltaic cell being manufactured by Sharp-ECD Solar, Inc. in Shinjo, Japan. Please note the intrinsic layers which I have indicated to you are critical and must be made with the lowest possible density of gap states in order to have the highest quality material. The cell has the following layer structure:

1. Thin steel substrate
2. Multi-layered photovoltaic amorphous silicon alloy layers (approximately 1μm thick; tandem cells have six layers)
3. ITO
4. Grid pattern
5. Encapsulant

Fig. 3 The Ovonic continuous roll-to-roll photovoltaic processor

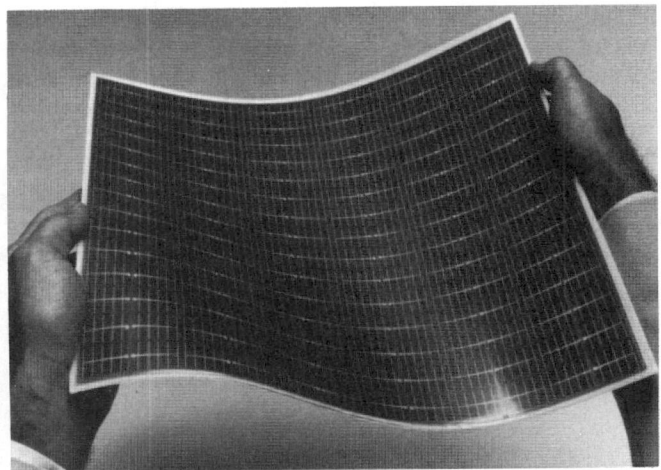

Fig. 5 A one square foot Ovonic amorphous cell

Figure 3 shows the Ovonic Continuous Roll-to-Roll Photovoltaic Processor. This plasma deposition machine for depositing multi-layered amorphous alloys has many technical advantages. It is approximately 35 feet long, has multiple deposition modules and continuously processes a 16-inch wide stainless steel substrate (approximately 8 mil thick). Amorphous photovoltaic thin films (less than $1\mu m$ thick) having a six-layered structure (PINPIN), as shown in Fig. 2, are deposited on a roll of 16-inch wide, 1000-foot long stainless steel continuously in single pass. Mass production of low-cost tandem amorphous solar cells utilizing a roll-to-roll process is now a commercial reality. Much credit for our progress in photovoltaics must be given to Dr. Masatsugu Izu, head of our Photovoltaics Project and to the staff at ECD.

Fig. 4 Continuous one foot wide rolls of mass-produced Ovonic solar cell material

The output of this machine is shown in Fig. 4.

Figure 5 shows a 1 square foot Ovonic amorphous solar cell. The efficiency of such a cell is in the 7% range while similar cells have been prepared in the laboratory with efficiencies of approximately 10%. In addition, much higher energy conversion efficiencies can be obtained with the same process by using the multi-cell layered or tandem thin-film solar cell structures shown in Fig. 2. These devices can exhibit enhanced efficiency by utilizing a wider range of the solar spectrum. Since the theoretical maximum efficiency for multi-cell structures is over 60%, one can certainly realistically anticipate the production of thin-film amorphous photovoltaic devices with efficiencies as high as 30%. Our production device is already a two-cell tandem cell and we have made prototype three-cell tandem cells. We have solved not only the problems of interfacing the individual cell components but also the difficulties associated with a 1-foot wide device deposited on a continuous web. It is the development of new alloys of lower band gap which is the area of greatest interest now since without materials with band gaps as low as 1.1 eV with low density of gap states, one cannot hope to achieve this very high efficiency goal.

We have discussed how a thin film is formed in a plasma by generating the most reactive species possible, i.e., free atoms of hydrogen and fluorine. The film is amorphous since the fullest expression of unusual three-dimensional bonding, charge configurations, and local order can only be fully realized in amorphous materials in which atoms have three-dimensional freedom and choices without the constraints of periodicity. In such a material, the notion of equilibrium positions of atoms loses its classical meaning, metastability takes on an entirely new meaning, and there is a unification of physics, chemistry and geometry so that, indeed, one can consider a new discipline has evolved. This scientific discipline is based on the technology of generating chemical bonding, plasmas being the "furnaces" for forming the necessary species and configurations. I have stated this thesis in the following manner: "Anyone who has worked with a highly excited plasma should become converted to the notion that...the radical chemistry that goes on transiently in such a plasma and in its interaction on surfaces makes for configurational combinations of such richness that even though the same elements are involved, the materials can differ substantially merely by altering the reactivity of the plasma. In other words, the elements have not changed, but their (chemical) interactions have. I call this radicalization of the plasma."(13)

In effect the plasma interacting with the solid creates a new solid state chemistry. Once we accept that nonequilibrium configurations can be of importance in amorphous materials, it becomes clear that bonding options afforded by excited atoms are associated with relaxation times and pathways that provide new bonding possibilities. This is why thermal chemistry is different from photochemistry, and why plasma chemistry plays such an important role.

When one discusses plasmas, great importance must be placed on generation rates, diffusion lengths, recombination kinetics, third body interactions and surfaces. Although a discussion of these topics is too large for this present paper, it should be clear that since they are preparation dependent, all of these factors have an effect upon the resulting film.

The concept that we are dealing with silicon alloys rather than elemental silicon is an important one since the addition of hydrogen and fluorine affects the three essential forces which operate in solid state chemistry whether in the crystalline or the amorphous form. They are:

1. the number of nearest neighhors; this is important for a tetrahedral material since it should be as close to four as possible;

2. the separation between atoms, i.e., the bond length; this is important in order to assure amorphicity rather than crystallinity; and

3. the angle between the two bonds emanating from a given atom; this is important again in terms of assuring the amorphous state through distortion.

If, for example, we relied only on the number of nearest silicon neighbors in tetrahedral materials, there could be the problem of the tendency toward crystallization, since the amoprhous state is more favored by low coordination numbers.(15) I proposed that the controlling influence in amorphous materials is their average coordination number and that divalent materials had the most tendency to form glasses. Alloying elements such as hydrogen and fluorine play a structural role in assuring amorphicity, not only separating the silicon atoms in such a manner as to prevent crystallization, but also generating the coordination required, precluding undercoordination, and relieving the strains as all the bonding orbitals are utilized. It should be clear that an alloy, rather than a single element, is the preferred method of making interesting amorphous materials.

This paper has concerned itself with free radicals which are instrumental in generating and controlling the local order and coordination number. These concepts have been most helpful in synthesizing new amorphous silicon alloys. Plasmas have been utilized containing silicon, fluorine and hydrogen in which the free radical nature of the hydrogen and fluorine atoms is the essential key to depositing an amorphous film with the lowest density of gap states and therefore the best electronic qualities.

We have theoretically simplified the multitude of events taking place in the plasma, on the surface, and in the film by utilizing this unifying concept. Our experiments support this premise and commercial photovoltaic devices using plasma decomposition are being manufactured. It is the activity of the free radicals in the plasma, on the surface, and in the solid, both by themselves and as intermediaries, which makes understandable the creation of an amorphous silicon alloy with such a low density of gap states that it is a superior electronic material.

References

1. M. Hirose, Growth and Characterization of Amorphous Hydrogenated silicon; Proc. 13th Conf. on Solid State Devices, Tokyo, 1981; Jap. Journal of Appl. Phys. 21 (1982) Supplement 21-1, pp. 275-281.

2. S.R. Ovshinsky, The Shape of Disorder, J. Non-Cryst. Solids 32 (1979) pp. 17-28. (Mott Festschrift.)

3. S.R. Ovshinsky, Amorphous Materials as Interactive Systems, in Proc. of the 6th International Conference on Amorphous and Liquid Semiconductors, Leningrad, November 18-24, 1975, pp. 426-436. Also see summary by H. Fritzsche, pp. 65-68.

4. S.R. Ovshinsky and A. Madan, A New Amorphous Silicon-Based Alloy for Electronic Applications, Nature 276 (1978) pp. 482-484.

5. C.H. Hyun, M.S. Shur and A. Madan, Determination of the Density of Localized States in Fluorinated a-Si Using Deep Level Transient Spectroscopy, Appl. Phys. Lett. 41, No. 2 (1982) pp. 178-180.

6. S. Guha, Tata Institute of Fundamental Research, Bombay, India and Energy Conversion Devices, Inc., Troy, MI, private communication.

7. H.U. Lee, J. deNeufville and S.R. Ovshinsky, Laser-Induced Fluorescence Detection of Reactive Intermediates in Diffusion Flames and in Glow-Discharge Deposition Reactors, presented at the 10th International Conference on Amorphous and Liquid Semiconductors, Tokyo, August 22-26, 1983, to be published.

8. S.R. Ovshinsky, Chemical Modification of Amorphous Chalcogenides, in Proceedings of the 7th International Conference on Amorphous and Liquid Semiconductors, Edinburgh, Scotland, June 27-July 1, 1977, pp. 519-523.

9. S.R. Ovshinsky, Localized States in the Gap of Amorphous Semiconductors, Phys. Rev. Lett. 36, No. 24, (1976) pp. 1469-1472.

10. R. Tsu, J. Gonzalez-Hernandez, J. Doehler and S.R. Ovshinsky, Order Parameters in a-Si Systems, Solid State Commun. 46 (1983) pp. 79-82.

11. R. Tsu, S.S. Chao, M. Izu, S.R. Ovshinsky, G.J. Jan and F.H. Pollak, The Nature of Intermediate Range Order in Si:F:H:(P) Alloy Systems, presented at the 9th International Conference on Amorphous and Liquid Semiconductors, Grenoble,

France July 2-8, 1981. Also published in J. de Physique, Colloque C4, supplement au no. 10, Tome 42, (1981), pp. C4-269-C4-272.

12. R.A. Flasck, M. Izu, K. Sapru, T. Anderson, S.R. Ovshinsky and H. Fritzsche, Optical and Electronic Properties of Modified Amorphous Materials, in Proceedings of the 7th International Conference on Amorphous and Liquid Semiconductors, Edinburgh, Scotland, June 27-July 1, 1977, pp. 524-528.

13. S.R. Ovshinsky, The Chemical Basis of Amorphicity--Structure and Function, Revue Roumaine de Physique 26, Nos. 8-9 (1981) pp. 893-903. (Grigorovici Festschrift.)

14. P. Hey and B.O. Seraphin, The Role of Hydrogen in Amorphous Silicon Films Deposited by the Pyrolytic Decomposition of Silane, Solar Energy Mater. 8 (1982) pp. 215-230. (Ovshinsky Festschrift.)

15. S.R. Ovshinsky, Lone-Pair Relationships and the Origin of Excited States in Amorphous Chalcogenides, in the AIP Conference Proceedings No. 31, Williamsburg, Virginia, March 25-27, 1976, pp. 31-36.

LASER-INDUCED FLUORESCENCE DETECTION OF REACTIVE INTERMEDIATES IN DIFFUSION FLAMES AND IN GLOW-DISCHARGE DEPOSITION REACTORS

Henry U. Lee, John P. deNeufville, and Stanford R. Ovshinsky

Materials Development Laboratory, Energy Conversion Devices, Inc., P.O. Box 5357, North Branch, New Jersey 08876

1. INTRODUCTION

Photovoltaic quality films of a–Si:F:H prepared by the plasma-induced reaction of, for example, SiF_4 and H_2 appear to contain unusual bonding configurations involving the intimate association of silicon with both fluorine and hydrogen atoms.[1] The creation of radical sub-fluorides of silicon has been suggested as a mechanism for the deposition of such films.[2] Indeed, the presence of various excited-state neutral and ionic silicon radical species has been demonstrated in the optical emission of plasmas containing SiH_4, SiF_4 and H_2,[3,4] and the silylene radical, SiH_2, has been proposed to play a dominant role in the deposition of a-Si:H from the HOMOCVD of SiH_4.[5]

The reaction of SiH_4 with F_2 at about 1 Torr produces chemiluminescence which reveals the presence of the excited-state radicals SiF, SiH and SiF_3 in such diffusion flames.[6] We have recently reported the presence of ground-state radicals HSiF and SiF in these flames by detecting their laser-induced fluorescence (LIF).[7] The detection of HSiF in the gas phase had not been previously reported. The silane flame work has been extended in the present study to reveal the spatial dependence of the relative HSiF and SiF concentrations. We also report here on the detection of the ground-state radical SiF by LIF in the plasma decomposition of SiF_4.

2. EXPERIMENTAL

The set-up used in our LIF studies of flames has been described elsewhere.[7] Briefly, the reagents silane and fluorine are mixed via a concentric nozzle assembly in a flow reactor at pressure well below 1 Torr. The SiH_4/F_2 flow rate ratio varied from 3 to 5. A turnable dye laser beam ($< 30\mu$ J/pulse) enters the reactor perpendicular to the axis of the nozzle and can be positioned to probe the chemical intermediates at a distance up to 50 mm from the nozzle throat. The ensuing laser-induced fluorescence is focused onto a photomultiplier tube which is gated by a boxcar integrator.

The same apparatus is easily modified for the LIF detection of radical species in an RF glow discharge. As shown in Fig. 1, the electrodes consist of two parallel (4 x 25 cm) stainless steel plates, one of which is connected to a variable-frequency RF power supply. The discharge is typically operated at 100 KHz and 10 W. The laser beam is directed parallel to the electrodes and the fluorescence passes through a hole (1-cm diameter) on the upper electrode. Suitable band pass filters are used to block the discharge emission occurring outside the region of interest (400 – 450 nm).

3. RESULTS AND DISCUSSION

In the LIF excitation spectrum obtained from the flow reaction of silane with fluorine, shown in Fig. 2, we have identified two band systems belonging to the species SiF and HSiF. Details on the assignments of these LIF bands have been presented elsewhere.[7] Although the HSiF absorption and thus its absolute concentration are unknown, we suspect that the concentrations of SiF and HSiF in this reaction are comparable, judging from the relative intensities of the LIF bands.

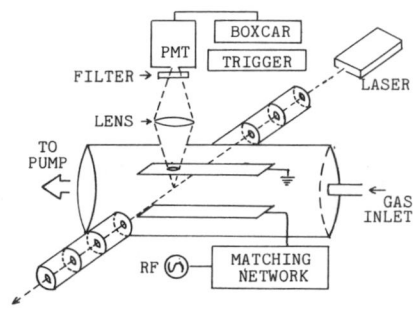

Fig. 1   Set-up for LIF probe of species in a glow discharge.

Reprinted by permission from *Journal of Non-Crystalline Solids*, Vols. 59/60, pp. 671–674 (1983).

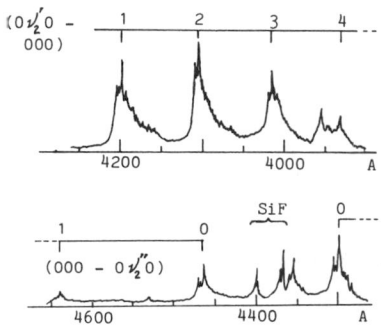

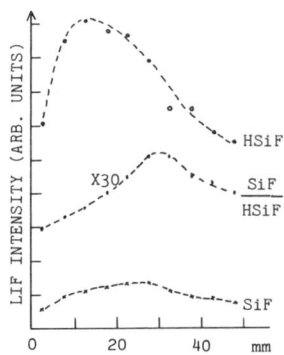

Fig. 2   LIF excitation spectrum of HSiF and SiF from the reaction $SiH_4 + F_2$.

Fig. 3   LIF intensity of SiF and HSiF vs. distance from the nozzle throat.

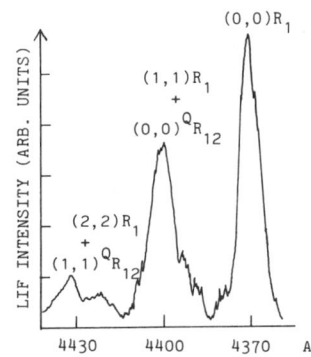

Fig. 4   LIF excitation spectrum of SiF from the RF plasma decomposition of SiF_4.

The detailed mechanisms of this and other analogous flow reactions are not fully understood. It seems plausible that the initial steps entail the stripping of the H atoms on silane by fluorine. Subsequent reactions generate species like HSiF, SiF, SiF_2, SiF_3 and SiF_4. There are approximately 75 conceivable exothermic bimolecular reactions of this sort, counting only those yielding two or more products. Assuming such a series of chain reactions is operative in our flow reactor, we would expect the relative concentrations of SiF and HSiF to vary with distance from the nozzle throat. Several factors can contribute to this variation, including a) the lateral pressure gradient, which is unknown; b) the number of reactions required to form HSiF (2-4, or an average of ~ 2.5) and SiF (2-5, or an average of ~ 3.5); c) the ratio of the reactants, SiH_4 and F_2, which is presumed to increase away from the nozzle due to the presence of excess SiH_4; and d) the possible reactions of SiF and HSiF with other intermediate species and with silane. The observation in Fig. 3 that the concentration of HSiF

peaks "earlier" than that of SiF is thus subject to multiple interpretations.

The successful flame measurements suggested that LIF could be applied to the detection of chemical intermediates present in glow-discharge deposition reactors. In our initial studies we have successfully detected SiF in the discharge decomposition of SiF_4 at a pressure of 0.5 Torr. Although the LIF signals in this case are far weaker than in the flame reaction, the SiF species shows up clearly in the excitation spectrum, as displayed in Fig. 4. A crude analysis of these LIF spectra (Figs. 2 and 4) indicates a vibrational temperature of about 800 ± 100 K for the SiF and HSiF products.

REFERENCES

1. S.R. Ovshinsky, J. Non-Cryst. Solids 32, 17 (1979).

2. S.R. Ovshinsky and A. Madan, U.S. Patent No. 4,226,898.

3. A. Matsuda and K. Tanaka, Thin Solid Films 92, 171 (1982).

4. F.J. Kampas and R.W. Griffith, Solar Cells 2, 385 (1980).

5. B.A. Scott, R.M. Plecenik and E.E. Simonyi, Appl. Phys. Lett. 39, 73 (1981).

6. C.P. Conner, G.W. Stewart, D.M. Lindsay and J.L. Gole, J. Amer. Chem. Soc. 99, 2540 (1977).

7. H.U. Lee and J.P. deNeufville, Chem. Phys. Lett. (in press).

ORDER PARAMETERS IN a-Si SYSTEMS

R. Tsu, J. Gonzalez-Hernandez, J. Doehler, and S.R. Ovshinsky

Energy Conversion Devices, Inc., Troy, Michigan 48084

(Received 10 January 1983 by J. Tauc)

1. INTRODUCTION

THE AMORPHOUS-CRYSTALLINE transition in a-Si and a-Ge has been studied for many years [1, 2]. However, unlike electrical and optical properties, no substantial modifications on the RDF (radial distribution function) for the amorphous phases have been reported for different methods of preparation or annealing conditions short of crystallization. More recently, Barna et al. [3] have found a small shift of the second peak in the RDF towards smaller r, and a more pronounced peak at 5 Å for the a-Si:H prepared by glow discharge technique. These authors concluded that a higher degree of local order is present in the glow discharge a-Si:H although they stated that there exist only minor differences in amorphous silicon samples with different preparations. Earlier Raman measurements [4] have not been shown to be sensitive to preparation conditions, however, recent results indicated that Raman measurements can indeed differentiate various types of amorphous silicon samples. Tsu et al. [5] have pointed out the differences in the Raman spectra of a-Si:F:H (fluorinated), a-Si:H (hydrogenated) and the sputtered a-Si; and Kshirsager and Lannin [6], in the Raman spectra of CVD a-Si compared to sputtered a-Si. To what extent the structural order in a purely amorphous phase influences the optical and electrical properties is an important question. Once a set of ordering parameters are established, it may then be possible to examine the role of structural defects on the basic properties of amorphous materials prepared by various deposition techniques.

In this paper we present the results of Raman scattering in terms of the frequencies, ω_{TO}; and ω_{TA} (TO- and TA-like phonons); the linewidth of the TO-phonon, $\Delta\omega_{TO}$; and the intensity ratio I_{TO}/I_{TA} for various a-Si systems heat-treated at different temperatures. Our results suggest that there is an increase in the local order for the hydrogenated a-Si:H and particularly for the fluorinated a-Si:F:H as compared to sputtered a-Si (sput). This increase in local order in the purely amorphous phase is related to the reduction in the distribution of bond-angle distortions compared to the perfect tetrahedrally bonded crystalline Si and Ge.

2. EXPERIMENTS

We have concentrated our investigation on three types of samples:

(1) a-Si:F:H, glow discharge deposition of amorphous silicon from RF decomposition of SiF_4 and H_2;
(2) a-Si:H, similar to (1) but SiF_4 is replaced by SiH_4 and;
(3) a-Si (sput), sputtered a-Si.

Fused quartz substrates have been employed for all samples in order to allow high temperature heat-treatment. Typically, sample type (1) contains 8% H, and 4% F, while sample type (2) contains 8% H. Heat-treatment was performed in a tube furnace with flowing nitrogen maintained at a pressure slightly above 1 atm. The annealing time was generally fixed at 6 hr. All Raman measurements were performed at room temperature with the 488 nm Ar-ion laser focussed by a cylindrical lens. The usual Raman set-up having a SPEX double monocromator with photon counting and thermoelectric cooler has been used throughout our investigations.

Reprinted by permission from *Solid State Communications*, Vol. 46, No. 1, pp. 79–82 (1983).

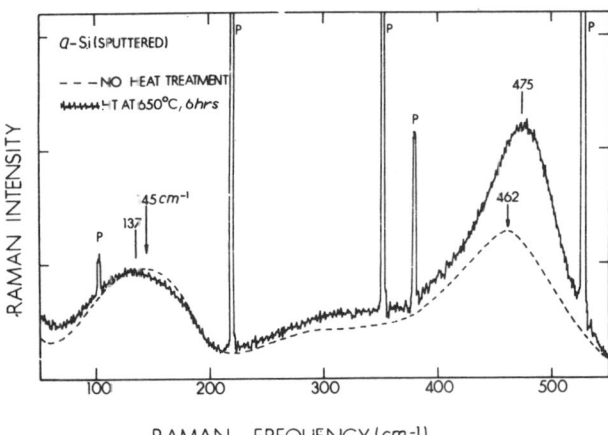

Fig. 1. Raman scattering intensity in arbitrary unit vs the Raman frequencies in cm^{-1} for a sputtered sample before and after heat-treatment at 650°C for 6 h. The noise and plasma line for the sample before heat-treatment have been removed for clarity.

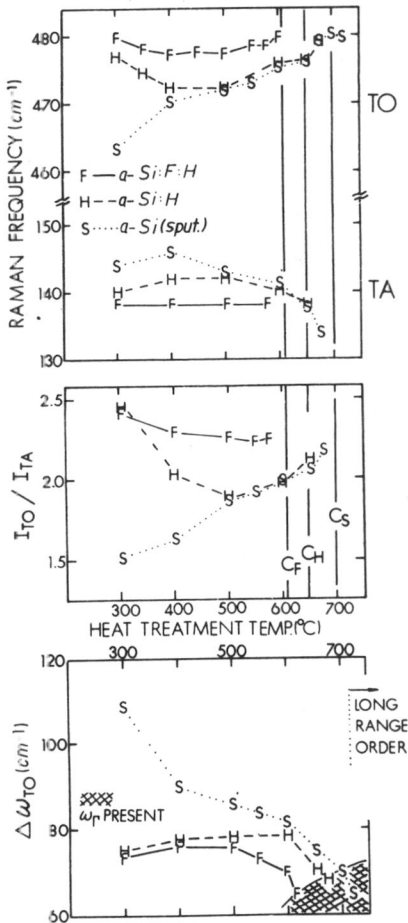

Fig. 2. The TO- and TA-frequencies (top), the intensity ration I_{TO} normalized to I_{TA} (center), and the TO-linewidth (bottom) vs the heat-treatment temperatures. The symbols F, H, and S are for the a-Si:F:H, a-Si:H and a-Si (sput) respectively. The cross-hatched region indicates the presence of crystallites.

Figure 1 shows the typical spectra for a sputtered a-Si sample before and after heat-treatment. The plasma lines of the Ar-ion-laser were left in purposely for an exact calibration of the Raman frequencies. Since we were interested in small frequency shifts, the superimposed plasma lines insured repeatability of our measurements. Note that ω_{TO} shifts from 462 cm^{-1} before heat-treatment to 475 cm^{-1} after annealing at 650°C for 6 h, while ω_{TA} shifts down from 145 to 137 cm^{-1}. There is very substantial change in the linewidth of the TO-phonons and the intensity ratio I_{TO}/I_{TA}. The summary of our measurements is shown in Fig. 2. The top figure gives ω_{TO} and ω_{TA} for various heat-treatment temperatures. Three types of symbols are used: F for a-Si:F:H; H for a-Si:H; and S for a-Si (sput). For the sputtered case, ω_{TO} increases rapidly as the annealing temperature is raised between 300 and 400°C, while ω_{TA} decreases rapidly at the high end of the heat-treatment temperature. The fluorinated and hydrogenated samples have high ω_{TO} before annealing, dropping as annealing temperature is raised, perhaps due to the evolution of hydrogen. Note that the fluorinated samples have a lesser effect from heating. The middle part of Fig. 2 gives the ratio of the intensity of the TO-phonon peak to that of the TA-phonon peak. Again the same behaviour has been obtained as ω_{TO}. The bottom part of Fig. 2 gives the linewidth of the TO-phonon, $\Delta\omega_{TO}$, obtained from doubling the value measured from the peak to the half-intensity point at the high frequency side of the TO-phonon. This was done to avoid complications arising from the asymmetry of the spectrum. The region of those temperatures where the ω_Γ (zone-center phonon) are present, indicating the presence of crystallinity, has been cross-hatched. Generally it is possible to detect the presence of microcrystallites whenever the volume fraction of microcrystallinity, X, exceeds approximately 3% [7, 8]. Therefore we designate the temperature at which $X > 0.05$ by C_F, C_H and C_S for the fluorinated, hydrogenated, and sputtered samples respectively. Obviously there is no well defined crystallization temperature because the logarithm of $(1/1 - X)$ follows an Avrami expression [8]. Nevertheless it is convenient later discussion to call C_F or C_S, a "crystallization temperature" for the particular material involved. For the sputtered samples, a vertical line is drawn at 700°C. The long-range order has been restored at temperatures above this line. The hydrogenated and particularly the fluorinated a-Si showed little change in $\Delta\omega_{TO}$ compared to the sputtered throughout most of the range of temperature of heat-treatment.

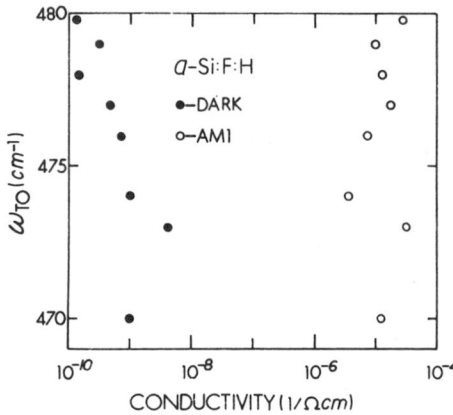

Fig. 3. Dark and photoconductivities (under AM1) vs the TO-frequency for a-Si:F:H.

In order to correlate our Raman measurements with electronic properties, we have shown in Fig. 3, the dark conductivity and the photoconductivity under AM1 excitation, vs the position of the TO-phonon frequency ω_{TO}. The higher ω_{TO}, the lower is the measured dark-conductivity. However, no trend may be seen with photoconductivity. Our results suggest that higher ω_{TO} does correlate with lower electronic defect density, but photoconductivity may involve trapping states with more complicated effects on the structure. In the next section we shall discuss the physical significance of these measurements.

3. DISCUSSION

As pointed out in the introduction section, direct structural investigation such as the RDF is not too sensitive to structural variation in amorphous silicon prepared by different methods, while Raman measurements, on the other hand, have demonstrated the capability of distinguishing various types of amorphous silicon more easily. Since Raman scattering is not influenced by impurities to any significant degree, it should serve as a powerful tool to gather structural information even though definite models are required to readily extract structional parameters such as the bond-angle deviations, the dihedral angle variations, etc. The tendency of an increase of ω_{TO} and decrease of ω_{TA} with improvement of local order is actually quite fundamental. Since the TO-phonon and TA-phonon are basically those phonons for c-Si near the zone-boundaries, the separation of ω_{TO} and ω_{TA} is due to the presence of two atoms per unit cell for c-Si. Naturally ω_{TO} and ω_{TA} should merge as disorder sets in to the point where one cannot even attempt to define the number of atoms per unit cell. Recalling

that [9] the center of gravity of phonons at X_4, W and L_3, for c-Si is located at 481 cm^{-1} and that for X_3 and L_3 is located at 132 cm^{-1} it is not surprising that the well annealed sputtered sample at a temperature below the crystallization temperature has $\omega_{TO} = 480$ cm^{-1} and $\omega_{TA} = 133$ cm^{-1}. As hydrogen is being evolved upon heating. ω_{TO} drops noticeably indicating that hydrogen not only serves to passivate dangling bonds, but also, to a certain extent, reduces internal strain, $\Delta\theta/\theta$, where θ is the bond-angle in a perfect tetrahedrally bonded system. The fact that fluorinated samples show the least effect of heat-treatment, and the highest ω_{TO} and lowest ω_{TA}, leads us to suggest that: (a) fluorinated samples have an overall better local order, and (b) since hydrogen is also present in our fluorinated samples, it is possible that fluorine prevents hydrogen form leaving.

It is significant to note that the quantity (I_{TO}/I_{TA}) $\Delta\omega_{TO}$ is almost constant for the sputtered a-Si over the entire range of heat-treatment. This indicates that the physical mechanism for the narrowing of the linewidth is directly operative on the intensity as well. As pointed out by Meek [10], the TO-phonon depends on both topology and bond-angle distribution of the network, where the latter is more affected by the bond-stretching motions which is sensitive to the arrangements of the bonds around each atom. On the other hand, the TA-phonon depends more directly on the topology. As shown in Fig. 1, the linewidth of the TA-phonons remain unaffected, therefore normalizing I_{TO} by I_{TA} is purely a way to account for the change in absorption leading to a different scattering depth.

If we assume that the lower limit of $\Delta\omega_{TO}$ is set by the linewidth of the crystalline density of states of the optical branch, for c-Si, $\Delta\omega_{TO}$ (no disorder) $\sim$ 30 cm^{-1} from neutron data. However, from careful annealing at small increment of heat-treatment temperature, we have established that 60 cm^{-1} is the lower limit. And this value is dependent on the types of a-Si. There is obviously a phase transition separating 60 cm^{-1} from 30 cm^{-1}. The "crystallization temperature" is also affected by local order, being 610, 650 and 700°C for a-Si:F:H, a-Si:H and a-Si (sput) respectively.

As discussed previously, ω_{TO} is related to the dark conductivity, and therefore, our best order in structure is reflected by the lowest electronic density of states in the gap.

In conclusion, the narrower linewidth, the lower "crystallization temperature", the higher ω_{TO} and the lower ω_{TA}, indicate an increase in local order in the amorphous phase for a-Si:H and particularly for a-Si:F:H. This improvement in structural order is consistent with the lower electronic gap-states. An attempt to extract structural parameter such as the bond-angle distribution from our Raman measurements is currently in progress. Preliminary work [11] suggests that $\Delta\omega_{TO}$ is proportional to the variance of the bond-angle $\overline{\Delta\theta^2}$.

Acknowledgement — This research is supported in part by The Standard Oil Company (Ohio).

REFERENCES

1. S.C. Moss & J.F. Graczyk, *Phys. Rev. Lett.* **23**, 1167 (1969).
2. W. Paul, G.A.N. Connell & R.J. Temkin, *Adv. Phys.* **22**, 529 (1973).
3. A. Barna, P.B. Barna, G. Radnoczi, L. Toth & P. Thomas, *Phys. Status Solidi* **41**, 81 (1977).
4. J.E. Smith, M.H. Brodsky, B.L. Crowder, M.I. Nathan & A. Pinczuk, *Phys. Rev. Lett.* **26**, 642 (1971).
5. R. Tsu, J.G. Hernandez, J. Doehler & S.R. Ovshinsky, *Bull. Amer. Phys. Soc.* **27**, 3, 207 (1982).
6. S.T. Kshirsagar & J.S. Lannin, *Proc. Int. Conf. Phonon Phys.* Bloomington (1981).
7. R. Tsu, J.G. Hernandez, S.S. Chao, S.C. Lee & K. Tanaka, *Appl. Phys. Lett.* **40**, 534 (1982).
8. J. Gonzalez-Hernandez & R. Tsu, *Appl. Phys. Lett.* **42**, 90 (1983).
9. P. A. Temple & C.E. Hathaway, *Phys. Rev.* **B7**, 3685 (1973).
10. P.E. Meek, *Phil. Mag.* **33**, 897 (1976).
11. M.F. Thorpe (private communication).

PASSIVATION OF DANGLING BONDS IN AMORPHOUS Si AND Ge BY GAS ADSORPTION

R. Tsu, D. Martin, J. Gonzalez-Hernandez, and S.R. Ovshinsky

Energy Conversion Devices, Inc., 1675 West Maple Road, Troy, Michigan 48084

(Received 28 August 1986)

I. INTRODUCTION

We have used molecular-beam deposition (MBD) with a base pressure of 10^{-10} torr to prepare a-Si and a-Ge at relatively low substrate temperatures. The films contain voids which can adsorb gases such as H_2, F_2, SiF_4, GeF_4, O_2, etc. Due to the high reactivity of the dangling bonds present, these molecules are broken down to their atomic constituents, thereby passivating the dangling bonds. Further annealing is required to form proper chemisorption and drive the passivating agents into smaller or isolated voids. Typically a decrease of dark conductivity by as much as 8 orders of magnitude is observed accompanied by the appearance of few decades of photocurrent above the dark current under (Air Mass 1) AM1. For a-Ge, using a similar procedure, fluorine is more effective as a passivating agent than hydrogen.

Due to the presence of dangling bonds, which are the major source of localized states in the gap, doping is not possible until these are removed or passivated. To date, the glow-discharge decomposition of silane[1] and a mixture of SiF_4 and H_2 (Ref. 2) results in an amorphous Si alloy having a sufficiently low concentration of gap states, of the order of 10^{16} cm^{-3} eV^{-1}, for efficient doping. There are a number of other techniques for introducing hydrogen into a-Si for the passivation of dangling bonds, such as sputtering in a hydrogen-argon mixture,[3] adding atomic hydrogen during evaporation of silicon,[4] post hydrogenation in a hydrogen plasma of a pure a-Si film prepared by UHV evaporation,[5] as well as hydrogenation of evaporated a-Si by diffusion,[6] and the evaporation of a-Si in a UHV system in a hydrogen environment.[7] To some degree, our approach to the problem of passivation for dangling bonds is similar to those in Refs. 6 and 7; however, we have not only studied the role of hydrogen over a wide range of substrate and annealing temperatures for both UHV-evaporated a-Si and a-Ge, but also for other gases such as O_2, F_2, SiF_4, GeF_4, and N_2 are included in this study. The need of maintaining an ultrahigh vacuum is particularly important for the hydrogenation of evaporated amorphous silicon as discussed by Ovshinsky and Izu.[8] As long as the base pressure is in the range of 10^{-10} torr, the effects of the substrate temperature manifest themselves as voids which determine the initial gas adsorption. Subsequent annealing not only affects diffusion but also modifies the void structure. It has been pointed out by Shevchik and Paul[9] that a lower bound on the density deficit due to voids, deposited at room temperatures, may be $\sim 5\%$. Substantial void density has also been reported for P-doped a-Si even after annealing at 950 °C.[10]

Since we are able to passivate the internal surfaces of the voids, we think that the void density after annealing is quite low. The high-void density reported in Ref. 10 may be due to bonding defects in the intergrain region. After annealed at a temperature above 350 °C, all connected voids apparently have been annealed out because no further adsorption of gases is possible. It is also possible that all dangling bonds in the connected voids have been reconstructed so that their reactivity has dropped below what is required to break any gas molecule. In hydrogen evolution, the first peak occurs around 320—350 °C,[11] and is attributed to the loosely bonded hydrogen on the connected voids. However, we found little variation of this temperature for a variety of gases, for example, hydrogen, oxygen, and fluorine. Therefore, the temperature of this first peak in evolution only characterizes the structure of voids rather than the strength of Si—H bond. Basically, it is not meaningful to think in terms of individual bonds when dealing with closely coupled adsorbates. For a-Si prepared at a substrate temperature of 300 °C or at a low temperature followed by annealing at a temperature above 300 °C in ultra-high-vacuum (UHV), it is difficult to introduce passivating elements such as H or F via adsorption; rather, a low-energy implantation is necessary. It has been pointed out for the chemisorption of hydrogen on silicon surface[12] that the Si(100) 2×1 surface involves the dimer model[13] while the Si(111)7×7 surface involves a vacancy model.[14] Since the internal void surface is more complex, it is reasonable to assume that both models apply to our situation. Our result that dangling bonds on internal surfaces of voids can break down molecular hydrogen does not conflict with Block's[15] contention that a freshly cleaved silicon surface cannot break down molecu-

Reprinted by permission from *Physical Review B*, Vol. 35, No. 5, pp. 2385–2390 (15 February 1987).

TABLE I. Enthalpy changes for various molecules reacting with dangling bonds for Si. The dot indicates a dangling bond. Crosses represent the breaking of bonds. Note that H_2 can be bonded only when dangling bonds are present because energetically it is favorable, but O_2, F_2, etc. can break Si—Si bond and be bonded. This is major difference between H_2 and other gas molecules under consideration.

Gas	Bond Strength	Si-site	Bond Configuration	Change (kCal/mole)
H_2	103	$-\overset{\bullet}{\underset{\|}{Si}} - \overset{\bullet}{\underset{\|}{Si}}-$	$-\overset{H}{\underset{\|}{Si}} - \overset{H}{\underset{\|}{Si}}-$	49
H_2	103	$-\underset{\|}{Si} * \underset{\|}{Si}-$	$-\underset{\|}{Si}-H\ H-\underset{\|}{Si}-$	-4
O_2	34	$-\underset{\|}{Si}\cdot\ \cdot\underset{\|}{Si}-$	$-\underset{\|}{Si}-O-\underset{\|}{Si}-$	398
O_2	34	$-\underset{\|}{Si} * \underset{\|}{Si}-$	$-\underset{\|}{Si}-O-\underset{\|}{Si}-$	292
F_2	37	$-\underset{\|}{Si} * \underset{\|}{Si}-$	$-\underset{\|}{Si}-F\ F-\underset{\|}{Si}-$	180
SiF_4		$\underset{\underset{\|}{-Si-}}{\overset{\overset{\|}{-Si-}}{Si\cdot\ \ \ \cdot Si}}$	$\underset{\underset{\|}{-Si-}}{\overset{\overset{\|}{-Si-}}{Si-F\ F-Si-F\ F-Si}}$	106

lar hydrogen without the presence of a high electric field. Fields on the order of 10^7 V/cm are present in amorphous silicon and germanium due to the bond-angle deviation from the tetrahedral angle.

II. EXPERIMENTAL PROCEDURES

Typically, we prepared the a-Si and a-Ge films by molecular-beam deposition in a UHV system having a base pressure of 10^{-10} torr, at a substrate temperature T_s varied between -120 and $300\,°C$. Gas in molecular form was introduced into the chamber for adsorption. The sample was then annealed after the gas was pumped out. During this procedure and subsequent annealing, conductivity was being monitored in situ. We shall depict the enthalpy changes using H_2 as an example.

$$H_2 + -\overset{\bullet}{\underset{\|}{Si}} - \overset{\bullet}{\underset{\|}{Si}} - = -\overset{H}{\underset{\|}{Si}} - \overset{H}{\underset{\|}{Si}} - + 49\ \text{kCal/mole}$$

Table I lists some of the reactions.

Figure 1 explains the experimental procedure in detail. We take the case for H_2 adsorption. An a-Si layer of 1000 Å thickness is evaporated at $T_s = -26\,°C$ using a 1000 Ω cm silicon wafer source in a typical molecular-beam deposition with a base pressure $\sim 10^{-10}$ torr in the growth chamber. The as-prepared sample has a room-temperature conductivity $\sim 10^{-3}$ Ω cm, measured in the introduction chamber. Hydrogen gas at 6 torr is let into the introduction chamber and is allowed to remain for 30 min before pump-out. Annealing is done back in the growth chamber. The reason for moving the sample back and forth between the introduction chamber and the growth chamber is to avoid the difficulties of pump-down after 6 torr of hydrogen has been introduced, and to exclude the possibility of the presence of atomic hydrogen. Conductivity in the dark and under AM1 light are measured in situ and the activation energy for the dark conductivity is measured during cool-down from the annealing temperature T_A and room temperature. Also shown in Fig. 1 is the result for hydrogen leaked into the chamber after annealed under UHV at 350°C for 1 h, which is not any different from the case without any gas present, indicating that there is no significant amount of connected voids to allow hydrogen molecules permeating the film. In an entirely different experiment,[16] we have shown that voids are indeed greatly reduced after annealing at 320°C. Therefore, we have discounted the possibility that the failure to show any effects of hydrogen adsorption may be caused by the loss of reactivity of the dangling bonds due to reconstruction after annealing.

Figure 2 shows results for hydrogen adsorption and subsequent heat treatment for the MBD sample deposited at $-26\,°C$. The dashed curve represents σ_1 under AM1 and the solid curve σ_d in the dark. T_c is the temperature at which crystallization occurs as determined by Raman scattering.[17] We note that there is no discontinuity in σ_d at T_c, and the UHV-annealed curve has no dip. The dip in σ is caused by the presence of hydrogen adsorption and subsequent passivation of dangling bonds. Figure 3 shows the temperature dependence of the dark conductivity for

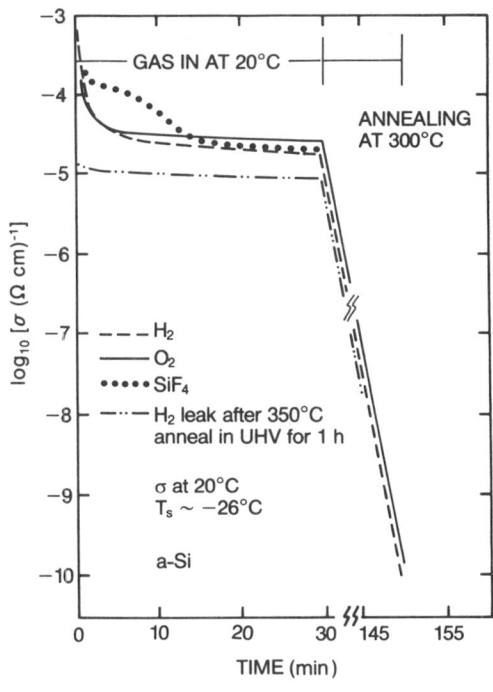

FIG. 1. Dark conductivity versus time.

127

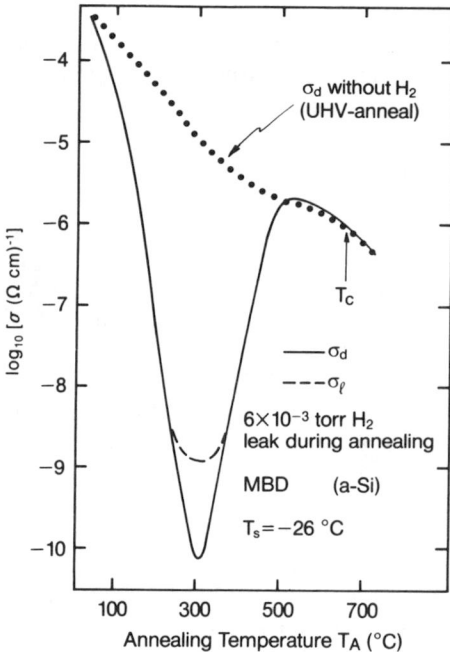

FIG. 2. Dark conductivity, σ_d and photoconductivity σ_1 under AM1 illumination versus annealing temperature with and without the presence of molecular hydrogen. The amorphous silicon was evaporated at $-26\,°C$ substrate temperature.

the same sample in Fig. 2. We note that annealing at $300\,°C$ for 1 h produces the highest activation energy. Comparing with the slope drawn for 0.8 eV, $\Delta E \sim 0.8$ eV is comparable to the good glow discharge a-Si:H. The UHV-annealed curve has a $\Delta E \sim 0.2$ eV. The effect of SiF_4 adsorption is shown in Fig. 4. The minimum has

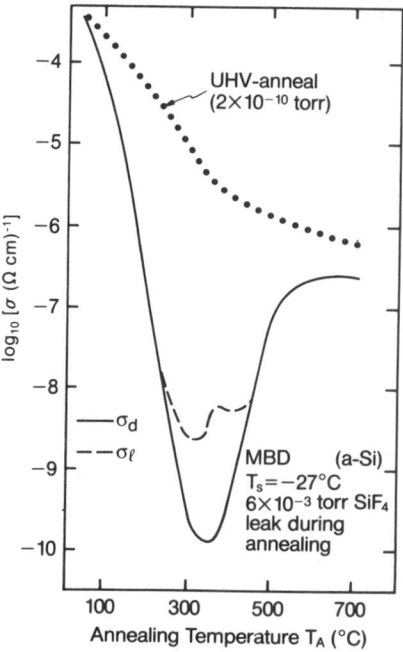

FIG. 4. Similar to Fig. 2, for SiF_4.

shifted to a slightly higher T_A, around $350\,°C$. The conductivity under AM1, σ_1 is almost 2 orders of magnitude higher. Of all the gas-adsorption methods of passivation, SiF_4 gives the best σ_1/σ_d ratio. Although the dark conductivity σ_d does not seem to be too different for oxygen adsorption shown in Fig. 5, σ_1 is definitely lower. We have thus far shown that the *dip in conductivity at a temperature of annealing around* $300-350\,°C$ *is due to adsorption of molecular gas followed by passivation of dangling bonds.* The slight inflection for the UHV-annealed curve is possibly due to the residue gas in the UHV chamber.

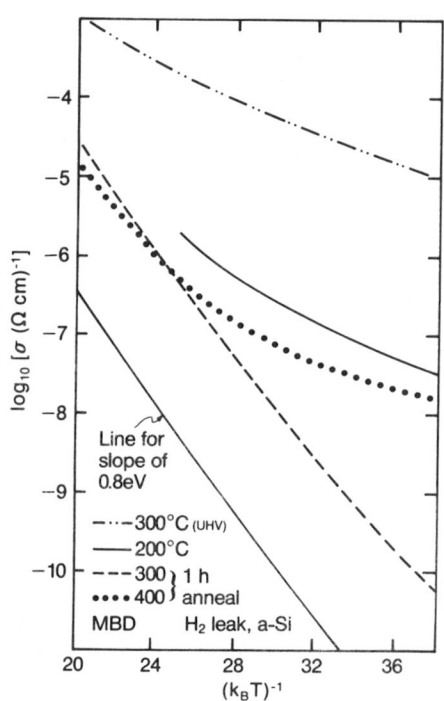

FIG. 3. Temperature dependence of the dark conductivity.

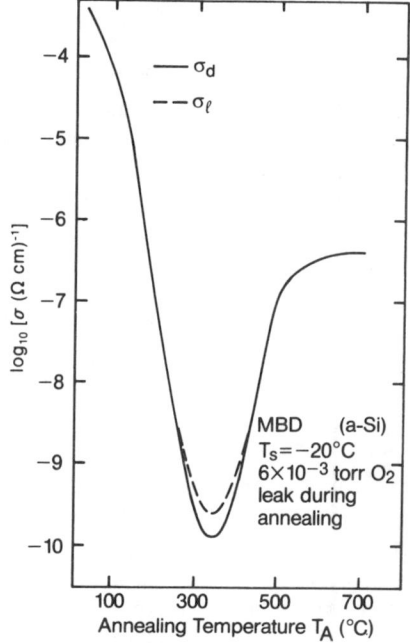

FIG. 5. Similar to Fig. 2, for O_2.

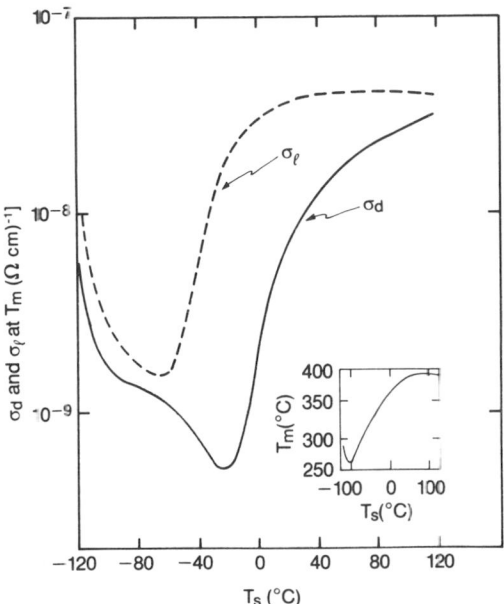

FIG. 6. σ_d and σ_1 at T_m versus the substrate temperature. T_m is defined by the annealing temperature T_A at which σ_d is minimum.

Since the triply bonded N_2 is difficult to break into atomic nitrogen, when we repeated this experiment with N_2 only a slight dip showed up due to a small amount of gas impurity in N_2. Figure 6 shows σ_d and σ_1 at T_m defined by the temperature T_A at which σ_d is minimum, versus the temperature of the substrate T_s for air. The insert shows T_m versus T_s. The dependence of T_m on T_s is largely due to the dependence of the connected voids on T_s. In other words, the lower the substrate temperature,

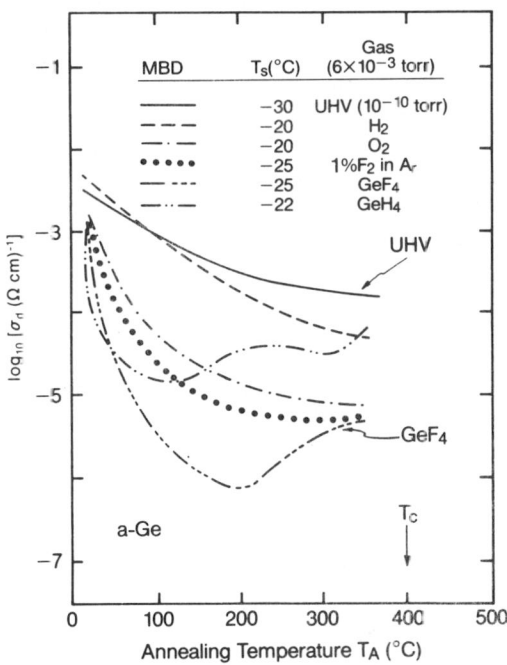

FIG. 7. Dark conductivity versus annealing temperature for a-Ge with various molecular-gas species.

the lower T_m is. Although Fig. 6 applies to the case of air, all other gas species follow nearly the same curve. Therefore we propose that T_m is more influenced by the structure rather than the bond strength of the adsorbates. It is important for us to emphasize that the experiment with air was conducted with annealing in a nitrogen furnace where there was no possibility of having atomic hydrogen. Thus we have shown that dangling bonds on void surface do catalyze molecular species into atomic species subsequently chemisorbed upon annealing. In this case, the active species in air is oxygen. Basically, the substrate temperature determines the connectivity of the voids and the void fraction, and annealing closes the channels as well as heals the surfaces both in terms of reconnecting the broken bonds and reconstructing the dangling bonds as occurring in surface reconstruction. Since closely coupled states of the chemisorbed atoms form a couple state, squeezing the channel results in a reduction of those states, thereby expelling the adsorbates.

Figure 7 shows the effects of annealed a-Ge samples with various adsorbates. Again, UHV-annealed samples show no passivation as expected. What is striking is in the comparison between GeF_4 and H_2. While H_2 serves well for the passivation of dangling bonds in a-Si, it does nearly nothing for a-Ge. Since the bond energy of Ge—H is 69 kcal/mol, only slightly lower than 76 kcal/mol for Si—H, the first case of Table I for Ge will give an enthalpy change of 35 instead of 49 which should be sufficient. Therefore, the failure of having a significant effect with H_2 indicates that there is a marked change in what characterizes the voids in Ge. The experimental results with H_2 or SiF_4 are similar for silicon and quite different for germanium indicating differences in reconstruction. From an experimental point of view, since there is no significant change after introduction of a gas leak, whether annealing is performed with the gas present or after pumping-out, we adopted the latter procedure. Without pumping out the gases such as F_2, GeF_4, etc., during annealing, the film will be etched away quickly.

Since we were quite puzzled by the observation that H_2 adsorption produced almost no effect for a-Ge, we decided to examine the morphology with TEM. Figure 8 shows the TEM results for three samples of a-Ge having $T_s = -25$, 100, and 200 °C. We note that the structure is very open for $T_s = -25$ °C. Our negative results suggested that either hydrogen does not passivate the germanium dangling bonds due to excessive surface reconstruction, or columnar structure masks any effects of passivation. In order to show that the poor result of passivation with hydrogen for a-Ge does not originate from any intrinsic problem in the Ge—H bond, we prepared an a-Ge film of 700 Å thick at $T_s \sim 180$ °C where the majority of the connected voids were absent and followed by implantation at 180 °C of hydrogen using a differentially pumped sputtering gun. Two voltages, 5 and 2 kV were used to produce a flatter hydrogen profile. The conductivities σ_d and σ_1 under AM1 are plotted versus temperature $(k_B T)^{-1}$ in Fig. 9. A typical a-Ge:H from glow discharge is also

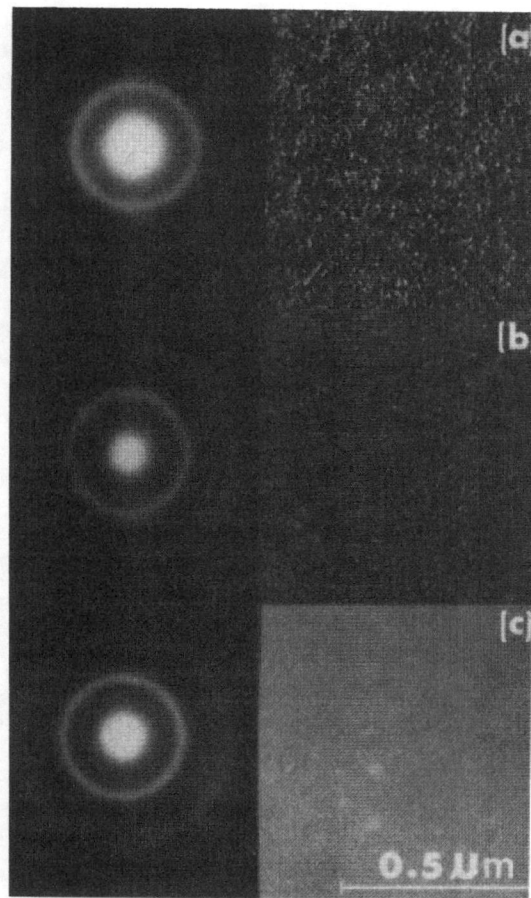

FIG. 8. TEM for three samples of evaporated a-Ge: (a) $T_s = -25\,°C$, (b) $T_s = 100\,°C$, and (c) $T_s = 200\,°C$.

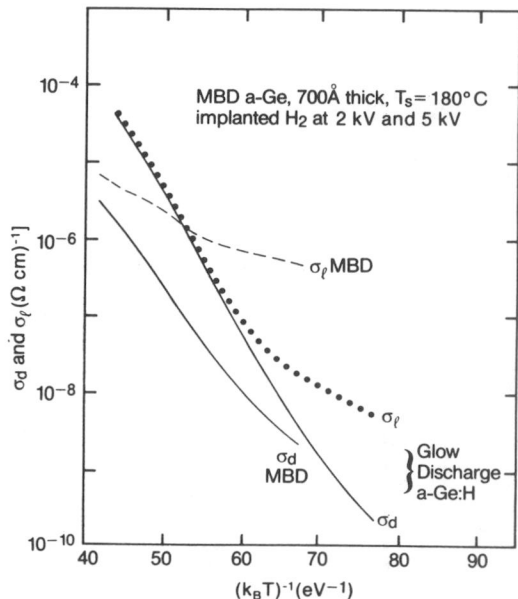

FIG. 9. Comparison of a-Ge:H by glow discharge with molecular-beam depositions followed by UHV hydrogen implantation.

shown for easy reference. We note the ratio of σ_1 to σ_d is much better for the hydrogen-implanted a-Ge. Therefore, this preliminary result indicates that the passivation of dangling bonds in a-Ge may depend on the success in controlling the surfaces of voids, and/or morphology.

III. DISCUSSIONS

A recent study in the hydrogen profiling using ^{15}N nuclear reaction on implanted samples by Woodyard et al.[18] indicated that for hydrogen concentrations of 5% and higher, a simple diffusion model does not apply for the redistribution of hydrogen in the annealed a-Si:H. This conclusion supports our present notion that the temperature 300—350 °C, marking the onset of hydrogen evolution, characterizes structure change in the form of closing the connected voids and restructuring the dangling bonds in a manner not dissimilar from surface reconstruction in crystalline solids. This model explains why the minimum conductivity at a given annealing temperature, T_m, is far more sensitive to the substrate temperature for the deposition of pure a-Si, and almost insensitive to the different gas species. The use of gas adsorption allows us to pas-

sivate effectively the dangling bonds in connected voids. For example, using the photoconductivity data, it is possible to estimate[19] the remaining dangling bond density. For $\sigma_1 \sim 10^{-8}(\Omega\,cm)^{-1}$, the estimated dangling bond density is $10^{18}\,cm^{-3}$, almost 4 orders of magnitude below the hydrogen concentrations in the film. These 4 orders of magnitude may very well represent the ratio of the connected voids to the isolated void fractions or vacancies. Having most of the defects passivated by the gas-adsorption method, the remaining defects may be further passivated with low-energy implantation of hydrogen with a differentially pumped sputtering gun operated in UHV. A brief account of this technique has been described in Ref. 18. The original concept was contained in Ref. 8. A detailed account of the combination of gas adsorption and implantation is to be found in a recent patent.[20] It is important to note that this method of producing a-Ge:H having a photoconductivity under AM1 twice as high as the dark conductivity may not be of technological importance, but certainly represents a step toward a better fundamental understanding. It was emphasized by Ovshinsky that atomic hydrogen and/or fluorine play decisive roles even in glow-discharge depositions.[21] We have also produced a-Si:H using the same implanting technique at a substrate temperature of 300 °C. The film produced has similar quality compared to the best a-Si:H by glow-discharge deposition and also exhibited light-induced degradation.[22] As far as the role of passivation of a-Ge by hydrogen, superficially it seems obvious that specific deposition techniques do play a role. From the fundamental point of view, however, we should place more emphasis on the overall relationship of voids, reconstructions, and passivating species rather than specific deposition techniques.

IV. CONCLUSIONS

Several results of this work seem worthy of note. Molecular species such as H_2, O_2, etc., can be chemisorbed on the internal void surfaces of amorphous silicon and germanium on account of the high reactivity of the dangling bonds. The primary process of gas evolution is governed by structural consideration rather than bond strengths. The role of annealing after gas adsorption is to convert the weakly physisorbed species to chemisorption. Once connected voids are annealed away, no more gas adsorption can take place. The remaining isolated voids may be passivated by further UHV implantation. The intrinsic random electric fields in tetrahedrally bonded amorphous material assist in surface reactions. The bottom line of all these experiments is the need of ultrahigh-vacuum, in the 10^{-10} torr regime. Taking advantage of the huge dynamic range of the *in-situ* conductivity measurement, it is possible to investigate the role of voids over a much wider range, thus opening a new avenue for the investigation of the passivation of defects in *a*-Si and *a*-Ge, as well as the mechanism of grain-boundary formation in microcrystalline materials, and the study of surface chemistry and surface physics in general.

ACKNOWLEDGMENTS

We are grateful to the Standard Oil Company (Ohio) for their support of this work, and to J. H. Block of Fritz-Haber-Institut der Max-Planck, Berlin for a discussion on the role of high electric fields in surface reactions.

[1] R. C. Chittick, J. H. Alexander, and H. F. Sterling, J. Electrochem. Soc. **116**, 77 (1969).

[2] S. R. Ovshinsky and A. Madan, Nature **276**, 482 (1978).

[3] W. Paul, A. J. Lewis, G. A. N. Connell, and T. D. Moustakes, Solid State Commun. **20**, 969 (1976).

[4] D. L. Miller, H. Lutz, H. Wiesmann, E. Rock, A. K. Ghosh, S. Ramamoorth, and M. Strongin, JAY, **49**, 6192 (1978).

[5] D. Kaplan, N. Sol, and G. Velesco, Appl. Phys. Lett. **33**, 440 (1978).

[6] J. Jang, J. H. Kang, and C. Lee, J. Non-Cryst. Solids **35 & 36**, 313 (1980).

[7] N. Kniffler, W. W. Múller, J. M. Pirrung, N. Hanisch, B. Schroder, and J. Geigu, J. Phys. (Paris), Colloq. **10**, C4-811 (1981).

[8] S. R. Ovshinsky and M. Izu, U.S. Patent No. 4 217 374 (1980).

[9] N. J. Shevchik and W. Paul, J. Non-Cryst. Solids **16**, 55 (1974). B. G. Bagley, D. E. Aspres, A. C. Adams, and C. J. Mogeb, Appl. Phys. Lett. **38**, 56 (1981).

[11] D. K. Biegelsen, R. A. Street, C. C. Tsai, and J. C. Knights, Phys. Rev. B **20**, 4839 (1979).

[12] H. H. Madden, Surf. Sci. **105**, 129 (1981).

[13] R. E. Schlier and H. E. Farnsworth, J. Chem. Phys. **30**, 917 (1959).

[14] J. J. Lander and J. Morrison, J. Chem. Phys. **37**, 727 (1962).

[15] The role of an electric field in the surface reactivity was pointed out by J. H. Block. See for example, T. Sakata and J. H. Block, Surf. Sci. **130**, 313 (1983).

[16] R. Tsu, J. Gonzalez-Hernandez, S. S. Chao, and D. Martin, Appl. Phys. Lett. **48**, 647 (1986).

[17] J. G. Hernandez, D. Martin, S. S. Chao, and R. Tsu, Appl. Phys. Lett. **45**, 101 (1984).

[18] J. R. Woodyard, D. R. Bowen, J. Gonzalez-Hernandez, S. C. Lee, D. Martin, and R. Tsu, J. Appl. Phys. **57**, 2243 (1985).

[19] L. Voget-Grote, W. Kimmorle, R. Fischer, and J. Stake, Philos. Mag. B **41**, 127 (1980).

[20] R. Tsu, J. Gonzalez-Hernandez, D. Martin, and S. R. Ovshinsky, U.S. Patent No. 4 569 697 (1986).

[21] S. R. Ovshinsky, Proceedings of the International Ion Auger Congress, ISIAT and IPAT, Kyoto, 1983 (unpublished), p. 817.

[22] R. Tsu, D. Martin, J. Gonzalez-Hernandez, and S. R. Ovshinsky (unpublished).

CHEMICAL BOND APPROACH TO THE STRUCTURES OF CHALCOGENIDE GLASSES WITH REVERSIBLE SWITCHING PROPERTIES

Josef Bicerano and Stanford R. Ovshinsky

Energy Conversion Devices, Inc., 1675 West Maple Road, Troy, Michigan 48084

(Received 6 December 1984)

1. Introduction

Since the publication of the paper announcing the discovery of reversible switching phenomena in certain types of chalcogenide glasses [1], much effort has been devoted towards the characterization, improvement, and understanding of the properties of these materials [2]. This interest has been stimulated both by the basic scientific questions that have to be answered in order to understand the structures and properties of these alloys, and by their obvious potential as reversible threshold and memory switching elements in a wide variety of electronic devices [2,3].

A considerable amount of additional theoretical work has been carried out to understand the properties of chalcogenide glasses in general [4], and glassy switching materials in particular (for a simple model of the electronic transformations in these materials see Ovshinsky and Sapru [5]; this model was further elaborated and the importance of lone pair orbitals was emphasized by Ovshinsky [5]). The relevance of the amount of connectivity of an alloy in determining whether it will have threshold or memory switching properties, was already pointed out in the original paper [1]. The differences between threshold and memory alloys have been explained structurally in terms of the relative rigidities of the atomic networks formed by different percentages of divalent, trivalent and tetravalent alloying elements [5]. To summarize, it has been suggested that the memory switches have lower energy barriers to reversible local order changes and amorphous–microcrystalline structural phase transitions due to the weaker bonds and a more flexible network containing an abundance of divalent elements, while the more rigid threshold switches can only undergo electronic transitions which are automatically reversed when the applied voltage causing the excitation is turned off.

A method which has been recently used to examine the structures and properties of various types of chalcogenide glasses is the chemical bond approach (see e.g. refs. [1,5,6]). Although this technique is qualitative and at best only semi-quantitative, it has provided many valuable insights which are all the more useful in view of the difficulty of treating these problems in an ab initio manner.

In section 2, a version of the chemical bond approach is described in detail and applied to some threshold and memory alloys. The threshold alloy chosen for examination is $Si_{18}Ge_7As_{35}Te_{40}$ which is the most extensively studied alloy of this type (denote by T for brevity) [7]. Two memory materials have been examined: $Ge_{15}Te_{81}Sb_2S_2$ (denote by M1, this alloy is especially useful for applications in microelectronics) [8] and $Ge_{24}Te_{72}Sb_2S_2$ (denote by M2) (Formigoni [8]).

In section 3, the results of the calculations are discussed in greater detail, and speculations are made concerning their implications in terms of the possible preparation of alloys with optimized properties.

2. Procedure and results

We will assume that: (1) Si and Ge bond tetravalently, As and Sb trivalently, and S and Te divalently. This is equivalent to neglecting dangling bonds and other valence defects as a first approximation. Van der Waals interactions, which can provide a means for further stabilization by the formation of links much weaker than regular covalent bonds, are also neglected. Then, the following average coordination numbers $\bar{C}$ are obtained for the three materials: $\bar{C}(T) = 2.85$, $\bar{C}(M1) = 2.32$ and $\bar{C}(M2) = 2.50$. (2) Atoms combine more favorably with atoms of different kinds than with the same kind (this assumption, which is generally found to be valid for glass structures, has been used as far back as Zachariasen [9] in his covalently bonded continuous random network model). This condition is equivalent to assuming the maximum amount of chemical ordering possible. It favors the formation of a glass structure by increasing the glass transition temperature. In using this assumption, we consider Si and Ge, or S and Te, which have identical valences, to be "of the same kind", and thus omit Si–Ge and S–Te bond pairs from further consideration. Bonds between like atoms will then only occur if there is an excess of a certain type of atom, so that it is not possible to satisfy its valence requirements by bonding it to atoms of different kinds alone (see the example of Te–Te bonds, in the discussion below). (3) Bonds are formed in the sequence of decreasing bond energy until all available valences of the atoms are saturated.

The bond energies $D(A–B)$ for heteronuclear bonds have been calculated by using the relation

$$D(A–B) = [D(A–A) \cdot D(B–B)]^{1/2} + 30(x_A - x_B)^2$$

proposed by Pauling [10], where $D(A–A)$ and $D(B–B)$ are the energies of the homonuclear bonds (ref. [10], table 3-4 (p. 85); the $D(A–A)$ values used are in units of kcal/mol.: 42.2 for Si, 37.6 for Ge, 32.1 for As, 33 for Te, 50.9 for S, 30.2 for Sb, and 44 for Se), and x_A and x_B are the electronegativities of the atoms involved. (ref. [11], the x_A values used are: 1.90 for Si, 2.01 for Ge, 2.18 for As, 2.10 for Te, 2.58 for S, 2.05 for Sb, and 2.55 for Se).

The types of bonds expected to occur in the T, M1 and M2-type materials to any appreciable extent are listed in table 1, together with their bond energies. In the memory materials, the fairly strong Sb–S($D = 47.6$ kcal/mol) and Ge–Sb($D = 33.7$ kcal/mol.) bonds are not expected to occur, since the Ge–S bonds are expected to saturate the valences of S, while the very abundant as well as stronger Ge–Te bonds are expected to saturate the valences of Ge.

Assumption (3) mentioned above can be applied directly and in its simplest form to the memory materials, where there is no ambiguity about the formal order in which the bonds are formed. The S atoms strongly bond to Ge. The remaining Ge atoms, are bonded to the Te atoms, which are present in large numbers. The Te atoms also fill the available valences of the Sb atoms. After

Table 1

Energies of bonds expected to occur in T, M1 and M2 Materials [a]

T		M1 and M2	
A–B	$D(A–B)$	A–B	$D(A–B)$
Si–As	39.2	Ge–S	53.5
Si–Te	38.5	Ge–Te	35.5
Ge–As	35.6	Sb–Te	31.6
Ge–Te	35.5	Te–Te	33.0
As–Te	32.7		
Te–Te	33.0		

[a] $D(A–B)$ values are given in kcal/mol.

Reprinted by permission from *Journal of Non-Crystalline Solids*, Vol. 74, pp. 75–84 (1985).

Table 2
Average expected numbers of neighbors

Central atom	Its neighbors	Average number for material		
		T	M1	M2
Si	As	2.98	–	–
	Te	1.02	–	–
Ge	As	2.24	–	–
	Te	1.76	3.73	3.83
	S	–	0.27	0.17
As	Si	1.53	–	–
	Ge	0.45	–	–
	Te	1.02	–	–
Te	Si	0.46	–	–
	Ge	0.31	0.69	1.28
	As	0.89	–	–
	Sb	–	0.075	0.08
	Te	0.34	1.235	0.64
Sb	Te	–	3.0	3.0
S	Ge	–	2.0	2.0

all these bonds are formed, there are still unsatisfied Te valences which must be satisfied by the formation of Te–Te bonds. This bond picture represents a system of Te chains (as in amorphous Te), heavily crosslinked by the tetravalent Ge and trivalent Sb atoms. The S atoms preferentially occupy bridging positions between pairs of second neighbor Ge sites.

On the other hand, assumption (3) must be refined in order to become applicable to the threshold material, since D(Si–As)-D(Si–Te) ≈ 0.64 and D(Ge–As)-D(Ge–Te) ≈ 0.14 kcal/mol. are of the same order of magnitude as $kT \approx 0.59$ kcal/mol. at room temperature ($T \approx 298.15$ K). This means that no one type of bond will completely dominate around Si and Ge atoms. After weighting by the Boltzmann factors $e^{-E/kT} = e^{D/kT}$, we find that the probability of Si–As bonds is approximately 74.6% and the probability of Si–Te bonds is approximately 25.4% around Si. Similarly, the approximate probabilities of Ge–As and Ge–Te bonds around Ge are 55.9% and 44.1% respectively. This more complicated bonding pattern helps the formation of a glass structure in which all the atoms are properly incorporated. Unlike the memory alloys, the bonding in the T alloy is not dominated by a template formed by the preponderance of atoms of a given valence.

Table 2 lists the average expected number of neighbors of each type for each central atom, calculated by using the methods described above. If we now further assume that the bond energies are additive, we can estimate the cohesive energy (CE), i.e., the stabilization energy of an infinitely large cluster of the material per atom, by summing the bond energies listed in table 1 over all the bonds expected in the material (in the proportions estimated by the values given in table 2). This is equivalent to assuming a simplified model consisting of noninteracting electron pair bonds highly localized between adjacent pairs of atoms. The results are: CE(T) = 2.35 eV/atom, CE(M1) = 1.75 eV/atom, and CE(M2) = 1.92 eV/atom (1 eV/atom = 23.05 kcal/mol.).

3. Discussion

The magnitudes of $\bar{C}$ and CE follow the same ordering as expected from qualitative arguments concerning relative rigidities of networks containing different percentages of divalent, trivalent and tetravalent atoms. They are also consistent with the general mechanism suggested [5] for the switching properties of these glasses, and with the fact that the thickness of films and the possibility of structural phase transitions are correlated (Formigoni [8]).

It is interesting to compare the calculated CE values with those obtained by assuming that the actual CE values are equal to weighted averages of the CE values of the elements (ref. [12], the CE values of the elements are listed here as being 4.63 eV/atom for Si, 3.85 for Ge, 2.96 for As, 2.23 for Te, 2.75 for Sb, and 2.85 for S) forming the alloys. This gives 3.03, 2.50 and 2.64 eV/atom for T, M1 and M2 respectively. Thus, our calculated CE values for T, M1 and M2 are 77.6%, 70.0% and 72.7% respectively, of those obtained by the simple averaging process. Two immediate observations are that (i) these CE values are all underestimated by comparable amounts; but (ii) the percentage by which they are underestimated varies in the order M1 > M2 > T, indicating that our simple analysis based on the assumption of the additivity of bond energies works better for larger $\bar{C}$ and fewer divalent atoms.

It is also instructive to compare the CE values obtained for the simple elemental solids by using the bond additivity assumption, with their experimental CE values [12]. This assumption, in conjunction with Pauling's D(A–A)

values (see previous note in text concerning ref. [10]), yields calculated CE values which are the following percentages of the experimental values for the elements forming the switching alloys: Si(79.1%), Ge(84.7%), As(70.6%), Sb(71.5%), S(77.5%) and Te(64.2%). With the exception of S, the trend clearly shows that the bond additivity assumption gives a better estimate of the CE for elements with larger coordination numbers. The percentage errors are comparable to those calculated above for the alloys. Since S in only present in the memory materials in a very small amount (2%), while Te is the most abundant element in all three alloys, the process of comparing the CE values obtained by the bond additivity assumption with those obtained by taking weighted averages of the CE values of the elements [12], and the resulting conclusions presented in the previous paragraph, remain valid.

The most important effect involved here is clearly electron correlation: A calculation that assumes simple additivity of noninteracting electron pair bonds completely neglects the stabilization caused by the correlations between the electrons and the resultant lowering of the total energy of the system. It is well-known that especially for the simplest small molecules the energy lowering caused by chemical bonding is underestimated if electron correlation is neglected. In fact, rigorous *ab initio* calculations indicate that the consequences of this neglect can be drastic enough to cause, for example, the diatomic fluorine molecule (F_2) to be unbound at the Near-Hartree–Fock level of computation, but to be bound with the expected dissociation energy once electron correlation is included by a technique such as configuration interaction [13].

Furthermore, since electron correlation effects are especially important for systems containing atoms such as Te with more than one lone pair highly localized on the same atom, the order by which the amount of the underestimation varies (M1 > M2 > T) can be rationalized.

The glass transition and crystallization temperatures (T_g and T_x respectively) have been measured for several related alloys [14]. They have the following values $Ge_{15}Te_{81}Sb_2S_2$ ($T_g = 123°C$, $T_x = 175°C$); $Ge_{33.3}Te_{66.6}$ (225, 256); and $Ge_{33.3}Sb_{33.3}Te_{33.3}$ (249, 263). Thus, both T_g and T_x increase in the same order as the CE predicted by our simple method for this series of materials. However, this apparent correspondence between CE and phase transition temperatures should not be used carelessly, since the transition temperatures depend on the product of the transition. For example, while the CE of elemental S(2.85 eV/atom) is much larger than that of elemental Te(2.23), the melting point of rhombic sulfur (112.8°C) is much lower than that of crystalline tellurium (449.5°C) [15]. This is because crystalline sulfur is a molecular solid made up of S_8 units. Therefore, while most of the CE comes from the strong S–S bonds within the S_8 units, it is sufficient to merely break the weak Van der Waals interactions holding different molecules together in order to melt the solid. The situation can also be quite complex and involve other factors in the determination of the T_g and T_x values for the amorphous alloys. For example, a model has been proposed for the relationship between T_g, the band gap E_g, and the connectivity $\bar{C}$ (deNeufville and Rockstad [6], Sarrach et al. [6]). Thermal crystallization has also been studied experimentally in considerable detail for the sputtered $Ge_{20}Se_{40}Te_{40}$ alloy [16].

Another observation about these alloys (from table 2) is that they all contain Te–Te bonds, at least at this level of analysis. While an appreciable amount of Te–Te bonding is unavoidable for the memory materials since Te atoms are present in large excesses, it is surprising to note the presence of an average number of 0.34 Te–Te bonds per Te atom in the threshold material. While this is much smaller than the amount of Te–Te bonding in the memory alloys and might even decrease further at a more accurate level of computation, it is too large to be an artifact of the simple chemical bond approach used here, and unlikely to go away at any level of accuracy of calculation.

The number of Te–Te bonds calculated here might also decrease by the formation of valence alternation pairs (VAPs) in which two neutral threefold-coordinated Te atoms in a defect configuration are replaced by a positively charged threefold-coordinated and negatively charged singly coordinated Te ion. This process, formally denoted by $2C_3^0 \rightleftharpoons C_3^+ + C_1^-$, has been proposed [4] as a possible mechanism to explain many of the structural and electronic transport properties of chalcogenide glasses.

The chemical bond approach can further help us to understand why these glasses all contain an abundance of Te. For example, if we were to assume that in the threshold glass all the Te atoms were replaced by Se, we would find a drastically different bond picture. D(Si–Se) = 55.8 and D(Ge–Se) = 49.4 kcal/mol. are far greater than D(Si–As) and D(Ge–As), so that there is no competition between Se and As in degree of preference for bonding to Si and Ge. Since D(As–Se) = 41.7 kcal/mol. As cannot compete with either Si or Ge in forming bonds to Se. Thus, the Se atoms would first completely saturate the 72 valences of the eighteen Si atoms, and then use up 8 of the 28 available valences of the seven Ge atoms. The remaining Ge valences would use up 20 of the 105 available valences on the 35 As atoms. Then, the As atoms would have

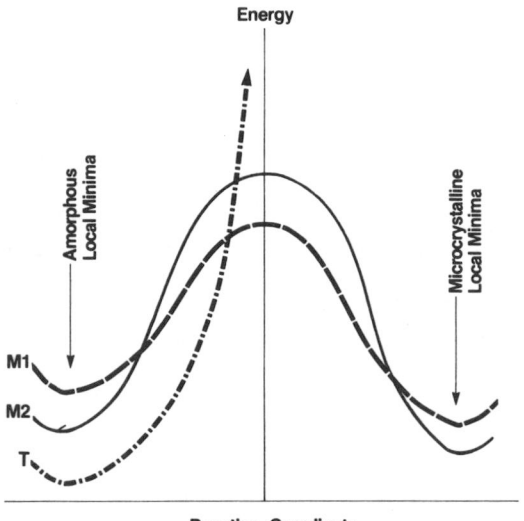

Energy

M1

M2

T

Reaction Coordinate

Fig. 1. Schematic illustration of the reversible amorphous–microcrystalline memory transition. The potential minima of memory alloy M1($\overline{C} = 2.32$) are the shallowest, and its barrier to rearrangement is the lowest. M2 (which has $\overline{C} = 2.50$) has deeper minima and a higher barrier towards the structural phase transition. The threshold material (T), which has $\overline{C} = 2.85$, does not undergo the transition. It is conjectured in the Discussion that a percolation threshold $\overline{C}^P$ might exist for this transition above which the energy barrier becomes infinite for the extended system as shown here for T. (The reaction coordinate and the energy axis are in arbitrary units.)

no other choice but to satisfy their remaining valence requirements (85 out of a total of 105) by bonding among themselves. In other words, they would not be incorporated into the glass structure at the right proportion, but probably form a segregated arsenic-rich phase. In fact, experiments have failed to produce a homogeneous solid analog of the T alloy with all Te atoms replaced by Se and still exhibiting a threshold switching effect (Formigoni [8]). However, under certain conditions, it is possible to produce such an alloy in the liquid phase (Ovshinsky [8]). In contrast, the heteronuclear bond strengths of the bonds to Te are of the right order of magnitude to allow these atoms to become incorporated properly.

The chemical bond approach can also help us to understand why Sb and S which are present in such small amounts (only 2% each) play a crucial role in the memory glasses by enhancing the reversibility of these films in memory processes (Moss and de Neufville [8]). We have seen (table 2) that the Sb atoms preferentially bond to Te, while the S atoms preferentially bond to Ge. Thus, the conjecture on the influence of Sb and S in inhibiting GeTe formation (Moss and de Neufville [8]) is probably correct, and likely to be related to the influence of these elements in disrupting GeTe crystallite growth by bonding to Ge and Te in amounts that are small, but significant for thin films. For thicker films, their disruptive effect would be less, since there would be three directions in which crystals could fully grow, overcoming the influence of Sb and S. In fact, for thicker quaternary films, such a significant inhibition of telluride formation has not been observed (Moss and de Neufville [8]).

It should also be noted that if the two atomic per cent of S atoms in the memory alloys are replaced by Se atoms, the chemical bond approach predicts that the Se atoms will also be bonded to Ge, although with a somewhat lower bond energy (49.4 kcal/mol.). Therefore, the general bonding pattern would remain the same, unlike the drastic changes caused by the Te → Se replacement in the T alloy.

As was mentioned previously, the threshold and memory-type properties of these materials have been explained structurally in terms of the relative amounts of steric hindrances and energy barriers to matrix rearrangement [5]. It is also possible to extend this model by using a bond percolation model [17], with a percolation threshold $\overline{C}^P$ for the average coordination number of the network of divalent, trivalent and tetravalent atoms. Below $\overline{C}^P$, the connectivity of the alloy would be low enough for the disturbance caused by the electrical impulse to result in the breakage of structural bonds, with the breakage then propagating throughout the material like a wave. This propagation could thus create nucleation sites for an extended structural phase transition of the type observed in the memory alloys. As $\overline{C}^P$ is approached from below, the energy barriers to such a percolative flowlike propagation of

structural change would gradually become greater, suppressing the structural phase transition. At first, the distance to which the structural change could propagate would gradually decrease. Then, as $\overline{C}^P$ is reached and the energy barrier becomes infinite, the local breakages or rearrangements of bonds caused by the motivating potential would not be able to propagate, leaving only the possibility of threshold switching via electronic excitation. This situation is illustrated by the schematic drawing in fig. 1.

It should be clear that we are considering here the possibility of a percolation threshold for the occurrence of a well-defined reversible phase transition. The statement that the energy barrier would go to infinity at $\overline{C}^P$ refers to this specific structural phase transition for an extended system. Otherwise, it is of course possible to destroy the structural integrity of the material completely by applying excessive amounts of energy. However, this might create another phase with an excessive amount of stability and would not be a reversible memory transition (Ovshinsky [7]).

The possibility that the average coordination number $\overline{C}$ of a network may be fundamentally related to its vitrification properties, with special values of $\overline{C}$ resulting in certain properties (such as a higher glass forming tendency) has been extensively discussed (de Neufville and Rockstad [6] and refs. [2,5,18]). If a $\overline{C}^P$ value exists as a specific percolation threshold in these materials, interesting effects may be observed as one approaches it very closely. From the fact that the T alloy has $\overline{C} = 2.85$ while the M2 alloy has $\overline{C} = 2.50$, it is clear that if it exists $2.50 < \overline{C}^P < 2.85$, i.e., it follows the general principles elucidated in ref. [5] and is only slightly above the $\overline{C} \approx 2.4$ value suggested by Phillips [19] for an ideal glass former which is nearly strain-free. (A more mathematical treatment of the subject of rigidity and floppiness in glass structures, in terms of percolation theory, has recently been provided by Thorpe [20].)

In view of these considerations, it should be interesting to investigate the $2.50 < \overline{C} < 2.85$ region by chemical modification [19], always keeping in mind that the chemical bond approach should be used to check that an alloy would have all of its atomic constituents well-integrated into the network. (Remember the example given above of what happens when Te is replaced by Se in the threshold alloy! A naive calculation of $\overline{C}$ would still give 2.85, but only as an average over a phase-segregated heterogeneous material.)

Yet another possibility that comes to mind if $\overline{C}^P$ exists is the effect of lowering system dimensionality. It is well-known [14] that percolation becomes more difficult when the dimensionality of a system is decreased. Therefore, if very smooth and very thin films could be made of a material with a $\overline{C}$ very close to but still below $\overline{C}^P$ and thus still exhibiting a memory transition, the value of $\overline{C}^P$ itself might begin to decrease as the film becomes thinner. At some critical thickness (t^C) the material need no longer exhibit memory switching but probably still have switching properties (Ovshinsky [8]). This supports the possibility of using layers of the same material for both threshold (thickness below t^c) and memory (thickness above t^c) switching action.

Another consequence of the considerations above is that a memory material with a large $\overline{C}$ is likely to be a more stable memory alloy. For example, M2 is likely to be better than M1, and a memory alloy with $\overline{C} > 2.50$ is likely to result in even further improvement. This is because it would be able to undergo the reversible structural phase transition that forms the basis of the memory action more effectively.

We would like to thank David Adler, Napo Formigoni and John DeNeufville for helpful discussions, and Steve Hudgens for a critical reading of the manuscript.

References

[1] S.R. Ovshinsky, Phys. Rev. Lett. 21 (1968) 1450.
[2] S.R. Ovshinsky, in: Physical Properties of Amorphous Materials, eds., D. Adler, B.B. Schwartz and M.C. Steele, Institute for Amorphous Studies Series, Vol. 1, Ch. 2 (Plenum, New York, 1985).
[3] S.R. Ovshinsky, J. de Phys., C4, Suppl. 10, 42 (1981) C4–1095.
[4] M. Kastner, D. Adler and H. Fritzsche, Phys. Rev. Lett. 37 (1976) 1504;
M. Kastner, J. Non-Crystalline Solids 31 (1978) 223;
D. Adler, J. de Phys. C4 Suppl. 10, 42 (1981) C4–3.
[5] S.R. Ovshinsky and K. Sapru, Proc. 5th Int. Conf. on Amorphous and Liquid Semiconductors, eds., J. Stuke and W. Brenig, (1974) p. 447; S.R. Ovshinsky, Phys. Rev. Lett. 36 (1976) 1469; S.R. Ovshinsky, Proc. Int. Topical Conf. on Structure and Excitation of Amorphous Solids, Williamsburg, VA, March 24–27, 1976, p. 31.
[6] J.P. deNeufville and H.K. Rockstad, Proc. 5th Int. Conf. on Amorphous and Liquid Semiconductors, eds., J. Stuke and W. Brenig (1974) p. 419; D.J. Sarrach, J.P. deNeufville and W.L. Haworth, J. Non-Crystalline Solids 22 (1976) 245; T. Shimizu, R. Negishi and N. Ishii, J. Non-Crystalline Solids 31 (1979) 287; N. Tohge, T. Minami and M. Tanaka, J. Non-Crystalline Solids 38&39 (1980) 283; see also ref. [18].
[7] D. Adler, H.K. Henisch and N. Mott, Rev. Mod. Phys. 50 (1978) 209; D. Adler, M.S. Shur, M. Silver and S.R. Ovshinsky, J. Appl. Phys. 51 (1980) 3289; S.R. Ovshinsky, J. Non-Crystalline Solids 2 (1970) 99.

[8] S.C. Moss and J.P. deNeufville, Mat. Res. Bull. 7 (1972) 423; H.K. Rockstad and R. Flasck, Proc. 5th Int. Conf. on Amorphous and Liquid Semiconductors, eds., J. Stuke and W. Brenig (1974) p. 1311; N. Formigoni, unpublished experimental results; S.R. Ovshinsky, unpublished exp. results.

[9] W.H. Zachariasen, J. Am. Chem. Soc. 54 (1932) 3841.

[10] L. Pauling, The Nature of the Chemical Bond, 3rd ed. (Cornell University Press, 1960) p. 91.

[11] A.L. Allred, J. Inorg. Nucl. Chem. 17 (1961) 215.

[12] C. Kittel, Introduction to Solid State Physics, 5th ed. (Wiley, New York, 1976) p. 74.

[13] H.F. Schaefer III, The Electronic Structure of Atoms and Molecules (Addison–Wesley, New York, 1972).

[14] Research on the Properties of Amorphous Semiconductors at High Temperature, ARPA Contract DAHC 15-70-C0187, Final Technical Report (1973).

[15] Handbook of Chemistry and Physics, 64th ed., R.C. Weast. ed. (CRC Press, 1983) pp. B-33 and B-34.

[16] D.J. Sarrach, J.P. DeNeufville and W.L. Haworth, J. Non-Crystalline Solids 27 (1978) 193.

[17] R. Zallen, The Physics of Amorphous Solids (Wiley, New York, 1983) Ch. 4, 5.

[18] S.R. Ovshinsky, Proc. 7th Int. Conf. on Amorphous and Liquid Semiconductors (1977), ed., W.E. Spear, p. 519; S.R. Ovshinsky and D. Adler, Contemp. Phys. 19 (1978) 109.

[19] J.C. Phillips, J. Non-Crystalline Solids 35&36 (1980) 1157 and refs. therein.

[20] M.F. Thorpe, J. Non-Crystalline Solids 57 (1983) 355.

CRITICAL MATERIALS PARAMETERS FOR THE DEVELOPMENT OF AMORPHOUS SILICON ALLOYS

Stanford R. Ovshinsky[1] and David Adler[2]

[1]Energy Conversion Devices, Inc., 1675 West Maple Road, Troy, Michigan 48084
[2]Department of Electrical Engineering and Computer Science, Massachusetts Institute of Technology, Cambridge, Massachusetts 02139

INTRODUCTION

Solar cells were invented more than 30 years ago, but have not yet achieved the commercial success predicted many times in the past. There is no question that they efficiently convert the eternally available sunlight directly to electricity, they have the reliability inherent in a system with no moving parts, they can be used as either distributed power sources or central power stations, and they yield minimal environmental contamination or safety hazards.(1) However, it is clear that additional criteria must be imposed to guarantee their large-scale commercial utilization, since they must compete successfully with conventional power sources such as coal, gas, oil, and nuclear energy. The desired properties are straightforward to identify:

(1) Efficiency. The solar cells must have a high initial conversion efficiency n_0, defined as the ratio of the electrical power generated to the power of the sunlight intercepted by the entire module. While it has been estimated that cells must exhibit at least 15% efficiency for use in central power stations,(2) a figure we are now approaching, there is an equally important market, viz. distributed power, where such an efficiency is not necessary. Distributed power has the advantage of eliminating the cost and losses associated with transmission lines.

(2) Stability. The efficiency must remain high after at least 20 years of continuous operation. If the stability, S, is defined as the ratio of the efficiency after 20 years of exposure to sunlight, n_f to n_0, i.e.,

$$S = n_f/n_0, \qquad (1)$$

then it is necessary to optimize the product $n_0 S$ rather than n_0. This arises because solar-cell degradation ordinarily occurs primarily over the first few months of operation, and $n_f = n_0 S$ is thus essentially the average efficiency over the entire lifetime of the device. It is extremely misleading to quote only n_0 in presenting solar-cell data if S is significantly less than unity.

(3) Cost. It is evident that the cost of the module, C, in, e.g., dollars per peak watt, is crucial. High cost has been a major reason that high-quality crystalline Si and GaAs solar cells have not achieved market penetration to date,(1) and C must be reduced sufficiently to compete with the available alternative sources of electricity. Of course, it is not only the cost of the modules that is important, but the cost of the delivered power, including the transportation and installation of the cells.

(4) Power Density. The possibility of distributed power via solar-cell use is very appealing as a means of reducing the economic and energy costs of electrical transmission over long distances. But transportation and installation of bulky solar cells are also expensive. In addition, solar cells have major value both in outer space and in remote areas. For all of these reasons, high power per unit weight, P, in peak watts per kilogram, is extremely desirable. In addition, if the cells are flexible, there are major advantages in transportation and storage.

(5) Abundance. Irrespective of cost, widespread conversion to solar-generated power requires the use of enormous quantities of material. Thus, even if, for example, GaAs solar cells could be made cost competitive using currently produced materials, the finite abundance of Ga would preclude their providing a major fraction of the world's electricity budget. In reality, partial depletion of the ores would begin to drive the cost up enormously.

(6) Toxicity. Especially in distributed residential and commercial applications, materials toxicity could have a deleterious effect on market penetration. For example, easily formed compounds of As and Cd can be lethal in relatively low concentrations, and completely reliable safety precautions in the event of fires, etc. could increase the cost of the solar power a great deal.

Reprinted by permission from *Materials Research Society Symposium Proceedings*, Vol. 49, pp. 251–264 (1985).

Although it is difficult to put the proper weight on all the above factors, especially considering the diverse potential applications, some simply calculated figure of merit would be useful to evaluate the relative merits of different solar-cell technologies. A linear coefficient has been proposed:(3)

$$F = \eta_0 \, S \, P \, / \, C \qquad (2)$$

which has the virtue of ease of estimation. We shall use this figure of merit subsequently to quantify some of the current progress being made in amorphous-silicon-alloy solar-cell technology.

ADVANTAGES OF AMORPHOUS SILICON ALLOYS FOR SOLAR-CELL APPLICATIONS

There are many reasons why amorphous silicon alloys are highly desirable for solar-cell applications. These include:

(1) Silicon is essentially infinitely abundant. It is the second most common element in the earth's crust and there is a sufficient quantity in a 1mm layer of sand over the Sahara desert to cover the entire earth's surface with amorphous solar cells.

(2) Amorphous silicon alloys are excellent absorbers of sunlight. A layer of only 500nm is sufficient to attain high solar energy conversion efficiencies. Consequently, thin-film solar cells on inexpensive substrates are feasible. This suggests both a very low cost for the starting materials and the possibility of high power densities using low-density substrates, in sharp contrast with single-crystalline or polycrystalline silicon solar cells which require more than 10 times the thickness to absorb the same fraction of visible light.

(3) High-quality amorphous silicon alloys can be prepared using inexpensive starting materials, such as silicon tetrafluoride gas, another reason why low-cost solar cells are feasible.

(4) Amorphous silicon alloys can be deposited very rapidly, further suggesting a low overall solar-cell cost. Recently, photoreceptor-grade amorphous silicon alloys have been grown at the extremely rapid rate of 36 μm/hour using a 2.45 GHz microwave glow discharge,(4) opening up the eventual possibility of ultrarapid deposition of solar-grade material as well.

(5) The lack of lattice mismatches in multilayered amorphous structures offers the possibility of very high solar energy conversion efficiencies via the fabrication of stacked cells of different band gaps using an array of alloys.

(6) With appropriate alloying, the range of band gaps achievable using these amorphous materials can cover the gamut from 1.2 eV to 2.2 eV, enabling the possibility of using multilayer devices to make efficient use of the entire solar spectrum in a stacked cell configuration.

(7) Amorphous silicon alloys do not contain any highly toxic components.

In brief, amorphous silicon alloys appear to possess all of the essential desirable characteristics for commercial sources of solar-generated electricity. The material is nontoxic, infinitely abundant, and capable of being made into stable low-cost, high-efficiency, high-power-density solar cells. In fact, they appear to be unique in this respect.

SELECTED PROPERTIES OF AMORPHOUS SILICON ALLOYS

Although amorphous silicon alloys have the impressive characteristics enumerated in the previous section, the development of solar cells based on their use is far from straightforward. In this section, we describe the origin of some of the problems that have had to be overcome to achieve commerciality.

(1) Amorphous silicon alloys form primarily tetrahedrally bonded networks, a reflection of the chemical nature of Si, which is in Column IV of the Periodic Table and bonds optimally using sp^3 orbitals. Because this represents the maximum possible number of bond per atom using only s and p electrons, no low-energy defects exist. Thus, only negligible defect concentrations are required from purely thermodynamic considerations, in sharp contrast with the case of amorphous chalcogenide alloys.(5) This opens up the possibility of depositing semiconductor-grade amorphous silicon alloys. However, strain-induced defects are present under ordinary circumstances and these degrade the materials and preclude their use in any conventional semiconductor device.

(2) The origin of the defects in amorphous silicon alloys is the intrinsic strains necessarily introduced on deposition. The fact that the optimal coordination number, Z, for Si is Z = 4 results in its forming an overconstrained network.(6,7) Most of this strain is relieved by bond-angle distortions which induce valence and conduction band tail states. Although these band tails limit somewhat the maximum open-circuit voltage, V_{OC}, in solar cells, they primarily act as shallow traps for photogenerated electrons and holes. Since the trapped carriers are almost always re-released before they recombine, their presence does not reduce the short-circuit current density, J_{sc}, or fill-factor, ff, significantly, and thus the band tails do not have a major effect on solar-cell efficiency. However, in certain regions of the film, the bond-angle distortions necessary to relieve the strain are sufficiently large that the resulting increase in total energy would exceed the energy reduction due to formation of the bond. In that case, the bond is unstable and a defect center appears. This could take the form of an isolated dangling bond, T_3, in the conventional notation,(5) a two-fold-coordinated Si atom, T_2, or even a defect complex.(8) Such defect centers introduce states deep in the gap, which act as recombination centers that degrade both J_{sc} and ff, and sharply reduce solar-cell efficiency.

(3) A partial solution to the problem of strain-induced recombination centers is to alloy the amorphous silicon with hydrogen. Since the

137

coordination number of hydrogen is Z = 1, the average coordination, $\bar{Z}$, of hydrogenated amorphous silicon, $a\text{-Si}_{1-x}H_x$, is

$$\bar{Z} = 4 - 3x, \qquad (3)$$

which is considerably lower than 4 for typical hydrogen concentrations. Thus, a-Si:H films are much less overconstrained than pure a-Si, and the defect concentration can be reduced from $\sim 10^{-3}$ to $\sim 10^{-8}$. In addition, the fact that the Si-H bond is approximately 40% stronger than the Si-Si bond results in the absence of fully bonded states in the gap resulting from Si-H bonds. The somewhat greater electronegativity of H relative to Si leads to a hydrogen-induced removal of states from near the top of the a-Si valence band to positions much deeper in the gap.(5) However, this results in a net increase of the gap with increasing hydrogen concentrations, thus inducing a decrease in the theoretical maximum solar-cell efficiency.

(4) The last-mentioned problem could be easily solved by the development of amorphous alloys with reduced band gaps. This would have the further desirable feature of providing the opportunity for fabrication of very high efficiency stacked cells with different gaps. The most obvious answer would be the use of Si-Ge alloys, since Ge is chemically similar to Si but has a considerably smaller energy gap. However, hydrogenated amorphous Si-Ge alloys (9) and even hydrogenated amorphous Ge itself (10) have sufficiently large defect concentrations to preclude their use in efficient solar cells. The origin of this is most likely the weaker Ge-H bond relative to Si-H taken together with the greater chemical tendency of Ge toward divalency and thus T_2 defects.(3,6)

In brief, although amorphous silicon alloys appear to be ideal for use in commercial solar cells, many problems have to be overcome. These primarily involve sharp reductions in the concentration of strain-induced defect centers.

CHEMISTRY AND PHYSICS OF AMORPHOUS SILICON ALLOYS

In this section, we elaborate some of the previous discussion in order to pinpoint the chemical and physical origin of the problems with conventional amorphous silicon alloys. In particular, we discuss the basic band structure, the nature of the traps and recombination centers, the problem of material stability, and the origin of the additional defects that characterize Si-Ge alloys.

The basic band structure of amorphous silicon alloys can be derived from their known local atomic structure.(5) The vast majority of Si atoms are surrounded by four other atoms bonding primarily covalently and forming approximately a regular tetrahedron around the central atom. Disorder enters both geometrically, with the introduction of $\pm 10^0$ distortions around the 109.5^0 tetrahedral bond angle, and chemically, with the introduction of bonded hydrogen or fluorine. A good approximation to the density of states of a-Si:H can be obtained by considering a small cluster, $Si_4H(sat)_9$, in which a central Si atom is surrounded by three other Si atoms and one H atom in a tetrahedral configura-

tion with the outer Si atoms saturated by a monovalent atom at the appropriate Si-Si separation of 2.35Å.(11) This calculation shows that the valence and conduction bands of a-Si:H are similar in character to those of c-Si, except for the disappearance of the sharp structure (van Hove singularities) that characterize systems with long-range order. Because of the fact that the Si-H bond is about 40% stronger than the Si-Si bond and H is somewhat more electronegative than Si, the presence of bonded hydrogen lowers the energy of states near the top of the valence band, thus increasing the energy gap of a-Si:H relative to that of pure a-Si. The presence of bond-angle and dihedral-angle disorder leads to the appearance of band tails, experimentally (12) of exponential form:

$$g_c(E) = g_0 \exp\left[-(E_c - E)/kT_c\right], \qquad (4)$$

with $T_c \approx 325K$ for the conduction band, and:

$$g_v(E) = g_0 \exp\left[(E_v - E)/kT_v\right], \qquad (5)$$

with $T_v \approx 550K$ for the valence band. As mentioned previously, the band-tail states limit V_{oc} in solar cells.

Since a-Si:H forms an overconstrained network, small concentrations of defect centers are also present. These include dangling bonds, T_3 centers, which introduce two defect states near midgap, and perhaps other centers such as two-fold-coordinated Si atoms, T_2, which introduce four states in the gap, or three-center bonds, T_{3c}, with bridging hydrogen atoms.(8,13) As discussed previously, the defect states near the Fermi energy increase the recombination current and thus reduce J_{sc} and ff in solar cells.

After exposure to intense light, e.g., several hours of sunlight, the concentration of neutral dangling bonds, T_3o, in a-Si:H films sharply rises,(14) increasing the recombination current and concomitantly degrading solar-cell efficiency. This photostructural change is known as the Staebler-Wronski effect (15) and has been shown to be an intrinsic property of a-Si:H.(16) The effect is driven by a particular recombination branch (17) and can be reversed by annealing,(15) so that it is self-limiting. However, the degradation in solar-cell efficiency can be severe,(18) yielding up to a 40% reduction in several days of operation prior to saturation. A partial solution of this problem can be achieved by the use of thinner films. Tandem devices even with the same band-gap alloy in both junctions have demonstrated improved stability, because of a reduction in the recombination current in very thin layers (since greater concentrations of photogenerated carriers are then collected).

Finally, it is also useful to analyze the source of the problems with Si-Ge alloys. We have previously pointed out (19,3,5,6) that Ge chemically has a much stronger tendency than Si toward divalency, a state in which an s lone pair (Sedgwick pair) remains nonbonded on the Ge atom. The resulting four states in the gap sharply increase $g(E_f)$ and thus the recombination current in alloys containing Ge. Clearly, the incorporation of hydrogen cannot of itself cure this problem. In addition, the Ge-H bond is somewhat weaker than the Si-H bond, which can have two deleterious consequences. First, H preferentially attaches to Si in Si-Ge alloys, thereby leading to relatively high concentrations of Ge dangling bonds. Second, if the bond is sufficiently weak, Ge-H antibonding states may contribute to the conduction band tail or even form defect states deeper in the gap. (Note that the chemical trend of Column IV elements toward divalency increases as the Row in the Periodic Table increases, resulting in still greater defect concentrations in, e.g., alloys containing Sn and Pb.(6))

To summarize, the origin of the problems in a-Si:H based solar cells can be traced to: (1) the inherent bond-angle disorder; (2) the strain-related defects; (3) the photostructural changes; and (4) the tendency of Ge toward divalency. Since these are all problems intrinsic to the material itself, it is clear that cures must involve new materials development.

ROAD TO UTOPIA

In order to solve the scientific and technological problems necessary for the development of amorphous silicon alloy solar cells of sufficient efficiency, stability, and cost to achieve commercial success vis-a-vis conventional power sources, the following four steps must be completed:

(1) Develop a very high quality material for single-cell devices. To achieve energy conversion efficiencies beyond 10% requires band tails with both T_c and T_v below 600K and $g(E_f) < 10^{16} cm^{-3} eV^{-1}$.

(2) Develop a process for rapid, continuous production of large-area, high-efficiency modules. This requires attainment of (a) high deposition rates, (b) excellent uniformity, and (c) little or no loss in efficiency in large-area compared to small-area devices.

(3) Produce cells which retain their initial efficiency after many years of exposure to sunlight. A projected 20 years of operation in relatively sunny areas requires no significant degradation for about 20,000 hours of operation at AM1 (100 mW/cm^2) conditions.

(4) Develop very high quality stable alloys with reduced band gaps for use in high-efficiency stacked configurations. Such alloys must also have $g(E_f) < 10^{16} cm^{-3} eV^{-1}$ to provide the necessary increase in efficiency over single-junction devices.

RESULTS ACHIEVED BY ECD AND SOVONICS

In this section, we show how our theoretical understanding of amorphous silicon alloys has allowed us to accomplish all of the goals enumerated in the previous section. We describe the progress attained to date in solar-cell development, and evaluate the achievements quantitatively in terms of the figure of merit of Eq. (2).

The first step of developing a high-quality material for use in a single-cell device has been

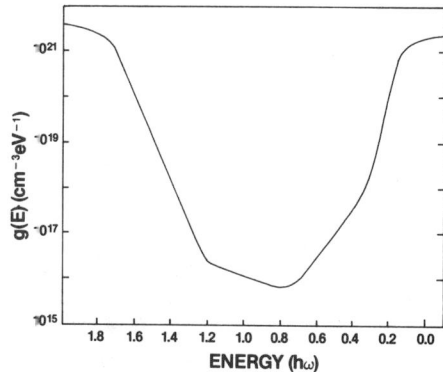

Fig. 1   Density of states of a high-quality amorphous silicon alloy film.

accomplished. Carefully produced a-Si:H films and especially our proprietary a-Si:F:H films with band gaps in the 1.7-1.8 eV range have been used to produce solar cells with energy conversion efficiencies in the 10% range. A typical density of states curve for such materials, as determined by a combination of optical absorption data, photothermal deflection spectroscopy, and transient and steady-state photoconductivity experiments (20) is shown in Fig. 1. Both valence and conduction band tails fall off with characteristic temperatures below 600K and the midgap density of states is less than $10^{16} cm^{-3} eV^{-1}$, as desired.

The second step, development of a high-speed continuous process for large-area production, is unique to ECD and its partners.(3) This required the bold step of designing and building a machine that had not previously been contemplated. The concept was to use rolls of an inexpensive substrate such as stainless steel to produce solar cells continuously in a completely automated manner. This eliminates both bulky glass modules and inefficient, space-consuming batch processing. Figure 2 shows the production of 16-inch wide, 1000-foot long rolls of solar cells via our proprietary continuous-web process at the Sharp-ECD plant in Shinjo, Japan where commercial cells for consumer applications have been produced for the past two years. Figure 3 demonstrates the high level of automation in this

Fig. 2   Continuous   one-foot-wide   rolls   of mass-produced Ovonic solar cell material.

Fig. 3   Interior of the Sharp-ECD Solar, Inc. plant.

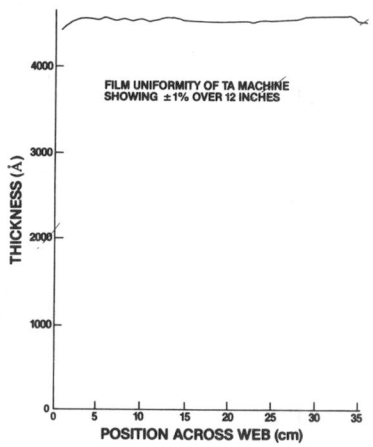

Fig. 5.   Film thicknesses as a function of position across a strip 40-cm wide, deposited by our continuous-web process.

plant, in which at most four employees are necessary during the production run. Figure 4 is a photograph of the latest version of our processor, now in operation at the Sovonics plant in Troy, Michigan.

There are several noteworthy factors in our approach.   The first was our choice of a one-square-foot format at a time when everyone was involved in small-area processing. This required solving the potential difficulties of adhesion of the film to the substrate, uniform large-area deposition, and maintenance of cell integrity on flexible substrates, all nontrivial problems. As an example of the uniformity achieved, Fig. 5 shows the film thickness as a function of position across a 30 cm (12 in) section of substrate,(21) and Fig. 6 shows the spatial distribution of solar-cell parameters over a one-square-foot tandem a-Si:F:H device with an average efficiency of 8%, in which 90% of the sub-cells exhibit efficiencies in the 7-9% range.(21,22)

The second factor was to process a 1000-foot roll continuously with high yield and high efficiency. To achieve this, we had to solve the problems of uniform gas flow, homogeneity of the gas mixture, and control of the speed of the web, among many others.   It is of great significance to our overall strategy that all this was achieved in a

tandem configuration.   Our approach has led to our being awarded the basic patents in the field, and we now have several years of production experience behind us.   The impressive results attained to date include a 9.57% average efficiency over a one-square-foot single-cell device which has a peak sub-cell   efficiency   of   10.5%,(23)   and   a continuous-web one-square-foot single-cell device fabricated on a production machine with an average efficiency of 8.2%.(22)   In addition, a fully encapsulated one-square-foot module with an active area efficiency of 8.3% has been produced with a measured output of 7.33 W_p.

As we discussed previously, none of the initial efficiency data have any significance if the cells degrade during operation. However, using our knowledge of the physics of amorphous silicon alloy materials and devices, we have designed and produced multijunction solar cells with efficiencies in the 8% range which retain over 99% of their initial efficiency after 2000 hours of AM 1 exposure, as is clear from Fig. 7.(24)   Extrapolation of these data indicates that such devices will retain over 95% of their initial efficiency after 20 years of actual operation in sunlight.   Also shown in Fig. 7 for

Fig. 4   View of a portion of the Sovonics production plant, showing the Sovonics Photovoltaic Processor.

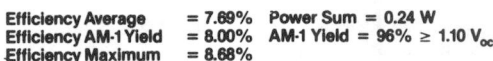

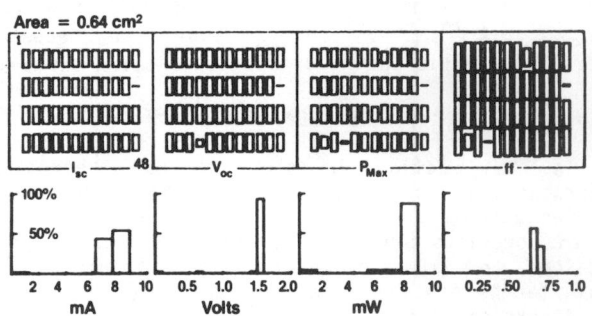

Fig. 6   Solar   cell   parameters   of   the individual 48 sub-cells on a one-square-foot solar cell.

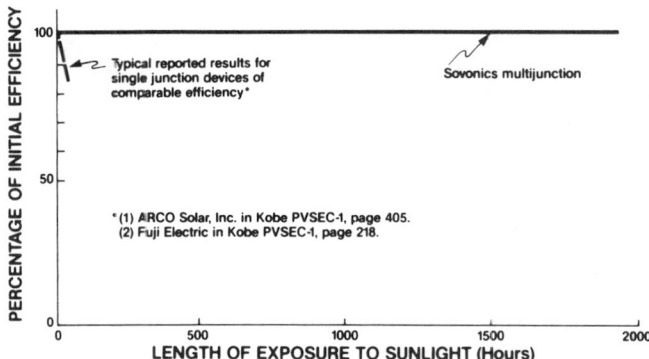

Fig. 7 Stability of a Sovonics multijunction solar cell after 2000 hours of AM1 exposure compared to recent results on typical single-junction a-Si:H cells.

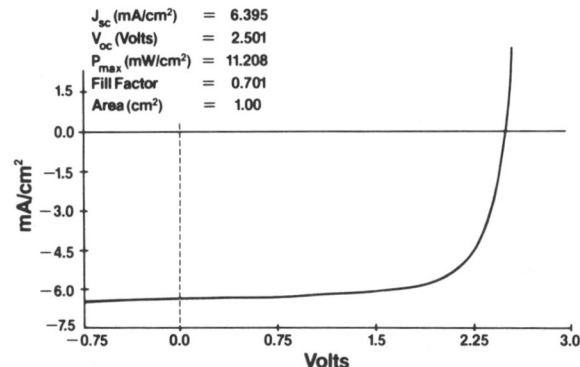

Fig. 9 J-V characteristic of a multiple band gap triple junction Ovonic solar cell with an efficiency of 11.2%.

comparison is the recent data reported by ARCO Solar and Fuji Electric on the degradation of single-junction devices of comparable efficiencies.(18)

Finally, it is of the utmost importance to develop smaller band-gap alloys of comparable quality and stability. As is evident from Fig. 8, our a-Si-Ge:F:H alloy has solved this problem. The only difference in the subgap optical absorption between the a-Si:F:H alloy and the a-Si-Ge:F:H alloy is the lower gap in the latter. A translation of the $\alpha(\omega)$ data of the a-Si-Ge:F:H film to higher photon energies by the 0.3 eV difference in band gaps results in complete superposition of the two curves. This demonstrates that $g(E_f)$ of the Si-Ge alloy is sufficiently low for solar-cell applications. Note that although the a-Si-Ge:H film shown for comparison in Fig. 8 has a similar exponential tail to the fluorinated films, the greater defect density in the absence of fluorination yields a considerably higher midgap absorption.

Using the a-Si-Ge:F:H alloy characterized in Fig. 8, we have been able to fabricate a single-junction solar cell with 9.0% efficiency.(20,24) More important, we have produced for the first time dual-band-gap tandem solar cells with efficiencies greater than those of single-junction cells. Figure 9 shows the J-V characteristic of a multijunction device with 11.2% efficiency.(24) We have also produced high-efficiency one-square-foot devices--tandem with 9.42% average efficiencies, triple with 9.83% average efficiencies, (23) and quadruple with 7.1% average efficiencies.(23,24)

Finally, we have also developed an ultralight module which produces power at the impressive density of 615 Wp/kg,(25) more than twice the previous record. The significance of this can best be appreciated by returning to the proposed figure of merit, Eq. (2), and noting that F ≈ 0.1 - 0.2 for conventional c-Si or a-Si:H solar cells, but F = 34 for our current prototype ultralight module. With further improvements in efficiency and substrate development, we are planning an ultralight module with F = 750. A photograph of a typical present-day ultralight device is shown in Fig. 10.

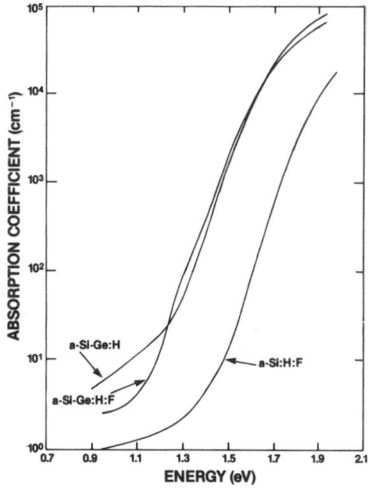

Fig. 8 Absorption coefficient as a function of photon energy for a-Si:Ge:F:H alloy films compared to that of an a-Si:F:H film with a 0.3 eV larger band gap.

Fig. 10 Rolls of one-foot wide ultralight Ovonic solar cell.

141

EFFECTS OF FLUORINE

It is worthwhile to emphasize that the achievement of each of the four goals discussed in the previous two sections relies on the incorporation of fluorine in the amorphous alloy. An understanding of the chemistry of amorphous silicon alloys has been clouded in the past. The high-quality "amorphous silicon" originally reported (26) turned out to be a hydrogenated alloy,(27) in fact with the hydrogen responsible for the quality of the material. However, a-Si:H still has residual defects, leading to $g(E_f) \sim 10^{15}$-10^{16}cm^{-3}eV^{-1}, even in the best films. Although the dispersion in material quality of both a-Si:H and a-Si:F:H is larger than the differences between the two types of alloys, there is evidence that the best a-Si:F:H films have lower defect concentrations than the best a-Si:H films. Field-effect measurements (28) indicate smaller values of $g(E_f)$ in the fluorinated films, and independent DLTS experiments (29) suggest that it isprimarily the density of interface states that is reduced by fluorination. Such states are particularly important in semiconductor devices such as solar cells.

The original report (28) of the resistance of a-Si:F:H alloys to photostructural effects has also been independently confirmed.(30) In fact, this property has been shown to increase the relative stability of fluorinated single-junction solar cells.(31) In addition, the use of fluorine results in striking increases in the achievable deposition rate,(4) a very important parameter in the final cost of the completed devices.

Finally, as we have already discussed (see Fig. 8), fluorination is essential in the production of high-quality amorphous Si-Ge alloys. As with all of the others, this result has been independently confirmed.(32,33) It is of major significance that the fluorinatd Si-Ge alloys are also much more stable than comparable unfluorinated alloys. Guha (20) has recently measured the photoconductivity of two Si-Ge alloys with the same optical gap, and found the initial value of the fluorinated material was twice that of the unfluorinated material. Moreover, after 16 hours of AM1 exposure in a coplanar configuration, the photoconductivity of the a-Si-Ge:H film had decreased by 34%, compared to only a 5% reduction in the a-Si-Ge:F:H film. Clearly, there is a real difference between the two alloys. In addition, through the years we have incorporated fluorine in a wide variety of amorphous alloys, including high-quality a-Si-C:F:H alloys.

All of the above materials differences can be understood theoretically from the following considerations:

(1) Si-F and Ge-F bond are primarily ionic, while Si-H and Ge-H bonds are primarily covalent. Since ionic bonds do not impose any bond-angle constraints, the use of fluorine results in films with lower overall strains.

(2) Si-F and Ge-F bonds are about half again as strong as Si-H and Ge-H bonds. There is thus no possibility that the antibonding states of the fluorine bonds are located either in the gap or in the conduction band tail.

(3) Fluorine is the most electronegative atom in the Periodic Table. The incorporation of fluorine thus reduces the concentration of negatively charged dangling bonds (T_3^- centers) in tetrahedrally bonded materials. Since the trapping of holes by T_3^- centers induces a Staebler-Wronski degradation, the fluorinated films exhibit increased stability.

(4) Fluorine expands the valency of both Si and Ge. The concomitant reduction in divalency, particularly in films containing Ge, lowers $g(E_f)$ sharply.

(5) Fluorine acts in the plasma, at the surface during growth, and in the bulk to induce unique local configurations that would otherwise not be present in the final film.(34)

CONCLUSIONS

We have shown how our proprietary approach which utilizes the incorporation of fluorine makes superior films of tetrahedral amorphous alloys, including amorphous silicon, silicon-germanium and silicon-carbon alloys. It is also useful in increasing the deposition rate and improving the stability of the material. These fluorinated alloys combined with our continuous-web production process have been used to fabricate high-efficiency stable multijunction solar cells. An ultralight module which delivers power at a density of 615 W$_p$/kg has also been produced.

Acknowledgments

We wish to acknowledge S.J. Hudgens, J. Yang, J. Hanak, S. Guha, J. Doehler, M. Izu, P. Nath, W. Czubatyj and many others at ECD whose contributions have made possible the devices discussed in this paper and thank them for making their data available to us prior to publication.

References

1. See D. Adler and S.R. Ovshinsky, Chemtech, in press, for a recent review.
2. M.K. Armstrong-Russell, W. Freedman and E.E. Spittes, S.P.I.E. Proc. 407, 132 (1983).
3. S.R. Ovshinsky, Tech. Digest Intern. PVSEC-1, (Kobe, Japan, 1984) p. 577. (As presented.)
4. S.J. Hudgens and A.G. Johncock, these proceedings.
5. D. Adler, in Physical Properties of Amorphous Materials, ed. D. Adler, B.B. Schwartz and M.C. Steele (Plenum Press, N.Y., 1985) p. 5.
6. S.R. Ovshinsky, in Physical Properties of Amorphous Materials, ed. D. Adler, B.B. Schwartz and M.C. Steele (Plenum Press, N.Y., 1985) p. 105.
7. S.R. Ovshinsky, A.I.P. Conf. Proc. 31, 67 (1976).
8. S.R. Ovshinsky and D. Adler, Contemp. Phys. 19, 109 (1978).
9. W. Paul, D.K. Paul, B. von Roedern, J. Blake and S. Oguz, Phys. Rev. Lett. 46, 1016 (1981).
10. J.A. Reimer, B.A. Scott, D.J. Wolford and J. Nijs, Appl. Phys. Lett. 46, 369 (1985).
11. K.H. Johnson, H.J. Kolasi, J.P. deNeufville and D.L. Morel, Phys. Rev. B 21, 643 (1980).

12. T. Tiedje, in Semiconductors and Semimetals, ed. by R.K. Willardson and A.C. Beer (Academic Press, N.Y., 1984) vol. 21C, p. 207.

13. S.R. Ovshinsky, in Amorphous and Liquid Semiconductors, ed. by W.E. Spear (C.I.C.L., U. of Edinburgh, 1977) p. 519.

14. H. Dersch, J. Stuke and J. Beichler, Phys. Stat. Sol. B 105, 265 (1981).

15. D.L. Staebler and C.R. Wronski, J. Appl. Phys. 51, 3262 (1980).

16. C.C. Tsai, J.C. Knights and M.J. Thompson, J. Non-Cryst. Solids 66, 45 (1984).

17. S. Guha, J. Yang, W. Czubatyj, S.J. Hudgens and M. Hack, Appl. Phys. Lett. 42, 5881 (1983).

18. D.P. Tanner and K.W. Mitchell, Tech. Digest Intern. PVSEC-1, (Kobe, Japan, 1984) p. 405. Y. Uchida, M. Kamiyama, Y. Ichikawa, T. Hama and H. Sakai, ibid, p. 217.

19. S.R. Ovshinsky and K. Sapru, in Amorphous and Liquid Semiconductors, ed. J. Stuke and W. Brenig (Taylor & Francis, London, 1974) p. 447.

20. S. Guha, unpublished data.

21. J. Doehler, unpublished data.

22. M. Izu and K. Hoffman, unpublished data.

23. P. Nath, unpublished data.

24. J. Yang, unpublished data.

25. J. Hanak, unpublished data.

26. W. E. Spear and P.G. Le Comber, Phil. Mag. 33, 935 (1976).

27. S.J. Hudgens, Phys. Rev. B 14, 1547 (1976).

28. S.R. Ovshinsky and A. Madan, Nature 276, 482 (1978).

29. C.H. Hyun, M.S. shur and A. Madan, Appl. Phys. Lett. 41, 178 (1982).

30. H. Matsumura and S. Furukawa, in Amorphous Semiconductor Technologies and Devices, ed. Y. Hamakawa (North-Holland, N.Y., 1982) p. 88.

31. Y. Kuwano, M. Ohnishi, H. Nishiwaki, S. Tsuda, H. Shibuya and S. Nakano, in Proc. 15th IEEE Photovoltaics Specialists Conf., (1981) p. 698.

32. K. Nozawa, Y. Yamaguchi, J. Hanna and I. Shimizu, J. Non-Cryst. Solids 59-60, 533 (1983).

33. S. Nakano, Y. Kishi, M. Ohnishi, S. Tsuda, H. Shibuya, N. Nakamura, Y. Hishikawa, H. Tarui, Y. Takahama and Y. Kawano, these proceedings.

34. S.R. Ovshinsky, in Proc. of the Intern. Ion Engineering Congress, ISIAT '83 & IPAT '83, (Kyoto, Japan, 1983).

LOW PRESSURE MICROWAVE GLOW DISCHARGE PROCESS FOR HIGH DEPOSITION RATE AMORPHOUS SILICON ALLOY

S.J. Hudgens, A.G. Johncock, and S.R. Ovshinsky

Energy Conversion Devices, Inc., 1675 West Maple Road, Troy, Michigan 48084

1. INTRODUCTION

As glow discharge amorphous silicon alloys become more important for commercial production of photovoltaic, xerographic, optical sensing, and microelectronic devices, economic considerations have motivated considerable research aimed at increasing deposition rate and gas utilization efficiency. Recently, a low pressure microwave glow discharge process was reported (1) which produces good quality a-Si:F:H films, with essentially 100% gas utilization at deposition rates > 100Å/sec.

This paper will compare plasma I-V and mass spectroscopic data from 13.56 MHz, rf, and 2.45 GHz, mw, glow discharge, SiH_4 plasmas produced with otherwise identical process parameters in an attempt to understand low pressure mw deposition. IR and electrical properties of mw deposited films will be analyzed in light of differences in these characteristics.

2. EXPERIMENTAL DETAILS

The deposition system,(1) shown in Fig. 1, incorporates a double floating probe system for measurement of plasma I-V characteristics and a residual gas analyzer (Inficon IQ 200) for analysis of plasma species. The rf antenna is a 5 cm square stainless electrode and the mw antenna is a rigid 0.64 cm diameter slotted coaxial cable, enclosed in 1.9 cm diameter fused silica shroud. The antenna to substrate distance is 3 cm when either antenna is inserted. The probes are 0.64 cm diameter stainless steel rods separated by 1.0 cm and located 1.0 cm from either antenna. The probe rods extend throughout the length of the tube. All depositions and plasma diagnostic measurements were made using 30 sccm flow of SiH_4, applied excitation power, P = 110 W, and substrate temperature, T_s = 250°C.

3. RESULTS

Electric probes (2) have previously been used in SiH_4 rf glow discharge plasmas to measure plasma electron temperature and ion concentration. Current versus differential probe voltage data are shown in Fig. 2 for rf and mw plasmas. Because the double diameters are equal, the I-V characteristics are symmetrical so only the first quadrant data are shown here. Using the analysis of Chen,(3) the results shown in Table I were obtained.

Although the rf plasma Te values are higher than previously reported,(2) the low pressure mw plasma is clearly much more efficiently excited than the rf, giving both higher relative ion concentration,

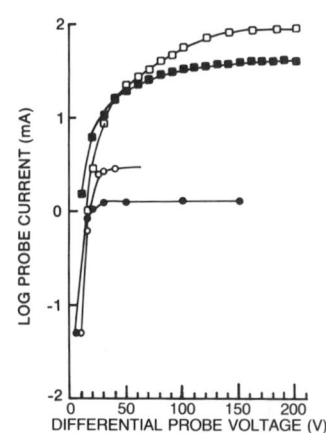

Fig. 2 Probe I-V characteristics, ● P_{rf}= 50mTorr, ○ P_{rf}=200mTorr, ■ P_{mw}= 50mTorr, □ P_{mw}= 200mTorr

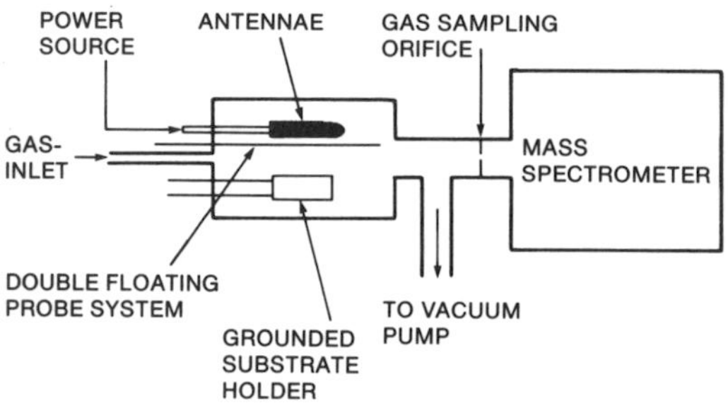

Fig. 1 Experimental Apparatus

(POWER SOURCE, ANTENNAE, GAS SAMPLING ORIFICE, GAS-INLET, MASS SPECTROMETER, DOUBLE FLOATING PROBE SYSTEM, GROUNDED SUBSTRATE HOLDER, TO VACUUM PUMP)

Reprinted by permission from *Journal of Non-Crystalline Solids*, Vols. 77/78, pp. 809–812 (1985).

<table>
Table I
</table>

Plasma	Pressure (mTorr)	kTe (eV)	$\dfrac{n_0(mw)}{n_0(rf)}$
mw	30	42	13
rf	30	9	
mw	200	61	11
rf	200	8	

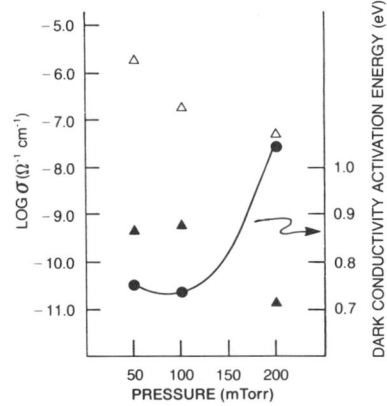

Fig. 4   mw   film   electrical   characteristics, σ_p= Δ, σ_d= $\blacktriangle$, ΔE = $\bullet$

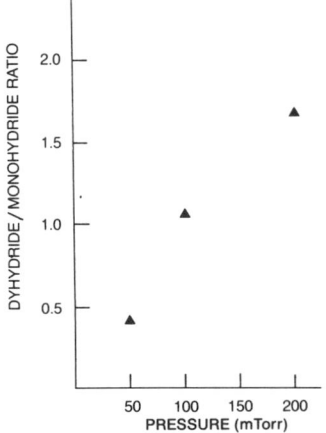

Fig. 3   Hydrogen Bonding Ratio for mw films from IR absorption

$n_0(mw)/n_0(rf)$, and apparently higher electron temperature, T_e, for the same applied power. This trend to higher electron concentration with increasing excitation frequency is predicted by theory;(4) however, since double probes sample only the high energy tail cf the electron distribution, kT_e may be overestimated for plasmas with excess, non-Maxwellian, high energy electrons.

Increasing pressure couples energy to the plasma more efficiently. The same trend with pressure is seen in the gas utilization efficiency, η. Using the mass spectrometer to measure the partial pressure of SiH_4 with and without the plasma ignited, we see: $\eta_{rf}(30mTorr) \lesssim 0.05$, $\eta_{rf}(200mTorr)=0.77$, $\eta_{mw}(30mTorr)=0.97$, $\eta_{mw}(200mTorr)=1.00$. Higher silanes were not observed in the mw plasmas at either pressure. The rf plasma, however, produced detectable Si_2H_6 at both pressures. For comparison, $\approx 0.2\%$ Si_2H_6 gas phase abundance has been reported (5) for SiH_4 plasmas at 500mTorr with approximately one tenth of the applied rf power density used here.

Infrared spectroscopy on 1 μm thick mw deposited films allows the calculation (6) of the ratio of hydrogen bonded as polyhydride to that bonded as monohydride, R, as a function of system pressure shown in Fig. 3. This same trend to increasing R with increasing operating pressure has been previously noted in rf deposited films.(6) Degradation in mw film's electronic quality, due to increased mid gap defect density accompanying the increased polyhydride

fraction, is seen in Fig. 4. Here, the co-planar dark conductivity and photoconductivity, measured at AM-1 illumination, are obtained after heat drying samples at 200°C in vacuum.

4. DISCUSSION

The degradation of high gas utilization, high deposition rate rf deposited films due to increased plasma polymerization is well documented.(7) Although mw films also appear to degrade for the same reason when deposited at higher pressures, high deposition rates and gas utilization can be achieved with mw plasmas at low pressure. The plasma chemical reactions,(8-10) which produce depositing species in rf glow discharge plasmas, are still the subject of ongoing research. Evidence indicates that the principal depositing species are neutral free radicals.(9,11) Since no variation of deposition rate is observed when a 100V bias voltage is applied between substrate and electrical probes in the mw system, we can conclude that, here too, neutral free radicals are the primary film forming species; however, the higher values of kT_e in the mw plasma permit the generation of species not found in rf plasmas. The identity of these new species and their influence on the properties of mw produced high deposition rate films is the subject of continuing research.

The authors wish to acknowledge the expert help of L.R. Peedin in sample preparation and M. Lipton and E. Norman for their assistance in preparation of the manuscript.

References

1. S.J. Hudgens and A.G. Johncock, Proc. MRS Spring Meeting (San Francisco, 1985), to be published.
2. E.R. Mosburg, R.C. Kerns and J.R. Abelson, J. Appl. Phys. 54 (1983) 4916.
3. F.F. Chen, in Plasma Diagnostic Techniques, edited by R.H. Huddlestone and S.L. Leonard (Academic Press, New York, 1965) Chap. 4.
4. C.M. Ferreira and J. Loureiro, J. Phys. D 17 (1984) 1175.
5. P.E. Vanier, F.J. Kampas, R.R. Corderman and G. Rajeswaran, J. Appl. Phys. 56 (1984) 1812.
6. R.C. Ross, I.S.T. Tsong, R. Messier, W.A. Lanford and C. Burman, J. Vac. Sci. Technol., 20 (1982) 406.
7. R.C. Ross and J. Jaklik Jr., J. Appl. Phys. 55 (1984) 3785.

8. J.P.M. Schmitt, J. Non-Cryst. Solids 59/60 (1983) 649.

9. G. Turban. Y. Catherine and B. Grolleau, Plasma Chemistry and Plasma Processing, 2 (1982) 61.

10. F. Kampas, Semiconductors and Semimetals, 21A (1984) 153.

11. S.R. Ovshinsky, Proc. Int'l. Ion Engineering Congress., Kyoto, Japan (1983) 817.

BASIC ANTICRYSTALLINE CHEMICAL BONDING CONFIGURATIONS AND THEIR STRUCTURAL AND PHYSICAL IMPLICATIONS

S.R. Ovshinsky

Energy Conversion Devices, Inc., 1675 West Maple Road, Troy, Michigan 48084

There are many dogmas which have been uncritically accepted in the amorphous field that are basically unsound, the chief one being that the short-range order surrounding each atom is the same as in the "corresponding" crystal and that all that is different is some fluctuations in bond lengths and angles. We propose to show that short-range order is not conserved in amorphous materials, since space is related to energy in a basically different manner. The consequences of this lead to entirely new local structures than are found in crystalline materials and to the fact that many of these structures have no analogs in crystals and are inherently anticrystalline in nature.

Arguments as to the similarities of x-ray diffraction and other structural measurements in amorphous and crystalline materials are based upon average nearest neighbor relationships, and disorder is ordinarily invoked to explain the discrepancies that exist; however, it is the local (several Å) relationships in three dimensions which are not susceptible to the present analyses, in which the uniqueness of amorphous materials is exhibited. For example, while chalcogenide atoms are primarily divalently bonded and normal structural bonding (NSB) (1) of silicon atoms is mainly tetrahedral, it is the Rosetta Stone of understanding amorphous materials that it is the deviant electronic configurations (DECs) (2) that control the important electronic transport properties.

The magnitude of the mobility gap (2) is affected by chemical composition, its sharpness by compositional disorder and the deposition techniques, and it is manipulatable by the bonding energies of the atoms involved. The NSBs determine the optical band gap, but the transport properties are determined by the states within the gap. How often has it been asked by workers in th field, "Why does hydrogenated amorphous germanium (or tin) have a higher density of gap states than hydrogenated amorphous silicon"? When I have suggested that it is due to lack of tetrahedralness, the answer has been "But we have checked, and 95% of the atoms are in tetrahedral relationships." It is the non-tetrahedral configurations which control the transport properties. The fact is that less than 0.1% of non-tetrahedral sites are all that is necessary to affect the density of gap states drastically, while well over 1% is needed to be detected in structural experiments.

The orbital interactions provided by the structural links of alloying elements take place in three-dimensional space and the changes in the environment produced by such interactions have significance. The total interactive environment (TIE) (4) must be taken into account to understand the special properties of amorphous solids for it reflects the redistribution of charges and relaxations as alloying takes place. In fact, the possibility of perturbing and altering a local environment is one of the great opportunities available to designers of amorphous materials.

Through the use of a stereochemical approach, we can explain some of the seeming mysteries associated with our field. To understand the various mechanisms involved, we emphasize that space is utilized in a unique manner in amorphous materials by noting that in a crystal, formation of the lattice permits virtually no degrees of freedom to the atoms. Once the first atom is in place and the orientation set by the next two, there are no remaining degrees of freedom and all the remaining atoms have predetermined positions. We can say that the space surrounding each atom is passive. In contrast, in an amorphous material, the environment is crucial, for in every direction of available space there are electrical (i.e., chemical) forces which affect the bonding options, and, therefore, the shape and occupied space of the ensuing molecular configuration. There are many degrees of freedom for an atom as it is incorporated into an amorphous material. Since these degrees of freedom are used up as bonds are formed, we can describe the environment as active. Unlike crystals, the local environment is never exactly the same and the atom making the choice not only adjusts itself to the material previously formed, but the atoms nearby also relax, thus adding an extra degree of dimensionality that can be looked at as a fourth dimension.(5,6). If all of the orbitals available are close in energy, it is the modulation, that is, the additional or subtractional energy of the fluctuating surrounding charges of the environment, which is decisive in determining the bonding

---

Reprinted by permission from *Journal of Non-Crystalline Solids*, Vol. 75, pp. 161–168 (1985).

choice. Understanding this permits us to emphasize that the total interactive environment (TIE) is an important parameter both in the understanding and in the design of new materials.

since there are many degrees of freedom for an atom in amorphous materials, atoms which present bonding options are of fundamental importance. This is because they utilize space to develop molecular complexity so that many new configurations are generated and the ensuing convoluted bonding/non-bonding arrangements are essentially anticrystalline in nature. Such structures are important for the special electron, phonon and chemical interactions that are the hallmarks of amorphous materials, and as such we can distinguish them from crystals by their different quantum interactions. For example, the phonon interactions with other quanta in amorphous materials are basically different than in crystals. I have been advocating this point of view since the early 1960's when I called one of my first threshold switches a Quantrol. The varying complex irregular structures in three-dimensional space suggest the potential for adapting fractal theory as a means of defining the nature of seemingly random disorder.(6)

In a crystalline solid, all bond angles are predetermined, spatial orientations are rigidly defined, and bond strengths are known. The very essence of crystallinity is an almost absolute determination in space of each atom. In an amorphous solid, with such constraints removed, we can think of the space not used as energy available. Crystalline physics is akin to a flatland, that is, of lower dimensionality, while amorphous materials are three-dimensional and even above. Multiparticle theory must be developed to take into account the new geometric interactions of carriers. Just as there is a type of dimensionality associated with energy, the three-dimensional options where all angles are available for orbital interactions add an extra dimensionality which provides for electronic and chemical actions simply not available in crystalline solids. Unlike a crystal, the freedom that remains after local bonding is optimized still has potential energy in the sense that until the entire three-dimensional space is utilized, the exact free energy is still not completely determined. In other words, amorphous materials are multiequilibrium materials. Until all the three-dimensional space is utilized, relaxations can take place so that it is possible to reduce the free energy further, creating configurations separated in space by energy barriers.

Filling space in an amorphous solid is not a random packing of particles, but on an atomic basis there is a predetermined force field in all directions where bonding choices are made, contingent upon the chemical and electrical constituents of this field. If one puts oneself in the place of an atom and looks around in all directions, one finds that the choice of nearest neighbors is dictated by the chemistry of the available atoms, coupling spatial arrangement and energy. (In the following paper,(7) we explain the specific chemical bonding approach which is most helpful in designing new materials.) One also finds that one would move in response to any perturbation in the force field. This means that excitation in amorphous solids can redistribute charges in a local

area and the conformational change can either relax back to its original position as in a unistable material or result in configurational change as in a bistable (i.e., memory) material. Such changes can either be of a subtle nature or be an amorphous-to-crystalline transformation.(8,9) Depending upon the material design, they can also be reversible. Photostructural changes are examples of the reversibility which depends upon the elasticity of the structures involved and are directly related to the coordination number;(10) for example, divalent materials are more reversible than tetrahedral materials, in which the lack of flexibility, that is, the stiffness of structure and the strength of the new configurational bonds, provides energy barriers to reversible changes.

The question of coordination is an important one. The NSBs of tellurium and selenium are divalent and the DECs utilize the lone pairs.(10,11) The intertwined, chain-like structures of tellurium can easily crystallize at $10^{\circ}C$, where the coordination then goes to trivalency; however, if arsenic is added, even in small amounts, an anticrystalline bonding configuration takes place and more three-dimensionality is present.(5,6)

Elemental amorphous silicon has bonding predispositions toward tetrahedralness which are perturbed by the positional displacement of the silicon atoms away from each other. Complete tetra-hedralness costs too much in strain energy (12,13) and, in fact, if sufficient heat is applied, crystallization occurs and the material becomes four-fold coordinated.

Elemental amorphous silicon has over $10^{19} cm^{-3}$ states in the gap because of its lack of bonding completion; therefore, the material has limited use. If we make an alloy of silicon with hydrogen and preferably also fluorine, then tetrahedralness can be completed without crystallization. To do so, we must decrease the overall coordination of the material through the use of low-coordination in local areas where the coordination goes from two or three to four. The silicon-hydrogen and/or silicon-fluorine bonding is another example of anticrystallinity. While this anticrystallinity would appear to be compositional rather than positional, the new configurations are forms of chemically caused spatial displacement in which the steric barriers of translational geometry inhibit crystallization.

As can be seen from the above, the coordination in an amorphous material inevitably fluctuates locally. This in itself becomes the basis for new structures and new geometries which are anticrystal-line in nature and which break up tendencies toward periodicity. This is the reason why boron, carbon, transition metals, and elements with lone pairs, such as chalcogens, are basic elements for anticrys-talline bonding, since they generate fluctuating coordination in three dimensions which disturbs lattice building and epitaxial replication. For example, boron with its three-center bonding possibilities, carbon with its multiorbital choices, the multichoices of d-orbitals of the transition metals, and the flexibility in space or new bonding configurations of, for example, tellurium, permit bonding options, especially in multielemental materials, which result in anticrystalline structures.

Unlike in crystalline lattices, one can accommodate and combine many different elements in amorphous materials. Three-element tetrahedral materials such as silicon-hydrogen-fluorine alloys are common, as are five and six-element chalcogenide alloys. It is not merely that the atoms interfere with each other's lattice-building tendencies by mechanically displacing atoms that would "match" each other, but the distortion and mismatch are frozen or fixed in space by new chemical connections in which the laws of periodicity cannot function.

What are the relationships of structure and function? First of all, small amounts of hydrogen and fluorine can cap dangling bonds. They can do so in crystalline materials as well; however, in amorphous materials, they also act structurally in that they generate new anticrystalline geometries for the hydrogen and fluorine act stereochemically to make new configurations intrinsically different from that of the silicon-silicon bond. These configurations act as steric hindrances, both in the strength of their chemical bonds and the new shapes they assume while relaxing from the plasma into the solid. They are distributed throughout the material and provide energy barriers to crystallization. The corresponding local structural changes are correlated and reflected in the electronic properties of the amorphous alloy. The sharpness of the mobility gap is also affected. The overall gap changes from 1.1 eV in the crystal to 1.5 eV and over in the amorphous material with the order of $10^{19} cm^{-3}$ states in the gap in the pure solid. In the alloys, the gap state concentration can be reduced by many orders of magnitude. These changes are of exceptional importance in device design. Such alloy materials should not be designated as "amorphous silicon."

The ability of an amorphous material to utilize its space in an adjustable manner different from a crystal is tied to its flexibility which is directly connected to the coordination number of the atoms involved. An increase of coordination adds stiffness to the material and the use of space is constrained as exemplified by elemental amorphous silicon. As we alter the coordination in amorphous silicon by alloying, we relieve strains, particularly through the use of fluorine. In pnictide and chalcogenide materials, we utilize the three-dimensionality that can be designed into such materials to stiffen the system by the use of bridges and crosslinks.

Three-dimensionality is thus intimately connected with flexibility. Chain- and ring-like materials such as chalcogens provide a whole spectrum of orbital angles that are positioned in various directions by the bends and folds of the structures so that they intersect in new orbital relationships, producing a TIE not found in crystalline materials.(4,10) Here also, the crosslinking atoms, i.e., the alloying atoms, alter the overall coordination of the material by increasing it, yet locally there are some states of decreased coordination. The use of bridges and crosslinks to lock in amorphicity performs the same role in altering both the layered materials of Group V and the tetrahedral materials such as silicon.

We can understand the principles of how structure and electronic compensation are connected by starting with an elemental material and then seeing how fluctuating coordination can provide the mechanism to accomplish compensation and also inhibit crystallization. Chalcogenide materials are compensated thermodynamically by the interactions of the lone-pair orbitals. Uncompensated orbitals, such as are prevalent in elemental amorphous silicon, can be compensated only by additional elements capping them and/or adding the necessary structural relaxations so that bond pairing can take place. In order to understand this basic factor of compensation, we review some fundamental chemical considerations. From the description we are about to give, we can see why and how amorphous materials are truly synthetic in nature.

If we were considering only single elemental materials, then we would note that for atoms in Group IV, all solids but lead ordinarily bond via strong sp^3 interactions. These hybrids give the maximum possible number of bonds per atom (without using d or f electrons), four, and usually form rigid three-dimensional structures. The Group V atoms have an s lone pair and bond p^3, primarily right-angle bonding (with some sp angular widening). This leads to either puckered layers or ribbon-like structures without the three-dimensional rigidity of the Group VI solids. The Group VI elements are unique in possessing an outer electron p lone pair, yet being able to bond p^2 and form solids. Since this type of bonding usually yields only two bonds per atom, these solids consist of either chain-like or ring-like structures, giving a maximum of one-dimensional rigidity. Group VII elements each have two p lone pairs, but usually can form only one bond, yielding diatomic molecules; Group VIII atoms ordinarily cannot bond at all.

It would be simple enough, therefore, to draw the conclusion that Group VI solids form helixes and chains, Group V solids form layered structures, and Group IV solids form three-dimensional structures. To synthesize the most interesting materials (for that is the beauty of breaking the symmetry and eliminating the lattice), the preferred materials are those that are not made of single elements but have atoms from other Groups added; therefore, if we add crosslinks such as germanium or arsenic from Groups IV or V to Group VI solids, we transform the material into a fully three-dimensional structure. If we bridge the layers of Group V solids with the elements from either IV or VI, we can transform the sheet-like structures into a three-dimensional solid. Since a Group IV element such as silicon cannot fully complete its four orbitals without introducing large strains and dangling bonds, we must add the previously discussed bridging, crosslinking, and chain-relieving elements from Group VII, like fluorine, or Group I, like hydrogen, to allow a more complete three-dimensional structure to be generated.(14)

As an example of how space-energy relationships are reflected in anticrystalline geometries in an amorphous solid, we consider the previously mentioned "problem" of making good quality intrinsic amorphous materials of germanium and tin. While normally tetrahedral, in the amorphous phase they have an increased tendency toward lower coordination bonding due to their "Sedgwick" pairs.(5,6) I have pointed out that this results in these elements being analogous to tellurium by virtue of their greater susceptibility to divalency, their flexibility and the use of their lone pairs for fluctuating bonding purposes.(5,6) To rectify this problem, we use fluorine in the plasma, on the surface and in the solid to chemically force tetrahedralness. We have used this to advantage with amorphous carbon as well.(5,6)

So far, we have discussed materials which are generally inhomogeneous in local bonding (~ 2Å) but homogeneous in overall structure (> 10Å). In elemental amorphous materials, I have called these materials structural alloys, since the single element provides different structures in the same material and its properties are affected accordingly. Through the years, we have developed concepts and techniques to deliberately make multilayered, thin film structures in which we can control the composition of a material so that it is either modulated or heterogeneous. This is made possible only by discarding the principles that have been sacred to crystalline materials, that is, equilibrium configurations, stoichiometry, lattice and symmetry rules, etc. In compositionally modulated and multilayered materials, we move from atoms, orbitals and molecular configurations to domains and clusters, but the principles are still the same. The domains and clusters have their own special geometries. They can be viewed as being a "macroscopic" DEC configuration rather than "microscopic" atoms and molecules, but still they share the same characteristics of presenting unique orientations, orbital bonding and active sites for interaction with their surrounding environment. I believe that this concept will allow us to utilize frontier orbital theory (15) in amorphous materials. On a structural basis, macrochemistries and geometries operate on the same principles that we have discussed herein, that is, such fluctuations, whether in layered or compositional form, interfere and prevent repetition of like atomic configurations from taking place. By definition, if there is no lattice, there is no lattice mismatch.

From this discussion, we can extract some basic rules of amorphicity.

1. The first rule: the basic chemical elements, molecules and clusters can generate topologies which are inherently noncrystalline when they offer multiorbital choices or bonding options to their nearest neighbors. Convoluted three-dimensional geometries prevent percolation paths which would link atoms together to form a lattice.

2. The second rule: the physical displacement of atoms, positionally, translationally and compositionally, is best achieved by fixing the spatial relationships through homopolar and heteropolar chemical bonding in three-dimensional space. This is the basis for steric hindrances involved in bonding and crosslinking.

3. The third rule: space not utilized provides an effective extra dimensionality which can be reflected in various relaxation modes and multiequilibrium structures.

4. The fourth rule: a total interactive environment (TIE) has to be taken into account in understanding both the normal structural bonds (NSBs) and the defect structures, that is, the deviant electronic configurations (DECs), in these materials.

5. The fifth rule: the bonding choices and bonding reconstructions, while inherently chemical in nature, are strongly topologically dependent. This is related to the three-dimensional spatial freedom in amorphous solids.

6. The sixth rule: the relaxation processes and phonon interactions are so basically different that we consider amorphous materials as having different quantum interactions than do crystalline materials.

7. The seventh rule: the fact that all of the above considerations are reflected in the electronic activity is the Rosetta Stone of amorphous materials, which can be stated: deviations from local order, that is, the DECs and TIEs, control the transport properties of the materials just as perturbations of periodicity control those of crystalline materials.

Just as there could not have been a crystalline transistor without the understanding of the importance of perturbations of periodicity, there can be no important amorphous devices without utilizing the last rule as a design tool.

In this short paper, we have provided the basic approach which has allowed us to make a working science out of what appeared to be an "amorphous" field.

ACKNOWLEDGMENTS

I want to thank David Adler for his helpful suggestions.

REFERENCES

1. S.R. Ovshinsky, "Chemical Modification of Amorphous Chalcogenides," Proc. of the 7th Intl. Conf. on Amorphous and Liquid Semiconductors, Edinburgh, Scotland,(1977) p. 73.

2. S.R. Ovshinsky and D. Adler, "Local Structure, Bonding, and Electronic Properties of Covalent Amorphous Semiconductors," Contemp. Phys. 19, 109 (1978).

3. M.H. Cohen, H. Fritzsche and S.R. Ovshinsky, "Simple Band Model for Amorphous Semiconducting Alloys," Phys. Rev. Lett. 22, 1065 (1969). See also N.F. Mott, Adv. Phys. 16, 49 (1967).

4. S.R. Ovshinsky, "The Chemical Basis of Amorphicity: Structure and Function," Revue Roumaine de Physique 26, 893 (1981). (Grigorovici Festschrift.)

5. S.R. Ovshinsy, "Fundamentals of Amorphous Materials," in Physical Properties of Amorphous Materials, eds. D. Adler, B.S. Schwartz and M.C. Steele, Institute for Amorphous Studies Series, vol. 1, Plenum Press, New York (1985) p. 105.

6. D. Adler, ed., Disordered Materials: Science and Technology, Selected Papers by S.R. Ovshinsky, Bloomfield Hills, Michigan, Amorphous Institute Press (1982).

7. S.R. Ovshinsky, "Chemistry and Structure in Amorphous Materials: The Shapes of Things to Come." in Physics of Disordered Materials, ed. D. Adler, H. Fritzsche and S.R. Ovshinsky, Institute for Amorphous Studies Series, Plenum Press, New York (1985) p. 37.

8. J. Bicerano and S.R. Ovshinsky, "Chemical Bond Approach to Glass Structure," J. Non-Cryst. Solids, in press.

9. S.R. Ovshinsky, "Reversible Electrical Switching Phenomena in Disordered Structures," Phys. Rev. Lett. 21, 1450 (1968).

10. S.R. Ovshinsky, "Electronic and Structural Changes in Amorphous Materials as a Means of Information Storage and Imaging," Proc. 4th Intl. Congress for Reprography and Information, Hanover, Germany 109 (1975).

11. S.R. Ovshinsky and H. Fritzsche, "Amorphous Semiconductors for Switching, Memory and Imaging Applications," IEEE Trans. Electron Devices ED-20, 91 (1973).

12. S.R. Ovshinsky and K. Sapru, "Three-Dimensional Model of Structure and Electronic Properties of Chalcogenide Glasses," Proc. 5th Intl. Conf. on Amorphous and Liquid Semiconductors, Garmisch-Partenkirchen, Germany 1973, ed. J. Stuke and W. Brenig, Taylor and Francis, London (1974).

13. M. Kastner, D. Adler and H. Fritzsche, "Valence-Alternation Model for Localized Gap States in Lone-Pair Semiconductors," Phys. Rev. Lett. 37, 1504 (1976).

14. S.R. Ovshinsky, "Amorphous Materials As Interactive Systems," Proc. 6th Intl. Conf. on Amorphous and Liquid Semiconductors, Leningrad, 1975: Structure and Properties of Non-Crystalline Semiconductors, ed. B. T. Kolomiets, Nauka, Leningrad, 426 (1976).

15. H. Fritzsche, "Summary Remarks," Proc. 6th Intl. Conf. on Amorphous and Liquid Semiconductors, Leningrad, 1975: Structure and Properties of Non-Crystalline Semiconductors, ed. B. T. Kolomiets, Nauka, Leningrad, 65 (1976).

16. S.R. Ovshinsky, "The Shape of Disorder," J. Non-Cryst. Solids 32, 17 (1979). (Mott Festschrift.)

17. K. Fukui, "Role of Frontier Orbitals in Chemical Reactions," (Nobel Lecture), Science 218, 747 (1982).

SUPERCONDUCTIVITY AT 155 K

S.R. Ovshinsky, R.T. Young, D.D. Allred, G. DeMaggio, and G.A. Van der Leeden

Energy Conversion Devices, Inc., Troy, Michigan 48084

(Received 22 May 1987)

The accomplishment of high-temperature superconductivity is of immense scientific and technological importance. Several critical transition-temperature barriers have recently been breached since the long-standing record temperature of 23.2 K for Nb_3Ge was exceeded. The most important milestones were the announcement of $T_c \approx 30$ K in lanthanum barium copper oxide by Bednorz and Müller,[1] whose work was based upon materials developed by Michel and Raveau,[2] and the work of Chu, Wu, and others,[3] based upon the replacement of lanthanum by yttrium, which resulted in superconductivity at temperatures of approximately 95 K.

Indirect measurement techniques[4,5] have been used[6] previously to infer the existence of regions of high-T_c phases in Y-Ba-Cu-O systems. We report here for the first time the direct measurement of a zero-resistance superconducting state at 155 K. This result was observed in a new chemical system, Y-Ba-Cu-F-O.

In the quest for even higher transition temperatures, variations and replacements in the metallic portion of the compounds have not been fruitful. It was found that yttrium could be replaced by most of the rare-earth metals with achievement of approximately the same T_c.[7] Our approach has been to synthesize a new five-element compound, which has resulted in the achievement of considerably higher T_c's. In one sample, the transition to zero resistance was observed at temperatures as high as 155 K. In another sample, an abnormally rapid decline of resistivity was observed starting at room temperature, which suggests the existence of phases exhibiting superconducting onset above room temperature. We have also observed anomalies in the magnetic properties, as well as weak flux trapping below 260 K. This material contains at least four presently identified structural phases. Work is continuing to identify unambiguously the structure of the high-T_c phases. In previous papers[8] we have discussed the significant role that fluorine plays in affecting

the electronic and structural properties in other multielemental materials, particularly in affecting orbital interactions. These factors[9] motivated us to synthesize the fluorinated Y-Ba-Cu-O compounds reported here.

Samples with nominal compositions $Y_1Ba_2Cu_3F_xO_y$ ($x=0$, 1, 2, 3, and 4) were prepared from two master compositions, $Y_1Ba_2Cu_3O_{6.5}$ and $Y_1Ba_2Cu_3F_4O_{4.5}$, which define the compositional extremes ($x=0$ and $x=4$). The starting reagents used to prepare the oxide and fluoroxide were (Y_2O_3, $BaCO_3$, and CuO) and (Y_2O_3, BaF_2, and CuO), respectively. Each master composition was prepared by a mixing of the sieved reagent powders, and grinding and firing of them in air in a Pt crucible at 950 °C for 8 h. After the initial firing the master compositions were reground. Samples were prepared from mixtures of the two master compositions which were pressed into $\frac{1}{2}$-in. pellets and sintered at 950 °C in flowing O_2 for 48 h, and then cooled to 200 °C over a period of 6 h. Samples were examined by x-ray powder diffraction and microprobe to determine their structure, phases, and composition. Table I summarizes the various phases and compositions which have been identified.

Electrical resistance was measured by means of a standard four-probe method on samples with silver-paint contacts. Rectangular bars $10 \times 2 \times 1$ mm^3 were cut from pellets for the measurements. The applied constant current ranged from 100 μA to 10 mA depending on the sample resistance. The dc magnetic susceptibility and flux trapping were studied with a SQUID magnetometer.

Of the five nominal compositions studied, only those with $x=0$, 1, and 2 show superconductivity, as measured by resistivity and magnetic susceptibility measurements. The room-temperature resistivity of samples with $x=3$ and $x=4$ is greater than 20 MΩ. The samples with $x=0$ show resistance-temperature behavior similar to that reported by others[7] for typical $YBa_2Cu_3O_{7-\Delta}$, with a modest decrease of resistivity from room temperature

Reprinted by permission from *Physical Review Letters*, Vol. 58, No. 24, pp. 2579–2581 (15 June 1987).

TABLE I. Various phases and compositions of $Y_1Ba_2Cu_3F_xO_y$ from x-ray diffraction and microprobe analysis.

x	$Y_1Ba_2Cu_3O_{7-\Delta}$	$Y_1Ba_2Cu_3O_y$:F	$Y_2Ba_1Cu_1O_y$:F	$Y_1Cu_2O_y$:F	BaF_2	CuO
$x=0$	×					
$x=1$		×	×		×	×
$x=2$		×	×		×	×
$x=3$			×	×	×	×
$x=4$				×	×	×

to the onset temperature of 95 K, reaching the zero-resistance state at 90–92 K. The $x=1$ sample showed a behavior essentially identical to that of the $x=0$ sample. Dramatically different and complex behavior was observed, however, in samples with $x=2$.

The temperature dependence of the resistance of a portion of the $x=2$ pellet is shown in Fig. 1. During the initial cooling, the sample (sample 1) completely lost its resistance at ≈ 168 K. After warming and measurement during second cooling, an increase of resistance was observed and the zero-resistance state was not reached until 148 K. An appreciable change in resistance was again observed during the second warming. The zero-resistance state remained (within our instrument noise level of $\approx 10^{-8}$ V) until 155 K. Several higher-temperature resistance transitions were also observed. The resistance anomalies seen upon warming and cooling are remarkable and may be connected with filamentary conduction.

Figure 2 shows a plot of the temperature dependence of the average resistivity for another sample (sample 2) with the $x=2$ composition. The average resistivity calculated with the assumption of uniform current density falls dramatically with temperature for this sample beginning at room temperature or above, and achieves resistivity values 5 times lower than that of single-crystal copper before the zero-resistance state is achieved at ≈ 91 K. We were able to fit the resistivity to a good approximation over the entire range by using a $T^{8.3}$ law. Since the resistivity became so low with decreasing temperature, a large current of 10 mA, or a current density of 0.5 A/cm^2, had to be applied during measurement. With the assumed multiphase filamentary type of conduction, this applied current density might have washed out more sharply defined high-T_c transitions. For comparison, the resistivity-temperature plot of pure copper[10]

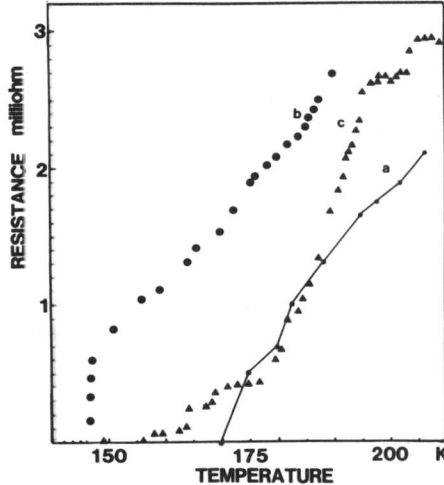

FIG. 1. Resistance vs temperature measured at constant current of 1 mA for sample 1 with nominal composition $YBa_2Cu_3F_2O_y$. Curve a shows the resistance upon initial cooling (line drawn to guide the eyes); curve b, data obtained upon second cooling; and curve c, data from warming after second cooling. In the superconducting state, voltage is less than 10^{-8} V, the noise level of the voltmeter. Sample dimensions are area $=1\times2$ mm^2, length $=10$ mm. This gives an average resistivity estimate of less than 2×10^{-7} Ω cm in the superconducting state.

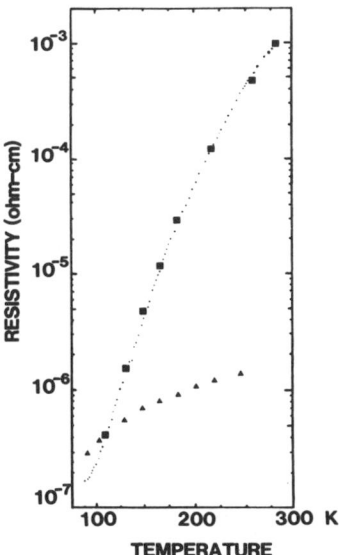

FIG. 2. The logarithm of the average resistivity of sample 2 of $YBa_2Cu_3F_2O_y$. Average resistivity was calculated on the assumption of uniform current density. Resistivity was found to follow T^n, where $n \approx 8.3$; these points are indicated by squares. The ideal resistivity of pure copper (Ref. 10) is also plotted (triangles).

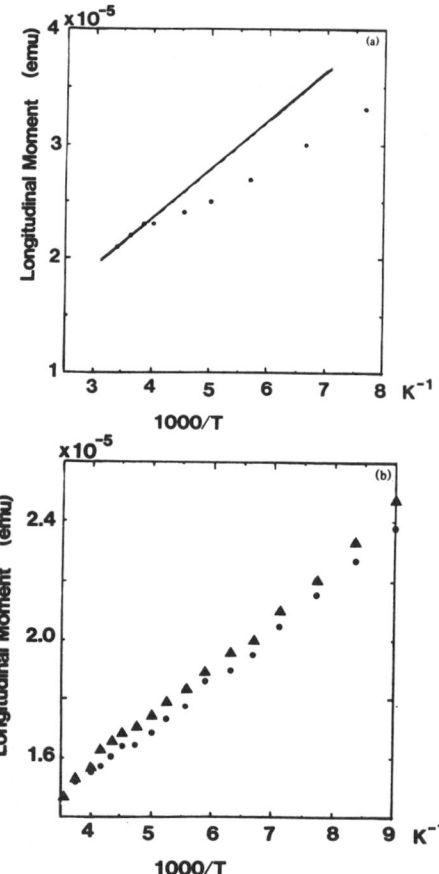

FIG. 3. Longitudinal moment vs $1000/T$ for $YBa_2Cu_3F_2O_y$. (a) Measurements made with use of 100-G field. Note the diamagnetic deviation from a Curie-Weiss law below 250 K. (b) Weak magnetic hysteresis at 20 G. Data for warming after zero-field cooling are indicated by circles and data from cooling with field applied are indicated by triangles.

is also included in Fig. 2. The actual resistivity of the current-carrying phases will certainly be much less than the calculated average. The fact that the average resistivity of the multiphase ceramic material at a temperature of 91 K is 5 times lower than copper provides another indication of the existence of higher-temperature superconducting phases.

The magnetic measurements in Fig. 3 suggest that only a very small volume fraction of the sample is superconducting at these high temperatures. Figure 3(a) plots the magnetic moment as a function of temperature in a magnetic field of 100 G. The diamagnetic deviation from the Curie-Weiss law at temperatures below 250 K is an indication of the Meissner effect in this small volume fraction. A large diamagnetic response below 90 K is also observed but not plotted here. Weak magnetic hysteresis is observed as shown in Fig. 3(b), where the data on warming after zero-field cooling are indicated by circles and the data on cooling with field applied, by triangles. The facts that (1) the data of warming and cooling measurements coincide from room temperature until 260 K and deviate below this temperature and (2) the paramagnetism is stronger in the cooling process provide additional verification of the existence of high-temperature superconducting phases.

In summary, a zero-resistance state at 155 K has been observed in the fluorinated Y-Ba-Cu-O system and phases with even higher T_c have also been observed. The abnormally rapid decline of resistivity starting at room temperature suggests the existence of phases exhibiting superconducting onset above room temperature. The high-temperature phases are in the process of being identified. In the samples reported here, the volume fraction of the high-T_c materia. is very small as indicated by the magnetic data. Since a zero-resistance state requires a percolation path, the small-volume-fraction superconducting phase appears to be in a filamentary form. Work continues both to identify clearly the structure of the high-T_c phase or phases and to optimize materials synthesis to increase their yield.

We wish to thank Dr. S. J. Hudgens, Professor J. T. Chen, Dr. E. Teller, and G. Wicker for stimulating discussions on various aspects of this work. We would also like to thank B. Chao, L. Contardi, and D. Pawlik for structure analysis and G. Fournier for technical assistance.

[1]J. G. Bednorz and K. A. Müller, Z. Phys. B **64**, 189 (1986).

[2]C. Michel and B. Raveau, J. Solid State Chem. **43**, 73 (1982); J. Provost, F. Studer, C. Michel, and B. Raveau, Synth. Met. **4**, 157 (1981).

[3]M. K. Wu, J. R. Ashburn, C. J. Tong, P. H. Hor, R. L. Wong, L. Gao, Z. J. Huang, Y. Q. Wang, and C. W. Chu, Phys. Rev. Lett. **58**, 908 (1987); P. H. Hor, L. Gao, R. L. Meng, Z. J. Huang, Y. O. Wang, K. Forster, J. Vassiliow, and C. W. Chu, Phys. Rev. Lett. **58**, 911 (1987).

[4]A. M. Saxena, J. E. Crow, and M. Strongin, Solid State Commun. **14**, 799 (1974).

[5]H. Sadate-Akhavi, J. T. Chen, A. M. Kadin, J. E. Keem, and S. R. Ovshinsky, Solid State Commun. **50**, 975 (1984).

[6]J. T. Chen, L. E. Wenger, C. J. McEwan, and E. M. Logothetis, Phys. Rev. Lett. **58**, 1972 (1987).

[7]See, for example, A. R. Moodenbaugh, M. Suenaga, T. Asano, R. N. Shelton, H. C. Ku, R. W. McCallum, and P. Klavins, Phys. Rev. Lett. **58**, 1885 (1987); P. H. Hor, R. L. Meng, Y. Q. Wang, L. Gao, Z. J. Huang, J. Bechtold, K. Forster, and C. W. Chu, Phys. Rev. Lett. **58**, 1891 (1987).

[8]S. R. Ovshinsky, in *Physical Properties of Amorphous Materials,* edited by D. Adler, B. B. Schwartz, and M. C. Steele (Plenum, New York, 1985), and Rev. Roum. Phys. **26**, 893 (1981), and J. Phys. (Paris), Colloq. **42**, C4-1095–C4-1104 (1981), and J. Non-Cryst. Solids **32**, 17 (1979).

[9]S. R. Ovshinsky, to be published.

[10]Data taken from G. K. White and S. B. Woods, Phil. Trans. Roy. Soc. London, Ser. A **251**, 272 (1959).

SUPERCONDUCTIVITY IN FLUORINATED COPPER OXIDE CERAMICS

S.R. Ovshinsky, R.T. Young, B.S. Chao, G. Fournier, and D.A. Pawlik

Energy Conversion Devices, Inc., 1675 West Maple Road, Troy, Michigan 48084

1. BACKGROUND

It is difficult to do justice to the scientific and technological impact of the recent discoveries in high T_c superconducting ceramic oxides. It is now clear that the oxides that were the basis of the lanthanum copper oxide ceramics which moved the superconducting temperature from 23 to 30 to 40K (1) have been around for some time.(2,3)

Substitution of elements which alter the stereo-chemistry of these materials was shown by Wu, Chu and associates (4) to have an important effect on zero resistance superconductivity, moving the range to approximately 95K.

We previously reported that through the use of fluorine,(5) we have been able to dramatically increase the zero resistance temperature by 60-70K--by far the highest increase--to 155-168K. (See Fig. 1.)

This result has been confirmed by several groups, for example, in China,(6) Taiwan (7) and Sweden.(8) The most recent confirmation was by a group at North American Philips in Physical Review Letters.(9) The Philips group carefully confirmed our fluorinated findings in various samples. They were able to increase T_c from 95K to a maximum of 110K by thermal cycle treatment of the conventional materials. Only with the addition of fluorine were they able to go up to 159K. The paper also confirms our results that the increase in T_c occurs systematically and monotonically with increasing fluorine addition. The 159K temperature was achieved by exactly following the published ECD recipe with two fluorines substituted for oxygen. In fact, they duplicated the solid state reaction synthesis utilizing BaF_2 as reported by us.

We also have been able to show (Fig. 2) an abnormally rapid decline of resistivity starting at room temperature which indicates the existence of a phase exhibiting superconducting onset above room temperature. It is also important to point out that the resistivity of this sample <u>above</u> T_c is four times lower than that of copper.

Since there is a small volume fraction of such high T_c material allowing for a conducting percolation path, we have utilized magnetic measurements to probe the high temperature phases that are dispersed through the material and have observed the diamagnetic signal as well as flux trapping below 280K as shown in Fig. 3. Indeed, we have observed similar phenomena at temperatures as high as 305K.

The early samples were made in a typical solid state reaction in pressed pellet form. In order to get better control of the fluorination reaction we have since utilized the plasma technique described in the next section.

X-ray diffraction and electron microprobe analyses indicate that the samples prepared from solid state reaction are multiphasic, containing at least four phases. Three of them, i.e., $Y_2Ba_1Cu_1O_y$:F, BaF_2, and CuO, are nonsuper-conducting. The fluorine content of the supercon-

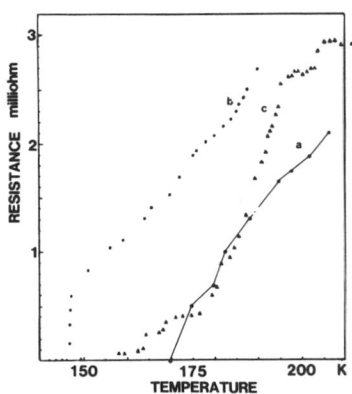

Fig. 1   Resistance vs temperature measured at constant current of 1mA for a sample with nominal composition $YBa_2Cu_3F_2O_y$. Curve a shows the resistance upon initial cooling (line drawn to guide the eyes); curve b, data obtained upon second cooling; and curve c, data from warming after second cooling.

Reprinted by permission from *Reviews of Solid State Science*, Vol. 1, No. 2, pp. 207–219 (1987).

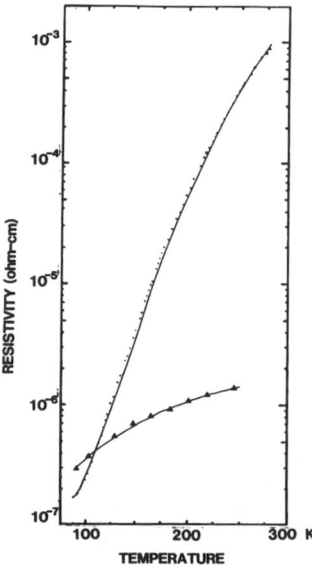

Fig. 2   The logarithm of the average of a YBCOF samples. Average resistivity was calculated on the assumption of uniform current density. Resistivity was found to follow T^n, where $n \simeq 8.3$. The ideal resistivity of pure copper is also plotted (triangles).

ducting $Y_1Ba_2Cu_3O_{7-x}$ (YBCO) phase is very small ($\simeq$ 0.2 at.%). The volume fraction of the high-temperature phase estimated from the strength of the diamagnetic signal was only about 1 part in 10^4. Because of the multiphases and because of such small volume fraction, we have not yet been able to identify the specific high T_c phases.

Various fluorides have been used in the starting reagent including BaF_2, CuF_2, NH_4HF_2. We found that regardless of what initial fluoride was used, the BaF_2 phase was always observed. Due to the strong affinity of barium to fluorine, barium will preferentially react with fluorine to form BaF_2 before the formation of YBCO. The amount of YBCO phase in the material is therefore critically dependent upon the amount of fluorine additive in the mixture. Since the majority of the fluorine reacted with the barium, the fluorine incorporation into the YBCO phase became very minimal. High fluorine content in the YBCO may only exist at the interface between YBCO and BaF_2. The role that fluorination plays with respect to surfaces, twin boundaries and/or grain boundaries still must be clarified.

In this paper, we report two low-temperature fluorination techniques with the aim of controlling the amount of fluorine incorporated into the YBCO structure without the formation of BaF_2.

II. GAS PHASE REACTION

In this approach, NH_4HF_2 was used as the fluorine source. At room temperature, NH_4HF_2 is in a solid form. It melts at $120^{\circ}C$ and subsequently decomposes to NH_3 and $2HF$.

Single phase fully reacted YBCO powder was first mixed with appropriate amounts of NH_4HF_2 to a nominal composition of $F_{0.2}$, $F_{0.5}$, and $F_{1.0}$. These mixed powders then reacted in the alumina boats in air at a temperature range of from 300 to $900^{\circ}C$. The X-ray powder data of the $F_{0.5}$ and $F_{1.0}$ samples revealed that BaF_2 forms at temperatures as low as $300^{\circ}C$ and its peak intensity increases linearly with increasing temperatures. The amount of fluorine incorporated into the YBCO powder is very low (< 1 at.%). These powders were then pressed into pellets and sintered under O_2 at $950^{\circ}C$ followed by slow cooling. All samples show typical 90K superconducting transition. However, magnetic measurements showed diamagnetic signals at high temperatures (250-300K) indicating other high temperature phases.

Since single crystal materials have little practical use, in order for technological progress to be made, reliance upon epitaxial high temperature crystalline formation must be overcome. Through the

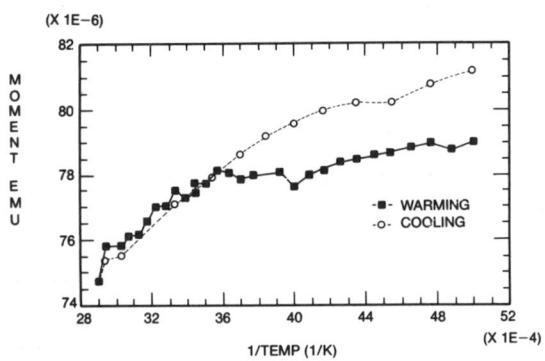

Fig. 3   Magnetic moment vs 1/T for a YBCOF sample. Measurements made with use of 40G field. Data for warming after zero-field cooling are indicated by "■" and data from cooling with field applied are indicated by "o".

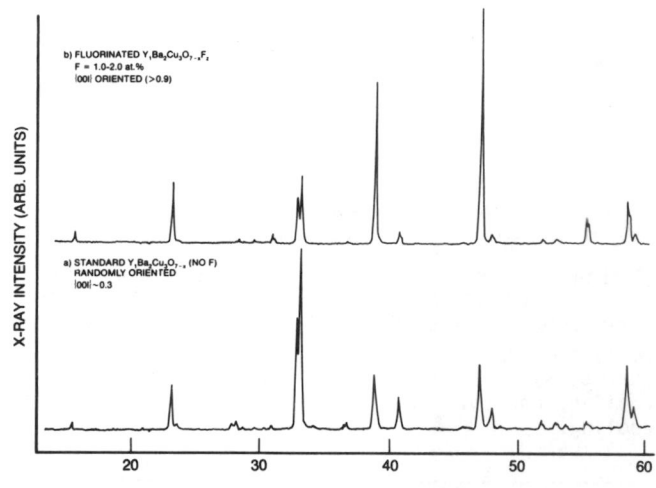

Fig. 4   X-ray diffraction showing the preferred orientation in the fluorinated material.

Region 1		Region 2		Region 3	
Y	7.1	Y	7.5	Y	6.6
Ea	14.3	Ba	15.3	Ba	15.9
Cu	21.1	Cu	22.6	Cu	22.7
O	34.6	O	46.9	O	51.9
F	22.9	F	7.7	F	2.9

Fig. 5   SEM   cross-sectional   micrograph (backscattered mode).

use of fluorine, we have been able to change the random crystalline structure of the conventional material and have been able to get preferred orientation.

With our approach, we observed a strong preferred orientation effect on these pellets, especially in the pellets with $F_{0.2}$. Figure 4 shows the comparison of the X-ray diffraction pattern of a standard YBCO pellet and a slightly fluorinated YBCO pellet. The much stronger [00ℓ] peaks in the fluorinated pellet clearly indicate such orientation effect. From the intensity ratio, we estimated that more than 90% of the crystalline grains are aligned with c-axis perpendicular to the sample surface. A dopant amount of fluorine, or perhaps HF, plays an important role in controlling the nucleation and promoting the preferred crystal growth. We anticipate that this same approach should be suitable to the growth of oriented polycrystalline films. Due to the very anisotropic nature of these materials, producing oriented microcrystalline films would be an essential step in achieving the high critical current needed for practical applications.

III. MICROWAVE PLASMA TREATMENT

The objective of this treatment is to use microwave to generate F^+ free radicals and subsequently to replace a controlled amount of oxygen in the YBCO at a temperature below the BaF_2 formation. The microwave plasma experiment is performed in a tubular reactor at a power level of 50-100W with a frequency of 2.45 GHz. The gas mixtures are Ar and NF_3. The operating pressure is $\simeq$ 20mτ. The NF_3 flow rate, sample temperature, reaction time, and microwave power level are important controlling parameters. Both orthorhombic $YBCO_7$ and tetragonal $YBCO_6$ (10) were fluorinated under similar conditions.

Figure 5 is a cross-sectional SEM micrograph of $YBCO_7$ pellet after being treated with fluorine at $400°C$ for 1 hour. Several important features can be seen from this figure:

(1) Fluorine diffuses rather fast in this material--at a given time and temperature, a penetration depth of $\simeq$ 50 um is obtained.

(2) The microprobe analysis indicates that (i) fluorine primarily substitutes for oxygen; (ii) the metal ratio remains unchanged; and (iii) no BaF_2 is observed. However, like any diffusion-controlled process, the F concentration gradient is seen.

(3) A layer type of structure is developed in the fluorinated region. Detailed microprobe analysis indicates that fluorine concentration is higher in the darker and closely layered region. We speculated that these layers are associated with twin boundaries. The fluorine may preferentially diffuse through the twin boundaries and then laterally diffuse to the grains. However, a correlation with TEM is needed before the model can be confirmed.

We have applied similar plasma treatment to treat the $YBCO_6$ pellets. The results are shown in Figure 6. The starting pellets are orthorhombic $YBCuO_{6.8}$. The sample is then converted to tetragonal with an expanded c-axis after annealing at $600°C$ for three hours in UHV (curve b). The microprobe analysis indicated that the composition

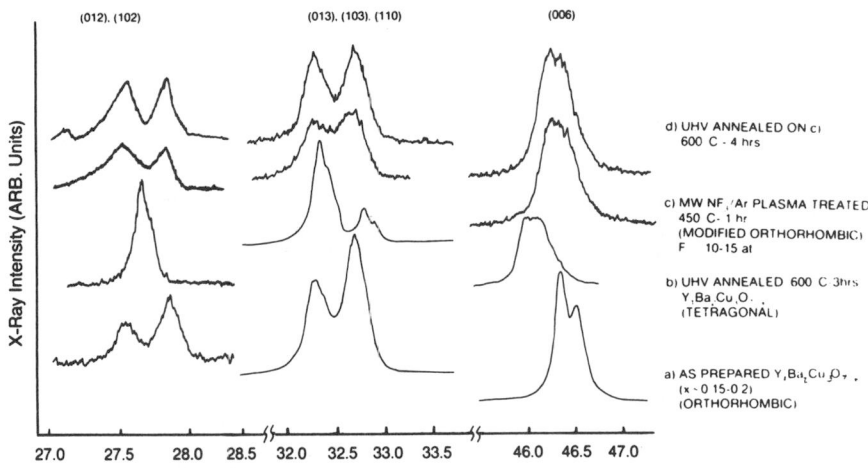

Fig. 6   X-ray diffraction showing the orthteta-orth phase transformation.

becomes $YBCO_{6.0}$. The lower Cu valence resulting from the reduction of oxygen is believed to be the cause of the expansion of the c-axis. After fluorination, the structure is converted into orthorhombic with the (006) peak shifting back to the initial position (curve c). This result suggests that the fluorine is not only incorporated into the structure but also located at the ordered Cu linear chain. Furthermore, the fluorine substitution restored the Cu valence. The microprobe data show that after fluorine treatment, the F and O contents are $\simeq$ 15 at.% and 41 at.%, respectively, which resulted in a nominal composition of $Y_1Ba_2Cu_3(OF)_{7.4}$. The 0.4 extra fluorine may occupy the ordered vacancy site. This may be the reason for the modified orthorhombic structure. The fluorinated sample was then reannealed in UHV at 600°C for 4 hours. There was no detection of fluorine evolution from the sample monitored from a quadropole mass spectometer. The X-ray diffraction data also confirm that the structure remains as the modified orthorhombic structure with no shift of the (006) peak. Furthermore, there is no indication of BaF_2 formation.

We should emphasize that the process has not yet been optimized. However, these data strongly indicate that once fluorine is bonded into the structure, it forms much stronger bonds and is much more stable than oxygen. The results also show that fluorine can be incorporated into the structure without reacting with Ba to form BaF_2. This finding is of scientific and technological importance. There has been a great fear that the oxygen diffusion problem might be the Achilles heel of high T_c superconductivity. We have shown that oxygen can be replaced by fluorine at which point neither the remaining oxygen nor the fluorine diffuses. The oxygen retention and stability problem is in principle solved. The fluorinated material is therefore qualitatively different from the conventional YBCO material.

The electrical properties of the microwave fluorinated samples can generally be classified into

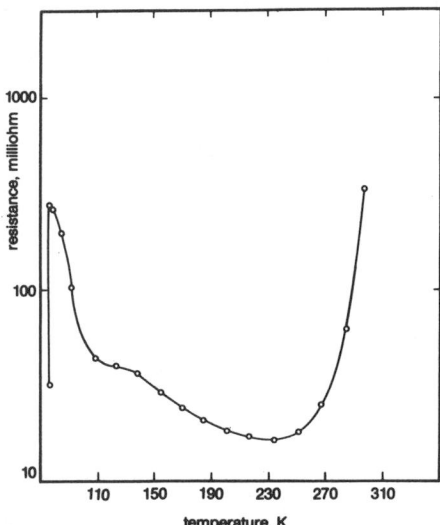

Fig. 7   Resistance vs temperature plot of an over-fluorinated sample.

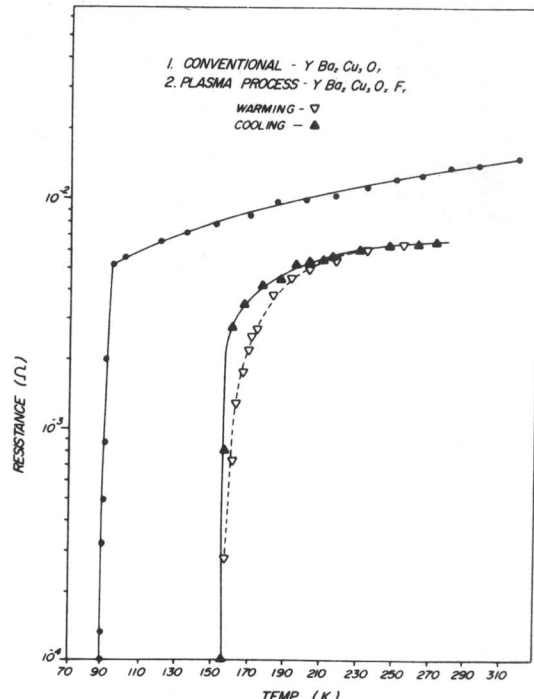

Fig. 8   Resistance vs temperature plot showing 154K zero resistance transition of a microwave treated YBaCuOF sample.

three categories. The over-fluorinated samples usually exhibit semiconductor behavior which are either superconducting at lower temperature or not superconducting at all (Fig.7). The lightly-fluorinated sample normally shows typical 90K transition. The samples which show zero resistance transition in the 155K-168K range are in a delicate balance of fluorination (Fig. 8). That such a balance can be achieved by the crude means of solid phase reactions and the more sophisticated means of plasma is of great importance.

Since microwave plasma reaction is a surface treatment, the fluorine incorporation is primarily through a diffusion controlled process. Therefore, the concentration gradient is present. Furthermore, the fluorine incorporation is very sensitive to local power fluctuation, gas depletion, temperature variation as a result of microwave heating, etc. Therefore, other methods are being developed so that the narrow window of reproducibility can be widened.

IV. SUMMARY

In summary, the superconducting zero resistance transitions at 155-168K have been observed in multiphase solid-state synthesized samples as well as low-temperature fluorine-treated samples. The amount of fluorine and its direct location have been generally but not specifically established. Due to the extremely reactive nature of fluorine, methods that will enhance reproducibility still require attention.

In addition to the very high temperature achieved, we showed that dopant amounts of fluorine promote the much-desired oriented crystal growth. We also found that the replacement of weakly-bonded oxygen by fluorine offers a solution to the material stability problem.

High T_c materials are critically dependent on the Cu-O chain, oxygen stoichiometry, ordered oxygen vacancies and the stereochemistry associated with mixed valency control.(11) The amount of fluorine substitution and its location affect all of these parameters and therefore are critical in the determination of T_c.

References

1. J.G. Bednorz and K.A. Muller, 2 Phys. B64, 289 (1986).
2. C. Michel and B. Raveau, J. Solid State Chem. 43 73, (1982); J. Provost, F. Studer, C. Michel and B. Raveau, Synth. Met. 4, 157, (1981).
3. I.S. Shaplygin, B.G. Kakhan and V.B. Lazarev, Russian Journal of Inorganic Chemistry 24, 1478 (1979).
4. M.K. Wu, J.R. Ashburn, C.J. Tong, P.H. Hor, R.L. Wong, L. Gao, Z.J. Huang, Y.Q. Wang, and C.W. Chu, Phys. Rev. Lett. 58, 908 (1987); P.H. Hor, L. Gao, R.L. Meng, Z.J. Huang, Y. Q. Wang, K. Forster, J. Vassiliow, and C.W. Chu, Phys. Rev. Lett. 58, 911 (1987).
5. S.R. Ovshinsky, R.T. Young, D.D. Allred, G. DeMaggio and G.A. Van der Leeden, Phys. Rev. Lett. 58, 2579 (1987).
6. Y.R. Meng, Y.R. Ren, M.Z. Lin, Q.Y. Tu, Z.J. Lin, L.H. Sang, W.Q. Bing, Proc. Int. Sym. on High T_c Superconductors, July 1, (1987) Beijing, China (to be published).
7. J.H. Kung, Proc. 1987 Sym. on Low Temp. Phys. Sept. 7-8, (1987) Hsin-Chu, Taiwan.
8. C. Krontiras, private communication.
9. R.N. Bhargava, S.P. Herko and W.N. Osborne, Phys. Rev. Lett. 59, 1468 (1987).
10. J.S. Swirnea, H. Steinfink, J. Mat. Res. 2, 424, (1987).
11. S.R. Ovshinsky, to be published.

A STRUCTURAL CHEMICAL MODEL FOR HIGH T_c CERAMIC SUPERCONDUCTORS

S.R. Ovshinsky,[1] S.J. Hudgens,[1] R.L. Lintvedt,[2] and D.B. Rorabacher[2]

[1]Energy Conversion Devices, Inc., 1675 West Maple Road, Troy, Michigan 48084
[2]Department of Chemistry, Wayne State University, Detroit, Michigan 48202

(Received 15 December 1987)

INTRODUCTION

During the past year a large number of models have been proposed to explain high T_c superconductivity in Cu-O ceramic compounds. A nearly equal number of novel charge carrier pairing mechanisms have been invoked in these models as alternatives to the conventional phonon mediated process described in the BCS theory. Some of these models have recently been reviewed by Rice.(1)

In the current paper we propose a model for high T_c superconductivity with a carrier pair mechanism based on antiferromagnetic spin coupling which explicitly focuses on the role of chemical bonding and the oxidation states of constituent copper atoms in the reported crystal structure of the Y-Ba-Cu-O ceramic superconductors. Although electron-pairing mechanisms which involve antiferromagnetic spin coupling have previously been discussed,(2-6) the current proposal uniquely interprets the manner in which this mechanism is manifested in the Cu-O ceramic superconductors. Specific attention is directed to $YBa_2Cu_3O_{7-x}$ as a model system. However, the proposed model has greater generality as will be discussed.

In $YBa_2Cu_3O_{7-x}$, each element plays a role--in some cases several roles--in creating conditions required for achieving high T_c superconductivity. For yttrium this role is not unique as experiments have shown that it can be substituted completely by most of the trivalent lanthanides with essentially unchanged superconducting properties.(7-10) As exceptions, substitution of Pr, Tb or Ce results in changes in T_c. However, it has been argued (11) that these elements, alone among the lanthanides, have stable tetravalent states, and that these states are likely to be present in materials prepared at high temperature in oxidizing environments. Thus, substitution of these elements in the Y-Ba-Cu-O system, unlike the other lanthanides, necessitates oxidation state changes elsewhere in the compound, similar to those produced by oxygen deficiency. The partial substitution of divalent Ba by Sr or Ca has also been shown to have little effect on T_c.(8) The small effects which have been observed could be attributed to an alteration in structural equilibrium, but might involve chemical or electronic effects as well.(11)

By contrast, substitution for Cu (12) and changes in O content have profound consequences. In particular, reduction of O content, through the formation of vacancies, is found (13-17) to have a significant effect on both the transition temperature and the crystal structure of the Y-Ba-Cu-O materials. Therefore, our consideration of the chemical basis for the existence of high T_c superconductivity in these materials will focus primarily on the Cu-O chemistry, structure, and magnetic properties. Generalization to other elements which might play analogous roles in new materials will then be considered.

PROPOSED STRUCTURAL CHEMICAL MODEL

Assignment of Oxidation States

Figure 1 presents a schematic diagram of the arrangement of copper and oxygen atoms in orthorhombic $YBa_2Cu_3O_7$ as revealed by crystallographic studies.(18) As illustrated, the copper atoms are arranged in two distinct patterns: (i) upper and lower, two-dimensional dimpled "sheets" in which the copper atoms are coordinated to five oxygen atoms of which four are nearly coplanar with the copper while the fifth is in an apical position resulting in an overall square pyramidal coordination sphere; (ii) central copper one-dimensional "chains" (represented in boldface in Fig. 1) in which the copper atoms are coordinated to four oxygens in a flat planar array. These differences in chemical bonding and the apparent formal oxidation states of the two types of Cu form the basis for the proposed high T_c superconducting model.

We construct a model for high T_c superconductivity by first conceptually treating the "sheets" and the "chains" as isolated systems. Interactions between these structures, which play a key role in the phenomenon, will be introduced later. We additionally simplify the initial description by postulating that, in each of the two structures, the Cu atoms are all in the same oxidation state. We

Reprinted by permission from *Modern Physics Letters B*, Vol. 1, Nos. 7/8, pp. 275–288 (1987).

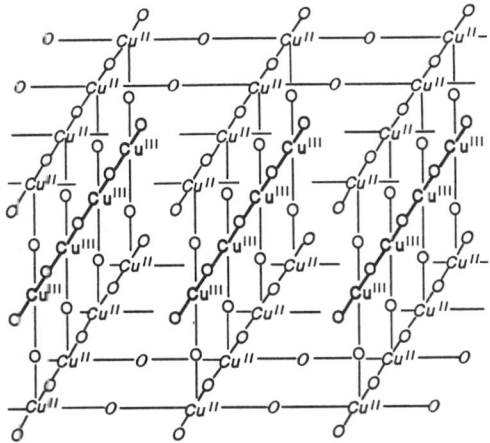

Fig. 1. Schematic representation of the Cu—O skeletal network in the $YBa_2Cu_3O_7$ superconductor. The 4-coordinate copper atoms in the "chains" (shown in boldface) are assigned a formal oxidation number of +3. The 5-coordinate copper atoms in the upper and lower "sheets" (shown in italics) are assigned a formal oxidation number of +2. (Note that some of the cross-linking oxygen atoms in the "sheets" are omitted for clarity.)

therefore start with $YBa_2Cu_3O_7$ by proposing that the 5-coordinate copper atoms in the upper and lower "sheets" are formally in the +2 oxidation state. Although Cu^{II} commonly exhibits a coordination number of 6, with the two donor atoms along one axis being elongated as a result of Jahn-Teller distortion, 5-coordinate Cu^{II} is well known.[19] If not otherwise constrained, 5-coordinate Cu^{II} is square pyramidal with the copper being displaced approximately 0.1-0.3Å above the basal plane,[20] this distance increasing as the strength of the apically coordinated donor atom increases. The Cu atoms in the "sheets" are, in fact, displaced from the basal oxygen plane by approximately 0.27Å,[18] creating the "dimples" referred to earlier. (Note that no attempt has been made to represent this latter feature in Fig. 1.)

We further propose that the copper atoms in the central "chains" are formally in the +3 oxidation state. Although reported occurrences of Cu^{III} are still relatively rare, there are a number of examples of this oxidation state in the literature. In particular, Margerum has shown [21] that Cu^{III} is readily obtained when copper is coordinated to "hard" donor atoms such as deprotonated amide nitrogens (or, presumably, oxide ions). In the one crystal structure of a Cu^{III} complex which has been reported to date,[22] Cu^{III} is shown to exhibit planar 4-coordination as expected for an internally spin-paired d^8 system.

These assignments for the copper atom valencies are consistent with the overall stoichiometry of $YBa_2Cu_3O_7$. Assuming the elements Y, Ba, and O have their expected valencies of +3, +2, and -2, respectively, the average oxidation state for Cu must be 2.33 which implies that two-thirds of the

copper atoms are in the +2 oxidation state while the remaining one-third are +3. Moreover, recent iodometric titrimetric analyses by Appelman et al. (23) of $YBa_2Cu_3O_{6.9}$ have established the presence of Cu^{III} in quantities which are consistent with the preceding oxidation state assignments.

Spin State Assignments

The foregoing oxidation state assignments allow us to formulate a hypothesis regarding the mechanism for superconductivity in $YBa_2Cu_3O_7$ materials which can account for the observed properties at various stages of deoxygenation. Furthermore, the proposed mechanism permits conjectures to be made regarding modifications in the crystal lattice which might lead to higher T_c values.

The Cu^{II} atoms in the upper and lower "sheets" have a d^9 electronic configuration and, thus, possess an unpaired spin. However, since we know from magnetic susceptibility measurements that carefully prepared samples of $YBa_2Cu_3O_7$ exhibit a zero Curie-Weiss term,[24] we must conclude that the majority of Cu^{II} spins in the "sheets" are coupled antiferromagnetically, resulting in no local magnetic moment. Such antiferromagnetic coupling comes about through an O intermediary in the superexchange process. The anti-parallel spin alignment utilizes the p-orbitals of the oxygen atom to form the coordinate bond and is favored by the large Cu-O-Cu bond angle of approximately $165°$.[18] This coupled spin state in the "sheets" need not result in a spatially extensive, time-independent, two-dimensional Néel antiferromagnet but may, in fact, manifest itself in the form of long range instantaneous antiferromagnetic spin correlations such as the "Quantum Spin Fluid" state found by Shirane et al. (25) to exist in the Cu-O "sheets" of La_2CuO_4.

In the "chain" structure, we note that the O-Cu-O bond angle along the "chain" direction is exactly $180°$ (18) with the Cu atoms sitting at the center of a square planar array of O atoms. Here again our identification of the oxidation state of these atoms as Cu^{III}, which are internally spin paired with a d^8 electronic configuration, is consistent with the observed coordination geometry and bond angles and is dictated by our previous assignment of Cu^{II} for the Cu atoms in the "sheets." The oxygen atoms perpendicular to the "chain" direction are, in fact, the apically-coordinated oxygen atoms of the Cu^{II} atoms in the upper and lower "sheets." Thus they provide a direct communication between the Cu atoms in the "chains" and those in the "sheets" and form the basis in our model for charge propagation in these materials as outlined later.

Structural Effects of Deoxygenation

Experiments have shown that oxygen is readily removed from $YBa_2Cu_3O_7$. At relatively low temperatures (400-500°C), the careful removal of

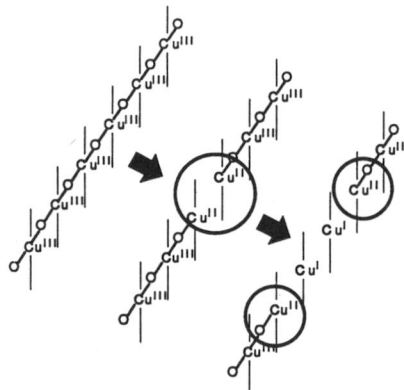

Fig. 2. Schematic representation of the proposed process of deoxygenation in a single "chain" (left) in which the loss of a single bridging oxygen (center) creates two Cu^{II} atoms (circled) adjacent to the vacated site. Such 3-coordinate Cu^{II} atoms are thermodynamically unstable, thereby destabilizing the remaining adjacent bridging oxygen atoms and leading to further oxygen loss to create stable 2-coordinate Cu^{I} (right) and two new unstable Cu^{II} atoms (circled) at the edge of the deoxygenated region. This process continues in a single chain until the entire chain is converted to disconnected Cu^{I} atoms.

oxygen produces a range of compositions of the general formulation $YBa_2Cu_3O_{7-x}$. A number of remarkable properties of these oxygen deficient compounds have been reported which must be addressed by any model for this overall system. First, neutron scattering experiments (17,26) indicate that, over the range of oxygen deficiency $0 \leq x \leq 1$, oxygen is preferentially removed from "chain" sites, producing at $x = 1$ "chains" which are totally free of bridging oxygen. Second, over this range of x, room temperature resistivity, magnetic susceptibility, and, particularly, superconductivity transition temperature reveal the occurrence of two clearly demarcated regions with distinctly different T_c.(24,27)

Based on our assignment of the oxidation states of the copper atoms in $YBa_2Cu_3O_7$, the removal of a single oxygen from a "chain" would result in the reduction of the two adjacent Cu^{III} atoms to Cu^{II} each of which will then be 3-coordinate (Fig. 2). Such a coordination number for Cu^{II} is virtually unknown and, presumably, unstable. As a result, these Cu^{II} atoms should readily undergo the loss of the other adjacent oxygen atom in the "chain" resulting in their further reduction to Cu^{I} with a coordination number of 2 (Fig. 2). (Such linearly coordinated Cu^{I} atoms are well known in other compounds (28) and are presumably stable.) At the same time, however, two new 3-coordinate Cu^{II} atoms are produced at the edge of the deoxygenated region. This process will result in a cascade in which a Cu-O-Cu "chain" will begin to "unzip" once a single bridging oxygen atom is removed. As a result, deoxygenation should continue in any single

"chain" until all Cu^{III} atoms have been reduced to linearly coordinate Cu^{I} atoms. This process accounts for both the preferential removal of the oxygen atoms in the "chains" and for the absence of local magnetic moments in partially deoxygenated material.(24)

When all copper "chains" have been fully reduced, the overall stoichiometry conforms to $YBa_2Cu_3O_6$ (i.e., $x = 1$) and the former "chains" will consist entirely of disconnected Cu^{I} atoms, a proposition which has been previously expressed by Cava et al.(27) However, according to the deoxygenation mechanism described above, partially deoxygenated systems ($0 < x < 1$) should consist of a mixture of Cu^{III}-O-Cu^{III} "chains" and oxygen-free linear arrays of Cu^{I} atoms as illustrated in Fig. 3.

Electrical Properties

The Y-Ba-Cu-O system, as described above, is an insulator. To obtain the observed electrical properties of the system, both above and below T_c, we must now consider charge transfer interactions between the "sheets" and the "chains." "Sheet"/ "chain" interaction allows the "sheets" to act as electron reservoirs which can transfer carriers to the "chains," providing the basis for electrical conduction and a means of establishing conditions to give high T_c superconductivity. The O atoms lying at the apex of the square pyramidal Cu atoms in the "sheet" provide an accessible pathway for such "sheet"/"chain" coupling. Each electron promoted

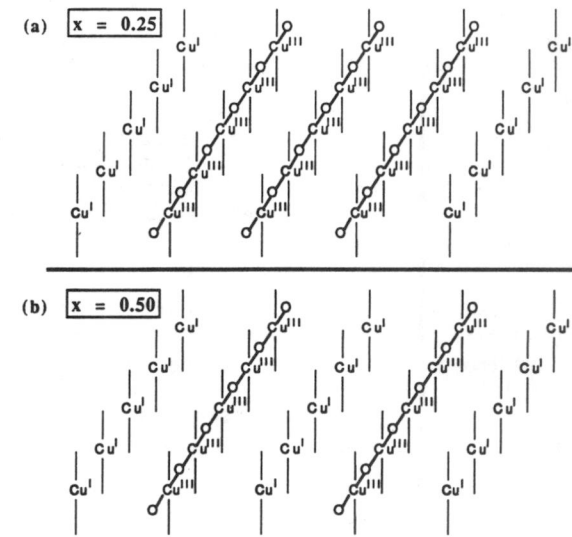

Fig. 3. Schematic representation of the predicted patterns of alternating "chains" of Cu^{III}-O-Cu^{III} and disconnected linear arrays of Cu^{I} atoms postulated to predominate at various stages of deoxygenation. At $x = 0.25$ (Fig. 3a), three adjacent Cu^{III}-O-Cu^{III} "chains" exist between linear arrays of disconnected Cu^{I} atoms. At $x = 0.5$ (Fig. 3b), each Cu^{III}-O-Cu^{III} "chain" is sandwiched between linear arrays of Cu^{I}, thereby eliminating cooperativity between the "chains."

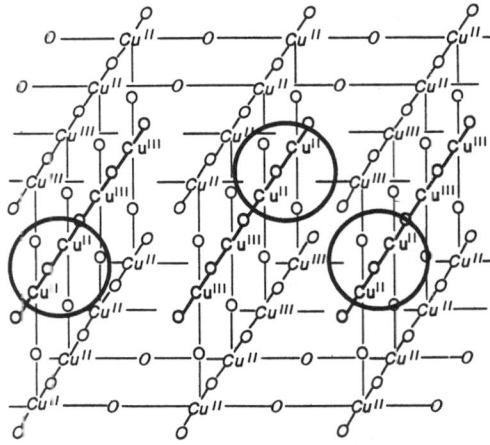

Fig. 4. Conceptualized representation showing the instantaneous generation of spin-paired Cu^{II} atoms (circled) in the "chains" and the corresponding generation of paired Cu^{III} "holes" in either the upper (left and right) or lower (center) "sheets." In reality, the spin-paired Cu^{II} atoms are delocalized.

from the "sheet" in effect leaves behind a Cu^{III} atom and converts one of the "original" Cu^{III} atoms in the "chain" to Cu^{II}. As electrons are transferred up and down from the adjacent "sheets" to the Cu-O-Cu "chains," a mixed Cu^{II}/Cu^{III} system will be formed in both the "sheets" and the "chains" as illustrated conceptually in Fig. 4.

This process does not, of course, produce localized Cu^{II} atoms on the "chains" nor does it leave Cu^{II} atoms with unpaired spins on the "sheets" since this would produce local magnetic moments. Rather the free carriers present in this mixed oxidation state are delocalized, resulting in the observed Pauli-Landau temperature-independent magnetic susceptibility and the metallic type electrical conductivity. In fact, the carrier delocalization and increase in entropy, which result from the "sheet" to "chain" charge transfer, provides the thermodynamic driving force to produce the mixed oxidation state. Although the holes and electrons created in this manner must exhibit a coulombic interaction, bound excitons are not formed because the screening length which will be present in this metallic system is shorter than the "sheet" to "chain" distance.

With about 2/3 of the Cu^{III} "chain" atoms converted to Cu^{II} and, correspondingly, 1/3 of the Cu^{II} atoms on each of the "sheets" converted to Cu^{III}, there will be a density of electrons and holes of 6×10^{21} cm^{-3}--a value approximately equal to the measured free carrier density above T_c. If we assume that the holes created on the "sheets" by this process have higher mobility than the electrons transferred to the "chains," we can also explain the p-type thermopower measured (29) for this system. The delocalized carriers on the "sheets" and

"chains" also interact themselves through the O atom intermediaries. Because of delocalization, however, the coupling for this interaction is weaker than the previously described antiferromagnetic coupling which exists between the localized spins on the Cu^{II} atoms in the "sheets." Despite the reduced strength of this free carrier antiferromagnetic coupling, at sufficiently low temperatures two spins on alternate sides of a bridging O atom can be in a favorable position, because of the large bond angle, to couple through the superexchange process to produce an anti-parallel spin pair.

The superexchange coupled spin pairs both on the "chains" and on the "sheets" are mobile, strongly bound, spinless composite particles which will obey Bose statistics. At any given concentration of these spin pairs, therefore, there will exist a Bose condensation temperature at which a transition to a superconducting ground state will occur. Thus, there are three important temperatures in the problem. The first two are the spin-pairing temperatures for the holes on the "sheets" and electrons in the "chains," respectively, and the third is the Bose condensation temperature.

Based on the foregoing concepts, the superconducting state can be achieved in three different ways. First, if both spin pairing temperatures are lower than the Bose condensation temperature, superconductivity will occur, in principle, at a temperature just below the highest spin pairing temperature. This situation is analogous to that found in conventional low temperature metallic superconductors where T_c is determined by the strength of the carrier pairing interaction.

In the second possibility, one spin pairing temperature could lie above the Bose condensation temperature and one below. In our model, the higher spin pairing temperature would likely be that of the "chain." This would give rise to a situation in which, above T_c, normal conduction would occur through both ordinary Fermi particles and uncondensed spin pairs. Since Hall Effect and thermopower measurements in Y-Ba-Cu-O show normal state p-type conductivity, this could be explained, as previously mentioned, by postulating a higher mobility for the holes on the "sheets," allowing them to dominate the electrical properties of the material, thus "hiding" the electrical consequences of the uncondensed spin pairs on the "chains" above T_c.

The third possibility is that the lowest temperature is the Bose condensation temperature. This novel situation would have dramatic consequences for electrical properties of the normal state, to say the least, resulting in the occurrence of charge transport exclusively through uncondensed spin pairs! It is not clear whether one could successfully explain all of the measured transport and magnetic properties of these materials in terms of such peculiar charge carriers.

To consider these possibilities, we obtain a simple estimate of the Bose condensation temperature for the model system in the following way. The volume occupied by the wavefunction of a Bose particle can be approximated as a cube with dimension

163

equal to the particle's de Broglie wavelength. The de Broglie wavelength, in turn, is determined by the momentum of the particle and, therefore, by its thermal energy. This temperature dependent Bose particle interaction volume increases with decreasing temperature. When the interaction volume grows to become equal to the volume available per Bose particle in the system, the Bose particles interact so as to bring about condensation. The condensation temperature, T_0, can be written as

$$T_0 = 3.31 \frac{h^2}{(2\pi)^2 k_B m} n^{2/3}$$

where h is Planck's constant; k_B is Boltzmann's constant; m is the effective mass of the Bose particle; and n is the number density. For $YBa_2Cu_3O_7$ the greatest density of Bose particles (Cu^{II}-O-Cu^{II} spin pairs) will occur when 2/3 of the "chain" Cu atoms are in the +2 oxidation state. If we use this density and take $m = 2m_e$, where m_e is the electron mass, we obtain for the Bose condensation temperature, $T_0 \approx 3000K$. We can obtain another estimate which takes into account the two-dimensionality of the "sheets" by equating the "sheet" area available per spin pair to the square of the de Broglie wavelength, but this also gives a value of T_0 equal to a few thousand Kelvin.

These results could be overestimated to some extent as a result of using too small an effective mass. However, an effective mass of $m > 20m_e$ would be required to obtain $T_0 = T_c$. Therefore, we should conclude that in these high T_c systems, as in the low temperature metallic superconductors, the transition to the normal state is likely to be determined by pair breaking and not by the Bose condensation.

Although it is difficult to obtain an estimate of the spin pairing temperatures in terms of this simple model, it is clear that, because of the increased orbital overlap between Cu d-electrons and O p-electrons which occur in the 180° bond angles along the "chains," we would expect spin pairs to be more strongly bound through superexchange on these structures than in the "sheets." Support for this is obtained from recent nuclear spin lattice relaxation experiments on the Y-Ba-Cu-O system reported by Warren et al.(30) which show a clear indication of the presence of two distinct pairing energies, with substantially larger energies for quasiparticle formation (pair breaking) on the "chains" than on the "sheets." Their data indicate spin pairing temperatures of approximately 60K and 200K when one assumes the standard BCS weak-coupling gap ratio of $2\Delta/kT_c \approx 4$.

Superconductivity Changes upon Deoxygenation

According to the recent studies of Cava et al.,(28) as illustrated in Fig. 5, partial deoxygenation of $YBa_2Cu_3O_{7-x}$ causes only minor changes in T_c up to $x \approx 0.25$ when the value of T_c suddenly drops from $\approx 90K$ to $\approx 60K$. Upon further removal of oxygen this latter T_c value is maintained until x

≈ 0.5 when the T_c value decreases again by over 30K.

In accordance with the model which we have proposed, x = 0.25 can be viewed as corresponding to the point at which one fourth of all "chains" have been reduced to oxygen-free linear arrays of Cu^I atoms resulting in an average of three adjacent Cu^{III}-O-Cu^{III} "chains" between Cu^I linear arrays (Fig. 3a). The sudden change in properties at this point (Fig. 5) suggests that the higher T_c value derives from extended three-dimensional communication between the coupled spin pairs in adjacent "chains." When x = 0.5, half of the "chains" have been reduced and, on the average, all Cu^{III}-O-Cu^{III} "chains" are sandwiched between linear arrays of Cu^I atoms (Fig. 3b). Since the Cu^I atoms are incapable of participating in the mixed valence charge propagation process, three-dimensional interaction between adjacent intact "chains" is virtually ruled out and a further significant reduction in T_c results (Fig. 5).

In this same range of the oxygen deficiency parameter, x, we see in Fig. 5 that an increase in room temperature resistivity occurs in the region of x < 0.25. Presumably this is due to elimination of conductive Cu-O-Cu mixed valence "chain" structures resulting in an inhibition in the formation of mixed oxidation states on the "sheets" and "chains." This reduction in free carrier density is also accompanied by a slight decrease in magnetic susceptibility as observed.

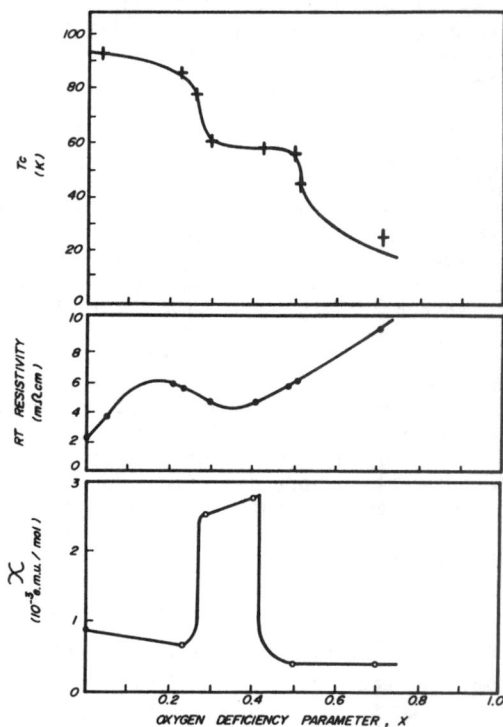

Fig. 5. Effects of partial deoxygenation in $YBa_2Cu_3O_{7-x}$ on the superconducting transition temperature, T_c (top), the room temperature resistivity (center), and the room temperature magnetic susceptibility, X (bottom). Data are from references 24 and 27.

Once the transition to the T_c = 60K phase has occurred, where $0.25 \leq x \leq 0.5$, the room temperature resistivity dips, and the magnetic susceptibility shows a dramatic increase (Fig. 5). These data suggest the formation of an ordered phase as has been pointed out by Cava et al.(28) Indeed, these authors have reported evidence, from electron diffraction measurements, which suggest vacancy ordering of some type, although the lattice still retains orthorhombic symmetry. In terms of our model, this ordering is predicted to be an arrangement of intact Cu^{III}-O-Cu^{III} "chains" and completely oxygen deficient linear Cu^I arrays. This stands in contrast to the ordered arrangement of empty and occupied oxygen sites on single "chains" as proposed by the previous authors. The increase in free carrier magnetic susceptibility in this range is remarkable and is much greater than would be associated with the slight increase in conductivity observed, suggesting a change in effective mass, which may be due to details of band structure related to the ordered "chains."

Two ordered arrangements of "chains" can occur in this range of x. At x = 0.25 groups of three intact "chains" can be separated by a single oxygen-free array of Cu^I atoms (Fig. 3a), while at x = 0.33 pairs of intact "chains" can be separated by an oxygen-free array. When x = 0.5, half of the "chains" have been reduced and, on the average, all intact Cu^{III}-O-Cu^{III} "chains" are sandwiched between oxygen-free Cu^I arrays (Fig. 3b). It is interesting to note that the T_c value reached for x > 0.5 is typical of the "sheet" type ceramic superconductors La_2CuO_4 and $La_{2-x}(Ba,Sr)_xCuO_4$, suggesting that, with the loss of all "chain"/"chain" interaction, the effective dimensionality of the system has become "sheet-like."

As $x \to 1.0$, the remaining Cu^{III}-O-Cu^{III} "chains" are reduced to Cu^I and at x = 1 all superconductivity is lost. This loss in superconductivity can be attributed to the decreasing possibility of creating the required mixed oxidation state either in the "sheets" or in the "chains" and, ultimately, to the transition from an orthorhombic to a tetragonal structure.

Clearly, a more thorough analysis is required to understand details both of the antiferromagnetic coupling between delocalized carriers and the proposed "chain"/"chain" coupling. The origin of the free carrier magnetic susceptibility enhancement in the ordered oxygen deficient phase will also require closer attention. Additionally, we must develop a better understanding of how the observed T_c in $YBa_2Cu_3O_7$ is arrived at through interaction of the "chain" and the "sheet" systems. The small spatial separation of these structures and the existence of interactions, including superconducting proximity effects, should result in both structures participating in the superconducting state.

Application of this simple model to the La-Cu-O system is also possible. In these structures, although the same antiferromagnetic spin pairing mechanism is presumably operative, no "chains" exist and the mixed valence Cu^{II}/Cu^{III} system is

established on the "sheets" either by doping with Sr or Ba or through the incorporation of excess O. As has been mentioned, the La-Cu-O systems exhibit transition temperatures which are near those found for the Cu-O "sheet" systems which remain after deoxygenation of Y-Ba-Cu-O. These systems, of course, will not exhibit the complex behavior with O removal seen in the Y-Ba-Cu-O system, nor will they ever likely exhibit the high T_c behavior, both as a result of the limited dimensionality of their structure.

Predicted Requirements for Achieving Enhanced T_c

The foregoing interpretation of the behavior of partially deoxygenated Y-Ba-Cu-O systems suggests that optimal T_c values are obtained when electron pairs in the "chains" interact in three-dimensions involving the adjacent "chains" as well as the adjacent "sheets." This suggests that T_c might be further enhanced by synthesizing new materials in which layers of "sheets" and "chains" are fully interleaved in an extended three-dimensional array. This would have the additional benefit of converting Cu atoms in the "sheets" to a 6-coordinate geometry, thereby removing "dimples" from the "sheets." By changing the 165° O-Cu-O bond angles to 180°, the spin pairing temperature on the sheets would be increased. In addition, the spin pairing temperature in the "chains" might be increased by shortening the distance between the Cu^{III} atoms by the substitution of more tightly bound bridging anions, thus increasing the strength of the antiferromagnetic coupling.

Such anion substitutions might require altering the charge on the other cations (i.e., making appropriate substitutions for Y and Ba) to achieve balance of the overall charges. Moreover, as some of the bond distances change in the overall crystal lattice, care needs to be taken to maintain optimal Cu-X-Cu bond angles, both for promoting charge propagation and to optimize antiferromagnetic coupling. We believe that such substitutions can be made in a fruitful manner as, for example, may be indicated in preparations of multi-phase samples of partially fluorinated $YBa_2Cu_3O_{7-x}F_y$ which were initially shown by Ovshinsky et al.,(31) and subsequently confirmed by others,(32) to contain some regions with T_c = 159K. The effect of fluorine on oxygen diffusion and microcrystal orientation have been discussed elsewhere.(33)

Finally, one can consider the possibility of synthesizing new mixed oxidation state systems in which other transition metal atoms with appropriate multiple oxidation states replace Cu. Very recent indications of high T_c superconductivity in La-Sr-Nb-O (34) and Y-Ba-Ag-O (35) ceramic films support our belief that the Cu-based systems may not be unique but rather only members of a larger family of antiferromagnetically spin coupled mixed oxidation state systems.

Acknowledgement

The authors would like to acknowledge stimulating discussions with Prof. H. Fritzsche.

References

1. T. M. Rice, Z. Physik $B67$, 141 (1987).
2. J. E. Hirsch, Phys. Rev. Lett. 59, 228 (1987).
3. L. Pauling, Phys. Rev. Lett. 59, 225 (1987).
4. P. W. Anderson, Science 235, 1196 (1987).
5. V. J. Emery, Phys. Rev. Lett. 57, 2794 (1987).
6. H. Kamimura, Jpn. J. Appl. Phys. 26, L627 (1987).
7. P. H. Hor, R. L. Meng, Y. Q. Wang, L. Gao, Z. J. Huang, J. Bechtold, K. Forster, C. W. Chu, Phys. Rev. Lett. 58, 1891 (1987).
8. E. M. Engler, V. Y. Lee, A. I. Nazzal, R. B. Beyers, G. Lim, P. M. Grant, S. S. P. Parkin, M. L. Ramirez, J. E. Vasquez, R. J. Savoy, J. Am. Chem. Soc. 109, 2848 (1987).
9. L. Soderholm, K. Zhang, D. G. Hinks, M. A. Beno, J. D. Jorgensen, C. U. Segre, I. K. Schuller, Nature 328, 604 (1987).
10. B. W. Veal, W. K. Kwok, A. Umezawa, G. W. Crabtree, J. D. Jorgensen, J. W. Downey, L. J. Nowicki, A. W. Mitchell, A. P. Paulikas, C. H. Sowers, Appl. Phys. Lett. 51, 229 (1987).
11. S. R. Ovshinsky, in Festschrift in Honor of Heinz Henisch, S.R. Ovshinsky, R.W. Pryor and B.B. Schwartz, Eds., Institute for Amorphous Studies Series (Plenum, New York, 1988), in press.
12. Y. Maeno, T. Tomita, M. Kyogoku, S. Awaji, Y. Aoki, K. Hoshino, A. Minami, T. Fujita, Nature 328, 512 (1987).
13. P. K. Gallagher, H. M. O'Bryan, S. A. Sunshine, D. W. Murphy, Mater. Res. Bull. 22, 995 (1987).
14. A. Santors, S. Miraglia, F. Beech, S. A. Sunshine, D. W. Murphy, L. F. Schneemeyer, J. V. Waszczak, Mater. Res. Bull. 22, 1007, (1987).
15. I. K. Schuller, D. G. Hinks, M. A. Beno, D. W. Capone, II, L. Soderholm, J. P. Locquet, Y. Bruynseraede, C. V. Segre, K. Zhang, Solid State Commun. 66, 385 (1987).
16. J. D. Jorgensen, B. W. Veal, W. K. Kwok, G. W. Crabtree, A. Umezawa, L. S. Nowicki, A. P. Paulikas, Phys Rev. $B36$, 5719 (1987).
17. R. Beyers, G. Lim, E. M. Engler, R. J. Savoy, T. M. Shaw, T.R. Dinger, W. T. Gallager, R. L. Sandstrom, Appl. Phys. Lett. 50, 1918 (1987).
18. M. A. Beno, L. Soderholm, D. W. Capone, II, D. G. Hinks, J.D. Jorgensen, I. K. Schuller, C. U. Segre, K. Zhang, J. D. Grace, Appl. Phys. Lett. 51, 57 (1987).
19. (a) B. J. Hathaway, Coord. Chem. Rev. 35, 211 (1981). (b) B. J. Hathaway, D. E. Billing, Coord. Chem. Rev. 5, 143 (1970).
20. See, e.g., (a) P. W. R. Corfield, C. Ceccarelli, M. D. Glick, I. W.-Y. Moy, L. A. Ochrymowycz, D. B. Rorabacher, J. Am. Chem. Soc. 107, 2399 (1985). (b) V. B. Pett, L. L. Diaddario, Jr., E. R. Dockal, P. W. R. Corfield, C. Ceccarelli, M. D. Glick, L. A. Ochrymowycz, D. B. Rorabacher, Inorg. Chem. 22, 3661 (1983); and references therein.
21. D. W. Margerum, Pure Appl. Chem. 55, 23 (1983).
22. L. L. Diaddario, W. R. Robinson, D. W. Margerum, Inorg. Chem. 22, 1021 (1983).
23. E. H. Appelman, L. R. Morss, A. M. Kini, V. Geiser, A. Umezawa, G. W. Crabtree, K. D. Carlson, Inorg. Chem. 26, 3237 (1987).
24. R. J. Cava, B. Batlogg, C. H. Chen, E. A. Rietman, S. M. Zahurak, D. Werder, Nature 329, 423 (1987).
25. G. Shirane, Y. Endoh, R. J. Birgeneau, M. Kastner, Y. Hidaka, M. Oda, M. Suzuki, T. Murakami, Phys. Rev. Lett. 59, 1613 (1987).
26. F. Beech, S. Miraglia, A. Santors, R. S. Roth, Phys. Rev. $B36$, 8778 (1987).
27. R. J. Cava, B. Batlogg, C. H. Chen, E. A. Rietman, S. M. Zahurak, D. Werder, Phys. Rev. $B36$, 5719 (1987).
28. H. Hope, P. P. Power, Inorg. Chem. 23, 936 (1984).
29. N. Mitra, J. Trefny, M. Young, B. Yarar, Phys. Rev. $B36$, 5581 (1987).
30. W. W. Warren, Jr., R. E. Walstedt, G. F. Brennert, G. P. Espinosa, J. P. Remeika, Phys. Rev. Lett. 59, 1860, (1987).
31. S. R. Ovshinsky, R. T. Young, D. D. Allred, G. DeMaggio, G. A. Van der Leeden, Phys Rev. Lett. 58, 2579 (1987).
32. R. N. Bharagawa, S. P. Heako, W. N. Osborne, Phys Rev. Lett. 59, 1468 (1987).
33. S. R. Ovshinsky, R. T. Young, B. S. Chao, G. Fournier, D. A. Pawlik, Proc. Int'l. Conf. on High Temperature Superconductivity, Drexel University (1987), (World Scientific, Singapore, 1988), to be published.
34. T. Ogushi, Y. Hakuraku, Y. Honjo, G. N. Suresha, S. Higo, Y. Ozono, I. Kawano, T. Numata, Low Temp. Phys. (Jan. 1988), to be published.
35. K. K. Pan, H. Mathias, C. M. Rey, W. G. Moulton, H. K. Ng, L. R. Testardi, Y. L. Wang, Phys. Lett. $A125$, 147 (1987).

PART II

DEVICE APPLICATIONS

REVERSIBLE STRUCTURAL TRANSFORMATIONS IN AMORPHOUS SEMICONDUCTORS FOR MEMORY AND LOGIC

S.R. Ovshinsky[1] and H. Fritzsche[2]

[1]Energy Conversion Devices, Inc., Troy, Michigan 48084
[2]James Franck Institute, University of Chicago, Chicago, Illinois, 60637

ORDER-DISORDER transformations such as ordering of magnetic or ferroelectric domains are used in many information storage devices. The structural transformation of amorphous semiconductors from the amorphous to a more ordered state can be used for the same purpose if the structure transformation can be achieved fast and reversibly.[1] Amorphous semiconductors have then the advantage over other information storage devices in that they can be cheaply produced and easily shaped in many different configurations. Furthermore, for semiconductors the difference between the physical properties of the amorphous and the crystalline state is particularly large. This enables one to retrieve and read the binary information, stored in the form of the structural state, with good signal to noise ratio despite extensive miniaturization. This paper describes a) a few examples of materials capable of reversible structure transformations, b) some methods for initiating the transformations, and c) a model in which the electrical or optical creation of excess electron-hole densities plays an important role in determining the crystallization rate.

MATERIALS

Several phenomena which are often labeled "switching" and "memory effects" have been reported[2] in a large variety of materials. Among these are thin films of the transition metal oxides, of silicon oxide and aluminum oxide, as well as amorphous layers of Si, Ge and B. The physical processes responsible for the switching and memory effects are likely to be different in the various material classes. Only a few have been positively identified such as the semiconductor-metal transition in certain transition metal oxides, which is associated with a transformation between two crystal

structures at a critical temperature. The switching effects observed in amorphous Si, Ge, and B may be similar to that of chalcogenide alloy semiconductors; the memory effects, however, must have a different origin, as will be discussed in a later section. The degree of reproducibility required for practical devices has been achieved, as far as is known, only with chalcogenide alloys. The results reported here were obtained with the system $Te_{81}Ge_{15}X_4$, where X represents one or two elements of group V or VI of the periodic table. This composition is close to $Te_{85}Ge_{15}$, the eutectic point of the Ge-Te binary. Glass formation is particularly favorable near the eutectic because of the low eutectic temperature (375°C) and relatively high viscosity in the melt.[3] The X additives to the pure eutectic composition were found to further stabilize the amorphous state without preventing crystallization to occur as a result of excitation by a light or current pulse. Se-Te alloys were also found to yield results of high reproducibility.

The effects of thermal cycling[4] on some properties of $T_{81}Ge_{15}X_4$ are illustrated in Fig. 1. In the amorphous state the material has a resistivity of about 10^7 ohm-cm at room temperature. Although the temperature is shown in centigrade, the abscissa of Fig. 1 is proportional to $1/T$, the reciprocal absolute temperature. This scale is chosen to display the exponential behavior of the resistivity ρ which follows

$$\rho = \rho_0 \exp(\Delta E/kT)$$

with $\Delta E \approx 0.5$ ev up to the glass transition temperature T_g. The change in heat capacity occurring near T_g is noticeable on the differential thermal analysis (DTA) graph labeled (a). Near T_1 the resistivity drops by several orders of magnitude while the DTA curve shows an exothermic peak which represents the heat of transformation as the material devitrifies. Since the composition is close to the eutectic the crystallization process is accompanied by phase separation predominantly into finely divided tellurium and GeTe regions.[5] Cooling the material from a temperature slightly above T_1 preserves the low resistive ordered state. Near $T_M = 375$°C the crystalline eutectic melts and the resistivity increases to a value which is approximately an extrapolation of the resistivity curve of the vitreous state. The material

---

S. R. OVSHINSKY is Founder and Chairman of the Board of Directors, Energy Conversion Devices, Inc., Troy Mich. H. FRITZSCHE is Professor of Physics, James Franck Institute, University of Chicago, Chicago, Ill.

This manuscript is based on a paper presented at the annual conference sponsored by the Electronic Materials Committee of the Institute of Metals Division of the Metallurgical Society of AIME and held August 30-September 2, 1970, in New York City.

Reprinted by permission from *Metallurgical Transactions*, Vol. 2, pp. 641–645 (1971).

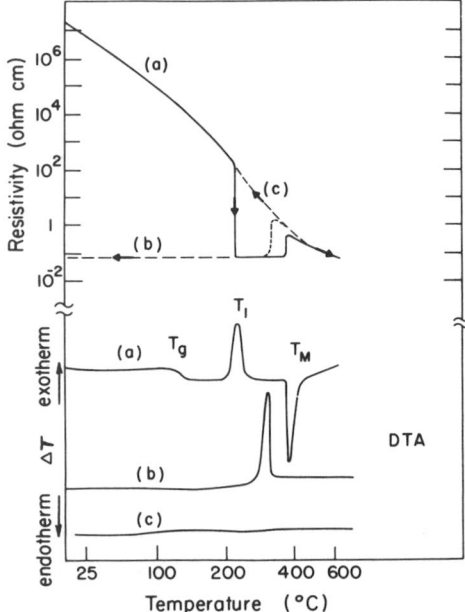

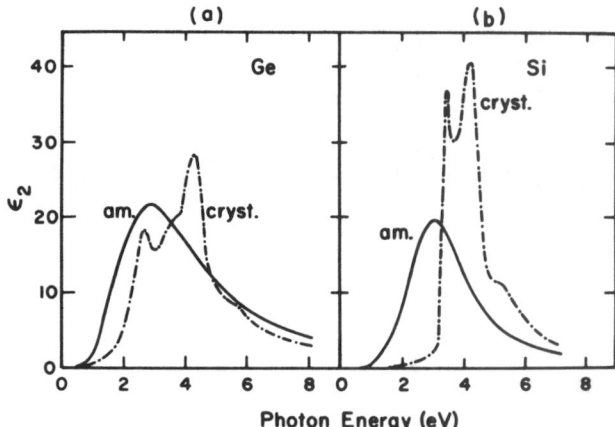

Fig. 2—Imaginary part of dielectric constant ϵ_2 for (a) germanium according to J. Tauc *et al*. (Proc. Conf. Physics of Non-Crystalline Solids, Delft 1964) and (b) silicon according to D. Beaglehole *et al*., J. Non-Crystalline Solids, 1970, vol. 4, p. 272 in the amorphous and crystalline state.

Fig. 1—Temperature dependence of the resistivity and the differential thermal analysis curves for a typical "memory type" material showing a reversible structural transformation. Curve (a): heating rate = 25°C per min. Curve (b): cooling rate less than 25°C per min. Curve (c): cooling rate faster than 25°C per min. The ordinate of the DTA graph represents the difference ΔT between the temperature of the sample and that of an inert reference material. Exothermic processes manifest themselves therefore as an upward peak in the DTA curve.

is again disordered and probably in structure similar to the vitreous state except of course for the larger thermal motion. The vitreous state and the original high resistivity can be restored by cooling the liquid faster than about 25°C per min (see curve c). Slow cooling yields solidification into the crystalline eutectic (after some supercooling), which manifests itself by an exo-

thermic solidification peak (see DTA curve b) and the low resistance state at room temperature.

The DTA curves of Fig. 1 show that the amorphous semiconductor material can, in principle, be brought reversibly into a crystalline state or an amorphous state by appropriate heat inputs and control of the quench rates. The amorphous and the crystalline state of this and many related materials differ strongly in their electrical and optical properties. Any of these properties can be used to read out the binary information, *i.e.*, the structural state of the material. The resistivity at room temperature, for instance, differs by six or seven orders of magnitude. The optical transmission is altered by many orders of magnitude for wavelengths close to the absorption edge and the reflectivity changes

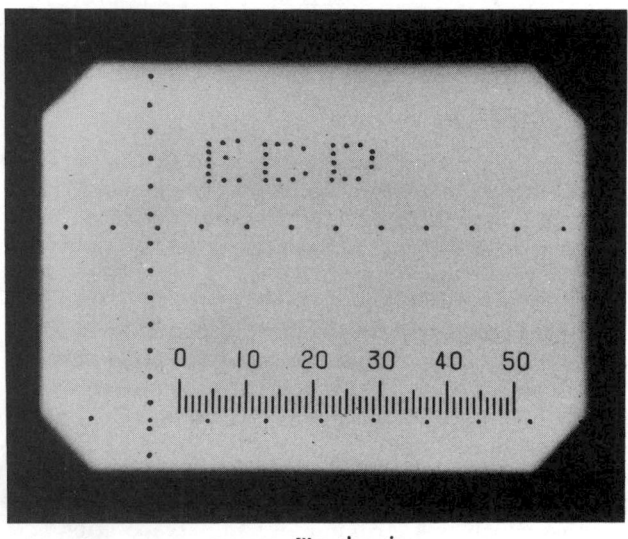

memory film showing
dots written by laser

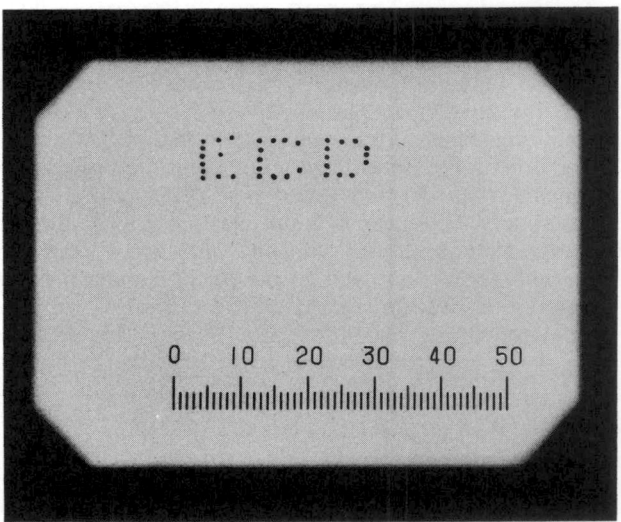

same film after erasing
some dots with same laser

Fig. 3—Microscope view of a glass substrate coated with an amorphous semiconductor film. The dots of high absorption are produced with a 1 μsec long laser pulse. The same laser can "erase" dots, as evidenced by the right hand side of this figure. Dot diameter is about 2 μm.

in tellurium-rich alloys by about a factor two.[6] Large changes in the optical properties of semiconductors occur because the lack of long range order in the amorphous state lifts the restriction, valid for crystals, that only those optical transitions can take place which conserve crystal momentum. As an example of the changes which typically occur, the absorptive part of the dielectric constant, ϵ_2, of the amorphous and crystalline state of germanium and silicon[7] are shown in Fig. 2. Similar curves for the materials discussed here are not available, although the same physical processes are operative, because of the finely polycrystalline and phase-separated nature of the eutectic composition.

From the above discussion it appears that the speed of transformation might be very limited and hence might constitute a major obstacle for any practical application. This, however, is not the case. The next section describes observations of structure transformations on time scales of microseconds. Indeed, these experiments suggest that other than thermal processes are operative which considerably accelerate the transformation rates.[8]

OPTICALLY AND ELECTRICALLY INDUCED STRUCTURE CHANGES

One method of producing a structure change uses a light beam from a mixed gas Argon-Krypton laser.[9] Fig. 3 shows a microscope slide coated by sputtering with a $2\,\mu$ thick vitreous film of a Se-Te compound. The dots, produced with 1 to $10\,\mu$ sec long laser pulses, look dark because the transmission of the more ordered material in the dots is decreased for red light by about a factor of one hundred. The laser beam intensity of between 10 to 100 mw was focused by a 10 power objective to a spot of about $3\,\mu m$ in diam. The smaller dots seen in Fig. 3 are $2\,\mu m$ in diam. The focal plane of the laser beam is placed at the interface of the amorphous layer and the glass substrate so that only the interface material was exposed and dust or surface irregularities on the top surface do not affect the resolution of "read" signal. If one decreases the intensity of the laser beam by a factor of four and moves the beam over the line of dots, they disappear as shown on the right hand side of Fig. 3. One can write and erase dots repeatedly. By using a 500 to 1000Å thick film of the same material or of sputtered $Te_{81}Ge_{15}X_4$ on a carbon coated 200 mesh electron microscope grid the change into the crystalline and back into the amorphous state has been monitored by electron diffraction.[8] The crystallite size was estimated to be less than a few hundred angstroms. Upon becoming more ordered, the transmission edge shifts to longer wavelengths, making the material in the dot more opaque to visible and red light. Fig. 4 shows an experimental arrangement suitable for a high density mass memory. The glass substrate in form of a disk carries the vitreous film and can be rotated by a stepping motor. The laser light beam is controlled by an electronic logic circuit and an electronic shutter. The red laser line at $\lambda = 6764$Å is used in the case of the selenium-base alloy to measure the transparency of the dot. This enables one to use the same laser and optical alignment for both writing and reading the binary information. The mechanical arrangement resembles that

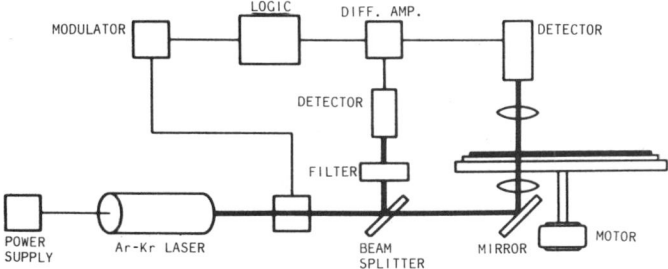

Fig. 4—Experimental arrangement for optical mass memory. The amorphous semiconductor film is a thin coat on the transparent disk carried by the motor.

used for magnetic storage disks. Instead of a magnetic read head, the moving arm carries here a tiny mirror which controls the radial distance of the read and write laser beam. A dot density of about 2×10^7 per sq cm is presently used, but this figure is far below the maximum practical storage bit density. This maximum density is probably determined by optical diffraction effects and the precision of mechanically relocating the laser beam on any given spot.

In view of the rather slow thermal crystallization rates of many of these materials and particularly of the observation that a cooling rate of slightly less than 25°C per min can prevent crystallization, it is remarkable to find crystallization in the chalcogenide film after less than $5\,\mu$sec exposure to a laser pulse. A possible solution to this problem will be discussed in the next section.

In the above case the transformation of the structural state was obtained with a pulse of light. The change in optical properties, the transmission or reflection, for instance, is used to read the binary information. The reversible structure change can also be produced by a current pulse if a memory type amorphous semiconductor*[4] is placed as a thin film between two electrodes.

An oscilloscope trace of the current-voltage characteristic of a memory switching device[11] is shown in Fig. 5. For describing the time sequence of the characteristic let us start with the semiconductor film in the amorphous state. Its resistance will be high (typically 10^6 to 10^7 ohms) and a breakdown process, similar to that observed in the threshold switch,[1,11] has to occur before sufficient current can flow to initiate the structural change along the current filament which forms during the breakdown process. The current has to flow for about 1 m-sec to "set" the memory switch in its "on" state. It remains on even without current flowing. The "reset" current pulse for returning the memory switch into the high resistance "off" state is larger in magnitude than the "set" pulse but only of μsec duration. The "read" operation can be very fast because the memory switch acts like a resistor in either the "off" or "on" state and hence can be read with a speed which is limited only by the lead inductances. More than 10^9 com-

---

*Memory type amorphous semiconductors exhibit an ordering transformation as evidenced by a DTA curve similar to the one shown in Fig. 1. Other amorphous semiconductors do not devitrify within the time span of the experiment before reaching the liquid state. Another class of amorphous semiconductors, such as germanium and silicon films, devitrify at a certain temperature but cannot be brought back to the amorphous state by quenching from the liquid state because their coordination number is different in the two states.

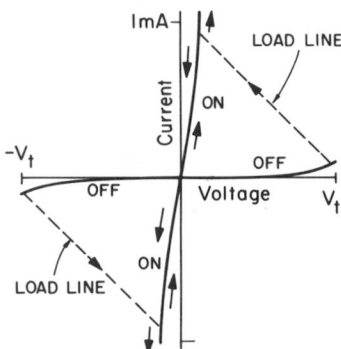

Fig. 5—In the Memory Switch the conducting state is "set" by a voltage pulse of m-sec duration and exceeding the threshold voltage, which in this case is about 10 v. The "on" state is permanently conducting, and the current can pass through it in either direction. A "reset" current pulse of μsec duration restores the "off" state.

plete memory changes are achieved without noticeable deterioration.

This has been described in greater detail before.[10] Of practical interest is the possibility of miniaturizing these switches and of integrating their thin film deposition and photolithographic processing technique with that of conventional silicon diodes. As an example, Fig. 6 shows a 256-bit array of amorphous semiconductor memory switches whose cross points are isolated with silicon diodes.

It is tempting to explain the operation of a memory switch on the basis of the information contained in Fig.

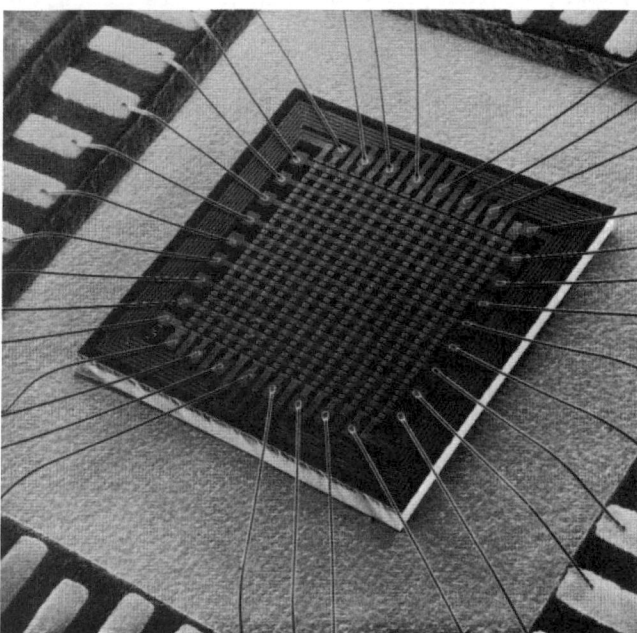

Fig. 6—A micrograph of a 256-bit memory array built By R. G. Neale. The array measured 0.3 cm on a side. Here amorphous semiconductor memory switches are integrated with conventional silicon diodes to form an orthogonally addressable matrix of memory cells. Each memory cell consists of a memory switch in series with an isolation diode and measures 0.01 by 0.012 sq cm.

1 in the following manner. Starting from the high resistance "off" state, the permanently conducting "on" state could be reached by a "set" pulse which raises the temperature of the breakdown path to somewhere between the transformation temperature T_1 and the melting temperature T_M. The "on" state would then be retained in this path until its temperature is raised above T_M by a "reset" pulse of sufficient magnitude and brevity so that the cooling rate is fast and the molten material in the path is quenched to its original high resistance amorphous state. It appears, however, that the dynamics of the transition is not adequately explained by this purely thermal description. This will be discussed more fully in the next section.

DYNAMICS OF THE STRUCTURE TRANSFORMATION

In the laser induced structure transformation crystallization was achieved in the irradiated spot in less than 5 μsec. In the electrical memory switch crystallization was achieved with a current pulse within millisecond. In contrast to these observations, samples of the same material have been annealed in the supercooled liquid region between T_g and the melting point T_M for minutes without crystallization.[8] Furthermore, the crystalline state avoided by quenching from the liquid at a minimum rate as low as 40°C per min. These observations raise the following question. If the thermal crystallization proceeds on the time scale of minutes, why then can the same material be transformed by a current pulse in less than milliseconds and by a laser pulse in a microsecond?

These observations may be understood if the crystallization rate is strongly increased by the presence of large excess concentrations of light-induced or current-induced electron-hole pairs. Dresner, Stringfellow,[12] and others[13] have indeed reported an increase in the growth rate of selenium crystals. An increased rate by a factor of 20 was measured in amorphous selenium at 84°C using an incident illumination intensity of about 3×10^{18} electron-hole pair producing light quanta per sq cm per sec. Even with this level of illumination the growth rate was found to be only 3×10^{-5} cm per sec which is very much slower than the transformation rates observed by us. Evans et al.[10] reported complete crystallization of a 2 μm thick sample of $Ge_{15}T_{81}X_4$ within 0.045 sec. This implies a minimum rate of 4×10^{-3} cm per sec. The crystallization produced by a laser pulse of a few microsecond duration must have proceeded at a rate of about 1 cm per sec. The crystallization of the conducting path in the 2 μm thick film of the memory switch described above required a setting time of a few milliseconds. From this one might estimate a rate of about 0.1 cm per sec. The transformation speeds in the optical and electrical memory units are in general appreciably higher than the rate of photo-crystallization in selenium. This is not implausible since the laser beam intensity is about 10^4 times stronger than that used in the selenium experiments. Furthermore the excess concentration of electron-hole pairs is likely to affect the nucleation rate in addition to the growth rate.

Why electron-hole pairs can increase these rates may be understood in the following manner. The amorphous semiconductors used for this study have a covalently bonded random network structure. The amorphous state is stabilized by the presence of the group V elements which crosslink the group IV and VI elements. In creating an electron-hole pair, a valence bond electron is lifted from a bonding state to an antibonding state. This weakens the structure at sufficiently large concentrations to facilitate diffusion and the formation of polycrystallites.

The reverse process, *i.e.*, the return to the amorphous state, is possible if the "erase" laser pulse in the optical memory or the "reset" current pulse in the memory switch brings the ordered region to a temperature above T_M where the material is disordered. Although upon subsequent cooling the material remains in the crystallization range between T_M and T_1 for a time which is comparable in length with the previously mentioned crystallization times, it fails to crystallize because during that time there are no excess electron-hole pairs. It passes through this range when the light is off and no current is flowing. As a result the disordered state is preserved and the original amorphous material is restored.

SUMMARY

Memory effects are observed in amorphous semiconductors which are associated with a reversible structural change. The structure change can be produced and reversed by an electrical or an optical pulse. The transformation rate into the more ordered state appears to be strongly increased by electrical or optical generation of excess electron-hole densities. Resistivity, optical transmission and reflection, and in general, any physical property strongly affected by the order-disorder transition can be used for reading the binary information.

The amorphous semiconductor materials are fabricated by cathode sputtering and shaped by photolithographic etching processes. The ease of manufacture, high speed, small size, and their inherent insensitivity to impurities make these materials attractive for many computer and control applications.

ACKNOWLEDGMENTS

It is a great pleasure to acknowledge the ingenuity, inventiveness, and invaluable help of R. G. Neale, D. C. Nelson, R. Fleming, and J. Feinleib in the development and characterization of the devices.

REFERENCES

1. S. R. Ovshinsky: *Phys. Rev. Letters,* 1968, vol. 21, p. 1450; Proc. of 1968 Elec. Comp. Conf., Wash., D.C., p. 313.
2. Many recent reports on this subject as well as earlier references can be found in the Proceedings of two conferences, "Semiconductor Effects in Amorphous Solids", New York, May 1969, and "Intl. Conf. on Amorphous and Liquid Semiconductors", Cambridge, England, Sept. 1969, published, respectively, as vol. 2 and 4 of the *J. Non-Cryst. Sol.,* 1970.
3. D. Turnbull and M. H. Cohen: in *Modern Aspects of the Vitreous State,* J. D. Mackenzie, ed., vol. 1, p. 38, Butterworth & Co., Ltd., London, 1960; D. Turnbull: *Phys. Non-Cryst. Sol.,* Proc. Intl. Conf., p. 41, North-Holland, Delft, 1965.
4. H. Fritzsche and S. R. Ovshinsky: *J. Non-Cryst. Sol.,* 1970, vol. 2, p. 148; H. Fritzsche: IBM J. Res. and Dev., 1969, vol. 13, p. 515; E. A. Fagen and H. Fritzsche: *J. Non-Cryst. Sol.,* 1970, vol. 2, p. 170.
5. A. Bienenstock, F. Betts, and S. R. Ovshinsky: *J. Non-Cryst. Sol.,* 1970, vol. 2, p. 347.
6. J. Feinleib and S. R. Ovshinsky: *J. Non-Cryst. Sol.,* 1970, vol. 4, p. 564.
7. J. Stuke: J. Non-Crsyt. Sol., 1970, vol. 4, p. 1.
8. J. Feinleib, J. de Neufville, S. C. Moss, and S. R. Ovshinsky: *Appl. Phys. Letters* (in print).
9. J. Feinleib, J. de Neufville, and S. C. Moss: Bull. Am. Phys. Soc. II, 1970, vol. 15, p. 245.
10. E. J. Evans, J. H. Helbers, and S. R. Ovshinsky: *J. Non-Cryst. Sol.,* 1970, vol. 2, p. 334.
11. The threshold switch is characterized in several articles in *J. Non-Cryst. Sol.,* 1970, vol. 2. See also H. K. Henisch: Sci. Ann., 1969, vol. 221, p. 30.
12. J. Dresner and G. B. Stringfellow: *J. Phys. Chem. Solids,* 1968, vol. 29, p. 303.
13. I. A. Paribok-Aleksandrovich: *Soviet Phys. Solid State,* 1970, vol. 11, p. 1631.

RADIATION HARDNESS OF OVONIC DEVICES

Stanford R. Ovshinsky,[1] E.J. Evans,[1] D.L. Nelson,[1] and H. Fritzsche[2]

[1]Energy Conversion Devices, Inc., Troy, Michigan 48084
[2]James Franck Institute, University of Chicago, Chicago, Illinois 60637

Introduction

If a semiconductor device is irradiated by, for instance, X-rays a large number of electron-hole pairs are created which cause a change in the equilibrium currents. Sometimes these changes are enhanced by appreciable gain factors of the device and remain long after the radiation pulse has ceased. Neutron radiation causes atomic displacements and lattice defects which, in turn, result in electrically active centers or traps in the gap and, thereby, alter the bulk and surface properties of the semiconducting material.

In contrast to this, one expects amorphous materials to be much less affected by radiation because of their intrinsic lack of long range order. Care has to be taken to prevent radiation-induced devitrification.[1] One of the reasons for their greater radiation tolerance is that the amorphous and vitreous semiconductor materials contain a high concentration of traps and recombination centers in the energy gap which minimize the effect of ionizing radiation and of local rearrangements of atoms due to neutron bombardment.

During the recent years several devices[2-6] have been developed at Energy Conversion Devices, Inc., in which the principal functional element is a thin film of amorphous semiconductor. Because of their inherent bistable characteristics, the Ovonic switches may be electrically biased to form circuit functions required for computers and logic circuits. Flip-flops, astable multi-vibrators, gates, shift registers, and oscillators are among many circuits which have been built using as active elements only Ovonic threshold switches (in the following abbreviated as OTS).[7,8] The Ovonic memory switch[2,3] (abbreviated as OMS) is another device based on the amorphous semiconductor technology. It maintains bit information without the requirement for power and allows for non-destructive readout.

Motivated by the simplicity of the basic structure of these switching devices--an amorphous film between two electrodes--which adds the advantage of integrated circuit fabrication and microminiaturization to the broad range of circuit functions realized by these devices, a program was undertaken to test their tolerance to ionizing radiation transients and to radiation damage by fast neutron irradiation.

This report will be confined to radiation tests performed on Ovonic threshold switches (OTS) with fast neutrons and with broad band X-rays from the FX-1 Superflash X-ray facility of Ion Physics Corporation in Burlington, Mass. Radiation tolerance experiments on Ovonic memory switches (OMS) are presently under way. They show equally promising results.

The novelty of the Ovonic device technology requires first a brief resume of the construction and the properties of the OTS devices. The experimental details and results of the radiation tests will then be presented. In the Conclusions, a brief theoretical discussion of the material properties will be given as they relate to the radiation testing.

OTS Device Construction and Performance

The reversible electrical switching phenomena described below have been ob-

Reprinted by permission from *IEEE Transactions on Nuclear Science*, Vol. 15, pp. 311–321 (1968).

Fig. 1. OTS device in DO-7 package. The thickness of the amorphous semiconductor film between the carbon electrodes A and B is greatly exaggerated. S = Spring.

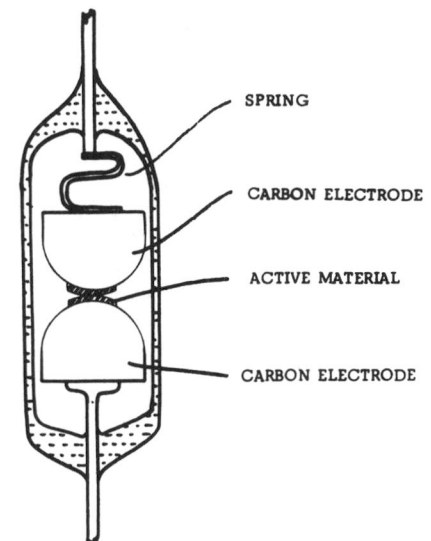

SPRING

CARBON ELECTRODE

ACTIVE MATERIAL

CARBON ELECTRODE

served in various types of disordered materials, particularly in inorganic glassy polymers, covering a large composition range. The switching devices used for the irradiation tests consisted of a vitreous chalcogenide material containing (in atom percent) 48°/₀ tellurium, 30°/₀ arsenic, 12°/₀ silicon, and 10°/₀ germanium. The composition is chosen such as to maximize the resistance against devitrification.

The material can be evaporated, cathode sputtered, or hot pressed so that it forms an approximately 10^{-4} cm thick film between the polished surfaces of the two carbon electrodes as shown in Fig. 1. For the devices tested the film was evaporated. The contact resistance between the electrodes and the film is low compared to the bulk resistance and no p-n junction exists in the amorphous film. The spring pressing the carbon electrodes together exerts a force of about 50 grams. The curvature of the carbon electrodes affects the area of contact which is typically 10^{-4} cm². The resistances of these devices were between 1 and 10 MΩ at low voltages and room temperature.

As the voltage V is increased, the current I begins to increase faster than ohmic and above 2 volts reaches a region in which

$$I = I_0 \exp(V/V_0) . \qquad (1)$$

As shown in Fig. 2, the OTS device switches from the highly resistive to a conducting state when the applied voltage

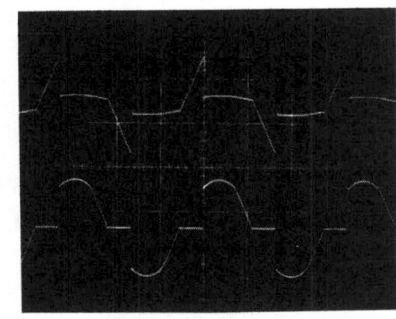

a) OTS Voltage
   10 Volts /div. -vert.
   5 ms./sec. - hor.

b) OTS Current
   10 ma./div.-vert.
   5 ms./sec.- hor.

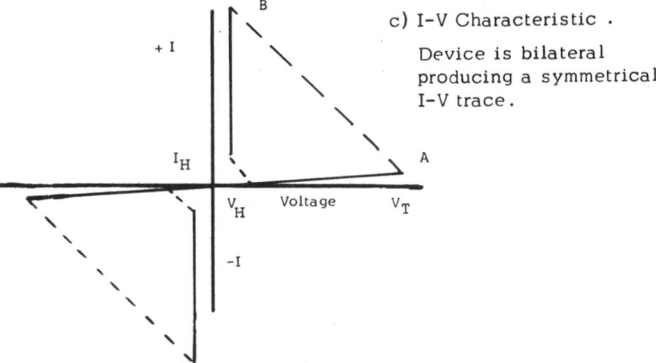

c) I-V Characteristic .

Device is bilateral producing a symmetrical I-V trace.

Ovonic Threshold Switch behavior in a series circuit consisting of an OTS ,a 2.2K load resistor, and a 60 Hz power supply.

Fig. 2. Response of OTS device to 60 Hz voltage:

Part (a): Voltage across the device. Vertical scale 10 Volts per division. Horizontal scale 5 msec per division.

Part (b): Current through the device. Vertical scale 10 mA per division. Horizontal same as above.

Part (c): I-V characteristic.

exceeds a threshold value V_T. Fig. 2 shows oscilloscope pictures of (a) the voltage across the device, and (b) the current passing through the device as a function of time. In this case 60 Hz a.c. voltage was applied across the device and a 2.2×10^3 ohm load resistor. Fig. 2(c) shows a redrawn curve of the I-V characteristics. The conducting state is characterized by a holding voltage V_H and a holding current I_H. When the current is decreased below I_H, the OTS device returns to its original highly resistive state. The switching time from point A to B can be as small as 10^{-10} seconds. It was less than 10^{-9} seconds for the devices tested.

The OTS devices are inherently symmetric with respect to the polarity of the applied voltage. It remains symmetrical even when the electrodes are of different contacting areas or of different materials. The switching process is repeatable. Alternating or full wave unfiltered rectified voltage have been applied continuously to devices over periods of many months without noticeable change in their characteristics.

The threshold voltage V_T increases nearly linearly with the thickness of the amorphous semiconducting film. The holding voltage V_H is very weakly dependent on film thickness. The test devices had approximately the following characteristics: $V_H = 1V$, $I_H = 0.5$ mA, $V_T = 10, 20, 30,$ or 40 V, and a dynamic resistance in the conducting state of 5 ohms.

Results of Neutron Irradiation Tests

Pulsed Neutrons

In April 1967, four OTS devices having V_T 30V were subjected to a fast neutron fluence of 1.8×10^{14} n/cm^2 (E > 10 KeV) at the Harry Diamond Laboratory Triga reactor.[9] When examined four weeks later the devices showed no change in threshold current I_T, voltage V_T, and switching time. Some statistical fluctuation of V_T was observed. It was about one percent of V_T before and after the test.

Five high current OTS devices, referred to as S-50 switches, which can carry a 7 Ampere pulse for 10 msec were exposed at the same facility to 6×10^{14} n/cm^2. These devices have metal electrodes for improved heat sinking but they show the same electrical characteristics. Post-radiation measurements indicated a drop of up to 20% in V_T and a small increase in V_H. The other OTS parameters

remained unchanged. Eight low current OTS devices having V_T between 10 and 30 V tested at that time showed no change in parameters. It was later discovered that mechanical shocks which occurred while mounting the experimental S-50 devices into the reactor could have caused the observed changes in V_T and V_H. It can be seen from Fig. 1 that the spring and the loose fit of the electrodes in the glass tube make this package quite sensitive to shock. This difficulty is avoided in all film units, in which the active material is sandwiched between evaporated metal electrodes. Such units are presently being made and will be tested in the near future.

14 MeV Neutrons

In August 1967 five OTS with V_T=20V and five S-50 devices were exposed[9] for 12 hours at the Cockroft-Walton 14 MeV neutron generator of General Atomic, La Jolla, California. A total fluence of 1.5×10^{13} n/cm^2 was obtained. The V_T=20V devices showed no change. The S-50 devices exhibited a threshold voltage V_T between 18 and 19 V but they recovered to their original V_T=35V on subsequent switching operations. Again the experimental S-50 devices were noted for some mechanical internal instability so that the observed effect cannot be attributed to radiation until shock resistant devices are tested.

Fast Neutron Fluence of 1.2×10^{17} n/cm^2

The most severe fast neutron radiation test[10] to date was conducted on July 14, 1968. A delay line stabilized relaxation oscillator circuit, consisting of one OTS, two carbon resistors, one ceramic capacitor, and an electromagnetic delay line, was inserted in the fast neutron flux of the water moderated steady state nuclear reactor at the University of Michigan. The circuit is discussed in detail in the Appendix. In order to accumulate a high fast neutron fluence it was necessary to locate the circuit in a high flux region which had a correspondingly high gamma rate. The resultant gamma heating caused the circuit which was potted in RTV rubber to rise to a temperature of about 100°C at package center. The circuit temperature was monitored with a copper-constantan thermocouple. The power supply voltage was adjusted for optimum operation after the circuit temperature had stabilized in the neutron flux.

The oscillator circuit was exposed for a period of 22.5 hours during which it accumulated a total integrated fast neutron fluence of 1.2×10^{17} n/cm^2. In addition to the measurement of the oscillation period, it was possible to determine at intervals during the exposure whether or not any changes in V_T occurred during the test. The methods used to determine these device parameters are discussed in the Appendix.

Up to 6×10^{16} n/cm^2, the threshold voltage was measured repeatedly. It varied between 6.5 and 8 Volts. The variations in V_T could be correlated with corresponding changes in the circuit temperature which varied between a low of 82°C and a high of 118°C. Since the temperature dependence of V_T is about 0.5°/₀ per degree, the temperature variations were sufficient to cause the observed V_T changes within the ±0.5 Volt error of measuring V_T. During this exposure the oscillation period remained at the value of 3.6 microseconds which is determined by the length of the delay line.

The next and final data were taken at a fluence of 1.2×10^{17} n/cm^2. The oscillator still functioned at this high fluence. Its performance deviated, however, from that observed on earlier data points in the following ways: (1) The threshold voltage had increased to 10 Volts (T = 105°C), (2) the oscillation period had decreased to approximately 3.4 microseconds, and (3) the output voltage had reduced in amplitude. Since the oscillator was well stabilized at this new frequency it can be concluded only that degradation in the delay line had occurred. Degradation of insulating materials is a prime suspect for this change since most insulators are significantly affected at these radiation levels. The apparent increase in threshold voltage accompanied by a reduction in output signal also suggests that passive components or the cable which shunts the OTS are probably the cause for the measured threshold voltage change since an increase in V_T normally results in this circuit in an increase in output signal.

Results of Flash X-Ray and 2.5 MeV Electron Tests

Several OTS devices were tested[11] in April 1968 at the FX-1 Superflash X-ray facility of Ion Physics Corporation in Burlington, Massachusetts. The purpose of the test was (1) to measure the photoresponse of an OTS as a function of gamma dose rate and (ii) to determine the minimum dose rate which initiates OTS switching for different preset bias voltages $V < V_T$.

The FX-1 facility consists of a 2.5 MeV Van de Graaf which can direct an electron beam either directly at the test sample or onto a heavy metal target. In the latter case a broad band Bremsstrahlen spectrum is emitted to which the sample is exposed. In either of the two modes the radiation pulse lasts for about 20 nsec.

Because the behavior of the OTS devices during the irradiation pulse was to be studied, the devices were tested in the circuit shown in Fig. 3. This circuit provides a constant bias voltage V when $V < V_T$, and serves as a relaxation oscillator when the device threshold voltage V_T is exceeded. Since the blocking resistance of the OTS is much larger than 5.7K, the voltage at the capacitor C yields the bias voltage V when $V < V_T$ and also the threshold voltage V_T as the amplitude of the relaxation oscillations.

Since the photo current response of the OTS device is expected to be small compared to other noise voltages occurring during the X-ray flash, it was found necessary to form a bridge with the additional resistors $R_3 = 4.7K$ and R_4 equal to the OTS blocking resistance under bias condition. A wide band differential amplifier (40 MHz) was used as the null instrument to detect a 20 nsec photo response pulse or the current pulse which would result from the discharging of C if the X-ray flash induced the OTS to switch.

The circuit with a standard TLD dosimetry capsule next to the OTS device was mounted on the pole face of the Superflash X-ray facility, which was operated in the Bremsstrahlen mode. All other test components were shielded with lead bricks.

Three different OTS switches with V_T = 10V, 20V, and 30V were subjected to a dose rate $\dot{\gamma} = 1.8 \times 10^{11}$ rad/sec on

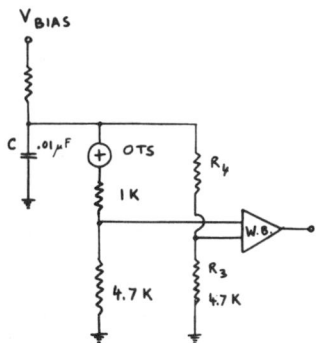

Fig. 3. Test circuit for OTS device used in superflash X-ray unit.

forty successive flashes. The results are the following.

No transient switching was ever observed although the bias voltage was pre-set to within 3% of V_T on 10 of the 40 gamma pulses. On all tests the values of V_T before and just after the pulses agreed. No photocurrent could be detected. There was no difference between the transient output of a circuit with or without the OTS device in place. Furthermore, a null experiment was performed with a shielded resistor of 100KΩ, the same as R_4, in place of the OTS device. From this we estimate that the photocurrent must have been less than 0.2 mA.

When the OTS device in the test circuit was placed directly in the electron beam, transient switching was observed in every case. This was found to be caused by voltages induced in the signal leads rather than by the direct effect of the impinging electrons because the transient switching continued even after a 1-inch thick aluminum shield was placed over the device. Rather small induced voltages caused the devices to trigger because these were biased near V_T.

It seems important to point out that no device degradation occurred from the direct exposure to more than five electron beam pulses and that the total energy deposition from the beam was not sufficient to cause thermo-mechanical damage to any of the mechanical packages (DO-7 packages).

Discussion and Conclusions

Ovonic threshold switches made of amorphous semiconducting materials have been shown to withstand fast neutrons in excess of nvt = 1.2×10^{17} n/cm^2 and up to 2×10^{11} rad/sec gamma pulses. Moreover, the devices appear not to be significantly affected by transient pulses of flash X-ray to levels of at least 2×10^{11} rads/sec. It was demonstrated that the devices can still function at a fast neutron fluence which is considered marginal for most passive components today and is well above the anticipated neutron fluence levels for most military applications of electronics.

The radiation insensitivity is not limited to this particular OTS device. It is characteristic of the amorphous semiconductor materials of which they are made. Therefore, it is reasonable to expect that a similar radiation tolerance will be observed with essentially all devices made of amorphous semiconductors.

Since the Ovonic technology began only recently after the discovery by Ovshinsky of the anomalous high field transport properties of certain materials, one can expect a family of devices with this radiation tolerance.

To demonstrate that it is the amorphous material to which the high tolerance is to be attributed let us explore to what extent our present understanding of these substances permits us to estimate the radiation tolerance.

Since the early work on the vitreous state by A. F. Ioffe[12] and B. T. Kolomiets,[13] the amorphous semiconductors have received increasing attention during recent years.[14-16] Although this field is still in its infancy several novel features begin to appear which set the disordered structures apart from the crystalline materials. In the following we shall discuss those features which seem important for the understanding of the reversible switching effects and the high radiation tolerance of the Ovonic devices.

The amorphous semiconductors fall into three major categories. These are characterized by the dominance or the combination of certain kinds of structural disorder as follows:

Elemental amorphous materials, such as amorphous Te, Se, Ge, or Si, are characterized by positional disorder of the atoms, or groups of atoms like tetrahedral groups or other structural elements. Structural faults, such as vacancies and voids, may give rise to acceptors or, less frequently, to donors. If the local bonding and the size of the atoms favor statistically some types of structural faults then localized states may be found concentrated in certain energy regions within the band gap.

Covalent amorphous alloys, such as the chalcogenide or boron glasses, mixed oxide glasses, and inorganic amorphous polymers. These materials possess compositional disorder in addition to positional disorder. They have band gaps less than 2 eV.

Molecular amorphous materials, such as oxides and nitrides, with band gaps larger than 2 eV. Examples are SiO, Al_2O_3, BN, Si_3N_4, and Ta_2O_5. Into this category also belong liquid insulators like liquid ammonia and anthracene. They contain positional disorder but because of the tight binding situation and the small overlap of neighboring orbitals, these materials may contain as an additional type of disorder foreign atoms or, as a result of deviations from stoichiometry, atoms in different valence states. These often act as donors or acceptors and form a narrow band of states within the band gap of the host material.

Because the materials of the second category are of greatest interest to Ovonic devices we restrict the following discussion to them. These materials remain intrinsic, in the sense of exhibiting a thermal activation energy of the resistivity which is nearly one-half the optical absorption edge energy, over a range of composition considerably greater than that produced by dopants or by departures from stoichiometry in crystalline semiconductors. This is the result of the positional and compositional disorder which allows most atoms to complete their valence state and thereby establishes a high degree of compensation. Since the energy gap of semiconductors is predominantly determined by the bond arrangement in the immediate vicinity of the atoms, one expects in these inorganic glassy polymers an energy gap with a high density of localized states tailing in from both the conduction band and the valence band.

Fig. 4a is a schematic drawing of the density of states against electron energies. Since the band widths normally increase with increasing energy one expects the conduction band edge to be more smeared out than the valence band edge.

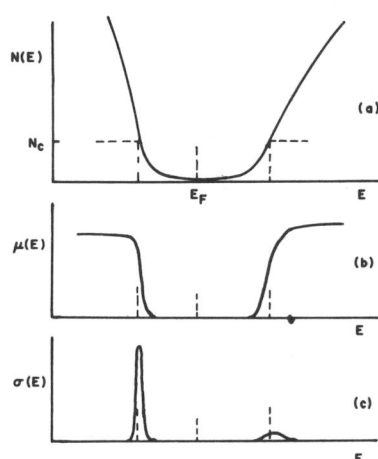

Fig. 4. Model of an amorphous semiconducting alloy of the second category (see text). Schematically drawn as a function of electron energy are in Part (a) the density of states $N(E)$, in Part (b) the mobility $\mu(E)$ of holes and electrons, respectively, and in Part (c) the differential conductivity $\sigma(E)$.

Approximately half of the gap states originate from either band. The Fermi level E_F will, therefore, be close to the center of the gap. Field effect measurements[17] indicate that near E_F the density of gap states lies between 10^{19} and 10^{20} cm^{-3} eV^{-1}. This agrees with the observation that composition changes of the order of one percent do not noticeably change the electrical characteristics of these materials.

Near the center of the gap the gap states are strongly localized. The small overlap of the wave functions of neighboring centers leads to a low probability for an electron exchange or hopping conduction.[18] The states become increasingly delocalized and conducting as one approaches the bands. The mobility $\mu(E)$ as a function of energy is sketched for this situation in Fig. 4 part b. Since fewer states lie within a given energy interval near the conduction band edge as compared to the valence band edge the mobility function rises less steeply at the conduction band edge. The leveling off of the mobility probably occurs near the critical concentration N_c, discussed by Mott,[15,19] which separates the region of localized states from the region of nonlocalized states. In the energy region for which $N(E) < N_c$ the mobility $\mu(E)$ is governed by hopping and decreases very rapidly with $N(E)$.

A differential conductivity $\sigma(E)$ due to electrons and holes may be defined by multiplying $N(E)$ $\mu(E)$ with the charge and the appropriate Fermi distribution function $f(E)$ and $[1-f(E)]$, respectively. Fig. 4, part c shows $\sigma(E)$. The area under these curves represents the total conductivity. The sudden drop of $\sigma(E)$ on the band gap sides is caused by the rapid drop in the mobility functions $\mu(E)$. The drop of $\sigma(E)$ on the outer sides is determined by the fact that the concentration of electrons and holes, respectively, falls off exponentially with the energy separation from the Fermi level. The peaked shape of $\sigma(E)$ explains why the temperature dependence of the conductivity of these materials exhibits a definite activation energy despite the large number of gap states. The conduction is predominantly by holes because the valence band is less smeared out as explained above.

The large concentration of gap states has several important consequences for the functioning of OTS devices and for their high radiation tolerance. It may be of interest to discuss some of these presently.

1. They reduce the thickness of the space charge layer near the electrodes to about 30Å which is sufficiently thin to assure low resistance contacts. The presence of a large density of gap states make it, therefore, possible to have a well-insulating material in which the conduction is bulk limited rather than contact limited.

2. The large overlap in energy of those levels originating from the valence band with those from the conduction band produces near the center of the gap a large density of charged traps. These traps, when occupied, can easily be ionized at moderate field by lowering the trap depth and reducing the capture cross section.[20] The close spacing of the localized states in both space and energy makes it likely that internal field ionization from these states play an important role in the carrier generation necessary to initiate the conducting state.

3. With the Fermi level locked by the high gap state density, changes in composition by as much as one percent produced, for example, by neutron capture and subsequent nuclear transmutation can be tolerated. Also, a concentration of radiation-induced atom displacements equal to the concentration of states in the gap is not likely to produce changes in the electrical characteristics.

4. A large fraction of the gap states near the Fermi level are expected to be charged because in this energy range the states stemming from the conduction and from the valence bands overlap. These states will act as efficient recombination centers. As a consequence the recombination lifetime is very short and the excess carrier concentration remains small even at high radiation dose rates.

Using the above model, let us estimate the tolerance to photo transients produced by a strong pulse of ionizing radiation.

Optical absorption data yield an energy gap of about $E_g = 0.9$ eV for the material used. The carrier concentration at 300°K is then about $n_o = 10^{12}$ cm^{-3} assuming a density of state mass equal to the free electron mass. With the resistivity of the material being $\rho = 2 \times 10^7$ ohm-cm the carrier mobility is of the order of 1 cm^2/V sec and the conductivity relaxation time is about 10^{-15} sec.

The carrier generation rate g may be estimated[21] by assuming that

$$g = \dot{\gamma} \, g_o Z/Z \, (Si) \, . \qquad (2)$$

Here $\dot{\gamma}$ is the dose rate in rad/sec referred to silicon, g_o is the electron-hole pair generation constant per rad, and $Z/Z \, (Si) = 3$ is the ratio of the average atomic weight of the amorphous material to the atomic weight of silicon. Because of the proportionality of g_o with the band gap, we can assume g_o to be that of silicon, $g_o \simeq 4 \times 10^{13}$ cm^{-3} rad^{-1}.

For the present case of $\dot{\gamma} = 1.8 \times 10^{11}$ rad/sec one obtains with Eq. (2) a generation rate $g = 2 \times 10^{25}$ pairs/cm^3 sec. The resulting excess carrier concentration, Δn, can be estimated using the usual rate equation at equilibrium

$$d(\Delta n)/dt = g - \Delta n/\tau = 0 \qquad (3)$$

which lead to

$$\Delta n = g\tau \, . \qquad (4)$$

The recombination lifetime τ can be estimated from N_t, the concentration of recombination centers, σ_t, their capture cross section, and $v = 10^7$ cm/sec, the thermal velocity of the carriers:

$$\tau = (N_t \, \sigma_t \, v)^{-1} \, . \qquad (5)$$

In accordance with our model we assume $N_t \simeq 10^{19}$ cm^{-3} charged recombination centers near the center of the gap. Their cross section will be limited by their average spacing (50Å) so that $\sigma_t \simeq 2.5 \times 10^{-13}$ cm^2. Substituting these values into Eqs. (4) and (5) one obtains $\tau = 4 \times 10^{-14}$ sec and $\Delta n = 8 \times 10^{11}$ cm^{-3}. This excess carrier concentration would have produced a photo current which is much smaller than the 0.2 mA detection limit of the flash X-ray test. It appears, therefore, plausible that the high tolerance to photo transients observed in these amorphous semiconductors is the result of the short recombination lifetime which in these materials is more than six orders of magnitude shorter than the lifetime of minority carriers in silicon transistors.

Acknowledgments

We gratefully acknowledge several illuminating discussions with M. H. Cohen and K. W. Böer, and the devoted help of I. M. Ovshinsky and of the staff of Energy Conversion Devices, Inc.

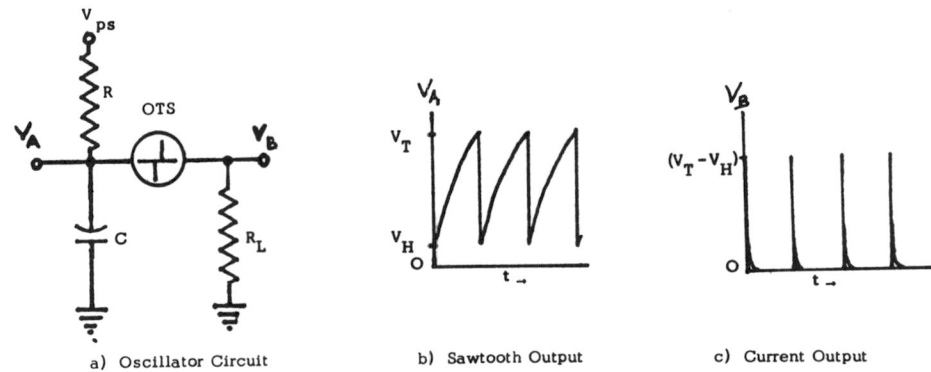

a) Oscillator Circuit　　　b) Sawtooth Output　　　c) Current Output

Fig. 5.　Simple relaxation oscillator
circuit with OTS device.　The wave forms
at points A and B are shown.

Appendix

Description of Delay Line Oscillator

The electrical operation of a delay
line stabilized oscillator incorporating
an Ovonic Threshold Switch (OTS) as the
active element will be described.　This
circuit was used for the neutron irradiation test at the University of Michigan.

The simplest form of oscillator using this device is the relaxation oscillator shown in Fig. 5.　Assume that
initially no power is applied to the
circuit and the OTS is in its nonconducting state.　When power is applied,
the capacitor, C, charges through resistance, R, toward the power supply
voltage at an exponential rate.　When the
voltage across the capacitor is equal to
the threshold voltage of the OTS, the OTS
fires discharging the capacitor through
R_L.　If R is selected to be large enough
so that it cannot supply a current exceeding the holding current I_H to the OTS,
the OTS will turn off and the cycle will
repeat as indicated in the wave form
figures of Fig. 5.　The resistance R_L is
usually chosen to be much less than R so
that the wave form at point A approximates a saw tooth wave and the wave form
at point B is a series of voltage spikes
with a decay time $R_L C$.　The frequency and
amplitude of this oscillator are strongly
dependent on both the OTS threshold voltage and the power supply voltage.　The
delay line stabilized oscillator discussed below alleviates these dependen-
cies.

The delay line stabilized oscillator
shown in Fig. 6 is a modified relaxation
oscillator which incorporates an electromagnetic delay line for frequency and
amplitude stabilization.　The simple
oscillator was modified first by substitution of a 1.8 microsecond delay line
for the load resistor R_L and, secondly,
by the addition of a small resistor in
series with the capacitor to monitor the
capacitor discharge current.　The delay
line was terminated in a short at one end
so that any pulse sent into the line will
return inverted after a period of twice
the delay line length (3.6 microseconds).
Upon application of power supply voltage
V_{PS} to this circuit, the capacitor, C, is
charged through resistor R until its
voltage exceeds the threshold voltage of
the OTS.　The OTS then fires discharging
the capacitor into the delay line generating a positive pulse at the delay
line input.　The positive pulse traverses
the delay line arriving at the shorted
end in 1.8 microseconds.　The pulse is
then inverted and reflected back to the
input arriving 3.6 microseconds after its
initiation.　The amplitude of this reflected pulse in a lossless delay line
would be double the input amplitude; however, due to the finite loss of the delay

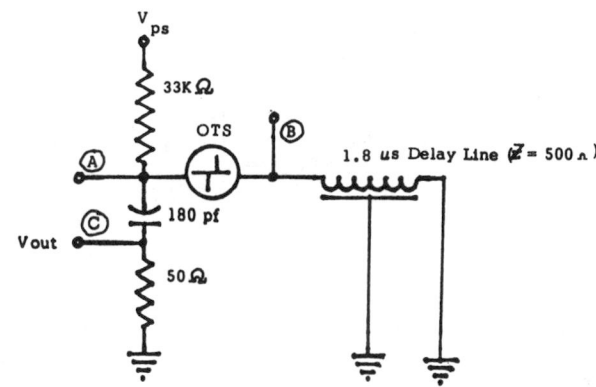

Fig. 6.　Delay line stabilized oscillator
with OTS device.　The wave forms at
points A, B, and C are shown in the next
figure.　V_{PS} = power supply voltage.

181

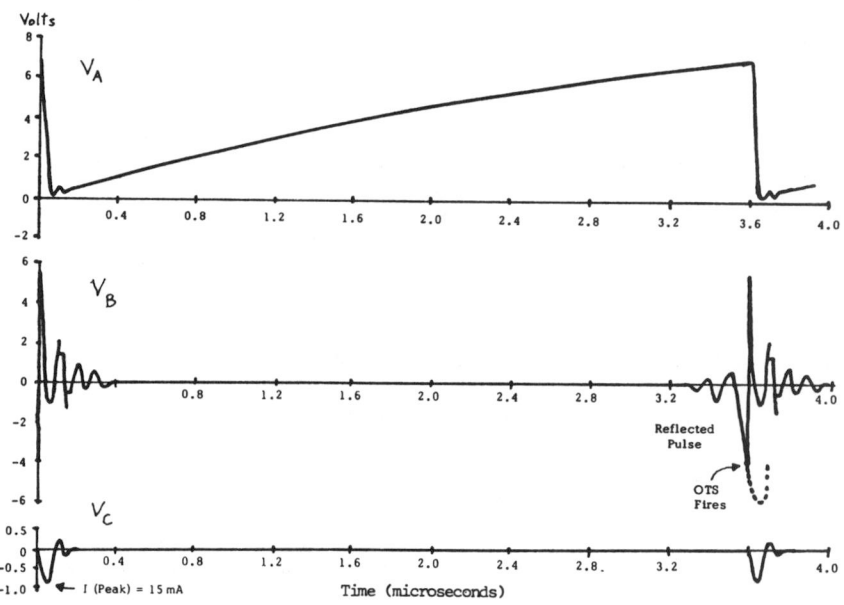

Fig. 7. Wave forms observed as a function of time at points A, B, and C of circuit shown in previous figure.

line the pulse amplitudes of the reflected pulse is approximately equal to the initiation pulse.

During this 3.6 microsecond period, capacitor C is again being charged through resistor R since the OTS is non-conducting. The time constant RC is selected so that the voltage across the capacitor, 3.6 microseconds after firing the OTS, is not yet equal to the threshold voltage V_T. With the arrival of the reflected inverted pulse, the voltage across the OTS exceeds its V_T and the OTS again fires generating a new positive pulse at the delay line input repeating the oscillation cycle. As mentioned above, the RC time constant must be long enough so that firing will be initiated by the reflected delay line pulse. If too long a time constant is used, it is possible for the delay line oscillator to lock on a subharmonic with one-third frequency since the voltage at the first pulse reflection is not sufficient to fire the OTS.

Fig. 7 shows typical wave forms at points A, B, and C of Fig. 6. As with the standard relaxation oscillator, the voltage at point A resembles a saw-tooth wave form. The second wave form shows the voltage at point B across the delay line input. As can be seen from this figure, considerable ringing occurs on the delay line immediately after the OTS fires. This ringing is used to advantage

since it provides a forced turnoff for the OTS thereby allowing the charge resistor R to be reduced since the requirement that it supply less than holding current is no longer valid. Also indicated on this figure (dotted curve) is the assumed shape of the reflected pulse if the OTS threshold voltage had not been exceeded. The third wave form of Fig. 7 is the voltage V_C which is proportional to the discharge current in the capacitor showing the 15 milliamp 100 nanosecond current pulse. This voltage was selected to monitor the oscillator performance during the reactor tests since long coaxial cables connected at this point did not disturb the circuit operation. The wave forms indicated in Fig. 7 were measured with a power supply voltage of approximately 18 Volts. This results in the capacitor charging to approximately 7 Volts prior to firing. This voltage added to the amplitude of the reflected delay line pulse defines the maximum OTS threshold voltage which can be tolerated. As the power supply voltage is reduced, the oscillator will unlock from the fundamental mode and begin to oscillate in the one-third subharmonic mode. This dependence on power supply voltage range can be used as an indicator of changes in threshold voltage. The actual OTS threshold voltage can also be determined by reducing the power supply voltage to zero and gradually increasing it until a first firing is observed. Since the right side of the OTS is at ground and the left side is approximately at power supply voltage prior to firing, the power supply voltage necessary for first firing is equivalent to the device threshold voltage.

When the temperature of the oscillator circuit is increased from room temperature to 100°C, the optimum power supply voltage reduces to about ten volts which is consistent with the expected reduction in threshold voltage of the OTS.

Small variations in oscillator frequency can occur due to distortion of the reflected pulse from the delay line and variation in the OTS turnon delay time. Typical frequency variations due to these effects are 1 KHz out of 280 KHz.

References

1. A. N. Bobrova and E. M. Lobanov, in Radiation Effects in Solids (Akad. Nauk SSSR, Moscow, 1963); J. T. Edmond, J. C. Male, and P. F. Chester, J. Sci. Instr. 1, 373 (1968).

2. S. R. Ovshinsky, U. S. Pat. No. 3,271,591.

3. S. R. Ovshinsky, Proceedings of the IV Symposium on Vitreous Chalcogenide Semiconductors, Rochester, 1967.

4. S. R. Ovshinsky, Proceedings of the International Colloquium on Amorphous and Liquid Semiconductors, Bucharest, Rumania, 1967.

5. S. R. Ovshinsky, Proceedings of the IV Symposium on Vitreous Chalcogenide Semiconductors, Leningrad, USSR, 1967.

6. G. Sideris, Electronics, Sept. 1966, pp. 191-195; J. D. Mackenzie, Electronics, Sept. 1966, pp. 129-136.

7. J. A. Perschy, Electronics, July 1967, pp. 74-84.

8. E. J. Evans, Picatinny Arsenal Technical Report 3698 (1968).

9. The tests were conducted by Picatinny Arsenal, Dover, New Jersey, and the results are summarized in Ref. 8.

10. Test conducted by D. L. Nelson, Energy Conversion Devices, Inc., Troy, Michigan.

11. Tests conducted by E. J. Evans, under the auspices of Picatinny Arsenal, Dover, New Jersey.

12. A. F. Ioffe and A. R. Regel, in Progress in Semiconductors, A. F. Gibson, Ed. 4, 239 (1960).

13. See review article by B. T. Kolomiets, Phys. Stat. Sol. 7, 359 (1964); ibid., 7, 713 (1964); and numerous later publications.

14. J. Tauc, Science 158, 1543 (1967).

15. See review article by N. F. Mott, Adv. in Phys. 16, 49 (1967).

16. J. T. Edmond, Brit. J. Appl. Phys. 17, 979 (1966).

17. H. Fritzsche, E. A. Fagen, and S. R. Ovshinsky, to be published.

18. N. F. Mott and W. D. Twose, Phil. Mag. 10, 107 (1961).

19. N. F. Mott, Phil. Mag. 17, 1259 (1968).

20. G. A. Dussel and R. H. Bube, J. Appl. Phys. 37, 2797 (1966).

21. Frank Larin, Radiation Effects in Semiconductor Devices, John Wiley and Sons, Inc. (1968), pp. 15-16.

22. C. A. Klein, J. Appl. Phys. 39, 2029 (1968).

REVERSIBLE HIGH-SPEED HIGH-RESOLUTION IMAGING IN AMORPHOUS SEMICONDUCTORS

S.R. Ovshinsky and P.H. Klose

Energy Conversion Devices, Inc., Troy, Michigan 48084

INTRODUCTION

Energetic processes have been described for fast and controllable ordering and disordering of amorphous semiconductors.(1,2,3,4,5) Concomitant with these changes are alterations of various physical parameters, all of which we use for the temporary or permanent storage and display of information (6). The present paper reports on some physical characteristics associated with photostructural changes reflecting imaging processes in amorphous materials.

MATERIAL PREPARATION

The preparation of the thin films (7) used for storage or display is usually done by vacuum deposition of the prereacted bulk material at a system pressure of less than 10^{-5} mm Hg. Single layer structures and multiple layer structures are deposited by this technique on a wide variety of substrates. Most commonly used are acetate, mylar, and glass. During the vacuum deposition the substrate temperature is kept below or above the glass transition temperature of the material depending on whether an amorphous or a crystalline film is preferred. The end use of the film determines its thickness which generally ranges from .1 micron to several micron. Aside from the preparation of the bulk material the vacuum deposition is the first and final step in the actual production of photostructural film.

BASIC CHARACTERISTICS OF PHOTO STRUCTURAL MATERIALS

The chemical composition of these materials is designed to fill the requirements for reversible light induced crystallization. In bulk form, one typical composition ($Te_{82}Ge_{16}Sb_2$) gives rise to the characteristic DTA (differential thermal analysis) (8) curves shown in Figure 1. Trace (a) shows a heating cycle. At the transition tempera-ture (T_1= 230°C) crystallization occurs as evidenced by the exotherm. The large endotherm at 370°C represents the extra energy required to melt the material. A relatively fast cooling cycle (50°C/min) shown in trace (b) "freezes in" the amorphous structure. A slow cooling cycle allows for crystallization to occur. (Note exotherm at 350°C trace (c). Trace (d) is an example of a fast heating cycle such that no crystallization occurs upon heating. Thus in such a material the amorphous or crystalline state can be obtained by the application of suitable shaped pulses of thermal energy. The slow time scale of the DTA measurement (such measurements are usually carried out over a period of several minutes) permits quasi equilibrium conditions to exist within the material and no unusually high crystallization rates are observed. However, if a material such as the one

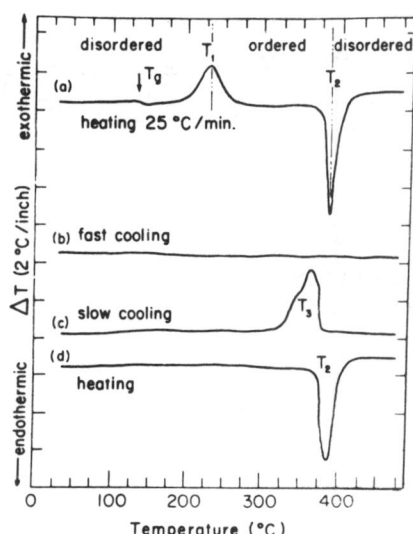

Figure 1. Differential thermal analysis of a chalcogenide alloy suitable for memory applications. Traces (a) and (d) are heating curves, (b) and (c) are cooling curves.

Reprinted by permission from *Society for Information Display, 1971 International Symposium, Digest of Technical Papers*, pp. 58–61 (May 1971).

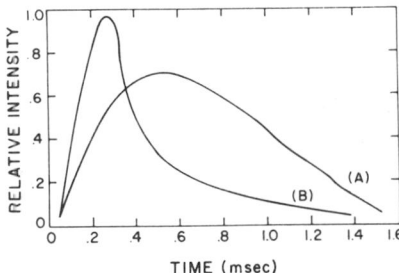

Figure 2. Energy profile of Xenon flash used to crystallite the memory film (a) and to restore the original amorphous structure (b).

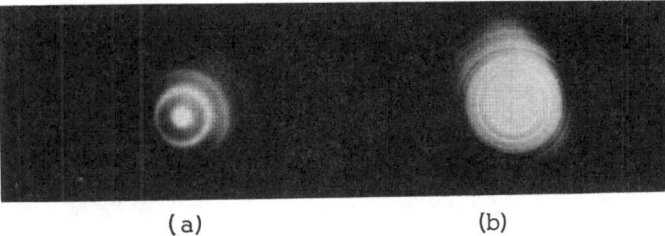

(a)   (b)

Figure 3. Electron diffraction pattern of amorphous film before exposure (a) and in the crystalline state after exposure to light and heat (b).

described above is prepared as a thin film, it is found to crystallize within a very short period of time. An example of the pulse shape suitable to "write" (crystallize) and "erase"(amorphousize) is given in Figure 2. As predicted by the DTA result the erase cycle requires a larger amplitude over a shorter period of time than the write cycle. In Figure 3 electron diffraction carried out on such a film clearly indicates the structural change from the amorphous state (3a) to the crystalline state (3b).

IMAGE & DISPLAY PROPERTIES

As indicated above one of the materials developed at ECD is sensitized, exposed and developed in a single step by a flash exposure of about 1 millisecond. Assuming a rather conservative bit density of $10^7 /in^2$ the rate at which bits are recorded is 10^{10} bits/sec in^2. Another material system can be exposed by laser pulses as short as .5 microsecond. At the other end of the spectrum materials have been developed at ECD in which the photostructural change is produced by ordinary ambient light energy (10^{-5} joule/cm^2 or less. Optical as well as electrical contrast are produced in the exposure. The optical contract arises from the difference in the reflectivity in the amorphous vs the crystalline phase. The contrast being about 2:1. The high electrical contrast (100:1) arises from the change in the conductivity and may be used to "read out" and transmit the image electronically in serial form. An example of the high electrical con-

trast is shown in Figure 4. Figure 4 shows a part of a copy of a frame of microfiche reproduced with an amorphous photo structural film as it appears in a scanning electron microscope (SEM). Very high emission of secondary electrons is observed in the crystallized image area. (Magnification 90X). Various energy sources such as an electron beam, a laser or an electronic flash have been used to temporarily or permanently store information in these materials in either a vacuum or normal ambient environment (9).

The range of materials under development at ECD extends in sensitivity from high (10^0 joule/cm^2 to rather low illumination levels 10^{-5} joule/cm^2. In the more sensitive materials the photonucleation process can occur independently from the crystallization process so that the development is carried out separately if desired. Again the parallel bit rate is very high and both electrical and optical contrast are available for read out or display. No fixing step is required for either material system since the amorphous as well as the crystalline state are stable at ordinary storage temperatures, and can be designed for other selected temperatures. This property permits "ad on writing" without the need for erase of a large data block. As discussed earlier (4) (SEAS Conf. SRO), the presence of the crosslinks can restore the amorphous state of the material during the "erase cycle". In a reversible system locally the chemical composition might be altered during the recording cycle, however, since the cross-linking additives are selected so as to not form compounds with the crystalline phase they become available again as crosslinking agents during the "erase" cycle. Chemical catalysts by reducing the energy required to break the bonds of the crosslinked polymer, can facilitate the formation of heterogeneous nuclei in the illuminated area. This part of the imaging process which we call photonucleation does not have to be concurrent with the imaging part involving crystallization of the amorphous matrix, since the crystallization process can be either thermally activated and/or photon activated. An example of photonucleation is in Figure 5 (micrograph 100X).

Figure 4. Electrical contrast of exposed photo structural film. Contrast (about 100:1), arises from the greatly increased emission of secondary electrons in the image area. (SEM magnification 90X).

Figure 5. Example of photonucleation in Ovonic photostructural films. The number of lights nucleated, thermally developed grains in the image area is $5 \times 10^6/cm^2$ vs 10^4 grains/cm^2 in the non image area (micrograph 100X).

Figure 6. Example of film showing practically no grains developed in the non image area (micrograph 100X).

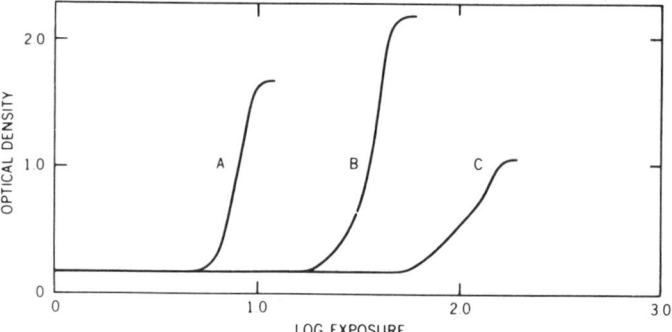

Figure 7. Photographic characteristics of three Ovonic photostructural films. The optical density and gamma are controlled by physical and compositional parameters.

The number of thermally developed grains in the image area is $5 \times 10^6/cm^2$ while the number of grains in the non image area is only $10^4 cm^2$. With suitable additives the number of grains can be reduced to practically zero as shown in Figure 6 (micrograph 100X). The tonal range presently obtainable with these films depends to a large extent on the type of exposure. A 2 millisec flash exposure generally produces a large gamma while a longer exposure with a low intensity light source results in a gamma of 1.5 to 2.0. Three typical curves showing optical density as a function of relative exposure as seen in Figure 7. Most of the Photosensitive materials under develpment at ECD are self fixing i.e., they cannot be exposed or

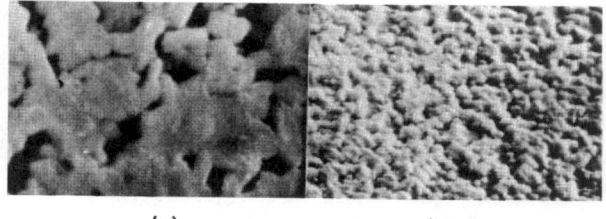

(a)   (b)

Figure 8. SEM micrograph (5000X) of exposed frilm showing the large range of grain size obtainable with these films. (8a) coarse grain 1-10 micron, (8b) fine grain .1-1.0 micron.

erased even with large amounts of radiation, e.g., bright sun light, without an appropriate bias temperature. At elevated temperatures, however, photo enhanced crystallization (10) occurs in the area where light is absorbed.

The sensitivity and the image contrast generated on this basis are relatively small. However, if both the photonucleation and the growth rate are light controlled, very high contrast and sensitivity can be achieved in this medium.

Contrast in all of the photosensitive amorphous materials described here implies both optical contrast as well as electrical contrast. The optical contrast arises partly from the large number of small crystallites $(10^7 = 10^{10}/cm^2)$ in the image area which act as scattering centers in transmission as well as reflection and partly from the change in reflectivity. As a result, the material appears to be negative working in transmission and positive working in reflection. Figure 8 is an example of the range of grain sizes available in these films as seen with the SEM at a magnification of 5000X.

SUMMARY

From our design model and experimentation we feel that the speed and sensitivity of these films can be comparable to silver halide. The general concept is that a relatively small number of light quanta can develop the nucleating point for crystal growth. The growth process, which can be combined with, but also can be independent of light, is independent of the original energy consideration of the nucleation process. This forms the basis of photographic gain in the medium The speed at which a nucleating point can occur can be the speed of a carrier excitation process. The experimental results presently obtainable with photostructural response of amorphous semiconductors as presented today in general terms, we feel, give a strong indication of the potential these materials have in imaging and communication.

REFERENCES

1) S. R. Ovshinsky, Ovonic Switching Devices, International Colloquium on Amorphous and Liquid Semiconductors, Bucharest, Rumania, 1967

2) S. R. Ovshinsky, Ovonic Switching Devices, Electronic Components Conference, Washington, D. C. 1968

3) S. R. Ovshinsky, Patent #3530441, 3271591

4) S. R. Ovshinsky, An Introduction to Ovonic Research, Journal of Noncrystalline Solids, 2, 99-106 (1970)

5) Stanford R. and Iris M. Ovshinsky, Analog Models For Information Storage and Transmission in Physiological Systems. Mat. Res. Bull. 5(1970) 681

6) Stanford R. and Iris M. Ovshinsky, Analog Models for Information Storage and Transmission in Physiological Systems. Mat. Res. Bull. 5(1970)681

E. A. Fagen and H. Fritzsche, Electrical Conductivity of Amorphous Chalcogenide Alloy Films, Journal of Non-Crystalline Solids 2(1970)170

E. J. Evans, J. H. Helbers and S. R. Ovshinsky, Reversible Conductivity Transformation in Chalcogenide Alloy Films, Journal of Non-Crystalline Solids, 2(1970)334

J. Stuke, Review of Optical and Electrical Properties of Amorphous Semiconductors, Journal of Non-Crystalline Solids, 4(1970)1

J. Feinleib and S. R. Ovshinsky, Reflectivity Studies of the Te(Ge,As) based Amorphous Semiconductor in the Conducting and Insulating States. J. Non-Crystalline Solids 4(1970)564.

7) R. G. Neale, Device Structure and Fabrication Techniques for Amorphous Semiconductor Switching Devices. J. of Non-Crystalline Solids, 2, 558 (1970)

8) H. Fritzsche, Physics of Instabilities in Amorphous Semiconductors. IBM Journal of R & D, 13 (1969) 515

9) J. Feinleib, J. P. de Neufville, S. C. Moss and S. R. Ovshinsky, Rapid Reversible Light –Induced Crystallization of Amorphous Semiconductors, Applied Physics Letters, L1199 (to be published)

10) J. Dresner and G. B. Stringfellow, Electronic Processes in the Photo-Crystallization of Vitreous Selenium, J. Phys. Chem. Solids, 303-311 (1968).

REVERSIBLE OPTICAL EFFECTS IN AMORPHOUS SEMICONDUCTORS

J. Feinleib,* S. Iwasa,* S.C. Moss, J.P. deNeufville, and S.R. Ovshinsky

Energy Conversion Devices, Inc., 1675 W. Maple Road, Troy, Michigan 48084

In a previous paper,(1) we described the optical changes that can occur in thin films of Te based amorphous semiconductors under the influence of sharply focused, short pulses of laser radiation. Optical and electron microscopic examination of the transformed areas showed convincingly that a single source of laser radiation of appropriate intensity, wavelength and duration could transform those amorphous films to a crystalline state in the exposed area, and that this change could be reversed by reexposure at an altered intensity. In the course of that investigation, similar reversible optical effects were observed in other amorphous chalcogenide systems, based on long chain polymeric glasses, which differ from the amorphous-crystalline transformation.

The observations we will describe here cannot be classed as bulk properties of the amorphous material, but are highly dependent on the specific experimental conditions. The effects were produced in the following manner. The 6471Å line from a CW Krypton laser was passed through a modulator to produce pulses of variable duration including operation in the continuous mode. This light was then focused to spot size of 1-5μm by appropriate optics. The samples were prepared by either evaporation or sputtering onto clean Corning glass substrates, and the active thicknesses used were between 0.5 and 5 μm. In the experimental set-up, the laser spot was focused through the glass substrate onto the interface between the active chalcogenide material and the glass substrate. The substrate was rotated so that the laser radiation would effect concentric circular tracks on the samples. In this manner, when the laser-modulator operated in a pulsed mode, discrete spots could be produced on the tracks, whereas in the CW mode of operation, the spots within the tracks could be exposed to laser radiation without requiring a high positioning accuracy. At sufficiently high intensities, the CW operation would produce a visible track that was found to be caused by physical flow of the active material through heating

in the track. This effect enabled easy identification of the exposed areas and comes about because of the decrease in surface tension with increasing temperature which causes the material at either side of the warm track to pull the track material apart leaving small depressions or grooves at the track edges.

A typical experiment proceeded as follows. The laser intensity or pulse duration was adjusted to produce crisp circular dark areas when viewed through a microscope of about 200x magnification. The material typically had a transmission edge in the red wavelength of the laser so that with the active material thicknesses mentioned, sufficient tungsten light was transmitted for visual observation. When these areas were produced, the laser was then operated in a CW mode with the sample rotating. The intensity was adjusted until it was observed that the dark spots were just reversed so as to be optically indistinguishable from the background within the track. This process could be repeated to observe the effects of multiple reversals. Various analytic experiments were then performed.

The top surface of the active area was examined with a scanning electron microscope (SEM). Figure 1 shows a sequence of typical tracks that have been cycled a few times where the last written spots were not reversed. The spots appear here to produce a near-spherical bulge on this top surface, while the track itself, caused by the CW reversing cycle, apparently has the material flowing into ridges on the sides much as we noted earlier. The reversed spots were no longer identifiable in the tracks by this examination.

In another test, the active material was slowly etched in CS_2, which is sensitive to structural alterations in these polymeric glasses,(2) until the glass substrate was exposed. In Fig. 2, a sample containing such residual tracks is viewed in the SEM. In the top track, there were three or four successive cycles of spot formation and reversal where successive spots are not on top of each other. Although the optical examination could not resolve spots that had been erased, the chemical etch appeared to leave a residue of material that

---

*   Present address: ITEK, Physics Department, Lexington Research Labs., Lexington, Massachusetts 02173, U.S.A.

Reprinted by permission from *Journal of Non-Crystalline Solids*, Vols. 8–10, pp. 909–916 (1972).

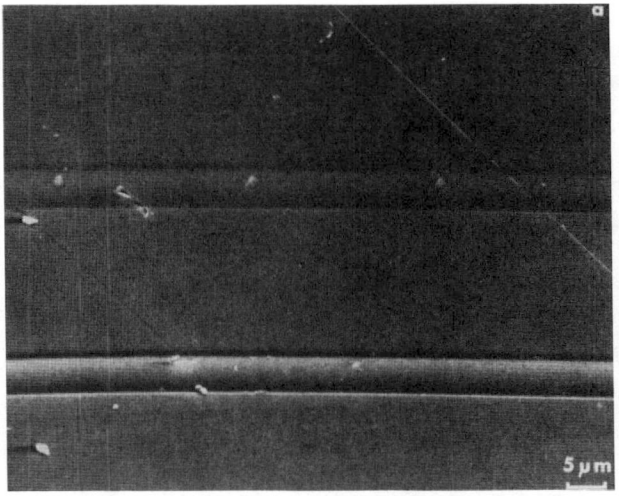

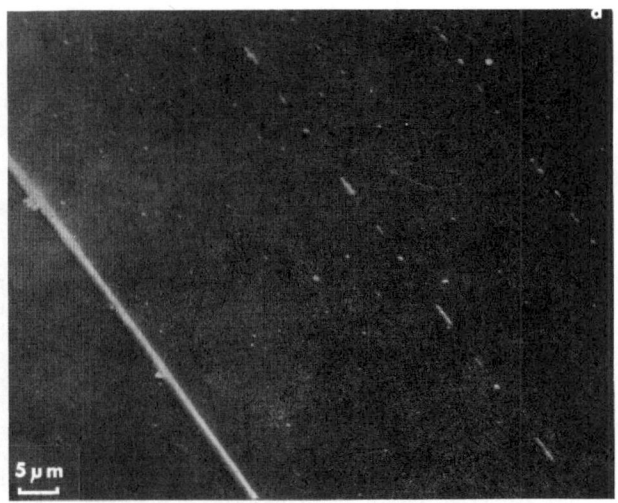

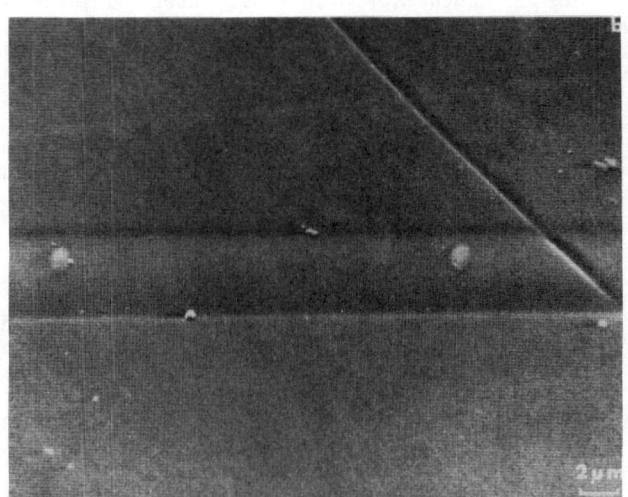

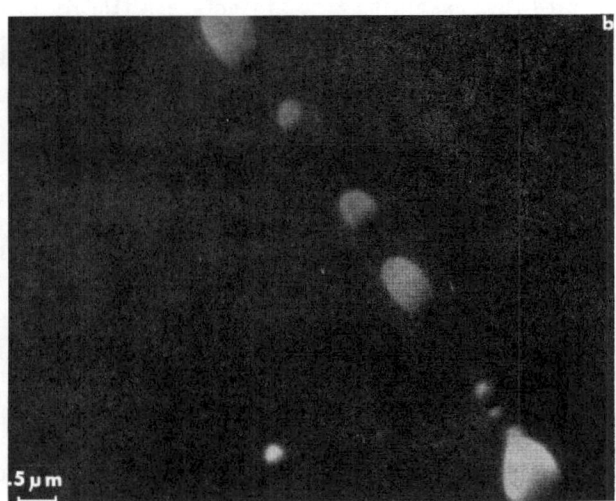

Fig. 2 SEM photomicrographs of material residue on the glass substrate after prolonged etching in CS_2. The films had previously undergone a series of write and erase cycles with a final erase which left them, prior to etching, optically and visually as in the lower track of Fig. 1a. (b) is a magnified view of the residual erased spots of the track in the upper right corner of (a) which had been written and erased three times. The central track in (a) underwent five reversals.

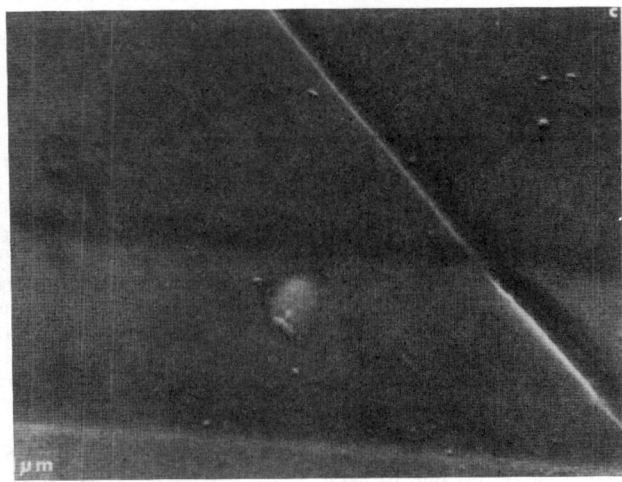

Fig. 1 SEM photomicrographs of a series of laser written spots in a track that has undergone several write-erase cycles. In (a) the lower track is left erased while the upper track is left in the written condition. In the magnifications of (b) and (c) can be seen a scratch at the right which reveals the profile of the erase track. Occasional debris and torn pieces result from mishandling of the samples.

was differentiable from the unwritten background material of the tracks. It is important to note that the first spot of a sequence suffers more reversals (n of them) than the nth where there have been n formation-reversal cycles. Yet it apparently remains nearly as distinct as the spot which underwent only a single reversal. This residual material is thus more resistant to the etch than the background material since electron probe microanalysis showed that the spots visible in the photograph are indeed composed of active material of the original composition while the surrounding active material has been entirely etched away to the bare substrate. Many repeated cyclings left, after etching, the continuous track shown in the bottom of the figure. This may be interpreted as either the consequence of an eventual juxtaposition of written spots, or possibly as a slow alteration in the material caused by the CW reversal exposure. This condition of continuously transformed material did not, however, signal the end of our ability to perform formation-reversal cycles.

In other experiments, a parting layer was placed between the active material and the substrate. Even though the interface was therefore not identical with those of the other samples, the spot writing and reversals appeared to behave similarly. After cycling tracks, the active film was floated off the substrate and the interface surface examined directly with the SEM. These experiments indicated that the spherical bulge observed on the top surface was not always accompanied by a corresponding deformation at the interface, i.e., the interface in the track was often more or less smooth with and without written spots. This latter observation was verified by examining the interface without total absorption in the film. Thus, only interface reflection was observed. Here again, the written spots could not be observed on the interface.

The ability to observe the results described depended critically on certain experimental parameters. If the light spot was larger than a few microns, the sharply defined spots would break up in mottled areas, some of which were more difficult to reverse. Also, the effect seemed to reach the interface and leave a deformation there, observable in our reflection experiments.

The wavelength of exposing light was important. The krypton red line is closely matched to the absorption edge of the amorphous chalcogenide film so that in these thicknesses up to 50% of the incident light is transmitted. An attempt to produce similar effects using krypton yellow or shorter wavelength light, which is nearly totally absorbed, did not produce the effect observed. In the latter case, the absorbed energy appeared to affect principally the interface material leaving a deformation there.

There was also a marked correspondence between thickness of the amorphous film and both the pulse energy required to form a spot and the ability to achieve a clean reversal. These observations, though qualitative, have led us to a model of the reversible optical effects which appear consistent with the data and are related to the physical properties of the long chain, amorphous, chalcogenide semiconductors.

The spot formation appears to occur in the following manner. The focused laser light is absorbed throughout the thickness of the film exposed by it. In the chalcogenide film, the glass transition temperature, or temperature at which there is a marked drop in viscosity of the glassy material, is only some tens of degrees above ambient while the volatilization temperature is several hundred degrees higher. Thus the exposed material quickly becomes very fluid with a vapor pressure which increases with the continued absorption of light until either a vapor bubble nucleates or thermal expansion produces a bulge. If, at this instant the light is removed, there is then a very rapid cooling and quenching of the material around the bubble or the ensuing void. If the cooling is

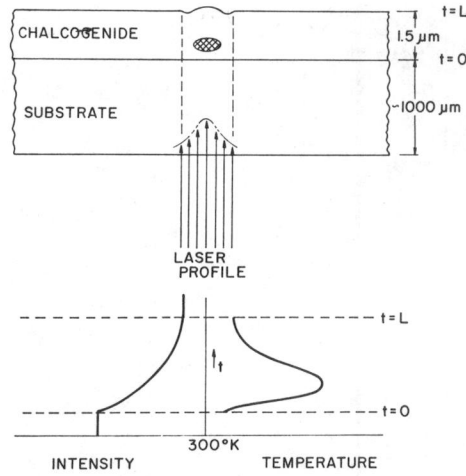

Fig. 3 Schematic of the laser interaction with a long chain chalcogenide film illuminated through a transport substrate where the film is partially absorbing to the laser beam. Included is an approximate spatial profile of the beam, a cross section of the written spot including the (cross-hatched) bubble and qualitative profiles of both laser intensity and temperature through the film at about the peak in the laser pulse.

rapid enough the bubble deformation remains frozen in and the interior of the bubble or void remains more or less empty with the deformation frozen at the top surface. At first thought, we might expect a bubble to form as a hemispherical surface moving up from the interface. However, since the substrate is transparent to the laser light, this substrate interface surface can remain relatively cold and so can cool the immediately adjacent active material. Thus we can expect that, under appropriate focusing conditions, the hottest region of the active film is some distance above the interface and the bubble or void pocket is therefore formed in the interior. The light intensity and temperature profiles responsible for this effect are sketched qualitatively in Fig. 3. The observed optical effect of this embedded bubble in transmission is a dark spot, principally due to the refraction and scattering of the light. To reverse this optical effect, it would appear that all that is required is for the material surrounding the bubble to absorb enough light to reach a temperature well above the glass transition temperature but not so hot as in the formation pulse. At this point, the bubble collapses and the material continues to flow to form a more or less smooth track.

Although this explanation is consistent with the visual observation of the film, we must explain the apparent reemergence on etching of the spot that has been erased. It is well known (2) that CS_2 has a different etching rate for the monomer components of this glassy material in comparison with the more polymerized components, the latter having the slower rate. It would thus appear that in the spot formation, the heating plus photo-effect is sufficient to change the ratio of the two components. The increase in polymer ratio is most apparent in the interface material, so that it remains behind in the areas previously written while the background is entirely etched away. If a spot is written and subsequently reversed so as to be optically indistinguishable from the unwritten material, it appears that the monomer/polymer ratio is not also reversed. While this is consistent with the known long annealing times required for this process,(3) comparably long times also prevail at the higher temperatures at which the polymerization originally took place. In other words, it seems that, if all that were available to us at the formation or write temperature were heat for a few μsec, there would be far insufficient time to accomplish the polymerization that the CS_2 etch reveals. We are thus nearly forced to the conclusion that we are observing a photoenhancement of polymerization. In a manner identical to that described in ref. 1, the incident photons on absorption promote electrons out of bonding orbitals, thereby weakening the bonds and enormously enhancing the molecular mobility. This leads to a large increase in the polymerization rate. Accompanying this change, however, we observe no change in intrinsic optical properties (monomer versus polymer) and therefore the bubble can be reversed leaving no optical residue while the etch can detect the residual monomer/polymer ratio changes. This polymeric transformation is only secondarily related to the optical effects observed which appear primarily related to the flow properties of the chalcogenide glass. It does, however, contribute to the thermo-mechanical stability of the laser spot because of the increased viscosity and hardness associated with polymerization. The reversible optical characteristics of these amorphous chalcogenides are thus physical phenomena which are coupled to the unique glass forming and visco-elastic properties of the materials combined with their characteristic semiconductor optical properties.

We wish to thank Professor A. Eisenberg for several interesting discussions on the visco-elastic properties of inorganic polymers.

References

1. J. Feinleib, J. deNeufville, S.C. Moss and S.R. Ovshinsky, Appl. Phys. Letters, 18 (1971) 254.
2. G. Briegleb, Z. Physik. Chem. A, 144 (1929) 321.
3. A. Eisenberg, private communication.

IMAGING IN AMORPHOUS MATERIALS BY STRUCTURAL ALTERATION

S.R. Ovshinsky and P.H. Klose

Energy Conversion Devices, Inc., 1675 W. Maple Road, Troy, Michigan 48084

One of our major approaches to information storage and control through the years (1-5) has been to have external information alter the physical structure of amorphous material and have the change of structure detected unambiguously by the concomitant electrical, chemical, optical and other changes that accompany the structural transformation. In this manner, external information is encoded into an internal structural change.

Among others the rheological qualities of polymeric material are an important factor that can be used for information storage and control. This is true especially where fast high energy inputs are available which can initiate elastomeric changes. Various mechanical and deformable alterations of the material can be frozen in depending upon relaxation and quenching rates and the free energy configurations available in a viscous medium. This is particularly true of long chain polymeric materials. In short chain polymeric materials such as tellurium, crystallization is one of the best available options. In long chain amorphous polymers, various other configurational options are available. A memory utilizing the flow characteristics of long chain polymers is described by Feinleib et al. in another paper at this meeting.(6)

This paper will concern itself with the crystallization phenomenon in both long chain and short chain polymers; in particular, the use of light to initiate the ordering process in amorphous materials so that a clearly defined photographic image is achieved.

Figure 1 gives a listing of the major physical and/or chemical changes which can be observed as a result of the phase transition. In many instances the properties of the amorphous phase differ by many orders of magnitude from those of the crystalline phase thereby leading to recording or storage media of unusually high signal-to-noise ratio. The transition from the amorphous to the crystalline phase may be effected by several different means independently or in combination. We here illustrate the various forms of energy which can be used to generate nuclei which are then developed into larger crystallites by thermal or other means.

A large portion of the energy spectrum can be used to overcome the energy barriers. For example, at the high energy side of the spectrum the absorption of light of a certain frequency can break specific bonds thus providing the building blocks for the formation of nucleation sites. At the low end of the energy spectrum thermal breaking of bonds together with movement of molecular segments aid configurational and structural transitions. We have observed the initiation of crystallization by a range of energy sources. L. Pellier of our laboratory observed, early in the 60's, electron beam crystallization.(7) We have also reported (1,8-11) the effects of heat, light, electric field, laser and pressure.

The energy concept as shown in Fig. 1 in conjunction with our experimental results serves to illustrate that it is possible to bias our materials with one form of energy and then have another

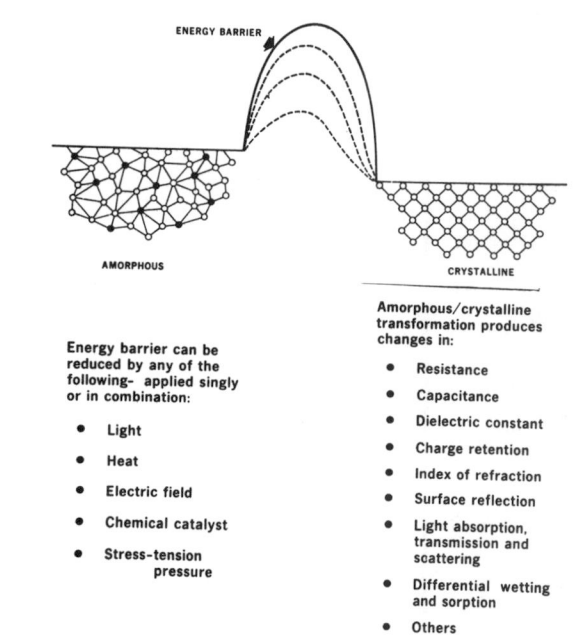

Energy barrier can be reduced by any of the following- applied singly or in combination:

- Light
- Heat
- Electric field
- Chemical catalyst
- Stress-tension pressure

Amorphous/crystalline transformation produces changes in:

- Resistance
- Capacitance
- Dielectric constant
- Charge retention
- Index of refraction
- Surface reflection
- Light absorption, transmission and scattering
- Differential wetting and sorption
- Others

Fig. 1   Mechanism for information retrieval and display by structural transformation.

Reprinted by permission from *Journal of Non-Crystalline Solids*, Vols. 8–10, pp. 892–898 (1972).

Fig. 2   Example of a high contrast image made
with a 0.4 μm selenium base film. The image is
shown in the reflection mode. Resolution ~ 100
lines/mm, contrast ~ 20/1, gamma ~ 2, exposure ~
10^{-4} J/cm^2.

trigger the reaction. Obviously, irrespective of
how the nuclei are initiated, thermal energy is very
useful for the dry development of crystal size. In
the short chain polymer for the most part cross
linking additives are used to form the desired
amorphous structural configuration while the
mechanical entanglement of the ring and long chain

fig. 3a.

fig. 3b.

Fig. 3   Reflection (a) and transmission (b)
grey scale reproduced with amorphous selenium base
films. Film thickness 0.4 and 0.1 μm,
respectively. Exposure 10^{-3} J/cm^2.

polymeric structure can by itself be sufficient to
act as energy barriers although even in these
materials additives can be helpful.

The following figures are examples of images and
various forms of readout obtainable with two of the
many types of amorphous imaging films developed at
ECD.

In the selenium base system developed in our
laboratory the light-initiated nucleation process
can occur at room temperature independent of the
subsequent thermal development.

Figure 2 is an actual size photograph of such a
film viewed in the reflection mode. The high
resolution of which this system is capable is, for
example, suitable for applications such as
microfiche duplication. Negative contrast is
observed if the film is viewed in transmission.
Since the crystallites in the imaged area scatter
the light more efficiently than the unexposed
portion of the film, the imaged area appears dark.
In Fig. 3, a standard step wedge is used to produce
the grey scale in reflection (a) and transmission
(b). As above the material is positive working in
reflection and negative in transmission. The
optical density differential $D_{max} - D_{min}$ is usually of
the order of 1. Typical gamma values range from 1.5
to 6.

In Fig. 4 a cross-sectional view of the
amorphous film illustrates three steps that we
believe are involved in producing an image. In the
first step photons are absorbed in the amorphous
material which produce local alteration in structure
leading in step two to nucleation. These nuclei
also represent the latent image. In the final step
the nuclei which can act as trapping sites for
carriers are thermally developed into crystallites.
Obviously, we are oversimplifying. A detailed
description of the various photochemical steps
involved in the image formation in amorphous films
will have to be the subject of another paper.
Excitation as well as inhibition of crystallization
can be light induced as we have been able to
demonstrate in our laboratory, that is where light
has been absorbed, crystalline growth has been
inhibited (negative image) or excited (positive
image).

Another important parameter to be considered in
imaging in amorphous films is the number and size of
crystallites which, of course, depend on the number
of activator sites and the film thickness.

The estimated quantum yield of the selenium
system based on the number of photons necessary to
initiate complete crystallization is at least of the

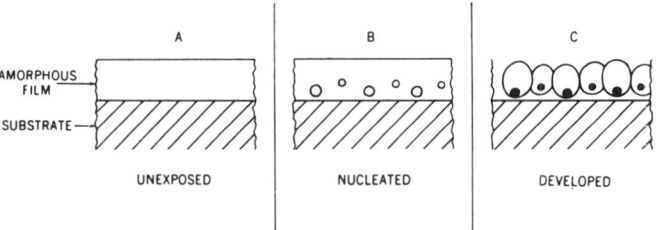

Fig. 4   Schematic diagram of photonucleation
and development in Ovonic photostructural films.

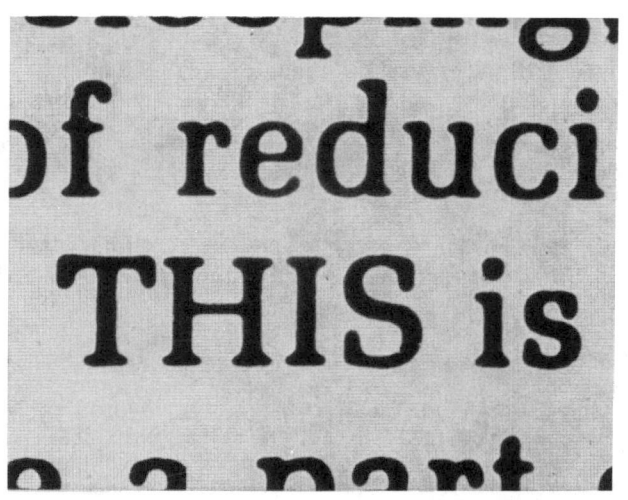

Fig. 5 High contrast high resolution (Te base) film. Resolution 500 lines/mm. Optical density 3/1. Exposure 10^{-1} J/cm^2 (magnification 100x).

order of 100. That is for every incident photon 100 atoms of crystalline selenium are produced. We feel on the basis of our model and our presently available experimental results that the light-initiated structural changes in amorphous systems can have very large total quantum yields.

The second system we will discuss is tellurium based. The very much smaller grain size of the tellurium base films quite naturally lends itself to very high resolution imaging as can be seen in Fig. 5. After exposure this film has an electrical contrast of about 1000 to 1 and an optical contrast of about 2 to 1. After a very brief processing step the optical contrast is 1000 to 1. This figure shows the film after processing. The resolution of this example is 500 lines per mm. In another example of electrical contrast obtained with a film crystallized upon exposure to light and heat we have found that if an imaged film is bombarded with electrons a much larger number of secondary electrons can be collected in the imaged area than in the non-imaged area. A still different form of read-out can be obtained if the imaging medium is adapted to electrostatic printing and copying requirements. In one application the generally crystalline, low resistivity film is switched by a laser beam into the amorphous state with its resultant high resistivity. The amorphous high resistivity areas are then charged and processed by the conventional techniques of electrostatic printing.(10)

To demonstrate the very rapid structural transformations which we have discussed above, a sample of a 1000Å tellurium base film vacuum deposited on a regular mylar substrate is placed in contact with the negative and exposed with a 2 ms xenon flash. The total exposure is between 10^{-1} and 1 J/cm^2. The film is reversed to its original condition, i.e., ready to receive new information, by briefly heating it to 170°C. Thus, we have developed a reversible imaging system based on the amorphous-crystalline phase transition without sacrificing parameters such as resolution and image stability. With the addition of a brief processing step, we obtain maximum contrast. This film may be stored at 100°C without damage in either the imaged or unimaged state.

The reproduction of information from an external source incorporated in structural changes in amorphous materials can be accomplished in a fast and sensitive as well as a reversible manner.

Acknowledgment

The authors would like to gratefully acknowledge some experimental work of R. Hallman and G. Simpson which was incorporated in this paper. Our special thanks go to I. Scislowicz and J. Sivec for their part in the preparation of the experimental samples.

References

1. S.R. Ovshinsky and I.M. Ovshinsky, Mater. Res. Bull. 5 (1970) 681. See also for earlier references.
2. J. Feinleib and S. Ovshinsky, J. Non-Crystalline Solids 4 (1970) 564.
3. J. Feinleib, J. deNeufville, S.C. Moss and S.R. Ovshinsky, Appl. Phys. Letters 18 (1971) 254.
4. S.R. Ovshinsky and P.H. Klose, 1971 Intern. Symp. Digest of Technical Papers, May 1971, Philadelphia, PA., pp. 58-61.
5. S.R. Ovshinsky, in: proc. 1968 Electronic Components Conf., May 1968, Washington, D.C., pp. 313-317.
6. J. Feinleib, S. Iwasa, S.C. Moss, J. deNeufville and S.R. Ovshinsky, J. Non-Crystalline Solids 8-10 (1972) 909.
7. L. Pellier and S.R. Ovshinsky, unpublished data.
8. S.R. Ovshinsky, J. Non-Crystalline Solids 2 (1970) 99.
9. S.R. Ovshinsky and H. Fritzsche, Met. Trans. 2 (1971) 641.
10. H.K. Henisch, Sc. Am. 221 (1969) 30.
11. M.P. Southworth, Control Engineering 11 (August 1964) 69.

NEW THIN-FILM TUNNEL TRIODE USING AMORPHOUS SEMICONDUCTORS

R.F. Shaw,[1] H. Fritzsche,[2] M. Silver,[3] P. Smejtek,[3] S. Holmberg,[4] and S.R. Ovshinsky[4]

[1]Esso Research and Engineering Company, Linden, New Jersey 07036
[2]James Franck Institute, University of Chicago, Chicago, Illinois 60637
[3]University of North Carolina, Chapel Hill, North Carolina 27514
[4]Energy Conversion Devices, Troy, Michigan 48084

(Received 2 December 1971)

A tunnel triode was fabricated using Al_2O_3 dielectric to form the hot-electron tunnel cathode and an amorphous film of $As_{34}Te_{28}Ge_{16}S_{21}Se_1$ to separate the base and collector. The structure of the device is shown in Fig. 1. The electrodes were all aluminum. A 2000-Å emitter electrode was evaporated onto a glass substrate and anodized to form an Al_2O_3 layer 125 Å thick as described by Onn and Silver.[1] This was then coated with a 125-Å Al layer to form the base electrode followed by a 1500-Å sputtered layer of the amorphous material over which was subsequently deposited a 2000-Å collector electrode.

The vacuum emission characteristics of the uncoated tunnel cathodes were similar to those prepared by Onn and Silver.[1] The thick Al_2O_3 layer permitted operation at applied voltages in excess of the Al work function, thus permitting the tunneling carriers to escape the electrode surface and be emitted as hot electrons. The forward-current transfer ratio α for vacuum emission was on the order of 10^{-4}.

The triodes were first made with neutral contacts to the amorphous layer, i.e., the conduction was bulk limited with a characteristic activation energy for conduction of 0.58 eV and a room-temperature resistivity of 10^8 Ω cm. With neutral contacts to the amorphous layer, low-temperature operation was necessitated as the Ohmic current at low collector-base voltages dominated effects due to injection from the tunnel cathode.

Operation at 77 °K with and without a collector field showed the injected hot electrons thermalized and formed an equilibrium space charge in the amorphous layer adjacent to the base electrode causing further injected electrons to be accelerated back into the base electrode. An increase in the emitter-base current resulting in an increase in injected current caused a momentary increase in the short-circuit collector current which decayed back to zero as the space charge equilibrated. Termination of the emitter-base current

caused a large reverse current which delayed back to zero with a 22-sec time constant at 77 °K. The magnitude of the reverse current was proportional to the injection level prior to its cessation.

In order to obtain room-temperature operation, a blocking contact was made to the amorphous layer. This was accomplished by exposing the amorphous layer to a 50% relative humidity atmosphere prior to the collection electrode deposition. Photovoltage and capacitance as a function of bias voltage indicated the presence of an accumulation layer resulting in a barrier of 0.4 eV. Current-voltage measurements on the collector-film-base sandwich confirmed this value and showed the conductance to be electrode limited with an increase in resistance of 5×10^3.

Measured values of the short-circuit collector current with forward and reverse emitter-base currents are shown in Fig. 2. The values of α at zero collection bias vary from device to device but are in the range 10^{-2}–10^{-1}.

The lack of injection with the reverse emitter-base currents is contrary to the findings of Delord,[2] who suggested that electrons could be photoemitted by luminescence in the oxide layer. The higher values of α result

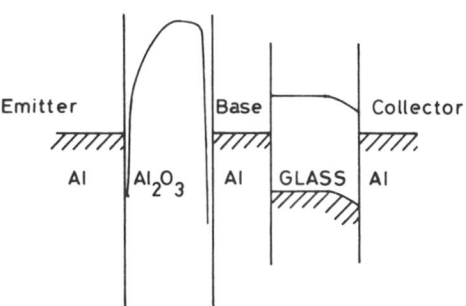

FIG. 1. Band diagram of tunnel triode.

Reprinted by permission from *Applied Physics Letters*, Vol. 20, No. 7, pp. 241–243 (1 April 1972).

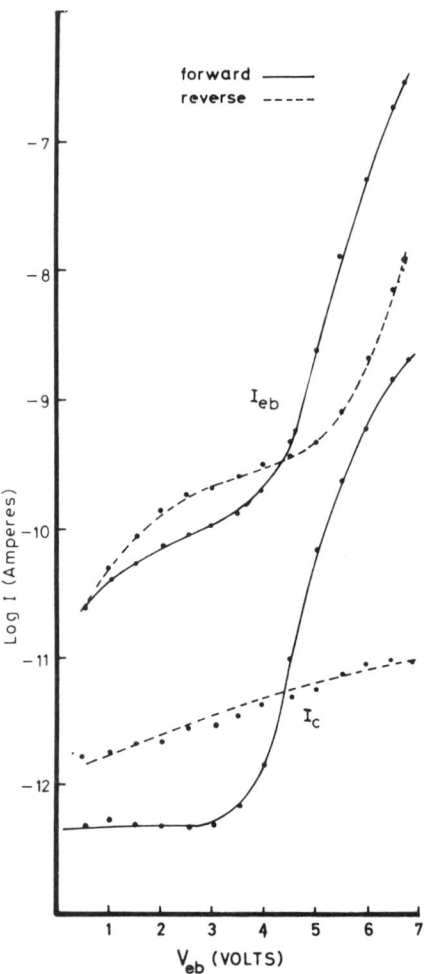

FIG. 2. Short-circuit collector current and base current as a function of forward and reverse emitter bias.

from the lowering of the effective work function by the electron affinity of the amorphous layer.

Prior investigations of tunnel-triode structures by Mead[3] and others[4, 5] involved the injection of majority carriers. In this structure the injected electrons are minority carriers: The material is p type by thermopower measurements. The injection of minority carriers gives rise to the possibility of gain.

If charge neutrality overides recombination, the injection and thermalization of hot electrons results in an increase in the conduction-band population above the mobility edge. If the Fermi level is pinned, then an equal increase in the valence-band hole population would result.

Excepting deep trapping and recombination through deep sites, the electron current would be

$$I_e = eF\tau_e / T_e \tag{1}$$

and for hole current

$$I_h = eF\tau_h / T_h, \tag{2}$$

where τ is the free-carrier lifetime, F is the genera-

tion rate, i.e., the injection current, and T is the transit time. If, as assumed, the generation rates for electrons and holes are equal, then the current amplification will be

$$G = I_h / I_e = T_e \tau_h / T_h \tau_e. \tag{3}$$

If the mobilities are field independent, then G would be field and geometry independent. However, since some moderately deep trapping occurs this is not to be expected.

Figure 3 shows that $G > 1$ may be possible. Curve A is the normal collector-current–collector voltage characteristics of the sandwich with no injection. Curve B is the collector current with a constant emitter-base current of 10^{-7} A. At zero collector voltage the collector current was 8.15×10^{-10} A. Curve C is the sum of the "dark" current plus the collector current at zero volts. Assuming the injected current is independent of collector voltage, the ratio of curves B to C would represent the amplification of injected current as a function of collector voltage. This ratio peaks at a collector voltage of 0.65 V with a value of 3.5. Ratios as large as 12 have been obtained. This decrease in the ratio as the field is further increased may be explained by the transit time approaching the dielectric relaxation time

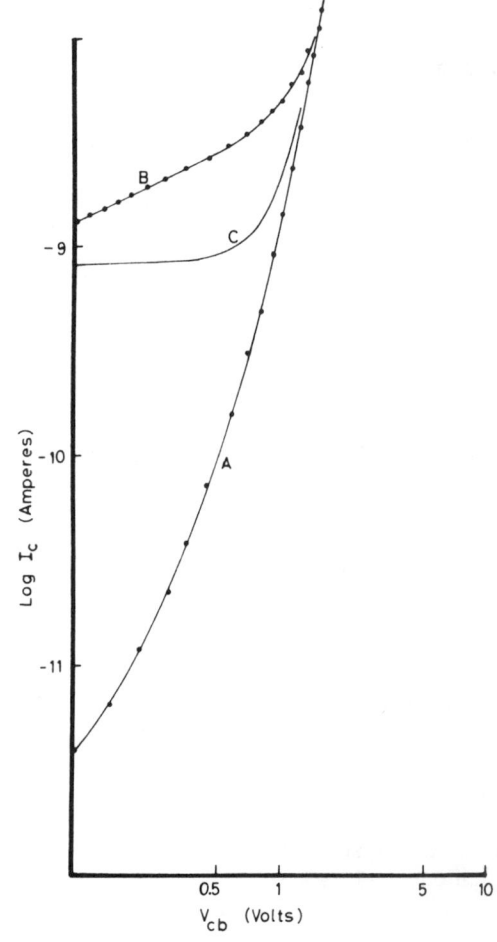

FIG. 3. Collector current as a function of collector bias with and without injection.

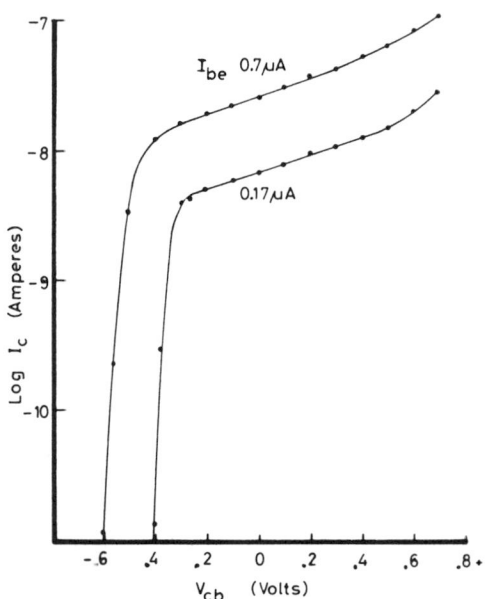

FIG. 4. Collector current as a function of collector bias polarity for two injection levels.

resulting in the onset of space-charge-limited current flow.

These results indicate that the introduction of an amorphous p-type material permitting the injection of minority carriers forms the basis of a new all-thin-film device, the advantages of which would be simpler fabrication and low power consumption. Also, assuming a device size similar to present integrated MOS's and geometrically extrapolating capacitances from the present device, switching speeds of 10^{-8} sec should be obtainable. This arrangement would also be useful for many physical-property measurements such as electron transit time without the complications normally encountered in photogenerated carrier measurements. The vacuum equivalent of photoemitted electron-retarding potential measurements may also be made, as shown in Fig. 4, to study electron energy distributions and thermalization times.

[1]D.G. Onn and M. Silver, Phys. Rev. 183, 295 (1969).
[2]J.F. Delord, Appl. Phys. Letters 11, 287 (1967).
[3]C.A. Mead, J. Appl. Phys. 32, 646 (1961).
[4]D.V. Geppert, Proc. IRE 48, 1644 (1960).
[5]R.N. Hall, Solid-State Electron. 3, 320 (1961).

MECHANISM OF REVERSIBLE OPTICAL STORAGE IN EVAPORATED AMORPHOUS AsSe AND $Ge_{10}As_{40}Se_{50}$

J.P. deNeufville, R. Seguin, S.C. Moss,[†] and S.R. Ovshinsky

Energy Conversion Devices, Inc., Troy, Michigan 48084

1. INTRODUCTION

The holographic response (Keneman, 1971; Ohmachi and Igo, 1972) of evaporated As_2S_3 at 4880Å has been ascribed to photo-polymerization of a hard-sphere As_4S_6 glass (deNeufville et al., 1973). The effect of illumination is to shift the location of the optical absorption edge ($\Delta E = -0.06eV$), and to increase the refractive index ($\Delta n = 0.08$). A comparable increase in refractive index occurs during annealing of evaporated As_2S_3 at 189°C, near its glass transition temperature, accompanied by a smaller edge shift ($\Delta n = -0.03eV$). While the optical absorption edge can be reversibly cycled by exposure at 25°C and annealing at 180°C ($\Delta E = -0.03eV$), the accompanying index changes are relatively small ($\Delta E = \leq 0.01$). Furthermore, the X-ray diffraction pattern is unaffected by cycling, suggesting that the optical effects arise from the creation (exposure) and elimination (annealing) of a nonequilibrium concentration of trapped charge carriers rather than from an innate structural change. The behavior of evaporated As_2Se_3 is entirely analogous to As_2S_3 with the exception that the reversible optical edge shift is much smaller ($\Delta E = -0.01eV$).

In contrast to these essentially irreversible holographic media, certain ternary (e.g., $Ge_{10}As_{40}Se_{50}$) memory alloys (Ovshinsky, 1970) exhibit relatively larger reversible changes in refractive index ($\Delta n = 0.03-0.04$) when exposed at 25°C

and annealed in the vicinity of T_g (Igo and Toyoshima, 1973). In this investigation we have reproduced these results for $Ge_{10}As_{40}Se_{50}$ and examined the resultant structural transformations by X-ray diffraction. We have also examined the energy dependence of the refractive index for this alloy and a variety of other alloys, including As_2S_3 and As_2Se_3, in the virgin, annealed, and exposed conditions. By fitting these data to the Wemple-DiDomenico (1971) dispersion relationship we have calculated the shifts of the lowest interband optical transition energy E_0 as a function of exposure and annealing, and thus related these changes to the structural state of the films.

2. EXPERIMENTAL

Thin (2-5μm) films of As_2S_3, As_2Se_3, AsSe and $Ge_{40}As_{40}Se_{50}$ were evaporated onto microscope slides at a residual pressure ca. 5×10^{-8} torr from a constant temperature source. Congruent evaporation was indicated by a constant evaporation rate of ~ 1000Å/min. The illuminations were performed under UHV conditions using a 5 watt microscope illuminator and a heat filter. Annealing was performed in a nitrogen atmosphere in the dark. X-ray diffraction experiments were performed with CuK_α radiation (deNeufville et al., 1973).

The optical parameter of interest was the refractive index n as a function of photon energy between 0.62 and 1.4eV. In this non-absorbing portion of the optical spectrum $n^2(\omega) = \epsilon_1(\omega)$. $n(\omega)$ was determined from the wavelengths of

[†]Present Address: University of Houston, Houston, Texas, U.S.A.

Reprinted by permission from the *Proceedings of the Fifth International Amorphous and Liquid Semiconductor Conference*, Garmisch-Partenkirchen, Germany (1974), pp. 737–743.

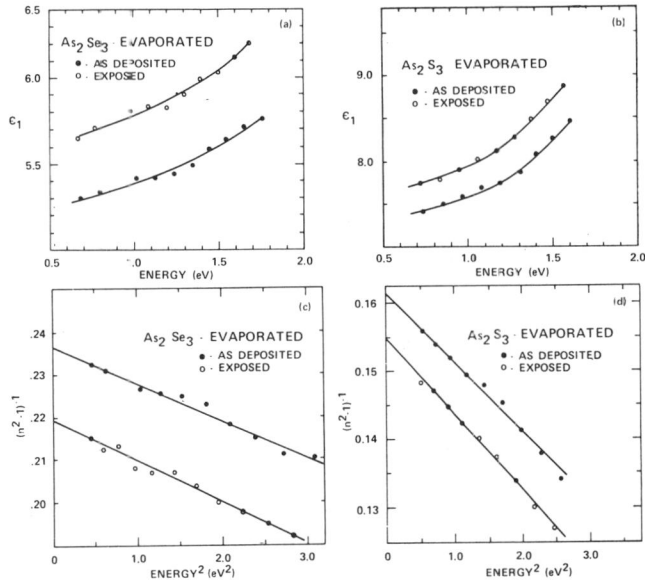

Fig. 1   (a) $\epsilon_1(E)=n^2(E)$ plotted   versus photon energy E for evaporated 2.36 μm As_2S_3 film before and after 17 hours tungsten illumination. (b) $\epsilon_1(E)$ versus photon energy for evaporated 1.86μm As_2Se_3 film before and after 17 hours tungsten illumination. (c) $(n^2-1)^{-1}$ versus E^2 for n(E) data in (a). (d) $(n^2-1)^{-1}$ versus E^2 for n(E) data in (b).

interference maxima in transmission, using the relationship $m\lambda_{max}=2nd$, where m is an integer and d is the film thickness. The values of λ_{max} could be determined to ±0.5%, but the values of d are uncertain to ±2%. Therefore, incremental changes in n for a given film are as precise as λ_{max}, if d is unchanged, whereas absolute values of n are no more precise than those of d. We observed that d was unaffected by exposure ($\Delta d < 0.5\%$) in all cases, but we could not directly determine the effect of annealing upon d to equivalent precision.

3. RESULTS

The refractive index behavior for these materials is typified by the data for As_2S_3 and As_2Se_3 in Fig. 1(a) and 1(b). n(E) can be fitted by the Wemple-DiDomenico (1971) dispersion relationship:

$$\epsilon_1(\omega) = n^2(\omega) = 1 - E_dE_0/[E_0^2 - (\hbar\omega)^2] \qquad (1)$$

where E_0 and E_d are single oscillator fitting constants which measure the oscillator energy and strength respectively. E_0 for these chalcogenides closely approximates the peak in $\epsilon_1(\omega)$ (Wemple, 1973) corresponding to the mean energy of transitions from valence band (lone pair) states to

Table 1. Index of refraction of amorphous films measured at $(\hbar\omega)=1eV$. See text for description of annealing and exposure conditions.

Composition	Film thickness (μm)	Condition	n	Δn
$As_{40}S_{60}$	2.36	as evaporated	2.331	
		exposed	2.415	+0.084
	2.33	as evaporated	2.384	
		annealed at 180°C	2.457	+0.073
$As_{40}Se_{60}$	1.86	as evaporated	2.785	
		exposed	2.848	+0.063
	1.92	as evaporated	2.785	
		annealed at 180°C	2.842	+0.057
$As_{50}Se_{50}$	4.07	as evaporated	2.632	
		annealed at 150°C	2.700	+0.068
		exposed after anneal	2.733	+0.033
$Ge_{10}As_{40}Se_{50}$	2.30	as evaporated	2.700	
		annealed at 180°C	2.662	−0.038
		exposed after anneal	2.696	+0.034

Table 2. Oscillator parameters E_o and E_d of amorphous films.
See text for definitions of E_o (oscillator energy) and E_d
(oscillator strength).

Sample	Condition	E_o, eV	E_d, eV
$As_{40}S_{60}$	as evaporated	5.20	22.0
	exposed only	4.76	21.7
	annealed only	4.78	22.0
$As_{40}Se_{60}$	as evaporated	3.99	24.7
	exposed only	3.73	24.0
	annealed only	3.85	25.3
$As_{50}Se_{50}$	as evaporated	4.75	26.9
	annealed only	4.43	26.5
	exposed after anneal	4.07	24.7
$Ge_{10}As_{40}Se_{50}$	as evaporated	4.02	23.8
	annealed only	4.33	25.1
	exposed after anneal	4.03	23.8

conduction band states (Kastner, 1972; Drews et al., 1972). By plotting $(n^2-1)^{-1}$ versus $(\hbar\omega)^2$ as in Fig. 1(c) and 1(d) and fitting the data by a straight line, E_d and E_o can be directly determined from the intercept (E_o/E_d) and the slope $(E_d E_o)$.

Comparable data were obtained for $Ge_{10}As_{40}Se_{50}$ and $As_{50}Se_{50}$. The values of n and Δn at 1eV for all four alloys in the evaporated, annealed and exposed conditions are listed in Table 1. Note that the effects of annealing and exposure are additive for As_2S_3, As_2Se_3 and AsSe, whereas annealing of $Ge_{10}As_{40}Se_{50}$ lowers n. The change of n between the annealed and the exposed states, which indicates relative performance as a reusable phase holographic medium, is larger for $Ge_{10}As_{40}Se_{50}$ (Igo and Toyoshima, 1973) and AsSe than for As_2S_3 (Ohmachi and Igo, 1972) or As_2Se_3 (deNeufville et al, 1973).

The $n(\omega)$ data for As_2S_3, As_2Se_3, AsSe and $Ge_{10}As_{40}Se_{50}$ have been fitted to $(n^2-1)^{-1}$ versus E^2 plots to determine the values of E_d and E_o which are listed in Table 2 as a function of annealing and exposure treatments. Each grouping of data corresponds to a single sample. We estimate the absolute uncertainty of the E_o and E_d values in Table 2 as ±3% and the relative uncertainty for a given film as ±1.5%. In spite of these relatively large uncertainties, it is clear that annealing of

As_2S_3, As_2Se_3 and AsSe produces the increases in n listed in Table 1 by a reduction of E_o rather than be an increase in E_d. Furthermore, the increase in n accompanying exposure of annealed AsSe and $Ge_{10}As_{40}Se_{50}$ is predominantly associated with a decrease in E_o.

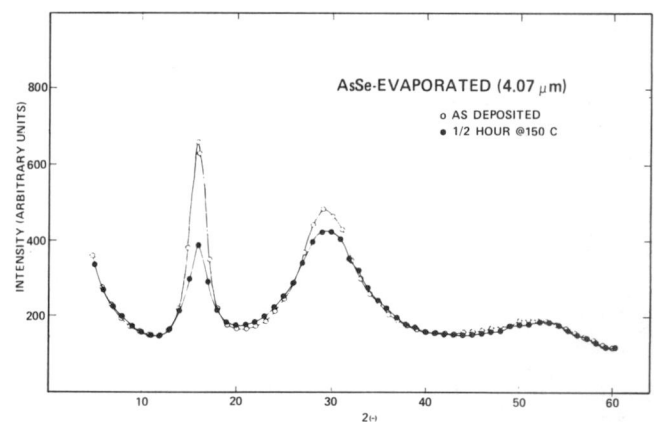

Fig. 2   X-ray diffraction profiles before and after annealing at 150°C of AsSe film vapor deposited over a microscope slide. The radiation is CuK_α (λ=1.54Å). A subsequent diffraction profile obtained after exposing the annealed film to tungsten illumination for 17 hours in UHV was identical to the profile obtained prior to illumination.

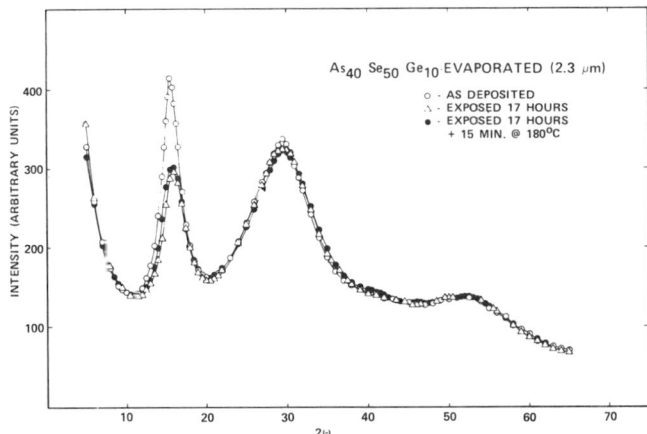

Fig. 3 Sequence of X-ray (CuK$_\alpha$) diffraction profiles of a 2.30μm Ge$_{10}$As$_{40}$Se$_{50}$ film after vapor deposition, after incandescent exposure, and after subsequent annealing at 180°C.

To determine whether the relatively large differences in n between the annealed and the annealed-exposed conditions for AsSe and Ge$_{10}$As$_{40}$Se$_{50}$ correspond to different structural states, we performed the X-ray diffraction experiments shown in Figs. 2 and 3. For AsSe a relatively large structural change accompanied annealing, and no diffraction change was detected as the result of subsequent exposure. For Ge$_{10}$As$_{40}$Se$_{50}$, a comparable structural change accompanied exposure, and an insignificant further change accompanied subsequent annealing. In both cases, the change from the virgin patterns mostly involves a weakening and a slight broadening of the diffraction maximum at 2θ=15.9° corresponding to an equivalent crystalline interplanar spacing of 5.57Å. Thus, the increase in n of ~ 0.035 which accompanies exposure of AsSe and Ge$_{10}$As$_{40}$Se$_{50}$ is not the result of a detectable perturbation of the radial distribution function as indicated by the diffraction patterns. However, it can be inferred from Figs. 2 and 3 that a detectable perturbation of the R.D.F. must occur in the vicinity of 6.85Å, the inferred hard-sphere diameter, as the result of annealing of the virgin films.

4. DISCUSSION

The interpretation of these results involves an extension of the model previously proposed (deNeufville et al., 1973) to account for the photo-structural effects in As$_2$S$_3$ and As$_2$Se$_3$. Thus, we conclude that virgin films of AsSe and Ge$_{10}$As$_{40}$Se$_{50}$ contain largely molecular units (As$_4$Se$_4$ and/or As$_4$Se$_6$), the hard-sphere correlations of which account for the strong and narrow X-ray diffraction peaks at k≈1.13Å^{-1} for a sphere radius r=3.42Å. This structural interpretation is not confirmed by mass spectrometry of the evaporation products of AsSe, As$_2$Se$_3$ and As$_2$S$_3$ (Agarwal, 1973), and in no case is a significant fraction of the expected eight or ten atom molecules obtained. The negative result may arise from the disintegration of the larger molecules into smaller fragments by ionization in the mass spectrometer, so that a direct determination of the molecular constituency of these films by Raman spectroscopy will be required.

Annealing of the AsSe and Ge$_{10}$As$_{40}$Se$_{50}$ films results in a weakened and slightly broadened peak at k≈1.13Å^{-1}. However, this first peak is not displaced and is much less smeared out by annealing for these two alloys than in the cases of As$_2$S$_3$ and As$_2$Se$_3$ (deNeufville et al., 1973). We conclude that a significant intermolecular correlation is retained in such films. Furthermore, we assume that this correlation signifies that a substantial equilibrium concentration of a large monomeric molecular species, presumably As$_4$Se$_4$, is present in these films under the annealing conditions.

In prior work we tentatively ascribed the irreversible change in the refractive indices of evaporated As$_2$S$_3$ and As$_2$Se$_3$ films with exposure or annealing to a shift of the first interband transition E$_0$ to lower energies (deNeufville et al., 1973). This interpretation is supported by the E$_0$ and E$_d$ data presented in Table 2 for both of these alloys and for AsSe as well. For all three alloys the decrease in E$_0$ with annealing is significant. This shift in E$_0$ accompanying partial polymerization in the case of AsSe and essentially complete polymerization in the cases of As$_2$S$_3$ and As$_2$Se$_3$ is not unexpected, since the first peak in ε$_2$(ω) is progressively red shifted for the various phases of selenium as the monomer (Se$_8$) to polymer (Se$_n$) ratio is decreased (Dalrymple and Spear, 1972; Stuke, 1970).

However, the changes in E$_0$ which result from exposure of annealed AsSe and Ge$_{10}$As$_{40}$Se$_{50}$ films, but which are unaccompanied by any significant changes in structure as revealed by X-ray diffraction, presents a greater problem in interpretation. We suggest that the non-equilibrium distribution of carriers which arises from exposure can significantly perturb the $\epsilon_2(\omega)$ spectrum without any concomitant structural rearrangements for alloys which contain a significant equilibrium concentration of a monomeric species. Therefore, the requirements for a reversible holographic medium of this sort appear to include: (a) the presence of monomeric constituents; (b) a suitable optical gap so that a significant increase in carriers accompanies exposure; and (c) a suitable glass transition temperature so that annealing of the disequilibrium carrier distribution can only occur well above the exposure temperature.

These requirements are met by evaporated amorphous AsSe and $Ge_{10}As_{40}Se_{60}$ films. Furthermore, the analogies between these two materials suggests that the ternary alloy may contain As_4Se_4 molecules in a heavily cross-linked polymeric $GeSe_2$-rich matrix, while the binary alloy contains coexisting monomeric and polymeric species of comparable composition analogous to amorphous Se. The presence of excess trapped charges after exposure apparently perturbs the electronic interaction between the chalcogen lone pair electrons of the polymer and monomer constituents (Ovshinsky, 1973) without causing gross structural changes.

ACKNOWLEDGMENTS

We acknowledge with thanks many discussions of the storage mechanisms with H. Fritzsche. The mass spectrometric analyses were performed by S.C. Agarwal of the University of Chicago, whose help we gratefully acknowledge.

REFERENCES

S.C. Agarwal, 1973, (private communication).

R.J. F. Dalrymple and W.E. Spear, 1972, J. Phys. Chem. Solids 33, 1071.

J.P. deNeufville, S.C. Moss and S.R. Ovshinsky, 1973 (submitted to J. of Non-Cryst. Solids).

R.E. Drews, R.L. Emerald, M.L. Slade and R. Zallen, 1972, Solid State Commun. 10, 293.

T. Igo and Y. Toyoshima, 1973, J. Non-Cryst. Solids 11, 304.

M. Kastner, Phys. Rev., 1972, Lett. 28, 355.

S.A. Keneman, 1971, Appl. Phys. Lett. 19, 205.

Y. Ohmachi and T. Igo, 1972, App. Phys. Lett. 20, 506.

S.R. Ovshinsky, 1973, presented at Topical Meeting on Optical Storage of Digital Data, Aspen, Colorado, (to be published).

S.R. Ovshinsky, 1970, U.S. Patent 3,530,441.

J. Stuke, 1970, J. Non-Cryst. Solids 4, 1.

S.H. Wemple, 1973, Phys. Rev. B7, 3767.

S.H. Wemple and M. DiDomenico, 1971, Jr., Phys. Rev. B3, 1338.

SOLAR ELECTRICITY SPEEDS DOWN TO EARTH

S.R. Ovshinsky

Energy Conversion Devices, Inc., Troy, Michigan 48084

The US Department of Energy is probably the best known body to have claimed that photovoltaic conversion devices cannot compete with fossil fuels or uranium until at the earliest 1986, or more likely 2000. Such statements are based on the known current technological problems of three types of silicon—single crystal, polycrystalline, and amorphous. With amorphous silicon the chief difficulty has been the high density of states (as explained by Taylor on p 672) which restricts the necessary current flow.

This problem is now solved by creating new silicon based materials in which fluorine plays an important role. Using this material our company is now in a position to construct effective p-n junctions from amorphous silicon. For designers of electronics devices, the implications of this development spread beyond solar cells into the computer and semiconductor industries.

Any suggestion that silicon could be the natural choice for photovoltaic devices is not, unfortunately, correct for its single crystal, polycrystalline, or elemental amorphous forms. This may sound surprising. While single-crystal silicon is not a likely candidate—because over 95 per cent of a single crystal is wasted in producing a photovoltaic device—such devices have done yeoman service in space and isolated installations with efficiencies ranging from 10 to 18 per cent. Their cost, however, is prohibitive for general use. Unlike transistors, which depend upon high density in small areas, solar cells for the most part require large areas if power systems are to be economically viable.

There are some exceptions, where higher temperature materials such as gallium arsenide can be used in conjunction with concentrators of various sorts. But small-area devices, while promising, are not likely to see widespread application. Polycrystalline materials are now also stronger candidates, but they suffer from disorder and grain boundary problems which give them all the disadvantages of both disorder and the physics connected with crystallinity. Grain boundaries limit the current flow, while polycrystalline as well as single-crystal silicon are "indirect band gap" materials: over 100-microns thickness is needed to absorb the sunlight effectively.

Technologically useless material

Elemental amorphous silicon is a technologically useless material (see Figure 1). Being tetrahedral in structure, it does not have the flexibility to eliminate voids or compensate for the opposite effects of dangling bonds. These defects control its conductivity and insert such a large density of states in the band gap as to make it practically immeasurable. The economic significance of the scientific term "band gap", as shown in Figure 1, can be put into context by regarding the reduction of the density of states in the gap as equatable with efficiency. This means that elemental amorphous silicon can be ruled out as a possibility for photovoltaic devices.

How does this statement agree with the results achieved by W. E. Spear and P. G. Le Comber at the University of Dundee? Their elemental amorphous silicon, material announced in 1975, had a much improved density of states over earlier efforts and their pioneering efforts to control amorphous silicon have led to a $5 \cdot 5$ per cent efficient solar cell. The answer is that despite the original assumption, their material was not amorphous silicon.

This takes us back to an ideological attitude in science which assumed that elemental amorphous semiconductors such as germanium and silicon were simple systems which could be understood by comparing them with their much studied and well-known crystalline counterparts. This is an attractive view, but it is like looking for the coin under

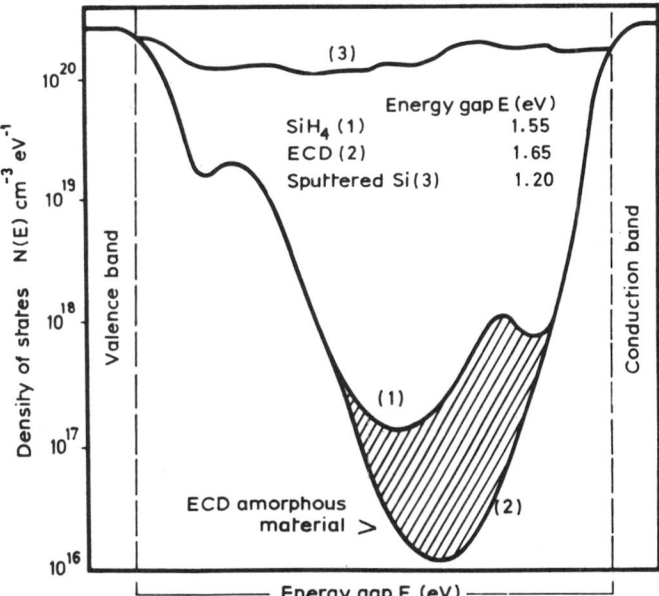

Figure 1   Density of states curves for three materials—1. glow discharge silicon made by Spear and Le Comber; 2. ECD's amorphous material and 3. sputtered silicon

Reprinted by permission from *New Scientist*, Vol. 80, No. 1131, pp. 674–677 (30 November 1978).

Amorphous alloy	Si:F:H	Si:H
Density of undesirable localised states	Low	Too high for efficient solar cells
Doping efficiency	High	Low
Dependence of electrical parameters on deposition temperature	Nil	High
Band gap	1·65 eV	1·55 eV
Photoconductivity	Good	Good
Photostructural changes in operation	Absent	Present and major
Hydrogen content	~0·5%	>5%
Fluorine content	~4%	None

the lamp post instead of where it was lost. Multicomponent amorphous materials such as the chalcogenides were considered to be basically different from elemental amorphous silicon. This view is correct but irrelevant, for to make a useful silicon material, a multicomponent amorphous material must be formed, just as in the chalcogenides. In the case of the Dundee work, a gas, silane, was decomposed by a glow discharge method and it seemed that silicon was deposited on the substrate. In any case, the silane materials as shown in Figure 1 had superior photoconductivity based on a lower density of states. Soon afterwards, H. Fritzsche at Chicago, J. Knights at Xerox (formerly of the Cavendish), and Bill Paul at Harvard University found that the hydrogen component could be as high as 30 per cent and that the material therefore was a silicon-hydrogen alloy.

The development of such a material by the Dundee group, was, and is, an important contribution to the science of amorphous semiconductors. David Carlson and Chris Wronski at RCA quickly built a 5·5 per cent efficient cell based upon the use of a Schottky barrier diode, and John Wilson and his colleagues at Heriot-Watt University, Edinburgh, developed an MIS (metal-insulator-semiconductor) configuration which has possibly even more interesting ramifications for large-area photovoltaic cells. However, the overall materials problem had not been solved, because the silane-derived alloy still had too many states in the gap, which reduced the general performance of the material.

An impenetrable barrier soon became apparent. Despite the fact Dr Spear and his group showed that a p-n junction was possible, its effectiveness as a junction was limited because of the large density of states. This was reflected in the fact that the Fermi level could not be moved any closer to the conduction band than two-tenths of an electron volt upon using relatively large amounts of dopant. In other words, the conductivity of the material could not be totally controlled. An effective depletion layer like that which exists in crystalline silicon could not be produced. This led to a cul-de-sac, as efficiencies have not been improved since the original announcements. It was not at all apparent how other useful semiconductor devices such as those of the transistor industry could be developed.

We at Energy Conversion Devices (ECD) have spent nearly 20 years inventing, developing, and understanding amorphous materials. We work with a large number of different types of amorphous materials including silicon. But until our announcement at the Seventh International Amorphous and Liquid Semiconductors meeting in Edinburgh in 1977, we were mostly identified with multicomponent chalcogenide and related materials. These have proved themselves in computers, memories, and imaging applications.

But with photovoltaic devices, we took a different view from that associated with the silane-decomposed materials. Arun Madan and I, working on the principles I have been proposing that useful amorphous materials containing silicon (which obviously as an element does have many advantages as we have noted) must be made in a multi-component manner since the concepts that are useful to chalcogenides are also applicable to the tetrahedral materials. In Edinburgh, I demonstrated that a great many materials as diversified as those based on group III (such as boron), group V (such as silicon), group V (such as arsenic) and group VI (such as tellurium) could be synthesised and their optical band gaps and electrical activation energies independently controlled. We considered the universality of amorphous materials, stressing their similarities rather than their differences. While such materials can sustain very high temperatures and are almost utopian for thermoelectric and other applications, they would not be chosen for near-term photovoltaic uses.

By making a new silicon alloy based on fluorine, which can also contain other elements such as hydrogen, along our steric design principles, we were able to demonstrate for the first time that a material could be made with over an order of magnitude less density of states than the best silane material (Figure 1). The unusually low density of states is an inherent property of the new alloy. This has led to a dramatic improvement in the effectiveness of the doping of amorphous semiconductors (Figure 2). Thus a problem with hydrogenated silicon was that it failed not as an elemental material, but as a multicomponent material, because it was not "multi" enough.

The favourable economics that are a feature of the manufacture of amorphous materials, can now be realised with multicomponent silicon-based materials, whose efficiencies can be analogous to those of crystalline photovoltaic cells. But even this year, at the International Solar Society meeting in Denver, several speakers said that the problem with amorphous silicon materials was the inability to reduce the

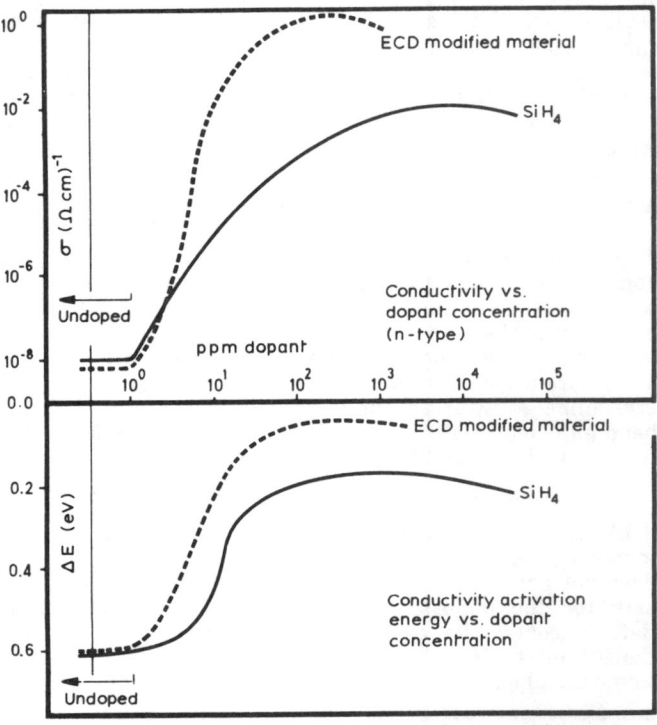

Figure 2 How doping concentration affects two important parameters of amorphous materials

Amorphous materials are creeping into the semiconductor business. This is a 1024 bit integrated circuit memory

remaining density of states, which restricts the necessary current flow.

Figures 1 and 2 show how this problem can be overcome. These new materials have a band gap different from either elemental amorphous silicon or silane-decomposed silicon, and a different internal structure. This difference in bonding structure is shown not only by the inherently fewer localised states in the gap but also, from a practical point of view, by the fact that these materials do not show the photostructural changes that have been demonstrated by David Staebler and Chris Wronski at RCA. Because ours is a new, exceptionally stable alloy, its physical characteristics differ from the silicon-hydrogen alloys. Its electrical properties offer—for the first time in amorphous materials—a density of states low enough to permit doping in the efficient parts-per-million range. The unique structural and transport properties of these materials overcome the scientific barriers that have limited the widespread use of silicon-based amorphous materials for photovoltaics. We are only awaiting installation of modern fabrication equipment before we can manufacture effective p-n junctions from amorphous materials.

I had been working on the assumption that elemental materials were too limiting in that one had to live with their very specific characteristics. We, therefore, have been considering what kind of bonding arrangements are possible within an amorphous solid that would allow molecular structures to be designed in which such tailor-made materials could possess properties superior to, and not as limiting as, elemental amorphous materials. This approach uses steric considerations which involve electronegativity, atomic size and reactivity resulting in preferred bonding and charge configurations. I have been proposing that it is necessary to have deviant as well as primary bonding choices so that an amorphous structure can be complete and stable. We have explored concepts such as three-centre bonds, coordination bonds, multiorbital bonding, and others into a field in which one thought strictly

in terms of silicon being a tetrahedral material and tellurium being a divalent material. We showed that silicon's tetrahedral nature was not to be taken for granted; that tellurium's divalency was not all that made it interesting; and that selenium's properties, even in its elemental form, depended on several steric structures.

New and important device properties

It appears from our modification of the chalcogenides, arsenides, oxides, nitrides, carbides, borides, and silicides, and now the multicomponent light element alloys of silicon, that we have proved that multi-elements in amorphous materials are needed to give new and important device properties—as well as to enhance our fundamental understanding of chemical bonding in amorphous materials. It is this understanding that allows us to tailor-make the band gap and to eliminate and control the density of states in the gap of amorphous solids. The parentage of such states differs qualitatively and quantitatively from those found in crystalline solids and in some simple elemental materials. The science is at its beginning, not at its end, and the technology moves in many directions besides energy conversion.

Many commercial companies are now working in the field. Exxon is now the home of Wronski, previously with RCA. And in Japan, where the universities have taken an early interest in amorphous materials, Hamakawa and his colleagues have built a device of different configuration based on silane materials whose efficiency is close to the figures of the RCA group. There is hardly a company, either in energy or in semiconductors, or a university, that is not now involved in amorphous work.

We believe that new materials that are not restricted to those made by silane deposition offer an opportunity for the field to be freed not only from the accidental results of a decomposition process, but from the limitations of the use of a single element. Xerox built an entire industry on amorphous selenium. The semiconductor industry was founded on crystalline silicon. And we can now synthesise new alloys where we can design in the best features of selenium or silicon and yet control their properties independently of the limitation of the nature-given character of the single element. Freeing ourselves from the single-element concept also discloses the exciting part of the science of amorphous materials because the differences from crystalline materials offer us many new insights. It is possible that amorphous materials may become as ubiquitous as crystalline materials in the coming years because of their economic advantage.

We believe, and have experimental evidence, that low-cost amorphous materials can be useful in areas of energy as diverse as thermoelectricity, electrochemistry, catalysis, superconductivity, energy storage and control photovoltaics. We have also just begun to explore the optical interactions in amorphous materials.

Freeing ourselves from not only the restraints of crystalline symmetry and stoichiometry but also from the limitations of elemental amorphous structures has allowed us to design materials rather than merely accept nature's dictates. We synthesise new materials, like plastics, that were not considered before. We can look forward to an exciting time in science and technology.

AMORPHOUS PHOTOVOLTAIC CELLS

Stanford R. Ovshinsky and Arun Madan

Energy Conversion Devices, Inc., 1675 West Maple Road, Troy, Michigan 48084

1. INTRODUCTION

The position of the Department of Energy (DoE) reiterated specifically at this meeting (1) is that photovoltaic solar energy conversion will not begin to make a large-scale contribution to our overall energy production until 1986 at the earliest, and more probably the year 2000. It is the theme of this paper that the basic materials problems which have led to such pessimistic projections have already been solved and that, given an intensive program of development, economically competitive photovoltaic cells can be mass produced in the near future.

Single-crystal solar cells work well but are more than an order of magnitude too expensive. Polycrystalline materials are still too costly and suffer from low efficiencies resulting from defects such as grain boundaries. It is now generally accepted that amorphous materials have the intrinsic low cost enabling the fabrication of solar cells that are economically competitive with fossil fuel and nuclear power. This is because amorphous semiconductors can be fabricated in large-area, thin-film configurations using inexpensive starting materials and a low-cost deposition process. However, the silane-decomposed amorphous films proposed for photovoltaic use have too large a density of localized states in the gap to allow doping, p/n junction formation, or the high carrier mobility necessary for efficient devices. As pointed out at this meeting, the DoE is presently spending a considerable sum of money in attempts to find amorphous materials, particularly silicon, with a reduced density of localized states (1).

In this paper we shall present experimental evidence that shows that a new Ovonic material has fewer localized states in the gap by more than an order of magnitude compared with previously reported amorphous semiconductors. It also shows a dramatic increase in doping efficiency so that higher carrier mobilities, and more efficient solar cells can be obtained.

2. MATERIALS FOR SOLAR PHOTOVOLTAIC ENERGY CONVERSION

Solar cells made of single-crystalline silicon, while relatively efficient (10-18%) have serious economic disadvantages and are presently about a factor of 20 too expensive for large-scale commercial use. The material has an indirect optical absorption edge, requiring the use of 200 μ m thick active layers, and this, together with the necessity of using ultra-pure solar-grade silicon and the fact that only 5% of the initial material appears in the final device, makes the starting material already too expensive. In addition, since crystals cannot be grown with cross-sectional areas in excess of about 10 cm, cell fabrication and assembly costs are prohibitive. Unlike transistors, wherein major cost reductions have been achieved via higher device densities, solar cells require the use of large-area arrays for significant energy conversion. (We are not ignoring the possibility of use of small-area cells with concentrators and tracking systems. We have developed high-temperature amorphous materials which remain extrinsic to 400°C and are promising in this regard.)

Polycrystalline materials, while potentially less expensive than single crystals, still require crystal-growth processes, and also suffer from the necessary deleterious appearance of large densities of grain boundaries, particularly in the less expensive films. These grain boundaries act as carrier traps and greatly reduce the solar cell efficiency. Polycrystalline devices thus appear to combine the disadvantages of both periodicity and disorder. At present, the CdS-based cells developed by Böer and coworkers (2) are the

Reprinted by permission from *Solar Energy Symposia of the 1978 Annual Meeting,* Denver, Colorado, 28–31 August 1978, pp. 69–73.

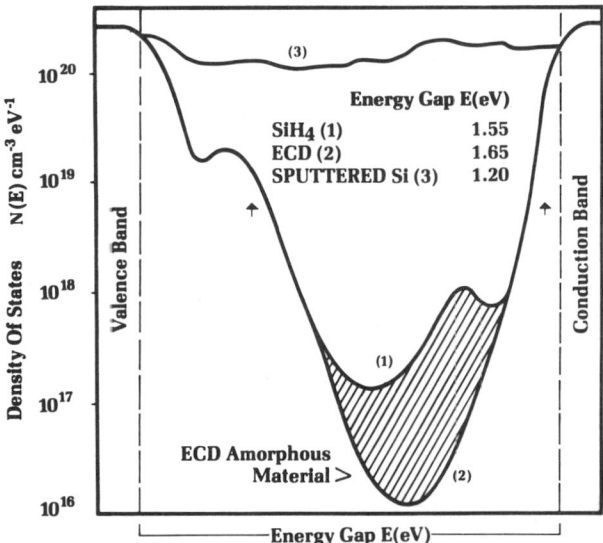

Fig. 1. Density of states in the gap. Density of localized states deduced from the field effect experiment. Arrows indicate extent of ECD experimental data.

leading contenders in this field and show much promise.

Amorphous films do not possess any long-range order, and therefore require no crystal growth. Their disorder precludes the possibility of any indirect optical-absorption edges, so that they can absorb almost all the incident solar energy in about one micron. However, the problem of large densities of localized states makes them unsuitable for solar cells. Then Carlson and Wronski (3) announced that they had fabricated an amorphous silicon Schottky-barrier device with a conversion efficiency of 5.5%. Later, Wilson et al. (4), using an MIS configuration, and a Japanese group which included Hamakawa (5), using a p-i-n geometry, also obtained approximately 5% conversion efficiencies with amorphous materials. All of these solar cells use silicon-hydrogen alloys produced from the glow-discharge decomposition of silane (SiH_4) gas, materials discovered by Chittick et al. (6) and pioneered and developed by Spear and Le Comber (7). These materials, although originally thought to be pure amorphous silicon, have recently been shown to contain 10-35% residual hydrogen (8-12). Since the density of states is roughly a measurement of the efficiency potential of a material, we present Fig. 1 to be used as a guide for the relative densities of localized states in: the top curve representing density of localized states in evaporated or sputtered samples of elemental amorphous silicon whose densities of states are so high that they are virtually immeasurable, the middle curve representing the best amorphous silicon-hydrogen

alloys, and the lower curve representing our Ovonic materials.

There are several problems associated with even the best silane-decomposed films insofar as their use in solar cells is concerned. Firstly, as is evident from Fig. 1, the density of localized states in the gap of the best silane material (middle curve) is still too high. This results in inefficient doping and also these states reduce the carrier mobilities and recombination times of the free carriers. In addition, the doping is inefficient since the Fermi level cannot be moved closer than 0.2 eV from the edge even with heavy doping. This has thus far precluded the fabrication of effective p/n junctions. The intrinsic properties of the materials are strongly dependent on substrate temperature and other deposition parameters, resulting in the necessity for careful preparation techniques. The materials show major photo-structural changes (13) which raise questions concerning device lifetimes. Despite major efforts at many laboratories, there has been no progress in solving these problems.

3. AMORPHOUS SEMICONDUCTOR MATERIALS

The problems that have affected the development of solar cells based on amorphous silicon-hydrogen alloys are related to ideology that expresses itself in the following manner: since elemental crystalline silicon is a well-studied and well-understood material, elemental amorphous silicon can be understood by a direct comparison to its crystalline counterpart. In this section, we emphasize the inaccuracies of such an approach. To begin with, it is clear that the supposed "simplicity" of elemental amorphous silicon is irrelevant since, as the top curve of Fig. 1 shows, its extremely high density of localized states makes it essentially useless as an electronic material. It is only when silicon is combined with other elements, either accidentally as in silane-decomposed films, or specifically designed as in the materials which we shall discuss, that it has important device possibilities. The multi-element approach has been the mainstay of our work in chalcogenides (14).

The problem of density of states in silicon-based materials has to do with the rigidity of the tetrahedral structure. In the amorphous state, the energy required to complete the structure and to have spin pairing is so great as to generate voids, cracks, and dangling bonds. Hydrogen acts as a polymeric structural element which can bond in several ways. Its small atomic size and its ability to bond specifically with only one other atom provides the necessary flexibility to a silicon-based amorphous material so as to produce a material which intrinsically has fewer states in the gap. However, hydrogen can introduce defects of its own, and, as we shall make clear in the next section, other elements can accomplish the same result in a different manner.

We have extensively worked with amorphous solids based on alloys of materials from Columns III-VI in the Periodic Table. In particular, the Column VI (or chalcogen) elements form the basic ingredients of the chalcogenide glasses which have found commercial applications via their switching, memory, and imaging effects (14-17). As we have pointed out (18,19), chalcogen atoms, through their primary divalency, lone-pair interactions, and ability to crosslink with other atoms, provide a more flexible structure than the tetrahedrally bonded materials. The energy-saving distortions connected with such flexibility encourage bonding and orbital relationships which result in spin pairing. Other states in the gap, including those low-energy bonding configurations, are ordinarily present in chalcogenide glasses (18,20,21). At approximately the same time as Spear and Le Comber showed that amorphous silicon-hydrogen alloys could be doped (7), we reported the first doping of chalcogenide glasses (22). Extension of the same principles to elements from Columns III, IV, and V of the Periodic Table led to the chemical modification and achievement of extrinsic conduction in a wide variety of amorphous semiconductors (21,23-25). Whatever the nature of the amorphous matrix, it is now possible to tailor-make materials analogous to the synthesis of new plastics by independently varying important parameters such as optical band gap and electrical activation energy using our principles of chemical modification (23). For example, we have been able to make a large-band-gap silicon-carbon alloy while reducing the electrical activation energy sufficiently that the conductivity is essentially metallic. These materials are particularly attractive for thermoelectric applications.

Ten years ago, the CFO (Cohen-Fritzsche-Ovshinsky) model (26) for multicomponent amorphous alloys was developed. Although it was never intended to describe elemental amorphous solids, the model was apparently able to explain the experimental results on amorphous silicon successfully (27). Placed in our perspective that amorphous silicon has now turned out to be an alloy which requires orbital distortions and flexible links affecting charge relationships and therefore having similarity to chalcogenide alloys, then this seemingly paradoxical situation is clarified. The proper approach to developing an amorphous silicon-based alloy with more desirable properties than silane-decomposed films is to first understand the nature of the problems in the latter. We pointed out that hydrogen in intrinsic as well as boron- and phosphorus-doped amorphous silicon-hydrogen alloys could exist in several different configurations, including as a bridging element in a three-center bond (21,23). These configurations will ordinarily produce unnecessary states in the gap. We have utilized the steric principles described more fully elsewhere (28) and have chosen elements based on their size, electronegativity, electronic configuration, and reactivity, for example, fluorine (28,29), as components in a new silicon-based amorphous alloy superior to silane-decomposed films for solar energy conversion. The electronegativity of fluorine is a strong organizer and inducer of nearest-neighbor structures. The charge relationships and equilibrium that ensue are possibly as important as the dangling bonds that are eliminated by its orbital reactivity.

4. A NEW AMORPHOUS ALLOY FOR SOLAR CELLS

As discussed previously, amorphous silicon-hydrogen alloys deposited from silane gas have several serious problems which have not been overcome. In this section, we present the results of a new approach using the principles discussed in the last section via which we have tailor-made a material with structural stability, with properties essentially independent of the deposition parameters, and with a much reduced density of localized states in the gap. This material has an optical band gap of 1.65 eV, as compared with 1.55 eV for the best silane-decomposed samples, and 1.1 eV for crystalline silicon, and is thus a completely new material with the optimal gap for solar photovoltaic energy conversion. It also, for the first time, permits efficient doping. As is clear from the lower curve of Fig. 1, the density of localized states of ECD's new semiconductor is drastically lower than that of silane-decomposed films. The high-energy peak that appears below the conduction-band edge in silane-decomposed films is absent. As is clear from Fig. 2, this results in much more efficient n-type doping, and, in fact, effective doping has been achieved using only 50 vppm AsH_3 in the premix. In addition, the previously discussed 0.2 eV activation-energy barrier in the silane-decomposed films has been overcome with the new material. Details of the chemistry of the latter will soon be published elsewhere (29), but we wish to point out here that these newly synthesized alloys based on silicon and fluorine can contain three or more elements (for example, other elements such as oxygen are sometimes utilized) and have properties all of which are superior to those of the best silane-

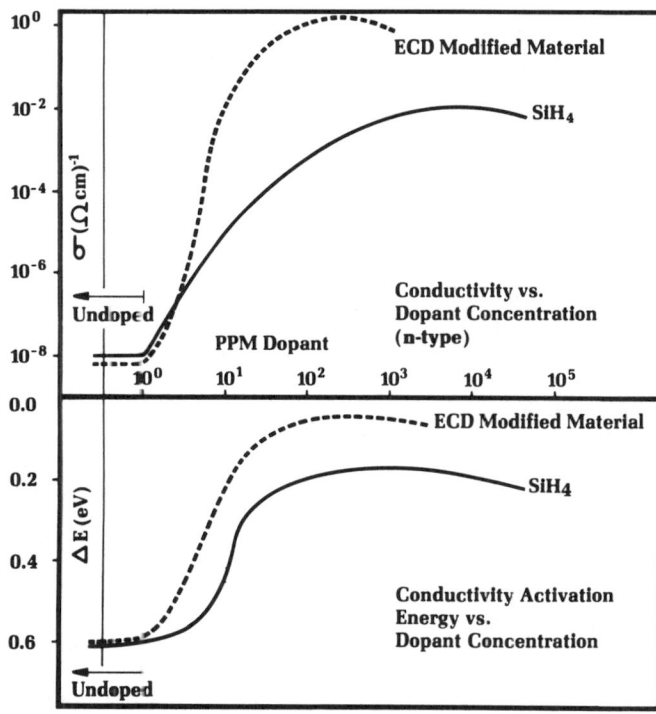

Fig. 2. Doping of amorphous semiconductors.

decomposed films for electronic applications. Their photoconductivity is equal. State-of-the-art photovoltaic devices utilizing Schottky barrier and other geometries have been fabricated. As shown in Fig. 3, the economics of a wide-area device based on our new material can be realistically estimated to be competitive with coal, gas, oil and nuclear power at the present time. Efficiencies analogous to crystalline materials are now possible and we are working on graded-gap devices with very high energy-conversion efficiencies.

5. SUMMARY

We have described a new amorphous semi-conductor alloy which has properties superior to

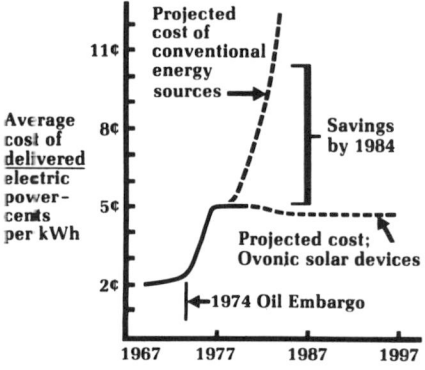

Fig. 3. Cost comparison of ovonic solar devices with conventional sources.

those of silane-decomposed films. The new material has the optimal gap for solar photovoltaic energy conversion, has physical properties essentially independent of the deposition procedure, shows no photostructural changes, and is unique in having a greatly reduced density of localized states as well as providing doping efficiencies usually associated with crystalline materials. The economics of solar cells based on thin films of this material deposited over wide areas can be estimated to be within the DoE's 1986 goal of $0.50 per peak Watt at the present time.

6. ACKNOWLEDGEMENT

We should like to thank David Adler for his many helpful suggestions.

7. REFERENCES

(1) Feucht, D.L. "Advanced Materials Photovoltaics R & D," presented at this conference.

(2) Böer, K.W. "The CdS/Cu$_2$S Solar Cell," Phys. Stat. Sol (a), *40*, 355 (1977).

(3) Carlson, D.E., and Wronski, C.R. "Solar Cells Using Discharge-Produced Amorphous Silicon," J. Electron. Mater., *6*, 95 (1977).

(4) Wilson, J.I.B, and McGill, J. "Amorphous Silicon MIS Solar Cells," Nature, *272*, 152 (1978).

(5) Okamoto, H., Nitta, Y., Adachi, T., and Hamakawa, Y. "Glow Discharge Produced Amorphous Silicon Solar Cells," presented at the ICSFS in Tokyo, Japan, July 5-8, 1978.

(6) Chittick, R.C., Alexander, J.H., and Sterling, H.F. "The Preparation and Properties of Amorphous Silicon," J. Electrochem. Soc., *116*, 77 (1969).

(7) Spear, W.E., and Le Comber, P.G. "Electronic Properties of Substitutionally Doped Amorphous Si and Ge," Phil. Mag., *33*, 935 (1976).

(8) Fritzsche, H. *Amorphous and Liquid Semiconductors,* edited by W.E. Spear (Edinburgh: Center for Industrial Consultancy and Liaison, University of Edinburgh, 1977) p.3.

(9) Fritzsche, H., Tsai, C.C., and Persans, P. "Amorphous Semiconducting Silicon-Hydrogen Alloys," Solid State Tech., January 1978, p. 55.

(10) Paul, W., Lewis, A.J., Connell, G.A.N., and Moustakas, T.D. "Doping, Schottky Barrier and p-n Junction Formation in Amorphous Germanium and Silicon by RF Sputtering," Solid State Commun., *20*, 969 (1976).

(11) Knights, J.C., Hayes, T.M., and Mikkelsen, Jr., J.C. "Coordination of Arsenic Impurities in Amorphous Silicon-Hydrogen Alloys," Phys. Rev. Lett., *39*, 712 (1977).

(12) Madan, A., Le Comber, P.G., and Spear, W.E. "Investigation of the Density of Localized States in a-Si Using the Field Effect Technique," J. Non-Cryst. Solids, *20*, 239 (1976).

(13) Staebler, D.L., and Wronski, C.R. "Reversible Conductivity Changes in Discharge Produced Amorphous Si," Appl. Phys. Lett., *31*, 292 (1977).

(14) Ovshinsky, S.R., and Fritzsche, H. "Amorphous Semiconductors for Switching, Memory, and Imaging Applications," IEEE Trans. Electron Devices, *ED-20,* 91 (1973).

(15) Ovshinsky, S.R. "Reversible Electrical Switching Phenomena in Disordered Structures," Phys. Rev. Lett., *21,* 1450 (1968).

(16) Ovshinsky, S.R., and Ovshinsky, I.M. "Analog Models for Information Storage and Transmission in Physiological Systems," Mat. Res. Bull., *5,* 681 (1970).

(17) Ovshinsky, S.R. "Amorphous Materials as Optical Information Media," J. Appl. Photo. Eng., *3,* 35 (1977).

(18) Ovshinsky, S.R. "Localized States in the Gap of Amorphous Semiconductors," Phys. Rev. Lett., *36,* 1469 (1976).

(19) Ovshinsky, S.R., and Sapru, K. *Amorphous and Liquid Semiconductors,* edited by J. Stuke and W. Brenig (London: Taylor & Francis, 1974) p. 447.

(20) Kastner, M., Adler, D., and Fritzsche, H. "Valence-Alternation Model for Localized Gap States in Lone-Pair Semiconductors," Phys. Rev. Lett., *37,* 1504 (1976).

(21) Ovshinsky, S.R., and Adler, D. "Local Structure, Bonding, and Electronic Properties of Covalent Amorphous Semiconductors," Contemp. Phys., *19*, 109 (1978).

(22) Ovshinsky, S.R. *Structure and Excitations of Amorphous Solids,* edited by G. Lucovsky and F. L. Galeener (New York, AIP, 1976) p. 31.

(23) Ovshinsky, S. R. *Amorphous and Liquid Semiconductors,* edited by W. E. Spear (Edinburgh: Center for Industrial Consultancy and Liaison, University of Edinburgh, 1977) p. 519.

(24) Flasck, R., Izu, M., Sapru, K., Anderson, T., Ovshinsky, S. R., and Fritzsche, H. *Amorphous and Liquid Semiconductors,* edited by W. E. Spear (Edinburgh: Center for Industrial Consultancy and Liaison, University of Edinburgh, 1977) p. 524.

(25) Ovshinsky, S. R. Sapru, K., and Dec, K. *The Physics of SiO_2 and its Interfaces,* edited by Sokrates T. Pantelides (New York: Pergamom Press, 1978) p. 304.

(26) Cohen, M. H., Fritzsche, H., and Ovshinsky, S. R. "Simple Band Model for Amorphous Semiconducting Alloys," Phys. Rev. Lett., *22,* 1065 (1969).

(27) Spear, W. E., and Le Comber, P.G. Phil. Mag. In press.

(28) Ovshinsky, S. R. "The Shape of Disorder," 1979 Mott Festschrift. To be published in J. Non-Cryst. Solids.

(29) Ovshinsky, S. R. and Madan, A. To be published.

METAL–INSULATOR–SEMICONDUCTOR SOLAR CELLS USING AMORPHOUS Si:F:H ALLOYS

A. Madan, J. McGill, W. Czubatyj, J. Yang, and S.R. Ovshinsky

Energy Conversion Devices, Inc., 1675 West Maple Road, Troy, Michigan 48084

(Received 5 May 1980; accepted for publication 1 August 1980)

Amorphous hydrogenated silicon (a-Si:H) alloys produced from the radio-frequency glow discharge of SiH_4 gas have been studied extensively in recent years[1]. Thin-film photovoltaic devices have been fabricated from these alloys in Schottky barrier[2] metal-insulator-semiconductor (MIS),[3] and p-i-n[4] configurations with conversion efficiencies of up to 5.5, 4.8, and 4.5%, respectively. *

We have reported[5–7] that a-Si:F:H alloys with a low density of gap states ($\sim 10^{16}$ cm^{-3} eV^{-1}) and possessing a high photoconductivity can be fabricated using the radio-frequency glow discharge of mixed SiF_4 and H_2 gases. We have also shown that this type of material can be doped easily by introducing small amounts of PH_3 or AsH_3 in the gas phase to obtain conductivities (for the n^+ layer) $\sim 10 \Omega^{-1}$ cm^{-1}, a three-orders-of-magnitude improvement over those reported for an a-Si:H alloy.[8] Recently, a conversion efficiency of 1.3% was reported[9] for an MIS-type device using a-Si:F:H as the active material. In this letter, we report a substantially higher conversion efficiency of 6.3%, which in fact exceeds the maximum efficiencies that have been reported for devices fabricated from the a-Si:H-type alloy.

A typical device structure is shown in the inset of Fig. 1. (Full details concerning the conditions required for deposition have been given elsewhere.[6]) A thin highly conducting n^+ layer (~ 800 Å) was deposited onto a reflecting Mo bottom contact. Next, ~ 5000 Å of active photoconductive a-Si:F:H was deposited using a volume gas ratio of $SiF_4/H_2 \simeq 5/1$, which, as we have previously demonstrated, possesses a relatively low density of states.[6] (The photoconductivity under AM-1 excitation of this component is typically in the range 10^{-4}–$10^{-3} \Omega^{-1}$ cm^{-1}, which provides for a low series resistance in operation.) Then, a 20-Å-thick layer of an oxide such as Nb_2O_5 was thermally evaporated and contact was made to the device using 70 Å of high-work-function Au:Pd (90:10) metal. Finally, a layer of 350-Å-thick ZnS served as an antireflection coating.

*At the time of publication, these efficiencies were 6.6 to 6.7%, and since then they have been moving up continuously.

Figure 1 shows a typical dark current density-voltage (J-V) characteristic for forward and reverse bias. We note that the rectification ratio at 0.5-V bias is about 10^5. The dark diode ideality factor for this device is $n = 1.2$. The departure from $n = 1.0$ can be explained on the basis of the oxide layer as indicated in the formulation by Card and Rhoderick.[10]

In Fig. 2, we show the fourth quadrant of the illuminated J-V characteristics as measured at AM incident power (83 mW cm^{-2}) using a commercially available solar simulator (Optical Radiation Co. Model No. SS-1000-20C). The characteristics of the device are open-circuit voltage $V_{OC} = 0.75$

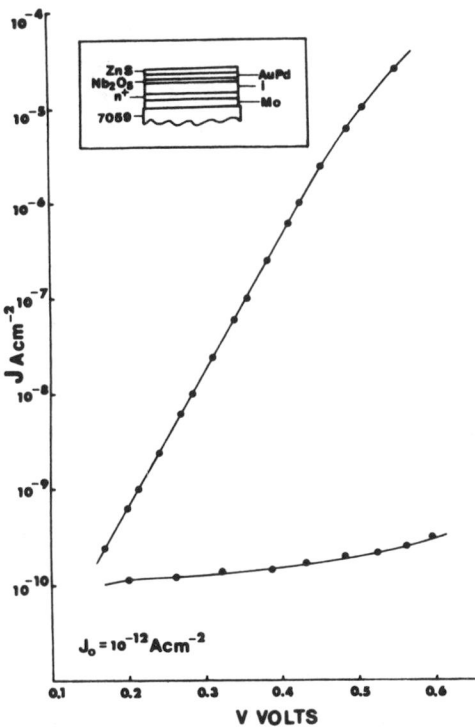

FIG. 1 Foreward and reverse J-V characteristics of a typical device. The insert shows a schematic of the device structure.

Reprinted by permission from *Applied Physics Letters*, Vol. 37, No. 9, pp. 826–828 (1 November 1980).

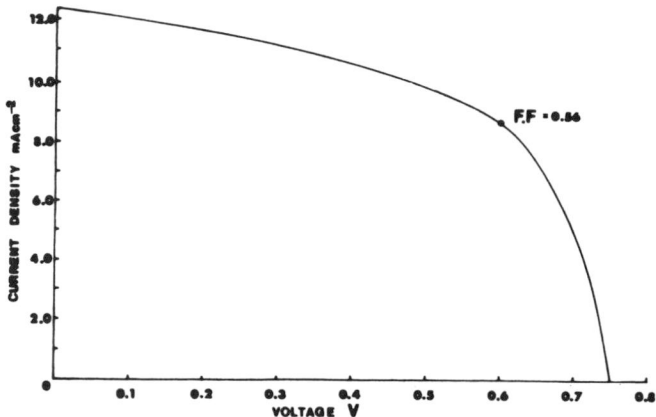

FIG. 2 The fourth quadrant of the current-voltage curve is plotted. The maximum power point and fill factor are indicated and give an efficiency of 6.2% under AM-1 illumination.

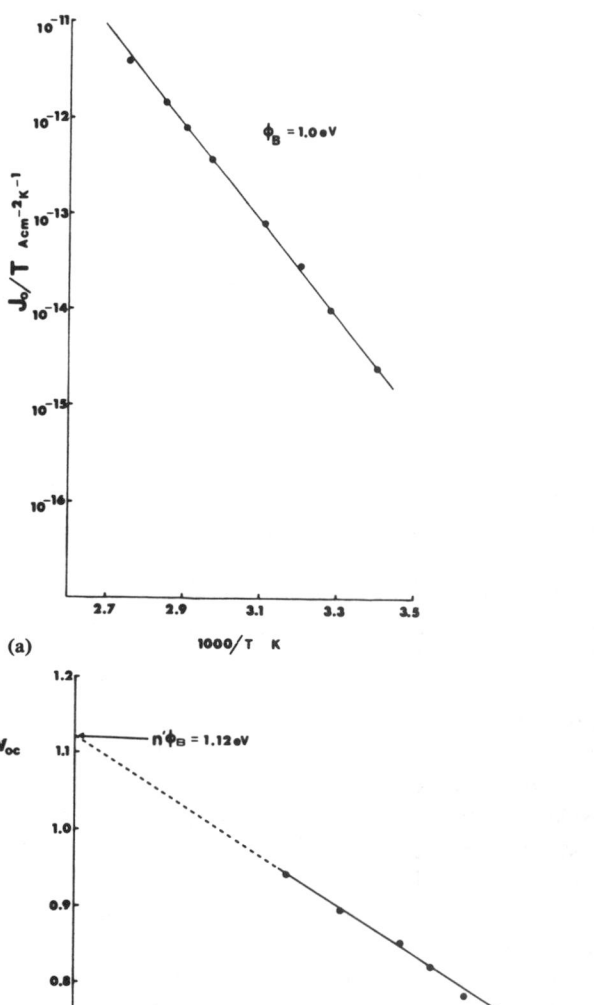

FIG. 3(a) plot of J_0/T against reciprocal temperature yeilds $\phi_B = 1.0$ eV[1].(b)plot of V_{OC} against temperature extrapolated to T = O gives a value for n' $\phi_B = 1.12$ eV.

V; short-circuit current density $J_{SC} = 12.24$ mA cm^{-2} fill factor of 0.56; and conversion efficiency of 6.2%. The area of the device is 4.2 mm^2. This efficiency has been reproduced in several cells. For example, another such cell exhibits $J_{SC} = 12.9$ mA/cm^2, $V_{OC} = 0.80$ V, and a fill factor of 0.61 at a power of 100 mW/cm^2, yielding a 6.3% overall conversion efficiency. Further improvements in the MIS cell can be achieved with optimization of the antireflection coating, the contacts, the intrinsic layer, and the n^+ layer. For example, by utilizing multilayer antireflection coatings such that reflection losses are minimized over the entire range from the UV to about $\lambda = 600$ nm, higher currents would result. Further improvement can be brought about by an optimum compromise between the conflicting requirements for maximum light transmission and minimum sheet resistance. Fine tuning of the intrinsic and the n^+ layer thickness could also lead to an improved performance of the current collection.

Since amorphous silicon-based alloys are low-mobility solids, diffusion theory[11] should be applicable for estimating effective barrier height ϕ_B. Then the saturation current J_0 can be written as,

$$J_o = e\mu N(E_c)kTE_{max} \exp(-e\phi_B/kT), \qquad (1)$$

where the electron mobility is assumed to be the same as in a-Si:H alloys, $\mu \sim 0.1$ cm^2 V^{-1}s^{-1},[1] $N(E_c) \sim 10^{21}$ cm^{-3} eV^{-1} is the density of states at the conduction band edge,[6] and $E_{max} \sim 10^5$ V cm^{-1} (Ref. 12) is the electric field at the metal-semiconductor interface. Using these values and the measured $J_0 = 10^{-12}$ A cm^{-2}, we estimate ϕ_B to be 1.0 eV. (Note that the value of μ may be both field-dependent and much higher near the top of the barrier, where the field has its maximum value. If we assume that the mobility increases by a factor of 100, an extreme situation, then the value of ϕ_B is increased only slightly, to 1.1eV.)

Alternatively, the slope of a plot of J_0/T as a function of reciprocal temperature yields ϕ_B without any assumptions about the preexponent in Eq.(1). Figure 3(a) shows such a plot, which gives $\phi_B = 1.0$ eV.

Finally, we have also obtained ϕ_B by plotting the open-circuit voltage V_{OC} against temperature, as shown in Fig. 3(b). Since

$$V_{OC} = (n'kt/q)\{\ln[J_L(V)/J_0] + 1\}; \qquad (2)$$

where $J_L(V)$ is the light-generated current and is a function of bias V, because the current is almost entirely due to drift,[13] and n' is the diode ideality factor under illumination. [The value of n' was measured to be temperature independent within 10% experimental error. Further, the variation of the short-circuit current over the temperature range (300–180 K) investigated was less than a factor of 2.] We note that $n' = 1.0$ is a substantial reduction from its dark value of 1.2. This could be due to the potential drop in the interfacial layer becoming negligibly small due to interface states that may take on a positive charge.[14] Using $n' = 1.0$, Fig. 3(b) shows

$\phi_B = 1.12$ eV, which is in good agreement with ϕ_B estimated from the diffusion theory (1.0 eV), and when measured from the slope of J_0/T versus reciprocal temperature (1.0 eV).

In summary, we have presented compelling evidence that MIS solar cells fabricated from a-Si:F:H alloys are clearly at least as, in our opinion, more efficient than those prepared from a-Si:H alloys. Various measurements of the barrier height yeild the same result: $\phi_B \cong 1.0$ eV for the structure shown in the inset of Fig. 1.

We should like to thank Ron Himmler, Lynn Bement, Robin Stiers, and Larry Christian for their expert assistance in the preparation and measurements of samples and diodes. We also thank Professor D. Adler, Professor H. Fritzsche, Professor M. Shaw, and Professor M. Shur for stimulating and useful discussions. We are indebted to ARCO for their assistance and cooperation and, in particular, to Dr. C. Gay for his helpful comments.

[1] See, e.g., W.E. Spear Adv. Phys. **6**, 811 (1977).

[2] D.E. Carlson and C.R. Wronski, J. Electron. Mater. **6**, 95 (1977).

[3] J.I.B. Wilson, J. McGill, and S. Kinmond, Nature **272**, 152 (1978).

[4] H. Okomoto, Y. Nitta, T. Asochi, and Y. Hamakawa, Surf. Sci. **86**, 486 (1979).

[5] S.R. Ovshinsky and A. Madan, Nature **276**, 482 (1978).

[6] A. Madan, S.R. Ovshinsky, and E. Benn, Philos. Mag. **40**, 259 (1979).

[7] A. Madan and S.R. Ovshinsky, J. Non-Cryst. Solids **35 36**, 171 (1980).

[8] W.E. Spear and P.G. LeComber, Philos. Mag. **33**, 935 (1976).

[9] M. Konagai and K Takahashi, Appl. Phys. Lett. **36**, 599 (1980).

[10] H.C. Card and E.H Rhoderick, J. Phys. D **4**, 1589 (1971).

[11] E.H. Rhoderick, *Metal-Semiconductor Contacts* (Clarendon, Oxford, 1978).

[12] M. Shur, W. Czubatyj, and A. Madan, J. Non-Cryst. Solids **35 36**, 731 (1980).

[13] W. Czubatyj, M.Shur, K. Ng, and A. Madan, Proceedings of the 14th IEEE Photovoltaics Conference, San Diego, 1980 (unpublished).

[14] H.C. Card and E.S. Yang, Appl. Phys. Lett. **29**, 57 (1976).

PROGRESS IN LARGE AREA PHOTOVOLTAIC DEVICES BASED ON AMORPHOUS SILICON ALLOYS

John P. deNeufville, Matsatsuga Izu, and Stanford R. Ovshinsky

Energy Conversion Devices, Inc., 1675 West Maple Road, Troy, Michigan 48084

Introduction

Photovoltaic devices based on amorphous alloys of silicon with H and/or F have recently generated substantial interest in the photovoltaic community and the larger solid state physics and electronics communities as well. These developments have been reviewed recently by D. Carlson of RCA Labs[1]. It was Carlson's early collaboration with C. Wronski[2] which served to rekindle scientific and technological interest in a-Si:H produced by the glow discharge induced decomposition of SiH_4 gas and the deposition of Si-rich fragments of this gas onto substrates held at 200-300°C[3-5].

The photovoltaic energy option has received keen attention at all levels as a potential panacea to the energy crisis facing the world today[6]. Not only is its fuel, sunlight, inexhaustable on any reasonable time scale, its energy output, electricity, nearly universally useable for any energy needs, but its capital costs can, in principle, be acceptably low[7]. Recently, however, the photovoltaic option has tended to be dismissed as a very long term solution to the energy crisis[8], and the photovoltaic industry, nurtured with major governmental investments, is meeting increasing economic uncertainties. The basic problem facing the industry is the absence of any major market for photovoltaic arrays at the current prices of $7-12 per peak Watt coupled with the realization that conventional photovoltaic devices, based on 250 μm slices of crystalline silicon, can never be produced at prices in the range of 35-50¢ per peak Watt which would be required to make the cells competitive with conventional sources of electrical power including such fuels as coal, oil, gas and uranium.

It is the position of ECD that the mass production of low cost photovoltaic cells and panels based on thin films will be required before the photovoltaic revolution can begin in earnest. Our economic analyses indicate that a thin film technology, capable of mass producing photovoltaic cells with active layers in the thickness range 0.5-1.5 μm, can provide the requisite systems cost breakthrough if the cells have an efficiency in the range of 7-10% and a useful life of 20 years. Such cells, produced at a cost of 35-50¢ per peak Watt, could compete effectively with conventionally generated electricity in certain markets within the U.S.

Efficiencies of photovoltaic devices based on a-silicon seemed bounded for several years by the 5.5% level originally reported by Wronski and Carlson. Furthermore, little progress was reported in the preparation of efficient larger area devices, even of the order of 1 cm^2, in the same time frame. Within the past year, however, research efforts have produced progress in device efficiency[9] and area[1] based on both metal/insulator/semiconductor/metal (MIS) and conductor/p-type semiconductor/intrinsic semiconductor/n-type semiconductor/conductor (PIN) sandwiches which can be illuminated by making one or more of the photovoltaic diode electrodes partially transparent. Indeed, the Sanyo Electric Company in Japan has introduced consumer products which are powered by a-Si:H photovoltaic cells. While such cells have relatively low conversion efficiency for sunlight, it is gratifying to note that a 6.6% active area AM1 (simulated sunlight) efficiency has recently been reported by Madan et al.[9] for a 0.73 cm^2 device based on an a-Si:F:H alloy.

In the following discussions several aspects of this new class of photovoltaic devices will be highlighted. First, the structure and the electronic properties of amorphous silicon alloys will be examined from a simplified but fundamental point of view, emphasizing those characteristics which make this material a viable candidate for photovoltaic applications. Next, the nature and characteristics of the deposition process will be examined to establish the plausibility of a low cost mass production

Reprinted by permission from the *Proceedings of the 16th Intersociety Energy Conversion Engineering Conference*, Atlanta, Georgia, 9–14 August 1981, Vol. 3, pp. 2217–2220.

method for manufacturing such cells. Finally, the unique position of ECD with respect to this goal will be examined in the light of recent laboratory results concerning the scale-up of those photovoltaic cells from 1-5 cm^2 to ~1000 cm^2.

Discussion

A. Amorphous Photovoltaic Materials

Since the early work of Sterling[3], Chittick[4] and Spear[5] it has been known that "amorphous silicon", produced by the glow discharge induced decomposition of silane (SiH$_4$) gas onto substrates held at 200-300°C, has interesting photoelectronic properties including a high photoconductivity and an optical band gap in the range of 1.7 eV. Subsequent analyses[10,11] revealed that such films contain substantial concentrations of H (10-20 at%) in addition to silicon, and are, thus, amorphous alloys based on silicon. Excellent accounts of the subsequent progress in the characterization of such a-Si:H films and the application of them to photovoltaic solar energy conversion are available in recent publications [1,12].

Most amorphous semiconductors, prepared either as bulk glasses or thin films deposited by sputtering, evaporation, etc., tend to have roughly symmetrical distributions of conduction band and valence band tail states, leading to a pinning of the Fermi level near the center of the forbidden gap. Furthermore, band bending at the contacts of such films with metals or other semiconducting layers tends to produce low barrier heights and narrow space-charge regions due to the relatively high density of states in the vicinity of the Fermi level. Materials such as a-GeTe$_2$, a-GeSe$_2$, a-As$_2$Te$_3$, etc. contain large concentrations of a chalcogen atom (S, Se, or Te) which is normally in two-fold coordination. It now appears that low coordination numbers can lead to a multiplicity of bonding configurations[13] and thus to a broadened density of states function (both valence band and conduction band) within the forbidden band gap.

Pure amorphous silicon, on the other hand, tends to have a high concentration of actual structural defects, including microvoids and dangling bonds, when it is prepared at temperatures below 300-400°C or so. When amorphous silicon is deposited at temperatures between 500-600°C, such as by silane pyrolysis[14], it has no microvoids, and essentially no hydrogen, but still retains a significant concentration of intrinsic dangling bonds[12], presumably arising from the topological constraints of constructing a non-crystalline tetrahedrally bonded network[15]. Such films have sharp optical absorption edges and a band gap of roughly 1.6eV, but have a sufficiently high level of mid-gap states associated with these dangling bonds to seriously limit the width of the space charge region and the lifetimes of carriers, leading to unacceptable characteristics for photovoltaic device applications.

In contrast to the underconstrained covalent networks of chalcogenide glasses on the one hand, and to the overconstrained network of pure amorphous silicon on the other hand, the a-Si:H and a-Si:F:H alloys appear to represent nearly optimally constrained networks. For example, the presence of singly coordinated hydrogen atoms lowers the average number of network coordination points per silicon atom from nearly 4 in pure a-Si to about 3.8 for a-Si$_{85}$H$_{15}$. Optimal network constraint manifests itself not only via the elimination of the presence of intrinsic dangling bonds which occur in pure a-Si, but also via the reduction in bond angle strain, thus minimizing the undesirable tailing of conduction band and valence band states into the forbidden gap[16].

As a final point with regard to amorphous silicon-based alloys for photovoltaic devices, it is appropriate to emphasize the characteristics of the optimized alloys themselves and to indicate the role of "fine tuning" of alloy chemistry upon device performance and stability. The decomposition of silane via a glow discharge is an irreversible reaction resulting in the production of a-Si:H and H$_2$. The substrate temperature affects simultaneously the degree of hydrogen incorporation in the growing film and the concentration of dangling bonds as indicated by electron spin resonance[14]. In such circumstances the simultaneous optimization of both sources of mid-gap states, network strain and point defects, cannot be achieved. ECD has approached this problem from the view point of introducing the constituent, F, into the film using a reagent, gaseous SiF$_4$, whose reaction with hydrogen to produce silicon is reversible at elevated electron temperatures. This reversibility permits the simultaneous deposition and etching of Si by SiF$_4$-bearing plasmas. The resulting films, in addition to containing F, have a lower density of mid-gap states and, thus, a wider space charge region in photovoltaic cells, than films made in the a-Si:H system. The enhanced photo-stability of the fluorinated ECD films[17] may arise from the reduced hydrogen levels compared to a-Si:H.

B. Economics of Amorphous Photovoltaic Devices

The economics of photovoltaic solar energy conversion are determined by three crucial parameters: the cost of the collector, its lifetime, and its efficiency. Conventional photovoltaics based on crystalline silicon have adequate efficiency and lifetimes, if properly packaged, but have a cost of $7-12 per peak Watt, based on current methods and rates of production. While 10-fold reductions in these costs have been projected[7] based on greatly expanded volume, such projections may be unrealistic[8], particularly in light of significant contributions of electrical energy to these costs. Crystalline silicon, with its high melting point, limited ductility, great hardness and extreme purity requirements for photovoltaic cell manufacture, is an intrinsically costly material to handle and process. Furthermore, for efficient solar energy absorption, a crystalline silicon photovoltaic cell must be 10-50 μm thick and is normally 250 μm thick.

By contrast, amorphous silicon alloys can be deposited at reasonable temperatures (200-300°C) from gases such as SiH_4, SiF_4 and H_2 which are relatively inexpensive and easy to handle. Furthermore, due to a step-like optical absorption edge, amorphous silicon alloys with an optical band gap of 1.7 eV can, in a 1 μm thickness, absorb roughly as many solar photons as a 10 μm slice of crystalline silicon which has a band gap of 1.1 eV but a shallow absorption edge. Thus the fabrication costs of amorphous cells can be much less than those of crystalline cells, and become strongly controlled by such factors as the cost of the substrate and the module used to support and connect the individual cells.

Optimization of the efficiency of photovoltaic cells using an amorphous absorber is an active area of current research as suggested in the introduction. The sharp absorption edge of a-Si-based alloys favors a higher open circuit and operating voltage than for crystalline Si cells. Photovoltaic device efficiency, however, is a product of open circuit voltage (OCV), short circuit current (SCC) and "fill factor" (FF) which represents the fraction of the (OCV)(SCC) product available at the point of maximum power output of the device. SCC and FF of the amorphous devices are limited by the minority carrier diffusion length of ~0.2 μm versus ~50 μm in crystalline Si. Based on our experience with a-Si:F:H alloys at ECD, device efficiencies of 8-10% (based on active area) should be possible with optimized large area cells in the near future. Such efficiencies, corresponding to 7-8.5% efficiency averaged over the entire surface of a photovoltaic array including frame and grid areas, are comparable to those attainable with commercially available crystalline silicon PV modules measured on the same basis.

Given these summaries of cost and efficiency factors for photovoltaic cells based on amorphous silicon alloys, it is appropriate now to outline the methods which have been devised to prepare such large-area amorphous cells at a low cost. The large area (~1000 cm^2) structures we have evaluated at ECD involve a stainless steel substrate which is overcoated with p (B-doped), i and n-type (P-doped) a-Si:F:H layers, an indium-tin oxide semitransparent electrode which also functions as an anti-reflection coating, and a current collecting grid. Passivation and cell protection is provided by a glass cover although this layer can be an integral part of the cell itself. Eight such cells would be combined in a 2' x 4' module. A ptototype array consisting of three modules undergoing roof-top environmental testing is shown in Fig. 1. For a high volume (20 x 10^6 ft^2 per year)

Fig. 1: 24 ft^2 operating photovoltaic power supply. Each 1ft^2 element is a single photovoltaic cell, all 24 of which are connected in series to produce a useful supply of electrical power.

manufacturing operation we project initial costs of 44¢ per peak Watt, based on an 8.5% overall conversion efficiency and an array to be mounted on a new roof, oriented and sloping at the appropriate collection attitude.

SUMMARY

An overview of photovoltaic devices based on amorphous silicon alloys provides encouraging signs that this option has a good chance to contribute to energy generation in the U.S. in the next few years. ECD's success in producing a ft^2 cells based on an a-Si:F:H alloy suggests that the glow discharge deposition technology can be scaled up for high-volume production. ECD is now constructing a pilot facility to further evaluate the feasibility of achieving this objective and to manufacture prototype cells using a continuous automated method of manufacturing.

ACKNOWLEDGEMENTS

We acknowledge the many contributions of our colleagues Vin Cannella, Steve Hudgens and Herb Ovshinsky to the ideas and new developments presented here.

REFERENCES

1. D. E. Carlson, Solar Energy Materials 3 (1980) 503-518.

2. D. E. Carlson and C. R. Wronski, Appl. Phys. Lett. 28 (1976) 671.

3. H. F. Sterling and R. C. G. Swann, Solid State Electron. 8 (1965) 653.

4. R. C. Chittick, J. H. Alexander and H. F. Sterling, J. Electrochem. Soc. 116 (1969) 77.

5. W. E. Spear and P. G. LeComber, J. Non-Crystalline Solids 8-10 (1972) 727.

6. P. D. Maycock and E. N. Stirewalt, "Photovoltaics", (Brick House Publishing Co., Andover, Massachusetts, 1981).

7. U.S. Department of Energy, Assistant Secretary for Energy Technology, Division of Solar Technology, "National Photovoltaics Program: Multi-Year Program Plan" (draft) (Washington, D.C.: U.S. Department of Energy, 1980).

8. "Solar Photovoltaic Energy Conversion" (American Physical Society, New York, 1979).

9. A. Madan, W. Czubatyj, J. Yang, J. McGill and S. R. Ovshinsky, 9th Int'l. Conf. on Amorphous and Liquid Semiconductors, Grenoble, France (1981) (to be published).

10. M. H. Brodsky, M. A. Frisch, J. F. Ziegler and W. A. Lanford, Appl. Phys. Lett. 30 (1977) 561.

11. C. C. Tsai, et al., Proc. 7th Int'l. Conf. on Amorphous and Liquid Semiconductors, ed. W. E. Spear (CICL, University of Edinburgh, 1977) 339.

12. H. Fritzsche, Solar Energy Materials 3 (1980) 447-501.

13. S. R. Ovshinsky and K. Sapru "Amorphous and Liquid Semiconductors", ed. J. Stuke and W. Brenig (London: Taylor and Francis, 1974) 447.

14. M. Janai, D. D. Allred, B. O. Seraphin and H. S. Gurev, Solar Energy Materials 1 (1979) 11.

15. J. C. Phillips, J. Non-Crystalline Solids 34 (1979) 153; J. C. Phillips, Phys, Rev. Lett. 42 (1979) 1151.

16. J. D. Joannopoulos, Phys. Rev. B16 (1977) 2764.

17. S. R. Ovshinsky and A. Madan, Nature 276 (1978) 482; A. Madan, S. R. Ovshinsky and E. Benn, Phil. Mag. 40 (1979) 259.

AMORPHOUS PHOTOVOLTAICS

David Adler[1] and Stanford R. Ovshinsky[2]

[1]Department of Electrical Engineering, Massachusetts Institute of Technology, Cambridge, Massachusetts 02139
[2]Energy Conversion Devices, Inc., 1675 West Maple Road, Troy, Michigan 48084

The only obstacle to the immediate widespread use of solar cells is the enormous cost of conventional crystalline-silicon solar cells. The problem lies in the production of crystalline-silicon arrays:

- Crystalline silicon must be 100 μm thick. This intrinsic requirement cannot be overcome.
- Crystals now are pulled from the melt, which limits the diameter of single solar cells to 10 cm.
- More than half of the material is lost during cutting and polishing.

A concomitant problem with crystalline-silicon solar cells is the energy payback period, i.e., the time of operation needed before the energy used to fabricate the cell is returned in useful power. At present, crystal growth, diffusion, encapsulation, and assembly require so much energy that the payback period probably exceeds the assumed 20-year lifetime of the cells. Even with projected cost reductions resulting from the development of semicrystalline-silicon (ribbon) solar cells, the energy payback period will still remain at 5–10 years. We should not underestimate the importance of this fact. If we suddenly were to find ourselves in a crisis in which only five years' worth of fossil fuel reserves remained, a crash program to convert to crystalline solar cells could not be successful, no matter how many people and resources were committed to the project. We cannot repeal the laws of physics, and conservation of energy is a basic physical principle. Today a solar panel on the roof of a house with sufficient output to deliver about 1500 kWh/month would cost $100,000 without storage facilities. If such a panel were to last 20 years and all electricity generated were used efficiently, it would deliver power at 30¢/kWh, not including maintenance or interest costs. Because this price is about five times greater than that of today's electricity, cost reduction is essential before large-scale terrestrial use of solar cells can begin.

Let us dispel at the outset several myths of solar cell development. First, it is often pointed out that other crystalline-silicon-based devices, such as hand calculators, have undergone price decreases of greater than a factor of 100 over the past decade and, by analogy, similar progress will make crystalline-silicon solar cells competitive by 1987. However, cost reductions in the electronics industry resulted almost entirely from our progressive ability to grow more devices on the same size chip; because the devices are small and extremely large numbers of products are made from the same masks, the major cost of crystalline-silicon circuits is determined only by the number of processing steps. Unfortunately, the very high density of devices now possible and the potential density increases, which will soon result from submicrometer technology, will not help the solar cell cost problem in the least. This is because solar energy impinges on the earth at an average rate of only 1 kW/m^2. To collect sufficient energy to alleviate the world's electrical problems requires large-area devices; the density of devices is thus irrelevant. Although concentration is technically possible, the minimum projected cost of concentration and tracking systems is approximately a factor of five too large. In addition, the diffuse light from the sun, representing up to 50% of the total solar flux in many areas, is lost in concentration schemes.

A related myth concerns cost reductions that follow the development of mass production techniques in a new area. However, crystalline-silicon p–n junctions have been fabricated in bulk quantities for the electronics industry for many years and have already produced the savings that are expected.

Finally, it is sometimes stated that because the important parameter is cost per watt, then low-cost solar arrays can be made using inexpensive, low-efficiency glassy or polymeric semiconductors. This is misleading because efficiency is important in most built-up areas. A cheap cell with an efficiency of 0.001% would require one km^2 to meet the average electrical requirements of a modest-sized single-family house. Only in the most remote areas would such an inefficient use of land be justified.

Two routes lead to cost-effective solar cells: We can try to reduce the cost of crystalline-silicon solar cells, or we can try to increase the efficiency of inherently low-cost cells based on amorphous (noncrystalline) materials. About 90% of the U.S. government-supported solar-cell program prior

Reprinted by permission from *Chemtech*, Vol. 15, pp. 538–546 (1985).

to 1980 took the former path. However, progress was extremely slow. On the other hand, we at Energy Conversion Devices (ECD) initiated a large research effort to develop high-efficiency amorphous solar cells (1). A few other companies, primarily in Japan, followed our lead and began studies of amorphous solar cells, but it was not until efficiencies above 5% were reported that general attention was given to such devices (2). Here we describe the enormous progress in amorphous solar cells over the past few years.

Solar-cell physics

To understand the origin of the problem, we must review the physics of solar cells. This enables us to focus on the fundamental materials parameters, which must be optimized and made cost effective in any type of solar cell, crystalline or amorphous.

To produce electricity from sunlight, we must fabricate a device incorporating

- a material that absorbs a large fraction of the incident light;
- a material in which the light-absorbing process creates potential carriers of electricity, i.e., mobile pairs of (negatively charged) electrons and (positively charged) holes;
- an internal electric field that removes the electron from the vicinity of the holes or vice versa (otherwise, the potential carriers will simply recombine where they were created, and the light-absorbing process will serve only to heat up the material); and
- a geometry in which essentially all of the separated carriers reach the external circuit without recombining anywhere in the material.

The first two requirements restrict us to the use of semiconductors as the active material. Metals reflect sunlight; insulators either transmit the light or absorb it but transfer it to heat directly without the creation of potential carriers of electricity.

In metals the energy gap between the most energetic electrons and the lowest energy band available for excited electrons is small enough that electrons can move freely into the higher (conduction) band. In insulators this energy band gap is too wide for electrons to traverse; under ordinary conditions all electrons remain in the lower (valence) band.

On the other hand, semiconductors are characterized by an energy gap that is larger than that in metals and smaller than in insulators. The material becomes a conductor when electrons have moved from the valence band into the conduction band. The conduction band then has an excess negative charge, and the valence band contains positively charged holes.

To traverse the energy gap, E_g, between the two bands electrons must have a certain amount of energy. This energy can be supplied by heat so that semiconductors become more conductive with increased temperature. The energy can also be supplied by light that has sufficiently energetic photons, hence the use of semiconductors in photovoltaic cells.

The sunlight reaching the earth consists of a broad spectrum peaking near 5000 Å. It is most intense in the visible region (4000–7000 Å) because atmospheric O_2 and O_3 absorb light in the UV below 4000 Å, and CO_2 and H_2O vapor absorb in the IR, above 7000 Å.

Not all crystalline semiconductors absorb this light equally well. Some materials, such as crystalline silicon, absorb light poorly down to a certain wavelength, beyond which light absorption typically increases 100-fold. Materials having this property are said to possess an indirect edge. This property has practical consequences: To absorb most of the incident sunlight, the semiconductor thickness in a solar cell must be 100 times greater in materials with an indirect edge than in those with a direct edge. Typically, crystalline silicon must be approximately 100 μm thick.

The wavelength at which semiconductors absorb light is an internal characteristic, a function of E_g. Note that semiconductors with a large energy gap waste most incident sunlight so that if $E_g > 2.5$ electron-volts (eV), the solar cell will be inefficient.

On the other hand, the energy gap also represents an upper limit on the size of the internal electric field that can be supported by the material. This means that light of a lower wavelength than that corresponding to E_g is absorbed, but its maximum output is degraded in proportion to the size of the gap. Thus, small-gap materials are also inefficient because they can supply power only at small voltage. For example, the maximum voltage attainable through the use of a 1.0-eV gap semiconductor is less than 1.0 V. Clearly, there must be an optimal energy gap for solar cells, and that is about 1.5 eV. Because of the losses at both high and low wavelengths, the maximum efficiency associated with any semiconductor is about 30%. This relatively low theoretical maximum follows only from the broad nature of the solar spectrum; however, we cannot change the solar spectrum.

We could improve this maximum theoretical efficiency by using a composite consisting of semiconductors that have different energy gaps. Such a tandem or graded-gap device could yield theoretical efficiencies up to 60% but cannot ordinarily be fabricated from crystalline semiconductors because the mismatch in lattice constants introduces large regions in which the photogenerated carriers recombine. Even in the rare cases in which the two lattice constants match, such a structure would be inordinately more expensive using crystalline semiconductors because of the increased number of processing steps per cell.

The requirement of an internal electric field to separate the photoexcited carriers can be met in a straightforward manner. The technique used is diffusion, similar to coloring water by adding a small amount of ink: Although it may be added in one region, the ink diffuses and ultimately darkens the water uniformly. Thus we can introduce excess electrons into semiconductors by doping them with atoms that have more electrons than the host material. An example of such an n-type semiconductor is germanium doped with arsenic. When an n-type semiconductor is placed in contact with the pure material, some of the electrons diffuse into the latter. Because electrons are negatively charged, this diffusion creates an internal electric field. We can also dope semiconductors with atoms that have fewer electrons than the host material. For example, germanium doped with indium results in a p-type semiconductor, which has a deficiency of electrons, or positive holes. When we place a p-type next to an n-type semiconductor, diffusion is increased. The practical result is a p–n junction solar cell.

Other configurations are possible, but the principle is the same: The diffusion of charged particles from the side of the junction where there is an excess of one type to the other side creates the internal field. The two important factors for device efficiency are the strength of the internal field and its spatial extent. Although the maximum strength is determined by the band gap, the complete geometry can often result in attaining a larger or smaller fraction of the theoretical maximum. However, the extent of the field is determined by a quantity known as the density of states in the gap. This quantity arises because semiconductors have defects that result in states in which there are electronic energies that ordinarily are forbidden, i.e., within the energy gap of the material. When the charged carriers (the electrons or holes) diffuse, they can be trapped by states in the gap, and these trapped carriers contribute to the internal field. Thus in a material with a small density of states in the gap, the carriers diffuse farther before being trapped, and the result is a more extensive internal field.

The major problems in solar cell technology arise because photoexcited, separated carriers must reach the external circuit. Just as states in the gap trap the carriers, which diffuse in the dark after the junction is established, they also trap photogenerated carriers trying to produce an external current. But every trapped carrier leads to reduced efficiency of the solar cell. Thus, perhaps the most important reason for the efficiencies not approaching the theoretical maximum is a large density of states in the gap.

Because these gap states result from defects in the material, crystalline semiconductors must be ultrapure and carefully grown. Most impurities introduce states in the gap because they ordinarily have different chemical valence, and the crystalline constraints force a defective configuration. In addition, grain boundaries that exist in polycrystalline material and structural defects such as vacancies, interstitials, dislocations, and interface mismatches lead to states in the gap. The need to grow ultrapure, high-quality single crystals leads to the inordinately high cost of conventional solar cells.

To summarize, the important factors in determining the efficiency of a potential semiconductor for solar cell applications are

- the value and nature of the energy gap (a 1.5-eV direct gap is optimal) and
- the density of states in the gap, which should be as small as possible.

To these, we add the other requirements for large-scale terrestrial use: A solar cell must
- be composed of low-cost, abundant chemicals,
- not degrade in operation over 20 years of use, and
- be capable of being made into an operating cell at little additional cost.

Amorphous semiconductor solar cells

The use of amorphous thin films in solar cell applications offers several advantages:

- They can be deposited over extremely large areas (compared with the maximum 20-cm wafer size for growth of single crystals).
- They can be tailored so that the gap is the optimal 1.5-eV value using straightforward compositional changes.
- They have low material requirements, due to strong absorption of sunlight by an amorphous semiconductor, which results in efficient cells that are only 1 μm thick as opposed to the 100-μm thickness required by crystalline-silicon cells.
- They do not require large densities of interface states near the contacts.
- They offer the possibility of fabricating high-efficiency tandem-type devices at small additional costs.
- They allow low-cost deposition, typically less than 5% of the processing cost of crystalline silicon.

Nevertheless, except for our work at ECD, amorphous semiconductors generally were dismissed as a serious possibility before 1975. This was the result of the folklore that has plagued the field of amorphous solids since the very beginning. In denigrating amorphous solar cells, it was noted that the dark conductivity of amorphous semiconductors is very small, the density of states in the gap is enormous, the free-carrier diffusion lengths are extremely small, and doping is impossible.

Therefore, it was concluded that solar-cell efficiencies with these materials could never exceed 1%, much less reach the minimum 7% needed for commercial application. Fortunately, this logic did not deter Carlson et al. (2), who announced the development of an amorphous-silicon-based solar cell with an efficiency in excess of 5%.

What was wrong with the reasoning that showed that such performance was impossible? As is common in this kind of technological prognostication, it was based on a combination of a lack of imagination and a surfeit of irrelevancies. First, the low dark conductivity of amorphous semiconductors is completely irrelevant to their use as solar cells; only their photoconductivity is important, as Xerox has amply demonstrated with its electrostatic copying machine, which uses amorphous selenium–arsenic alloys as the active material, and as Hitachi proved with its Saticon TV pick-up tube, which uses amorphous selenium–arsenic–tellurium alloys in a tandem geometry. These materials are very efficient photoreceptors.

Second, the density of states in the gap of many amorphous films is not an intrinsic phenomenon. In amorphous solids, impurities do not ordinarily enter with the wrong chemical valence because all crystalline constraints are absent. In addition, vacancies, interstitials, and dislocations do not exist. Even at interfaces, the lack of lattice mismatches sharply reduces the density of states in the gap.

Gap states have two origins: The first is from disorder, due primarily to bond-angle distortion and other strains of

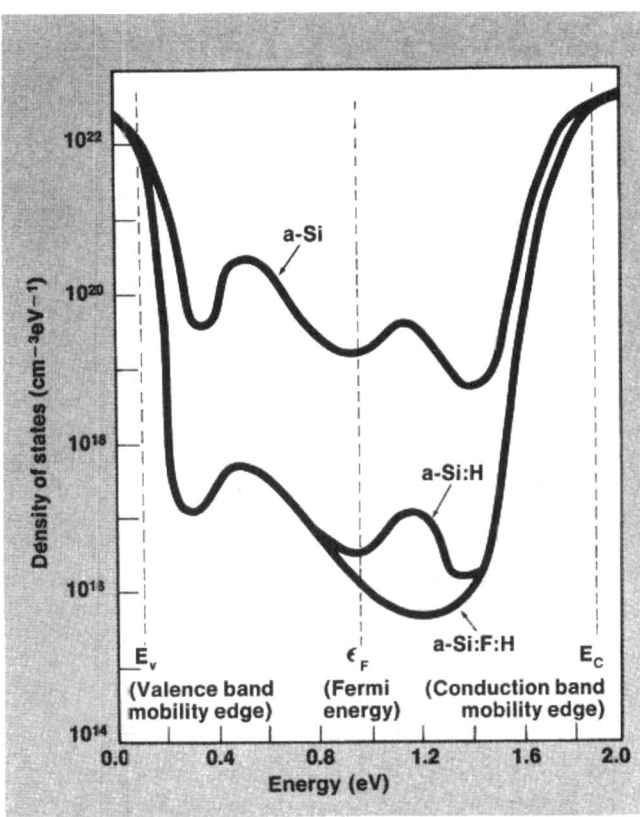

Figure 1. Density of states in the gap of amorphous silicon alloys. Pure amorphous silicon (a-Si), a-silicon–hydrogen alloys, and a-silicon–fluorine–hydrogen alloys are all shown (8). These data do not reflect further improvements in the past five years

a more long-range nature. The other is from incorrect chemical bonding, resulting from such causes as network constraints and the nonequilibrium deposition procedures commonly employed (3). We can sharply reduce the former by lowering the rigidity of the network by chemical alloying (4). We can minimize the latter, for example, by adding significant densities of lower valence atoms to a tetrahedrally bonded matrix or by reducing the average coordination number in general.

Figure 1 shows the density of states in the gap for amorphous-silicon (a-Si) films. Pure a-Si contains large densities of strains and dangling bonds (silicon with coordination less than the optimal four nearest neighbors), and the resulting films are useless in electronic devices. However, when a-Si is decomposed from silane (SiH_4) gas, the density of localized states is reduced by about a factor of 1000 (5). The reason for this is the incorporation of residual hydrogen into the silicon matrix. Hydrogen ordinarily bonds covalently only with one nearest neighbor, thus reducing the average coordination of the material and relieving strains. In addition, hydrogen easily ties up most of the residual dangling bonds that might remain in the films. Following Chittick et al. (6), Spear and Le Comber (7) were able to show that silane-decomposed films could easily be doped by adding either phosphine (PH_3) gas to the silane for n-type doping or diborane (B_2H_6) gas for p-type doping.

These results appear paradoxical to anyone familiar only with crystalline semiconductors. In the case of a-Si alloys, the useful electronic material is not pure but highly impure. In fact, a wide class of chemical impurities does not reduce but rather enhances film quality. Of course, the great advantage is that we can use much less expensive starting materials because amorphous films experience no constraints from long-range periodicity; most impurities are optimally bonded in the matrix, thus reducing the density of states in the gap. In addition, no expensive crystal growth is necessary and, because doping is accomplished by a simple mixture of different gases, no high-temperature, energy-intensive diffusion steps are required.

In principle, the low free-carrier diffusion lengths of amorphous semiconductors could present a problem. However, because Yang has already achieved solar energy conversion efficiencies in excess of 11% (8), it is clear that this problem is not serious. The fact is that efficient amorphous solar cells can be designed to work primarily by drift (the motion of charged particles in an electric field) rather than by diffusion. For efficient solar cell operation, all that is essential is that the photoexcited carriers reach the contacts before recombining. This is expressed mathematically by $t < \tau$, where t is the transit time and τ is the carrier lifetime.

In conventional solar cells, t depends primarily on the speed with which the carriers diffuse to the region of the internal electric field. However, amorphous cells can be fabricated so that most carriers are photogenerated within the electric field region and no initial diffusion is necessary. If the field is sufficiently strong that $t < \tau$, efficient cells are possible. For amorphous solar cells, this requires fields of only about 1000 V/cm (9); the internal electric fields in a-Si alloy cells are typically more then 10 times larger than this value over a region of up to half the thickness of the semiconductor. The reason such a low internal field is required to attain $t < \tau$—despite the fact that the average carrier mobility (velocity per unit field) is so low in amorphous silicon—is that recombination is also small. The free electron must find a hole with which to recombine, and its low mobility does not allow it to reach the hole rapidly. Consequently, it does not matter that it takes a long time for a photoexcited carrier to reach the contact as long as the average recombination time is still longer.

We can use many different geometries to make efficient a-Si alloy solar cells. However, because the undoped material appears to have a lower density of states in the gap than either p- or n-doped materials, it is better to have the bulk of the cell intrinsic (i). On the other hand, interfaces are needed both to attain the required internal electric field and to ensure against carrier losses at the contacts. Consequently, cells deposited on glass substrates are based on a p–i–n geometry, usually with a transparent conductor either on the top or as the substrate. This material, which is typically based on indium or tin oxide, lets the incident sunlight in and extracts either the photoexcited electrons or the holes. The p–i and the i–n interfaces induce internal electric fields within the intrinsic material, and essentially all the free carriers that are photoexcited within the regions of internal fields contribute to the solar cell current.

Thus, the most efficient solar cells are those with the most extensive internal fields. Because the extent of the internal field is determined by the distance carriers from the *n* and *p* regions diffuse into the intrinsic material before achieving equilibrium, and because equilibrium requires the diffusion of a fixed total number of electrons across the junction, the field will penetrate the longest distance in those materials with the lowest density of localized states. Thus, the most important aspect of the materials in an amorphous solar cell is the density of states in the gap.

Amorphous-silicon–hydrogen solar cells

As we have seen (Figure 1) the large defect densities of pure a-Si films can be enormously reduced through the incorporation of hydrogen. This can be done by decomposition of silane (SiH_4) gas,

$$SiH_4 \rightarrow Si + 2H_2$$

or by adding hydrogen to the atmosphere of other deposition processes for pure a-Si. Yet residual defects remain in resulting films because hydrogen cannot compensate for a small fraction (perhaps 1 in 1000) of the dangling bonds in pure film. In this small fraction the necessary bond length (1.05 Å) and bond angles (109.5°) with the other three nearest neighboring atoms cannot be achieved within the geometry determined by the previously deposited layers (10). This residual density of localized states is a major obstacle to improving the efficiency of silane-decomposed solar cells.

Since the work of Spear and Le Comber (7), some clever techniques have been used to increase overall efficiency. Using a heterojunction configuration in which the *p* material is an amorphous-silicon–carbon–hydrogen alloy, Catalano et al. achieved an overall conversion efficiency in excess of 10% on small ($\sim$1 cm^2) devices (11). Furthermore, a-Si–H solar cells have been produced commercially in Japan both by batch processing (12) and by a continuous-web, roll-to-roll process (13).

Unfortunately, a major problem has emerged with regard to the efficiencies of a-Si–H solar cells. In the presence of intense light, these cells degrade rather quickly, and a 10% device becomes only a 5% device within days of operation in actual sunlight. The origin of this degradation is a photostructural change (14, 15) caused by the recombination of photoexcited carriers (16). ECD has solved this problem through a combination of chemical and configurational means. Their devices will retain more than 95% of initial efficiency after 20 years of operation (17).

Amorphous-silicon–fluorine–hydrogen solar cells

Ovshinsky and co-workers have developed a new a-Si alloy that contains hydrogen and fluorine (14, 18). It is easily prepared through the decomposition of silicon-tetrafluoride–hydrogen gas mixtures

$$SiF_4 + 2H_2 \rightarrow Si + 4HF$$

The density of states in the gap of the resulting a-Si–F–H alloy (Figure 1) is further reduced in the localized state density, partly because fluorine forms an ionic rather than a covalent bond with silicon (10). Because ionic bonds do not require a precise bond angle, fluorine has more space available to compensate for residual defects than does hydrogen. The lower density of localized states in a-Si–F–H alloys is reflected in the more extensive internal fields obtained in devices based on these materials (19).

Amorphous-Si–F–H alloys have many other advantages for solar cell applications. SiF_4 is much less expensive than SiH_4, a fact that should lower the ultimate cost of devices based on SiF_4, and a-Si–F–H alloys are much less resistant to scratching and other mechanical damage. The much higher doping efficiency of a-Si–F–H alloys provides a large advantage in contact technology. Furthermore, a-Si–F–H alloys exhibit significantly fewer of the disastrous photoinduced effects that characterize a-Si–H alloys (14, 15).

Another major advantage of the use of fluorination in amorphous solar cells is in improving the quality of lower band gap materials for use in tandem solar cells. Hydrogenated amorphous germanium (a-Ge–H) and germanium–silicon alloys (a-Ge$_x$Si$_{1-x}$–H) have proved inferior because of the large residual concentration of defect states in the gap (20). These very likely result from the chemical predilection for Ge to exhibit twofold as well as tetrahedral coordination (4, 21). Incorporation of fluorine expands the valence of Ge, essentially eliminating this defect (4). Using a-Ge$_x$Si$_{1-x}$–F–H materials, ECD has achieved a record 12% efficiency in 1-cm^2, triple-layered devices (8).

Roll-to-roll processed solar cells

ECD has developed a means of mass producing amorphous-silicon-based solar cells of arbitrary width using a continuous roll-to-roll process. Single and tandem cells have been produced. The present production machines have the following structure:

- a thin stainless steel substrate,
- multilayered a-Si alloy layers (about 1 μm thick) and tandem cells with six layers,

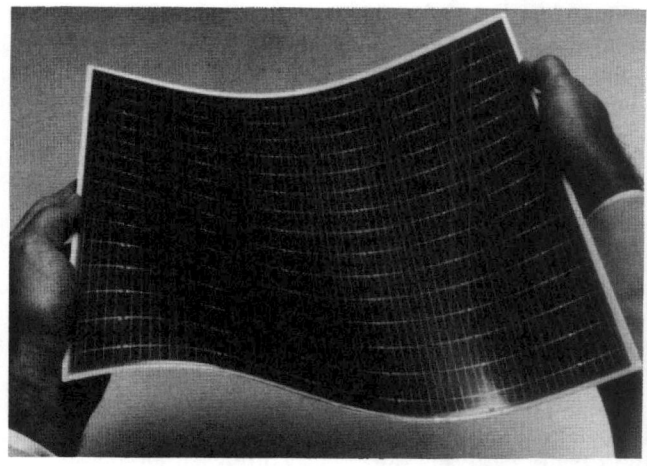

Figure 2. A 1-ft^2 amorphous solar cell

- a transparent conducting layer (indium tin oxide),
- a grid pattern, and
- an encapsulant.

Deposition of high-quality a-Si alloy layers in a uniform high-speed process is the key step technologically.

Such wide continuously deposited cells will supply electricity at the lowest rate because

- the total thickness of active material is less than 1 μm, and the material usage is small;
- silicon, fluorine, hydrogen, and other materials used in the device are abundant and inexpensive;
- a thin, low-cost substrate is used;
- product yield is high; and
- development of high-efficiency cells in the future will further reduce the material cost.

The labor cost associated with solar cell production is low because the process uses simple, high-rate, highly automated processing for the complete fabrication of photovoltaic modules. Specifically, six layers of tandem a-Si alloy solar cells are plasma-deposited continuously on a roll of wide stainless steel substrate in a single pass. The development of microwave techniques, which have already been used to achieve deposition speeds of 36 μm/h (22), show that the line speed can be increased by several orders of magnitude. Other downstream processing steps for the fabrication of photovoltaic modules also use simple, high-rate, highly automated machinery.

An important aspect is the continuous roll-to-roll process, which is more cost effective than a process using rigid sheet substrates. ECD has demonstrated several advantages:

- An automated mass production line was readily adapted at minimal cost.
- Once the technology was established, the scale-up to a higher volume and a higher production rate was straightforward.
- The use of a thin substrate in a roll form reduced the cost of substrate and material handling. The amount of edge loss was also minimized; of course, a wide range of materials can be used for substrates.
- Solar cells have been cut to various sizes from a roll of the film. This provided flexibility in product size and shape.

Engineering for the scale-up to develop a wide, continuous roll-to-roll (plasma deposition) process for a-Si alloys has proceeded stepwise according to the following sequence.

The first step was to develop a large-area plasma deposition machine capable of uniform 1-ft^2 a-Si deposition. We have confirmed the following points:

- Deposition thickness and photovoltaic quality are highly uniform over more than one square foot (1-ft^2 cells have been fabricated) (Figure 2).

Figure 3. Continuous roll-to-roll photovoltaic processor for depositing six layers of amorphous silicon on 1000-ft-long rolls of stainless steel. The steel roll is loaded into the overhead compartment and fed through six processors, which are supplied with deposition gases

Figure 4. Sharp–ECD production plant in Shinjo, Japan. Amorphous-silicon deposition process is at far left. Clean-room conditions prevail. Amorphous-silicon-covered stainless steel roll is about to be covered with indium tin oxide to reduce reflection and to complete the solar cell. Unit at right is for circuit imprinting

Figure 5. Continuous 1-ft-wide rolls of mass-produced solar cell material

- Solar cells from this large-area plasma deposition machine are almost as efficient as those from small R&D deposition machines. Present efficiencies of 1-ft^2 cells exceed 10% in laboratory devices and 8% in devices fabricated with production machines.
- The product yield is high.
- Low-cost, large-area substrates have been identified and used.
- Downstream cell fabrication processes are compatible with high-volume production.

The second step was to develop a small-scale, continuous roll-to-roll a-Si plasma deposition machine. The following results have been obtained:

- Solar cells produced by the continuous processing machine are now nearly as efficient as those produced on small R&D equipment.
- Uniformity of the deposition is excellent—superior to that of a batch 1-ft^2 system.
- The production yield is high.
- Winding the stainless steel substrate that is coated with a-Si alloy into a roll did not impair solar cell performance.
- A high-quality substrate that can be used in roll form (16 in. wide and 1000 ft long) was developed.

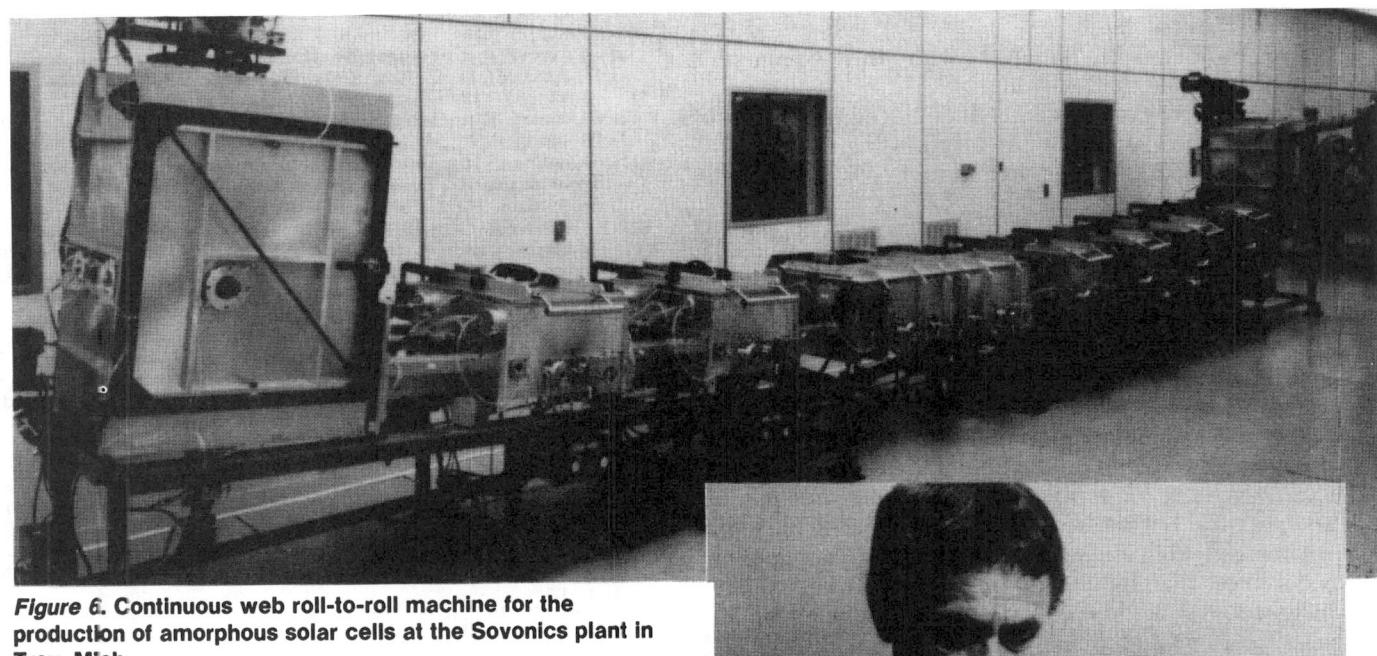

Figure 6. Continuous web roll-to-roll machine for the production of amorphous solar cells at the Sovonics plant in Troy, Mich.

The third step was to develop a commercial-scale, roll-to-roll plasma deposition machine for depositing multilayered a-Si alloys. This machine is approximately 35 ft long (Figure 3), has multiple deposition areas, and processes 16-in.-wide stainless steel substrates continuously. Amorphous-Si photovoltaic thin films (less than 1 μm thick) with a six-layer structure (p–i–n–p–i–n) are deposited continuously on a roll of 16-in.-wide, 100-ft-long stainless steel substrate in one pass.

The commercial machine has been in operation for several years at Sharp–ECD Solar, Inc. (SESI) in Japan, a joint business venture of Sharp Corporation and the technology partnership of ECD and the Standard Oil Company (Ohio) (Figure 4).

Reproducible results

Important results obtained with this commercial machine include the following:

- The efficiency of the solar cells produced is as high as or higher than those from batch processing.
- Product yield is higher than that of cells produced by batch processing.
- The product is uniform from the beginning to the end of a 1000-ft roll.
- It is possible to reproduce the production line.
- For several years, 1000-ft substrate rolls have been processed continuously on a commercial basis.
- Tandem cells have been produced commercially for the first time.

Figure 5 shows the amorphous film going through the production line. Mass production of low-cost tandem a-Si solar cells using roll-to-roll processing is not only possible and practical but is already taking place at SESI, where commercial production has been under way for two years. So far, more than five million Sharp calculators have been

Figure 7. Joseph Hanak of ECD demonstrates an ultralightweight 1-ft² solar cell

manufactured with the solar cells produced by this machine.

The success of the SESI machine led to the design and construction of new generations of roll-to-roll solar cell manufacturing facilities. Figure 6 shows a continuous web machine in its start-up phase in the Sovonics plant in Michigan. It was constructed under a joint venture between ECD and the Standard Oil Company (Ohio). The output of this machine will be used for the direct generation of electric power. In addition, large-area ultralightweight solar cells have been deposited on thin substrates (Figure 7). They have converted sunlight to electricity at a rate of 1620 peak W/kg, more than six times

NASA's previous record. New machines have produced 1-ft^2 devices with efficiencies greater than 10%. Continuous web tandem solar cells have been fabricated with total efficiencies beyond 8%. Ultrastable devices with 7.5% efficiencies have been tested and shown to maintain more than 99% of their initial power after 1800 h of continuous illumination, indicating their ability to provide more than 95% of their original output after 20 years of actual operation. The heralded new generation of solar electricity has finally arrived.

Acknowledgment

We thank Masat Izu, Stephen Hudgens, Jeff Yang, Prem Nath, Joe Doehler, Wally Czubatyj, Subhendu Guha, Joe Hanak, and many others for providing us with their data prior to publication.

References

(1) Ovshinsky, S. R. In "Amorphous and Liquid Semiconductors"; Spear W. E., Ed.; C.I.C.L.: University of Edinburgh, U.K., 1977; p. 519.
(2) Carlson, D. E.; Wronski, C. R.; Pankove, J. I.; Zanzucchi, P. J.; Staebler, D. L. *RCA Rev.* 1977, *38*, 211.
(3) Ovshinsky S. R.; Adler, D. *Contemp. Phys.* 1978, *19*, 109.
(4) Ovshinsky, S. R. In "Physical Properties of Amorphous Materials"; Adler, D.; Schwartz, B. B.; Steele, M. C., Eds.; Plenum Press: New York, N.Y., 1985; p. 105.
(5) Madan, A.; Le Comber, P. G.; Spear, W. E. *J. Non-Cryst. Solids* 1976, *20*, 239.
(6) Chittick, R. C.; Alexander, J. H.; Sterling, H. F., *J. Electrochem. Soc.* 1969, *116*, 77.
(7) Spear, W. E.; Le Comber, P. G. *Solid State Commun.* 1975, *17*, 9.
(8) Yang, J.; Mohr, R.; Ross, R. *Tech. Dig. Int. PVSEC-1, Kobe, Jpn.* 1984, p. 1.
(9) Adler, D.; Silver, M.; Madan, A.; Czubatyj, W. *J. Appl. Phys.* 1980, *51*, 6429.
(10) Adler, D. In "Semiconductors and Semimetals"; Willardson, R. K.; Beer, A. C., Eds.; Academic Press: New York, N.Y., 1984; Vol. 21A, p. 291.
(11) Catalano, A.; D'Aiello, R. V.; Dresner, J.; Faughnan, B.; Firester, A.; Kane, J.; Shade, H.; Smith, Z. E.; Swartz, G.; Triano, A. In "Proceedings of the 16th IEEE Photovoltaic Specialists Conference"; IEEE: New York, N.Y., 1982; p. 1421.
(12) Kuwano, Y. *Oyo Denshi Bussei Bunkakai* 1980, *385*, 17.
(13) Ovshinsky, S. R.; Adler, D. *Nikkei Sci.* 1983, *13*(9), 60.
(14) Ovshinsky, S. R.; Madan, A. *Nature* 1978, *276*, 482.
(15) Matsumura, H.; Furukawa, S. In "Amorphous Semiconductor Technologies and Devices"; Hamakawa, Y., Ed.; North-Holland: Amsterdam, The Netherlands, 1980; p. 88.
(16) Adler, D. *Sol. Cells* 1983, *9*, 133.
(17) Ovshinsky, S. R. *Tech. Dig. Int. PVSEC-1, Kobe, Jpn.* 1984, p. 577.
(18) Madan, A.; Ovshinsky, S. R.; Benn, E. *Philos. Mag.* 1979, *40*, 259.
(19) Shur, M.; Czubatyj, W.; Madan, A. *J. Non-Cryst. Solids* 1980, *35–36*, 731.
(20) Oda, S.; Yamaguchi, M.; Hanna, J.; Ishihara, S.; Fujiwara, R.; Kawate, S.; Shimizu, I. *Tech. Dig. Int. PVSEC-1, Kobe, Jpn.* 1984, p. 429.
(21) Adler, D. In "Physical Properties of Amorphous Materials"; Adler, D.; Schwartz, B. B.; Steele, M. C., Eds.; Plenum Press: New York, N.Y., 1985; p. 5.
(22) Hudgens, S. J.; Johncock, A. G. In "Materials Issues in Applications of Amorphous Silicon Technology"; Adler, D.; Madan, A.; Thompson, M. J., Eds.; MRS: Pittsburgh, Pa., 1985; p. 368.

THE BREAKING OF THE EFFICIENCY–STABILITY–PRODUCTION BARRIER IN AMORPHOUS PHOTOVOLTAICS

Stanford R. Ovshinsky and Jeffrey Yang

Energy Conversion Devices, Inc., 1675 West Maple Road, Troy, Michigan 48084

INTRODUCTION

There are three key issues that challenge amorphous photovoltaics, namely, efficiency, stability, and production. In this paper we will review the current status of our amorphous photovoltaics at ECD in terms of these three issues and demonstrate how we have broken the efficiency-stability-production barrier.

Single-junction PIN devices using conventional hydrogenated amorphous silicon alloy suffer severe degradation in their performance after prolonged exposure to sunlight. Laboratory-scale devices with initial efficiencies as high as 10% rapidly degrade to 5-6%; hence making claims of initial efficiencies insignificant. The relatively low value of the stabilized efficiency coupled with the low throughput of the conventional batch manufacturing process form an insurmountable barrier to low-cost power usage.

We have recognized the problems associated with hydrogenated amorphous silicon alloys, and have developed a new fluorinated material which addresses the problems described above.(1-4) We have also recognized the limitations of single-junction solar cells and put into practice, not only in research and development but in production, the tandem structure which minimizes recombination losses and reduces degradation.(5) The Sharp-ECD solar plant established in 1981 in Shinjo, Japan is a good example. The amorphous silicon processor employs the roll-to-roll process from which a roll of 1000-foot long, 16-inch wide stainless steel substrate is coated by six layers of amorphous silicon materials in a continuous fashion and produces a roll of lightweight and flexible tandem cells. Millions of calculator cells have been manufactured.

Using fluorinated amorphous silicon and silicon-germanium alloys, we have already achieved a record 13% conversion efficiency in our laboratory.(4,6) This device has a triple-junction stacked-cell configuration, a configuration that offers exceedingly good stability. We have successfully developed new narrow band gap fluorinated materials with excellent sub-band gap absorption properties.

These new materials can lead to 20% conversion efficiency. The fluorinated alloys, the improved stability with tandem structure, and the unique manufacturing technology based on our roll-to-roll process have allowed us to break the efficiency-stability-production barrier.

CURRENT STATUS OF FLUORINATED AMORPHOUS SILICON ALLOY DEVICES

In order to achieve high efficiencies in a tandem structure, one must first develop high-quality materials from single junction devices. For a PIN configuration, this necessarily means that not only does one need high-quality intrinsic material but also high-quality n^+ and p^+ layers.

We have previously reported on the development of high-quality intrinsic and n^+ materials with the incorporation of fluorine.(1,7,8) We have recently reported on the development of a fluorinated microcrystalline p^+ silicon alloy which has high dark conductivity and low optical loss.(9) Using these fluorinated materials, we have fabricated single-junction PIN devices on stainless steel substrate with light entering the p^+ layer and achieved an 11.3% conversion efficiency for a 1 cm^2 active area device. The J-V characteristic is shown in Table I. Two-cell tandem devices and three-cell triple devices have conversion efficiencies of 11.4% and 12.0%, respectively, as shown in Table I. It should be pointed out that although one-cell, two-cell, and three-cell devices have similar conversion efficiencies, the stability of multi-junction devices is much superior to that of the single-junction devices.

CURRENT STATUS OF FLUORINATED AMORPHOUS SILICON-GERMANIUM ALLOY DEVICES

While the device efficiency using amorphous silicon alloy with 1.7 eV band gap approaches its theoretical limit, one must develop narrow band gap materials to broaden the spectral response and increase the efficiency.

Reprinted by permission from the *Proceedings of the SPIE Symposium on Photovoltaics for Commercial Solar Power Application*, Cambridge, Massachusetts, 18–19 September 1986, Vol. 706, pp. 88–93.

TABLE I

Photovoltaic characteristics of various device structures under AM1 (100 mW/cm^2) illumination at 25°C. Devices have an active area of 1 cm^2.

Structure	V$_{oc}$(volts)	Jsc(mA/cm^2)	FF	η(%)
SS/NIP/ITO	0.93	17.8	0.68	11.3
SS/NIPNIP/ITO	1.82	9.5	0.66	11.4
SS/NIPNIPNIP/ITO	2.70	6.42	0.69	12.0
SS/NI'P/ITO	0.78	22.1	0.58	10.0
SS/NI'PNIP/ITO	1.69	10.9	0.68	12.5
SS/NI'PNIPNIP/ITO	2.58	7.0	0.72	13.0
SS/NI'PNI'PNIPNIP/ITO	3.20	5.69	0.64	11.7

I   a-Si:F:H
I'   a-Si:Ge:F:H

We have developed a 1.5 eV fluorinated amorphous silicon-germanium alloy that exhibits very low sub-band gap absorptions.(10) Figure 1 plots absorption coefficient versus energy for a-Si:F:H, a-Si:Ge:F:H, and a-Si:Ge:H made at ECD and conventional a-Si:Ge:H. It should be clear that the fluorinated materials are of higher quality. It is also shown by Guha (10) that fluorinated material exhibited much reduced light-induced effect.

Using the 1.5 eV fluorinated amorphous silicon-germanium alloy, we have achieved a 10.0% efficiency in a single-junction device. The J-V characteristic is shown in Table I. Dual band gap two-cell tandem, three-cell triple, and four-cell quadruple devices were fabricated using the fluorinated 1.5 eV and 1.7 eV materials; efficiencies of 12.5%, 13.0%, and 11.7% were achieved (6) for their respective structures; the J-V characteristics are shown in Table I. The 13% efficiency represents the highest value reported for amorphous solar cells. Figure 2 shows the J-V characteristic of this device.

In terms of stability, we have shown (6) that a dual band gap triple device with initial efficiency of 11.2% retained 90% of its initial value after 2500 hours of continuous AM1 exposure, as shown in Fig. 3. The 10% stabilized efficiency value is the highest achieved to date. Our best stability data on quadruple devices show no loss in efficiency after 2500 hours of continuous light exposure, as shown in Fig. 4.

To further increase device efficiencies beyond 13%, one must obtain high-quality materials with band gap narrower than 1.5 eV. We have successfully developed fluorinated materials that have band gaps of 1.40 eV, 1.34 eV, 1.25 eV, and 1.15 eV. These materials exhibit excellent sub-band gap absorption properties as shown in Fig. 5. It should be pointed out that curves 1 and 2 represent 1.7 eV and 1.5 eV materials that allowed us to obtain the 13% efficiency. These two curves have very similar slopes as well as low defect densities. It is further noted that curves 3, 4, 5, and 6 also have similar slopes and low defect densities. As these new fluorinated materials are fully incorporated into multi-junction device configurations, efficiencies approaching 20% or even beyond can be

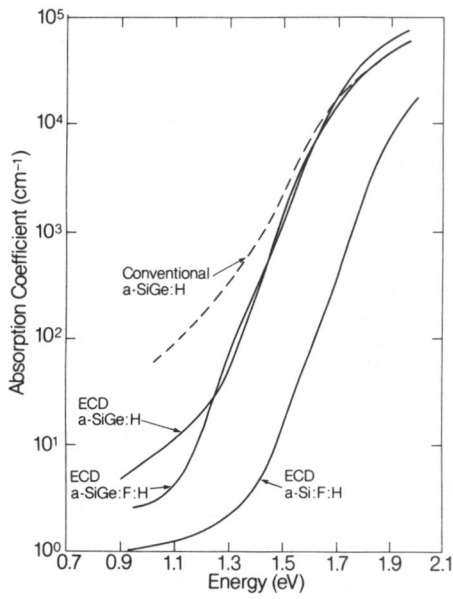

Fig. 1 Absorption coefficient as a function of photon energy for a-Si:Ge:F:H and a-Si:Ge:H alloy films compared to that of an a-Si:F:H film with a 0.3 eV larger band gap. Also shown is a conventional a-Si:Ge:H curve.

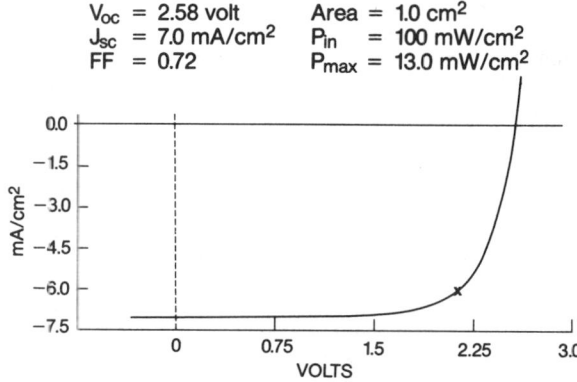

V_{OC} = 2.58 volt   Area = 1.0 cm^2
J_{SC} = 7.0 mA/cm^2 P_{in} = 100 mW/cm^2
FF = 0.72 P_{max} = 13.0 mW/cm^2

Fig. 2   Current-voltage characteristic of a triple-junction solar cell with an efficiency of 13%.

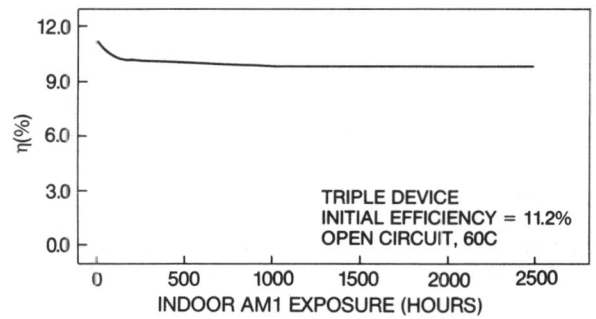

Fig. 3.   Conversion efficiency versus light-soaking time for a dual-band gap triple device.

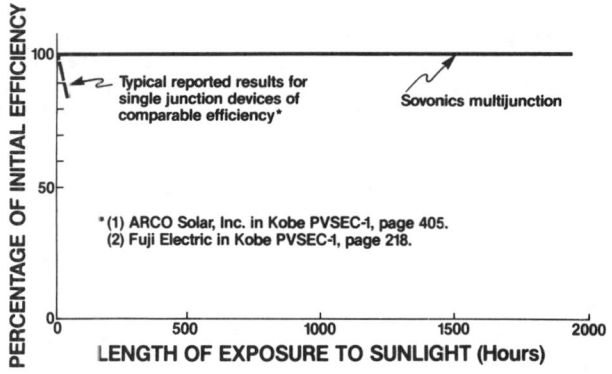

Fig. 4   Stability of a Sovonics multi-junction solar cell after 2500 hours of continuous AM1 exposure compared to recent results on typical single-junction a-Si:H cells.

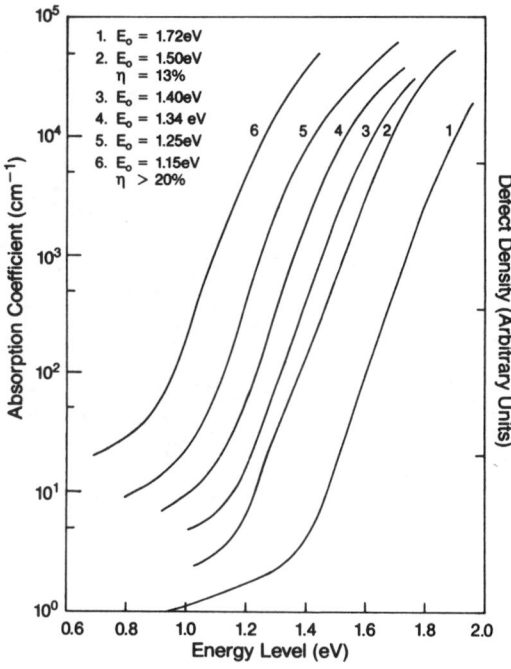

Fig. 5   Absorption coefficient as a function of photon energy for a-Si:Ge:F:H alloy films with various band gaps.

Fig. 6   Both newsprint, above, and advanced thin-film Ovonic photovoltaic devices, below, use low-cost roll-to-roll manufacturing process.

anticipated. It is important to note that such a device will be made in a single monolithic process, the process that we currently employ for manufacturing.

STATUS OF PRODUCTION

While there is no question that amorphous photovoltaics can provide a nondepletable,

229

Fig. 7 View of a portion of the Sovonics production plant, showing the Sovonics Photovoltaic Processor.

pollution-free, worldwide attainable source of energy, a low-cost mass production technology must be developed to serve this purpose. As early as the 1970's, ECD recognized this need and has since designed and built four generations of the roll-to-roll amorphous silicon processors. This efficient and continuous process can be compared with the printing of newspapers, as illustrated in Fig. 6. In Fig. 7, we show a portion of our present production facility located in Troy, Michigan.

Since the roll-to-roll process is a continuous deposition process, ECD had designed a proprietary means to effectively minimize dopant contamination. This is best illustrated by SIMS data on the dopant concentrations in the intrinsic layer obtained from

Sharp's roll-to-roll processor built by ECD, as reported by Hirobe et al.(11) They found that the boron and phosphorous concentrations were less than 5 x 1016 atoms/cc in the intrinsic materials, indicating the excellent isolation between deposition regions.

The uniformity of the deposition across the web is also a crucial parameter in ensuring the product uniformity. Figure 8 shows the uniformity of the film across the web obtained from our roll-to-roll production machine. Our production machine is currently producing two-cell tandem devices using fluorinated amorphous silicon alloys. Figure 9 illustrates the excellent yield and consistency in efficiency of several consecutive production runs.

To address the issue of power delivered per unit weight, we have developed an ultralight module (12) which produces power at the impressive density of 2,418 W_p/kg, more than 10 times the previous record and nearly twice NASA's 1995 goal.

We have also studied the effect of air mass on various device structures.(13) Our outdoor performance data indicate that multi-junction devices deliver higher power output than single-junction devices.

Our tandem modules also received much favorable evaluation in terms of accelerated stress testing than single-junction modules in the market as reported by an independent study.(14) Our performance data have also been validated in Florida and India.

SUMMARY

We have reviewed the current status of amorphous photovoltaics at ECD in terms of the three key issues: efficiency, stability, and production. Using fluorinated materials and multi-junction approach, we have achieved a 13% conversion efficiency with exceedingly good stability. Our high yield and consistently good efficiency in production coupled with rapidly advancing science and technology can soon bring amorphous

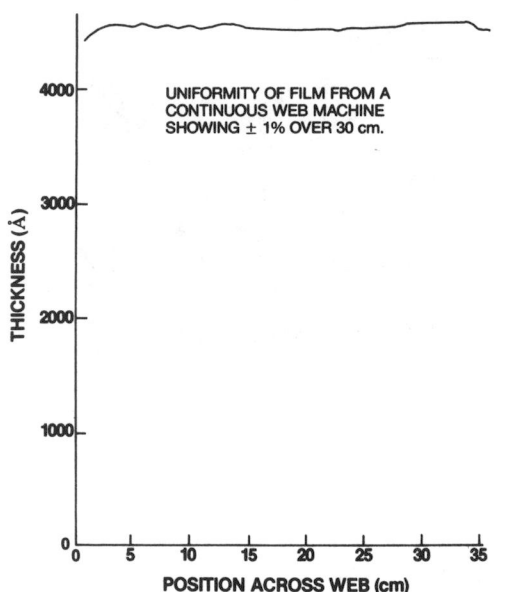

Fig. 8 Film thickness as a function of position across a strip 35-cm wide, deposited by our continuous web process.

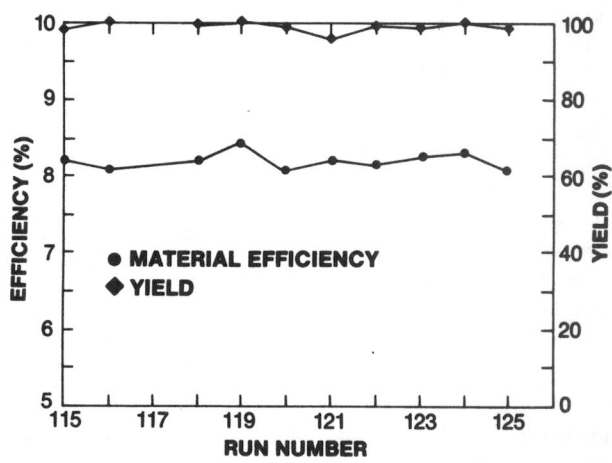

Fig. 9 Efficiency and yield as a function of run number for a typical set of runs on a roll-to-roll production machine.

photovoltaics competitive to conventional fuels with less than one dollar per peak watt.

Acknowledgements

The authors wish to acknowledge the great team effort of our photovoltaic group, especially the contributions of Steve Hudgens, Masat Izu, Subhendu Guha, Prem Nath, Rosa Young, Joe Doehler, and Joe Hanak. We wish to thank Ghazaleh Koefod and Melissa Lipton for their assistance in preparing this manuscript.

References

1. S.R. Ovshinsky and A. Madan, Nature, 276, 482 (1978).
2. S.R. Ovshinsky, in Proceedings of the International Ion Engineering Congress (ISIAT '83 & IPAT '83), p. 817, Kyoto, Japan (1983).
3. S.R. Ovshinsky, in Physical Properties of Amorphous Materials, ed. D. Adler, B.B. Schwartz, and M.C. Steele, p. 105, New York: Plenum Press (1985).
4. S.R. Ovshinsky, in Proceedings of the 18th IEEE PV Specialists Conference, p. 1365, Las Vegas (1985).
5. S. Guha, J. Yang, W. Czubatyj, S.J. Hudgens, and M. Hack, Appl. Phys. Lett. 42, 588 (1983).
6. J. Yang, R. Ross, R. Mohr, and J.P. Fournier, in Materials Research Society Symposia Proceedings, Palo Alto, CA (1986). In press.
7. R. Tsu, M. Izu, V. Cannella, S.R. Ovshinsky, G.J. Jan, and F.H. Pollak, J. Phys. Soc. Jpn. 49, Suppl. A, 1249 (1980).
8. R. Tsu, S.S. Chao, M. Izu, S.R. Ovshinsky, G.J. Jan, and F.H. Pollak, in Proceedings of 9th International Conference on Amorphous and Liquid Semiconductors, Grenoble, France, 1981; J. de Physique, Colloque C4, Supplement au no. 010, 42, C4-269 (1981).
9. S. Guha, J.Yang, P. Nath, and M. Hack, Appl. Phys. Lett. 49, 218 (1986).
10. S. Guha, in Proceedings of the 11th International Conference on Amorphous and Liquid Semiconductors, Rome, Italy, Sept. 2-6, 1985; J. Non-Cryst. Solids 77 & 78, 1451 (1985).
11. T. Hirobe, M. Katayama, Y. Shimada, T. Nagayasu, H. Oka, H. Izawa, H. Nanbu, N. Shiozaki, H. Morimoto, T. Takemoto, and N. Nakajima, in Proceedings of the 2nd International PVSEC, p. 471, Beijing, China, August 19-22 (1986).
12. J. Hanak, in Proceedings of the 18th IEEE PV Specialists Conference, p. 89, Las Vegas (1985).
13. J. Yang, T. Glatfelter, J. Burdick, J.P. Fournier, L. Boman, R. Ross, and R. Mohr, in Proceedings of the 2nd International PVSEC, p. 361, Beijing, China, August 19-22 (1986).
14. J. Lathrop and E. Royal, in Proceedings of the 2nd International PVSEC, p. 386, Beijing, China, August 19-22 (1986).

AMORPHOUS SEMICONDUCTORS FOR MICROELECTRONICS

Stanford R. Ovshinsky

Energy Conversion Devices, Inc., 1675 West Maple Road, Troy, Michigan 48084

The solid-state revolution, which began in 1947 with the invention of the transistor, was made possible by the ability to make crystalline materials (at that time germanium) sufficiently free of defects that substitutional dopants could overcome the background noise of other defects and control the electronic transport properties of the semiconductor. Since the early 1930's Bloch, Wilson, and others had laid a sufficient theoretical groundwork in semiconductors so that transistor action could be predicted, demonstrated, and understood even though the point contact transistor had many mysteries associated with it. Figure 1 shows that historical lever that moved the world, the first transistor.

How disorder on crystalline surfaces, which I have called the progenitor of amorphicity, played an important role in the development of the transistor can be illustrated by the fact that the inventors of the transistor sought to make a field-effect device before a bipolar one, but failed because the defect density near the surface swamped out the transistor action. It was clear, even in those early days, that it was not only crystalline periodicity that was the important factor, but also the chemistry of the material was critical. That is why chemists made such important contributions to the development of the transistor. For example, although silicon was preferable because of its larger band gap, it stubbornly resisted the taming of defects and impurities which had been so successfully achieved in germanium.

Despite its accomplishments, it soon became evident that the germanium transistor had serious limitations. With its relatively small band gap, it did not have the temperature range required for many applications. In any case, the invention of the transistor stimulated a flood of materials research, and every conceivable type of crystalline structure was investigated, for example, gallium arsenide, indium antimonide, lead sulfide, mercury telluride, etc. Fortunately, silicon was ultimately conquered, primarily by chemists and materials scientists rather than by the electronics experts of the time, many of whom were still serving out their indenture to vacuum tubes and gaseous electronics which was considered to be the electronics of that time.

Recall that the announcement of the invention of the transistor was buried in a back page of the New York Times, together with the daily radio broadcast schedule.

With the adoption of silicon, all the other semiconductors became footnotes in the pages of technological history until recently, when after many, many years, longer than the entire history of amorphous semiconductors and after incredibly more investment, gallium arsenide has, phoenix-like, risen once more to command attention. Silicon is a high-temperature material and fortunately, unlike germanium, it has an easily grown oxide available for electrical isolation, so that the next great advance after the original invention of the transistor, the integrated circuit, was made possible. If the developers had stuck with germanium, this may never have happened as germanium oxide has an array of processing problems. Again, chemistry was the necessary handmaiden of physics.

Since silicon is an excellent semiconductor, is abundant and inexpensive, is capable of performing at high temperatures, and has an easily grown oxide

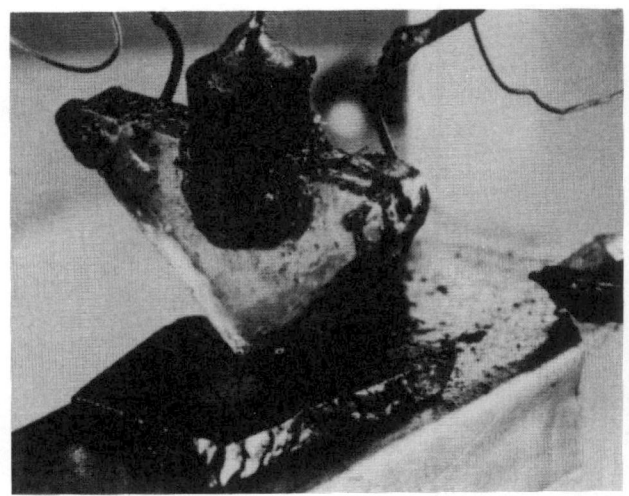

Fig. 1   The First Transistor

Reprinted by permission from the *Proceedings of the SPIE Symposium on Amorphous Semiconductors for Microelectronics*, Los Angeles, California, 21–22 January 1986, Vol. 617, pp. 2–9.

that enabled the development of integrated circuits, it appeared that progress in crystalline silicon devices would be virtually endless. It is interesting to note that once silicon oxide was used, amorphicity raised its head in a beneficial way. It was found that crystalline-silicon/amorphous-silicon-oxide interfaces have extremely small interface defect densities, and MOS structures became a reality.

More and more, research in transistors, and therefore in microelectronics, became a battle to develop higher device densities, therefore much smaller circuitry and lower costs. The number of devices on a chip was reflected qualitatively through many new complex circuits that expanded computer applications enormously. Obviously, these advances were not trivial; indeed, they spawned a whole new era of electronics. However, disquieting facts began to intrude. It became increasingly difficult and expensive to grow larger and larger crystals. Getting to the position of growing a six-inch crystal was an arduous task, and getting beyond it still more difficult. Therefore, advances had to take place in finer lithography. Micron and submicron lithography involve expensive and difficult procedures. One is faced now with crystalline silicon plants costing in the 200-300 million-dollar range, compared to only about 10 million dollars in the early days. In addition, the high density of circuit components is increasingly susceptible to stray and unwelcome disruptive charges. Heat dissipation is more and more of a problem. Furthermore, everyone is making virtually the same thing. There is very little proprietary position in the field, and Japan with its higher quality and lower costs is turning Silicon Valley into "Death Valley."

The microelectronics industry is now taking out the cannons instead of the rifles, figuring that the excruciating effort to get incrementally larger wafer size and finer photolithography can overcome all problems. This brute force approach means a tremendous expenditure of energy and money to accomplish relatively small gains, albeit the devices that will result will certainly be of importance to the electronics industry. According to a prominent figure in the semiconducting industry, "You've got an industry that has pushed its technology to the fundamental limits of nature." (1) This is both a crisis and an opportunity. As Yukawa (2) has said in reference to another area of science, "The fault surely lies in an excessive conservatism in the realm of ideas, retaining the same preconceptions and pursuing the same lines of development—an unfortunate and inefficient process. Even while keeping to one particular line of development, the objective will rarely be attained without some radical leap forward along the way."

We will show in this paper that we are taking such a leap forward. Based upon our experience and on the products we have developed, we can make the competitive battle taking place between device manufacturers making similar products throughout the world irrelevant by an entirely different approach—the use of proprietary amorphous materials, devices, and circuits.

Why amorphous rather than crystalline? Does not gallium arsenide, our old acquaintance and sometime friend, hold the greatest promise? After all, "speed is the name of the game." A factor of 10 increase in mobility over silicon is of great importance. Unfortunately, mobility does not always reflect itself in the final circuit speed since the difficulty of processing large, high-quality gallium arsenide crystals has not been solved. The typical gallium arsenide crystal wafer is only 2-3 inches in diameter, and, because of the high defect densities, epitaxial techniques are essential.

All the above appear to be crushing problems preventing mass production. Is there any solution à la crystalline silicon in sight? Gallium arsenide has chemical problems similar to those of germanium in that it cannot be easily oxidized for electrical isolation and is not readily amenable to the photolithographic techniques that have proven so rewarding in crystalline silicon technology. Can't all of these problems be solved? Can't we spend tremendous sums of money and resources to somehow reduce the density of defects, increase the size of the wafer, and develop simple isolation techniques? Obviously, some will go this route, but I submit that the answer to the question is why do it, for there is a basic approach which solves a great many of the problems outlined above, one on which we have been working for a great many years and which has now reached the point of fruition.

Let us compare gallium arsenide, crystalline silicon, and our amorphous silicon alloy devices. The problem which presently limits the gate densities available on gallium-arsenide chips is the large defect concentration. Presently, the best crystalline-silicon chips available provide 40,000 gates, while the best gallium arsenide chips provide about 800 gates. Memory devices show an even greater spread: crystalline silicon RAM chips of 1 million bits are available in sample quantities, while gallium arsenide chips of 1 thousand bits are in the process of development. Thus, crystalline silicon logic is about 50 times better in packaging density than gallium arsenide and crystalline silicon memory about 1000 times better than gallium arsenide. This margin is due primarily to the much lower defect densities possible with crystalline silicon.

Gallium arsenide circuits are of interest for their high-speed capabilities which arise because the field-effect mobility in gallium arsenide is higher than that of crystalline silicon by a factor of 5 to 10. Thus, individual gallium arsenide chips can be much faster than crystalline silicon chips.

The problem for computer designers, however, is that large numbers even of crystalline silicon chips are needed to make the sophisticated fast computers of interest. When gallium arsenide chips are used, still greater numbers of chips are required. The technical difficulty is that the time delays necessary to interconnect such enormous numbers of gallium arsenide chips prevent the systems designer from achieving the desired speed potential of the faster gallium arsenide chips. In fast crystalline-silicon computers, packaging actually accounts for most of the time delays. Clearly, there is little advantage to converting large-scale electronics to gallium arsenide.

Faced with packaging limitations, the optimal design goal is to get everything onto a single silicon substrate; this is one of the objectives of Wafer Scale Integration. But even using a high density substrate, electrical signals propagate with the very undesirable property that the signal delay is proportional to the square of the distance. Substrate area is even more important on high-density substrates than it is on (low chip density) printed circuit boards where signals propagate with delays proportional to distance.

The only answer to this problem appears to be amorphous silicon alloy devices. In our amorphous silicon alloys, "macroscopic" defect densities are conservatively about 100 times less than those in crystalline silicon, so that even greater gate densities can be achieved with our amorphous silicon alloy "chips" than with crystalline silicon chips. The principal factor limiting gate density, however, is line resolution through photolithography. Our amorphous silicon alloys offer two practical avenues of escape from this constraint: (i) large-area substrates for those applications where low cost rather than speed is the objective, and (ii) three-dimensional multilevel depositions for the highest speed (vertical) interconnects and the smallest horizontal sizes.

Viewed from this impact of defect density on chip size, and thus on system speed, our amorphous silicon alloys are conservatively about 100 times better than crystalline silicon, which is itself at least 40 times better than gallium arsenide in defect density. This "packaging" advantage accrues simultaneously in both cost and speed. The additional challenge for us has been to come up with a new high-speed transistor whose gate delay could be fast enough to make use of the 4000:1 advantage we have in packaging. Present gallium arsenide transistor gate delays are approximately 50 picoseconds, while conventional amorphous silicon thin-film transistors have gate delays of the order of 1 microsecond.

The conventional amorphous TFT approach has a field-effect mobility of 1 cm^2/Vsec which we have attained.(3,4) Praiseworthy as is this achievement, we were still limited to circuits which did not allow us to penetrate the high speed computer applications important for the future. This mobility problem has appeared to be basic and inherent in amorphous materials. But as can be seen in Fig. 2, there is reason to doubt the low mobility nature of amorphous solids in the presence of a high electric field. Figure 2 shows my original Ovonic threshold switch,(5) to this day the fastest room-temperature switch. One must use a Josephson-effect device operating at liquid-helium temperature to get similar speeds. The principle of operation of the Ovonic threshold switch involves double injection filling the large concentration of positively and negatively charged traps, inducing a rapid transition to a state in which conduction takes place via a plasma propagating through a quasi-neutral background.(6,7)

What has been needed is a transistor based on new physics; one that avoids the limitations of conventional bipolar and field-effect devices. Such a historical advance has now been made. We have developed a new concept for transistors (8)--perhaps the first new principle with broad device applications since 1947. The work of Hack,

Czubatyj, and Shur of our laboratory in creating this new transistor results in many new important advantages and applications, not the least being the reduction of the gate delay time to about 50 nanoseconds, only a factor of 1000 slower than that of gallium arsenide. Thus, our factor of 4000 advantage in gate density more than overcomes the disadvantage in delay time for larger assemblage of electronics, resulting in a <u>net figure of merit four times larger than that of gallium arsenide for our amorphous devices</u>. Similarly, in comparison with crystalline silicon, the amorphous silicon packaging advantage of 100:1 offsets the factor of 50 difference in gate delay, giving a <u>net figure of merit advantage of two to our amorphous devices over crystalline silicon technology</u>. Further reduction of the delay time using our new transistor concept should enhance this advantage considerably.

In terms of ultra-high speed devices, we will be announcing shortly a new amorphous thin-film transistor with switching times as fast as any of those reported in the most advanced crystalline materials.(9)

Of crucial importance to my point of view is that neither gallium arsenide nor crystalline silicon can be made in thin-film, large-area configurations, and that not only can neither compete with the dramatic increase in density or cost that we achieve in two dimensions with our large-area vertical structures, but also they fail absolutely as a realistic possibility for the necessary new generations of <u>three-dimensional</u> circuits. Our ability to make three-dimensional circuits is the sine qua non of our nth-generation computer approach and it completely separates amorphous from crystalline materials of any sort. For no matter what kind of crystalline geometries have been or will be attempted, the physics of crystallinity is constrained by the lattice; therefore, superlattices or strained-layer lattices cannot compete with our three-dimensional structures. Our approach is the only one that can realistically break the two-dimensional barrier, in the same way that the transistor broke the vacuum barrier and made two-dimensional solid-state devices feasible.

Gordon Moore of Intel has humorously used a cardboard model of a crystal wafer to show the size

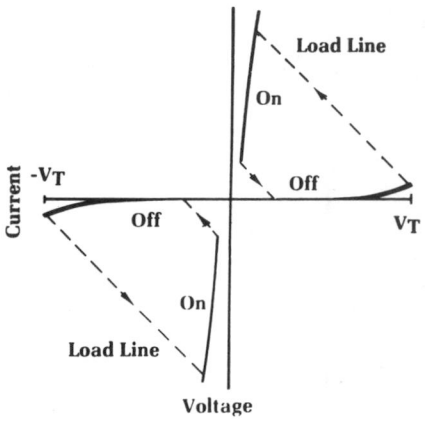

Fig. 2   IV Characteristics of Ovonic Threshold Switch

Fig. 3 Cardboard model of an imaginary "crystal wafer" on the left; a typical real wafer is shown in the inset; an "infinite crystal" on the right

Memory	Logic/ Control	Imaging Films	Photo Receptors
● **Semi-Conductor Memory**	● **Ovonic Threshold Switch**	● **Updatable** (Instant Dry) **Microfiche**	● **Copier**
■ **Non-Volatile (EEPROM)**		■ **MicrOvonic File**	● **Printer**
■ **High Density ROM**	● **Thin Film Transistor**	■ **MicrOvonic Terminal**	● **Instant Camera**
■ **High Density PROM**	● **Thin Film Diodes**		● **Video Camera**
■ **RAM**		● **Graphic Arts Film**	● **Sensors & Transducers**
● **Beam Addressable Memory**	● **Large Area Integrated Display**		● **Photo Transistors**
■ **Computer Mass Memory**		● **Instant Dry Photographic Film**	
■ **Video Disc**			

Fig. 5 ECD's information strategy of the 1960's and 1970's

that would be required in the late 1980's and 1990's to perform the tasks of the new electronics age. Figure 3 shows this next to our "infinite crystal," a 1000-foot long, 16-inch wide roll of our proprietary tandem amorphous silicon alloy on a stainless steel substrate. We have turned the ridiculous into the sublime. Crystals have a defect density of 1-5 defects per square centimeter, while our large-area, thin-film amorphous alloys have 1-3 defects per square foot. We therefore can make an amorphous analog of an infinitely large crystal with improved properties.

FROM: **"The Ovshinsky Switch"** by Stanford R. Ovshinsky
PROCEEDINGS OF THE FIFTH ANNUAL NATIONAL CONFERENCE ON INDUSTRIAL RESEARCH, CHICAGO, ILLINOIS, SEPTEMBER 18-19, 1969, 86-90.

The next slide shows a block diagram of a computer which many of you might have drawn differently. However, as presented here it emphasizes certain aspects that we want to point out. I look upon this computer block diagram as a kind of evolutionary tree

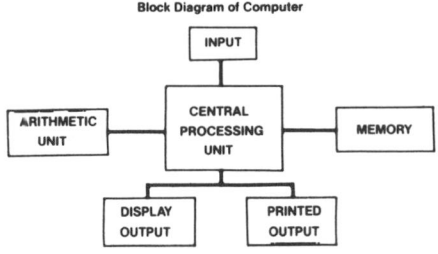

Block Diagram of Computer

because it allows us to clearly see the historical evolution of electronic technology. You can see the evolutionary branches just by examining what a computer is made of. For example, it still uses electromechanical input and output devices. It uses both vacuum and gas tubes for displays—such as CRTs and nixies. Then there are magnetic devices used in the memories and solid state devices in the arithmetic and central processing units. Here we see the evolution of the electronics industry.

What we have in mind is very ambitious and can't be done at once. Our idea is quite simple and there is no secret about it. What we intend to do for the first time is to find a way of performing all of these functions with film devices; all of them out of one area of science; all of them out of one area of technology; with unusual electrical results and at low cost.

Fig. 4 ECD's strategy for developing the all-thin-film computer--1969

Through the 1960's and 1970's I planned our semiconducting strategy based upon our position that we could carry out in amorphous thin-film form all the fundamental device functions that were performed by crystalline materials as well as show new phenomena and devices.(9-11) Figure 4 illustrates my strategy in 1969.(12) Right from the beginning we had shown thin-film possibilities for our devices, including not only the Ovonic switches and memories but also the necessary conductors, capacitors, dielectrics, diodes, and a rudimentary transistor. It was therefore possible to combine all this in an entire circuit on a single substrate (even a flexible one). We also demonstrated thin-film displays (12) and the first laser printers and optical memories.(12-14) Figure 5 shows our information strategy as updated in the mid 1970's. Much of that strategy has been accomplished. Figure 6 shows our continuing progress in making a complete computer with all of its functions in thin-film form. This computer we call the nth-generation computer because it is that radical leap forward that Yukawa spoke about.

Our continuing progress in developing the component parts for this computer is represented in part by other papers presented at this symposium by my collaborators at ECD.(15-18) Not only can we make microelectronic circuits in two dimensions using a continuous, high-speed, proprietary process, but since we can make vertical structures, we can also make multilayers so as to get three-dimensional circuits of unparalleled density as well as short propagation times. Since the motivation of my work in 1955 was to create a nerve cell model that had realistic dimensions and functions,(6,19) I am particularly pleased that we are moving ever closer to making new models of information processing that more and more approximate how the brain functions.

Putting all these factors together, we will be able to introduce more chemistry via the possibility of unlimited alloying, new physical principles, and three-dimensional structures to increase the

FROM: **"The Ovshinsky Switch"** by Stanford R. Ovshinsky
PROCEEDINGS OF THE FIFTH ANNUAL NATIONAL CONFERENCE ON
INDUSTRIAL RESEARCH, CHICAGO, ILLINOIS, SEPTEMBER 18-19, 1969, 86-90.

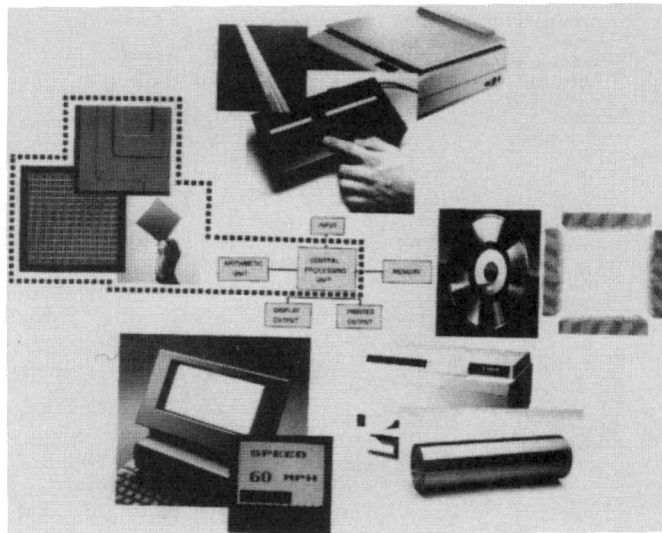

**What we intend to do for the first time is to find a way of
performing all of these functions with film devices; all of
them out of one area of science; all of them out of one area
of technology; with unusual electrical results and at low cost.**

Fig. 6   ECD's   progress   in   developing   the
all-thin-film computer—1985

microelectronic   capabilities   far   beyond   what
previously   was   thought   possible.   With   our   three-
dimensional   structures,   for   example,   a   10-layer
2.5 μm device deposited over a 1000 cm^2 substrate
would   yield   a   memory   capacity   of   about   2 billion
bits   of   information.   By   comparing   our   present
approach   to   what   is   expected   in   the   1990's   from   the
most   advanced   crystalline   extrapolations,(1)   we   note
that   we   would   have   100   times   more   capacity   than
crystalline   devices.   With   that   perspective,   one   can
appreciate   the   importance   of   the   new   advances
discussed here.

Since 1955 I personally, and since 1960 I and
the   other   researchers   at   ECD,   have   been   working   to
make   amorphous   materials   the   leading   materials   for
future   generations   of   the   information   field.   In
1960   I   predicted   that   telecommunications,   the
computer,   and   the   semiconducting   industry   would   all
be united.   While   it   appears   to   be   a   truism   now,   I
can   assure   you   that   it   was   not   taken   seriously
then.   A   new   era   means   new   inventions,   new
materials,   and   new   technology.   We   now   have   the
inventions,   we   have   the   materials,   and   we   have   the
technology.(8,20,21)   It   is   only   a   matter   of   time
and,   of   course,   money   to   fully   implement   our
strategy.   We   have   already   built   three-dimensional
devices   consisting   of   65   separate   layers,   totalling
less than 3000Å   in   thickness.   In   addition,   we   have
built   and   demonstrated   thin-film   memory   and   logic
devices   as   well   as   typical   computer   logic   blocks   and
ring   oscillators.   We   have   already   developed   several
entirely   new   thin-film   transistor   devices,   and
demonstrated   their   capabilities.   We   will   optimize
all   of   these   devices   over   the   next   few   years   by
application   of   the   chemical   and   physical   flexibility
inherent in our overall approach.

A   computer   is   an   entire   system.   It   includes
input devices that can be amorphous sensors, such as
digitizers,   and   output   devices,   such   as   laser
printers   utilizing   amorphous   drums.   In   parallel
with   our   development   of   three-dimensional   computers,
we   will   also   be   increasing   the   use   of   optics   in   our
electronic   computers,   first   as   hybrid   structures,
and   then,   in   the   future,   by   building   a   complete
optical computer.

Optical   interaction   with   amorphous   solids   has
been   a   major   theme   of   our   company   since   its
inception   in   1960.   An   important   new   ingredient   in
making   the   nth-generation   three-dimensional   computer
is   the   use   of   optics   to   connect   circuits.   One   of
the   most   serious   barriers   to   progress   in   all
information   technologies   has   been   the   packaging
problem: how   to   reduce   the   number   of   connectors   and
how   to   connect   high-density   circuits.   There   has   to
be   a   fundamental   new   approach.   We   believe   we   now
have it.   All   that   we   can   say   here   is   that   we   will
be   reporting   on   optical   interlayer   transmission   of
data   using   thin-film   light   emitting   and   detecting
devices   which   will   allow   complex   computers   to   be
fabricated   with   high-yield   device   processing
techniques.

We   have   shown   that   optical   devices   are   practical
using   our   amorphous   materials,   and   in   the   late
1960's   we   demonstrated   our   optical   memories   (21)
which   are   now   gaining   widespread   commercial
acceptance.   Our   licensees   to   date   include
Matsushita,   Hitachi,   Sony,   Asahi   Chemical   working
with   Fujitsu,   IBM,   and   Beijing   Institute   for
Aeronautics   and   Astronautics.   Recall   that   the
advantage   of   optical   memories   is   the   huge   density   of
data   they   can   store.   An   optical   disk   can   store   up
to   50   times   as   much   as   the   state-of-the-art   "hard"
magnetic   disks   and   up   to   1,000   times   as   much   as   a
floppy   disk   of   the   same   size.   Currently   available
nonmagnetic   optical   disks   are   "read-only;"   that   is,
data   can   be   written   on   them   just   once.   ECD's
technology   also   makes   possible   an   erasable   optical
disk,(14,21,22)   one   on   which   data   can   be   written,
erased,   and   rewritten,   similar   to   floppy   or   hard
disks,   but   with   the   critical   advantages   of   much
greater   information   density   and   of   reduced   risk   that
the valuable data can be lost.

In   summary,   we   have   described   an   approach   that
permits   us   to   design   amorphous   semiconductors   for
microelectronics   offering   an   escape   from   the
constraints   of   the   crystalline   lattice,   the   size   of
the   crystal,   the   limited   possibilities   of   chemical
alloying   in   crystals,   the   necessity   for
high-temperature   ultra-high-vacuum   processing,   the
lattice-mismatch   problem   in   heterojunctions,   and   a
multitude   of   other   problems   of   crystalline
materials.   We   are   synthesizing   chemistry,   physics,
and   materials   science   into   a   new   discipline   that
will   enable   us   to   build   three-dimensional   electronic
structures   and   integrated   optical   devices,   thereby
ushering   in   a   new   era   that   will   be   as   exciting   as
the   original   solid-state   revolution.   In   the   course
of   doing   this,   we   have   demonstrated   the   importance
of   invention   in   the   accomplishment   of   the   "radical
leap forward."

Acknowledgments

I   wish   to   thank   Drs.   David   Adler,   Robert
Johnson, and Stephen Hudgens for helpful discussions

and all my colleagues at ECD for their many contributions.

References

1. G. Dan Hutcheson, "Superchips: The New Frontier," Business Week, June 10, 1985, p. 84.
2. H. Yukawa, Creativity and Intuition: A Physicist Looks at East and West, (Kodansha International Ltd., Tokyo, 1973).
3. Z. Yaniv, H. Hansell, M. Vijan, and V. Cannella, "A 1 Micrometer Channel Length Amorphous-Silicon Alloy Thin-Film Field-Effect Transistor," Proceedings of Mat. Res. Soc. Symp. 33, 293 (1984).
4. Z. Yaniv, V. Cannella, J. Hansell, C. Wilner, and M. Vijan, "Novel Thin-Film Amorphous Silicon Alloy Approach to Drive Active Matrices Displays," Mol. Cryst. Liq. Cryst. 129, 149 (1985).
5. S.R. Ovshinsky, "Reversible Electrical Switching Phenomena in Disordered Structures," Phys. Rev. Lett. 21, 1450-1453 (1968).
6. S.R. Ovshinsky, "An Introduction to Ovonic Research," J. Non-Cryst. Solids 2, 99-106 (1970).
7. D. Adler, M.S. Shur, M. Silver, and S.R. Ovshinsky, "Threshold Switching in Chalcogenide-Glass Thin Films," J. Appl. Phys. 51, 3289-3309 (1980).
8. M. Hack, W. Czubatyj, and M. Shur, to be published.
9. S.R. Ovshinsky, to be published.
10. J.A. Perschy, "On the Threshold of Success: Glass Semiconductor Circuits," Electronics, July 24, 1967, p. 74.
11. R.G. Neale, D.L. Nelson, and G.E. Moore, "Nonvolatile and Reprogrammable, the Read-Mostly Memory is Here," Electronics, September 28, 1970, p. 56.
12. S.R. Ovshinsky, "The Ovshinsky Switch," in Proc. of the Fifth Annual National Conf. on Industrial Research, Chicago, (1969), 86-90.
13. L. Lessing, "Great Hopes from Ovshinsky's Little Switches Grow," Fortune, April 1970, p. 110.
14. "The Printed Word Goes Electronic," Fortune, September 1969, p. 116.
15. S.J. Hudgens, "Amorphous Silicon for Electrophotography," paper presented at this conference.
16. M. Shur, C. Hyun, M. Hack, and W. Czubatyj, "Amorphous Silicon Alloy Thin-Film Transistor Operation with High Field-Effect Mobility," paper presented at this conference.
17. Z. Yaniv, V. Cannella, A. Lien, J. McGill, and W. denBoer, "Progress in Two- and Three-Terminal Amorphous Silicon Switching Devices for Matrix Addressed LCDs," paper presented at this conference.
18. V. Cannella, J. McGill, and Z. Yaniv, "Bulk Limitation Effects in Amorphous Silicon Alloy Diodes," paper presented at this conference.
19. S.R. Ovshinsky and I.M. Ovshinsky, "Analog Models for Information Storage and Transmission in Physiological Systems," Mat. Res. Bull. 5, 681-690 (1970). (Mott Festschrift.)
20. S.R. Ovshinsky, "New Amorphous Materials for Computer Use," Digest of Papers, Spring CompCon 1979, 18th IEEE Computer Society International Congress, San Francisco, CA, February 26-March 1, 158-161-C.
21. For early references, see S.R. Ovshinsky, "Fundamentals of Amorphous Materials," in Physical Properties of Amorphous Materials, ed. by David Adler, Brian B. Schwartz, and Martin C. Steele, (Institute for Amorphous Studies Series, Plenum Publishing Corporation, 1985), 105-155. See also Fundamentals of Amorphous Semiconductors, Report of the Ad Hoc Committee on the Fundamentals of Amorphous Semiconductors, (National Academy of Sciences, 1972).
22. See, for example, U.S. Patent # 3,530,441, "Method and Apparatus for Storing and Retrieving Information."

EFFECTS OF TRANSITION-METAL ELEMENTS ON TELLURIUM ALLOYS FOR REVERSIBLE OPTICAL-DATA STORAGE

R.T. Young, D. Strand, J. Gonzalez-Hernandez, and S.R. Ovshinsky

Energy Conversion Devices, Inc., 1675 West Maple Road, Troy, Michigan 48084

(Received 15 May 1986; accepted for publication 11 August 1986)

The use of chalcogenide alloys for optical data storage was first reported by Ovshinsky in 1970.[1] He and co-workers discovered that reversible phase-change phenomena can be accomplished in several chalcogenide films using a focused laser beam.[2,3] In phase-change reversible data recording, the material requirements are intimately related to various parameters associated with system design such as available power from the semiconductor laser, spot sizes used for recording and erasing, disk-rotation speed, etc. Clearly, the material should be sensitive enough that the laser-power density and beam-dwell times are sufficient to complete the amorphous-crystalline–phase transformation. Furthermore, the optical properties of the recorded (amorphous) and erased (crystalline) areas should remain invariant upon cycling and be stable upon long-term storage.

Because of the appropriate optical and thermal properties, and the readily crystalizing behavior, tellurium alloys have been considered as the most attractive material for the phase-change, reversible optical-data storage. It is known that pure Te crystallizes rapidly at ~ 10 °C.[4] In order to stabilize the amorphous phase, tellurium alloys containing certain amounts of Ge, Sn, Sb, As, Se, S, O, etc., have been investigated.[5–15] The major difficulty encountered with this approach is that two contradictory characteristics of the material, i.e., the amorphous-phase thermal stability versus crystallization rate (τ_x) are involved. Finding a method to enhance τ_x while maintaining the amorphous-phase stability of the Te-based chalcogenide alloys is the purpose of the work. In this communication, we used a chemical-modification approach[16] to study the effect of various metal elements in Te-alloy films. We found that transition-metal elements such as Ni, Pd, Pt, etc., in a given Te-alloy film can improve the rate of crystallization substantially without any reduction in film stability.

Thin films of tellurium alloys were prepared by rf-diode sputtering from a target with compositions of $Te_{89}Ge_2Sn_9$. Metal elements such as Ni, Pd, and Pt were introduced into the film by co-sputtering from a target on which thin (1 cm^2) pieces of metal were uniformly distributed. Films were deposited on silicon substrates for analysis of chemical composition and depth profile. Films were deposited on glass substrates for crystallization-temperature (T_x) measurements, x-ray diffraction analysis, and phase-change static test experiments. The film thickness for static test experiments was chosen to optimize the contrast between the reflectivity signals for the amorphous and crystalline phases at 830 nm by utilizing interference effects,[9] which resulted in a film thickness of ~ 1000 Å.

Large-area phase transformations needed for experiments such as x-ray diffraction analysis, T_x measurement, etc., were carried out by using a XeCl excimer laser ($\lambda = 308$ nm, $\tau = 45$ ns) and a Xe flash lamp ($\tau \sim 50 \, \mu$s) for amorphization and crystallization, respectively. The τ_x of the films was measured using a static tester designed to study various characteristics of reversible-phase-change phenomena. A diode laser ($\lambda = 830$ nm), used for recording, erasing, and reading, was modulated to produce pulses of various intensities and durations. A HeNe laser was used for closed loop focus control. An objective lens with a numerical aperature of 0.5 was used, resulting in a spot size (FWHM) of $\sim 0.8 \, \mu$m. The sample was mounted on a stage driven by a stepping motor. To avoid film ablation and oxidation during cycling, samples for static testing were overcoated with a dielectric layer. The illumination is through the glass substrate.

Chemical compositions of the films, determined from x-ray energy-dispersive spectroscopy, are presented in Table I. Auger electron spectroscopy measurements indicated a uniform composition throughout the film thickness. Also shown in the table are crystallization temperatures T_x measured for films in the as-deposited state and after various numbers of cycles of large-area phase transformation. A typical plot showing T measurement of films at two different conditions is given in Fig. 1, where T_x is defined as the temperature at the middle of the sharp drop in the transmission of the HeNe probe beam. The data in Table I. shows that

Reprinted by permission from *Journal of Applied Physics*, Vol. 60, No. 12, pp. 4319–4322 (15 December 1986).

TABLE I. Chemical composition and crystallization temperature of TeGeSn films with various transition-metal additives.

Target	Film composition	Crystallization temperatures (°C)			
		as-deposited	1 cycle	5 cycles	10 cycles
$Te_{89}Ge_2Sn_9$	$Te_{88}Ge_2Sn_{10}$	101	93	86	87
$Te_{89}Ge_2Sn_9 + Ni$	$Te_{84}Ge_2Sn_{10}Ni_4$	102	86	82	82
$Te_{89}Ge_2Sn_9 + Pd$	$Te_{83}Ge_4Sn_{10}Pd_3$	102	92	89	90
$Te_{89}Ge_2Sn_9 + Pt$	$Te_{84}Ge_3Sn_{10}Pt_3$	125	116	90	91

there is a pronounced (10–15 °C) decrease of T_x for all films after the first phase transformation cycle; and T_x is stabilized after a small number of cycles. The addition of Ni and Pd to the films does not appear to have any influence on T_x before or after cycling. A small amount of Pt in the film raises T_x in the as-deposited film, then, after cycling, the T_x also stabilizes to the same value as that of the films with the unmodified composition.

The shift of T_x toward a lower temperature after cycling is evidence of phase separation; the separation of Te crystallities from the crosslinking and/or alloying elements results in the formation of a small amount of high-temperature phases such as GeTe, SnTe, etc. The subsequent laser-vitrification process is unable to melt and therefore, remix these phases to form the original "homogeneous" alloy. As a consequence, the crosslinking strength is reduced and a lower T_x is observed. The reduction in T_x is dependent on the degree of phase separation during crystallization and on the extent of remixing during vitrification.

The behavior of T_x suggest that the transition-metal elements neither chemically bond strongly with Te to stabilize the film nor react with the additive elements to destabilize the film; however, as discussed next, they are apparently effective nucleating catalysts that enhance the crystallization rate.

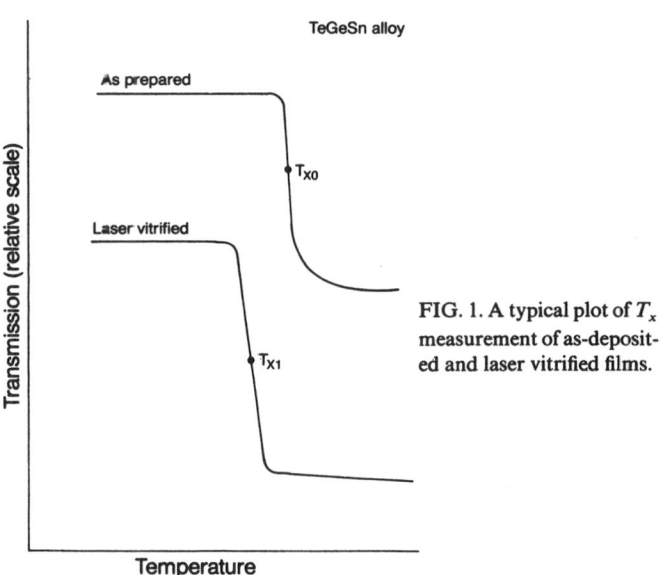

FIG. 1. A typical plot of T_x measurement of as-deposited and laser vitrified films.

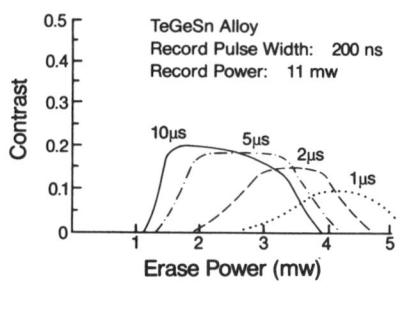

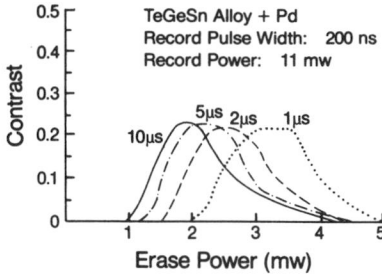

FIG. 2. Static tester results showing the effect of Pd on the crystallization rate in the TeGeSn films.

An example of the effect of these metal elements on the crystallization rate is shown in Fig. 2. These plots of optical contrast versus erase power using various erase-pulse widths are generated from static test experiments on TeGeSn films with and without Pd. The optical contrast is defined as $(R_c - R_a)/R_{a0}$, where R_c, R_a and R_{a0} are reflectivities of the test spot at its crystalline, laser-vitrified amorphous, and as-deposited amorphous state, respectively. The detailed procedure to generate each curve is described elsewhere.[17]

The results shown in Fig. 2 were obtained using four different erase-pulse widths within the erase-power interval of 1–5 mW. The power increment was 0.25 mW. It is clearly seen from the figure that a 5-μs laser pulse was required for the spot to reach the *maximum* contrast or to completely crystallize in the TeGeSn film; whereas, it took only a 1-μs laser pulse at a substantially lower power to achieve complete erasure when a few percent of Pd was added to the film. It is also seen from the figure that for a given erase-pulse width, there is an optimum power (or an optimum temperature) at which the film gives the best contrast. The results suggested that, for a given laser-pulse duration, there is a temperature for the film at which the rate of crystallization

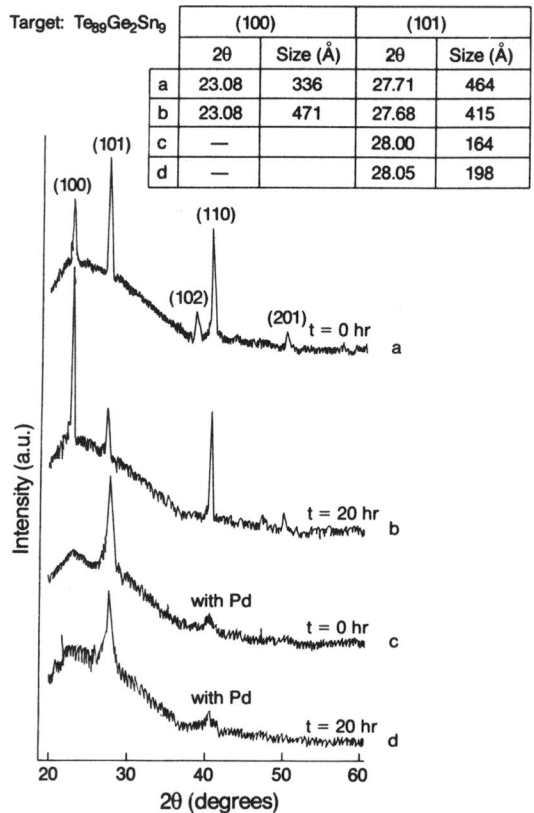

Target: Te$_{89}$Ge$_2$Sn$_9$

	(100)		(101)	
	2θ	Size (Å)	2θ	Size (Å)
a	23.08	336	27.71	464
b	23.08	471	27.68	415
c	—		28.00	164
d	—		28.05	198

FIG. 3. X-ray diffraction pattern of TeGeSn films with and without Pd. The films were crystallized after different storage times in the laser-vitrified amorphous phase.

reaches a maximum, which is not inconsistent with the prediction from the kinetic theory of nucleation and growth.[18] However, at sufficiently high powers, the center of the spots could melt and solidify to the amorphous phase which also reduces the contrast. To what extent the reduction of the contrast is due to incomplete crystallization generated by decreasing the crystallization rate or due to melting is currently under investigation.

In addition to the increase in crystallization rate, we also found that these metal elements reduce crystallite size and retain crystallite orientation with storage time. Figure 3 shows a comparison of the x-ray diffraction patterns of TeGeSn films with and without Pd. The as-deposited film was first crystallized with the flash lamp followed by large-area laser vitrification. For each film composition, x-ray diffraction traces for samples crystallized after different storage times in the laser-vitrified amorphous phase are shown. In one trace, the film was recrystallized immediately after the vitrification. In the other trace, the laser-vitrified film was stored at room temperature for 20 h prior to crystallization.

The x-ray diffraction pattern of all traces is that of polycrystalline Te but the ratios of the intensities of the peaks are different from those expected with randomly oriented polycrystalline Te. The nucleation mechanism of the TeGeSn films has not yet been ascertained; however, it is clearly indicated from the traces that the crystallite orientations of the

TeGeSn film changes with amorphous-phase storage time (see trace a and b). The addition of a few percent of Pd in the film leads to retention of crystallite orientation with storage as seen in trace c and d. Furthermore, from the angular breadth of the (101)-diffraction peak derived from a slow scan, we found that the crystallite size was reduced by more than a factor of two with the Pd in the film. The (100) and (101) peak position and the calculated crystallite sizes are shown in the figure.

Because of the highly anisotropic nature of the Te crystal, the optical properties of the crystalline phase are very sensitive to the crystallite orientations.[19] The change of crystallite orientation can result in a change of optical reflectivity. This condition could lead to a situation where the erased spot has a different reflectivity from the rest of the erased background, thus, an incomplete erasure of the recorded signal is detected, even though they were all fully crystallized. Apparently, this problem can be minimized through the proper control of nucleations.

The reduction of particle size, the retention of crystalline orientation, and the enhancement of the crystallization rate provided evidence that a small amount of Pd in the Te-alloy films is an effective nucleation catalyst. Similar effects are also observed with Ni and Pt added to the films. Evidently, these nuclei are present at such high concentrations and are distributed with sufficient uniformity that the crystallite size is significantly reduced and the crystallization rate is thus enhanced. Very likely, because of the high melting temperature, these nuclei may not even dissolve in the melt during the vitrification process. They form stable nuclei throughout the phase transformation process which retains the crystallite orientation. These metal elements may also have a catalytic effect on crystal growth, but further work is required before any conclusion can be drawn.

In addition to the TeGeSn alloys, the same technique has also been applied to TeSeSb alloys with similar or slightly higher T_x. An improvement in τ_x from 20 to 3 μs was observed.

Finally, we would like to point out that any glass-forming material would crystallize at a temperature above T_g, the glass-transition temperature. The T_x is therefore a parameter which is critically dependent on the measurement technique and heating rate. From differential thermal analysis, we found that T_x is approaching T_g for fast crystallizing material because the time scale of the experiment (e.g., ~1 °C/s) is too slow relative to the rate of crystallization (~μs). Effectively then, the T_x values reported in this work are equivalent or very close to T_g, which is recognized as the most important parameter used to predict the thermal stability of glasses.

Phase-change material offers several potential advantages over magneto-optic for reversible optical data recording, including higher carrier-to-noise ratio, better archival life, real-time erasability, etc. Moreover, the optical system in the disk drive requires no external magnetic field, and phase-change discs are more compatible with the read-only and write-once disk drives already appearing on the market. Material requirements, are, however, more stringent. For example, one of the major concerns of phase-change optical-disk media in the past was the tradeoff between erase speed and data-retention time. We have domonstrated in this work that with an appropriate approach and further research and development, this compromise can be eventually circumvented.

The authors would like to acknowledge H. Situ for the technical assistance, and D. Pawlik for x-ray diffraction measurements. We also express our gratitude to Dr. A. Bienenstock, Dr. H. Fritzsche, and Dr. R. F. Wood for many useful discussions. Finally, we thank M. Lipton and E. M. Norman for their assistance in the preparation of the manuscript.

[1] S. R. Ovshinsky, J. Non-Cryst. Solids 2, 99 (1970).

[2] J. Feinlieb, J. P. deNeufville, S. C. Moss, and S. R. Ovshinsky, Appl. Phys. Lett. 18, 254 (1971).

[3] J. Feinleib, S. Isann, S. C. Moss, J. P. deNeufville, and S. R. Ovshinsky, J. Non-Cryst. Solids 8–10, 909 (1972).

[4] H. Keller and J. Stuke, Phys. Status Solidi 8, 831 (1965).

[5] S. R. Ovshinsky, U. S. Patent No. 3530441 (1970).

[6] R. J. Von Gutfeld and P. Chandhari, J. Appl. Phys. 43, 4688 (1972).

[7] K. Weiser, R. J. Gambino, and J. A. Reinhold, Appl. Phys. Lett. 22, 48 (1973).

[8] R. J. Von Gutfeld, Appl. Phys. Lett. 22, 257 (1973).

[9] Mutsuo Takenaga, Noboru Yamada, Nenichi Nishiuchi, Nobuo Akahira, Takeo Ohta, Sugura Nakamura, and Tadaoki Yamashita, J. Appl. Phys. 54, 5376 (1983).

[10] Mutsuo Takenaga, Noboru Yamada, Shunji Ohara, Kenichi Nishiuchi, Michiyoshi Nagashima, Toshiaki Kashihara, Suguru Nakamura, and Tadaoki Yamashita, Proc. Soc. Photo-Opt. Instrum. Eng. 420, 173 (1983).

[11] A. E. Bell and F. W. Spong, Appl. Phys. Lett. 38, 920 (1981).

[12] D. Strand and D. Adler, Proc. Soc. Photo-Opt. Instrum. Eng. 420, 200 (1983).

[13] P. C. Clemens, Appl. Opt. 22, 3165 (1983).

[14] C. M. J. Van Uijen, Proc. Soc. Photo-Opt. Instrum. Eng. 529, 1 (1985).

[15] M. Chen, K. A. Rubin, V. Marrello, U. G. Gerber, and V. B. Jipson, Appl. Phys. Lett. 46, 734 (1985).

[16] R. Flasck, M. Izu, K. Sapru, T. Anderson, S. R. Ovshinsky, and H. Fritzche, Proceedings of the 7th International Conference on Amorphous and Liquid Semiconductors (Edinburgh, Scotland, 1977), pp. 524–528.

[17] R. T. Young, D. Strand, J. Gonzales-Hernandez, and S. R. Ovshinsky, Materials Research Society Proceedings, Palo Alto, CA, 1986, edited by D. Adler, Y. Hamakawa, and A. Madan (Materials Research Society, Pittsburgh, PA, 1986), Vol. 70, p. 697.

[18] See, for example, H. Rawson, Inorganic Glass-Forming Systems (Academic, New York, 1967), Chap. 3.

[19] See, for example, D. M. Chizhikov and V. P. Shchastlivyi, Tellurium and Tellurides (Collet's, London, 1970).

PART III

REVIEW ARTICLES

THE OVSHINSKY SWITCH

Stanford R. Ovshinsky

Energy Conversion Devices, Inc., 1675 West Maple Road, Troy, Michigan 48084

It is very difficult in 10 minutes to give you more than an overview of what is involved in the field of amorphous semiconductors. Also, I think you probably are more interested in applications at this conference rather than the physics involved.

The first slide gives an indication of the difference between crystalline and amorphous, or

Typical Elements and Characteristics

Tellurium · Arsenic · Silicon · Germanium

Fractured Ingot

Resistivity = 5 x 10⁶ ohm-cm at 300°K

Activation Energy (ΔE) = 0.5 eV

(Formula: $\rho = \rho_0 e^{\Delta E/kT}$)

glassy, materials. I am sure you are all familiar with how crystals for semiconductor devices must be very carefully grown with a minimum of imperfections and then doped with precise and minute amounts of impurities.

By contrast, to prepare an amorphous semiconductor material one takes some rather common elements and melts them. When the mixture cools it takes on a solid structure commonly called "glass" having no periodicity or long-range order. As you can see from the slide, the normal fracture for semiconducting glass is conchoidal and x-ray defraction analysis would further confirm its glassy state.

The next slide please. This slide gives a pictorial representation of what I have been discussing. On the left is a pure crystalline structure. One of the atoms has been replaced by a donor impurity that has resulted in an excess carrier. On the right is illustrated the non-periodic structure of a glass. Within the yellow area we illustrate the fact that excess carriers are not really excess in that they are

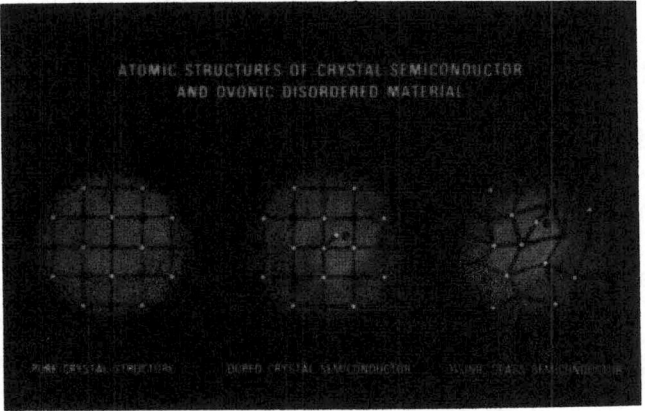

ATOMIC STRUCTURES OF CRYSTAL SEMICONDUCTOR AND OVONIC DISORDERED MATERIAL

readily bound by the large number of traps and recombination centers that exist in these semiconductors. For this reason, the composition of an amorphous semiconductor can vary substantially from its nominal composition without a corresponding change in electrical characteristics.

The next slide shows the electrical characteristics of one of our amorphous switches—the Ovonic Threshold Switch (OTS). The OTS can be made with a switching, or threshold, voltage over the range from a few volts to 300 V and higher, depending upon the requirements of the applications from which it is intended. When the threshold voltage is exceeded,

Reprinted by permission from the *Proceedings of the Fifth Annual National Conference on Industrial Research*, Chicago, Illinois, 18–19 September 1969, pp. 86–90.

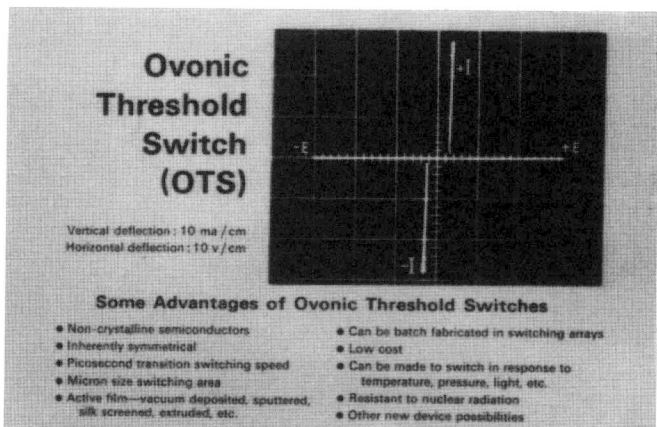

Ovonic Threshold Switch (OTS)

Vertical deflection: 10 ma/cm
Horizontal deflection: 10 v/cm

Some Advantages of Ovonic Threshold Switches

- Non-crystalline semiconductors
- Inherently symmetrical
- Picosecond transition switching speed
- Micron size switching area
- Active film—vacuum deposited, sputtered, silk screened, extruded, etc.
- Can be batch fabricated in switching arrays
- Low cost
- Can be made to switch in response to temperature, pressure, light, etc.
- Resistant to nuclear radiation
- Other new device possibilities

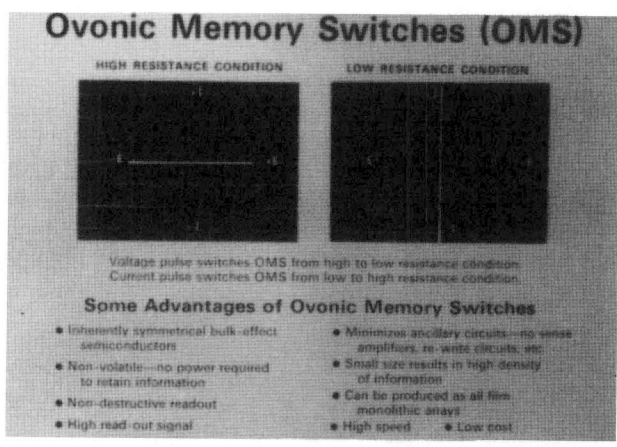

Ovonic Memory Switches (OMS)

HIGH RESISTANCE CONDITION   LOW RESISTANCE CONDITION

Voltage pulse switches OMS from high to low resistance condition
Current pulse switches OMS from low to high resistance condition

Some Advantages of Ovonic Memory Switches

- Inherently symmetrical bulk-effect semiconductors
- Non-volatile—no power required to retain information
- Non-destructive readout
- High read-out signal
- Minimizes ancillary circuits—no sense amplifiers, re-write circuits, etc.
- Small size results in high density of information
- Can be produced as all film monolithic arrays
- High speed
- Low cost

the OTS switches to the highly conductive state shown by the nearly vertical branch of the I-E characterization. Thus it switches from a blocking state, where the resistance is measured in megohms to the conductive state which can have a dynamic resistance of less than an ohm. While conducting, the voltage drop across the OTS is typically 1 V. When the device switches it does so with a transition time of about 150 picoseconds.

Notice the symmetry of the electrical characteristic with respect to polarity. Keep in mind that we are dealing here with a switch that does not have PN junctions or other rectifying elements. Consequently, the OTS provides a symmetrical response to input voltage.

The OTS consists of an active film of amorphous semiconductor between two electrodes. Typically active film thickness is 1 micron and a typical active region is 5 to 50 microns in diameter.

By making the electrodes as thin metal films, one has an all thin-film active component capable of batch fabrication by well-known and inexpensive methods. Of course, it is well-known in the electronics industry how to make passive elements—conductors, resistors and capacitors—of thin films so the emergence of a thin-film active device is an important step in achieving all thin film integrated circuits.

In the OTS you will note that there is a small amount of holding current necessary to keep it in the conducting state. When you take away the holding current, or minimum energy necessary to maintain the conductive state, then the OTS reverts to the blocking state.

The next slide shows the electrical characteristics of the bi-stable Ovonic Memory Switch (OMS). The scope photograph on the left

shows the blocking state; on the right is the conducting state. The OMS can be switched reversibly between these two states by appropriate electrical signals. Notice that the conducting state does not have a holding current so that the current can pass through zero without the device switching off. The OMS will remain in either impedance state definitely without power. We have conducted shelf-life tests in this manner for over 9 years. On the other hand, the device can be switched, repetitively, for hundreds of millions of times in applications where it is desired to frequently change stored information.

The OMS has the desirable characteristics of non-volatility (that is, no power required to retain data), non-destructive readout, and high output signal. Power consumption is low since no power is required except when the memory is switched or interrogated. Readout is very fast because the memory is essentially a resistor in both impedance states. An X-Y array utilizing OMS's, is capable of random access for writing and reading. It is important to point out that interrogation, even at megahertz rates, does not affect the memory switch. This is not always true of magnetic memory elements which are subject to creep induced by repetitive interrogations.

The next slide gives an indication of the state-of-the-art of thin film Ovonic switches. The photograph shows portions of 1-in.-square substrates containing 2,500 thin film switches. On the right is a sandwich-type construction which we call a "pore" device. It is essentially a capacitor in which a very small circular hole, or "pore", in the dielectric is filled with active material. On the left are shown "gap" devices in which the electrodes are coplanar, separated by an etched gap, with active material filling the gap.

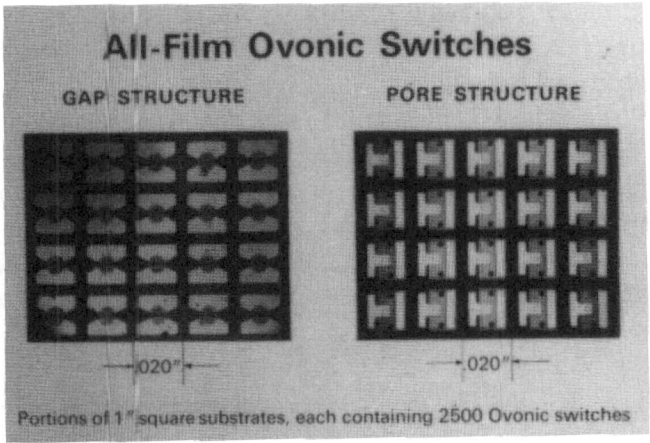

All-Film Ovonic Switches

GAP STRUCTURE   PORE STRUCTURE

—.020"—   —.020"—

Portions of 1" square substrates, each containing 2500 Ovonic switches

Block Diagram of Computer

The "pore" structure is best for low-voltage applications; the gap structure for high-voltage applications.

As you can see, much of the area allocated for each device is for contacts so that in interconnected arrays the packing density could be much higher than the 2,500 per square inch shown here. The nice thing about bulk devices of this kind is that there are no junctions or similar limitations to making the switching element as small as you like. Experimentally, we have employed active films as thin as 500 angstroms. Thus, we have here a fundamentally inexpensive technique for batch fabrication of thin film active components.

What does one do with these new components? It has been a popular and natural tendency to place them in competition with transistors. Of course, such reports make both transistor companies and ourselves unhappy because what we are really trying to do from the point of view of application and product development is to accent the differences between Ovonic switches and transistors. We are not trying to compete with the old established, sophisticated, and excellent technology that has developed in the transistor field. There are many uses of Ovonic switches that do not compete with transistors. These applications take advantage of the high-density and the low-cost possibilities of Ovonic switches as well as their unusual electrical and optical characteristics.

The next slide shows a block diagram of a computer which many of you might have drawn differently. However, as presented here it emphasizes certain aspects that we want to point out. I look upon this computer block diagram as a kind of evolutionary tree because it allows us to clearly see the historical evolution of electronic

technology. You can see the evolutionary branches just by examining what a computer is made of. For example, it still uses electro-mechanical input and output devices. It uses both vacuum and gas tubes for displays—such as CRTs and nixies. Then there are magnetic devices used in the memories and solid state devices in the arithmetic and central processing units. Here we see the evolution of the electronics industry.

What we have in mind is very ambitious and can't be done at once. Our idea is quite simple and there is no secret about it. What we intend to do for the first time is to find a way of performing all of these functions with film devices; all of them out of one area of science; all of them out of one area of technology; with unusual electrical results and at low cost. For a small company like ours which consists of only 100 people, to try to do this sort of job seems unrealistic. So let's see how we do approach it.

May I have the next slide please. Here we show a number of possible applications within a com-

Potential Ovonic Applications in Computers

INPUT	Keyboard Interactive Input/Output Systems
CENTRAL PROCESSING UNIT	Logic Micro-programmers Emulation
ARITHMETIC UNIT	Logic Registers
MEMORY	Main Memory Bulk Memory
DISPLAY OUTPUT	Numerics Alpha-Numerics Graphics Character Generator
PRINTED OUTPUT	Line Printer Page Printer

puter for amorphous semiconductors. The next slide shows those particular application areas for which our company is presently developing products.

The memory array we are developing is an electrically alterable read-only memory. We call it a

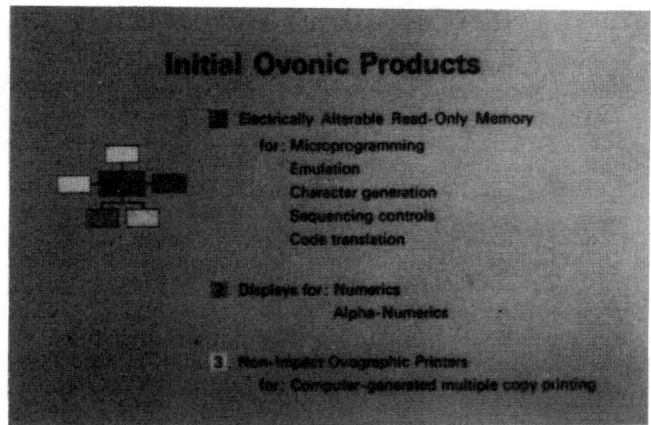

"Read Mostly Memory". Read-only memories are hard-wired; that is, their data is permanent so that a change of data requires reworking the memory or physically replacing at least a part of it. What is needed in the computer field is a read-only memory in which the data can be electrically altered; that is, a read/write memory that is non-volatile and has a fast read speed. Such a memory would provide a great deal of flexibility in computer systems, in performing functions such as micro-programing, emulation, character generation, code translation and sequencing.

The next slide shows our seven segment numeric displays. We use our threshold switches to control electroluminescent lamps of our own design. These have longer life and brighter outputs than lamps that are generally considered

state-of-the-art. By integrating OTS's and EL lamps the objective is to obtain a flat panel display with inherent memory. Such a display is advantageous from the point of view of smaller size, reduced circuitry, improved reliability and lower cost. The character height of the lamp we are developing is 0.6 inch, although, of course, since both electroluminescence and Ovonics are film technologies there is a great deal of flexibility

in making displays in various sizes and configurations. As an example, if a multi-digit display were required it could be made readily as a single strip rather than built up of a series of single-digit modules.

In the next slide we show a schematic diagram of a non-impact printing system we are developing using amorphous semiconductors. We use a computer-controlled laser to "write" on a drum coated with Ovonic memory material—similar to the material we use for our memory switches.

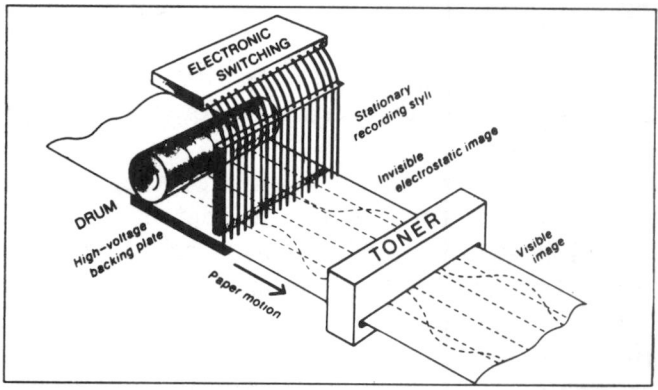

With the memory film initially in the ordered or conductive state the laser energy converts the material, where it impinges, to the high-impedance state. Electrostatic printing then is used to covert the image on the drum into hard copy—as many copies as are desired from the single image. Printing speed can be that of a printing press. Of particular importance is the fact that the material can be switched back to the original ordered state if it is desired to do so since the memory effect is reversible. Consequently, one can generate hard copy for proofreading. It can be edited, corrected, a new copy obtained, and so on until the copy is perfect, at which time final copies can be printed in large quantities.

Features of Ovography

1. Image Generation
   a. direct computer control
   b. other forms of modulation, e.g. vidicon, instruments
2. Permanent, but alterable, image
   a. multiple copies
   b. selective erasures and corrections (editing)
3. Printing on ordinary paper
4. Gray-scale recording
5. Multi-color recording
6. Audio-visual recording

Applications for Ovography

1. Computer Printer

2. Publishing (books, magazines, newspapers)
   Editing Printer
   Final Printer (Electronic Type Composition)

3. Copier

4. Home information set combining visual display, hard copy, computer and control terminal

5. Bulk memory

The next slide shows the state-of-the-art in terms of the quality of printing we are getting from our present model. We are very pleased with the results to date since we believe it compares favorably with other new developments in electronic printing with which we are familiar. By the way, this approach is not limited to electrostatic printing.

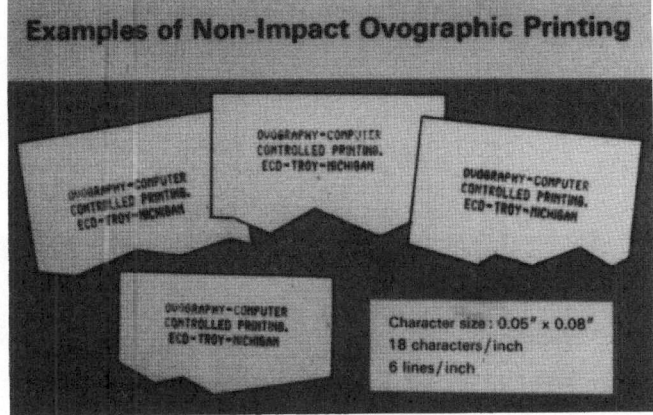

Examples of Non-Impact Ovographic Printing

It is obvious that the drum we have just described for our printer is a mass storage medium. Using the optical changes which occur simultaneously with the changes of electrical resistance when the material is switched, we have the basis for an optical mass memory. Densely packed data dots less than 2 microns in diameter are written with a finely-focused laser beam and can then be read out optically, such as by sensing the change in transmissivity or reflectivity as well as scattering that occurs in the material when it is switched. Because memory switching is reversible, the data can be erased at will and new data written in the same location. In other words, the memory film is a beam-addressable, beam-alterable mass memory medium.

I wish to emphasize that the threshold switch is an electronic device without structural change. The memory switch on the other hand utilized structural change as a means of preserving its information.

In this presentation it has been very difficult to do more than give you a superficial view of the applications that our materials, components, and systems can have. Important concepts to keep in mind are that these materials and devices are easy to manufacture and have unique electrical properties that enable us to do things that are either very difficult, expensive, or in some cases impossible to do with conventional devices and approaches.

Our work also represents an exciting new area in physics that will lead to many more applications and to a new area of technology. Our objective is to concentrate on specific products and bring them to the market place while continuing development. We are very proud of our organization and the very fine group of scientists associated with us. I see in the audience one of them, Dr. Hellmut Fritzsche from the University of Chicago, who has been an important contributor to our scientific developments and to our company.

I hope I have stimulated your interest in the field of amorphous semiconductors and their application possibilities. I thank you very much for your kind attention.

AN INTRODUCTION TO OVONIC RESEARCH

S.R. Ovshinsky

Energy Conversion Devices, Inc., 1675 West Maple Road, Troy, Michigan 48084

This paper will briefly review our past work in amorphous switching and discuss some of the operational concepts underlying our continued effort. In the past, periodicity of atomic structure has been the basis of the understanding of solid state physics. Part of the problem in understanding the amorphous state is the terms themselves, for the words "disordered" and "amorphous" for non-crystalline materials have misleading connotations as they project an image of a material with lack of specific structure. In this paper we will describe some of our working hypotheses which pay particular attention to spatial relationships of the localized structural units as a means of developing unusual electronic and electrical changes in amorphous materials.

In the late 1950's, the well-known Soviet physicist, A.F. Ioffe, considered the possibility that short-range order (1) could have a direct bearing on semiconducting properties and he, in conjunction with a notable group of scientists led by B.T. Kolomiets, started investigations into the characteristics of amorphous materials.(2,3) This Soviet group studied amorphous materials, measured their properties at low electric fields, and found them, in fact, to be rather uninteresting in the sense that they behaved like intrinsic crystalline semiconductors. Since intrinsic semiconductivity is common, one would not expect exciting new phenomena from such materials. Other conventional measurements important in crystalline materials were also discouraging when made on amorphous materials. For example, the Hall mobility is very low, almost non-existent, in most amorphous semiconductors. There appeared to be very few available carriers with little ability to move freely in an electric field.(4)

At approximately the same time, I independently started an investigation of the electrical properties of disordered materials with the avowed purpose of finding switching effects. My interest in disordered materials arose from an interest in neurophysiology. I had posed to myself the question of what is the physical basis of intelligence, that is, how do nerve cells store, switch, alter and transfer information in an organism.

We divided the problem into two parts, one having to do with the traffic problems of communication, that is, threshold switching events, and the other, the storage and alteration of information, which is memory. By building models which simulated biological action we could develop insights into possible biological mechanisms and at the same time develop a superior switch for technological use. This bionic approach has been a particular interest of both my wife's, who is a biochemist, and mine. We, therefore, worked on changing a film reversibly between a high and a low impedance (the threshold) and on storing and interrogating an energy event in a film by altering the atomic configuration of the material (the memory). Regarding the first, we originally expected a strongly disordered material to have many traps and recombination centers utilized for conduction processes by a sufficiently high electric field. As to the second type of switching, we viewed the expected change of atomic configuration as a change in the degree of local order in the disordered polymeric material. Many disordered systems were explored. Both threshold and memory switching were found and we proceeded with our original purpose of making nerve models and developing commercially usable switches.

In June of 1958, I succeeded in making a switch based on an amorphous material. I used a valve metal, the transition metal tantalum, which had an amorphous layer of tantalum oxide about 900Å thick, and an electrolyte solution formed a portion of the control electrode. Switching occurred when the film was sufficiently polarized. This work was widely reported in technical journals such as Electronics (5) and Control Engineering (6) in the summer of 1959, and discussed in the book Anodic Oxide Films by Young.(7) A suitable metallized electrode led to a memory version of this switch which was demonstrated at a lecture given at the Detroit Physiological Society Meeting in 1959.(8) A pulse of one polarity set the memory, and a pulse of the opposite polarity shut it off.

We have been demonstrating switching in amorphous materials since 1958 and our group at Energy Conversion Laboratories (now Energy

Reprinted by permission from *Journal of Non-Crystalline Solids*, Vol. 2, pp. 99–106 (1970).

Conversion Devices, Inc.) has been working continuously since 1960 with great commitment to glassy semiconducting switching. Our work, originally kept secret for commercial reasons, has led to many patents, granted and in process. The switches (9) that were developed can be divided into three types, all of them based on disordered materials (10) and fabricated by a common technology. The devices commercially utilized were made of chalcogenide materials. The electrical characteristics of three types of devices (threshold, memory and adaptive memory) will be described in detail in subsequent papers.

In the early 1960's, Dewald, Northover and Pearson (11) at Bell Labs were working with glasses. They also found amorphous materials which had memory action, and could be kept in either a conducting or a non-conducting condition by pulses of opposite polarity. They encountered problems of material stability and reproducibility, and their original project was eventually dropped. Movies of the memory device of the Bell Telephone Labs group recently shown (12) and a description of materials and electrodes used by them clearly show chemically unstable thermodynamic processes taking place.

It is important to caution that there are artifact switching and memory actions which are neither reproducible nor stable, nor long-lived, and can cause confusion among the uninitiated. These artifact switching effects can be summarized briefly. Placing two narrowly separated electrodes in a high electric field can cause the growth or throwing of a metallic bridge from the electrodes that can be destroyed by passing a high enough electrical current through it. This is an artifact because it uses up the electrodes and it is certainly not a reproducible or long-lived element. Even if the metal throwing takes place penetrating a solid film, it will still fall into the same category. Another common artifact is the use of chemically unstable material between two electrodes. Here chemical decomposition takes place, resulting in irreproducible and irreversible phenomena. This type can be characterized as using up the material. A third type draws upon the environment for its mechanism and concerns itself with forms of oxidation and reduction which are thermally induced. A form of this occurs with some oxides wherein a destructive dielectric breakdown can sometimes be healed by utilizing the oxygen either inside or surrounding the material to repair the oxide. Another type of artifact has to do with contacts that can react with the switching materials used. Nonstable switching can be seen as alloying takes place.

In contrast to this, ovonic threshold and memory switches are completely reversible. Their mechanisms do not involve chemical reactions with electrodes, materials or environment.

We will now briefly describe ovonic switching. The Ovonic Threshold Switch is a device which requires a sustaining current to keep it in the conducting condition. When the sustaining current is not present, the device reverts to its high resistance state. These devices have been cycled 8×10^{12} times without significant change of parameters, at which point the test was discontinued.

The mechanism is considered to be primarily an electronic one and does not involve structural change. Thermal effects can be cooperative with the electronic mechanism.(13-15)

The Ovonic Memory Switches can remain indefinitely either in the ON or OFF state without any sustaining energy input, that is, the information is stored in an available stable structure of the material and one must alter the structure of the material in order to change its electrical resistance. Electronic threshold switching initiates the change in conductance from high to low. The change of conductivity, however, is stored in the structure. In essence, this switch is based on reversible phase changes. A recent test run was stopped after 3×10^8 switching operations were observed without failure.

Let us now differentiate between the Ovonic Threshold Switch and the Ovonic Memory Switch. The threshold switch is made up of materials that are specifically chosen not to have structural changes, either at elevated temperatures or when slowly cooled. However, the memory material is chosen so that the cross-link elements, which shorten the polymeric chains and tie the bonding sites, allow for some polymeric movement that can include bond rotation and switching, chain shortening and lengthening and very limited spatial diffusion of atoms. Such atomic movement need not extend further than a few angstroms. The new structural configuration develops towards a different stable state which involves more order, therefore it tends toward crystallinity. This more conducting configuration is connected with lowering the local satisfaction of valence requirements of the amorphous state.

There is the possibility that as soon as positional and compositional disorder starts to change towards more order, the Fermi level moves away from the center of the gap and the material becomes a low resistivity semiconductor with perhaps a smaller band gap. Since this involves the changing of atomic overlap, it may be related to a Mott transition. It was Mott who also conceived of the ability of amorphous materials to accommodate changes in local structure which result in satisfying bonding requirements.(16)

Memory action involves the transformation of a material from one stable state to another. This transformation is affected by either the taking in or giving up of energy in the transitions. Fritzsche, who has been a most valued and early collaborator, has been utilizing the DTA to elucidate the properties of memory materials (17,18) and will report some of his findings at this meeting.(19) Electronic processes initiated by the electric field create the energetic conditions which provide for the limited atomic motion that allows the structural reorganization to take place. The changes of electrical resistance are, therefore, associated with ordering and disordering processes. Electric field, current density and thermal gradients may affect the filamentary ordering processes since they are anisotropic in nature.(9,13-15) Ridley has shown that in S-shape breakdown, filamentary conduction should be expected.(20)

A. Bienenstock, at this meeting, will show that, by using X-ray diffraction and spectrographic techniques, the transitions identified with DTA are associated with ordering through the formation of small crystallites.(21)

A third device which has interesting possibilities is a variation of the ovonic memory device called the Ovonic Adaptive Memory. Some aspects of this action will be discussed by Evans and Helbers.(22) The adaptive memory is deliberately designed so as to have a very large number or continuous range of resistance values between the OFF and ON conditions. It is energy-controlled in that the amount of energy put into the materials determines the resistive value which then retains the information state without the need for a holding energy and yet it can be altered. We consider this effect to be associated with a varying degree of structuring related to energy deposited in the material.

Amorphous materials are based upon positional and, in the case of alloys, also compositional disorder, However, paradoxically, it is the understanding of the role of short-range order in the form of local bonds and the steric architecture involved which is essential for both the understanding and the design of amorphous switching devices.

Various single elements can be condensed on cold substrates as amorphous films and yet they are usually not preferable for device usage. The spatial relationship of one atom to another can be altered by the increase in mobility of the atoms with temperature, for example, and therefore new local structures can be formed with different relative positions and changes of overlapping orbitals.(23) This is most pronounced in chain forming materials, but holds true for most polymer-forming single elements. Obviously, this is a form of memory action, for change of structure can be reflected in change of electrical resistance, but unfortunately single element memory is not very stable. For most memory work, one finds that once a memory made only of a single element crystallizes, the energy necessary to disrupt the crystal and revert it back to its amorphous state becomes high and difficult to control. This leads to destructive breakdown. It is useful to add atoms anchoring single element structures to prevent creep and aging, which are the reflections of configurational changes.

For improved memory action, material compositions are chosen which are allotropic in that they can have an ordered or disordered phase, that is, they can be either crystalline or amorphous, for example, germanium telluride.

The introduction of suitable crosslinking materials such as arsenic, phosphorus, etc. provides an additional chemical bonding distortion which favors the amorphous state. Under suitable energy conditions, at least one of the species in the mixture can become ordered and can become a nucleation point for other such arrangements. Disordering is accomplished by passing sufficient energies through the structure, causing randomization once more. Amorphousness can be assured by chemical bonding of crosslinking atoms.

The threshold materials are chosen with different structural principles in mind. The preferred materials are stabilized by alloying, so as to prevent atoms from moving and changing their spatial relationships. This is done by fixing the local order desired through the addition of compatible elements that can bind all available sites. These elements are chosen for two reasons: one, to provide interfering additives to inhibit crystallization, and two, to provide, through the processes of chemical bonds of different energies, a large density and a wide spectrum of localized states which bridge the valence and conduction bands. The band gap model of Cohen, Fritzsche and Ovshinsky (24-26) takes into account the importance of the localized states for the establishment of a high density of charged and neutral trapping centers. Fagen has reported,(27,28) and will report further at this meeting,(29,30) the large density of recombination centers that do exist at the Fermi energy. In our model, the almost complete local satisfaction of valence requirements yields the observed intrinsic behavior of the conductivity. Another possible explanation of intrinsic behavior will be described by our colleague, Böer.(31) In a material which has a high density of localized states, injecting processes can be considered to be a dominating factor in the transformation of the material into the conducting state, for if traps can be made to be inoperative, then clearly the material can become highly conducting until the current is decreased beneath a minimum holding current which is necessary to sustain the injecting process. Henisch is considering space charge effects in this kind of model.(32)

The field of disordered materials is an exciting one since it is a relatively new and expanding area in physics. Conventional concepts are going to be challenged and new techniques will be developed that will allow us to understand electronic processes better. As with the early transistor, one wipes out the distinction between pure and applied physics, for in order to see high field phenomena, one must really make a device. The device and the phenomena interesting to physicists become indistinguishable. Neale, an early collaborator, will describe the development of the manufacturing techniques that mate science and technology.(33)

What are some of the problems that we face as we develop amorphous switching? There is continuous material development, a portion of which is devoted to raising the glass transition temperature. This will minimize the problem of straight dc operation associated with some threshold switching. Another important part of our effort is to make sure that catalytic agents which enhance structural change are not introduced either during manufacturing or operation. We are also concerned with possible reactions between contact materials and semiconducting switching materials. Fleming has contributed greatly to our development with his materials work as well as directing the electroluminescent activities reported at this Conference.(34)

There is much work being done on circuit applications by the group at ECD under the direction of Nelson.(35) Various aspects of this work have been reported.(18,36,37) Programs in display and memory circuits are being actively pursued with some aspects of memory utilization being investigated by Sie. Electrical characterization of devices continues as a major project as exemplified by Shanks' paper at this meeting.(38)

The size of this meeting illustrates the great interest in amorphous switching and gives us great satisfaction. It has been a long journey from 1958 to the present. We have been fortunate to witness the increase of understanding in this area of physics and look forward to a rapid growth of this field. The work that will be reported here by our group reflects the progress being made.

References

1. This idea guided Mooser and Pearson to predict semiconducting behavior in a large variety of crystalline materials. See E. Mooser and W.B. Pearson, Can. J. Phys. 34 (1956) 1369.
2. For references to their work see A.F. Ioffe and A.R. Regel, Non-Crystalline, Amorphous, and Liquid Electronic Semiconductors, in: Progress in Semiconductors, Vol. 4, Ed. A.F. Gibson (Wiley, New York, 1960) pp. 237-291.
3. For references to their work see B.T. Kolomiets, Phys. Status Solidi 7 (1964) 359, 713.
4. For a discussion of electronic properties of amorphous materials see J. Tauc, Science 158 (1967) 1543.
5. S.R. Ovshinsky, Electronics 32 (1959) 76.
6. S.R. Ovshinsky, Control Engineering 6 (1959) 121.
7. L. Young, Anodic Oxide Films (Academic Press, New York, 1961) pp. 147-149.
8. S.R. Ovshinsky, The Physical Base of Intelligence--Model Studies, presented at Detroit Physiological Society, Dec. 17, 1959.
9. S.R. Ovshinsky, Symmetrical Current Controlling Device, U.S. Patent No. 3,271,591.
10. S.R. Ovshinsky, Phys. Rev. Letters 21 (1968) 1450.
11. A.D. Pearson, W.R. Northover, J.F. Dewald and W.F. Peck, Jr., Chemical, Physical and Electrical Properties of Some Unusual Inorganic Glasses, in: Advances in Glass Technology (Plenum Press, New York, 1962) pp. 357-365.
12. A.D. Pearson, in: Proc. Symp. on Instabilities in Semiconductors, IBM Watson Research Center, Yorktown Heights, New York, March 1969.
13. S.R. Ovshinsky, The Ovonic Switch as an Amorphous Switching Device, presented at IV Symposium on Vitreous Chalcogenide Semiconductors, Acad. Sci., USSR, Leningrad, 1967.
14. S.R. Ovshinsky, Ovonic Switching Devices, presented at Intern. Colloq. on Amorphous and Liquid Semiconductors, Acad. Soc. Repub. Rumania, Bucharest, 1967.
15. S.R. Ovshinsky, Ovonic Switching Devices, Proc. of 1968 Electronic Components Conf., Washington, D.C., 1968, p. 313.
16. N.F. Mott, Advan. Phys. 16 (1967) 49.
17. H. Fritzsche, Physics of Instabilities in Amorphous Semiconductors, Symp. on Instabilities in Semiconductors, IBM Watson Research Laboratory, Yorktown Heights, New York, March 1969.
18. H. Fritzsche, Bull. Am. Phys. Soc. [II] 14 (1969) 342.
19. H. Fritzsche and S.R. Ovshinsky, J. Non-Crystalline Solids 2 (1970) 148.
20. B.K. Ridley, Proc. Phys. Soc. (London) 82 (1963) 996.
21. A. Bienenstock, F. Betts and S.R. Ovshinsky, J. Non-Crystalline Solids 2 (1970) 347.
22. E.J. Evans, J.H. Helbers and S.R. Ovshinsky, J. Non-Crystalline Solids 2 (1970) 334.
23. Effects of this nature were observed in the form of an increase in the resistivity of evaporated films of Ge and Si as a result of heat treatment. For a review see P.A. Walley and A.K. Jonscher, Thin Solid Films 1 (1968) 367.
24. M.H. Cohen, H. Fritzsche and S.R. Ovshinsky, Phys. Rev. Letters 22 (1969) 1065.
25. H. Fritzsche and S.R. Ovshinsky, J. Non-Crystalline Solids 2 (1970) 393.
26. M.H. Cohen, J. Non-Crystalline Solids 2 (1970) 432.
27. E.A. Fagen, S.R. Ovshinsky and H. Fritzsche, Bull. Am. Phys. Soc. [II] 14 (1969) 311.
28. H. Fritzsche, E.A. Fagen and S.R. Ovshinsky, Bull. Am. Phys. Soc. [II] 14 (1969) 311.
29. E.A. Fagen and H. Fritzsche, J. Non-Crystalline Solids 2 (1970) 170.
30. E.A. Fagen and H. Fritzsche, J. Non-Crystalline Solids 2 (1970) 180.
31. K.W. Böer, J. Non-Crystalline Solids 2 (1970) 444.
32. H.K. Henisch, personal communication.
33. R.G. Neale, J. Non-Crystalline Solids 2 (1970) 558.
34. G.R. Fleming, J. Non-Crystalline Solids 2 (1970) 540.
35. D.L. Nelson, J. Non-Crystalline Solids 2 (1970) 528.
36. S.R. Ovshinsky and D.L. Nelson, Ovonic Switches and Their Application, Proc. IEEE Intern. Convention, New York, March 1969.
37. S.R. Ovshinsky, E.J. Evans, D.L. Nelson and H. Fritzsche, Radiation Hardness of Ovonic Devices, IEEE Trans. Nuclear Science (Dec. 1968) p. 304.
38. R.R. Shanks, J. Non-Crystalline Solids 2 (1970) 504.

AMORPHOUS SEMICONDUCTORS FOR SWITCHING, MEMORY, AND IMAGING APPLICATIONS

Stanford R. Ovhinsky[1] and Hellmut Fritzsche[2]

[1]Energy Conversion Devices, Inc., Troy, Michigan 48084
[2]James Franck Institute and Department of Physics, University of Chicago, Chicago, Illinois 60637

I. INTRODUCTION

THE UNIQUE properties of amorphous semiconductors and the advantages of producing them by means of thin-film processes, which do not limit their size and make them adaptable to integration with other solid-state technologies, have made these materials the base for a new area of science and technology. Applications of amorphous semiconductors either realized or anticipated include a very wide spectrum such as switching and memory devices, continuous dynode electron multipliers (channeltron), optical mass memories, phase contrast holograms, high-energy particle detectors, infrared lenses, ultrasonic delay lines, and microfiche transparencies. This is an impressive list in view of the fact that the scientific understanding of amorphous semiconductors today compares with that of crystalline semiconductors in the early 1950's.

In this paper we intend to discuss the amorphous semiconductor device and phenomena that deal with the handling of information in the form of switching, modulation, storage, and display. We try to summarize the available information concerning performance and reliability. The understanding of the physical processes leading to the various phenomena is rapidly growing; we shall present here the important observations on which our present understanding rests.

We start out by emphasizing some of the characteristic differences between amorphous and crystalline semiconductors that appear particularly important for understanding a variety of applications and phenomena.

II. CHARACTERISTICS OF AMORPHOUS SEMICONDUCTORS

What have amorphous semiconductors in common with crystalline semiconductors? Except for the fact that they are not metals and not insulators, very little. The differences from rather than the similarities to crystals turn out to be the exciting attributes of amorphous semiconductors. Two important aspects of crystalline semiconductors are missing in their vitreous counterparts. The first is the possibility of changing by impurity doping the conductivity of crystalline semiconductors by many orders of magnitude. The second is the possibility of forming p-n junctions by choosing different doping elements.

The reason for this is simply the following. In a crystalline semiconductor a donor or acceptor doping element acts as such because it is forced to take the place of a crystalline host atom and hence has either an excess or a deficiency of a valence electron. In an amorphous semiconductor, on the other hand, the local order is not forced to be the same everywhere; as a result each atom can satisfy its valence requirements and hence does not act as a conventional donor or acceptor as in crystalline semiconductors [1], [2]. This different response to chemical doping explains some characteristics that have important consequences for devices.

1) *Amorphous semiconductors behave similarly to intrinsic semiconductors. Their low conductivity enables one to observe high-field effects without excess heating.*

2) *Contacts to many amorphous semiconductors are nonrectifying and nonblocking essentially because these materials behave like intrinsic semiconductors. The majority carrier current equilibrates via generation and recombination with the relatively large minority current in the barrier region.*

3) A very important distinction concerns the chemical bonding. The most common crystalline semiconductors are the Group IV elements Si and Ge, and the III–V compounds GaAs, GaP, InSb, etc. These are tetrahedral semiconductors whose valence band is formed by the covalent bonding states and the conduction band by the antibonding state. Most amorphous semiconductors considered here have in contrast a Group VI element S, Se, or Te as a major constituent. These are normally in two-fold coordination. Consequently, the highest occupied band is formed by the two nonbonding p-electrons of these chalcogen elements

Reprinted by permission from *IEEE Transactions on Electron Devices*, Vol. ED-20, No. 2, pp. 91–105 (February 1973).

[3]. This makes for a very major difference. *In the chalcogenide semiconductors the unshared p-electrons of the Group VI elements form the valence band and the antibonding states the conduction band.*

4) A disordered semiconductor can exist in several structural states [4]. This structural factor constitutes a new variable with which the physical properties of the semiconductor can be controlled. Furthermore, structural disorder opens up the possibility to prepare in a metastable state new compositions and mixtures that far exceed the limits of thermodynamic equilibrium. Hence we note the following as a further distinguishing feature. *In many disordered semiconductors it is possible to control the short-range order parameter and thereby achieve drastic changes in the physical properties of these materials,* including forcing new coordination numbers for elements. This means that both the nonchalcogens, such as germanium and silicon, and the chalcogens, including tellurium, can have different coordination numbers than in the crystalline materials. This can lead to the possibility of coordinate or dative bonding in the spectrum of bonds available in the amorphous matrix that in a single sample can, in fact, contain all the various bond types from covalent to metallic. In the same sample an element can, in fact, have several coordinate numbers.

5) *The polymeric basis of amorphous materials permits morphological changes relating to the temporary plasticity of the material under various forms of energy.* These rheological effects can be semipermanently (reversibly) or permanently preserved by a quenching. Various disequilibria, especially in long chain polymers, result in conformational as well as configurational changes. All of these alterations to the material can be detected as information.

III. Materials and Applications

A. Structure Stable and Structure Reversible Materials

The number of possible amorphous semiconductors is immense because the structural disorder allows the existence in a metastable state of nonstoichiometric compositions and mixtures that have no crystalline equivalent.

Table I presents a classification and some examples of amorphous semiconductors. Using the structure parameter as a new variable one can distinguish two extreme cases and divide the materials useful for devices into two categories: type A, those whose structure does not change during the device operation; and type B, those whose structure can be changed in a controlled and reproducible manner.

Stable vitreous semiconductors of type A are found in Table I among 1c) the three-dimensionally cross-linked chalcogenide alloy glasses; 2) the semiconducting oxide glasses; and 3) the dielectric films. The latter are used as dielectrics in metal film capacitors and in the

TABLE I

Classification and Examples of Non-crystalline Semiconductors

1) Covalent noncrystalline solids
   a) Tetrahedral noncrystalline solids
      Si, Ge, SiC, InSb, GaAs, GaSb, . . .

   b) Two-fold coordinated and two-dimensionally bonded noncrystalline solids
      Se, S, Te, As_2Se_3, As_2S_3, . . .

   c) Cross-linked network noncrystalline solids
      Ge–Sb–Se, Ge–As–Se,
      Si–Ge–As–Te, As–Se–Te,
      As_2Se_3–As_2Te_3, Tl_2Se–As_2Te_3, . . .

2) Semiconducting oxide glasses
   V_2O_5–P_2O_5 MnO–Al_2O_3–SiO_2
   V_2O_5–P_2O_5–BaO CoO–Al_2O_3–SiO_2
   V_2O_5–GeO_2–BaO FeO–Al_2O_3–SiO_2
   V_2O_5–PbO–Fe_2O_3 TiO_2–B_2O_3–BaO
   . . .   . . .

3) Dielectric films
   SiO_x, Al_2O_3, ZrO_2, Ta_2O_3, Si_3N_4, BN, . . .

MOS technology. Ovshinsky [5] observed switching and memory effects in valve metal oxide films in 1958. These and related effects have been reviewed recently by Chopra [6] and Dearnaley *et al.* [7]. It appears that space-charge effects, ion drift, and the large bandgap of these materials limit the reproducibility of the observed effects and often lead to destructive breakdown [8]. These dielectric and anodic oxide films owe their conduction properties to the presence of donors (or acceptors). These result from defects or deviations from stoichiometry that are difficult to control during preparation or to maintain in device operation.

The chalcogenide alloy glasses, in contrast, behave like intrinsic semiconductors rather independent of trace impurities and annealing effects. There appear to be specific compositions in each glass-forming region of ternary or quasiternary systems that are particularly

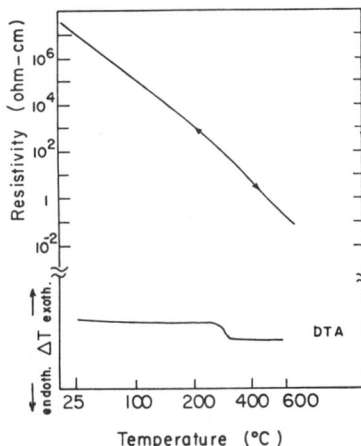

Fig. 1. Temperature dependence of the resistivity and differential thermal analysis (DTA) curve for a stable amorphous semiconductor (type A). The increase in heat capacity at the glass transition temperature $T_g \sim 300°C$ causes the step in the DTA curve. The T scale is linear in $1/T$.

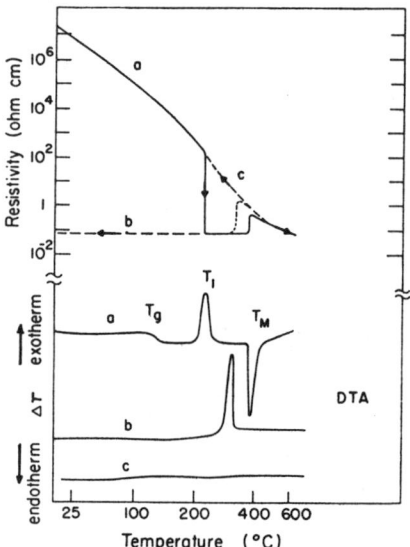

Fig. 2.   Resistivity and DTA curves for a type B material showing a reversible structural transformation. Curve (a): heating rate = 25°C/min. Curve (b): cooling rate less than 25°C/min. Curve (c): cooling rate faster than 25°C/min. The T scale is again linear in $1/T$. The ordinate of the DTA graph represents the T difference ΔT between the sample and an inert reference material as both are heated and cooled at the same rate. Exothermic processes occurring in the sample manifest themselves as an upward peak and endothermic processes as a downward peak in the DTA curve.

stable. These may approach Turnbull's concept [9] of an ideal metastable state by having an optimum number of stable bonds and cross links. They can be heated to the molten state and cooled slowly without devitrification. Fig. 1 shows as an example the resistivity curve of such a stable material and a trace of the differential thermal analysis (DTA). The absence of any sharp peaks in the DTA curve is a good indicator for the absence of structural changes.

Examples of semiconductors of type B whose structura can be changed reversibly between two structural states are compositions of the type $Te_{81}Ge_{15}X_4$, where X represents one or two elements of Group V or VI of the periodic table. This composition is close to $Te_{83}Ge_{17}$, the eutectic point of the Ge–Te binary. This point is particularly favorable for glass formation because of the relatively low eutectic temperature (375°C) and the high viscosity of the melt. The X additives to the pure eutectic composition stabilize the amorphous state without preventing crystallization from occurring as a result of excitation by a light pulse or a current pulse. The reversible structure transformation can be followed in slow motion by thermally cycling the material [10]. Fig. 2 shows the resistivity and DTA curves, respectively, of $Te_{81}Ge_{15}Sb_4$. In both Figs. 1 and 2 the abscissa, although shown in degrees centigrade, is proportional to $1/T$ to display the exponential behavior of the conductivity σ, which follows

$$\sigma = \sigma_0 \exp\left(-\Delta E/kT\right) \qquad (1)$$

with $\Delta E \sim 0.5$ eV and $\sigma_0 \sim 10^3$ $\Omega^{-1} \cdot cm^{-1}$ up to the glass

transition temperature T_g. This is the temperature at which a change in heat capacity is noticeable as a step in the DTA curve.

In Fig. 2 a structure change occurs near T_1. The DTA curve shows an exothermic peak, which is caused by the heat of transformation as the material devitrifies. Structure analysis reveals that the material above this temperature consists of finely divided crystalline Te and glassy GeTe regions. The former, being a degenerate semiconductor, causes the resistivity to drop at this temperature by many orders of magnitude. At the same time, many other physical properties undergo a drastic change, for example, reflectivity, optical transmission, density, wettability, dielectric relaxation, and secondary electron emission efficiency.

Cooling the material from a temperature slightly above T_1 preserves this structurally ordered state. The eutectic melts near $T_M = 375°C$ and the resistivity increases slightly. In the liquid the structure is similar to that of the vitreous state except for the larger thermal motion. Hence the original vitreous and high-resistivity state can be restored by rapid cooling (see DTA curve of Fig. 2). Slow cooling on the other hand leads to the crystalline eutectic (after some supercooling). The crystallization manifests itself by the exothermic solidification peak (see DTA curve b) and the low resistivity at room temperature. The structure transformations, controlled in this example by heating and different rates of cooling, are likely to be initiated or accelerated by light-generated or field-induced excess carriers in some of the mechanisms and devices described below.

Materials exhibiting reversible structure transformation are usually found in binary chalcogenide systems containing additives in low concentration to modify the rate processes. Ge–Te and Se–Te alloys [11]–[13] as well as sulphur compounds [14] were studied for light-induced structure transformations. These structure changes may be initiated by photodecomposition and ead to a separation or precipitation of another vitreous or crystalline phase or they might result in the formation of microbubbles and other elastomeric changes through the intermediary of light initiated or enhanced viscosity changes.

In any case, the structure change, and its reversal, is initiated by a certain amount of energy. This can be transmitted locally by a current pulse, a light pulse, an electron beam or, over a larger area, by exposing the semiconducting film to an optical image of visible or ultraviolet light. Either state is stable after the structural change has occurred.

In all cases, whatever type of structural change has occurred, whether it be crystallization, bubbles, or other kinds of elastomeric changes, a nucleation process can be operant and energy barriers can be lowered by catalytic processes.

The large differences in a number of physical properties of the two states allows the utilization of the re-

TABLE II

APPLICATIONS OF AMORPHOUS SEMICONDUCTORS

Energy Transfer Means	Property Affected	Application
	Without Change of Structure	
voltage	conductance	threshold switch, 2 and 4 terminals, and 3-terminal analog device
	With Change of Structure	
electrical pulse	conductance	memory switch
laser scanning	transmission	optical mass memory
		holography
	reflectance	photography
		microfiche
	chemical reactivity	
electron beam	wettability	printing
optical image exposure, e.g., room light, flash, UV, etc.	conductivity	electroluminescent display
	dielectric relaxation	electrostatic printing
	density	gravure plate
	secondary electron emission	electron beam memory

versible structure change in a variety of applications as illustrated in Table II. Although there are many more applications of stable vitreous semiconductors, only the threshold switch is included in Table II because its two distinct states of conduction that do not require a structure change will be contrasted with those of the memory switch.

In the following we shall try to summarize the present state of device development in the various categories of Table II.

B. Threshold Switch and Memory Switch

Fig. 3(a) and (b) show the current–voltage characteristics and the time response to an applied voltage of the threshold [5], [15]–[18] and the memory switch [11], [12], [19], respectively. The material for the threshold switch was a stable glass of type A of composition $Te_{50}As_{30}Si_{10}Ge_{10}$ and that for the memory switch a structure reversible film of type B similar to that shown in Fig. 2. In both cases a sputtered film of approximately 1-μm thickness is placed between carbon or metal electrodes of about 5×10^{-6} cm^2 cross-sectional area. Other types of nonoxide glasses such as As glasses and amorphous Ge and Si [20] show some kind of switching and memory effects but none with the reliability and stability of chalcogenide glasses.

Common to both types of devices is the switching process that occurs when the voltage drop exceeds a threshold voltage V_t or after a delay time t_D when a voltage pulse $V_p > V_t$ is applied. Once switching has commenced it proceeds very rapidly within a switching time $t_s < 10^{-10}$ s along the load line to the conducting branch of the characteristic. In this conducting state

the voltage drop is about $V_h \sim 1$ V and nearly independent of the current. The principal difference between the two devices is the existence in the threshold switch of a holding current value I_h below which the conductive state cannot be maintained, i.e., when the current falls below this value the threshold switch returns to its original high-resistance "off" state. In contrast, the memory switch retains its conductive state even when the current approaches and reaches zero or is reversed.

The memory switch retains its conductive state only after it has been kept in this state for a sufficient period of time, the lock-on time t_{LO}, to "set" the memory state. If the set pulse V_p is shorter than $t_D + t_{LO}$ then the memory switch reverts back to its high-resistance state as if it were a threshold switch. However, after "lock on" the memory switch is in a permanent conductive state as can be verified by an interrogating read pulse of either polarity as shown in the lower portion of Fig. 3(b). The lock-on period represents the time needed to allow the material in the current channel to devitrify. A brief reset pulse causes the memory switch to return to its

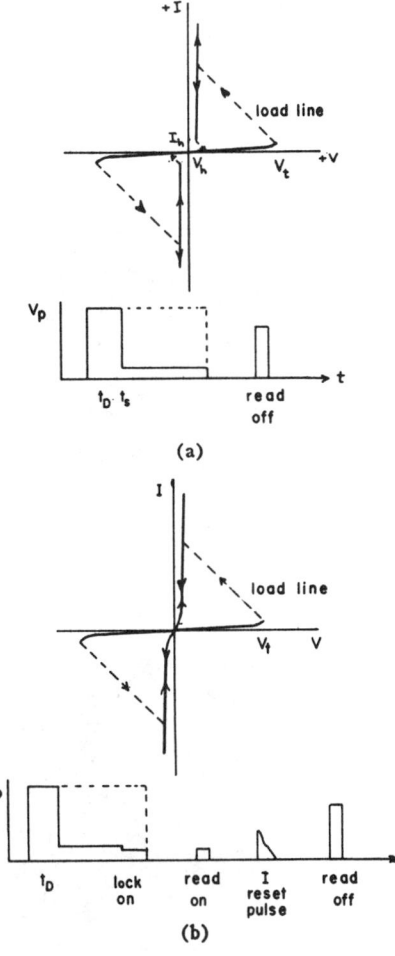

Fig. 3. Sketch of the current–voltage curve and of the time response to a voltage pulse $V_p > V_t$ and to interrogating voltage pulses $V < V_t$. (a) For a threshold switch. (b) For a memory switch.

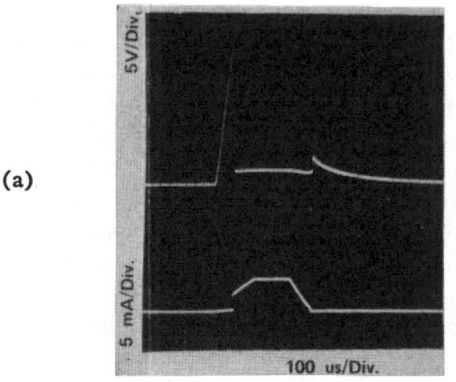

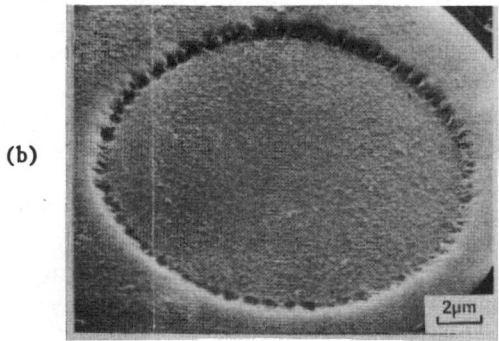

Fig. 4. (a) Time response of a memory switch to a voltage pulse $V_p > V_t$ sufficient to cause switching but of insufficient duration to produce the "lock-on" high-conductance state. (b) Shows an SEM picture of the type B material in the 20-μm-diameter pore of a memory device after 20 switching operations. The top electrode was removed by etching. No structural change is visible.

high-resistive (off) state. This is verified by the high voltage level of the interrogating "read-off" pulse as shown in the lower half of Fig. 3. The reset pulse pours sufficient energy into the crystallized area so as to both break the bonds and allow sufficient diffusion of the atoms forming the crystallized material. The subsequent rapid cooling of the material restores the original noncrystalline state. This is not only a "melt" condition, but one in which the chemical affinities of the cross-linking atoms aid in the establishment of the amorphous state.

Fig. 4 shows that a type B amorphous film of a memory switch remains unperturbed after being switched 20 times into the conducting state but without allowing sufficient time for lock on to occur. Fig. 4(a) shows the trapezoidal voltage and current wave forms experienced by the memory switch. During each operation a current of 4 mA flowed in the "on" state for 120 μs, which is insufficient for the permanent memory to form. Fig. 4(b) shows a scanning electron micrograph (SEM) of the amorphous film after the 20 switching operations. The upper electrode of the sandwich device, which consists of a 2000-Å layer of molybdenum covered by 1.5 μm sputtered film of aluminum, was removed chemically and the amorphous film was etched lightly for 20 s before this picture was taken. The latter etching process is known to enhance any structural differences. No damage, alloying, or structural changes resulting from

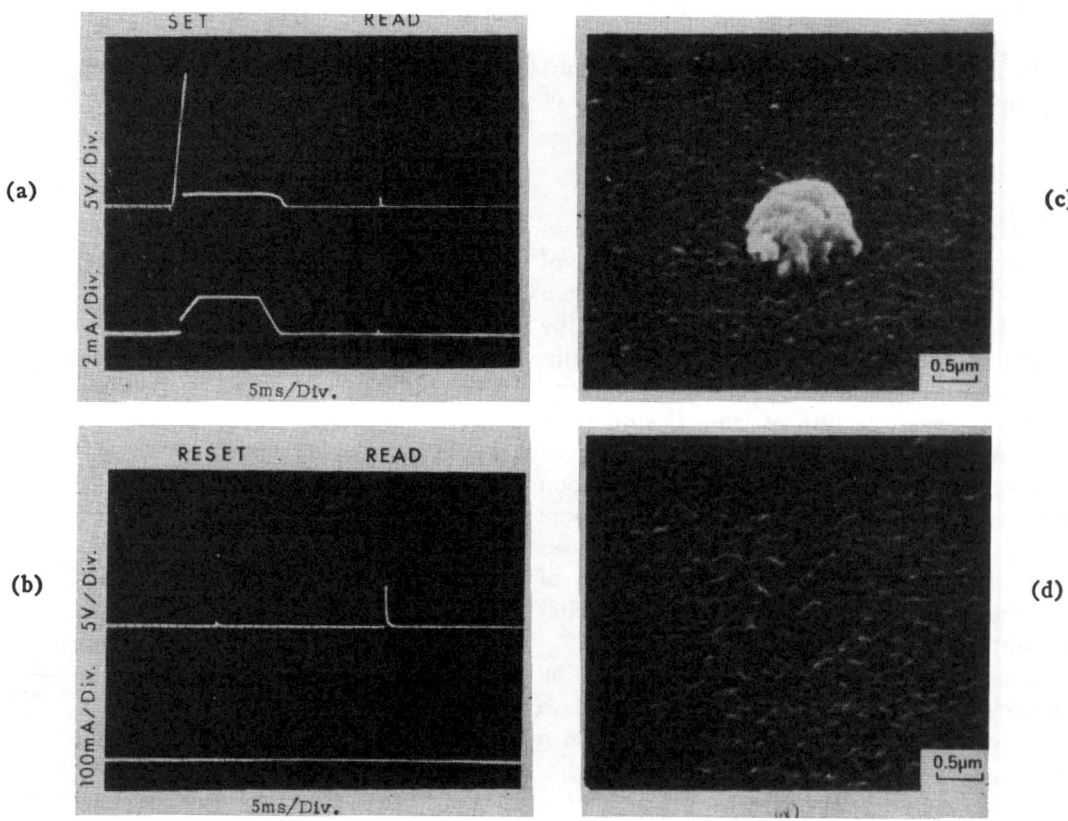

Fig. 5. (a) and (b) show the "set" and "reset" pulses applied to a memory device; (c) and (d) show SEM pictures of the etched amorphous type B material of memory devices left in the "set" and "reset" state, respectively.

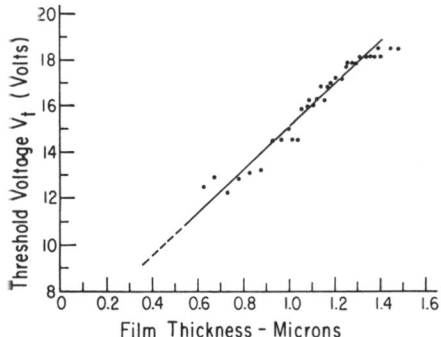

Fig. 6. Threshold voltage V_t as a function of electrode separation for material $Te_{81}Ge_{15}Sb_2S_2$.

the switching operations are observed. Type A stable amorphous films also show, of course, no structure change when the threshold switch is operated within specifications.

In contrast, there occurs a controlled structure change in the memory switch operation after lock on. For a memory switch a material of type B is used. Fig. 5(a) and (b) shows the voltage and current waveforms of a memory device during set (4 mA, 10 ms) and reset (120 mA, 5 μs) operations, respectively. The "lock-on" process, although verified to have taken place by the read pulse, is not discernible as a drop in holding voltage in this case [see Fig. 3(b)]. Fig. 5(c) and (d) shows SEM pictures of the etched amorphous layers of devices left in the "set" and "reset" states, respectively [21]. These pictures demonstrate that controlled crystalline filament growth during the lock-on process yields the high-conductance memory state while revitrification of this polycrystalline region occurs during the reset process [22]–[24]. The morphology of the polycrystalline region is visible in Fig. 5(d) because the etchant attacks the amorphous material more rapidly than the crystalline phase. The etchant was allowed to reduce the thickness of the amorphous layer by about 30 percent. The memory-set areas of devices after 30 set–reset operations did not vary appreciably from that of Fig. 5(c) and (d). The polycrystalline cylinder was found to consist predominantly of dendritic Te crystallites of 300–500 Å diameter pointing circumferentially outward to the cooler zone of surrounding amorphous material [21]. The reset pulse effects the elimination of the crystalline conducting paths and homogenizes the Te rich and Ge rich regions. The heating produced by the reset pulse and its duration might be insufficient to ensure complete mixing. Incomplete homogenization reduces the threshold voltage for subsequent set operations slightly. Moreover, it stabilizes the filament location which is desirable for reproducible device operations [24]. Stabilization of the filament location can be achieved more effectively by proper electrode shaping.

1) Threshold Voltage V_t: For a given amorphous semiconductor material the threshold voltage increases linearly with electrode separation as shown in Fig. 6 for a memory material $Te_{81}Ge_{15}Sb_2S_2$. The finite intercept of this curve at zero thickness indicates that the field at threshold is not uniform. From the slope of the curve in Fig. 6 one obtains a threshold field of about 10^5 V/cm. Average fields between $1–5 \times 10^5$ V/cm are usually found. Near the contacts the field is probably higher. The average threshold field remains independent of film thickness up to about 10 μm and then decreases as shown [16] in Fig. 7. No unique functional dependence can be given for large thicknesses since it is influenced by the thermal boundary conditions set by the shape and contacts. Temperature may be considered as a means of establishing the proper electric field for breakdown in thick films.

The magnitude of V_t decreases with increasing temperature as shown in Fig. 8 for $Te_{40}Ge_7As_{35}Si_{18}$, which is a threshold material, and for $Te_{81}Ge_{15}Sb_2S_2$, which is a memory material. The upper temperature of the memory device is determined by the thermal crystallization rate, which we estimated to be 10^{-8} cm·s^{-1} at 150°C. The upper limit for particular threshold switches depends among other factors on alloying reactions between the electrodes and the amorphous semiconductor. We have reported that switching can occur in the liquid state.[1] Busch *et al.* and Regel [25] also report threshold

[1] S. R. Ovshinsky, U. S. Patent 3 271 591.

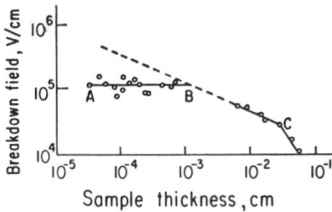

Fig. 7. Average switching field as a function of electrode separation. (After Kolomiets, Lebedev, and Taksami [16].)

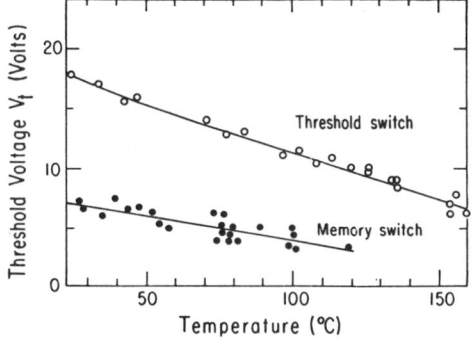

Fig. 8. Temperature dependence of threshold voltage of a threshold and a memory device, respectively. The memory device was reset after each measurement. Type A material: $Te_{40}Ge_7As_{35}Si_{18}$. Type B material: $Te_{81}Ge_{15}Sb_2S_2$.

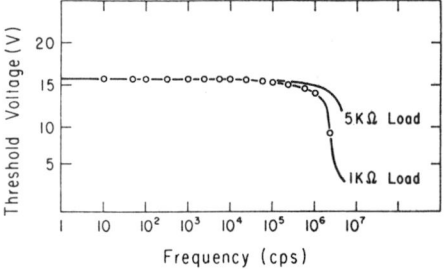

Fig. 9. Frequency dependence of V_t for threshold device having carbon electrodes. Electrode separation about 1 μm. The threshold device can be operated with dc voltages. A sinusoidal voltage was applied. The high-frequency limit depends on the current limiting load resistor. No drop in V_t is observed for first half-wave of applied voltage.

switching in semiconducting alloys of Te with Se and S some hundred degrees above the melting point. Carbon contacts must be used at such high temperatures.

The symmetry of the $I-V$ characteristic, including the magnitude of V_t with respect to the polarity of the applied voltage, has always been a distinct feature of amorphous semiconductor switches. This is found even when the two metal electrodes are of different material. Strong asymmetries in the switching characteristics have recently been observed [26] when n-type or p-type crystalline Ge was chosen as electrode material. These asymmetries cannot be explained by merely assuming that a rectifier is placed in series with the switch. The observations indicate that the injection properties of electrodes are essentially the same for different metals but different for n-type and p-type semiconductors.

Furthermore, Henisch and Vendura [26] showed that it is possible to initiate switching by a light pulse directed at the crystalline semiconductor electrode when the threshold switch is suitably biased.

The frequency dependence of V_t of a threshold switch is shown in Fig. 9. At high frequencies V_t drops because of the finite recovery time t_R discussed in the next section. At low frequencies the rate of change of the applied voltage is very slow so that quasi-equilibrium is maintained and switching is not influenced by rate processes. The low-frequency region and dc is at the present time accessible only for switches that have carbon electrodes and whose amorphous semiconductor films are free of loose ions that would be able to electromigrate.

2) Recovery Time: It is important to note that both devices recover their original switching voltage V_t and their high resistance only after a recovery time t_R, which is of order 10^{-6} s. This recovery time is believed to be in large part the time needed to dissipate the Joule heat accumulated during the conducting or reset period and to sweep out accumulated space charges. Using a rare double-pulse technique Henisch and Pryor [27] investigated the threshold voltage needed by a second pulse to reinstate the conducting state a period τ after the end of the first pulse. Their results are shown in Fig. 10. If τ is very small ($\tau < 10^{-7}$ s) no new switching process

is needed. For longer τ the second switching process occurs with a lower than normal threshold voltage V_t. They found that the recovery curve remains essentially unchanged as the current of the first pulse is increased by a factor of 10. This indicates that processes other than cooling influence the equilibration time.

The sum of delay time, recovery time, and, in the case of the memory switch, also the lock-on time determines the time response of these devices when the conducting state has been reached. However, bias-dependent relaxation-type oscillations having frequencies between 10^7 and 10^9 Hz have been produced with threshold switches using high-current-limiting resistors. In this case the device does not stay in or does not reach the fully conductive state so that Joule heating is limited to the discharge of the capacitively stored energy.

3) Delay Time t_D: As shown in Fig. 3 the delay time t_D measures the interval between the application of a voltage pulse V_p and the switching process. When V_p is only slightly larger than V_t then t_D is subject to appreciable statistical fluctuations. The magnitude of t_D as well as the statistical fluctuations decrease rapidly with increasing overvoltage $V_p - V_t$. As soon as the overvoltage exceeds V_t by about 20 percent the delay time is represented [15], [17] by the relation

$$t_D = t_{D0} \exp - [(V_p - V_t)/V_0]. \qquad (2)$$

The prefactor t_{D0} decreases with increasing temperature T and decreasing electrode separation d. For $T = 300$ K and $d = 1$ μm, t_D decreases typically from 10^{-6} s to 10^{-8} s as $(V_p - V_t)/V_t$ is raised from 0.2 to 1.0. Lee, Henisch, and Burgess [28] studied the statistical spread of t_D. They found relatively large statistical fluctuations of t_D for $(V_p - V_t)/V_t < 0.2$ and a rather abrupt transition to a regime at larger overvoltages in which the fluctuations are less than $10^{-3} t_D$. These small fluctuations of t_D enabled one to measure the time dependence of I and V during the switching process (switching time is less than 10^{-10} s) by means of a sampling scope that analyzes the data from a large number of successive switching processes.

The effect of reversing the sign of the switching pulse V_p on the delay time was studied by Shanks [17] and Balberg [29]. This was thought to be a sensitive test as to whether or not charge carrier injection plays a major

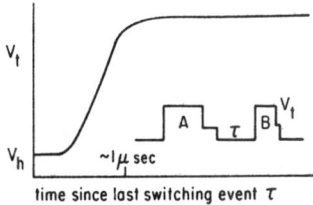

Fig. 10. Typical recovery curve of threshold voltage after switching process. (After Henisch and Pryor [27].)

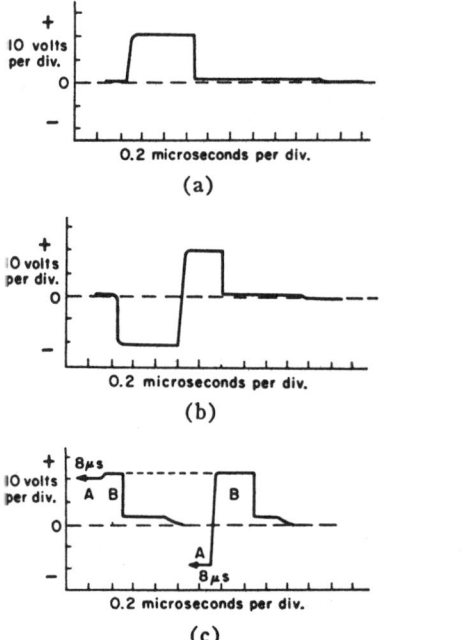

Fig. 11. Effect of preswitching pulse reversal and of a nonswitching A pulse on the delay time of a switching pulse. (a) No preswitching reversal: $t_D \sim 0.6$ μs. (b) Lengthening of delay time due to pulse reversal: $t_D \sim 0.9$ μs. (c) A symmetric shortening of a switching B pulse by a nonswitching A pulse for two relative polarities of A and B. (After Henisch and Pryor [27].)

role in the switching process and in determining t_D. They found t_D to remain the same even though V_p was reversed during the delay period. These experiments were recently repeated by Henisch and Pryor [27]. They extended the measurements to lower T and reduced the internal heating, which occurs while the device is in the on state, by using a low repetition rate. Furthermore, the overvoltage must be sufficiently high not to lie in the statistical fluctuation range of t_D. Fig. 11 shows their results obtained with a threshold switch at −70°C. In Fig. 11(a) one observes $t_D \sim 0.6$ μs. This value is independent of polarity. In Fig. 11(b) the pulse is reversed before it has time to cause switching. The reversal is found to lengthen total delay to about 0.9 μs. The longer the reversal is postponed, the larger is the lengthening of t_D.

Although the total delay is lengthened by the reversal to about 0.9 μs, the delay of the positive pulse is somewhat shortened (to about 0.4 μs) by the presence of the negative pulse. It appears, therefore, that the conditions established by one polarity are not totally opposed to those established by the other polarity. If that were the case the negative pulse condition would have to be undone before the positive pulse condition could cause switching; this would have resulted in a lengthening of the positive pulse delay in Fig. 11(b) beyond the original 0.6 μs observed in the absence of the negative pulse. In Fig. 11(c) one observed that a nonswitching pulse A shortens the delay of a switching pulse B more when A

and B have the same polarity than when they have opposite polarity. A shortening, however, occurs in both cases, which indicates again that only part of the effect of the A pulse on the delay time is polarity dependent. These polarity effects are enhanced at low temperatures, probably because the observation of space-charge effects requires that the dielectric relaxation time is sufficiently long compared to the time scale of the experiments.

4) Voltage Drop of Conducting State V_h: As shown in Fig. 3, the conducting branch of the I-V characteristic of both the threshold and the memory switch is nearly vertical. The current can be changed at virtually constant voltage V_h except at very small currents. In the case of some devices, the strip resistance of the thin film metal electrodes has to be subtracted to obtain V_h. The value of V_h for 1-μm-thick chalcogenide films is typically $V_h \sim 1$ V with electrodes of Mo, W, Ta, and nichrome and $V_h \sim 1.5$ V for carbon electrodes. A small magnitude of V_h is desirable because it governs for a certain current value the Joule heating generated. This small voltage drop in the conducting state and the large resistance ratio between the "off" and "on" state make these materials particularly suitable for switching and memory applications.

The reason for the near identity of V_h in threshold and memory switches is not yet understood as there is no coherent quantitative description of the conductive state at present. One observes that V_h is practically independent of: 1) temperature; 2) electrode separation for $0.4 < d < 5$ μm; 3) electrode area; and 4) current. We conclude from the second point that the potential drop V_h occurs predominantly near the electrodes. The third point tells us that the current flow is not uniform over the contact area but instead restricted to a region whose diameter is independent of the area. Fig. 5(c) shows that the conducting region had in that particular memory switch a diameter of about 2 μm.

It is well known [30] that filament formation is a common feature of current controlled switching as well as of negative resistance devices quite independent of the physical mechanism involved. The conditions which lead to a value of V_h, which is independent of current, have not yet been elucidated.

5) Stability, Life, and Reproducibility: The threshold voltage V_t can be determined with a voltage that increases linearly with time. For rise times Δt very long compared to t_{D0} of (2), V_t is essentially independent of Δt. The statistical fluctuation of V_t of a typical threshold switch is about a few percent of V_t. In a memory device a new switching operation can commence only after the off state is reestablished with a reset pulse. Fig. 12 shows a histogram of V_t values of a memory switch obtained from 900 consecutive set and reset cycles after the switch had been cycled 2700 times [24]. One observes fluctuations in V_t of roughly ±5 percent.

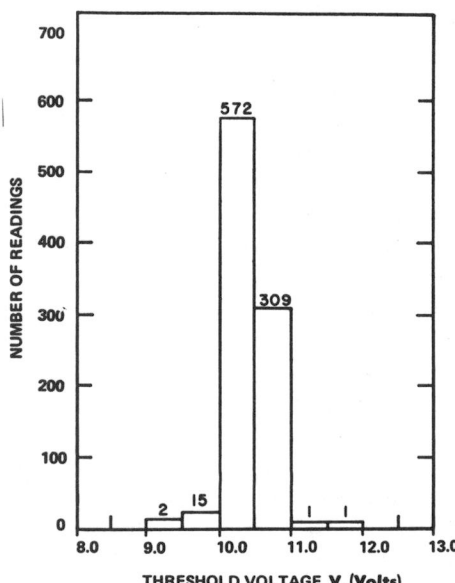

Fig. 12. Histogram of threshold voltage for a single memory device obtained from 900 consecutive set and reset cycles after cycling the device 2700 times. (After Cohen, Neale, and Paskin [24].)

TABLE III

HIGH TEMPERATURE STORAGE TEST OF MEMORY DEVICES

Number of 256-Bit Arrays Tested	Storage Temperature T_s(°C)	V_t/V_t (initial) After 1000 h Storage at T_s
3	70	1.00
1	80	1.12
1	80	1.12
1	80	0.91
1	90	0.97
1	90	0.95
1	90	1.03

The reproducibility of the characteristic properties of individual devices in a 256-bit array [31] or of different arrays depends predominantly on the clean-room standards in the production area and on the uniformity of the etching and film deposition processes. Although not altogether trivial, these problems are well known and solutions are available in the growing field of thin-film technology.

Threshold switches have been kept continuously switching for longer than five years using 60 Hz ac voltage and thin-film devices have been switched over 10^{13} cycles using unipolar pulses. A more severe testing procedure consists of subjecting the threshold switch 1) to variable load conditions and 2) to voltages of a single polarity. In case 1) the different load changes the internal current density and temperature distribution in the conducting state. After each change it is probable that some relaxation takes place on account of the viscoelastic properties of the amorphous materials and the internal strains produced by the nonuniform current flow. Therefore, devices must be designed with these factors in mind. Test 2) is used to differentiate between materials that are stable under conditions of high field and current densities and others that exhibit strong electromigration effects. Although a vitreous alloy might be stable when it is cycled to higher temperatures it still might suffer composition changes in strong thermal gradients and in high fields. Since some of these conditions cannot be easily reproduced in an experimental chamber, it was found convenient to test materials in a device configuration. For successful operation of threshold switching devices under dc and unipolar ac conditions the following precautions were found to be necessary or advantageous. 1) Reduction of gases trapped during the sputtering process in the amorphous layer. 2) Selection of a nearly ideally cross-linked network glass. This reduces the number of diffusing species and their diffusion constant. 3) The use of carbon electrodes. The large electronegativity difference between chalcogen and metal atoms sometimes leads to a slow alloying reaction that appears to become enhanced in the field at the anode. If these basic factors are ignored, some sort of switching and memory process can be observed even with alloying contacts but reliability and life of such devices are poor [32].

Individual memory switches have been subjected to more than 10^8 set–reset pulse trains as those shown in Fig. 3(b) without failure. However, subjecting a statistically large number of 256-bit memory arrays consisting of a 16×16 matrix of thin-film devices [31] to testing, one finds the first failure after about $1-2 \times 10^3$ set–reset cycles and an average failure free life of 2×10^4 set–reset cycles. Each failure case investigated so far by scanning beam electromicroscopy, after removing the top electrode by etching, indicated that the failure was related to solvable processing problems and not to the fundamental mechanism. The memory devices were found to remain in the set or in the reset condition for longer than one year (end of test) while being subjected to 10^{12} read cycles without change in condition.

Storage tests have been used to evaluate the long term stability of memory arrays at temperatures between -55°C and $+90$°C. Some test results of the room-temperature threshold voltage of RM-256 arrays [31] stored at elevated temperatures are shown in Table III. When lock on occurs, the bit must be reset, which essentially returns the memory device to its unstored condition. To allow the continuous accumulation of storage test hours for the threshold voltage, the following test procedure was adopted. V_t measurements were made at 25°C on two lines or 32 bits of an array after which the array is returned to its temperature chamber. This procedure is repeated at intervals of 10, 100, and 1000 h, each time on two new lines of the array. The last column of Table III shows that after 1000 h storage at T_s no significant changes occurred in V_t. Here V_t (initial) stands for the initial V_t. The ratios represent averages of measurements on at least 32 memory devices.

6) Radiation Resistance of Switching and Memory Devices: Several circuits containing standard threshold switches were chosen to test the feasibility of operating amorphous semiconductor devices in a radiation environment [33], [34]. Among these circuits were a relaxation oscillator [33], an astable multivibrator, an AND/OR gate with complementary outputs, and a *J-K* flip-flop [34]. All circuits exhibited a transient ionizing radiation tolerance of at least 10^{11} rad/s (referred to silicon) and a neutron radiation tolerance of at least 10^{16} n/cm² (1 MeV equivalent) without circuit malfunction. Although some malfunction did occur between 1×10^{11} and 7×10^{11} rad/s due to gamma induced noise, no permanent damage to any of the devices occurred, and in all cases normal operation was restored after the transient. Delamination of the glass-epoxy printed-circuit boards occurred at neutron fluences above 10^{16} n/cm² (1 MeV equivalent). A number of circuits survived a total fluence of 4.2×10^{16} n/cm² without any significant performance degradation.

Memory devices were tested [35] in the set and reset state. As a result of neutron exposure of 10^{16} n/cm² typical changes of the low "set" resistance were less than 50 percent and of the high "reset" resistance about 25 percent. These changes are insignificant in view of the fact that the ratio of reset to set resistance is larger than 10^3. Ionizing radiation in excess of 10^6 rad at rates of about 10^{11} rad/s did not cause any significant effects.

We conclude from these studies that certainly threshold and memory switches and perhaps other amorphous semiconductor devices [36] offer distinct potential for radiation hardened circuits and applications.

7) Four-Terminal Threshold Switch: The temperature dependence of V_t shown in Fig. 8 enables one to initiate switching by an external heat source when the amorphous semiconductor switch is properly biased. Assume for instance that a threshold switch, having at room temperature a threshold voltage $V_t = 100$ V, is in a load circuit with an ac or dc voltage insufficient for switching, say 90 V. Reducing the threshold voltage by heating to less than 90 V turns on the device.

Even when the external heat source is removed, the threshold switch can stay on and continue switching if the threshold voltage remains reduced. This occurs either when the applied frequency $\omega > t_R^{-1}$ is larger than the inverse recovery time ($\omega \geq 10^7$ Hz), see Fig. 10, or when the Joule heat generated raises the device temperature. In these cases the device switches off as soon as the current in the load circuit is interrupted for $t_R \sim 10^{-6}$ s or longer.

The small heat capacity of thin-film threshold switches enables one to apply the external heat by short pulse. A four-terminal switch is constructed by placing a thin carbon film (thickness about 0.15 μm) between the substrate and a coplanar threshold switch. The width of the carbon film is strongly reduced just below

the location where switching occurs between the thin carbon film electrodes of the device. The heater film is isolated from the switch by a 1-μm-thick film of silica. The isolation resistance is greater than 10^9 Ω; the important output coupling capacitance is of the order of 0.01 pF. The carbon film is used as a heater to control the V_t of the threshold switch. When the switch is biased with an ac voltage below threshold, the electrical pulse to the heater triggers switching. The present design permits one to switch the device with 10-μs pulses of digital circuit voltage level applied to the control resistor.

8) Three-Terminal Amorphous Semiconductor Device: By placing an amorphous semiconductor layer and a collector electrode on top of a tunnel cathode emitter, one obtains a thin-film tunnel triode. In this new device the current through the amorphous semiconductor layer can be modulated by controlling the injection of electrons from the tunnel cathode. The properties of this triode are described in a separate paper [37]. The injection current can also be used to decrease or increase V_t of the amorphous layer depending on the relative voltage polarity.

C. Reversible Optical Effects in Amorphous Semiconductors

There is hardly any physical property of a semiconductor that is not altered by a change in structure, as for instance a change from an amorphous to a polycrystalline state. In this section those applications are reviewed that are based on the change of the optical properties.

Whereas for memory switches materials are chosen that experience a large conductivity change between the structural disordered and the ordered state as shown in Fig. 2, the materials for most optical applications are selected to have the optical absorption edge at a desired wavelength region and a large change in the position of the absorption edge and in reflectivity [34]. The fact that most properties of amorphous semiconductors change continuously with composition provides a large margin of choice.

Another important difference between electrical memory devices and the optical effects discussed here is the following. In order to obtain the low-conductance state in a memory switch, the structure change has to occur in a region that extends from one electrode to the other. In contrast, optical effects are already achieved when a spot confined to a thin surface region is transformed. Furthermore, the structure change may but does not have to involve crystallization. Feinleib *et al.* [35] observed reversible optical effects caused by scattering centers that were induced by flow in the amorphous phase. These will be discussed in more detail below.

As is illustrated by Table II, structure changes can be produced either by the scanning and pulsed light pencil of a laser or by image exposure of a larger area. These

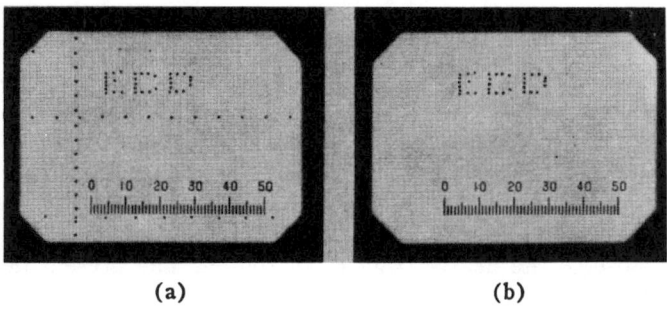

(a)   (b)

Fig. 13. Microscopic view in transmission of a glass substrate coated with an amorphous semiconductor film. (a) The dark dots of about 2-μm diameter are "written" with a laser pulse within 10^{-7} s. The same laser can "erase" dots, as evidenced by (b). (After Ovshinsky [11].)

two methods can be viewed as parallel and series writing of information [40]. The information thus imprinted locally on the amorphous film can in turn be read out by various means using one of the physical properties that have been altered.

1) Optical Mass Memory: One method of producing a local structure change [11], [13], [38] uses short pulses from a mixed gas argon–krypton laser directed against an amorphous film of a Ge–Te or Se–Te alloy, for instance, which is supported by a glass plate.

Fig. 13 shows an example of spots that were produced by laser pulses, 10^{-6} s long and each carrying an energy of 10^{-8} J. The spot size is about 2 μm in diameter. The structure change is confined to a small fraction of the film thickness where most of the light is absorbed. By measuring the light transmission of the spot during the pulse length, it was found that the structure change was actually completed in less than 10^{-7} s.

The spots shown in Fig. 13 were photographed on a microscope stage in transmission using an infrared image converter for the Ge–Te alloy (absorption edge ∼1 eV) and red light for the Te–Se alloy (absorption edge ∼2 eV).

In the amorphous–crystalline transformation the optical transmission is altered by several orders of magnitude for wavelengths close to the absorption edge and the reflectivity changes in Te rich alloys by roughly a factor of 2 [39]. One of the reasons why large changes in the optical properties of semiconductors occur is that the lack of long-range order in the amorphous state lifts the restriction valid in crystals, that only those optical transitions can take place that conserve crystal momentum. Furthermore, the polycrystalline structure of transformed material in the spot scatters an appreciable amount of light out of the beam.

In the case of Se and Se alloy films scattering appears to be the dominant effect. The scattering centers can be small voids formed in a pocket within the interior of the film [38]. It is believed that the temperature reached a maximum in this small region. As the laser pulse strikes the semiconductor film from the substrate side,

most light is absorbed in the lower half of the film thickness because the absorption coefficient for the laser wavelength is large. The effective cooling of the substrate causes the maximum temperature rise to occur somewhat inside the film rather than at the interface. At the place of maximum temperature a void pocket is produced and frozen in as the viscosity suddenly drops and the vapor pressure increases for about 10^{-6} s, the duration of the pulse.

The spots can be removed as shown in Fig. 13 by less intense laser radiation. This is plausible in the last case of void pockets because the voids collapse due to surface tension when the viscosity is reduced at a temperature somewhat above the glass transition temperature where the vapor pressure is negligible. The revitrification of the microcrystalline spots of the Ge–Te alloys probably involves brief local heating above the melting point. For this and other reasons it is advantageous to use the protected interface between the transparent substrate and the amorphous film.

It is important to find out whether the laser beam only acts as a heat pencil or whether the laser light excites a sufficient density of electron–hole pairs in excess of the thermal equilibrium concentration to significantly increase the rate of the structure transformation. In producing by means of a laser beam the structure transformation in VO_2, which is associated with the metal–nonmetal transition at 68°C, Balberg and Roach [41] found that the laser indeed acted only as a heat pencil. In several amorphous semiconductors, on the other hand, light-accelerated crystallization [42] has been observed at light intensities that are more than 10^4 times less than those used for obtaining the results shown in Fig. 13. The extent to which nonthermal effects play a role here might differ greatly in various classes of materials [12], [13], [23], [43].

A prediction by Ovshinsky concerning optical alteration of chain–ring ratios and possibly chain lengths was confirmed by Feinleib *et al.* [38].

A practical application of this effect for an optical mass memory might resemble the magnetic binary storage disk with a tiny mirror directing the laser beam in place of a magnetic reading and writing head as shown schematically in Fig. 14. The small size of the dot allows

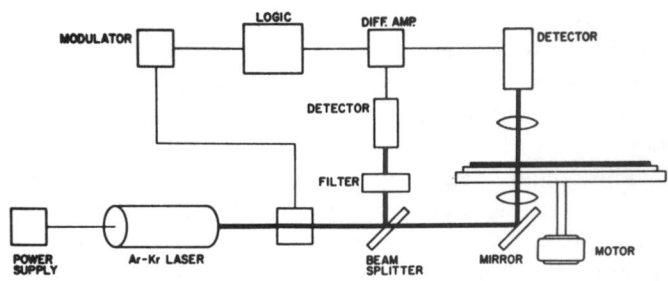

Fig. 14. Experimental arrangement for optical mass memory. The amorphous semiconductor film is a thin coat on the transparent disk rotated by the motor.

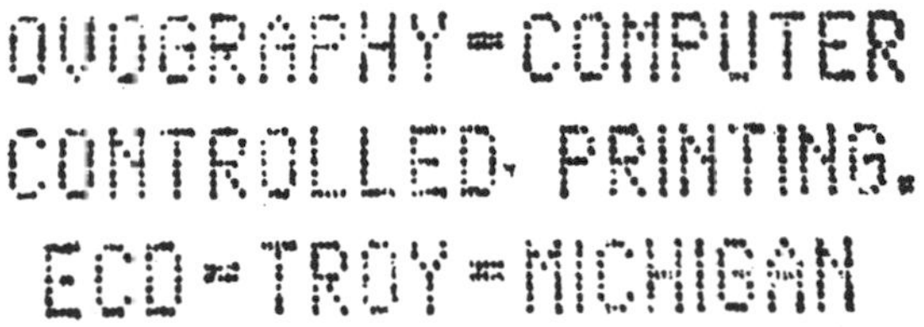

OVOGRAPHY - COMPUTER
CONTROLLED, PRINTING,
ECD - TROY - MICHIGAN

Fig. 15. Electrostatic printout obtained with ECD photostructural film printer. Each of the typewriter-size characters are generated on a 5×7 matrix by computer tape. (After Ovshinsky and Klose [11], [44].)

a twenty times larger bit density for the optically transformed amorphous layer than for the magnetic disk.

Although the repeatability of the write–erase–rewrite cycle is not of foremost importance in archival optical mass memories, at the present state of development reversibility in the thousands has already been achieved. It should be noted, of course, that the read process does not affect the structural memory state at all.

2) Computer-Controlled Multiple Copy Printing: The large difference in electrical resistivity of the vitreous and polycrystalline states makes a new approach to printing and plate making possible [11]. The printing plate may consist of a conducting drum onto which, for instance, a Se–Te alloy is deposited in a fine-grain polycrystalline state. Its resistivity in this state is about 10^6 $\Omega \cdot$cm. The position and triggering of a laser beam can be controlled by the digital logic through a shutter and a mirror. Fig. 15 shows an example of the electrostatic printout obtained with a prototype ECD printer. Each typewriter size character is computer tape generated on a 5×7 matrix. Where the laser pulse hits, the material becomes vitreous with a resistivity of about 10^{12} $\Omega \cdot$cm. An electric charge, deposited by a corona discharge onto the film, will quickly flow to ground at places of low resistivity. The charge will remain much longer on the high-resistivity image written by the laser. From here on, the methods of transferring and fixing the image onto paper can be the same as those used for the photoconductive image in the xerographic process. However, interesting applications are opened up by the fact that the amorphous image on the polycrystalline drum does not disappear after printing. Many copies can be made without the need for new image formation. Erasure of a word can be done with the same laser and erasure and reuse of the whole drum are simply accomplished by heat because the Se–Te film recrystallizes quickly near 120°C.

D. Imaging, Display, and Photography

A large area image can be produced by projecting an image onto the surface or by exposing an amorphous semiconducting layer through an appropriate template [11], [12]. The amorphous layer can be deposited on glass or on flexible substrates such as paper and mylar. It is convenient to distinguish 1) processes that yield an image directly after exposure from 2) those in which a latent image is produced (in the form of photonucleation, for instance), which then at the desired time is developed by heat or uniform radiation [44]. No "fixing" step is required for either of these processes. In some cases, as for instance the production of transparencies, slides, and microfiche copies, it is desirable to further process the picture by etching or adhesive lifting. Some of these will be reviewed briefly in the following.

Electroluminescent Display: When a 0.5–1 μm thick film of a structure reversible amorphous semiconductor is exposed through a contact template to a photoflash the exposed areas contain a large number of fine crystallites. The image is shown in Fig. 16(a) as viewed in reflected light. Fig. 16(b) demonstrates an electroluminescent display of such an image [44]. This engineering prototype of a display device consists of an interdigited electrode pattern covered with a thin layer containing electroluminescent phosphor. An amorphous semiconductor film is deposited on top of this layer. The display remains dark when an appropriate ac voltage is applied to the bottom electrode pattern because the electrode spacing is too large for the voltage chosen to produce the excitation field. As soon as certain areas of the amorphous film are transformed by photoflash into the polycrystalline state, the high conductance of this state capacitively couples the bottom electrodes and raises the field in these areas above threshold.

The device behaves somewhat like an electroluminescence–photoconductor storage panel, except that the image does not decay and is nonvolatile until deliberately erased. The conducting area can of course be "written" by a laser or an electron beam.

The sensitivity of an amorphous photostructural film can be greatly enhanced: 1) by first forming a latent image by photonucleation at room temperature and subsequent dry development for a few seconds by heat;

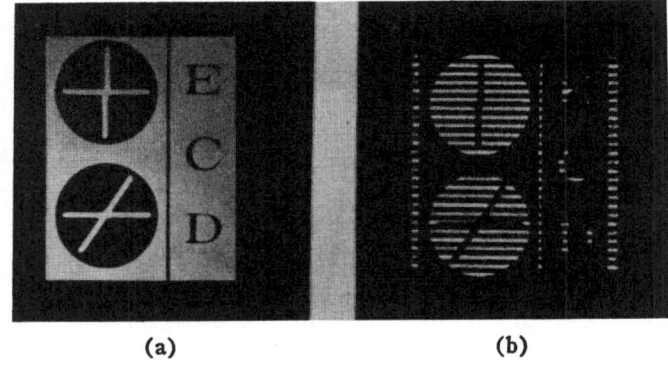

(a)   **(b)**

Fig. 16. (a) Example of a high-contrast image made by exposing a 0.4-μm-thick selenium-base film to a light flash. The image is shown in reflection. Resolution: 100 lines/mm, contrast 20 to 1. (b) Experimental model of electroluminescent display. The high conductivity of the image formed on the Ovonic film causes the lamp to light up in the imaged areas. (After R. Fleming, Energy Conversion Devices, Inc.)

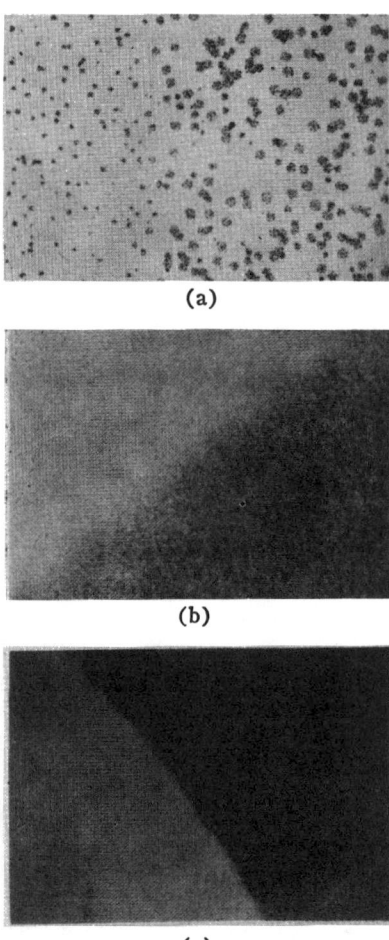

(a)

(b)

(c)

Fig. 17. (a) Example of photocrystallization. The number of crystallites in the exposed and dark areas is the same, their size is enlarged in the exposed area. (b) Example of photonucleation in Ovonic photostructural films. The number of light-nucleated, thermally developed grains in the image area is $5\times10^6/cm^2$ compared to 10^4 grains/cm^2 in the unexposed area (micrograph $50\times$, after Ovshinsky and Klose). (c) Example of high-contrast film showing practically no grains developed in the unexposed area (micrograph $50\times$, after Ovshinsky and Klose).

or 2) by exposing a film held at elevated temperature. The thermal bias temperature in the latter process is chosen so that crystal growth in the unexposed areas proceeds very slowly whereas growth in the exposed areas is accelerated by photocrystallization. Fig. 17 shows examples of low- and high-resolution films and of photonucleation and photocrystallization.

A gray scale is provided by the variation of grain size with exposure. Ovshinsky and Klose [45] also observed that light can inhibit crystalline growth, which then leads to a negative image.

The resolution is determined by the number and size of the crystallites in the exposed area. These depend on film thickness and the film composition. High-resolution films have typically 10^7 cm^{-2} crystallite grains about 1–10 μm in diameter and a resolution of 500 lines/mm [45].

The optical contrast after exposure is about 2:1. The crystallites in the imaged area scatter and reflect the light. The image appears, therefore, as a negative in transmission and as a positive in reflection. The optical contrast can be increased to 1000:1 by a very brief processing step. This step may be mechanical, such as dry stripping, which utilizes the different adhesivity of exposed and unexposed areas, or chemical etching or toning. An example of the latter is shown in Fig. 18. This picture of Sir Nevill Mott was taken with an exposure of 10^{-2} J/cm^2 followed by a brief etch for 0.5 s.

The high contrast and resolution and the dry and quick processing of amorphous photostructural films are great advantages for producing and copying projection material such as slides and microfiche. As an example a small area of a microfiche copy magnified $57\times$ is shown in Fig. 19.

Much greater sensitivity than the typical 10^{-2} J/cm^2 has been accomplished. There is reason to believe that photographic speeds may be achieved.

IV. CONCLUSIONS

The examples discussed here demonstrate the variety of ways by which the physical structure of amorphous

Fig. 18. Picture of a contact print exposed with 10^{-2} J/cm^2 followed by 0.5 s contrast etch.

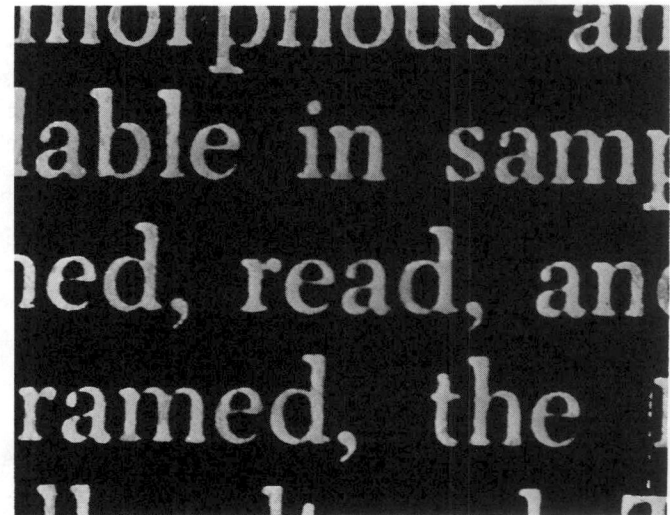

Fig. 19. Area of high-contrast high-resolution microfiche copy consisting of Ovonic photostructural film, Te-alloy, on mylar. $57\times$ magnification. Resolution: 500 lines/mm. Optical contrast 30 to 1 after brief etch.

materials can be altered and the concomitant changes in the electrical, optical, mechanical, and chemical properties that accompany the structural transformation. The results show the usefulness of amorphous semiconductors for information storage, control, and display.

One extends the use of the term "semiconductor" from its classical meaning when one utilizes semiconducting effects to initiate structural changes as we have done in some of our memory-type devices. The interaction of electronic activity with chemical bond changes is a most promising area of research.

Unusual electronic responses to high-field effects as well as the pertinence of excited carriers in amorphous chalcogenide semiconductors without structural change will, we believe, add to our knowledge of new type semiconducting action as well as giving us both digital and analog control.

For example, we can consider as a possibility that the electronic threshold action in chalcogenide amorphous materials is associated with the differing three-dimensional orbitals of lone pair atoms, such as tellurium, provided by the positional and compositional disorder and cross-linking elements making up the amorphous matrix. Such spatial arrangements and bunching up of lone pair orbitals in repulsive proximities to other tellurium atoms as well as the ability of the lone pairs to combine with and be in unusual coordination with alloying atoms could under the influence of an electric field or light have important consequences for both types of switching. Since many lone pairs would be nonbonding and biased upwards in energy by the lone pair electron interactions and affected and trapped by the variable coulombic interactions provided by the surrounding atoms, they could be excited and recombine without structural alteration of the material. The large number of groups of these lone pair electron configurations could thus be a source of the many carriers needed for the drastic change in conduction.

These electronic excitations can also initiate subsequent structural changes in memory-type materials where lone pairs can also provide coordinate bonding, making for weak cross links important for structural change.

There are exciting potentials of the relatively unexplored field of amorphous semiconductors. In view of the complexities of disordered systems it is not surprising that at times the theoretical understanding of these phenomena lags behind the technical developments.

Photonucleation and photocrystallization in these materials present a large new field for scientific research. They have received little attention before this work.

Space does not permit us to review the large amount of work on switching and memory effects in amorphous semiconductors that commenced after ECD's description of these effects and demonstration of the feasibility of commercial applications. The reader is referred to the proceedings of recent conferences [46], and a recent review by Adler on amorphous semiconductors [47]. The high-field switching and memory effects have been discussed recently by Mott [48].

ACKNOWLEDGMENT
The authors gratefully acknowledge the devoted help of I. M. Ovshinsky and of the staff of Energy Conversion Devices, Inc.

REFERENCES
[1] N. F. Mott, *Advan. Phys.*, vol. 16, p. 49, 1967.
[2] M. H. Cohen, H. Fritzsche, and S. R. Ovshinsky, *Phys. Rev. Lett.*, vol. 22, p. 1065, 1969.
[3] M. Kastner, *Phys. Rev. Lett.*, vol. 28, p. 355, 1972.
[4] We intend to include among *disordered* semiconductors not only homogeneous amorphous states, but also heterogeneous and morphologically complicated systems that are separated on a microscopic scale into two or more phases, some of which may be crystalline.
[5] S. R. Ovshinsky, *Electronics*, vol. 32, p. 76, 1959; *Contr. Eng.*, vol. 6, p. 121, 1959; L. Young, *Anodic Oxide Films*. New York: Academic, 1961, pp. 147–149.
[6] K. L. Chopra, in *Proc. Int. Congr. Thin Films* (Cannes, France), 1970, p. 351.
[7] G. Dearnaley, A. M. Stoneham, and D. V. Morgan, *Rep. Progr. Phys.*, vol. 33, p. 1129, 1970.
[8] N. Klein, *Thin Solid Films*, vol. 7, 149, 1971.
[9] D. Turnbull and D. E. Polk, in *Proc. 4th Int. Conf. Amorphous and Liquid Semiconductors*, 1971, see *J. Non-Cryst. Solids*, vol. 8–10, p. 19, 1972.
[10] H. Fritzsche and S. R. Ovshinsky, *J. Non-Cryst. Solids*, vol. 2, p. 148, 1970.
[11] S. R. Ovshinsky, in *Proc. 5th Annu. Nat. Conf. Industrial Research.* Beverly Shores, Ind.: Industrial Research, Inc., 1969, p. 86.
[12] E. J. Evans, J. H. Helbers, and S. R. Ovshinsky, *J. Non-Cryst. Solids*, vol. 2, p. 334, 1970.
[13] J. Feinleib, J. de Neufville, S. C. Moss, and S. R. Ovshinsky, *Appo. Phys. Lett.*, vol. 18, p. 254, 1971.
[14] R. G. Brandes, F. P. Laming, and A. D. Pearson, *Appl. Opt.*, vol. 9, p. 1712, 1970; A. D. Pearson and B. G. Bagley, *Mater. Res. Bull.*, vol. 6, p. 1041, 1971.
[15] S. R. Ovshinsky, *Automation (Cleveland)*, vol. 10, p. 45, 1963; *Contr. Eng.*, vol. 11, p. 69, Apr. 1964; in *Bull. Sci. USSR*, B. Kolomiets, Ed., vol. 9, p. 91, 1967; *Phys. Rev. Lett.*, vol. 21, p. 1450, 1968.
[16] B. Kolomiets, E. A. Lebedev, and I. A. Taksami, *Sov. Phys.—Semicond.*, vol. 3, p. 267, 1969.
[17] R. R. Shanks, *J. Non-Cryst. Solids*, vol. 2, p. 504, 1970; H. K. Henisch, *Sci. Amer.*, vol. 221, p. 30, 1969.
[18] D. R. Haberland, *Solid-State Electron.*, vol. 13, p. 207, 1970; *Frequenz*, vol. 24, p. 185, 1970; D. R. Haberland and H. Steigler in *Proc. 4th Int. Conf. Amorphous and Liquid Semiconductors*, 1971, see *J. Non-Cryst. Solids*, vol. 8–10, p. 408, 1972.
[19] S. R. Ovshinsky, presented at the Int. Conf. Amorphous and Liquid Semiconductors, Bucharest, Rumania, 1967; *Sci. J.*, vol. 5A, p. 73, 1969.
[20] J. E. Fulenwider and G. J. Herskowitz, *Phys. Rev. Lett.*, vol. 25, p. 292, 1970.
[21] C. H. Sie, M. P. Dugan, and S. C. Moss, in *Proc. 4th Int. Conf. Amorphous and Liquid Semiconductors*, 1971, see *J. Non-Cryst. Solids*, vol. 8–10, p. 877, 1972.
[22] H. Fritzsche and S. R. Ovshinsky, *J. Non-Cryst. Solids*, vol. 2, p. 393, 1970; vol. 4, p. 464, 1970; H. Fritzsche, *IBM J. Res. Develop.*, vol. 13, p. 515, 1969.
[23] S. R. Ovshinsky and H. Fritzsche, *Met. Trans.*, vol. 2, p. 641, 1971.
[24] M. H. Cohen, R. G. Neale, and A. Paskin, in *Proc. 4th Int. Conf. Amorphous and Liquid Semiconductors*, 1971, see *J. Non-Cryst. Solids*, vol. 8–10, p. 885, 1972.
[25] G. Busch, J. H. Güntherodt, H. U. Künzi, and A. Schwiger, *Phys. Rev. Lett.*, vol. 33A, p. 64, 1970; A. R. Regel, A. A. Andreev, and M. Mamadaliev, in *Proc. 4th Int. Conf. Amorphous and Liquid Semiconductors*, 1971, see *J. Non-Cryst. Solids*, vol. 8–10, p. 455, 1972.
[26] H. K. Henisch and G. J. Vendura, Jr., *Appl. Phys. Lett.*, vol. 19, p. 363, 1971.

267

[27] H. K. Henisch and R. W. Pryor, *Solid-State Electron.*, vol. 14, p. 765, 1971.

[28] S. H. Lee, H. K. Henisch, and W. D. Burgess, in *Proc. 4th Int. Amorphous and Liquid Semiconductors*, 1971, see *J. Non-Cryst. Solids*, vol. 8–10, p. 422, 1972.

[29] I. Balberg, *Appl. Phys. Lett.*, vol. 16, p. 491, 1970.

[30] B. K. Ridley, *Proc. Phys. Soc. (London)*, vol. 82, p. 954, 1963.

[31] R. G. Neale, D. L. Helson, and G. E. Moore, *Electronics*, vol. 43, p. 56, 1970.

[32] G. V. Bunton, S. C. M. Day, R. M. Quilliam, and P. H. Wisbey, *J. Non-Cryst. Solids*, vol. 6, p. 251, 1971.

[33] S. R. Ovshinsky, E. J. Evans, D. L. Nelson, and H. Fritzsche, "Radiation hardness of Ovonic devices," *IEEE Trans. Nucl. Sci. (Annual Conference on Nuclear and Space Radiation Effects)*, vol. NS-15, pp. 311–321, Dec. 1968.

[34] R. R. Shanks, D. L. Nelson, R. L. Fowler, H. C. Chambers, and D. J. Niehaus, Energy Conversion Devices, Inc., Troy, Mich., Final Tech. Rep. AFAL-TR-70-15, Mar. 1970.

[35] R. A. Smith, R. Sanford, and F. E. Warnock, in *Proc. 4th Int. Conf. Amorphous and Liquid Semiconductors*, 1971, also see *J. Non-Cryst. Solids*, vol. 8–10, p. 862, 1972.

[36] A. N. Bobrova and E. M. Lobenov, in *Radiation Effects in Solids.* Moscow: Akad. Nauk. U.S.S.R., 1963; J. T. Edmond, J. C. Male, and P. F. Chester, *J. Sci. Instrum.*, vol. 1, p. 373, 1968.

[37] R. F. Shaw, H. Fritzsche, M. Silver, P. Smejtek, S. Holmberg, and S. R. Ovshinsky, *Appl. Phys. Lett.*, vol. 20, p. 241, 1972.

[38] J. Feinleib, S. Iwasa, S. C. Moss, J. P. de Neufville, and S. R. Ovshinsky, in *Proc. 4th Int. Conf. Amorphous and Liquid Semiconductors*, 1971, see *J. Non-Cryst. Solids*, vol. 8–10, p. 909, 1972.

[39] J. Feinleib and S. R. Ovshinsky, *J. Non-Cryst. Solids*, vol. 4, p. 564, 1970.

[40] S. R. Ovshinsky, presented at the Gordon Conf. Chemistry and Metallurgy of Semiconductors, Andover, N. H., July 14–18, 1969.

[41] I. Balberg and W. R. Roach, in *Proc. Int. Conduction in Low-Mobility Materials* (Eilat, Israel), N. Klein, D. S. Tannhauser, and M. Pollak, Ed. London: Taylor and Francis, 1971, p. 77.

[42] J. Dresner and G. B. Stringfellow, *J. Phys. Chem. Solids*, vol. 29, p. 303, 1968.

[43] A. Hamada, T. Kurosu, M. Saito, and M. Kikuchi, *Appl. Phys. Lett.*, vol. 20, p. 9, 1972.

[44] G. R. Fleming, Meeting 27 of SAE Subcommittee A-20A, Flight Crew Station, Denver, Colo., Apr. 7, 1971; U. S. Patent application pending.

[45] S. R. Ovshinsky and P. Klose, in *Proc. 4th Int. Conf. Amorphous and Liquid-Semiconductors*, 1971, also see *J. Non-Cryst. Solids*, vol. 8–10, p. 892, 1972.

[46] *J. Non-Cryst. Solids*, vol. 2 and 4, 1970; *Proc. 4th Int. Conf. Amorphous and Liquid Semiconductors*, 1971 (*J. Non-Cryst. Solids*, vol. 8–10, 1972).

[47] D. Adler, *Critical Reviews in Solid State Sciences*, vol. 2. Cleveland, Ohio: Chemical Rubber Publishing Company, 1971, p. 317.

[48] N. P. Mott, *Phil. Mag.*, vol. 24, p. 911, 1971.

AMORPHOUS MATERIALS AS INTERACTIVE SYSTEMS

Stanford R. Ovshinsky

Energy Conversion Devices, Inc., Troy, Michigan 48084

The electronic nature of Ovonic threshold switching (1) has been definitively established.(2-8) Electronic action is also the initiating factor in our memories and the electronic concept has proven useful in establishing information encoding in an imaging manner as well.(9-22) It is now appropriate to elucidate the physical and chemical basis for such action.

Our work has entered a new stage technologically as well as scientifically since the Ovonic memory is now under the sponsorship of a leading computer manufacturer.(23) In the optical storage and imaging field, some of the world's largest manufacturers are involved with these concepts,(24) and there is now renewed interest in the use of the threshold switch and its variations. I believe that the amorphous materials utilized are a prime example of how one can relate structure and function and, therefore, will discuss how and why chemical bonds, molecular configurations and conformational relationships interact with electronic configurations and excitations in these materials. The mechanisms involved result in switching, memory effects, carrier modulation control and new areas of interest that follow from understanding the above.

Amorphous tetrahedral materials have basically two forms, both of which depend greatly upon the methods of material preparation. In the first form, in typical materials such as germanium and silicon, localized states can exist due to dangling bonds and voids. In the second, where dangling bonds and voids are minimized by, for example, compensation, the material is in its most stable amorphous form. One can now ask the basic question what does a particular atom see in 3-dimensional space in these materials? The answer is: an atomic configuration like itself with bonding distortions due to positional disorder reflecting the amorphous nature of the solid. Therefore, as one prepares increasingly better tetrahedral amorphous materials of the elemental type, they will resemble the corresponding crystalline materials except for differences in thermal conductivity.

Already the possibility of doping such materials is shown in the important work of Spear and LeComber.(25) Once localized states arising from structural defects are controlled and minimized, the possibility of ordinary doping exists, provided that the dopant atoms can be placed in tetrahedral environments. Such a substitutional impurity does not fulfill its local bonding requirements and consequently can act as a shallow donor or acceptor, depending on the electronic structure of the atom. There will still be some additional weak bonds that are the result of the stress caused by the addition of disorder. Under certain energy conditions, these distortions, that is, the energy barriers to the crystalline state, will be relaxed and the crystalline state becomes inevitable.

Let us now look at lone pair (26) amorphous semiconductors of which the chalcogenides are the prime examples. There are amorphous semiconductors other than chalcogenide glasses in which there can be lone pair participation, for example, glasses containing large amounts of arsenic have s-electron lone pairs and, while lower in energy than the chalcogenide p-electrons, can be affected by the disordered state, pushed up in energy and contribute to the switching process.(22,27) Stable reversible switching, however, requires that the carriers affected by the field come from primarily nonbonded electronic configurations, that is, lone pairs which can be freed by a relatively small electric field ($\sim 10^5$V/cm). While the chemical and structural criteria for designing unistable and bistable switching material have been given elsewhere,(14,27) because of their importance more details will be given later.

Let us consider how amorphous chalcogenide switching materials are intrinsically compensated, that is, do not have dangling bonds as do the more rigid tetrahedral amorphous materials. The divalent chalcogen atoms allow some degree of structural freedom since both chain and ring configurations are favored. Helical segments make for flexibility and easy bond distortions without bond breakage. Twists not only provide structural flexibility but, especially when stabilized by cross linking, introduce unique 3-dimensional electronic orbital relationships not available in tetrahedral materials. Therefore, considering these materials as polymers permits us to utilize the concept of cross linking (28) which, by the introduction of atoms

Reprinted by permission from the *Proceedings of the Sixth International Conference on Amorphous and Liquid Semiconductors*, Leningrad, USSR, 18–24 November 1975, pp. 426–436.

such as those in Groups IV and V to form chalcogenide glasses, not only provides structural stability but at the same time allows local valence satisfaction (compensation) by minimizing the concentration of dangling bonds. Multidirectional possibilities for bonding are accented since any two of the four outer p-electrons of the chalcogen can participate in the bonding. The other two are spin compensated as they are nonbonded lone pairs. Therefore, the chalcogen atoms serve several functions. They act as flexible links that covalently pick up, that is, compensate potential dangling bonds, e.g., cross links, via their divalency, and position and distribute in energy the remaining nonbonded lone pairs.

We would like to develop further concepts based on our earlier work (15,27,29) which seek to qualitatively and quantitatively distinguish the several types of localized states in amorphous chalcogenides. If one first decouples the structural skeletal matrix (lattice) which is strongly covalently bonded from the nonbonded lone pair electronic configurations which reside in this matrix, and then studies their recoupling through their interactions with each other, one can understand in a consistent manner both threshold and memory switching and thus how to make good switches and memories. This approach has been useful in conceiving new applications and in understanding the nature of the localized states in such materials, their parentage and their significance.

We now ask what a chalcogen atom sees as nearest neighbors. The answer is very different from the previously described situation for tetrahedrally bonded amorphous materials for geometric reasons. The chalcogen atoms have a richness of positional choice of atomic and electronic nearest neighbors. The interaction of the lone pairs with their polymeric environment involves 3 factors: electronic configuration, structural configuration and structural conformation (a change of structural configuration involves bond breakage, of conformation only bending). Relative to the latter, the presence or movement of lone-pair electrons has a profound effect in determining molecular shapes.(30) An electric field by repositioning the repulsive lone pairs would cause structural relaxation changing the angles of the bonds just as vibrational effects in the bonds could have some feedback on the lone pairs. It should be appreciated that the localized state distribution is, therefore, manipulatable under various nonequilibrium conditions. Because lone-pair electrons are nonbonded, they have an extra degree of freedom and rapidly rearrange themselves in a vibrating system in the unexcited state at room temperature.

Conformational localized changes related to excitation lead to the concept of metastability in our materials in the following manner: we consider two different time constants, one relating to electronic excitation and recombination, and the other to structural relaxation. An electronic excitation process that persists beyond the vibrational time constant would result in the formation of a metastable hole since the excitation changes local conformation. An excited lone pair creates a hole, in fact by being a radical two holes, that is, an empty orbital which was the originating point and an empty orbital at its extended excited state. During excitation in unstable materials there is a change of charge distribution which is reflected by a hole raised in energy which reverts back to its original position at the end of excitation. Such a hole can be compensated by further electronic processes. At room temperature, the metastability of the hole would only be seen under nonequilibrium conditions. At low temperatures, thermal fluctuations would be exponentially smaller and the metastability could persist for a very long time.

Is there experimental evidence for this? One should be able to determine single or double occupancy of localized states by studying paramagnetism of amorphous materials. For example, in noncompensated tetrahedral amorphous materials such as germanium and silicon, one would expect voids and dangling bonds to be reflected in singly occupied states. The experiment of Agarwal (31) confirms the lack of singly occupied centers in amorphous chalcogenides and their existence in tetrahedral amorphous materials. Let us now look at the important experiment of Bishop et al.(32) While no paramagnetism is observed in the chalcogenide samples whose electron distribution is at equilibrium, they found an ESR signal at $6-15^{\circ}K$ after excitation with a sub-band gap illumination. They interpret the spins as associated with holes localized on the chalcogen atoms. The nonequilibrium state brought about by illumination produces in addition a broad optical absorption band below gap energies. This nonequilibrium state decays slowly at low temperatures over a period of hours.

We believe that at room temperature thermal fluctuations mask the participation of the transient hole and the effect of stabilizing distortions around the hole, and rapid equilibration of the carriers occurs. Thus, the paramagnetism cannot be easily detected except under ultrafast and strongly nonequilibrium conditions such as in Ovonic switches or under conditions where thermal fluctuations are minimized and the distorted orbitals created by nearest neighbor and lone-pair interactions are frozen in position sufficiently long enough to be detected.

At the temperatures used by Bishop et al., thermal fluctuations are very small and, therefore, the local charge distribution associated with various configurations can be considered. We believe that the experiment can be understood by our proposed metastable localized hole and that the new electronic configuration interacts with and reflects the nearest neighbor environment so that the nonequilibrium holes or electrons or both are stabilized and prevented from recombination by bonding distortions around the carriers. Relaxation processes reestablish the original electronic configurations. Either thermal or infrared optical excitation in the new absorption band can cause equilibration by exciting electrons from valence states into stabilized holes or exciting stabilized electrons into the conduction band. This would explain the photobleaching effect observed by Bishop et al. with low energy photons.

There is a spectrum of localized states tailing into the gap from the valence band which is introduced by the interaction of lone pairs with each other and with their environment. Most of these are in paired configuration and thus diamagnetic; however, there are some singly occupied paramagnetic centers in which a localized hole exists on the chalcogen atom. Excitation of the material by either an electric field or light leads to a greatly augmented density of such holes. All of this is related to switching for the transient appearance of localized states associated with a disequilibration of the carrier distribution should also occur at room temperature under proper excitation conditions. This has been the leit motif of our work on switching in chalcogenide glasses.(33)

Our model extends the various band-mobility gap models (34,35) and makes a major distinction between two different kinds of electronic configurations specific to unstable amorphous chalcogenide materials.(14,27) The first is the arrangement of bonding electrons (ordinarily lying much lower in energy than the nonbonded lone pairs) which are responsible for the structural integrity of the material, even when a very large density of lone-pair electrons is excited. The second is the distribution of lone-pair electrons spread over a large energy range, but not primarily responsible for either the cohesive energy or structural stability of the material. It is among these nonbonded lone-pair configurations that low energy excitation processes play a role. These configurations, due to their varying environment in the disordered solid, have interesting and unique orbital relationships such as:

(i) repulsive interactions of lone pairs with nearby filled orbits, including other lone pairs, which spread the density of valence band states into the gap,

(ii) interaction of lone pairs with orbitals of nearest neighbors forming new bonding possibilities which can include the coordinate or dative bond,(27)

(iii) interaction of lone pairs with each other also forming new bond configurations. Like (ii), this has a net attractive force since lone pair orbitals are pushed so high above the valence band due to nearest neighbor interactions that new bond configurations are energetically favored.(15,27,29) As a variation of this electronic configuration, we suggest that lone pairs that have lost their electrons through nearest neighbor interactions can be compensated through the replenishment of electrons coming from other clusters of lone pairs so that there is a continuous mixing in which the parentage of the electrons may be lost.

A change in electronic configurations is accompanied by atomic conformational rearrangements of the neighboring atoms in polymeric materials. Thus, we see that matrix (lattice) and electronic configurations are interrelated. One, therefore, cannot separate electronic excitations from such rearrangements and a freezing in of metastable configurations at low temperatures becomes plausible.

The density of localized states is so great, $10^{19} eV^{-1} cm^{-3}$; that their excitation can lead to a sufficient concentration of free carriers to provide a metallic-like conducting state in what previously was a high resistance, nonconducting material. One has then, in principle, a way of explaining the nondestructive reversible breakdown and the high resistance as well as the highly conductive state in Ovonic threshold switches.(1) It is clear that, while breakdown is a bulk phenomenon, the redistribution of charge resulting from the availability and stability of a large density of localized holes together with the excitation of a large concentration of free electrons, opens up the further possibility of strong large scale injection processes.(27,36-39) Recombination kinetics are also affected by such excitation and, while rapid at room temperature, are orders of magnitude slower at liquid nitrogen temperature as shown, for example, by the experiments of Fagen and Fritzsche.(40) Of interest is that any change of position of the hole would affect the recombination rate, that is, recombination should be much slower once a metastable state is produced. This could be part of the mechanism which results in the breakback in the I-V curve of the Ovonic threshold switches.

In Ovonic memory-type materials, and there is a spectrum of them (just as there is of threshold materials), which are designed for bistability, the excited nonbonding carriers interact with the more flexible and the more weakly designed structural bonds and the stabilizing distortions caused by electrically or optically initiated excitation around the localized holes become nucleating centers for structural transformation.(15,27) Although the excitation processes in such materials are similar to those in threshold switches, structural configurational rearrangements can occur because this class of materials contains much less cross linking, weaker bonds, including Van der Waals, and more lone pair-lone pair interactions and thus is designed to be much less structurally stable in the amorphous state, especially to electronic excitation.

We now relate all of this to actual devices. Figure 1 represents the latest Ovonic memory technology. This is a 1024 bit, random access, static memory chip containing full decoding for a 256 word by four bit organization. Density of the chip shown is 5×10^4 bits per square inch. In addition, a 2048 bit chip has been designed with further improvements in performance and bit density.

Figure 2 illustrates density improvements by comparing an earlier version of a 256 bit undecoded

Fig. 1   Decoded 256x4 bit Ovonic memory chip. Magnification 10x.

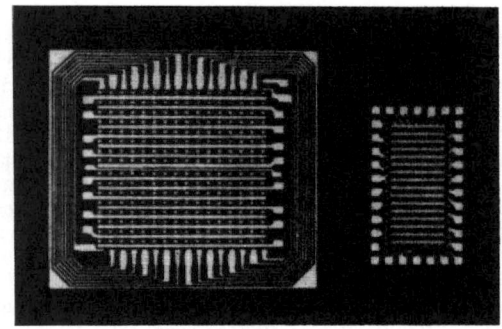

Fig. 2 Size comparison photo showing early and present versions of 256 bit undecoded Ovonic memory chip. Magnification 10x.

memory chip with the current functionally equivalent version. The obvious decreases in memory cell size was obtained by decreasing the voltage-current requirements of the Ovonic memory.

By rigorously following the principles so important in amorphous technology (14) and yet so rarely used, we have developed similar techniques for the production of integrated all-film Ovonic threshold switches which are DC stable devices whose testing was stopped, with the devices unchanged, at 10^{13} cycles.

In Figures 3-5 we illustrate the relation between density of states and the selection of materials for optimizing switching and memory effects in chalcogenides. The relative placement of the lone pair and the bonding bands plays a major role. There are different groupings of lone pairs, some of which are highly localized. Examining the density of states for the Ovonic memory material sketched in

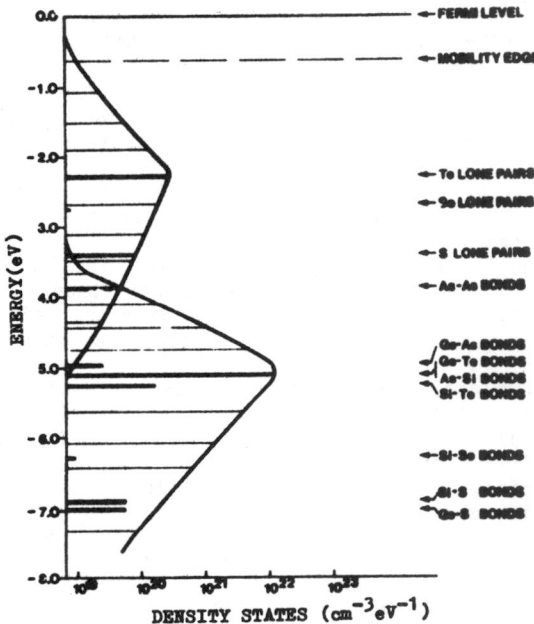

Fig. 4 Calculated density of states showing less overlap and stronger bond spectrum for an Ovonic threshold material.

Figure 3, the tailing of the bands comes from both positional and compositional disorder. It is evident that substantial overlap between the lone pair and the bonding band exists which allows a number of weak bonds as well as most of the lone pairs to be affected by the electric field. Note the proximity of the Te and the S lone-pair bands, the Te-Te bonding band and the Sb cross link which results in a structural instability under excitation. The lone pairs are structural buffers (27) and a large ratio of lone pairs to bonding pairs weakly stabilizes the structure of memory materials. When the lone pairs are excited, they therefore affect the surrounding weak bonds in various ways, one of which is that the overall electronic state connected with the large number of carriers excited during switching results in a shifting of charge density. One could expect charge compensation coming from the deeper electronic states, filling in the empty states and, therefore, aiding in the collapse of local structure in memory materials.

The overlap of bonds is related to the structural reversibility of memory switching. Amorphization is a local order condition in which atoms are apposed by energetic consideration in such a manner that there is bond preference for a nonperiodic structure. In terms of crystallinity of the memory materials, because of their longer chain and ring structures, excitation results in groups of tellurium atoms joining together and becoming nucleating sites. Any thermal process will aid in the crystal building phase. In summary, a large number of lone pairs are excited into antibonding orbitals in a material that is designed not to contain such excitation. All the above produces fast switching as well as bond breakage and an enhanced atomic mobility. Structural transformations therefore take place easily in such materials.

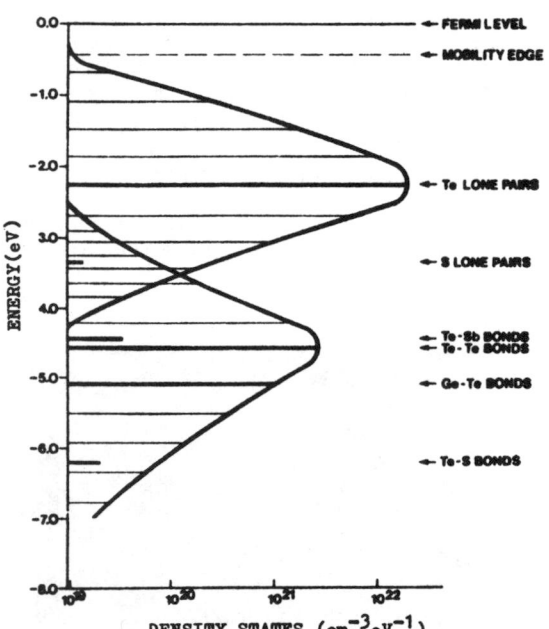

Fig. 3 Calculated density of states showing large band overlap for an Ovonic memory material.

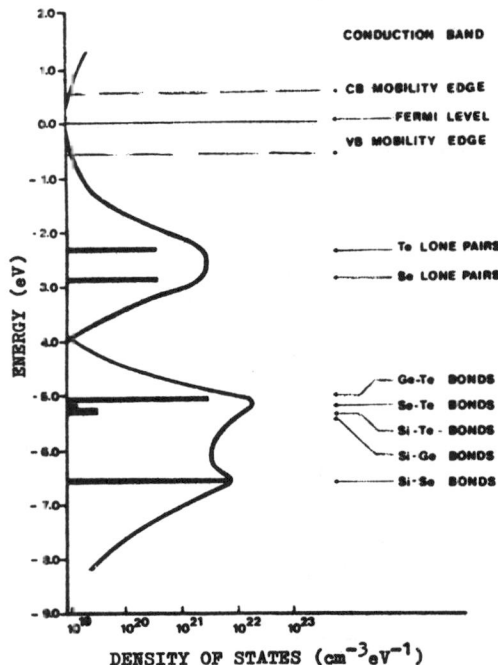

Fig. 5 Density of states of an Ovonic threshold material illustrating substantial isolation between the lone pair band and the bonding band.

In the threshold switch, however, we wish to isolate the most available lone pairs from the bonding band. Figures 4 and 5 illustrate this. In these materials, the two bands are energetically more separated, the ratio of the bonding to lone-pair states is increased, and the structural bonds are well below the Fermi energy. Switching results from the excitation of a spectrum of lone pairs from Te through Se and S, to As shielding the bonding electrons and thus maintaining the structural integrity of the glass. Being more heavily and strongly cross linked, threshold materials are less flexible than memory materials. The conformational changes spoken about earlier are very local and associated with electronic, not structural, changes.

Using these concepts one can develop optimum switching and memory materials since one can calculate and combine the photoemission spectrum of the material, density of states in the gap, electrical activation energy, material connectedness, glass forming ability, and glass transition temperature. However, unless logical fabrication procedures are used, the observation of switching will indicate little since any semiconductor has breakdown effects under sufficiently high field. We reiterate that the essentials for producing good reversible switches are: the use of nonreactive electrodes such as carbon for threshold switches and stress-free molybdenum for memories and the exclusion of catalytic impurities (which, of course, include moisture, unwanted oxygen, etc.) in all steps, from the preparation of the materials to their deposition and electrode attachment. We are amazed that the literature still reports the use of gold and other reactive electrodes, for even

refractory materials such as molybdenum have to be carefully made and still do not have the superior properties of carbon. Particular care must also be paid to interfacial problems. It is obvious that the integrity of the material must be protected at all times.

Other activities that we are presently engaged in, some of which are reported on by our associates here, relate to 3 and 4 terminal devices, heterojunction devices and all-amorphous junctions, all of which make this field a particularly promising one for the future.

Time does not permit a thorough discussion of our work related to optical properties in chalcogenide-type materials. However, there is one point that deserves attention here. The diffusion of atoms and ions through a chalcogenide polymeric material should be no mystery. Changes of conformation and configuration initiated by excitation processes can be viewed as providing pores which can be expanded and contracted, allowing or prohibiting diffusion. Especially important in aiding this process is the change in local order charge distribution by excitation.(15)

Figure 6 is an example of the photographic properties of one of our organo-chalcogen films sensitized to the visible part of the spectrum. The system has a large dynamic range and continuous tone gray scale. The resolution is over 600 line pairs/mm. We have experimentally attained an optical density of 1.0 with an exposure of 10^3 ergs/cm^2. At this sensitivity the amplification factor of the system is about 1000. While much work remains to be done, we see no theoretical barrier to achieving silver halide sensitivities.

The field of amorphous semiconductors is ideal for the encoding of information by excitation with

Fig. 6 Example of ECD dry process film showing continuous tone gray scale.

various forms of energy, especially electric field or light. Chaudhari and his group have not only written in the nanosecond range with our materials,(41) but they have also shown the feasibility of amorphous magnetic devices with superior qualities to those of crystalline materials.(42) There will be a trend toward optical computers, and the optical memories that we have described (9,14) may be a transition between conventional storage of information and computers of the future. We have shown that the field of amorphous semiconductors includes photographic as well as electronic control devices. The ability to have information detected unambiguously by various means will, I believe, make amorphous materials as ubiquitous as crystalline materials in the not too distant future.

I wish to acknowledge with thanks discussions with David Adler, Hellmut Fritzsche and Iris Ovshinsky.

References

1. S.R. Ovshinsky, Phys. Rev. Lett. 21, 1450 (1968).
2. H.K. Henisch and W.R. Smith, Phys. Status Solidi A 17, K81 (1973).
3. M.P. Shaw, S.H. Holmberg and S.A. Kostylev, Phys. Rev. Lett. 31, 542 (1973).
4. S.H. Holmberg and M.P. Shaw, 5th Intl. Conf. on Amorphous and Liquid Semiconductors, Garmisch-Partenkirchen, 687 (1973).
5. K.E. Petersen, D. Adler and M.P. Shaw, Appl. Phys. Lett. 25, 1585 (1974).
6. W.D. Buckley and S.H. Holmberg, Solid State Electron 18, 127-147 (1975).
7. D.M. Kroll, Phys. Rev. B 11, 3814-3821 (1975).
8. K.E. Petersen and D. Adler, Appl. Phys. Lett. 27, 625-627 (1975).
9. S.R. Ovshinsky, Industrial Research Conf., Chicago, September 1969, 86-89.
10. J. Feinleib, J.P. deNeufville, S.C. Moss and S.R. Ovshinsky, Appl. Phys. Lett. 18, No. 6, 254-257 (1971).
11. J. Feinleib, S. Iwasa, S.C. Moss, J.P. deNeufville and S.R. Ovshinsky, J. Non-Cryst. Solids 8-10, 909-916 (1972).
12. S.R. Ovshinsky and P.H. Klose, Proc. of the SID 13, 188-192 (1972).
13. S.R. Ovshinsky and P.H. Klose, J. Non-Cryst. Solids 8-10, 892-898 (1972).
14. S.R. Ovshinsky and H. Fritzsche, IEEE Trans. on Electron Devices ED-20, 91 (1973).
15. S.R. Ovshinsky, Invited Presentation at the Topical Meeting on Optical Storage of Digital Data, Aspen, Colorado (1973).
16. J.P. deNeufville, 5th Intl. Conf. on Amorphous and Liquid Semiconductors, Garmisch-Partenkirchen, 1352-1360 (1973).
17. J.P. deNeufville, R. Seguin, S.C. Moss and S.R. Ovshinsky, 5th Intl. Conf. on Amorphous and Liquid Semiconductors, Garmisch-Partenkirchen, 737-743 (1973).
18. J.P. deNeufville, S.C. Moss and S.R. Ovshinsky, J. Non-Cryst. Solids 13, 191-223 (1973/74).
19. H. Fritzsche, Proc. of the 5th Conf. (1973 International) on Solid State Devices, Tokyo.
20. S.R. Ovshinsky and P.H. Klose, Proc. of Conf. on Nonsilver Photographic Processes, Oxford, September 1973.
21. S.R. Ovshinsky and I.M. Ovshinsky, Invited paper, 1974 Society of Photographic Scientists and Engineers Fall Symposium, Washington.
22. S.R. Ovshinsky, Proc. Intl. Congress on Reprography and Information 1975, 109-114, Hannover, Germany.
23. Burroughs Corporation, Detroit, Michigan.
24. Agfa-Gevaert, N.V., Belgium, Asahi Chemical Industry co., Ltd., Japan, Fuji Photo Film Co., Ltd., Japan, IBM and Xerox.
25. W.E. Spear and P.G. LeComber, Solid State Comm. 17, 1193-1196 (1975).
26. M. Kastner, Phys. Rev. Lett. 28, 355 (1972).
27. S.R. Ovshinsky and K. Sapru, 5th Intl. Conf. on Amorphous and Liquid Semiconductors, Garmisch-Partenkirchen, 447 (1973).
28. S.R. Ovshinsky, J. Non-Cryst. Solids 2, 99 (1970).
29. S.R. Ovshinsky, submitted to Phys. Rev. Lett., 1975.
30. W.J. Orville-Thomas, Structure of Small Molecules, 50-51, Elsevier, Amsterdam, 1966.
31. S.C. Agarwal, Phys. Rev. B 7, 685 (1973).
32. S.G. Bishop, G. Strom and P.C. Taylor, Phys. Rev. Lett. 34, 1346 (1975).
33. For complete references see ref. 14.
34. N.F. Mott, Adv. In Phys. 16, 49 (1967).
35. M.H. Cohen, H. Fritzsche and S.R. Ovshinsky, Phys. Rev. Lett. 22, 1065 (1969).
36. S.R. Ovshinsky, Proc. 1968 Elect. Components Conf., Washington, D.C., 313 (1968).
37. H.K. Henisch, Sci. Am. 221, 30-41 (1969).
38. N.F. Mott, The Phil. Mag. 32, 159 (1975).
39. N.F. Mott, Nature 257, No. 5521, 15-18 (1975).
40. E.A. Fagen and H. Fritzsche, J. Non-Cryst. Solids 2, 480 (1970).
41. R.J. von Gutfeld and P. Chaudhari, J. Appl. Phys. 43, No. 11, 4688-4693 (1972).
42. P. Chaudhari, J.J. Cuomo and R.J. Gambino, IBM J. Res. Develop. 17, No. 1, 66-68 (1973).

AMORPHOUS MATERIALS AS OPTICAL INFORMATION MEDIA

Stanford R. Ovshinsky

Energy Conversion Devices, Inc., 1675 West Maple Road, Troy, Michigan 48084

(Received 18 February 1976)

We view amorphous materials as being ideal matrices for the encoding of information by optical or other means.[1-10] The result can be the storage and reproduction of information serially or in parallel. The laser is ideal for the digital storage of information. One can achieve an exceedingly high density (one micron spots closely spaced) with a high signal to noise ratio, and utilize the same laser for reading, writing, and erasing. For higher densities, scanning electron beams are used.[8]

This paper will concern itself with basic physical and chemical bonding concepts which we have developed and found operant in amorphous materials. Since this field is not yet widely known, we will review the various types of amorphous materials and put them in a systematic arrangement so that the physical and chemical arguments that will be used in elucidating mechanisms can be followed.

Amorphous materials are noncrystalline solids which can be made of inorganic materials, organic materials, or mixtures of the two. Their properties are affected by several factors: their local chemical bonding, that is, their short-range order determines the covalency and strength of the skeletal matrix structure. Also by removing the crystalline symmetry, one opens up many varying 3-dimensional new orbital relationships. Especially, as we shall see with the chalcogenides, this will affect electronic processes, charge distribution, and structural transformations.

Amorphous or glassy materials can be made as metals by exceptionally rapid quenching from a liquid, they can be semiconductors of various kinds, depending upon the type and strength of their bonds, or they can be large band-gap dielectrics. They can be single elements such as silicon or selenium, but often have most interesting properties in their alloy or nonstoichiometric form. While some of these materials can exist either in amorphous or crystalline form, many new compounds without crystalline counterparts are possible. Amorphous materials may also be combined with crystalline materials for interesting effects.

The semiconductors can be divided into the tetrahedrally bonded type such as germanium or silicon, or the lone-pair semiconductors where the valence band is made up of nonbonding lone pairs.[11] We are primarily interested in the latter for switching and imaging use. There are two important types of lone-pair amorphous semiconductor materials,[12] those without structural change and those with structural change. One has to understand the design parameters of the first in order to develop the second. It is the latter which are most useful for photoinitiated structural change, adding a new dimension to imaging possibilities.

The methods of making films of any of these materials range from sputtering and evaporation to hot pressing, casting and coating techniques. The films can be deposited on almost any substrate, including paper, or be free standing. It is the ability of ovonic films to be laid down in a grainless, tough, nongelatin form which can selectively absorb optical energy in certain molecular structures that also makes them interesting photographic media. Structural change type amorphous materials have free-energy considerations which can lead to amplification of the nucleating step, very important for increasing sensitivity. The amplification procedures can be along different energy paths than the original photochemical event, e.g., thermal energy can be used for the development step in a dry manner. The image thus developed is neither a dye nor a bubble, but of metallic appearance and nature, with excellent optical and archival properties.

There is a range of polymeric amorphous chalcogenide materials—there are those that are heavily, 3-dimensionally crosslinked with strong covalent bonds which are resistant to structural change, and where excitation interactions are transient and do not leave any configurational changes in the matrix. There are others in which the chain lengths are longer, crosslinks fewer, and which have a diversity of bond types and strengths allowing alterations in molecular geometry in response to excitation processes. Amorphous materials have more "room" for structural changes than crystalline materials.

The composition of a particular film depends upon several parameters—the choice of elements which provide the bonding and electronic configuration which can absorb and respond to the frequency of light utilized, and the design of the material so that structural change can take place as a result of the excitation. The choice of materials for actual products depends not only upon the absorption characteristics, but also the kind of optical contrast that is required, if indeed it is, since one can read the information not only optically, but electrically, or reproduce it onto other media.

In Figure 1, we illustrate our concept of structural memory wherein a material has bistability (the ability to coexist in at least two phases at room temperature). The excitation processes interact with the matrix causing structural changes in the matrix material which can be preserved. It is this type of material that is most suitable for optical encoding.

The ovonic memory materials, e.g., $Te_{81}Ge_{15}Sb_2S_2$, depend

Reprinted by permission from *Journal of Applied Photographic Engineering*, Vol. 3, No. 1, pp. 35–39 (Winter 1977).

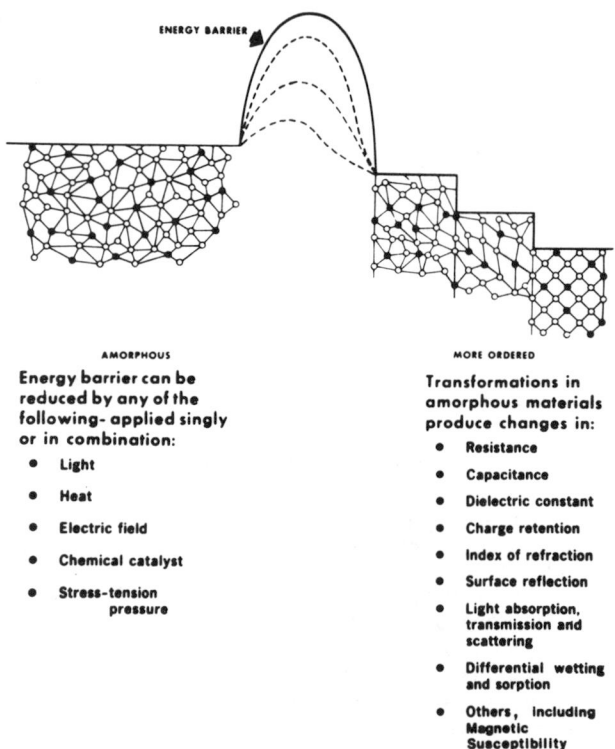

Figure 1. Mechanism for information retrieval and display by structural transformation.

AMORPHOUS

Energy barrier can be reduced by any of the following- applied singly or in combination:

- Light
- Heat
- Electric field
- Chemical catalyst
- Stress-tension pressure

MORE ORDERED

Transformations in amorphous materials produce changes in:

- Resistance
- Capacitance
- Dielectric constant
- Charge retention
- Index of refraction
- Surface reflection
- Light absorption, transmission and scattering
- Differential wetting and sorption
- Others, Including Magnetic Susceptibility

upon such principles. Figure 2[12] shows their typical structure. As opposed to the ovonic threshold (unistable) materials described elsewhere,[4,12] these memory materials show fewer crosslinks, more tellurium, and therefore more lone-pair interactions with each other, with other bonds and with the longer, more flexible chain structures. These materials are weaker and have varied bonds, including various lone-pair bonding possibilities; therefore, there is more molecular flexibility and less 3-dimensional stability. Especially the nonbonded lone pairs have a freedom of movement not permissible in bonded structures. They act as buffers and as they are shifted around, they can affect bond angles, relaxation modes and configurations of molecular structures and, therefore, affect their own charge environment. The environment, if altered, in turn will reposition the lone pairs. Lone pairs can also be bonded and, for example, a coordinate bond can be formed where the lone pairs act as donors to a vacant orbital. This could be one of the weak bonds designed into a matrix which enhances structural change.[5] The electronic excitation process is similar in both threshold and memory materials since it depends only on the amorphous state. However, in memory materials the amorphous state is designed to be much weaker and the energy generated by excitation can result in molecular alterations. Therefore, the electronic transition is structurally reactive. Its high-energy content and the hole's attraction to nearby negative-charge configurations and broken bonds are of great structural importance. The unsatisfied orbital can be utilized by whatever bond scissions and structural relaxation have been generated by the excitation process, including any concomitant thermal excitation which may occur. Since any increase of thermal energy raises the overall energy of the environment, additional nearby bonds can be broken. Increased temperature also changes viscosity, increases vibrational, translational, and

rotational movement and permits atomic diffusion. The reactivity of tellurium can be utilized by these processes to form new localized structures, i.e., nucleating points. Tellurium can join with other tellurium atoms and additional crystal growth can be thermally controlled around the site. With the appropriate laser, electronic and thermal action can be combined in a single-controlled energy pulse. The choise of the chalcogen used in structural change material depends on the type of changes required for detection of the encoded information ranging from subtle structural changes to complete crystallization.

There are many proven; as well as potentially useful photoconductive and photoemissive materials, among those without structural change and care is usually taken that structural change not take place. For example, xerography is based upon the transient photoconductive properties of amorphous selenium. Those that work with selenium know that, while it can remain amorphous up to certain temperatures because of the entanglements of the rings and chains on a substrate, if the useful photoconductive properties are to be preserved, the temperature of the environment must not be allowed to be high enough to cause crystallization. Selenium is a classic example of how crosslinking elements such as arsenic can give 3-dimensional stability so that crystallization is prohibited. To test this concept, all one need do is to add about 5% arsenic to selenium, and then try to crystallize it.[13]

Our concept of information encoding, however, is based upon structural change. This is best illustrated by referring to Fig. 2 again. Here we look upon information in the physical sense as the plastic ability of a matrix material to have impressed upon it by some energy source or sources a localized change which can be differentiated from the surrounding matrix by various means with the possibility of a very high signal to noise ratio. The term plasticity is meaningful since amorphous materials are supercooled liquids with a high viscosity which can be altered. We have, in fact, built a laser-controlled optical memory based upon the rheological qualities of the material.[14]

An important consideration to be learned from the above figure is that one need not have a complete amorphous to

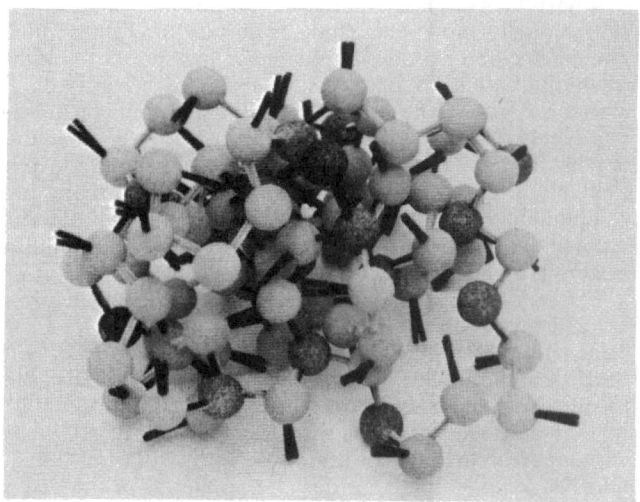

Figure 2. Photograph of a 3-dimensional model of a memory material. The lower part is shown in a relaxed position in order to clearly see how with subsequent chain folding and cross linking lone pair interaction takes place. Light balls represent Te atoms and dark sticks represent their lone pairs. Dark balls are Ge atoms. The 4 darkest balls are Sb and S.

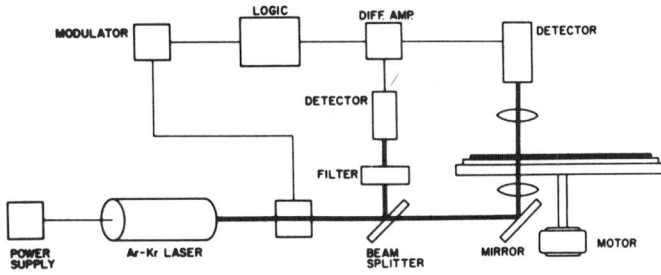

Figure 5. Experimental arrangement for optical mass memory. The amorphous semiconductor film is a thin coat on the transparent disk rotated by the motor.

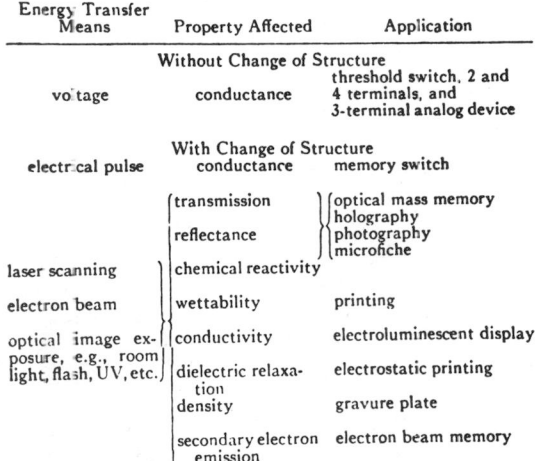

Energy Transfer Means	Property Affected	Application
Without Change of Structure		
voltage	conductance	threshold switch, 2 and 4 terminals, and 3-terminal analog device
With Change of Structure		
electrical pulse	conductance	memory switch
laser scanning	transmission	optical mass memory, holography, photography, microfiche
	reflectance	
	chemical reactivity	
electron beam	wettability	printing
optical image exposure, e.g., room light, flash, UV, etc.	conductivity	electroluminescent display
	dielectric relaxation	electrostatic printing
	density	gravure plate
	secondary electron emission	electron beam memory

Figure 3. Applications of amorphous semiconductors.

crystalline transformation, but only initiate it by subtle changes of bonding which then become points of lower free energy in the matrix for further structural change. It is in a real sense a thermodynamic entropy relationship in that the degree of structural randomness is decreased and the informational content is related to the degree of ordering that takes place.[6] There is a spectrum of subtle changes of bonding which can lead to gross changes such as the amorphous to crystalline transformation. The figure also indicates the various types of energy that can be used singly or in combination to transform a bistable material showing large detectable changes often of many orders of magnitude.

There is hardly any physical property of a semiconductor that is not altered by a change in structure, as for instance a change from an amorphous to a polycrystalline state. We review here those applications that are based on changes of optical properties.

Whereas for memory switches materials are chosen that experience a large conductivity change between the structurally disordered and the ordered state, the materials for most optical applications are selected to have the optical absorption edge at a desired wavelength region and a large change in the position of the absorption edge and in reflectivity. The fact that most properties of amorphous semiconductors change continuously with composition provides a large margin of choice.

As is illustrated by Fig. 3, structural changes can be produced either by the scanning and pulsed light pencil of a laser, or by image exposure of a larger area. These two methods can be viewed as parallel and series writing of information. The

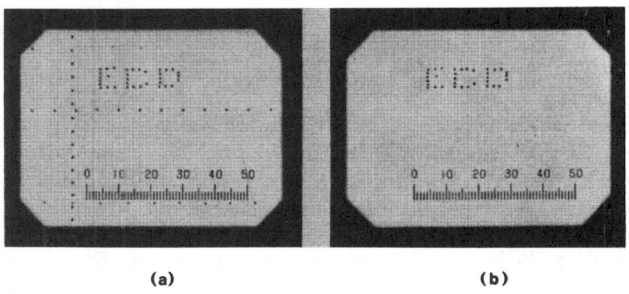

(a) (b)

Figure 4. Microscopic view in transmission of a glass substrate coated with an amorphous semiconductor film. (a) The dark dots of about 2 μm diameter are "written" with a laser pulse within 10^{-7}s. The same laser can "erase" dots, as evidenced by (b).

information thus imprinted locally on the amorphous film can in turn be read out by various means using one of the physical properties that have been altered.

(1) *Optical Mass Memory:* One method of producing a local structural change[1,14,15] uses short pulses from a mixed gas argon-krypton laser directed against an amorphous film of a Ge–Te or Se–Te alloy, for instance, which is supported by a glass plate.

Figure 4 shows an example of spots that were produced by laser pulses, 10^{-6}s long and each carrying an energy of 10^{-8}J. The spot size is about 2 μm in diameter. The structural change is confined to a small fraction of the film thickness where most of the light is absorbed. By measuring the light transmission of the spot during the pulse length, it was found that the structural change was actually completed in less than 10^{-7}s.[16] Chaudhari's group working on ovonic materials has reported amorphizing pulses in the nanosecond range.[17] Smith discusses power densities ranging from .1 to .3 nJ/μ^2.[18] This IBM research has also led to some important work on amorphous magnetic bubbles.[19]

The spots shown in Fig. 4 were photographed on a microscope stage in transmission using an infrared image converter for the Ge–Te alloy (absorption edge ~1 eV) and red light for the Te–Se alloy (absorption edge ~2 eV). In the amorphous-crystalline transformation the optical transmission is altered by several orders of magnitude close to the absorption edge, and the reflectivity changes in Te-rich alloys by roughly a factor of 2.[20] One of the reasons why large changes in the optical properties of semiconductors occur is that the lack of long-range order in the amorphous state lifts the restriction valid in crystals that only those optical transitions can take place that conserve crystal momentum. Furthermore, the polycrystalline structure of transformed material in the spot scatters an appreciable amount of light out of the beam. In the case of Se and Se alloy films, scattering can be made to be the dominant effect. The scattering centers can be small voids formed in a pocket within the interior of the film.[14] Either type of spot can be removed by less intense laser radiation. This is plausible in the case of void pockets because the voids collapse due to surface tension when the viscosity is reduced at a temperature somewhat above the glass transition temperature where the vapor pressure is negligible. The re-vitrification of the microcrystalline spots of the Ge–Te alloys can involve brief local heating. For this and other reasons, it is advantageous to use the protected interface between the transparent substrate and the amorphous film. The laser beam can act as a heat pencil, as well as excite a sufficient density of electron-hole pairs in excess of the thermal equilibrium concentration.

A practical application of this effect for an optical mass memory resembles the magnetic binary storage disk with a

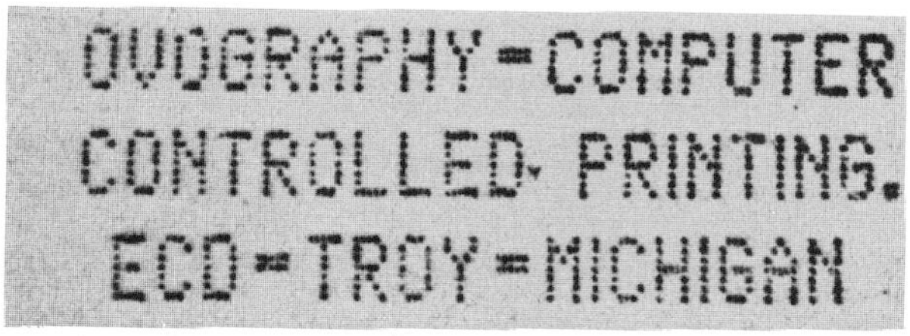

Figure 6. Electrostatic printout obtained with ECD photostructural film printer. Each of the typewriter-size characters are generated on a 5 × 7 matrix by computer tape.

tiny mirror directing the laser beam in place of a magnetic reading and writing head as shown schematically in Fig. 5. The small size of the dot allows a twenty times larger bit density for the optically transformed amorphous layer than for the magnetic disk.

Although the repeatability of the write-erase-rewrite cycle is not of foremost importance in archival optical mass memories, at the present state of development reversibility in the thousands has already been achieved. It should be noted, of course, that the read process does not affect the structural memory state at all.

(2) *Computer-Controlled Multiple Copy Printing:* The large difference in electrical resistivity of the vitreous and polycrystalline states makes a new approach to printing and plate-making possible.[1] The printing plate may consist of a conducting drum onto which, for instance, a Se–Te alloy is deposited in a fine-grain polycrystalline state. Its resistivity in this state is about 10^6 Ω cm. The position and triggering of a laser beam can be controlled by the digital logic through a shutter and a mirror. Figure 6 shows an example of the electrostatic printout obtained with a prototype ECD printer. Each typewriter size character is computer-tape generated on a 5 × 7 matrix. Where the laser pulse hits, the material becomes vitreous with a resistivity of about 10^{12} Ω cm. An electric charge, deposited by a corona discharge onto the film, will quickly flow to ground at places of low resistivity. The charge will remain much longer on the high-resistivity image written by the laser. From here on, the methods of transferring and fixing the image onto paper can be the same as those used for the photoconductive image in the xerographic process. However, interesting applications are opened up by the fact that the amorphous image on the polycrystalline drum does not disappear after printing. Many copies can be made without the need for new image formation. Erasure of a word can be done with the same laser and erasure and reuse of the whole drum are simply accomplished by heat because the Se–Te film recrystallizes quickly near 120°C.

The use of the laser in holography has underlined the need for a reversible medium. We have investigated various forms of holography utilizing our materials and have explained the work of others[5,21–26] who have used materials like ours and obtained similar results. We have been able to show that materials such as As_2S_3, As_2Se_3, and $Ge_{10}As_{40}Se_{50}$, which experience a refractive index change, have two different effects: one, the result of an irreversible polymerization of large monomeric species in which the structure changes from dense random packing into a continuous network; and the second, a reversible refractive index effect which does not involve gross structural changes, but rather a rearrangement of local coulombic relationships involving weaker bonds of ring- and particularly chain-segments based upon lone-pair configurations.

Photonucleation and thermal-crystallization around the

Figure 7. Example of ECD dry process film, showing continuous tone gray scale.

nucleating point provide an opportunity to extend the sensitivity of films which usually are in the 10^4 to 10^6 ergs/cm^2 range. Since reversibility is not a necessary feature for many photographic uses, we have chosen for its contrast and durability a nonreversible organochalcogen which can be incorporated in various amorphous matrices.[27,28] The material can be sensitized to the visible part of the spectrum. The system has a large dynamic range and continuous tone gray scale. The resolution is over 600 line pairs/mm. We have experimentally obtained an optical density of 1.0 with an exposure of 10^3 ergs/cm^2. At this sensitivity the amplification factor of the system is about 1000. While much work remains to be done, we see no theoretical barrier to achieving silver halide sensitivities. One can see the photographic properties of such a film in Fig. 7.

We design our films according to their commercial applications. For example, we have utilized our knowledge of materials and thin film technology to achieve image fidelity, edge acutance, contrast and process automation not presently attainable in graphic arts, and sensitivities similar to silver halide film. We have also designed microfilm systems that allow add-on, deletion, and annotation. Since we can write with electrical styli as well as light, we have shown feasibility for facsimile and convenience copying using our special paper. These are all active areas of commercialization for us.

We are dedicated to thin film imaging and have utilized our knowledge of materials and technology to understand how energy can interact with films of all sorts.[29] This broad approach has permitted us to investigate energy- and especially photo-initiated changes of morphology, rheology, and local order bonding in many materials, of which we do not demand either a completely ordered or disordered structure. While we are known primarily for our chalcogenide work, we utilize whatever films have desirable qualities for imaging purposes. Our understanding of the types of structural changes and of local order bonding relationships that can lead to long-range order has been a most useful tool for us to expand our area of science and technology. The laser has been one of our most productive tools for both scientific investigation and new product development.

References and Footnotes

1. S. R. Ovshinsky, "The Ovshinsky Switch," presented at the 5th Annual National Conference on Industrial Research, Chicago, September 19, 1969; *Proceedings* pp. 86–90.
2. S. R. Ovshinsky, "An Introduction to Ovonic Research," *J. Non-Cryst. Solids* 2: 99–106 (1970).
3. S. R. Ovshinsky and I. M. Ovshinsky, "Analog Models for Information Storage and Transmission in Physiological Systems," *Mat. Res. Bull.* 5: 681–690 (1970).
4. S. R. Ovshinsky and H. Fritzsche, "Amorphous Semiconductors for Switching, Memory, and Imaging Applications," *IEEE Trans. on Electron Devices* ED-20: 91–105 (1973).
5. S. R. Ovshinsky, "Optical Information Encoding in Amorphous Semiconductors," presented at the Topical Meeting on Optical Storage of Digital Data, Aspen, Colorado, March 1973.
6. S. R. Ovshinsky and I. M. Ovshinsky, "Optical Information Encloding in Amorphous Semiconductors," presented at the SPSE 14th Annual Fall Symposium, Washington, DC, October, 1974; *Proceedings* pp. 37–39.
7. S. R. Ovshinsky and P. H. Klose, "Reversible, High-Speed, High-Resolution Imaging in Amorphous Semiconductors," in 1971 SID International Symposium, Digest of Technical Papers, pp. 58–61; also see *Proceedings of the SID* 13: 188–192 (1972).
8. S. R. Ovshinsky and P. H. Klose, "Imaging in Amorphous Materials by Structural Alteration," *J. Non-Cryst. Solids* 8–10: 892–898 (1972).
9. S. R. Ovshinsky and P. H. Klose, "Imaging by Photostructural Changes," in Proceedings of the Symposium on Nonsilver Photographic Processes, New College, Oxford, September 1973; edited by R. J. Cox, Academic Press, London 1975, pp. 61–70.
10. S. R. Ovshinsky, "Electronic and Structural Changes in Amorphous Materials as a Means of Information Storage and Imaging," presented at the 4th International Congress for Reprography and Information 1975, Hanover, Germany, April 13–17, 1975; *Proceedings* pp. 109–114.
11. M. Kastner, "Bonding Bands, Lone-Pair Bands, and Impurity States in Chalcogenide Semiconductors," *Phys. Rev. Lett.* 28: 355–357 (1972).
12. S. R. Ovshinsky and K. Sapru, "Three-Dimensional Model of Structure and Electronic Properties of Chalcogenide Glasses," in Proceedings of the 5th International Conference on Amorphous and Liquid Semiconductors, Garmisch-Partenkirchen, Germany, September 3–8, 1973, edited by J. Stuke and W. Brenig, Taylor & Francis, London, 1974, Volume 1, pp. 447–452.
13. The introduction of disorder into periodicity is a common link between crystals and amorphous material. Surfaces even of single crystals are disordered and single crystals cannot be made into large area films or tapes for imaging purposes. Polycrystalline materials introduce new dimensions of disorder. Therefore, to study the spectrum of disorder is to begin the study of amorphous materials.
14. J. Feinleib, S. Iwasa, S. C. Moss, J. P. deNeufville, and S. R. Ovshinsky, "Reversible Optical Effects in Amorphous Semiconductors, *J. Non-Cryst. Solids* 8–10: 909–916 (1972).
15. J. Feinleib, J. P. deNeufville, S. C. Moss, and S. R. Ovshinsky, "Rapid Reversible Light-Induced Crystallization of Amorphous Semiconductors," *Appl. Phys. Lett.* 18: 254–257 (1971).
16. R. G. Neale and J. A. Aseltine, "The Application of Amorphous Materials to Computer Memories," *IEEE Trans. on Electron Devices* ED-20: 195–205 (1973).
17. R. J. von Gutfeld and P. Chaudhari, "Laser Writing and Erasing on Chalcogenide Films," *J. Appl. Phys.* 43: 4688–4693 (1972).
18. A. W. Smith, "Injection Laser Writing on Chalcogenide Films," *Appl. Optics* 13: 795 (1974).
19. P. Chaudhari, J. J. Cuomo, and R. J. Gambino, "Amorphous Metallic Films for Magneto-Optic Applications," *Appl. Phys. Lett.* 22: 337–339 (1973).
20. J. Feinleib and S. R. Ovshinsky, "Reflectivity Studies of the Te (Ge, As)-Based Amorphous Semiconductor in the Conducting and Insulating States," *J. Non-Cryst. Solids* 4: 564–572 (1970).
21. J. P. deNeufville, "Optical Information Storage," in Proceedings of the 5th International Conference on Amorphous and Liquid Semiconductors, Garmisch-Partenkirchen, Germany, September 3–8, 1973, edited by J. Stuke and W. Brenig, Taylor & Francis, London, 1974, Volume II, pp. 1351–1360.
22. J. P. deNeufville, S. C. Moss, and S. R. Ovshinsky "Photostructural Transformations in Amorphous As_2Se_3 And As_2S_3 Films," *J. Non-Cryst. Solids* 13: 191–223 (1973/74).
23. J. P. deNeufville, R. Seguin, S. C. Moss, and S. R. Ovshinsky, "Mechanism of Reversible Optical Storage in Evaporated Amorphous AsSe and $Ge_{10}As_{40}Se_{50}$," in Proceedings of the 5th International Conference on Amorphous and Liquid Semiconductors, Garmisch-Partenkirchen, Germany, September 3–8, 1973, edited by J. Stuke and W. Brenig, Taylor & Francis, London, 1974, Vol. II. pp. 737–743.
24. T. Igo and Y. Toyoshima, "A Reversible Optical Change in the As-Se-Ge Glass," *J. Non-Cryst. Solids* 11: 304–308 (1973).
25. S. A. Keneman, "Hologram Storage in Arsenic Trisulfide Thin Films," *Appl. Phys. Lett.* 19: 205–207 (1971).
26. K. Tanaka and M. Kikuchi, "On the Interpretation of Photographic Effects in Amorphous As-S Films," *Solid State Communications* 13: 669 (1973).
27. S. R. Ovshinsky, Y. Chang, and A. Eisenberg, to be published.
28. Y. Chang and S. R. Ovshinsky, 1973, unpublished.
29. H. Fritzsche, *Amorphous and Liquid Semiconductors*, edited by J. Tauc, Plenum Press, London, 1974, pp. 221–312.

THE SHAPE OF DISORDER

Stanford R. Ovshinsky

Energy Conversion Devices, Inc., 1675 West Maple Road, Troy, Michigan 48084

(Received 8 September 1978)

I feel that the best way for me to write a paper that honors Sir Nevill Mott is to continue a discussion on points of our mutual concern for Mott is not one who rests on his laurels but is in ceaseless struggle with nature, using the power of his mind as a searchlight into the unknown; in the case of amorphous materials, into the world of disorder. It is unfortunate that this negative emotive word, disorder, is used to describe amorphous materials since I believe that the operative description of the uniqueness of amorphous materials is really "freedom"--freedom from restrictions of crystalline symmetry, freedom from stoichiometry, and especially freedom of choices of bonding and nearest neighbor relationships. The multiple bonding and orbital relationships permitted by the removal of crystalline restrictions, while numerous, are specific and, in the manner and number in which they co-exist in an amorphous material, are definable as well as being unique.

This paper deals with the concepts which have been a personal guide for our work in amorphous semiconductors. Its theme is that in amorphous semiconductors we are involved with stereochemistry. The defect structure of these materials is connected with the orbital orientations, atomic spatial arrangements and geometric structures which interact in many ways in the varying three-dimensional space of an amorphous solid. This has been a recurrent theme in my work. "Amorphous materials are based upon positional and, in the case of alloys, also compositional disorder. However, paradoxically, it is the understanding of the role of short-range order in the form of local bonds and the steric architecture involved which is essential for both the understanding and the design of amorphous switching devices."(1) Since my theme is that the physical properties of amorphous semiconductors are determined by topology, that is, by the shape of local structure that are dependent on nearest neighbor relationships, it should be clear that we are dealing with microscopic polymorphic and allotropic arrangements where the term isomerism can describe many configurations in amorphous semiconductors.

It was considered to be self evident (2) that the field of amorphous semiconductors should be divided into tetrahedral and chalcogenide materials on the basis that amorphous silicon would have very little in common with the chalcogenides. This paper will concern itself with the opposite view; that is, that similar mechanisms involving steric design can be utilized in making both classes of materials into useful semiconductors. The molecular shapes must have built-in flexibility allowing the completion of local structures. In the case of the chalcogenides, it is their primary divalency and use of lone pairs together with other crosslinks that provide flexibility and determine structural configurations. The bond completion of the much more rigid tetrahedral structures requires the addition of other elements to provide the needed organization, distortion and flexibility. Hydrogen in silane-decomposed films has been a puzzle to many investigators since far more hydrogen is incorporated than is indicated by the density of defect states that it is supposed to be compensating. When looked at from our point of view, that hydrogen is also acting as a structural bridging and crosslinking atom with steric similarities to the chalcogenides and that its bonding and antibonding states make for additional states in the gap,(3) then it becomes clear that there is a need for a different structural and steric element to induce new bonding and charge configurations. An amorphous silicon-based material whose primary alloying element is not hydrogen is needed to provide the structural solution for a low density of states. The density of states and doping characteristics of such a unique material utilizing fluorine as a steric organizer will be described.

The elements which perform the tasks of organizing and completing the rigid tetrahedral structure of an element such as silicon can act in several ways--not only by the use of bridging crosslinks, and multi-orbital bonding choices which hydrogen achieves, but particularly fluorine due to its specificity of bonding, small size, electronegativity and reactivity can control the bonding and coordination of silicon as well as of other elements present, including hydrogen. Fluorine performs its steric role to perfection.

It is now evident that silicon does not have useful properties in its elemental amorphous form. Indeed, as in chalcogenides, other elements are

Reprinted by permission from *Journal of Non-Crystalline Solids*, Vol. 32, pp. 17–28 (1979).

needed for the required steric configurations to be present in the material, assuring compensation which gives amorphous silicon-based materials useful properties and allows for extrinsic silicon materials to be prepared. Inherent in amorphous semiconductors are several steric structures that may be required as energy-saving mechanisms to assure compensation. Grasping the steric requirements provides the understanding that amorphous semiconductors have unifying principles as well as fundamental differences. The choice of elements to affect and complete the bonding, coordination and crosslinking of chalcogenides or silicon atoms becomes a matter of steric design.

We have had several choices in terms of materials work--to attempt to visualize the bonding and non-bonding possibilities in these materials, to measure the deviations that we think occur, or to actually build materials which carry out the principles which guide us. The first and the last are the realistic choices available to us since the richness and sophistication of the field of amorphous materials is often obscured by the inability to measure directly the deviation of structure in the Angstrom range, necessary to correlate deviant configurations with various properties. As part of our efforts to understand the structural influences, we used the concept of lone-pair orbitals, their non-bonding influence on local order as well as their unusual configurations, including hybridization and deviant bonds.(4) We tried to identify the structural basis of defect formation by exaggerating the situation through optical excitation, hoping that this would make identification of the various states easier. We utilized methods which would detect the differences between electronic and thermal effects. Our optical work showed that there were several such defects. For example: "the optical effects in As_2Se_3 and As_2S_3 can be entirely accounted for in terms of defect formation (photodarkening) and polymerization (photostructural and thermostructural changes)....(5) In photodarkening, "we tentatively associate this effect with a non-equilibrium excess of trapped holes and electrons which are incapable of re-equilibrating at the temperature at which they are produced. Such trapped charges must lie within the mobility edges in sites which are sufficiently localized and deep-lying to inhibit recombination or re-equilibration with thermally generated carriers. Presumably, the electrostatic attraction of excess trapped charges constitutes a local deformation which could affect the joint density of lone-pair and conduction band states at energies approaching the optical gap energy."(5) Regarding the problem of identification of the defect states, we postulated a spectrum of such states associated with various types of lone-pair interactions, among which we discussed the possibilities of one- and three-electron bonds (6) and also stated: "the placement and interactions of the lone-pair configuration differ in space and energy as a function of material design. Excitonic type states should also result from some of these considerations. We suggest that they are possibly involved with the mechanism resulting in the critical electric field effect as well as the optical interactions observed."(6) Since subtle differences are important, the analytical tools that are used, whether X-ray diffraction, neutron scattering, Raman, or even EXAFS, give only average data not exact enough to clarify the defect state chemistry so important in these materials.* These techniques only show average bonding configurations; they are not as yet useful to describe the deviations in structure and the isomeric quality of the bonding structures involved in the defect states.

One of the obstacles in the field has been the word amorphous itself, which has been accepted as a sufficient scientific description of various films. A film can be amorphous to X-ray diffraction, electron microscopy, etc.; yet, the number and type of its defect states are not fixed since the material itself is not at its lowest free energy.(1) The temperature of crystallization is not the only measure of stability. Not only in chalcogenides but also in Group IV materials, atoms must be fixed in space and in energy by other atoms which lock them into place. Therefore, the ability to make stable amorphous materials even when new structures are created in a non-equilibrium manner (3) depends upon isomerism, steric hindrances, and preferred steric architecture.

Rather than the formlessness suggested by the term amorphous, we emphasize the importance of geometric configurations. The same elements in one configuration can result in neutral states and in another present a reactive or charged site. This means that the important concept of deNeufville regarding ordering (8) that occurs in amorphous semiconductors, such as in some of the chalcogenides, requires extension since the materials can still have _differing_ bonding configurations to one and the same element. Therefore, it is not that germanium and tellurium are attached to one another in an ordered manner, rather than germanium to germanium, or tellurium to tellurium, that is of importance, but that germanium and tellurium can be bonded to themselves and to each other in various ways and be in differing positions relative to their nearest neighbors. The nearby structures provide a local environment which can affect the distribution of localized states which is the basis of some of the defect chemistry of these materials. The same is true of silicon. The silicon-silicon bond is not as significant as the configurations permitted to silicon by the linking of small atom species such as hydrogen. I take as support for this contention that Paul and his group have shown the important result that some of the physical properties of silane-decomposed materials could be duplicated in sputtered films(9).

If the biological axiom that "structure is the basis of function" is applied to amorphous materials, then the crucial point is that while amorphous semiconductors have various categorizations such as tetrahedral, pnictide, and chalcogenide, there are universal principles which apply to them all. For example, the interesting factor in elemental tetrahedral amorphous materials is not that they are considered fourfold coordinated, but that, because they are amorphous, they cannot be completely fourfold coordinated, and therefore it is the deviations from such coordination that control the conductivity of the material. We are concerned with the charge configurations in all amorphous

---

*New Mossbauer techniques may be of help according to ref. (7).

semiconductors that are created or negated by such deviant bonding.(3,10) Electro-positive and electro-negative influences have an importance beyond bonding. They align, affect, and distort, that is, organize, nearby structures, including non-bonded configurations. The added elements can impose local-order charge conditions of both positive and negative nature which can affect the pinning of the Fermi level and are involved with carrier recombination kinetics. In a field where donors and acceptors have a physical meaning, we must now concern ourselves with their basic chemical significance. Coordinate bonding is a form of chemical donation and acceptance and has a spectrum of strengths of interaction. Complexing, chelation, and overcoordination, while new concepts to solid state physicists involved in crystalline semiconductor technology, have important implications in the amorphous field.(3,10) A satisfied spin orbital can still have charge activity. Even though lone pairs are spin compensated, they are negatively charged. In silicon-based materials, a dangling bond need not be the only form of an undercoordinated charge condition. Whether there are positive, negative, or neutral states in the gap depends on the charge condition which is associated with specific local-order bonding and structural configurations, associated with the type of added element.

The completion of orbitals requires a steric factor. It is probably not possible, or at least appears not practical, to make a tetrahedral amorphous semiconductor such as silicon intrinsic; that is, complete all of its bonding orbitals, without the addition of other elements acting as flexible links and local order templates that permit the three-dimensional rearrangement of bonding configurations. It simply costs too much energy for silicon to be fourfold coordinated with other silicon atoms in the amorphous state. Besides crosslinks, other factors can affect charge, compensate and control the conduction in chalcogenide materials, and can be operative in silicon-based amorphous materials as well as in many other materials with a wide variety of band gaps and compositions ranging from Groups III through VI.(3,11) We have clearly shown through a large number of experiments that chalcogenides as well as tetrahedral materials can have their Fermi level manipulated by the addition of the same elements. For example, we converted materials as diversified as germanium telluride and silicon into extrinsic materials through the addition of an alkali metal such as lithium (10,12) and silicon, silicon carbide, silicon nitride, silicon oxide, boron nitride and various arsenides and chalcogenides through the addition of transition metals. The chemical thinking behind this work has been covered in several papers.(3,10,13) However, for the purposes of this paper, it is sufficient to note that local order is sensitive to the number and arrangement of various orbital configurations such as d-orbitals. By various techniques new bonding environments can be created.

The steric influences of lone pairs and d-band orbitals are only some of the factors that can affect local configurations. Others are atoms which can provide flexible links, multi-orbital connections, promote π and σ bonding and hybridization, or through their charge affect the shape of local structures. In considering how to fit these chemical concepts into solid state physics (3) I collaborated with Adler,(10) whom I consider to be an outstanding figure in our field, to explain how the various bonding and antibonding arrangements could affect the states in the gap of amorphous semiconductors. This brings us to hydrogen which has proven to be a useful additive for silicon compensation.(9,14,15) As I pointed out,(3,10) silicon in conjunction with hydrogen can have varied type bonding including three-center bonds. Hydrogen, in compensating silicon, can act in a sense divalently permitting polymeric flexibility of structural rearrangements so that the completion of bonding can take place. Various free energy structural choices are available for silicon-hydrogen bonding depending on temperature. It is interesting that voids and other structural defects that are present in deposited amorphous silicon are now being found in silane-decomposed amorphous silicon alloys by Knights, Lucovsky and their colleagues who are connecting these voids with a specific silicon-hydrogen species.(16,17) Fritzsche, who has made outstanding contributions to this field, with his group has shown that there is an inhomogeneity indicated by the difference between the bulk and surface properties of silane-decomposed materials, and that surface states cannot be neglected.(18,19) This points out to us that because of their residual chemical activity, although better than elemental amorphous silicon which is not tetrahedral enough, these silane-decomposed materials would not be capable of completing tetrahedral structures to sufficiently reduce the state in the gap so that substitutional doping could be truly effective. Because hydrogen also acts structurally, it provides states in the gap.

In our prior work, we emphasized that multiple choice bonding leads to varying coordination in amorphous materials:

"A disordered semiconductor can exist in several structural states. This structural factor constitutes a new variable with which the physical properties of the semiconductor can be controlled. Furthermore, structural disorder opens up the possibility to prepare in a metastable state new compositions and mixtures that far exceed the limits of thermodynamic equilibrium. Hence, we note the following as a further distinguishing feature. In many disordered semiconductors it is possible to control the short-range order parameter and thereby achieve drastic changes in the physical properties of these materials, including forcing new coordination numbers for elements. This means that both the non-chalcogens, such as germanium and silicon, and the chalcogens, including tellurium, can have different coordination numbers than in the crystalline materials. This can lead to the possibility of coordinate or dative bonding in the spectrum of bonds available in the amorphous matrix that in a single sample can, in fact, contain all the various bond types from covalent to metallic. In the same sample an element can, in fact, have several coordinate numbers."(20)

This premise assumes that amorphous materials can be viewed as a matrix whose structural bonds have a normal coordination,(3,21,22) which is responsible for the cohesive energy,(6) but that there are several deviations from these normal bondings which can be "analog bonding where two local structures composed of different atoms can mimic one another....Possibly, tellurium can be forced by surrounding elements to take on the bonding appearance of polonium, while under certain conditions in non-chalcogenide glasses, arsenic can be distorted by its surrounding elements to resemble tellurium,"(4) or in lone-pair materials, there can be lone-pair interactions ranging from the coordinate bond to clusters of one- and three-electron bonds to coulombic interactions of a repulsive nature.(6) We considered that while in lone-pair materials the spins were paired and all bonds satisfied, there would be "in some regions charge imbalance and polarization from the lone pairs' interaction with nearest neighbors."(6) Our reasoning for bond fulfillment and spin pairing (6,12,21,22) can be extended to tetrahedral materials as explained here.

The removal of crystalline restraints provides the structural flexibility which encourages multi-choice bonding as the energy-saving mechanism for structural completion. The manner in which various elements fill the volume of an amorphous solid is related therefore to the size and charge of the atoms and their ability to link with their nearest neighbors. Related to orbital choice is the forcing of a desired configuration, also connected with deformability. The manner in which stiffness is built into an amorphous material not only determines the shape of local configurations, but adds the necessary orbitals which remove dangling bonds, balance or eliminate charge states, etc. This is why we emphasize the importance of organizers of local structures as well as bridging elements and crosslinks. Whether attached to a chalcogen such as tellurium or a Group IV element such as silicon, the degree of charge, size, number and strength of the linking atoms determine whether the distortions and relaxations of structure can accommodate the charge relationships in a manner that encourages bond completion and therefore pairing.

As can be seen from the above, we are suggesting that phonon relationships are not simple in amorphous materials. They depend upon the type of amorphous materials involved. Despite the commonality of principles that I have been proposing, I wish to emphasize that there is a broad spectrum of amorphous materials, especially connected to the shapes of the local structures and the degree of stiffness in the material. By dividing the primary structural bonding into an interactive system composed of the matrix and the electronic configurations within it, (6,12,21,22) it is possible to relate the strength of the electron-phonon coupling to the relaxations, distortions, deformability and chargedness of local structures, and therefore, deformability can be seen as the important compensating principle for the elimination of dangling bonds and the intermediary for spin pairing.

Taking the concepts of deformability, relaxation and stiffness to their logical conclusion, the most polymeric of the amorphous structures would have the least tendency for dangling bonds. Depending upon the number and strength of crosslinks, their chain and ring configurations are able to change shape and respond to charge relationships so that separations between the dangling bonds and opposite charges are small compared to the deformability inherent in the structural matrix. Because of the closeness in

energy of some of the configurations and conformations involved in many of these materials, there can be a conversion of one shape to another with little expenditure of energy.

We differ from Anderson in rejecting disorder as a sufficient explanation for the phenomenon of spin pairing in amorphous materials.(23) Anderson considers disorder as the basis of the phonon interaction which gives rise to an effective negative correlation energy for electrons leading to pairing. Pairing is not a result of disorder; in fact, in many films, especially of the tetrahedral type, it is not the rule. It is my position (3,6,21,22) that pairing in chalcogenides is connected with the primary divalency of the chalcogen atom and the various bonding and other interactions offered to lone pairs through the flexibility of local structure.(22) Some specific defects associated with chalcogenides which help explain the negative correlation energy have been pointed out by Mott and his colleagues (24-26) and Kastner et al.(27) In this paper, we wish to emphasize that pairing is an option to an amorphous tetrahedral material only if the matrix structure is deformable and flexible; that is, if the materials are designed through the principles enunciated here. For example, the small hydrogen atom acts as a pivot for linking and at times adding charge to the silicon atoms. This can lead to configurations which have many chain and ring characteristics similar to the chalcogenides. Fluorine as a structural unit can also be involved in chain forming and provides charge conditions of a negative nature. The juxtaposition of electrons so that spin pairing can take place is mediated through energy-saving matrix distortions. The potential bonding or pairing energy must be large enough to overcome the energy stored in matrix distortion. Fluorine is exemplary in this role. It is important to re-emphasize that the principles of bonding and pairing are true for amorphous silicon-based materials as well as chalcogenides. Utilizing amorphous silicon as a reference, we know that without additives it is subject to meandering annealing behavior without its structure being fully stabilized or dangling bonds eradicated. Adding

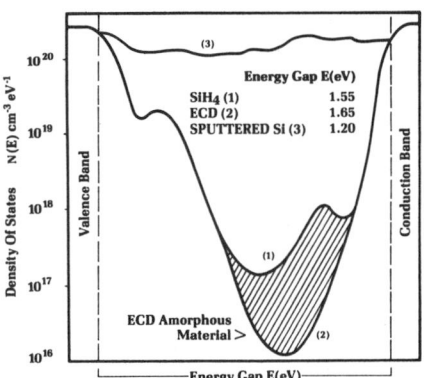

Fig. 1 Density of localized states deduced from the field effect experiment. Curve (1) refers to an amorphous film deposited from SiH_4, with substrate temperature held at 570K (Madan et al., J. Non-Cryst. Solids 20, 239 (1976)). Curve (2) shows $N(E)$ cm^{-3} eV^{-1} for the ECD materials. Curve (3) visualizes the density of states for sputtered silicon.

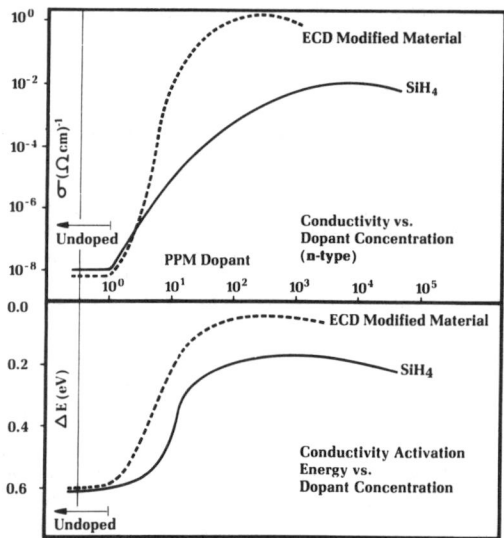

Fig. 2   Doping of amorphous semiconductors.

hydrogen in various manners, but especially through glow discharge, allows for the use of bridging atoms and varied local structures which utilize more of the silicon orbitals. However, the density of states of such a material is still quite high as evidenced by the inability to move the Fermi level closer than 0.2eV from the conduction band (see Fig. 1).

We have developed new silicon materials not based on silane decomposition which have unique properties based upon fluorine. The figures that follow show two important new results. Figure 1 shows that the density of states has been decreased by more than an order of magnitude from those materials produced by silane decomposition, but, most importantly, Fig. 2 is clear evidence that, for the first time, effective doping has been achieved in a silicon-based material. How this was accomplished, the measurements, and what materials and techniques were utilized, is the subject of other papers in collaboration with Arun Madan.(28-30) My premise which motivated our work was that films of elemental amorphous silicon made by various deposition means do not have three-dimensional stability, and that the structure of silane-decomposed amorphous silicon alloys is somewhat, but not fully, stabilized. The photostructural results of Staebler and Wronski,(31) support the argument of the similarities between chalcogens and tetrahedral materials. Our photostructural studies of the former (5,32-35) showed the polymeric base for optical changes whose principles we feel are applicable to the latter. Despite Carlson and Wronski's important device work (36) and Spear and LeComber's basic materials studies,(37) there is an inherent conductivity barrier in these materials associated with the remaining states in the gap which results in the need for an inordinate amount of doping. This raises serious doubts as to whether these materials can ever be utilized for efficient PN junctions. The remaining states in the gap would not result in the kind of depletion region needed for such devices.

Our material was designed to utilize a maximum of bonding orbitals of silicon by providing new bonding arrangements through other elements besides hydrogen. This resulted in a stable, amorphous, silicon-based material which is not affected by photostructural changes, and in which, for the first time, the density of states is so low that it can be doped in the effective parts per million range as in crystalline materials, resulting in conductivities greater than $1(\Omega\,cm)^{-1}$. Because of its analogies to crystalline materials in terms of density of states and doping levels, it has interesting device possibilities.

The question of common defect states in amorphous and crystalline materials has been dealt with lately by Kastner (38) whose chemical thinking has contributed to an understanding of the chalcogens (39) and whose continuous contributions have been of great value. He and Fritzsche have equated the defect states of crystalline and amorphous chalcogenides, as well as discussing the nature of defect states in lone pairs and other semiconductors.(40)

In terms of their crystalline-amorphous analogy, we take a different view by pointing out that while the bonding configurations can be similar, especially in crystalline chalcogenide materials which are so polymeric as to have a great degree of disorder as well as flexibility, by our criteria of stereochemistry, the transport properties must be different between amorphous and crystalline chalcogenides. The optical properties, of course, depend on the overall structural bonds and there should be a great deal of similarity between some amorphous and crystalline chalcogenides, but the defect states are not just single-electron or three-electron bonds, but a total configuration where the geometry of other atoms including that of the nearest neighbors is interactive. An active site depends upon not just the chemical bonds but on the shape of the local structure, which cannot be separated from the surrounding local environment. Because the local environment of a crystalline material is different from that of an amorphous material, the deviant bonds should have dissimilar effects upon the conductivity.

We have taken the position (21,22) that in amorphous chalcogenide materials there are many more states than D^+'s and D^-'s (24-26) or C_1^-'s and C_3^+'s, (27) and they would appear in space, and therefore in energy, in a different local environment than they would in a crystal. Since periodicity imposes spatial and energetic constraints even in chalcogenide crystals, the local order conditions are reflected both quantitatively and qualitatively in the defect states, which would result in different band gap pictures. Because of its polymeric nature, however, a chalcogenide crystal can be considered a transition between the rigid crystalline and the stabilized amorphous state.

Just as the understanding of impurities has become the hallmark of crystalline semiconductors, so too the various notations of Street and Mott (24) and Kastner et al. (27) are important to describe some kinds of defects in chalcogenide materials. Their work, as well as the pioneering efforts of Spear and LeComber,(37) to tame and control the conductivity of amorphous silicon, are landmarks in the process of understanding and utilizing amorphous semiconductors. The concepts of complexing, coordinate and dative bonding, multi-center bonding, multi-orbital configurations, overcoordination, induction and the influence of electronegativity on local structural configurations, d-band interactions and expansions, and lone-pair interactions, all embodied in the stereochemical approach, have been the basis on which we have developed families of amorphous devices ranging from switching,(41) memory,(42) and imaging (20) to energy conversion, superconductivity, electrochemistry and catalysis.

Neither disorder nor chemical order is sufficient to describe the interesting phenomena in amorphous semiconductors. What, we believe, gives the field its importance are the steric relationships involved both with chemical bonding and nearest neighbor relationships. These interactions can be as subtle as the perturbating and inducing influences of fluorine on the surrounding atoms or as drastic as the generation of new bonding structures by alloying.

Sir Nevill Mott, as an exemplar for all of us, by his many dominating contributions to making this last frontier of solid state physics understandable and an accepted and exciting area of science, has provided through his creativity the insight and leadership which have already received the highest scientific accolades. His contributions have not only been through his own work but by his developing a talented group of colleagues such as Davis and Street. What we must do to truly honor him is not stop now. In our young field we have already seen earlier assumptions overturned. Amorphous materials including chalcogenides _can be_ modified and doped.(3,12,37) Hopping conduction _can_ occur in chalcogenides,(3,11) and switching _is_ electronic.(41,43,44) There is much yet to be done, and Mott, as always will certainly be leading, discussing, clarifying and discovering.

It is not possible in a scientific paper to convey my personal feelings concerning Nevill Mott. He has been a most valued friend and advisor whose encouragement, help and support to both Iris and me have been deeply appreciated. There is a nobility to Mott that no prize can enhance. We give him not only honor but affection.

REFERENCES

1. S.R. Ovshinsky, J. Non-Cryst. Solids 2, 99 (1970).
2. Tetrahedrally Bonded Amorphous Semiconductors, eds. M. H. Brodsky, S. Kirkpatrick and D. Weaire (New York, AIP, 1974).
3. S.R. Ovshinsky, Amorphous and Liquid Semiconductors, ed. W.E. Spear (Edinburgh, CICL, 1977) p. 519.
4. S.R. Ovshinsky, Electronic-Structural Transformations in Amorphous Materials--A Conceptual Model, submitted to Nature--July 13, 1972, unpublished.
5. J.P. deNeufville, S.C. Moss and S.R. Ovshinsky, J.Non-Cryst. Solids 13, 191 (1973/74).
6. S.R. Ovshinsky and K. Sapru, Amorphous and Liquid Semiconductors, ed. J. Stuke and W.

Brenig (London, Taylor and Francis, 1974) p. 447.
7. Boolchand, personal communication.
8. J.P. deNeufville, J. Non-Cryst. Solids 8-10, 85 (1972).
9. W. Paul, A.J. Lewis, G.A.N. Connel and T.D. Moustakis, Solid S. Commun. 20, 969 (1976).
10. S.R. Ovshinsky and D. Adler, Contemp. Phys. 19, 109 (1978).
11. R. Flasck, M. Izu, K. Sapru, T. Anderson, S.R. Ovshinsky and H. Fritzsche, Amorphous and Liquid Semiconductors, ed. W.E. Spear (Edinburgh, CICL, 1977) p. 524.
12. S.R. Ovshinsky, Structure and Excitations of Amorphous Solids, eds. G. Lucovsky and F.L. Galeener (New York, AIP, 1976) p. 31.
13. S.R. Ovshinsky, K. Sapru and K. Dec, in Proc. Intl. Topical Conf., Yorktown Heights, New York, March 22-24, 1978, ed. S.T. Pantelides, p. 304.
14. H. Fritzsche, Amorphous and Liquid Semiconductors, ed. W.E. Spear (Edinburgh, CICL, 1977) p. 3.
15. J.C.Knights, Amorphous and Liquid Semiconductors, ed. W.E. Spear (Edinburgh, CICL, 1977) p. 433.
16. J.C. Knights, Tenth Solid State Device Conf. (Tokyo, Japan, August 29, 1978).
17. G. Lucovsky, R.J. Namanich and J.C. Knights, Phys. Rev., to be published.
18. M. Tanielian, H. Fritzsche, C.C. Tsai and E. Symbalisty, Appl. Phys. Lett., to be published.
19. C.C. Tsai and H. Fritzsche, Solar Energy Mater. 1, 29 (1979).
20. S.R. Ovshinsky and H. Fritzsche, IEEE Trans. on Electron Devices ED-20, 91 (1973).
21. S.R. Ovshinsky, Structure and Properties of Non-Crystalline Semiconductors, ed. B.T. Kolomiets (Leningrad, Nauka, 1976) p. 426.
22. S.R. Ovshinsky, Phys. Rev. Lett. 36, 1469 (1976).
23. P.W. Anderson, Phys. Rev. Lett. 34, 953 (1975).
24. R.A. Street and N.F. Mott, Phys. Rev. Lett. 35, 1293 (1975).
25. N.F. Mott, E.A. Davis and R.A. Street, Phil. Mag. 32, 961 (1975).
26. N.F. Mott and R.A. Street, Phil. Mag. 36, 33 (1977).
27. M. Kastner, D. Adler and H. Fritzsche, Phys. Rev. Lett. 37, 1504 (1976).
28. S.R. Ovshinsky and A. Madan, in: 1978 Ann. Meeting of the American Section of the Intl. Solar Energy Society (August 28-31, 1978, Denver, USA), to be published.
29. S.R. Ovshinsky and A. Madan, Nature 276, 482 (1978).
30. S.R. Ovshinsky and A. Madan, Phil. Mag., to be published.
31. B.L. Staebler and C.R. Wronski, Appl. Phys. Lett. 31, 292 (1977).
32. S.R. Ovshinsky, in: Digest of Technical Papers of the Topical Meeting on Optical Storage of Digital Data (March 19-21, 1973, Aspen, Colorado, USA) p. MB5-1.
33. S.R. Ovshinsky and P.H. Klose, J. Non-Cryst. Solids 8-10, 892 (1972).
34. S.R. Ovshinsky, in: Proc. of the 4th Intl. Congr. for Reprography and Information (April 13-17, 1975, Hannover, Germany) p. 109.
35. S.R. Ovshinsky, J. Appl. Photographic Eng. 3, 35 (1977).
36. D.E. Carlson and C.R. Wronski, Appl. Phys. Lett. 28, 671 (1976).
37. W.E. Spear and P.G. LeComber, Solid State Commun. 17, 1193 (1975).
38. M. Kastner, in: Topical Conf. on Atomic Scale

Structure of Amorphous Solids (Yorktown Heights, New York, April 3-5, 1978); publ. in J. Non-Cryst. Solids 31, 223 (1978).

39. M. Kastner, Phys. Rev. Lett. 28, 355 (1972).

40. M. Kastner and H. Fritzsche, Phil. Mag. 37, 199 (1978).

41. S.R. Ovshinsky, Phys. Rev. Lett. 21, 1450 (1968).

42. S.R. Ovshinsky and I.M. Ovshinsky, Mat. Res. Bull. 5, 681 (1970).

43. D. Adler, H.K. Henisch and N.F. Mott, Rev. Mod. Phys. 50, 209 (1978).

44. M.P. Shaw, M.S. Shur, K.F. Subhani, D. Adler, M. Silver and S.R. Ovshinsky, J. Appl. Phys., to be published.

THE CHEMISTRY OF GLASSY MATERIALS AND THEIR RELEVANCE TO ENERGY CONVERSION

Stanford R. Ovshinsky

Energy Conversion Devices, Inc., Troy, Michigan 48084

The problems of our modern society are directly related to materials. Depletable fossil fuels, shrinking silver supplies, the high cost of gold, platinum and other precious metals all are interfering with industrial growth and creating societal problems. The fossil fuels and uranium supply us with energy without which no society or civilization can endure. However, there is a tendency to focus only on the fuel aspects of materials and ignore the larger materials picture, that is, that materials such as platinum, palladium and iridium, which are the core of the catalytic and electrochemical industries, affect not only the generation but the use of energy in a dramatic fashion.

Silver is the basis of the present photographic industry; gold is widely used as a contact material; further, many promising developments are held back for lack of high-temperature, durable, low-cost materials. The need for materials inert to chemical attack is exemplified by the fact that corrosion in the automotive industry wastes over fourteen billion dollars a year. Consider the energy loss from this item alone. Passive coatings have always been of interest but the only synthetic approach that has been widely used is based on organic material, that is, plastics. The prime materials need today is for synthetic inorganic materials which can be designed, that is, tailor made, not only for passive but for active use. These new materials must meet the seemingly utopian combination of needs of high-temperature stability, inertness and yet also be applicable to energy conversion and other energy related problems.

In the past, various important parameters of a material have been in lockstep with each other. If one had a good thermal conductor, one had a good electrical conductor; if one had a high-temperature, large band gap material, then it usually was a low mobility solid, that is, its dielectric properties were emphasized.

The success of the transistor based upon crystalline symmetry and the control of impurities blinded some to the need for information encoding by virtue of structural change, and to the need for the development of new semiconductor materials for electronic control purposes.

This paper will show that a new approach allows us to independently control the properties that were previously considered inherently linked in a material and permits us to design functional analogs of platinum and other materials such as gold, silver, crystalline silicon, etc. at substantial economic savings and superior performance. They are truly synthetic inorganic materials uniquely suitable for the twin major concerns of our society--information and energy.

It was this concern with information and energy which motivated me to investigate disordered and noncrystalline materials. Amorphous materials can easily be alloyed and we have shown through the years that a wide variety of materials can be made both structurally unistable and bistable.(1) These materials range from small band gap, good conductors to large band gap dielectrics with special emphasis on those that are semiconductors.

We showed that amorphous semiconductors could be doped,(2) then showed that we could take elements from throughout the periodic table and control their conductivity relatively independently of their optical band gap.(3,4) Figures 1-7 (5) indicate the

Host Material	Active Modifier
Ge Te Se As	Ni, Fe, Mo
As	Ni, W
B_4C	W
Si C	W
Si	Ni, B, C
Si_3N_4	W
BN	W
TeO_2	Ni
Ge	Ni
SiO_2	W
$GeSe_2$	Ni
$Se_{95}As_5$	Ni

Figure 1. Chemical modification of amorphous semiconductors.

Reprinted by permission from *Journal of Non-Crystalline Solids*, Vol. 42, pp. 335–344 (1980).

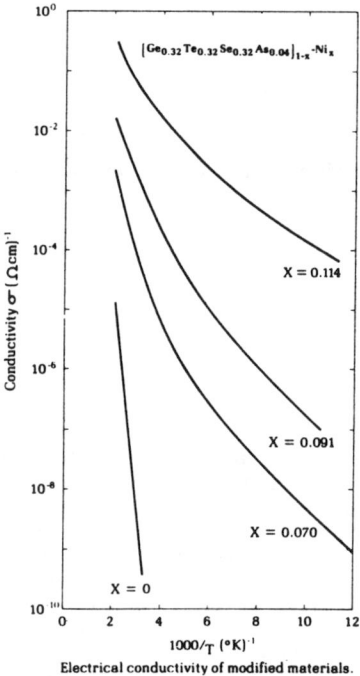

Electrical conductivity of modified materials.

Figure 2

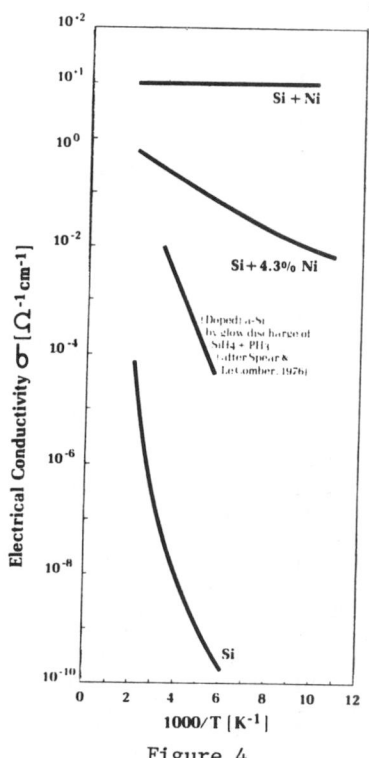

Figure 4

wide range of materials that can be designed and how the electrical activation energy can be changed at will.(3) In other words, we can synthesize and independently control all relevant parameters-- optical band gap, electrical activation energy, high melting temperatures, and even thermal conductivity. This synthesis and control are absolute necessities if the long dreamed of phenomenon of thermoelectricity is to make an impact on energy problems for we must be able to combine the conflicting requirements of poor thermal conductivity with good electrical conductivity and high Seebeck coefficient. Thermoelectricity has been considered the greatest fiasco in materials development; our new Ovonic modified materials change this picture drastically.(6,7)

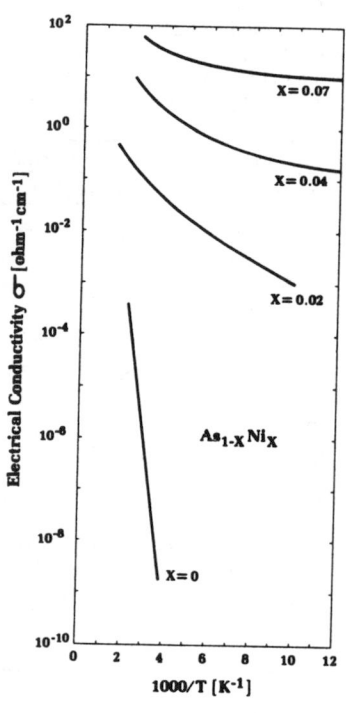

Figure 3

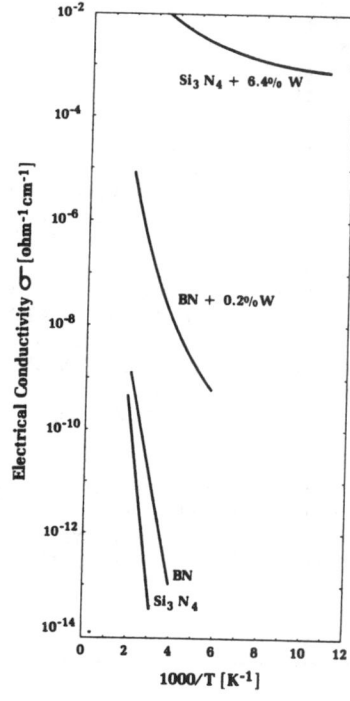

Figure 5

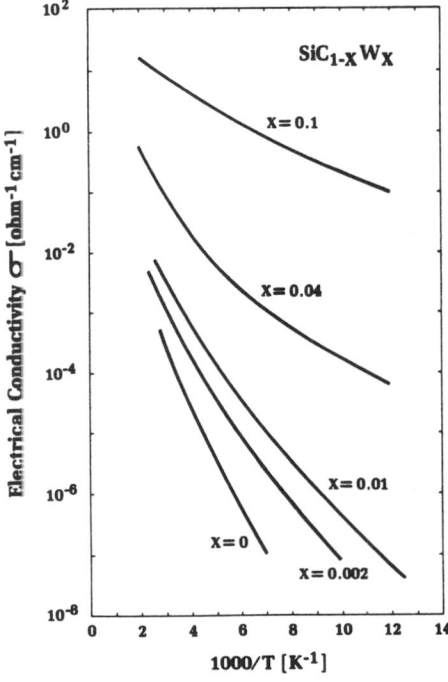

Figure 6

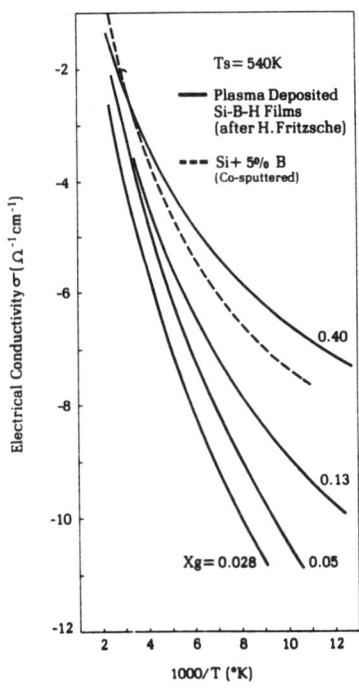

Figure 7

The ability to achieve these results was based upon the recognition that unusual bonding and nonbonding configurations could be defect atomic configurations controlling conductivity in both unstable and bistable amorphous materials.(8-12) Moreover, in the latter materials, the deviant configurations could become the nucleating points for structural change since the materials are deliberately made unstable to excitatory forces, opening up new vistas for memory storage and photographic systems.(13,14)

The key to the future is to be able to take elements from any part of the periodic table that we see fit and place their orbitals in bonding, nonbonding and antibonding positions that would be prohibited both in kind and in number in any substantial amount by the restraints of crystalline symmetry. This means that a new chemistry of amorphous solids is possible with potential for new devices which do not fit into neat categories of bonding orbital relationships and nonbonding configurations to which solid state physicists are accustomed. See Fig. 8.(3,15)

As Adler (in this issue) has described, crystalline and noncrystalline materials can have many similarities when having analogous chemical bonds. I wish to emphasize that with our approach we can design materials which have no relevant crystalline counterparts. These materials are possibly more important technologically and certainly scientifically.

I showed in Edinburgh, and Fig. 8 illustrates, the number of deviant configurations which control, increase and decrease states in the gap of amorphous semiconductors.(3) Everything from hopping conduction to band conduction can be designed into materials and Fermi levels unpinned and manipulated leading to important applications such as electrodes

with large band gaps, p or n type conductivity, chemical inertness and movable Fermi levels.

The manufacturing techniques include various methods of making solids, with cosputtering playing an important role. In no case need we be limited by the relaxation processes which have been restrictive factors even in amorphous materials.(10) We now can make stable, "nonequilibrium," electronic configurations which affect not only the band and mobility gaps,(16) but the states in the gap which can now be man-made. By uncoupling the individual parameters, we design materials to fit specific needs.

The structural aspects of compensating and doping additives are still not appreciated by many

---

Conventional DEC's

(1) Intrinsic: Vacancies, Interstitials, etc. (Crystal Only)
(2) Extrinsic: Substitutional Impurities

DEC's Which Characterize Amorphous Solids

(1) Intrinsic
  (A) Dangling Bonds (Tetrahedral Materials)
  (B) Lone-Pair Interactions (Chalcogenides)
  (C) Valence-Alternation Pairs (Non-Tetrahedral Materials)

(2) Extrinsic
  (A) Charge Compensation (e.g. Alkali Atoms)
  (B) Polyvalency and Coordinate Bonding (e.g.Transition metals, rare earths and electron deficient atoms)
  (C) Reduction of Lone-Pair Density (e.g. Oxygen)
  (D) Hybridization with Empty d-Orbitals (e.g. Silicon)
  (E) Multi-Center Bonding (e.g. Boron)
  (F) Multi-Electron Bonding (e.g. Carbon)

Figure 8. Deviant electronic configurations (DECs).

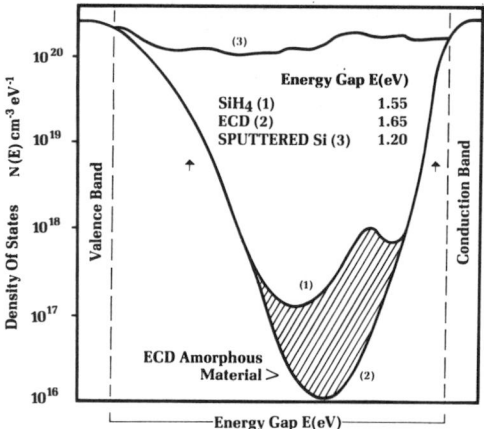

Density of localized states deduced from the field effect experiment.
Curve (1) refers to an amorphous film deposited from SiH₄ with substrate temperature held at 570K (Madan, LeComber, Spear, J. Non. Cryst. Solids. 20, 239, 1976).
Curve (2) shows N(E) cm-3ₑV-1 for the ECD material.
Curve (3) visualizes the density of states for sputtered Silicon.

	Energy Gap E(eV)
SiH₄ (1)	1.55
ECD (2)	1.65
SPUTTERED Si (3)	1.20

Arrows indicate extent of ECD experimental data.

Figure 9. Density of states
in the gap.

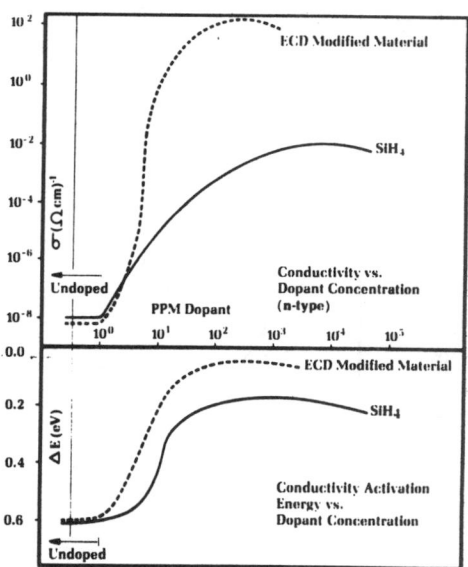

Figure 10. Doping of amorphous
semiconductors.

people working in the amorphous area. For example, astonishment is still expressed at the amount of hydrogen incorporated in silane-based amorphous silicon materials. The genesis of this is the error that was made in originally thinking that the materials produced from silane were elemental amorphous silicon rather than an alloy containing hydrogen.(17,18) Fritzsche (19) especially pointed out this error. Our conclusions are quite simple. Hydrogen is not only a compensating element but a structural element in an alloying system.(3) It does not simply tie up a dangling bond but enters the amorphous matrix much as bridges and crosslinks do in chalcogenide or even oxide materials. As such, it contributes to compositional and positional disorder which affect the band gap and band states of these materials. The various configurations that it can enter into in its bonding and nonbonding form not only affect the topological, geometric and steric relationships, but also affect charge relationships in the gap of materials.(3,20)

Figure 9 shows the density of states of three different materials--sputtered amorphous silicon which has such a high density of states that it is useless as a photoreceptor, the silane-derived silicon-hydrogen alloy showing a reduced number of states, and our new silicon-fluorine-hydrogen alloy which addresses itself to the multi-state and structural problems which I have mentioned.(21-23) We have designed an amorphous matrix which inherently has many fewer states in the gap by combining silicon, fluorine and hydrogen in a three-dimensional space so that dangling bonds, voids and other possible low free energy configurations are eliminated. As I have pointed out,(20) fluorine is not only a terminator of dangling bonds, but an organizer of local environment whose electronegativity can have a drastic inductive nearest neighbor effect. Bond angles and other distortions inherent in a tetrahedrally bonded amorphous material are certainly affected by the electronegative influence of fluorine. Our fluorine-based material allows the completion of tetrahedral structure whereas in amorphous silicon

or silane derived materials, the lowest free energy modes are dangling bonds, voids and other mismatched bonding configurations. Fluorine is an expander of silicon orbitals, probably utilizing even d-bands. Its strong bonding characteristics and its small size and specificity allow its atoms to become links in a matrix which combined with its polarizing force makes for a much tougher, unscratchable solid. The porosity and photostructural changes of silane materials are eliminated as well. Our alloy is a superior coating physically as well as being an excellent photoreceptor.

Figure 10 shows that the more complete structural matrix of our alloy can incorporate phosphorus and arsenic in a manner which provides superior n-doping. They are efficiently incorporated structurally as well as substitutionally because unlike the silane materials shown, the "noise level" of chemical mismatching is so low that true substitution can be picked up. This is not true in the p-type material. Boron, the classic dopant in

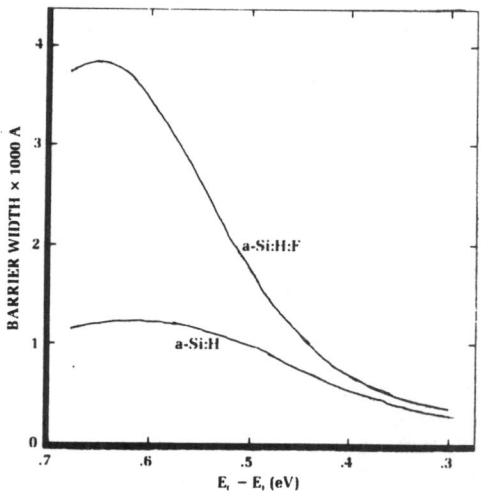

Figure 11. Depletion width comparison.

Figure 12

Figure 13

crystalline materials, due to its three-center bonds, adds states in the gap and therefore amorphous materials made with boron tend not to be doped but to be alloyed.(3) This results in inherently p-type materials rather than materials that are p-doped. Again, chemistry provides the understanding and therefore the answer.

Our new material and its characterization have been described fully elsewhere.(24-26) Of importance is that the well over an order of magnitude lower density of states and the orders of magnitude higher doping efficiencies result in a wider depletion width so necessary for the efficiency of an amorphous semiconductor.

Figure 11 (25) confirms this important prediction and since drift, not diffusion, is the vital parameter in amorphous silicon-based materials, then it should also follow that higher efficiencies can be obtained. Fortunately, this is so.

After many years of effort, 5 1/2% is still the published efficiency figure for silane-based devices. This efficiency barrier has now been broken in unoptimized devices composed of our silicon-fluorine-hydrogen alloy. We report 6 1/2% under AM-1 illumination with an M-I-S structure and an area of 4.2 square millimeters.(27) A tuning up of our process so as to minimize grid resistances should put us in the 7-8% range. While we are aiming toward 10% which the fluorine-based material should inherently be capable of, our present efficiency figures are already commercial since our devices can be made at an economic advantage to coal, gas, oil and uranium. Such a cost parity has been considered to be an impossible goal. The costs of single crystal and polycrystalline devices including ribbon are inherently too high. Our amorphous films not only can have a systems efficiency equivalent to a crystalline material through more effective use of space, but also their manufacturing costs are the lowest of any of the above technologies. Of course, their thicknesses are much smaller since amorphous materials are direct band gap materials and therefore much less material is used.

Figure 12 shows an experimental photovoltaic configuration of 21 arrays consisting of eight cells each.

Figure 13 shows a one square foot prototype panel of the same intrinsic material which is used in the 6.3% efficient devices. Note the uniformity of deposition.

Amorphous materials made in accordance with the chemical principles we have elucidated above are already being tested for a wide variety of other energy uses ranging from catalysis to superconductivity.

I would like to close on a personal note. I believe that amorphous materials will be ubiquitous and will help make the future energy rich. All past extrapolations have failed to take into account the new inventions in disordered, amorphous and glassy materials.

REFERENCES

1. S.R. Ovshinsky, J. Non-Cryst. Solids 2, 99 (1970).
2. S.R. Ovshinsky, Structure and Excitations of Amorphous Solids, ed. G. Lucovsky and F.L. Galeener (New York, AIP, 1976) 31.
3. S.R. Ovshinsky, Amorphous and Liquid Semiconductors, ed. W.E. Spear (Edinburgh, CICL, 1977) 519.
4. R. Flasck, M. Izu, K. Sapru, T. Anderson, S.R. Ovshinsky and H. Fritzsche, Amorphous and Liquid Semiconductors, ed. W.E. Spear (Edinburgh, CICL, 1977) 524.
5. These figures were first shown at the Proc. of the 7th Internat. Conf. on Amorphous & Liquid Semiconductors, Edinburgh, Scotland, June 27-July 1, 1977.
6. S.R. Ovshinsky, Amorphous Materials As Solar and Thermal Energy Convertors; SERI Research Seminar Series (15 March 1978).
7. S.R. Ovshinsky, Proc. of 14th Intersociety Energy Conversion Conference 2 (August 5-10, 1979) 1809.
8. S.R. Ovshinsky and H. Fritzsche IEEE Trans. on Electron Devices ED-20, 91 (1973).

9. S.R. Ovshinsky, Structure and Properties of Non-Crystalline Semiconductors, ed. B.T. Kolomiets (Leningrad, Nauka, 1976) 426.
10. S.R. Ovshinsky, Phys. Rev. Lett. 36, 1469 (1976).
11. S.R. Ovshinsky and K. Sapru, Amorphous and Liquid Semiconductors, ed. J. Stuke and W. Brenig (London, Taylor and Francis, 1974) 447.
12. J.P. deNeufville, S.C. Moss and S.R. Ovshinsky, J.Non-Cryst. Solids 13, 191 (1973/74).
13. S.R. Ovshinsky, in Proc. 5th Annu. Mat. Conf. Industrial Research, Beverly Shores, Ind.: Industrial Research, Inc. (1969) 86.
14. S.R Ovshinsky, in Proc. of the 4th Intl. Congr. for Reprography and Information (April 13-17, 1975, Hannover, Germany) p. 109.
15. S.R. Ovshinsky and D. Adler, Contemp. Phys. 19, 109 (1978).
16. M.H. Cohen, H. Fritzsche and S.R. Ovshinsky, Phys. Rev. Lett. 22, 1065 (1969).
17. W.E. Spear and P.G. LeComber, Solid State Commun. 17, 1193 (1975).
18. D.E. Carlson and C.R. Wronski, Appl. Phys. Lett. 28, 671 (1976).
19. H. Fritzsche, C.C. Tsai and P. Persans, Solid State Tech. 55, 482 (1978).
20. S.R. Ovshinsky, J. Non-Cryst. Solids 32, 17 (1979).
21. S.R. Ovshinsky and A. Madan, in 1978 Ann. Meeting of the American Section of the Intl. Solar Energy Society (August 28-31, 1978, Denver, USA) 69.
22. S.R. Ovshinsky and A. Madan, Nature 276, 482 (1978).
23. S.R. Ovshinsky, New Scientist 80, 674 (30 November 1978).
24. A. Madan, S.R. Ovshinsky and E. Benn, Phil. Mag. B40, 259 (1979).
25. A. Madan and S.R. Ovshinsky, 8th Intl. Conf. on Amorphous and Liquid Semiconductors, August 27-31, 1979, Cambridge, MA; J. Non-Cryst. Solid, 171 (Jan/Feb 1980).
26. M. Shur, W. Czubatyj and A. Madan, 8th Intl. Conf. on Amorphous and Liquid Semiconductors, August 27-31, 1979, Cambridge, MA; J. Non-Cryst. Solid, 731 (Jan/Feb 1980).
27. A. Madan, J. McGill, W. Czubatyj, J. Yang and S.R. Ovshinsky, Appl. Phys. Lett., in press.

PRINCIPLES AND APPLICATIONS OF AMORPHICITY, STRUCTURAL CHANGE, AND OPTICAL INFORMATION ENCODING

S.R. Ovshinsky

Energy Conversion Devices, Inc., 1675 West Maple Road, Troy, Michigan 48084

INTRODUCTION

Materials have always been the basis for new advances in civilization. The use of materials such as stone, iron, bronze, and the tools made from them are the result of the inventive process, and have been used to define the ages of humankind.

We are now faced with momentous problems stemming from the depletability of our natural resources. The challenges can and will be met by the invention of new synthetic materials that can interface with nature and generate the energy so necessary for future progress. The invention of these materials, which can convert various forms of energy directly into electricity, is already taking place.

The key issues of our present industrial society are energy and information. They are actually opposite sides of the same coin. Our energy problems are, in reality, difficulties involving the conversion of one form of energy (light, chemical, etc. to another (mechanical, electrical, etc.). Storage and processing of information also necessarily involve energy conversion and dissipation.(1,2) The transformation that take place in an amorphous solid have to do with both its informational and energy content. Figure 1 illustrates our concept of energy conversion which we approach in two ways--one, as a process by which light, heat, or chemical energy can be transformed into electrical energy through a suitable material; the other, where energy is used as a means of encoding and expressing information through the transformation that takes place in materials so that an input of one form of energy, such as light, electricity, or heat, an be recorded, stored, retrieved, transmitted, that is, communicated or switched into another intelligible form of information, such as computer memories, imaging films, or switching matrices. The common denominator is that the interactions and conversions take place in amorphous materials where the indivisibility of energy and information is uniquely expressed.

When I started work in this area in the 1950's, amorphous materials were not considered scientifically understandable nor to have any particular use. The selenium drum which gained popularity in the 1960's in xerography had evolved empirically and its amorphicity was considered incidental.(3)

We founded ECD in 1960 to develop a broad and basic approach to amorphous materials with applications such as various types of switching and control devices, memories, imaging films, diversified coatings, and energy conversion devices based upon principles described herein:

PRINCIPLES

1. Amorphicity is a generic term referring to lack of X-ray diffraction evidence of long-range periodicity and is not a sufficient description of a material. To understand amorphous materials, there are several important factors to be considered: the type of chemical bonding, the number of bonds

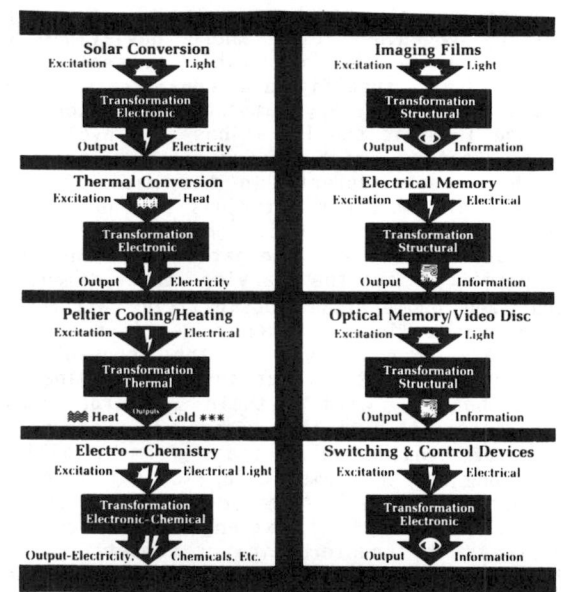

Fig. 1   ECD's energy conversion processes for energy and information.

Reprinted by permission from *Journal de Physique,* Vol. 42, Suppl. to No. 10, C4, p. 42 (1981).

generated by the local order, that is, its coordination, and the influence of the entire local environment, both chemical and geometrical, upon the resulting varied configurations. Amorphicity is not determined by random packing of atoms viewed as hard spheres nor is the amorphous solid merely a host with atoms imbedded at random. Amorphous materials should be viewed as being composed of an interactive matrix whose electronic configurations are generated by free energy forces and can be specifically defined by the chemical nature and coordination of the constituent atoms. Utilizing multiorbital elements and various preparation techniques, one can outwit the normal relaxations that reflect equilibrium conditions and, due to the three-dimensional freedom of the amorphous state, make entirely new types of amorphous materials--chemically modified materials.(4-8)

There are at least two systems operating in amorphous materials: the normal structural bonding which makes up the majority of the bonding configurations of the solid and controls its structural integrity and its optical energy gap, and the deviant electronic configurations (4,7) which are generated by the three-dimensional spatial freedom of individual atoms counter-balanced by the chemical forces surrounding them, their environment.(4,5,7,9-15)

The creation of deviant electronic configurations can be understood simply by viewing even an elemental amorphous material as having several different bonding configurations as available energetic options. It is vital to understand that the same atoms can be found in the same material in different configurations. For example, elemental amorphous silicon, while normally tetrahedrally bonded, has some atoms in which the coordination is not tetrahedral. The same is true of non-tetrahedral amorphous materials exemplified by the chalcogenides (14) where deviations from primary divalency are inexorably present. Local order is always specific and coexists in several configurations in every amorphous semiconductor. Steric and isomeric considerations are involved both with the factors which encourage amorphicity and those that create defects in the materials. The constraints in amorphous materials are not those of crystalline symmetry but are involved with asymmetrical spatial and energetic relationships of atoms permitted by the varying three-dimensional chemical and geometrical possibilities afforded by the amorphous solid. In such a solid, the bonding options are not only of the conventionally considered covalent type, but involve coordinate, or dative bonds with their charge-transfer characteristics.(4,9,11-14) There is not only a spectrum of bonding which spans from metallic to ionic in one and the same solid (10) but a spectrum of bonding strengths. A major factor involved in the spectrum of bond strengths in amorphous materials is the counteracting or competitive force of the chemical environment which acts to influence and alter the bond energy. We therefore can have a greater number of weaker bonds in an amorphous semiconductor than one would find in a crystalline solid.

Physicists have taken for granted that there is a thermodynamic drive toward crystallinity. We emphasize that there is an equally important energetic process that leads to amorphicity, that is, the preferred chemical bonding of atoms and the charge field produced by nonbonding electrons can alter a molecular structure so that it has an anticrystalline state. Crystals by definition have geometries that allow for repetition of the basic cell structure. The shapes that I am discussing are not rigid spherical balls but complex distorted shapes formed by localized pressures, repulsions, and attractions of surrounding forces, compressed here, elongated there, twisted along another axis, the very antithesis of a crystal cell model. These tangled networks are further inhibited from crystallinity by crosslinks and bridging atoms. They are constrained by virtue of the electron orbital relationships, including those of the lone-pair electrons, the chemical influence, the mechanical presence, and the spatial relationships of their nearest neighbors. Such complex three-dimensional forms favor chain and ring formation and are chemically characterized by fluctuations of valency and coordination and can have varied charge conditions. The electron-electron and electron-phonon interactions, therefore, cannot be characterized in a simple manner and the energetic considerations required to complete coordination depend upon the ability to spatially and energetically mate bonding positions.(11-14,16) This becomes increasingly difficult as one goes to the tetrahedral condition which is the reason why elemental amorphous silicon does not have completed structures but has weakened bonds, dangling bonds, and voids. In such cases, other elements with the proper size and charge are necessary, such as fluorine and hydrogen to make up the structural units that complete its bonding and provide stabilizing forces. While the bonding is more complete in materials made of silicon and hydrogen than in elemental silicon, the addition and substitution of fluorine for hydrogen is necessary to have an amorphous silicon-based alloy of optimum stability.(9,17-19) However, once this is accomplished, one no longer has elemental amorphous silicon, but an alloy.

2. Amorphous materials can be separated into two types: unstable, in which the materials' local order does not alter within a given range of temperature or excitation, and bistable, in which structural rearrangements and relaxations occur.(10,20) These can be of a reversible nature and result in a spectrum of changes from subtle nearest neighbor changes to those where many bonds are broken, resulting in various new types of short-range, intermediate and long-range order. Excitation causes conformational changes which result in structural changes so important for the information side of our work. Depending upon the design of the materials, there is again a range of these structural changes that can occur in amorphous materials. These can be detected chemically, for example, by etching, electronically by changes of resistance, as well as optically. All these effects can be made reversible, especially in the nontetrahedrally bonded materials such as the chalcogenides (see Fig. 2).(21-24)

3. The structural configurations in both unstable and bistable materials play a role in establishing the physical, chemical, and electrical properties of the materials. Crosslinks and bridging atoms are not only structural factors

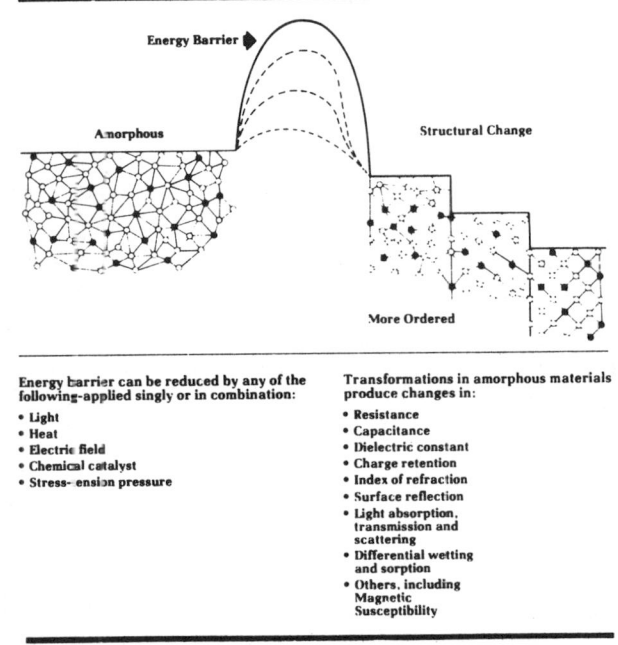

Energy Barrier

Amorphous

Structural Change

More Ordered

Energy barrier can be reduced by any of the following-applied singly or in combination:
• Light
• Heat
• Electric field
• Chemical catalyst
• Stress-tension pressure

Transformations in amorphous materials produce changes in:
• Resistance
• Capacitance
• Dielectric constant
• Charge retention
• Index of refraction
• Surface reflection
• Light absorption, transmission and scattering
• Differential wetting and sorption
• Others, including Magnetic Susceptibility

Fig. 2   Information storage/retrieval and display by structural transformation.

stabilizing amorphicity, but also contribute to the distribution of band states.(16,25) The density of states in the gap of an amorphous material controls its transport properties. The nature and number of the states not only affect the electrical conductivity but are also related to the chemical reactivity of the material.(16,20)

4. In amorphous solids, internal topology can result in another physical parameter, porosity, which results from all of the energetic and three-dimensional spatial factors discussed above, and which can be an important design factor not found in crystalline materials.(26)

5. Even elemental amorphous materials such as selenium, silicon, boron, etc. do not have uniform internal topology as do their crystalline counterparts but have various structural configurations such as rings, chains, voids, etc. I consider elemental materials to be in a sense structural alloys. Compositional alloys have diversity of structure based upon composition; elemental amorphous materials have diversity of structure based upon positional and translational relationships of similar atoms. The local environment in amorphous single elemental materials differs in important ways from that of their crystalline counterparts, and even though both types of materials may have common deviant bond structures, the number and type of such defect states are different in amorphous materials than in their crystalline counterparts. The total environment must be taken into account in the amorphous case in order to properly define the defect states.(9,12)

6. The major thrust of ECD's activities is centered on our conviction that amorphous materials are most interesting in their non-elemental form such as alloys or modified materials. The latter refers to materials in which the normal equilibrium bonding is disturbed by creating new configurations through the insertion of a perturbing element or elements with multi-orbital possibilities.(4,7) For example, alloying allows the optical band gap to be designed at will and yet chemical modification or doping by affecting gap states can independently affect and control the electrical conduction process. Alloys or modified materials can be made using many of the elements in the periodic table and the possibilities can be expanded by introducing as modifiers multi-orbital elements such as the d-band elements. These "pin cushion" orbitals (4,6) interact with the matrix and each other in ways that are unique. They can enter the bonding matrix but, most importantly, these multidirectional orbitals by their interaction can eliminate or create states in the gap. When they generate states, they can be of a much larger number than is allowable by alloying or even doping. Bonding possibilities in the amorphous state include coordinate bonds, three-center bonds, and the multi-orbital options of carbon.(4,7)

7. The elimination of the crystalline constraints and the consequent additional three-dimensional freedom afforded by the amorphous state permit the widest variety of bonding and antibonding orbital relationships that can be found in a solid and are our key to the ability to synthesize new inorganic materials not found in nature.

8. The understanding of spin pairing is involved with the structural flexibility of the amorphous material and the nature of the surrounding environment. Electron-phonon interactions in amorphous materials can be varied by changing either the composition or the preparation conditions. Spin pairing is not an inherent feature of all amorphous materials as has been suggested by Anderson.(27) As I previously pointed out,(11,28) it takes too much energy to pair spins in materials with a rigid matrix, for example, there is not enough bonding flexibility in sputtered amorphous tetrahedral materials to overcome the energy barrier between different configurations. However, the flexibility afforded by, for example, primarily divalent chalcogenide (lone pair) materials encourages the necessary electron-phonon interactions to drive the bonding deformation necessary to pair all the spins.(9,11-13) The energy barriers between such rearrangements are minimized by the effective electron-phonon coupling permitted through the relaxation modes available in amorphous materials designed with a flexible matrix. This is analogous to the competition between crystal-field effects and magnetic interactions in transition-metal compounds, leading to a high-spin state in some materials and a low-spin state in others.

In summary, the transport properties of amorphous materials can be understood from their deviant bonding character just as those of crystalline materials are understood and controlled by the deviations from long-range order. In amorphous materials, these deviations are directly involved with fluctuations of coordination and directionality of the bonding and nonbonding orbitals. The intersection of these spatially and energetically varied orbitals in three-dimensional space provides the opportunity for unusual

295

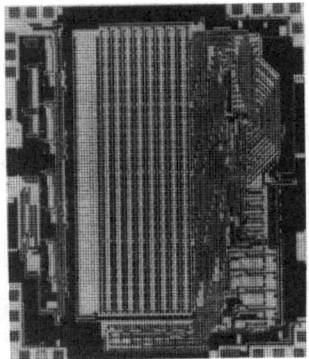

Fig. 3   1024 bit current mode logic Ovonic EEPROM measuring 136 mil by 160 mil.

electronic excitation and recombination mechanisms as well as the possibility for structural change if bistability is desired. In either case, bistable or unstable, deviant short-range order viewed against a background of the more commonly occurring normal short-range order is the Rosetta Stone for the understanding of electronic activity in amorphous materials.(4,7,14)

INFORMATION

As we have indicated above, the amount of stiffness in an amorphous matrix is a designable parameter. The most flexible structures are the reversible chalcogenide bistable materials which have an inherent plasticity in which information can be encoded by virtue of a structural change--from subtle bonding changes in the amorphous phase to an amorphous-to-crystalline transition. The alteration in the material reflects the informational content of the energy put in in either digital or analog form, that is, either in series or in parallel, and can be detected and read in any number of ways (see Fig. 2). As one adds stronger bonds and more crosslinks in building a more rigid material, reversibility of the amorphous-to-crystalline phase is diminished; for example, in amorphous silicon-based materials. These nonreversible materials can have device usages as well.

1. <u>Ovonic memory switches</u>. The reversible amorphous-to-crystalline transformation has become increasingly important as microprocessors have gained utilization in the office, factory and business environment. There is a great need for a

	1975	Now
$V_{Threshold}$	22V	8V
I_{Write}	150ma	5ma
I_{Read}	1ma	1ma
Write Time	10ms	1ms
Read Time	45ns	15ns
Processing Temperature (Max)	80℃	200℃
Storage Temperature (Max)	80℃	175℃
Operating Temperature (Max)	80℃	110℃
Cell Size	4mil²	0.75mil²

Fig. 4   Recent improvements in operating and manufacturing parameters of the Ovonic EEPROM device technology.

high-density, short-access-time, nonvolatile random access memory. The new improvements in the Ovonic EEPROM technology are most timely in this context.

The basic Ovonic memory switch can be used in a memory matrix to achieve a fast, completely nonvolatile EEPROM (electrically erasable programmable read-only-memory) device. In conjunction with Burroughs, and now with Sharp Corporation, a 1024 bit BCML (Burroughs Current Mode Logic) Ovonic EEPROM has been produced with access times of less than 15 nanoseconds (Fig. 3). Further materials development has subsequently improved the basic cell characteristics. The performance improvements are summarized in Fig. 4. These performance advantages are achieved by grading from tellurium-rich alloys on the top to a germanium-rich alloy on the bottom.(29-31)

2. <u>Optical memories</u>. Since the 1960's,(22-24,32-38) ECD has pioneered in and developed various types of optical memories based upon the principles of bistability. Ovonic materials have characteristics which make them ideally suited for optical information storage requiring high-density recording and high signal-to-noise ratios. The materials utilized exhibit a bistability in which both the amorphous and crystalline state can co-exist at ambient conditions. There are many properties which exhibit a dramatic difference between the two states. For an optical memory the characteristics that are important are the index of refraction and the absorptivity. The change in these two provides a film that can be read in either the reflective or transmissive mode. The materials are reversible (erasable) and have superior image quality.

The change in state of Ovonic material coated on an optical memory disc can be accomplished by a variety of methods. For example, a laser beam focussed to a sub-micron spot can be used to both crystallize and revitrify a single track or portion thereof on a spinning disc. A laser pulse that will vitrify the crystalline portion of the material can be used to recrystallize the same material by simply lowering the intensity by two thirds. Therefore, depending upon the laser intensity, the same spot can be repeatedly switched. Bulk erasure can also be accomplished by several means.

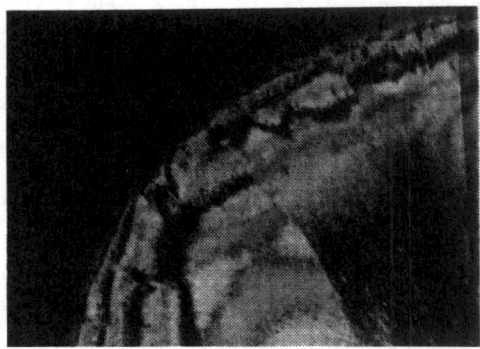

Fig. 5   Transmission electron photograph of a section of a crystalline dot on an amorphous background. The featureless area is the amorphous region.

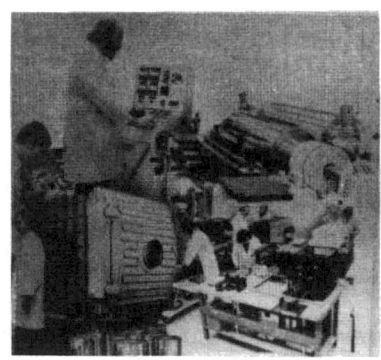

Fig. 6 Ovonic nonsilver photo-duplication film.

Fig. 7 400x enlargement of half-tone dots illustrating ECD's graphic arts film's ability to dot etch while holding a sharp border and maintaining dot hardness.

The application of light is all that is required to complete the change of state. No further processing is necessary. Lasers whose outputs range from the ultraviolet to the infrared can be used. The energy required to switch one square micron of material is between 0.1 and 0.2 nanojoules. On a disc spinning at 1800 rpm, this corresponds to less than 10 milliwatts of light at the disc.

The image quality is excellent. Figure 5 shows the ability to achieve a dot with an edge sharpness of 100Å. The resolution of the medium exceeds one thousand line pairs per millimeter. The spot size is not diffraction limited. The threshold recording aspects of the material can be used to record a smaller than diffraction limited spot by utilizing the Gaussian distribution of the intensity. Applications include archival and updatable computer storage, large volume data storage, consumer and industrial video disc and video disc mastering.

3. _Instant Dry Process Imaging Film._ We have developed a silverless dry process imaging film based upon an organotellurium compound. This film also utilizes an amorphous-to-crystalline phase transition and offers continuous tone with exceptionally high resolution and radiometric speed.

The mechanism can be described as an initial event resulting in the electronic excitation of a photosensitive reactant to a highly reactive species which then, through chemical action, initiates a reducing step forming a latent image. Through either simultaneous or sequential heat treatment, the organotellurium compound undergoes autocatalytic decomposition to tellurium which in the latter stages of thermal development generates crystallites in the form of needles which constitute the visible image.(34,39) Figure 6 is the reproduction of such an image.

4. _Photostructural Films._(40) We early developed materials which can be affected by light so as to make them selectively etched.(10,33) One such film is in production in Europe by Agfa-Gevaert for application as a graphic arts film. This nonsilver film has very high contrast coupled with high resolution. It is a contact speed film, available in both reversal and nonreversal forms. It can be used either for photographic or mask-making purposes.

The dot edge sharpness (see Fig. 7) is unexcelled. Dot edge sharpness refers to the linear distance over which the film is able to go from its lowest to its highest density. Since the film has only two available densities, the transition distance becomes virtually zero. The resolution of this film exceeds 600 line pairs per millimeter.

5. _MicrOvonic File System._ (See Fig. 8.) The essence of this system is a low melting alloy coated on polyester to form a fiche card which displays photostructural changes upon exposure to a given threshold level from pulsed energy sources. Typically, xenon flash or laser sources are used with pulse durations of 1μ-$1ms$ and energy values on the order of 0.1 joule/cm^2 to 1 joule/cm^2.

6. _Unstable Materials._ Amorphous materials can be designed from the chalcogenide to the tetrahedral type so as to be structurally stable under the influence of excitation processes. The threshold type is exemplified by the Ovonic threshold switch which I invented in 1960 and which has been fully described elsewhere.(21,41-42) Its switching speed at room temperature is comparable to that of Josephson diodes which require cryogenic temperatures. As digital applications increase, it is anticipated that this switch will find increasing utility. Transistor-type action in our Si:F:H alloy

Fig. 8 MicrOvonic File: this system permits not only the recording of documents on fiche but also provides for deletion and addition of all information.

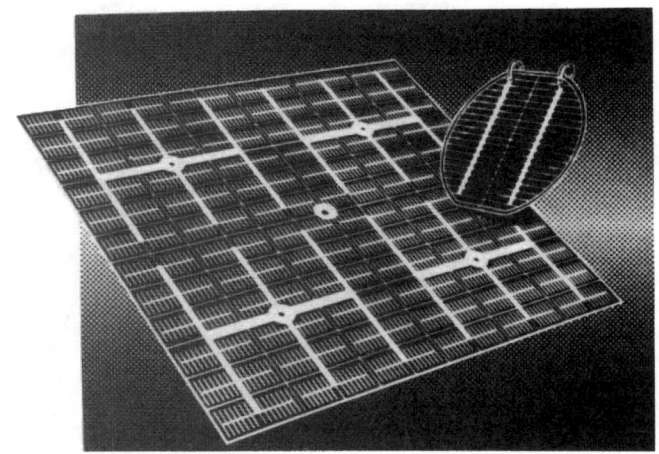

Fig. 9   Ovonic photovoltaic cell 1 ft. x 1 ft. compared to typical 4 in. diameter crystalline cell.

indicates that large-area thin-film transistors will soon see application.

ENERGY

Amorphous materials for energy usage are almost entirely of the unistable type. While there are many uses of amorphous materials utilizing the principles that we have outlined ranging from thermoelectricity and electrochemistry to superconductivity, the area of greatest immediate interest is photovoltaics.

We announced a silicon-fluorine-hydrogen alloy and described its material and photovoltaic device properties.(17-19,43-50) The efficiencies are very promising and Madan is reporting on them at this meeting. Figure 9 shows a one-square-foot array developed by Izu and his group at ECD. This device represents a great deal of attention to the production technology so necessary to utilize the inherent value of a large-area film. For the first time, alternative energy has the possibility of being competitive in cost to coal, gas, oil and uranium. Our targeted efficiencies for near-term production are 7-10% for these one-square-foot cells. We will report on newer alloys and configurations with even more optimal characteristics in the near future.

SUMMARY

There is a chemical and structural basis for amorphous materials which has guided our work from its very beginning. It has allowed us to alter band gaps at will through alloying, understand and define defect states and spin pairing, introduce doping in chalcogenides and modification across the periodic table. It has permitted us to utilize the super halogenicity of fluorine for its organizing (inductive) influence to manipulate size and charge of various atoms, and to design atomic and molecular configurations best suitable for specific tasks. The electronic excitatory processes in these materials have unique attributes that are reflected both in speed of switching and the highly conductive state seen in threshold switches. We have been able, through modification, to break the lockstep

which has connected the optical band gap, electrical activation energy and even the thermal conductivity of materials. By independently controlling and manipulating these important parameters, new areas of applications are possible. The ability to selectively retain or alter structure, the latter even in a reversible manner, has valuable device implications as well.

We believe the old debate whether or not one has to work with elemental amorphous semiconductors in order to understand the amorphous state has been resolved. Our view is that alloys and modified materials based upon structural and chemical considerations outlined herein are best suited not only to clarify the scientific understanding of amorphous materials but as a vehicle for new technology which will be beneficial to society.

REFERENCES

1. R. Landauer, IBM J. Res. Develop. 5 (1961) 183.
2 M. Tribus, Thermostatics and Thermodynamics (1961) xiii. (Published by D. Van Nostrand Company, Inc., New York.)
3. E.M. Pell, Xerography and Related Processes, ed. John H. Dessauer and Harold E. Clark (1965) 65. (Published by The Focal Press.)
4. S.R. Ovshinsky, Amorphous and Liquid Semiconductors, ed. W.E. Spear (1977) 519. (Edinburgh: Centre for Industrial Consultancy and Liaison.)
5. R. Flasck, M. Izu, K. Sapru, T. Anderson, S.R. Ovshinsky and H. Fritzsche, Amorphous and Liquid Semiconductors, ed. W.E. Spear (1977) 524. (Edinburgh, CICL.)
6. New Scientist 76 (1977) 491.
7. S.R. Ovshinsky and D. Adler, Contemp. Phys. 19 (1978) 109.
8. S.R. Ovshinsky, K. Sapru and K. Dec., in Proc. Intl. Topical Conf. on the Physics of SiO_2 and its Interfaces, Yorktown Heights, New York (1978).
9. S.R. Ovshinsky, J. Non-Cryst. Solids 32 (1979) 17.
10. S.R. Ovshinsky and H. Fritzsche, IEEE Trans. on Electron Devices ED-20 (1973) 91.
11. S.R. Ovshinsky, Structure and Properties of Non-Crystalline Semiconductors, ed. B.T. Kolomiets (1976) 426. (Leningrad: Nauka.)
12. S.R. Ovshinsky, Structure and Excitation of Amorphous Solids, ed. G. Lucovsky and F.L. Galeener (1976) 31. (New York: AIP.)
13. S.R. Ovshinsky, Phys. Rev. Lett. 36 (1976) 1469.
14. S.R. Ovshinsky and K. Sapru, Amorphous and Liquid Semiconductors, ed. J. Stuke and W. Brenig (1974) 447. (London: Taylor & Francis.)
15. For a fuller treatment see Grigorovici Festschrift to be published in the special issue of Revue Roumaine de Physique.
16. S.R. Ovshinsky, J. Non-Cryst. Solid 2 (1970) 99.
17. S.R. Ovshinsky and A. Madan, in Proc. 1978 Meeting of the American Section of the International Solar Energy Society, ed. K.W. Boer and A.F. Jenkins (1978) 69. (AS of ISES: Univ. of Delaware.)
18. S.R. Ovshinsky and A. Madan, Nature 276 (1978) 482.
19. S.R. Ovshinsky, New Scientist 80 (1978) 647.
20. S.R. Ovshinsky and I.M. Ovshinsky, Mat. Res. Bull. 5 (1970) 681.

21. S.R. Ovshinsky, Phys. Rev. Lett. 21 (1968) 1450.
22. S.R. Ovshinsky, J. Appl. Photog. Eng. 3 (1977) 35.
23. S.R. Ovshinsky, in Proc. 4th Intl. Congress for Reprography and Information, Hannover, Germany (1975) 109.
24. S.R. Ovshinsky, presented at the Topical Meeting on Optical Storage of Digital Data, Aspen, Colorado (1973).
25. M.H. Cohen, H. Fritzsche and S.R. Ovshinsky, Phys. Rev. Lett. 22 (1969) 1065.
26. S.R. Ovshinsky, presented at the Gordon Research Conference on Catalysis, New London, New Hampshire (1978).
27. P.W. Anderson, Phys. Rev. Lett. 34 (1975) 953.
28. H. Fritzsche, Electronic Phenomena in Non-Crystalline Semiconductors, ed. B.T. Kolomiets (1976) 65. (Leningrad: Nauka.)
29. S.R. Ovshinsky, U.S. Patent No. 3,271,591.
30. V.A. Bluhm, U.S. Patent No. 4,115,872.
31. S.H. Holmberg, U.S. Patent No. 4,177,475.
32. S.R. Ovshinsky, in Proc. of Industrial Research Conference, Chicago, Illinois (1969) 86.
33. S.R. Ovshinsky and P.H. Klose, J. Non-Cryst. Solids 8-10 (1972) 892.
34. S.R. Ovshinsky and P.H. Klose, Nonsilver Photographic Processes, ed. R.J. Cox (1975) 61. (Academic Press: London.)
35. J.P. deNeufville, Amorphous and Liquid Semiconductors, ed. J. Stuke and W. Brenig (1974) 1351. (London: Taylor & Francis.)
36. J. Feinleib and S.R. Ovshinsky, J. Non-Cryst. Solids 4 (1970) 564.
37. J. Feinleib, J.P. deNeufville, S.C. Moss and S.R. Ovshinsky, Appl. Phys. Lett. 18 (1974) 254.
38. S.R. Ovshinsky, U.S. Patent No. 3,530,441.
39. Y.C. Chang, and S.R. Ovshinsky, U.S. Patent No. 4,142,896.
40. The amount of disorder is a factor in many of these materials. Complete amorphicity is not always a requirement.
41. D. Adler, H.K. Henisch and N.F. Mott, Rev. Mod. Phys. 50 (1978) 209.
42. D. Adler, M.S. Shur, M. silver and S.R. Ovshinsky, J. Appl. Phys. 51 (1980) 3289.
43. A. Madan, S.R. Ovshinsky and E. Benn, Phil. Mag. B40 (1979) 259.
44. M.S. Shur, W. Czubatyj and A. Madan, Solar Energy Materials 2 (1980) 349.
45. A. Madan, W. Czubatyj, D. Adler and M. Silver, Phil. Mag. B42 (1980) 257.
46. A. Madan and S.R. Ovshinsky, J. Non-Cryst. Solids 35 & 36 (1980) 171.
47. M. Shur, W. Czubatyj and A. Madan, J. Non-Cryst. Solids 35 & 36 (1980) 731.
48. A. Madan, J. McGill, S.R. Ovshinsky, W. Czubatyj, J. Yang and M.S. Shur, in Proc. of the Society of Photo-Optical Instrumentation Engineers 248 (1980) 26.
49. A. Madan, J. McGill, W. Czubatyj, J. Yang and S.R. Ovshinsky, Appl. Phys. Lett. 37 (1980) 826.
50. D. Adler, M. Silver, A. Madan and W. Czubatyj, J. Appl. Phys. 51 (1980) 6429.

THE CHEMICAL BASIS OF AMORPHICITY
Structure and Function

Stanford R. Ovshinsky

Energy Conversion Devices, Inc., 1675 West Maple Road, Troy, Michigan 48084

Dedicated to Professor R. GRIGOROVICI
on the occasion of his 70th anniversary

Professor Grigorovici has made important contributions to the understanding and development of amorphous materials. In his long and productive career, he has been particularly involved with the structural basis underlying amorphicity.

As Grigorovici has pointed out, the prevailing thought in amorphous materials has been Joffe and Regel's empirical rule:(1) "that a molten or amorphous semiconductor retains its semiconducting properties in spite of the destruction of the long-range order if only the non-crystalline phase conserves the short-range order present in the related semiconducting crystal. The short-range order was characterized by the number of atoms in and the radius of the first coordination sphere."(2) He concluded that "There is at present not a single known exception to this rule."(2)

The conventional viewpoint is that this rule taken together with the microcrystalline approach, the perturbed crystal concept, and the continuous network approach completely account for the chemical and physical properties of amorphous solids.(3) With regard to the continuous network model, Professor Grigorovici has written, "In this approach the amorphous solid is supposed to consist of an infinite, non-periodical, three-dimensional array of interlinked atoms in which the short-range order about each atom is imposed by the same characteristics of the chemical bond as in the crystal."(4) However, just as Grigorovici implies his doubt in this statement by using the word "supposed," I have not believed that either the rule, the continuous network concept, or any of the other mentioned models were adequate to deal with the true nature of the amorphous state. While I cannot in this paper deal in depth with my objections to each of the models, I feel that by stating my basic premises, their important differences from the conventional approaches will become apparent.

A part of the overall thesis that I have developed since 1957 (5-7) is that: (1) the structural units that assure aperiodicity need not have any counterparts in an "analog" crystalline lattice; (2) the chemical bonds which are important in amorphicity can differ from those of related semiconducting crystals both in quantity and in quality; (3) the short-range order responsible for transport properties differs from that which results from the chemical bonds of the crystal; and (4) "the short-range order about each atom" is not imposed by "the same characteristics of the chemical bond as in the crystal." In other words, short-range order is not conserved in amorphous materials. Indeed, when crystalline symmetry is destroyed, it becomes impossible to retain the same short-range order. The reason for this is that the short-range order is controlled by the force fields of the electron orbitals and therefore the environment must be fundamentally different in corresponding crystalline and amorphous solids. In other words, it is the interaction of the local chemical bonds with their surrounding environment which determines the electrical, chemical, and physical properties of the material, and these can never be the same in amorphous materials as they are in crystalline materials.(8-14)

To begin with, deviant electronic configurations (DECs), which result from the interactive local environmental conditions, are dispersed throughout the normally structurally bonded amorphous matrix. Nonbonding electrons or empty orbitals can find themselves in local environments that are forbidden in a crystalline structure. I have characterized amorphous materials as being composed of normal structural bonds (NSB) as well as the deviant bonds (DECs) mentioned above.(12,13) The NSB, being in the vast majority in the amorphous matrix, are responsible for the cohesive energy of the material. Their bonding configurations can be chemically similar to those of crystalline materials; for example, amorphous chalcogenides are primarily divalent; amorphous silicon, primarily tetrahedral. However, although a material can be mostly divalent or mostly tetrahedral in its amorphous state, this does not mean that the material is analogous even in the single component phase to its crystalline counterpart. For example, neither the band gap of amorphous silicon nor of amorphous tellurium is similar in magnitude to its crystalline counterpart.

Reprinted by permission from *Revue Roumaine de Physique*, Vol. 26, Nos. 8/9, pp. 893–903 (1981).

To say that disorder is the cause of the difference is to beg the question. What is important is the local environment which interacts with and affects the bonding, generates DECs, and by its interaction with the bonding creates a different entity. My point is the _local environment_ of a crystal can never be truly the same as that of an amorphous material. The orbital relationships that can exist in three-dimensional space in amorphous but not crystalline materials are the basis for new geometries, many of which are inherently _anti-crystalline_ in nature. Distortion of bonds and displacement of atoms can be an adequate reason to cause amorphicity in single component materials. But to sufficiently understand amorphicity, one must understand the three-dimensional relationships inherent in the amorphous state, for it is they which generate internal topology incompatible with the translational symmetry of the crystalline lattice. Single component amorphous materials do not provide the clearest models of amorphicity. Although they resemble alloys in that they have a diversity of structural units such as, for example, the rings and chains in a-Se, they are pale images of compositional mixtures, since they do not have the richness of the three-dimensional relationships afforded by the addition of other elements and the multiple interactions that can result from them.

What is important in the amorphous state is the fact that one can make an infinity of materials that do not have any crystalline counterparts and that even the ones that do are similar primarily in chemical composition. The spatial and energetic relationships of these atoms can be entirely different in the amorphous and the crystalline forms, even though their chemical elements can be the same. The deviant bonds that control the electronic activity of amorphous materials and are therefore of prime importance in understanding the amorphous state, are made possible only by the elimination of the lattice and crystalline periodicity. Therefore, they have a configurational freedom, dictated by the energy of their environment, which simply is not available in periodic materials. _However, in terms of their activity, neither deviant bonds (DECs) nor the normal structural bonds (NSB) can be separated from their surrounding environment, for it is their interaction with it and the resulting feedback which create a combined new entity which I call the total interactive environment (TIE)._ TIEs are particularly important in representing the entire configuration of the DECs. The TIE has a localized geometry encompassing a total chemical site, which, if viewed as composed only of its parts, would not give the correct picture because the interaction of the parts creates a Gestalt different from and more than the sum of its parts.(14) There is in amorphous materials a broad spectrum of TIEs, and of particular relevance are the ones which incorporate DECs.

It follows that the effects of perturbating or varying the short-range order are very different from those that occur when the long-range order is affected since transient excitations are absorbed in specific localized configurations in an amorphous material and are dampened by the nearest-neighbor environment rather than perpetuated through the lattice as in crystalline materials. As amorphous solids become more rigid, the electron-phonon interactions change so that perturbations have more long-range consequences. I feel strongly that there are electron-phonon interactions in amorphous materials whose physical implications have not been analyzed even qualitatively. Since amorphous solids can have localized as well as extended phonons, electron-phonon scattering can be much different in the amorphous and crystalline states. The phonon modes are important since, when conduction occurs in amorphous materials, the normal scattering processes of the crystalline materials can be absent. This should be taken into account whenever the carrier mobility in amorphous materials is analyzed. It is important to determine how the transition from localized to extended states takes place under excitation conditions. In specific circumstances, these events do not happen only locally but can involve cooperative phenomena.

I believe there have been many qualitative errors in the analysis of the thermal properties of amorphous materials. An understanding of the physics of amorphous materials should begin by bearing in mind that thermal activation increases the overall energy throughout the material. Therefore, its activity is clear and rather trivial, for if one perturbs the vibrations between atoms throughout a material, there is no reason to consider any unique electronic responses. However, optical and electrical excitation can specifically couple to sites in the surrounding matrix in a manner much different from in crystalline materials. Since nonbonding electrons have little effect on the vibrations between atoms, lone-pair amorphous semiconductors can have particularly interesting high field effects.(8)

An electric field need not result in major effects on the primary matrix bonds (NSB) but can considerably affect the nonbonding charges and deviant local order composed of the charged TIE. I have been advocating for many years that an understanding of the chemistry and the structure of amorphous materials also requires a recognition of its photochemical and electrochemical reactivities. Basically, an excited amorphous material contains different chemical species than does its normal state. This factor has been an important tool for me in my work. I have pointed out that an amorphous matrix can have solvent and solute-like properties (15-18) since the absorption of a photon allows the activation of a specific 'solute' molecule in the presence of an excess of transparent 'solvent.' Therefore, light is a very selective tool in a glass, just as it is in liquids; in this respect, light is much different from heat, which affects the overall energy of the environment. Photoexcitation and recombination processes as well as photochemical reactions can be focussed into a small volume of amorphous material thereby changing local molecular configurations. Localized excitation and high electric fields can affect an amorphous material in various ways including inducing a cooperative delocalization of carriers. If the energy of excitation cannot be contained by the local environment, i.e., if the material is designed to be unstable to the energy of excitation, then conformational change will lead to configurational change, eventually resulting in localized phase changes and even crystallization. This type of amorphous material is called bistable, and is the basis of our Ovonic memory and imaging materials.(16-24)

Under the influence of a high electric field, sufficient free and nonbonding electrons can be excited in an amorphous chalcogenide so as to propagate as a plasma through the material. If this occurs in a glass in which the relaxation processes of the matrix can accommodate the excited conformational change by virtue of many strong bonds, then structural change will not occur and the amorphous material is unistable. Materials such as these are utilized in our Ovonic threshold switching.(19,21,22)

Electronic excitation of the proper energy and injection processes of the proper intensity can selectively affect those states created by the TIE. Charged sites can also act as efficient traps and recombination centers.

TIEs are reflected in the electronic density of states which controls the transport properties of the amorphous solid. If an amorphous material has a sufficiently low density of localized states, it can be utilized as a thin-film field-effect transistor (FET) in memory and circuit applications.

Understanding that displacement and redistribution of charge are inherent manipulatable parameters in amorphous materials has been an important factor in my thinking since the 1950's. I interacted amorphous materials with selected ions to alter their conductivities, their electronic, and, under specific conditions, their structural configurations, as well as restructuring local order with amphoteric atoms, and showed that dramatic electronic changes could result.(5,7,12,13,25,26) I utilized the ideas of both electronegative and electropositive charge in the design of materials in order to find ways of modifying the density of states of amorphous solids, and feel that the use of fluorine in some of our latest materials (27-29) exemplifies how not only electronegativity can affect bonding but can be fundamental controlling factor of the total interactive environment (TIE).

We take for granted that there are equilibrium structures representing the lowest energy configurations possible for any composition. But even the ground state is controllable in amorphous materials, since they generally have several bonding options. This means that an atom has different local configurations with which it can interact and that these choices can be close in energy. I have given elsewhere examples of how equilibrium can be outwitted to produce dramatic changes in, for example, electrical conductivity and have called it chemical modification.(12,13,26,29,30) This principle follows from the chemical compensation concepts that I proposed earlier.(8,31) The important point that I wish to emphasize here is how the concept of screening is connected with modification, compensation and the total interactive environment (TIE). I know the word "screening" has a precise meaning, that is, the effective reduction of the electrostatic force (either positive or negative) between two charged particles brought about by the displacement of any of the other charges in the material. I should like to expand now upon this definition by applying its principles to bonding and nonbonding configurations in amorphous materials.

Accepting the need to consider amorphous materials from a multi-particle point of view,

rather than from the two-dimensional flatland in which much of crystalline physics operates, helps in viewing a specific TIE. Let us consider a situation in which some atoms in a TIE are added, displaced, or replaced. It should be apparent that alterations of electrostatic forces will occur by virtue of the substitutions, and the individual bond energies will be modified. In other words, there can be a weakening of some of the bonds, a redistribution of charge, and even a change in sign of charge. It is in this sense that I use the term "screening," for not only are the overall atomic charges affected, but through polarization effects dipoles and higher order multipoles can vary the spatial charge distribution.

The charge effects inherent in amorphous materials are not just those resulting from dangling bonds, voids, charged pairs, etc., but can reflect a more subtle type of center, not simply connected with either under- or over-coordination, but rather representing a varying amount of effective charge or charge distribution. Only a part of a configuration need contain fluctuating charge. The various means of assuring overall charge neutrality is a subject in which I have been much interested,(32) and I believe that the chemical basis of the charge distribution resides not only in deviant bonds (DECs) but in the nature and shape of the TIEs. I believe that chemists are more prepared to understand the implications of these discussions since chemical donor and acceptor molecular interactions and the charge-transfer concepts involved with them have a relationship to the kinds of problems that chemists address. For example, a chemistry book that does not even mention amorphicity discusses donor and acceptor concepts and the relationship between equilibrium structures and different molecular environments, and states that "any charge density pattern is reflected in a characteristic structural pattern, which determines the reactivity pattern within the molecule."(33)

The charge-transfer characteristics of amorphous materials are tied very much to the TIE and are directly related to the spin pairing qualities of the materials, their corrosion resistance, their electrochemical and photochemical properties, their electric field effects, density of states and recombination processes. That is the reason that, while I had originally suggested unusual lone-pair orbitals, new bonding conditions and one- and three-electron bonds as vital factors in chalcogenides, I felt that the important KAF (34) model that developed from these thoughts did not sufficiently take into account the influence of the total interactive environment. The TIE not only is involved in the creation and expression of the valence alternation pairs (VAPs) but also shows that the VAPs should not be separated from their environment and indicates that there are also other charged centers that play a role in these materials. I feel that the suggestion that the concept of VAPs can be directly transposed to crystalline materials (35) is incorrect, since bonding should not be considered in isolation but as a part of the TIE.(14)

There is a spectrum of sites affecting electrical conductivity, trapping, and recombination processes associated with charge-transfer mechanisms inherent in the TIE principle. If an optical, electrical, chemical or even mechanical perturbating force is imposed, various transformations of the TIE can occur. I believe that the interconversion of such TIEs is the basis for the DC effect seen in some Ovonic threshold materials, and that such interconversions can also have dramatic influence on parameters such as conductivity and serve as nucleating points for crystallization in the less structurally stable memory materials.

Since Grigorovici introduced the valuable concept of amorphons, it is unusual to think of amorphous solids in terms of asymmetrical building-block clusters of atoms which fill space so that a crystalline lattice cannot be constructed. A more chemical view can be stated in the following manner: starting with the individual atom, the directionality of its outer orbitals is crucial, i.e., p orbitals, lone-pair orbitals, d orbitals, electron-deficient orbitals, and elements that exhibit multivalency all can be used to generate irregular configurations in three-dimensional space as they combine with other atoms into larger molecular structures. Therefore, in my opinion, there are configurations which are inherently anticrystalline. They have no crystalline analogs because the chemical forces that shape them have so twisted and deformed them that complex and distorted shapes are formed, and, therefore, it is impossible for them to be packed together to form a crystalline lattice. Some examples of this include the chemical forces that bind selenium atoms to arsenic atoms in the As-Se system as well as those that bind silicon and oxygen in amorphous SiO_x. The structural effects of crosslinks in amorphous materials much more resemble polymer chemistry than they do crystalline physics.(22,36)

I have discussed bulk amorphous materials as reflecting in a certain sense a continuous surface. One of the relevant aspects of this is that a surface need not just have dangling bonds, but because of the additional degree of freedom, reconstructions such as back bonding can take place. If one allows for the additional degrees of freedom that exist in the bulk of an amorphous solid due to the lack of crystalline constraints, then the analogy becomes even clearer; new bonding configurations become possible whose only restrictions are the energetic ones of the TIE. Physicists are accustomed to thinking in terms of donors and acceptors in band models. Those working in amorphous materials should think more chemically and recognize that, for example, electron-deficient bonds such as are associated with boron can act as acceptors (12,13) and that lone pairs can act as either acceptors or donors. I have stated on many occasions that the coordinate or dative bond that results from the interaction between a lone pair and an empty orbital is an important one in amorphous materials, and that there can be a spectrum of bonding, antibonding, and nonbonding configurations in one and the same amorphous material.(8-11,23) It is important to understand that the nonbonding orbitals, particularly lone pairs, can have a profound effect on the shape of local order configurations,(8-11,23) and are, therefore, an important factor in the TIE.

My arguments relative to spin pairing follow from the concept of a total interactive environment. The enhanced ability of such a configuration to "breathe" relative to a crystal allows strong electron-phonon interactions which in turn can induce spin pairing. This view differs sharply with that of Anderson who assumes that disorder itself is sufficient for a negative correlation energy.(37) My position is that if the energy barriers between different configurations are sufficiently large, e.g., in amorphous elemental tetrahedral materials, where the rigidity of the matrix does not permit much flexibility in the four bonding positions, then dangling bonds and voids ensue. On the other hand, the lower energy deformations of a more flexible matrix (lower coordination material) allow pairing to proceed and this is the physical basis of the negative correlation energy.(9-11,38)

The question of impurities and paramagnetism in amorphous materials deserves consideration. I feel that impurities are legitimate parts of the amorphous matrix and therefore can be responsible for paramagnetism. Paramagnetism can occur when an "impurity" atom (actually just a different atom in a multi-elemental compound or alloy) is chemically bonded and becomes an integral and inherent part of the matrix. If the additional electronic energy of such an atom is less than the energy necessary to deform the surrounding matrix, then paramagnetism remains. This is particularly favorable when there are so few of the "impurities" that they are spatially well separated. However, when I introduced transition metals which would normally be paramagnetic into an amorphous matrix in a chemically modified manner, diamagnetism rather than paramagnetism resulted.(12,29,39,40) There were sufficient centers within critical energy and spatial relationships so that the TIE could be affected by these orbitals and a redistribution of electrons could take place resulting in spin pairing, very much as crystal-field mechanisms can contribute to spin pairing.

In considering spin pairing in alloys and compounds rather than in elemental materials, the deformation potential is built into linking atoms, which can also act to eliminate dangling bonds. When added bridging and crosslinking elements are combined to make a new alloy, the necessary linkages for deformational purposes are introduced and pairing can be provided by the bonding and the charge influence of the specific bridging element.

When fluorine is used, its extreme electronegativity is also able to affect the TIE by removing an electron from one of the other components and opening up new bonding possibilities. This can induce pairing within a configuration in which paramagnetism might occur if the atom were neutral. The TIE is also affected by fluorine providing the energy necessary to alter the shape of the configuration so that nearby orbitals can be drawn together causing pairing. Its electronegativity can act not only to bond electrons but to attract them sufficiently so that they can pair other electrons in their proximal environment. It has been an important part of my thinking to consider the coordination expansion ability of fluorine which is so hungry for electrons that in silicon it can even induce six-fold coordination as it can act upon even the d orbitals of silicon and can increase silicon's coordination beyond tetrahedral. If the silicon is under-coordinated, fluorine can saturate it by forming an ionic bond or by interacting with its nearest neighbors. In addition, fluorine can be the intermediary for both structural and chemical changes, acting in a manner affecting nearby charge states, electrons and bonds. Its charge influences can even cause them to draw together, overcoming repulsive forces and effectively pairing their spins. The chemical combinations of fluorine also can affect spins in, for example, the following manner: two fluorine atoms can bridge spatially separated under-coordinated silicon atoms, thus eliminating their spins. The relatively small size of fluorine, its specific ability to bond to a single orbital, its extreme electronegativity and high inductive effect, and its ability to be a bridging structural unit as well as a bonding unit without adding new deleterious states in the gap, all combine to cause a-Si:F:H to be a superior electronic and physical material to a-Si:H.(14,27-29).

The greater hardness and density of a-Si:F:H over a-Si:H support these statements. Regarding a-Si:H alloys, I believe that my assertion that such materials are "two-step forward, one-step backward" materials has since been borne out. I suggested that hydrogen, although a small atom which ordinarily coordinates singly thus eliminating most but not all localized states, in addition to acting as a structural unit, has several different bonding possibilities with silicon, which could lead to the formation of new states in the gap.(12) For example, I have previously discussed the importance of three-center bonds in hydrogenated alloys.(12,13) The spectrum of weak bonding inherent in silicon-hydrogen configurations simply does not make a-Si:H a preferred material. Fluorine, in addition to its own strong bonding to silicon and silicon-hydrogen complexes, through its organizing action can, in fact, act to cause and specify certain silicon-silicon and silicon-hydrogen interactions and can therefore by affecting the TIE yield a new material with a lower density of localized states and much greater stability.

Anyone who has worked with a highly excited plasma should become converted to the notion that I have been advocating that the radical chemistry that goes on transiently in such a plasma and in its interaction on surfaces make for configurational combinations of such richness that even though the same elements are involved, the materials can differ substantially merely by altering the reactivity of the plasma. In other words, the elements have not changed, but their interactions have. I call this radicalization of the plasma.

My "bete noir" is continuously being told that such and such a material is amorphous silicon when it is truly a different material, i.e., an amorphous silicon-based alloy. The same is true for the chalcogenides. If an element is chemically incorporated into the matrix, whether or not it has unpaired spins, it becomes a new alloy. Even when a particular material has the same composition, but its properties are strongly dependent on the preparation conditions, the TIE is different and we have a different material. Isomerism and steric relationships must be taken into account when characterizing amorphous materials.

One can change compositions gradually or drastically in amorphous materials, making them more interesting and basically different from crystalline materials. There is a huge spectrum of amorphous materials that can be designed and tailor-made. This is why I call such materials synthetic—they do not depend upon what nature has provided us nor do they follow the dictates which tell us how to prepare crystalline alloys. Rather than being frightened by this proposition, and thinking of amorphous materials only in crystalline analogies, we should instead consider them to be like liquids where the character of the solution can be subtly or drastically changed by the addition of other chemical elements. Amorphous materials are not liquids, but they have a chemistry of their own which generates a myriad of internal shapes, bonding and electron distribution which control the physics involved. Equilibrium conditions either in a solid or a liquid are not very relevant in considering new material possibilities utilizing the synthetic approach.(41) I wish to repeat an old comment of mine—too many people entering the amorphous field gain comfort by looking for the lost coin under the lamppost rather than where they lost it. My position is that nature has provided us additional freedom in the amorphous state, and we should use it. As an inventor, I welcome it.

It is interesting to consider some basic electronic mechanisms that are different in amorphous and crystalline solids. The radiation hardness and recombination velocities of most amorphous solids are related to the many recombination centers and to relaxation modes that arise from the phonon relationships that I have been discussing. The ability not to propagate a disturbing event through a material combined with a built-in healing process and a rapid relaxation of the TIE make these materials intriguing, since they can be designed for either radiation selectivity or hardness.

Another point is that tunneling can occur in amorphous materials much more easily than in crystalline materials. Esaki has summarized the problem in crystalline materials quite well: "The Zener mechanism in dielectric breakdown has never been proved to be important in reality. If a high electric field is applied to the bulk crystal of a dielectric or a semiconductor, hot-electron effects such as impact ionization, avalanche, etc., precede tunneling, and thus the field never reaches a critical value for tunneling. In other words, observation of tunneling phenomena, as discussed here (band-to-band tunneling), has been limited to particular types of junctions: narrow p-n junctions, narrow Schottky junctions, conductor-thin-insulator junctions, etc."(42) I early noted in thin films of our chalcogenide threshold material (approximately several hundred angstroms thick) that the I-V characteristic of the threshold switch was transformed into one which had no "wings" (i.e., only the offset conducting state in the first and third quadrants remained) and was not temperature sensitive. I concluded that in an amorphous material, tunneling was the preferred mechanism, since free carriers in chalcogenides lack sufficient mean-free path to induce impact ionization and avalanche. It may very well be that the best examples of true Zener breakdown are in amorphous materials.(43)

In terms of devices, we will be reporting many new applications for amorphous materials in the near future. I believe that amorphous materials will not only add greatly to our scientific knowledge of many-body orbital interactions in three-dimensional space, but will help solve our societal and technological problems.

I have welcomed the opportunity to honor Professor Grigorovici. This paper represents a summary of my approach to amorphous materials, an approach which has allowed me to work as an inventor-scientist in this field since the 1950's. I know that Academician Grigorovici must be proud to think back to the time we had our international meeting in Rumania in 1967 with few people attending and no formal proceedings, and note the growth of the field by more than an order of magnitude, as indicated by this year's meeting in Grenoble, where there will probably be over 500 people with at least two bulky volumes of proceedings published.

It has been a great pleasure for Iris and me to have Radu Grigorovici as our friend and as a stimulating scientific colleague who has addressed basic and important questions in amorphicity and has greatly advanced the field.

I must add a personal note to this paper which deals with the intuitive areas of science. Hearing Radu Grigorovici play the piano at our home in Michigan and at his in Bucharest demonstrates the artistry of the man just as reading his scientific papers makes his scientific attributes clear. Let us make sure that as the science of amorphous materials becomes more institutionalized, it does not lose the fermenting creative spirit of its early days.

ACKNOWLEDGEMENTS

I wish to thank David Adler for his many helpful clarifications.

REFERENCES

1. A.F. Joffe and A.R. Regel, Progress in Semiconductors, ed. A.F. Gibson 4 (Heywood, London: 1960) p. 237.
2. R. Grigorovici, Amorphous and Liquid Semiconductors, ed. J. Tauc (Plenum Press, London & New York: 1974) p. 69.
3. R. Grigorovici, J. Non-Cryst. Solids 35 & 36 (1980) p. 1167.
4. R. Grigorovici, Amorphous and Liquid Semiconductors, ed. J. Tauc (Plenum Press, London & New York: 1974) p. 86.
5. J.D. Cooney, Control Engineering 6 (1959) p. 121.
6. L. Young, Anodic Oxide Films, (Academic Press, New York: 1961) p. 147.
7. S.R. Ovshinsky, The Physical Base of Intelligence-Model Studies, presented at Detroit Physiological Society, December 17, 1959.
8. S.R. Ovshinsky and K. Sapru, Amorphous and Liquid Semiconductors, ed. J. Stuke and W. Brenig (Taylor & Francis, London: 1974) p. 447.
9. S.R. Ovshinsky, Phys. Rev. Lett. 36 (1976) p. 1469.
10. S.R. Ovshinsky, Structure and Excitation of Amorphous Solids, ed. G. Lucovsky and F.L. Galeener (AIP, New York: 1976) p. 31.
11. S.R. Ovshinsky, Structure and Properties of Non-Crystalline Semiconductors, ed. B.T. Kolomiets (Nauka, Leningrad: 1976) p. 426.
12. S.R. Ovshinsky, Amorphous and Liquid Semiconductors, ed. W.E. Spear (Centre for Industrial Consultancy and Liaison, Edinburgh: 1977) p. 519.
13. S.R. Ovshinsky and D. Adler, Contemp. Phys. 19 (1978) p. 109.
14. S.R. Ovshinsky, J. Non-Cryst. Solids 32 (1979) p. 17. (Mott Festschrift.)
15. S.R. Ovshinsky, in Digest of Technical Papers, Topical Meeting on Optical Storage of Digital Data, Aspen, Colorado (1973), p. MB5-1.
16. S.R. Ovshinsky and P. Klose, Nonsilver Photographic Processes, ed. R.J. Cox (Academic Press, London: 1975) p. 61.
17. S.R. Ovshinsky, in Proceedings of the 4th International Congress for Reprography and Information, Hannover, Germany (1975) p. 109.
18. S.R. Ovshinsky, J. Appl. Photog. Eng. 3 (1977) p. 35.
19. M.P. Southworth, Control Engineering 11 (1964) p. 69.
20. S.R. Ovshinsky, presented at the IV Symposium on Vitreous Chalcogenide Semiconductors, Leningrad, USSR, May 1967.
21. S.R. Ovshinsky, Phys. Rev. Lett. 21 (1968) p. 1450.
22. S.R. Ovshinsky, J. Non-Cryst. Solids 2 (1970) p. 99.
23. S.R. Ovshinsky and H. Fritzsche, IEEE Trans. on Electron Devices ED-20 (1973) p. 91.
24. J.P. deNeufville, Amorphous and Liquid Semiconductors, ed. J. Stuke and W. Brenig (Taylor & Francis, London: 1974) p. 1351.
25. S.R. Ovshinsky and I.M. Ovshinsky, Mat. Res. Bull. 5 (1970) p. 681.
26. S.R. Ovshinsky, J. Non-Cryst. Solids 42 (1980) p. 335.
27. S.R. Ovshinsky and A. Madan, in Proceedings of 1978 Meeting of the American Section of the International Solar Energy Society, ed. K.W. Boer and A.F. Jenkins (AS of ISES, Univ. of Delaware: 1978) p. 69.
28. S.R. Ovshinsky and A. Madan, Nature 276 (1978) p. 482.
29. S.R. Ovshinsky, New Scientist 80 (1978) p. 647.
30. New Scientist 76 (1977) p. 491.
31. S.R. Ovshinsky, Electronic-Structural Transformations in Amorphous Materials, A Conceptual Model, submitted to Nature, July 13, 1972, unpublished.
32. M.H. Cohen, H. Fritzsche and S.R. Ovshinsky, Phys. Rev. Lett 22 (1969) p. 1065.
33. V. Gutmann, The Donor-Acceptor Approach to Molecular Interactions, (Plenum Press, New York & London: 1978) p. 6.
34. M. Kastner, D. Adler and H. Fritzsche, Phys. Rev. Lett. 37 (1976) p. 1504.
35. M. Kastner, J. Non-Cryst. Solids 31 (1978) p. 223.
36. S.R. Ovshinsky, Lecture Notes for Recent Advances in Polymeric Materials, March 1977, unpublished.
37. P.W. Anderson, Phys. Rev. Lett. 34 (1975) p. 953.
38. H. Fritzsche, Summary Remarks at the 5th International Conference on Amorphous and Liquid Semiconductors, Electronic Phenomena in Non-Crystalline Semiconductors, ed. B.T. Kolomiets (Nauka, Leningrad: 1976) p. 65.
39. R. Flasck, M. Izu, K. Sapru, T. Anderson, S.R. Ovshinsky and H. Fritzsche, Amorphous and Liquid Semiconductors, ed. W.E. Spear (CICL, Edinburgh: 1977) p. 524.
40. H. Fritzsche, unpublished communication.

41. S.R. Ovshinsky, to be published.
42. L. Esaki, <u>Tunnelling Phenomena in Solids,</u> ed. E. Burstein and S. Lundqvist (Plenum Press, New York: 1969) p. 50.

43. S.R. Ovshinsky, presented at the 1969 Gordon Conference on the Chemistry and Metallurgy of Semiconductors at Proctor Academy, Andover, New Hampshire, July 14-18.

306

FUNDAMENTALS OF AMORPHOUS MATERIALS

Stanford R. Ovshinsky

Energy Conversion Devices, Inc., 1675 West Maple Road, Troy, Michigan 48084

I.   INTRODUCTION

When I first began to study amorphous materials in the mid 1950's, the field appeared to be as mysterious as hieroglyphics had been to renaissance scholars. While it was taken for granted that amorphous materials had no real significance scientifically or technologically, it was clear to me, even then, that this was a rich, unexplored, and important area of science [for early references see 1-5]. Until then its major thrust was in the ancient art of glass making, and glass meetings devoted inordinate amounts of time to discussing "What is glass?"

Just as there was a Rosetta Stone which allowed the deciphering of the hieroglyphics of ancient civilizations, the following is the key I provided to make the nature of amorphous materials clear and to understand their physical properties.

It is the purpose of this paper to discuss how we broke the code, and how we have applied this insight to the development of an array of new devices, several of which will be described in detail. Such an understanding of our field is not yet widespread. For example, I recently received a book from Professors Yonezawa and Ninomiya,[6] both fine scientists. In discussing topologically disordered systems, i.e., amorphous materials, they state, "In this kind of disordered system, long-range order in the atomic distribution is completely broken while the short-range order (...referred to as SRO), is maintained in the sense that the coordination number of each atom remains the same as in the case of a corresponding ordered crystal, although bond lengths and angles in a disordered system fluctuate."

That statement is insufficient and can be misleading since the characteristics of amorphous materials are controlled not only by the fluctuations of bond lengths and bond angles with the consequent loss of periodicity and the establishment of chemical short-range order,[7] but also by the following interrelated factors which make up the Rosetta Stone for understanding amorphous materials. First, there is an <u>average coordination number</u> which defines the structural integrity of the material and its gap and is determined only by the chemistry of the constituent atoms; I have called this its <u>normal structural bonding (NSB)</u>. Second, it is the <u>deviations</u> from the optimal coordination number, the <u>deviant electronic configurations (DECs)</u>, that are essential to the understanding of the important phenomena in amorphous materials.[8,9] It is these DECs which determine the transport properties of amorphous materials and are responsible for the states in the gap. Third, there need not be "corresponding crystal structures," the central dogma of many working in the amorphous field, a leftover from crystalline physics with its inherent dependence on a lattice structure. The ability to <u>design</u> and <u>synthesize</u> a great variety of amorphous materials depends on the fact that many <u>do not</u> have corresponding crystal structures.

There is a subtle but important insight which should be kept in mind. It is that while short-range order and deviant electronic bonding represent distinct configurations whose total energy can be calculated, there is another distinction that reflects a localized region, the <u>total interactive environment (TIE)</u>. This TIE depends on a number of factors of which the nearest-neighbor bonding is but one; others include the effects of nearby chemical forces and of electrical charge distribution which are reflected in the overall three-dimensional topology and in the character of the states in the gap. Perturbations of the TIE can occur by excitational processes.[10]

It is difficult to understand now, but the absolutist belief of physicists in the dogma of the crystalline lattice as the basis of semiconductor science can be appreciated by tracing the attitude of Ziman, one of the leading figures in solid-state theory. In 1965 he wrote in his well-known introductory book on solid-state physics,[11] "A theory of the physical properties of solids would be practically impossible if the most stable structure for most solids were not a regular crystal lattice." Later, in 1969, at the Third International Conference on Amorphous and Liquid Semiconductors, he delivered a paper [12] entitled "How It It Possible To Have An Amorphous Semiconductor?" In this talk he proved that, since there is no regular lattice in amorphous

Reprinted by permission from D. Adler, B.B. Schwartz, and M.C. Steele, eds., *Physical Properties of Amorphous Materials,* Plenum Press, New York, 1985, pp. 105–155.

307

materials, there can be no band gap, and therefore these materials <u>cannot</u> be semiconductors. Of course, this misses the whole point of the CFO model with its concept of a mobility gap,[13] illustrated in Fig. 1. Finally, indicating how science progresses, or better, how scientists progress, Ziman later published another book, <u>Models of Disorder</u>,[14] in which he states, "Condensed-matter physics has expanded in recent years and shifted its centre of interest to encompass a whole new range of materials and phenomena. Fundamental investigations on the molecular structure of liquids, on amorphous semiconductors, on polymer solutions, on magnetic phase transitions, on the electrical and optical properties of liquid metals, on the glassy state, on metal ammonia solutions, on disordered alloys, on metallic vapours--and many other interesting systems--now constitute a significant proportion of the activity of innumerable physical and chemical laboratories around the world." He continues, "This research is not purely academic: disordered phases of condensed matter--steel and glass, earth and water, if not fire and air--are far more abundant, and of no less technological value, than the idealized single crystals that used to be the sole object of study of 'solid state physics.'" These contradictory quotes [15] suggest the climate in which we were living when I first discussed amorphous materials at scientific meetings. While the situation is much better these days, there are still remaining misconceptions which this paper will attempt to clarify. In so doing, we will address the fundamental principles of amorphous materials.

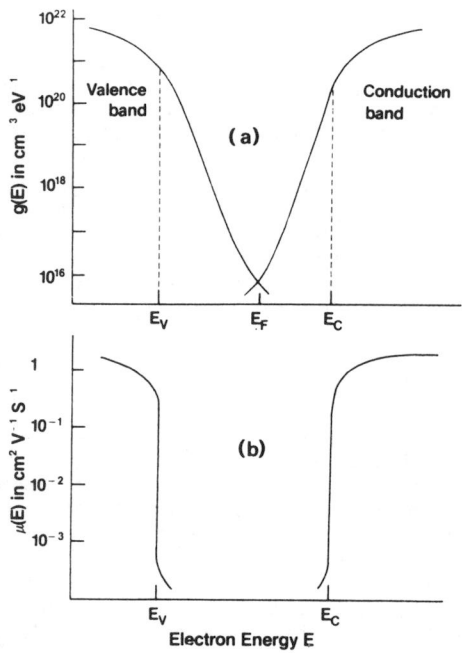

Fig. 1. Sketch of the densities of states of the valence and conduction bands and corresponding electron and hole mobilities. The magnitude of the mobilities should be regarded as approximate because no quantitative calculations have been made. States which are neutral when occupied are associated with the valence band, those neutral when empty with the conduction band; they overlap in the mobility gap. (Ref. 13.)

Kuhn's book [16] should be required reading for the historical and philosophical understanding of how scientific paradigms are developed. He discusses how anomalies appear in physical understanding, how the new solutions based upon original thinking unacceptable to the status quo physicists emerge, and then after a period of struggle, how a new mindset is generated and a new field is created. "More clearly than most other episodes in the history of at least the physical sciences, these display what all scientific revolutions are about. Each of them necessitated the community's rejection of one time-honored scientific theory in favor of another incompatible with it. Each produced a consequent shift in the problems available for scientific scrutiny and in the standards by which the profession determined what should count as an admissible problem or as a legitimate problem-solution. And each transformed the scientific imagination in ways that we shall ultimately need to describe as a transformation of the world within which scientific work was done. Such changes, together with controversies that almost always accompany them, are the defining characteristics of scientific revolutions." It is amazing how Kuhn's description applies to the development of amorphous materials in contradiction to the accepted and dogmatically defended crystal-line approach. I was particularly struck by his description of how scientists attempt to explain new phenomena by seeking to extend their conventional approaches to the point of irrationality. This can have unfortunate consequences. For example, what a waste of time it was for us to have to prove over and over that threshold switching was really electronic rather than thermal![17-32] Shaw [33], who for a time embraced the thermal theory, has recently put the final nail in its coffin.

II. CHEMICAL CONSIDERATIONS

It is important to discuss some of the still-remaining misconceptions. It appears puzzling to many that materials composed of exactly the same elements can have completely different structural and electronic properties, depending upon how they are processed. The reason that many amorphous materials are preparation-dependent is that the same elements can combine with each other in a number of different and distinct configurations. The local order actually chosen depends on the nature of the chemical bonding, which in turn is predicated on several factors, including dynamic considerations, for ours is not a chemistry of equilibrium states. The possibility of steric isomerism results in the same elements in different configurations displaying very different chemical reactivities and electronic properties. The internal freedom for placement of atoms in three-dimensional space without long-range order allows for new design possibilities not found in crystals. Indeed, it was stereo- and polymer chemistry that was my guide from the beginning.

Instead of a lattice of repetitive atoms, amorphous solids form a matrix where bonding and nonbonding orbitals with different energies interact in three-dimensional space, sometimes yielding charged centers and thus internal electric fields. The particular bonding option chosen by an atom as it seeks out an equilibrium position on the surface of a growing film is dictated by the kinetics, the orbital directionality, the state of excitation of the relevant atoms, and the temperature distribution during the deposition process. With the constraints of crystalline symmetry and lattice specificity lifted, new internal configurations can and do develop. Rheology plays a role, disclosing important differences between amorphous and crystalline materials, with the former exhibiting unique electron-phonon relaxation processes and pseudo-equilibria.

The chemical foundation of amorphous materials can be clarified by considering, e.g., why and how the carbon atom forms the basis of organic chemistry. For just as the varied bonding possibilities of carbon can generate many different configurations,[9,34] even though the same elements are involved, the multi-orbital choices of the elements in an amorphous material can lead to differing configurations. This is the basis for my broadly classifying our materials as synthetic. The difference between amorphous inorganic materials and synthetic organic materials is qualitatively important since we can make not only high temperature, chemically stable, passive materials which in themselves outperform plastics, but we can also make amorphous solids that are electronically active and can be used as switches, memories, transducers, photovoltaic cells, batteries, catalysts, superconductors, etc. Be reminded that there was no such list in 1960.

Directionality of bonding, multi-valences, and varied coordination possibilities, all of which are involved in the offering of multi-orbital choices, become the building blocks for amorphicity. It should not be a surprise that the temperature of the substrate, the state of excitation of the atoms, and the reequilibration kinetics all affect how orbital relationships are formed and how atoms select one another to make up a desired material. Therefore, substrate temperature, orbital directionality, multi-atomic interactions, sticking coefficients, free radical chemistry, and diffusion coefficients are important considerations in how atoms in amorphous materials relate to other atoms and build up their local geometries. These controllable parameters are important assets for they permit us to engineer many new and useful materials as well as being of great scientific value.

It is often asked if there is a fundamental difference between glasses and amorphous materials. The difference is simply that scientists who prepare materials by quenching from the melt make use of a longer time scale than those who deposit atoms on surfaces directly from vapor or plasma phases. Therefore, more equilibrium structures can be expected in glasses. The time and energy required for two atoms to bond to one another can be considered to be design parameters. For example, if there are four outer p electrons, as in a chalcogenide material, but only two bond in the NSB configuration, it is easier to prepare an amorphous material than if one has to bond four outer sp^3 electrons, as in elemental amorphous silicon. We can chemically aid the process by making it easier for atoms to bond to each other. How do we accomplish this? It is exceptionally difficult, if not impossible, to form amorphous silicon from the melt (except under laser energization), but it is easy to form amorphous selenium in this way. In the former

case, the liquid is not tetrahedrally coordinated and quickly crystallizes upon quenching. One has to add an interfering additive to prevent this crystallization and, more importantly, one has to bond all four outer electrons to obtain the tetrahedral structure. The rigidity of that structure can be understood by anyone who has tried to fit four surfaces together. In mechanics, one must insert a shim or a gib to do so. In stereo- and polymer chemistry, a "fitting link," either a crosslink or a bridge, is needed. In elemental amorphous silicon, it costs too much strain energy to try to bond all four orbitals [35-38] when all must be distorted to fit the local geometry. The result is that there are many strained bonds, dangling bonds are prevalent, and voids are formed in the solid. One would expect this from free energy considerations.

In contrast, in chalcogen elements, only two of the outer electrons need to be utilized for structural bonding. The remaining lone pair can assume a spectrum of nonbonding or bonding relationships.[39] Consequently, more flexible chain and ring structures result in the chalcogenides, more rigid structures in the tetrahedral materials. In both cases, I utilized stereo- and polymer chemistry concepts to control rigidity. In the tetrahedral materials, additional alloying elements are needed to reduce the strain and _lower_ the average coordination of the structure. They can also act in a bridging manner, like oxygen in fused silica. In the chalcogenide materials, alloying elements should be preferentially those that effectively crosslink the material, thus increasing the average coordination,[17,35,36,38-40] and making for more stable structures; i.e., they should _add_ rigidity. If these alloying and bridging rules are not followed, then the rapid quench rate that one achieves, e.g., by sputtering, only leads to the freezing in of local atomic mismatches and strains. Wherever there are strains or, more importantly, wherever there are _bonding options_, DECs are ordinarily created yielding large densities of localized states whose origin and significance need to be understood, especially if one wants to control or eliminate them. For example, the DECs in elemental silicon are generated by undercoordination; the DECs in chalcogenide materials arise from the various lone-pair configurations.[35,36,39] We can control them in the former by compensating the dangling bonds, e.g., with fluorine and hydrogen,[41] and in the latter by interacting the lone pairs with modifying elements.[8,38,42,43]

Chemical understanding must be translated into specific topological configurations, since the local geometries reflect the appropriate chemistry in amorphous materials and structure and function are indivisible. Grigorovici [44] was early interested in the structural configurations and internal topology of amorphous materials. Our work emphasizes the correlation of internal geometries with electronic properties. Surface topology in periodic materials is related to the lifting of restrictions in the free space above the surface. Therefore, the study of crystalline surfaces can be a useful first step in the study of bulk amorphous materials.[45,46] In fact, the unusual back-bondings at crystalline surfaces can provide clues of internal bulk configurations of amorphous materials. The TIE is different on the surface than in the bulk, for the third dimen-

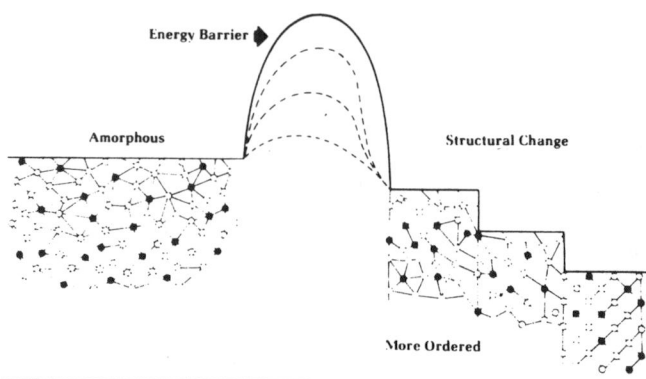

Energy Barrier

Amorphous

Structural Change

More Ordered

Energy barrier can be reduced by any of the following-applied singly or in combination:

- Light
- Heat
- Electric field
- Chemical catalyst
- Stress-tension pressure

Transformations in amorphous materials produce changes in:

- Resistance
- Capacitance
- Dielectric constant
- Charge retention
- Index of refraction
- Surface reflection
- Light absorption, transmission and scattering
- Differential wetting and sorption
- Others, including Magnetic Susceptibility

Fig. 2. Information storage/retrieval and display by structural transformation. (Ref. 50.)

sion in the bulk sets up its own chemical and electrical constraints.

Understanding that the types of defects available in amorphous materials are intimately related to the internal degrees of freedom unique to noncrystalline solids, one can appreciate that the defects are really part of the total interactive environment and part of the energy considerations therein. Defects need not be only dangling bonds, but can be very similar to the unusual bonding configurations that occur in amorphous chalcogenides or variations of the back-bonding that occur at surfaces. In the same amorphous material, there can be a whole spectrum of bonds including metallic, covalent, ionic and coordinate.[2] Whether they appear as defects or not depends upon the particular design of the material.

III. THERMODYNAMIC CONSIDERATIONS

Thermodynamically, if we have a system that has several possible configurations with essentially equal bulk energies open to it, depending, e.g., on the temperature distribution, we may well ask how the atoms developing into a solid choose between them. I would like to briefly discuss the meaning of metastability in amorphous solids. How often have we heard that amorphous materials are metastable? We should bear in mind that so is diamond! Should we consider tectites as unstable? Amorphous materials can be very stable indeed. When we want to utilize their metastability, we do so by design. The understanding of energy barriers on an atomic scale, as well as on a more macroscopic scale, is a crucial point. As we have shown, the barrier between the amorphous and crystalline phases can be controlled, as is sketched in Fig. 2. It is adjustable by altering the bond strengths of the

atoms involved, and it can be lowered or overcome by external energy sources. For crystallization to occur, there must be a cooperative action of a large cluster of atoms, but many subtle changes can occur first. Far more subtle barriers exist than the one between the amorphous and crystalline phases. Slight differences in energy can have important influences on the various conformations and configurations that are inherent in amorphous materials and the transformations available to them. One internal structure can be converted into another without affecting important properties of the material or, for that matter, without even breaking bonds, as indicated in Fig. 3. (However, the TIEs would be affected.) The closeness of energy of the various conformations and configurations can be masked by thermal vibrations (phonons) down to very low temperatures.[35] I interpret the so-called universal, low-temperature, two-level atomic tunneling systems seen in glasses and many amorphous materials as direct evidence of the multi-equilibrium possibilities that I have been describing. Other such evidence includes the photostructural changes that characterize both chalcogenide and tetrahedral alloys.

Rather than postulate that only bond switching is the source of the specific heat anomalies which have been viewed as atomic tunneling phenomena, larger-scale relaxations unique to the disordered and amorphous state could be the most accurate explanation, especially since these represent the conformational changes discussed in this paper. Such changes are directly related to variations in the TIE which reposition atoms, ions, and charged as well as neutral defects, to new positions related to the rest of their environment; i.e., there can be a new TIE as one changes the phonon concentration. This is reflected in the character and number of states in the gap. From the very beginning of my work in this field, I have been emphasizing that the coupling between electrons and phonons is, and must be, basically different in amorphous and crystalline materials. It is in pursuit of a direct demonstration of this concept that I have been actively working on superconductors since the early 1970's.[47,48] I am certain that investigations of the phonon spectra of amorphous material will some day be one of the most exciting new areas of scientific research.

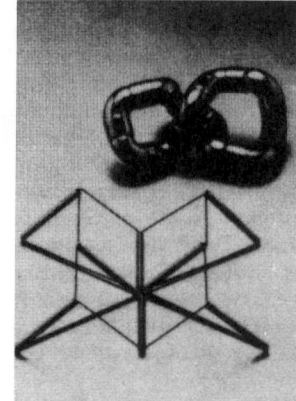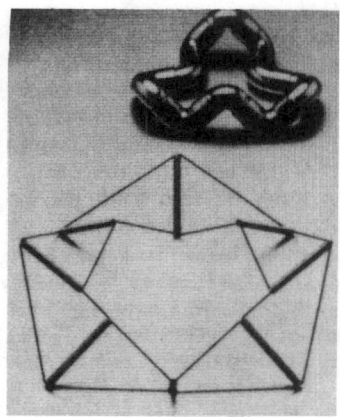

Fig. 3. Models illustrating conformational changes without bond breaking--the interconversion of one structural configuration into another.

Our concept of metastability begins on an atomic level, or, because atoms are not isolated in amorphous materials, rather on a molecular one. Let us assume that a local atomic cluster has been excited by inducing a transition from a low-energy molecular orbital to a higher-energy one, and ask what happens to the TIE? It must change, but how? It will change transiently if the local environment absorbs the excitation energy as it does in the Ovonic Threshold Switch; but if the added energy is dissipated through structural interactions that cannot contain the local conformational changes, as in the Ovonic Memory Switch, then the surrounding structure will disperse the energy in a manner which not only reshapes the conformation with an attendant redistribution of charge but also results in a configurational change, i.e., a breaking of bonds. These configurational changes can be designed to be reversible. There can be a whole spectrum of such changes, including the formation of crystallites. Whether the process is reversible or irreversible is basically a matter of the bond strengths, the size of the crystallites, and the topological and chemical environment. There are not only energy barriers in amorphous materials inhibiting crystallization but also many more subtle barriers involved with atomic and molecular scale changes which are part of the relaxation process unique in amorphous materials.

Reversible amorphization can be pictured as the dissolving of the periodic structure into the surrounding matrix.[49,50] This solute-solvent concept is an apt analogy, since it conjures up the picture of precipitating under certain sets of conditions and dissolving under others. Unlike the absorption of energy in a crystal which then propagates throughout the entire lattice, such events in amorphous materials can be very localized. That is why recombination of carriers has more important consequences in amorphous materials than in crystalline. A knowledge of the principles and processes of relaxation, nucleation, and of catalytic effects is necessary for the understanding of crystallization mechanisms in amorphous materials.

Not satisfied with the conventional wisdom that one had to have melting, i.e., a transition to the liquid phase, in order to reach the amorphous state, I proposed the concept of "amorphization" to describe the process of going from an ordered to a disordered system, and placed emphasis on this process occurring from chemical interactions without the temperature having to exceed the melting point although, of course, it may.[51] My theory, which has now been vindicated by many experiments, was that there is a dynamic chemical force tending to bring about the amorphous state, which can represent a configuration equally as attractive as the crystalline one under certain conditions. As an example of what we might call an "anticrystalline" configuration, let us consider a tellurium atom which is initially part of a chain as in crystalline tellurium but is also near an arsenic atom. If energy is supplied to the vicinity of this local area, the resultant displacement of the tellurium atom under consideration can cause it to align its orbitals within the chemical field of the arsenic atom and form a tellurium-arsenic bond. Since this is more stable than the tellurium-tellurium bond which it replaced, absorption of the energy in this case has led to a destruction of the crystal struc-

ture. The tellurium-arsenic configuration and, even more, the selenium-arsenic configuration are cross-linked, disordered ones, and can be thought of as anticrystalline. We have shown that there is an analog of the amorphization process in the mechanism of crystallization. If many free carriers are generated by light or electric field, then the relaxation processes favor ordering without the need for melting.

IV. CHEMISTRY AS A DESIGN TOOL

Right from the beginning of my work in amorphous materials, I have used a chemical approach as a basic design tool. The Periodic Chart of the Elements shown in Fig. 4 [52] has been for me primarily a means of deciding which elements could bond to each other in such a way as to control not only the shape and magnitude of the mobility gap but also the density of localized states in the gap. Since many of our materials are multi-component alloys, this concept can be illustrated by examples which will be detailed subsequently, but whose simple premises follow.

I have emphasized that it is not only the bond strengths but the type and number of crosslinks which control the barrier to crystallization. One can frustrate crystallization by steric hindrances. For a unistable material, we utilize maximum numbers of strongly bonded atoms and crosslinks, e.g., silicon, germanium, arsenic, and oxygen, as the crosslinks for a tellurium-based alloy. For a bistable material, we reduce the bond strengths of the alloying elements and also reduce their concentration, e.g., some or all of the arsenic can be replaced by antimony, which forms weaker bonds, and some or all of the silicon can be replaced by germanium, or even by tin or lead, for the same reasons. A glance at Fig. 4 shows the chemical logic in this method. It also follows that as we reduce the bond strengths, we concomitantly reduce the band gap of the material. The lone pairs in chalcogenides and the various configurations that they enter into control the transport properties of these materials.[39] We utilize small amounts of additional elements in our multi-component materials not only for their spatial, structural and chemical

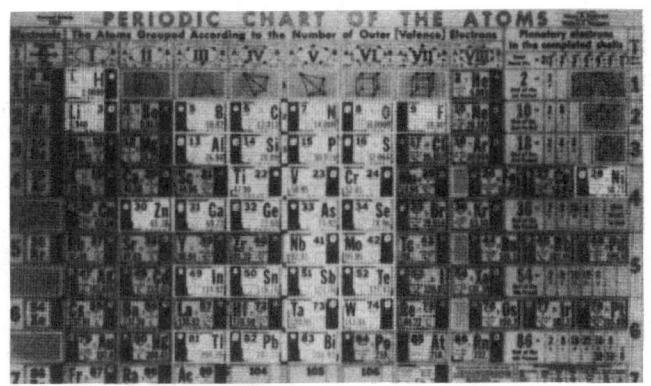

Fig. 4. Periodic chart of the elements with examples of the various elements that can be utilized to fabricate amorphous materials or to modify or dope them.

effects, but also for the influence they have on the electronic activity.

From the above, we can see that mere atomic displacements or simple distortions of a crystalline structure do not do the concept of amorphization justice. The two phases of our bistable materials can co-exist at room temperature. The balance can be shifted from one state to the other. For example, excitation in a memory material can result not only in crystallization but also in a tendency toward the amorphous state because of the chemical forces discussed above. We showed, e.g., that excitation could either inhibit crystallization or expedite it.[53] The outer electron lone pairs of the chalcogens are analogous to the double bonds of carbon in that their possible conversion into bonding-antibonding pairs opens up a host of different configurational structures with nearly the same energy. These new configurations fall into the category of DECs.

Fundamental to my way of thinking relative to the Periodic Chart has been the fact that low average coordination ordinarily favors the amorphous state. In multi-component alloys, the additional elements aid in assuring optimal coordination.(9,35] The balance between adding constraints, completing structures, and assuring rigidity becomes a chemical design parameter as will be seen from our subsequent discussion of tetrahedral materials.

Chemistry and structure are related through the concept of underlined connectivity,[38] for lattice constraints limit the ways in which atoms connect to each other, but the different possibilities in three-dimensional space in amorphous materials allow many new geometric configurations. The consequences of this concept, together with free energy considerations, is that there is not just one equilibrium but various equilibria. Structure and function are connected. If one wants to design and define a local order, then the entire local environment, the TIE, must be taken into account. Selective excitation can, in fact, add an important dimension in designing new configurations that would otherwise not be available through the usual thermodynamic considerations. We can also use such electronic pathways and their recombination events in amorphous materials for various memory and photographic applications.[2,45,46,49-51,55,56]

By perturbing the TIE, one also perturbs the density of states. It is no wonder that the Staebler-Wronski effect [54] can be understood as an example of a photostructural change [41] instead of appearing as some new esoteric phenomenon. If one excites carriers and recombination events occur in a material which has several different structural relaxations available to it, one can forget the conventional picture of a well-defined density of states;[55,57] one can readily see that there would be a redistribution of the localized states as a consequence of the redistribution of atomic configurations in three-dimensional space. From the beginning of our work we have used electro- and photostructural effects constructively for device applications. As has been shown in our laboratory by Guha et al.,[58] recombination is the mechanism which explains the worrisome Staebler-Wronski effect in hydrogenated amorphous silicon alloys.

Fritzsche's "hills and valleys" model [59] puts into perspective some of the consequences of the atomic fluctuations related to the density of states. The charge density of fluctuations in such situations is connected with positional relationships. Therefore, while local chemical bonding is of great importance because it has calculable bond strengths, and therefore short-range order, it does not adequately reflect the true spatial state of affairs of an amorphous solid. The overall positional charge-density fluctuations in three-dimensional space as well as the nature of the chemical bonds are an integral part of the total interactive environment.

It is important to reemphasize that in amorphous materials we do not use the concept of lattice but of matrix.[60] The normal structural bonding (NSB) makes up the great majority of bonding configurations, and therefore is responsible for the cohesive energy, the structural integrity, and the optical energy gap of the material. This gap, as we have discussed, can be adjusted by alloying, and is related to the bond strengths of the elements involved. Compositional, positional, and translational disorder inherent in amorphous materials are reflected in the shape and sharpness of the mobility edge and the density of states in its vicinity. The origin of the density of states in the gap as well as its control are also now quite clear. More subtle effects related to states near the mobility edge itself are still interesting areas of investigation, for these can act as traps and thus can have important device consequences. I am sure that as research progresses, we will be finding fine structure near and in the edge itself. We have already been successful in affecting the sharpness and steepness of the mobility edge by the choice of materials, the control of impurities and the generation of intermediate order.

The use of the term "disorder" is unfortunate since it ordinarily means deviations from periodic reference points, but if periodicity is not dominant, then we must substitute our own basic and specific noncrystalline principles. As pointed out previously, one can tailor the optical gap by the use of different covalently bonding elements which also affect the cohesive energy of the material. The alloying elements can further act as underlined structural crosslinks, assuring amorphicity. Following our rules, one can very specifically design materials: e.g., to increase the band gap of a tellurium alloy, add germanium; to increase it further, add stronger-bonding silicon. Similar increase of the band gap occurs if one substitutes arsenic for antimony, or adds selenium, sulfur, or oxygen. It is not unusual for amorphous materials to be multi-component alloys, with four or more elements. The bond strengths of all the elements affect and determine the overall gap. In terms of defects, DECs are generated by the three-dimensional spatial freedom of individual atoms counterbalanced by the chemical and electrical forces surrounding them, i.e., their environment. Therefore, a silicon alloy is primarily tetrahedral but its electronic properties, i.e., its transport properties, are controlled by the deviations from the NSB. These DECs are primarily responsible for the deep states in the gap of amorphous materials, and, depending upon their position in energy, can also play a role in the aforementioned shape of the

312

mobility edge. The matrix that we are discussing not only has relaxation modes which are different from a lattice structure, but has a degree of elasticity which becomes an exceedingly important parameter in material design.[35,36,38,39,61]

V. MECHANICAL PROPERTIES

The concept of elasticity is a common theme throughout this paper. For simplicity, consider the fact that as one changes the average coordination by replacing divalent materials in Group VI by tetrahedral materials in Group IV, e.g., silicon, the elasticity decreases. In order to attain necessary elasticity to make useful materials, atoms of lower valence are utilized. In contrast, if we start with divalent materials, we must add crosslinks to assure and control rigidity and stability. If we begin with tetrahedral materials, we add monovalent atoms such as hydrogen and fluorine to decrease the rigidity and to control and assure the tetrahedral structure of the Group IV atoms.

In order to understand how one goes from a flexible to a rigid structure, I proposed that the controlling influence was the network connectivity, which is characterized by a single parameter, the average coordination number C.[38] This average coordination number is related, of course, to the NSB. As one goes from primarily divalent materials, which have the greatest tendency for flexibility and the formation of glass, to tetrahedral materials with the greatest rigidity, the alloying and cross-linking elements that are added accomplish two purposes. They not only play a structural role, e.g., as can be seen in Fig. 5 [39] where the nonchalcogenide elements add rigidity to the solid, but they also provide an increase of the average coordination number. As the average coordination number is increased, the freedom in three-dimensional space is limited by placing a greater number of constraints on each atom; however, just as important, the freedom of chain and ring folding and

twisting is also controlled and inhibited. We need not go to more tetrahedral materials to increase the coordination of tellurium; oxygen and/or arsenic can increase both the average coordination and the size of the gap. In tetrahedral materials, there is much strain added as the bonding orbitals seek to complete their configurations. To relieve the strain, one alloys with atoms of a lower valency or with those which tend to form ionic bonds. As coordination is increased, e.g., in elemental amorphous silicon, the alloying atoms play the role of permitting completion of the tetrahedral structure by providing flexibility and electronic compensation. If they did not, DECs would be induced by virtue of the resulting under-coordination, and dangling bonds would be formed. Therefore, elasticity is intimately connected with coordination number: in chalcogenides, crosslinks and bridges play an important role; in a material such as amorphous silicon, alloying reduces the coordination. However, the average coordination of the silicon atoms themselves is increased, e.g., by the addition of fluorine, carbon, oxygen, nitrogen, etc. due to the reduction of the concentration of dangling bonds.

VI. PHONONS

I wish to emphasize that phonon activity in amorphous materials differs basically from that in crystalline materials, although there is a wide spectrum and in some materials similarities can exist. One should start with the simple premise that although crystalline solid only exhibit extended phonons, both localized and extended phonons characterize amorphous solids. There is a tendency for strong localized coupling in the flexible materials and weaker coupling in the tetrahedral materials. The matrix mediates the orbital energies in the divalent materials. The resulting spin pairing usually produces completely diamagnetic material. When the matrix is not deformable enough, such as in an as-deposited elemental amorphous silicon material, the electron-phonon interactions cannot provide the necessary pairing. The relaxations that are inherent in amorphous materials are, therefore, different from those of crystalline materials.

VII. MATERIALS SYNTHESIS FOR DEVICE APPLICATIONS

Let us see how these principles actually work in synthesizing materials for device purposes. We will start with the chalcogens and end with tetrahedral materials. As was pointed out earlier, amorphous devices fall into two categories.[17] The first are unstable materials, whose bond strengths and steric hindrances act to prevent crystallization; this class is illustrated in Fig. 5 by an Ovonic Threshold Switch. The crosslinks are numerous and the bonding is strong, and therefore structural changes such as crystallization do not occur within the device operating range. The second are bistable materials, in which there are fewer crosslinks and the bond strengths are weaker so that the barrier to crystallization can be overcome. An example is the Ovonic Memory Switch, shown in Fig. 6. Note how flexible and elastic the chalcogenide bistable memory material is compared to the unstable threshold switch. The average coordination for each is significantly different, C = 2.3 in the memory material while C = 2.9 in the threshold material.

Fig. 5. Model of an Ovonic Threshold Switch illustrating a large amount of strongly-bonded crosslinks assuring stability. The dark balls are Ge, Si, and As atoms. The light balls are Te atoms. (Ref. 39.)

Fig. 6. Model of an Ovonic Memory Switch showing fewer and weaker crosslinks and inherent flexibility which permit the reversible bistability. The light balls are Te atoms. The dark balls are Ge atoms. The darkest balls are Sb and S atoms. (Ref. 39.)

Figure 2 shows how the unique structural changes in amorphous materials, ranging from subtle relaxations to changes of phase including crystallization, become the basis of a whole new field of information and encoding devices, including new types of photography. Especially interesting is the fact that structural reversibility characterizes the more flexible materials so that one can cycle from, e.g., the crystalline state back to the amorphous. These changes of phase can be driven reversibly for more than hundreds of billions of cycles without degradation.

In all materials, we know the origin of the normal structural bonds. We already pointed out that the primary origin of the DECs in chalcogenides are the lone-pair electrons, either nonbonded or forced by the internal chemical and topological environment to assume a spectrum of bonding states, including one- and three-electron states.[39] In an important paper, Kastner, Adler, and Fritzsche [62] further developed this theme to explain the nature of the charged defects that result from these lone-pair interactions. They called the low energy one- and three-fold coordinated defect states valence alternation pairs (VAPs). It is interesting that these VAPs have the property suggested by the original CFO model,[13] large and equal concentration of positively and negatively charged centers which can act as efficient traps for excess electrons and holes. The elucidation of the lone-pair nature of the chalcogenides by Kastner [63] allowed us [35,39,64] to explain why there is no ESR signal in most chalcogenides despite the typical presence of a high density of states in the gap.[65] The fact that lone pairs are spin-compensated in all their variety of free or bonded conditions explains the above as well as how one can have a negative correlation energy. The difference between my explanation [35-37] and those of Street and Mott [66] and of Anderson [67] is that theirs are based upon disorder as sufficient for the negative correlation energy and allow unduly for dangling bonds, and therefore fail to distinguish between the chalcogenide and the tetrahedrally-bonded materials. My

position was, and is, that it is the inherent flexibility of the divalent state which permits the lone pairs to have the strong electron-phonon interactions that are the basis of the induced spin pairing.

Since switching and memory are such basic functions in our information-oriented society, it is important to note the special characteristics of the chalcogenide-based Ovonic devices and correlate them with the explanations given above. In the Ovonic Threshold Switch (see Fig. 7), we see a unique reversible transition between a high impedance and a low impedance state in less than 120 picoseconds at room temperature. (I have never understood the interest in Josephson Effect switches for computers, since they require liquid helium temperatures to achieve comparable switching times.) Such a device is completely independent of polarity and is made preferably in thin-film form, from less than 0.5 μm to many μm in thickness, depending upon the threshold voltage required.

When I first invented these devices, I called them Quantrols,[68,69] since I believed that the switching mechanism was electronic in nature and that such speeds could be observed only if there were a quantum basis for the electronic change of state. From 1960 on, I described the electrical characteristics of these devices.[70,71] In my 1968 paper,[17] I emphasized the electronic nature of the switching process, and explained the basis of the mechanisms of both threshold and memory switching phenomena. The application of a high electric field to specifically designed chalcogenide glasses induces a rapid switching process to a nonequilibrium conducting state followed by injection. The electronic basis of the process has been proven,[17-33] and it has many implications both to solid-state theory and to device potential. In the early 1960's, I performed a simple experiment to prove the electronic nature of the phenomenon by adding some selenium to the threshold materials (preserving the high-impedance state even in the liquid phase) and demonstrating switching above the melting point. Obviously, switching therefore was not based on a solid-to-liquid transition.

Now to discuss chemical-topological correlations. As pointed out, the Ovonic Threshold Switch is a heavily crosslinked material with strong bonds, and is therefore unistable; i.e., the electronic excitation does not change the basic structure. The Ovonic Memory Switch is deliberately made with fewer crosslinks and weaker bonds. Referring to Figs. 4-6, it can be seen that, e.g., if germanium is

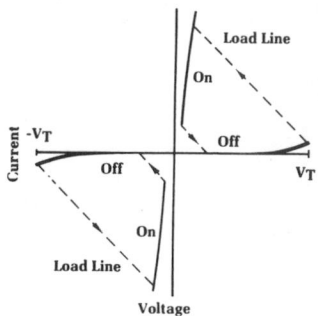

Fig. 7. Current-voltage characteristics of an Ovonic Threshold Switch. (Refs. 17 and 68.)

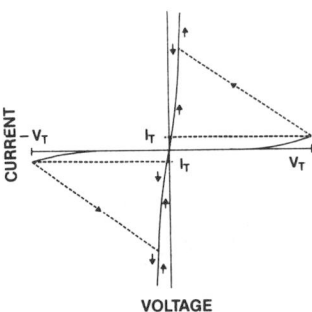

Fig. 8. Current-voltage characteristics of an Ovonic Memory Switch. (Refs. 17, 68 and 72.)

substituted for silicon, or antimony for arsenic, and the nonequilibrium threshold electronic switching effect is used to make the material reactive to electronic excitation, thus weakening or breaking the bonds, the subsequent thermal action which permits cooperative movement of atoms helps induce the memory state. Memory changes can be generated as well as accelerated by diffusion processes.[3] I have called materials which are based upon changes of local order bistable or phasechange materials. A typical current-voltage characteristic of an Ovonic Memory Switch is shown in Fig. 8.

Nowadays, the topic of artificial intelligence is of great interest. I feel that we are completing the grand circle that my wife and collaborator, Iris Ovshinsky, and I originally started in 1955 when we set out to understand the physical basis of intelligence, i.e., information, how it is encoded, switched and transmitted, and the energy transformations connected with it.[3,72] I proposed that this little-understood area of neurophysiology could be illuminated by considering that "disorder," i.e., local order, could play a crucial role. I felt that the energy transformations, excitations, and structural changes associated with amorphous materials would be valid models for nerve-cell action, and I built my first nerve-cell switching model and memory to prove the analogy.[72] I was particularly interested in the adaptive memory aspect of my model [73] and have continued with the adaptive memory concept as a learning "machine" in micron thicknesses ever since. To illustrate my work in this area since 1960, consider Fig. 9. We are also pursuing the three-dimensional circuit potential of amorphous materials.

We predicted and observed memory effects in amorphous semiconductors in response to electrical or optical pulses, and associated them with either irreversible changes if the materials were strongly bonded and did not have inherent flexibility, e.g.,

in amorphous silicon, or reversible changes in cases where more flexible structures were generated by utilizing, e.g., more weakly-bonded divalent alloys such as the chalcogenides. The adaptive memory, therefore, reflects the ratio of the amount of order to the energy input. Before 1960, I showed switching, memory, and adaptive memory action in transition metal oxides.[72]

I find it stimulating and fascinating to connect the new cosmological theories with the work being described here since they deal with the same types of problems, i.e., phase changes, supercooling, freezing-in of defects, nucleation, broken symmetry, etc., except that the time scale is a bit different when we are dealing with the origin of our universe (or universes)! Guth [74,75] assumes a liquid-to-crystal analogy whereas I would suggest that asymmetries of the amorphous state and the changes that can occur in it as described herein are more to the point; i.e., in the early transitional phase of the evolution of the universe, the theories of the amorphous phases discussed in this paper are more relevant than those of a crystalline phase. In fact, I believe that my multi-equilibria concept may have some connection and applicability to Guth's general theory. The unity of science is a marvel indeed!

As discussed previously, chalcogenide glasses have low values of the average coordination number. The network is not overconstrained and intermediate-range order is often observed. For nearly pure chalcogens, such as glasses in the Te-Se system, chemical crosslinking is very low. However, "mechanical" entanglements, especially as longer chains and rings are formed, serve as an energy barrier to crystallization, albeit a low one. Recall our rule that as one generates stronger chemical bonds, the band gap goes up as well; e.g., sulfur and oxygen both have stronger bonds to tellurium than does selenium, and thus the gap increases progressively as one replaces selenium by either sulfur or oxygen. If one combines several elements, then the bond strengths are averaged and the gap changes accordingly. Depending on the design of the material, especially the use of crosslinks involving particular bond strengths, crystallization can proceed at relatively rapid rates, especially in the presence of some activation such as increased temperature, incident light, or applied electric field. Similar energy input can be used in conjunction with rapid quench rates to return the material to the amorphous phase. Since the two phases are very distinct electrically and optically, and both phases are essentially completely stable at ambient conditions, such materials can be used as the basis for reversible nonvolatile memory systems. Either the crystalline or the amorphous phase can be used as the "zero" memory state. If the amorphous phase is considered the zero, writing can be accomplished by, e.g., applying a voltage pulse to crystallize a filament between two electrodes. As noted previously, chalcogenides ordinarily possess equal concentrations of positively and negatively charged defect centers which act as effective traps for injected free carriers. When an electric field is applied so that double injection takes place, the traps fill. Under these nonequilibrium conditions in an amorphous memory material, the large concentration of carriers

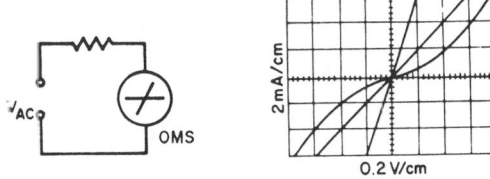

Fig. 9. Ovonic adaptive memory. (Ref. 72.)

weakens the structure, bond reconstructions take place, many covalent bonds are broken, and the rate of crystallization is enhanced (electrocrystallization). Typically, filaments of the order of 1 μm can be grown in times of less than 1 ms. The same results can be induced by optical excitation. It is important to point out that in the Ovonic Threshold Switch this filament is composed of carriers originating from nonbonding configurations, and therefore there is no structural change, i.e., no crystallization; in the Ovonic Memory Switch, the electronic threshold switching effect leads to desired structural changes.

A wide array of materials can be utilized for Ovonic memories. These include, but are not restricted to, chalcogenide alloys. Having worked on a particularly attractive chalcogenide system, i.e., tellurium-based materials, since 1960, we have been reporting on it in the scientific literature for many years. While multi-component alloys are ordinarily used, the memory mechanism can be understood by considering a simple example. For an alloy such as $Te_{83}Ge_{17}$, the eutectic composition for the Te-Ge system, application of a voltage pulse leads to a phase separation into Te-rich and GeTe-rich regions. Since tellurium crystallizes even at room temperatures, the Te-rich regions quickly form crystallites. Both Te and GeTe under somewhat nonstoichiometric conditions are semi-metals with conductivities over 10 billion times larger than the Te-Ge glass at room temperature. We have found that Te crystallites always grow when sufficient energy is coupled to virtually any Te-based memory glass, which can include O, As, Sb, Pb, etc. The differences in physical properties which serve as the memory mechanism are due to the properties of the Te crystallites on the one hand and the amorphous matrix on the other. In the case of electronic memories, the written filament is highly conductive whereas the unwritten glass is highly resistive. One can also write, and it is often preferable, by amorphizing an originally crystalline film. This has been accomplished in the nanosecond range.[76,77] The memory thus can be easily read by applying a small voltage across the contacts. In the Ovonic memory, we utilize other parameters such as large changes in reflectivity.[4,78] The amorphous-crystalline transition is a completely reversible one, a very important attribute.

To electrically erase, application of a sharp current pulse with a rapid trailing edge is all that is necessary. The electronic effects plus the consequent Joule heating are localized within the conducting filament, while the surrounding medium remains at room temperature, thus quenching the active material and reforming the nonconducting glass. I have found it helpful to consider both the precipitation of the crystallites from the matrix and their dissolving back in from a chemical point of view.[45,49] As I pointed out in 1973,[2] "This is not only a 'melt' condition, but one in which the chemical affinities of the crosslinking atoms aid in the establishment of the amorphous state."

A system designed on this principle acts as a nonvolatile electrically erasable programmable read-only memory (EEPROM), an important link between volatile random access memories (RAMs) and unalter-

Fig. 10. Ovonic high-speed, high-density Programmable Read-Only Memory (PROM) manufactured by Raytheon.

able read-only memories (ROMs). Ours were the first EEPROMs made, and were commercially available in the 1960's and 1970's. Their characteristics have been continually improved since then.

Because both the crystalline and amorphous phases of the material are completely stable at operating temperatures, it is evident that Ovonic memory switching can be used as the basis for ordinary ROMs and for archival applications by using a write-once mode. If one wants to assure irreversibility, it takes a change of chemistry, following the rules we have outlined, by utilizing stronger bonds than those in the reversible material. Figure 10 shows an Ovonic amorphous silicon-based, high-speed, high-density electrical PROM manufactured by Raytheon. Figure 11 is a commercially available (Panasonic) optical Ovonic memory based upon the chalcogenide crystalline-to-amorphous transition utilized as an optical PROM.

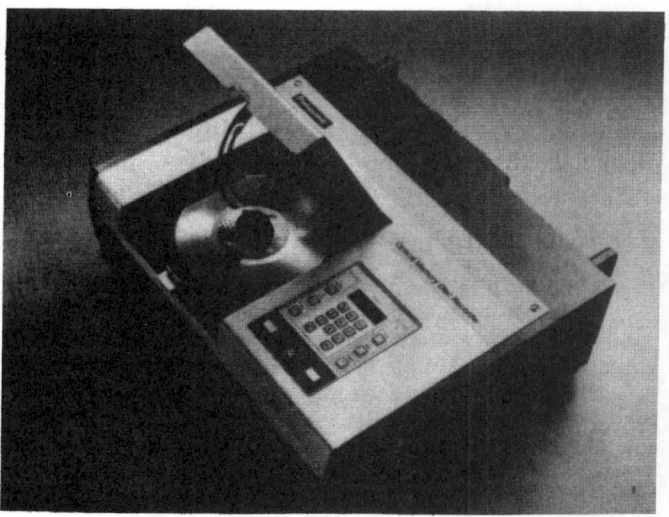

Fig. 11. Panasonic Optical Memory Disc Recorder.

The use of light to induce structural and phase changes has been very rewarding for us. Not wanting to hear again the dreary litany of thermal versus electronic models as the mechanism for changes in amorphous materials, I decided to utilize light to produce new types of optical recording and photographic imaging with unique properties. In addition, I showed that one could use lasers, electron beams, and ordinary light to create information encoding systems, both series and parallel. We were the first to accomplish the laser crystallization of amorphous materials, i.e., to cause crystallization to occur by utilizing a laser interacting with the materials.[4,79,80] I took the position that the simple explanation of melting and recrystallization was not adequate to describe fast laser crystallization. Melting in amorphous materials is the first refuge of ignorance. I proposed that in a material that is unstable to a large amount of excited carriers (initiated by light or electric field), changes of conformation occur and can result in exceedingly fast configurational changes such as crystallization.[3,80] In elemental tetrahedral materials, such relaxations are minor since crystallization requires very little more than eliminating the distortions, which are primarily bond angle changes. In the more flexible chalcogenide alloy materials, bond switching can also take place. This view was supported by much experimental evidence and in a 1971 paper in Applied Physics Letters,[81] we stated that "We have observed a high-speed crystallization of amorphous semiconductor films and the reversal of this crystallization back to the amorphous state using short pulses of laser light and evidenced by a sharp change in optical transmission and reflection. This optical switching behavior is analogous to the memory-type electrical switching effect in these materials which has received wide attention since the observation by S.R. Ovshinsky of both threshold and memory switching in amorphous semiconductors. ...we propose a model which closely relates the optical and electrical switching behavior, and shows that the phase change from amorphous to crystalline state is not only a thermal phenomenon but is directly influenced by the creation of excess electron-hole carriers by either the light, or, for the electrical device, by the electric field. The reversibility of the phenomenon in this model is obtained through the large difference in crystallization rates with the light on or off."

The idea of optical mass memory systems, for example, with the entire contents of a library stored on several disks, is a very appealing one, but was not seriously considered prior to our work on reversible phase changes, such as amorphous-to-crystalline or crystalline-to-amorphous transitions. That work opened up the possibility of optically writing, erasing, and reading via, e.g., the use of a laser which could be focused down to a 1-μm spot size. Resolution of this order of magnitude could provide information storage capacities of 10^8 bits/cm^2 and dramatically higher densities with the use of electron beams. Present-day technology allows the storage of about 250,000 pages of information on a single video disk, an even greater bit density. We showed that a number of multi-component alloys, such as glasses in the Te-Ge system, had very different reflectivities from the same material in its crystallized form, irrespective

of the particular components of the alloy. Similar properties can be attained with a multitude of other amorphous alloys as well. One can optimize specific properties by varying the composition. With the use of antireflection coatings tuned to either the crystalline or amorphous phase, the phase transition then provides many orders-of-magnitude changes in transmission upon writing and erasing.

My collaborators and I showed [81] that an ordinary laser pulse could both crystallize and amorphize a spot less than 1μm in diameter in under 1 μs, and that the same or another laser could be used to read in either a reflection or a transmission mode. We were operating optical disk systems based on these ideas in the 1960's and early 1970's. Exposure of the glass to a laser beam with characteristic frequency greater than the energy gap excites large concentrations of electron-hole pairs. These can have several effects including recombination and trapping by the charged defect centers, resulting in large densities of bond switching and broken bonds. Under these conditions, crystallization proceeds at an extremely enhanced rate (photocrystallization). This process is similar to the electric field-induced crystallization discussed previously. In the crystallized form, the material is more light absorbent than the glass. Exposure to the same laser beam thus can transfer an increased amount of energy to the written spot, returning it to the disordered state. Since the surrounding matrix is unaffected by the focused laser beam, it serves to provide the proper thermal as well as chemical environment, quenching in the anticrystalline configuration and thus reforming the glass. Consequently, either writing or erasing can be accomplished by the same laser pulse.

We have continually improved the parameters of reversible optical data storage disks, obtaining sharp increases in resolution, contract, and lifetime. In addition, optical memory techniques other than amorphous-to-crystalline transitions have been developed. One such technique [81] uses the self-focusing property of many glasses to rapidly nucleate a vapor bubble at the interface between the chalcogenide glass and an inert transparent layer. Self-focusing occurs whenever the energy gap of the glass decreases with increasing temperature, a common phenomenon. If the laser has a characteristic frequency very near the energy gap at room temperature, a small amount of laser-induced heating just below the interface will cause ever-increasing absorption in the same region, rapidly nucleating a small bubble. The bubble scatters light effectively, enabling the spot to be read easily. The entire memory can not only be laser-erased but can be block-erased by gentle heating with an infrared lamp. The advantages of such a system are smaller spot sizes (and thus higher resolution), faster write times, and lower energy cost per bit.

Other ideas conceived by us for optical memory applications include photostructural changes such as photodispersion (utilized in our MicrOvonic File,[60] photodoping, photodarkening, and holographic storage.[82] There is now no question that the much-needed mass memories of the near future will be optically written, erased, and read. Finally, we note the present-day importance of laser crystalliz-

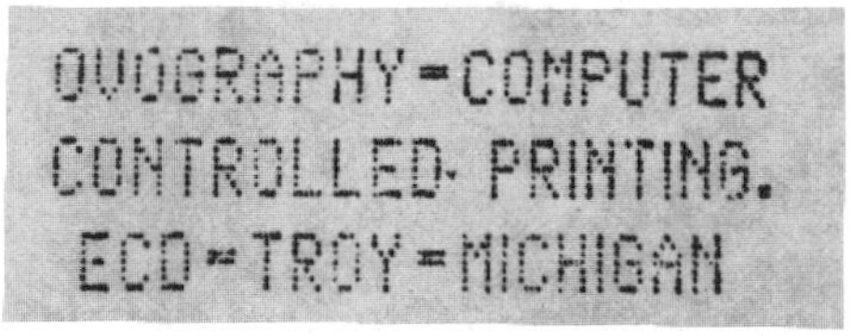

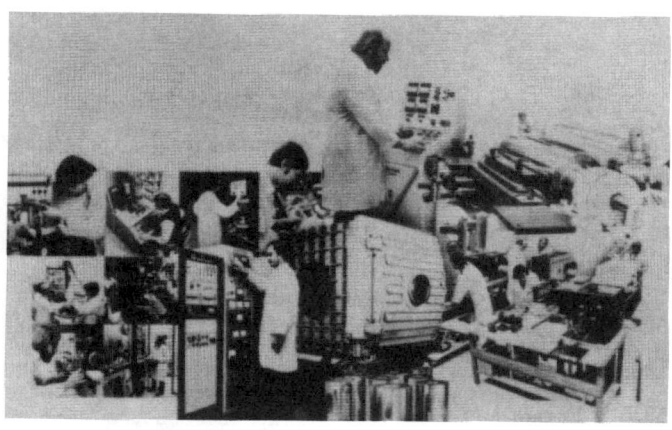

Fig. 12. Electrostatic printout obtained with ECD photostructural film printer. Each of the typewriter-size characters were generated on a 5x7 matrix by computer tape. (Ref. 4.)

ation of amorphous materials as an example of how new areas of technology can spring from basic scientific investigations. This use of amorphous solids as a vital step in preparing improved crystallites was not a subject of general scientific investigation until we demonstrated such phase changes. Laser printout is now accepted as a matter of course. We were the first to utilize lasers for such applications.[2,4,46,51,53] Figure 12 shows a printout of an early laser copying and printing demonstration. As the old Chinese proverb teaches, one should always leave a golden bridge of retreat so as not to humiliate one's opposition. In this vein, it is relevant to emphasize how the use of amorphous materials has proved to be crucial in the understanding, control, and operation of crystalline MOS devices. More and more crystalline scientists and technologists are appreciating the value of amorphous materials. To me, it has been a needless controversy since the understanding of disorder illuminates the inherent deviations from order in crystalline materials.

Looking farther in the future, still higher capacity memories will be essential. For such purposes, only x-rays, electron beams, or ion beams can yield the necessary resolution. The most promising technique at present involves the use of electron beams. Recent advances in electron optics suggest that 1000Å beams will soon be available, and even 100Å beams are a possibility. In the early 1960's, we showed that electron beams can be used to either crystallize or amorphize alloys. Furthermore, the crystalline and amorphous phases are quite distinct with regard to secondary-electron emission, so that the memory can be easily read by the electron beam. If 100Å resolution can be

Fig. 13. Ovonic nonsilver photo-duplication film. (Ref. 61.)

Fig. 14. Ovonic Continuous Tone Imaging Film exposed through a high resolution test mask, exhibiting a resolution in excess of 1200 line pairs per millimeter.

achieved, about 10^{15} bits of information can be sorted on a 30-cm disk, more information than is contained in the books in all the libraries of, e.g., a highly literate country such as Japan.

The Ovonic memory concept forms the basis for preparation of many types of instant, dry, stable, photographic films with unique amplification, high resolution, and gradation of tones. This is accomplished by varying the fraction of the glass which has been crystallized and the grain size of the crystallites.[45,46,55] ECD has produced an array of films with either ultra-high contrast or exceptional continuous tones for imaging applications. Additional flexibility arises from the fact that the image can be obtained either directly after exposure, as discussed previously, or in latent form, to be developed subsequently when desired. One mechanism for the latter approach is to use our proprietary organo-tellurides as the film material.[83] In this case, exposure to light induces nucleation centers which form the latent image. Subsequent annealing above the glass transition temperature then induces crystallization of the latent region which produces the desired image (see Fig. 13). Excitation also permits the diffusion of tellurium. Using these procedures, we have been able to attain significant amplification factors.

In addition to using the crystalline-to-amorphous/amorphous-to-crystalline transitions, we have developed materials in which local structural changes can be induced and detected optically. These have proven useful in updating or correcting images well after exposure. While the materials described here are of the instant dry development type, an exciting feature by itself, we have also designed materials which have excellent etching properties. These have been used for high resolution masks (see Fig. 14) and other photographic applications.[2]

VIII. CHEMICAL MODIFICATION

It was taken for granted that in amorphous materials certain important parameters were in lock step with each other, e.g., if one had a large band gap material, low electrical conductivity would necessarily result. I decided to challenge this dogma by showing that amorphous materials could be chemically modified, and that by controlling the states in the gap one could for the first time independently control the conductivity changes over many orders of magnitude (see Fig. 15). What was so exciting about these results was that we could

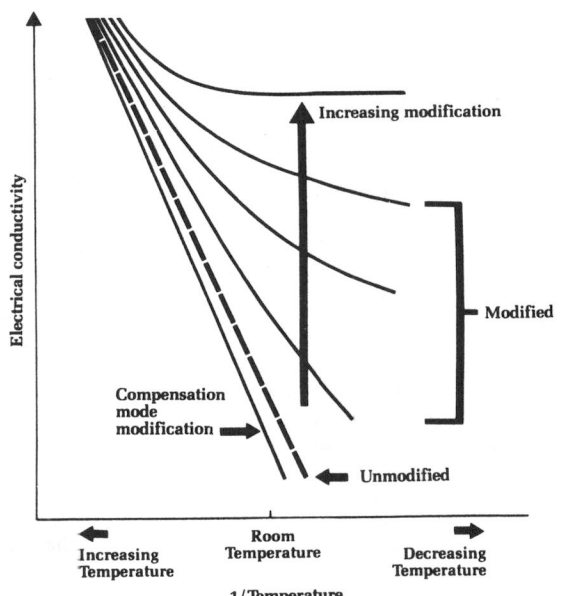

Fig. 15. Effect of chemical modification on the electrical properties of amorphous films. (Refs. 8 and 43.)

obtain large conductivity changes in elemental materials and in alloys containing elements from Group III through Group VI, including materials with drastically different band gaps.[8,9,42,43,84] (In the Periodic Chart of the Elements, Fig. 4, various atoms are darkened to show most of those used in the modification process.) In many cases, a small amount of modifier could _increase_ electrical resistance, while larger amounts _decrease_ it.

I had previously shown that lithium could achieve the same effect in chalcogenide glasses.[38] This was during the same period of time that, following the work of Chittick et al.,[85] Spear and LeComber [86] were demonstrating the possibility of substitutional doping in "amorphous silicon." (There still is a question about the effectiveness of p-doping in amorphous Si-H alloys.[8]) The fact that we could alter the conductivity of such a large variety of materials showed that we could outwit equilibrium and design a whole new family of materials with characteristics heretofore considered impossible. To put this in historical perspective, note the paper of Hamakawa,[87] which states, "the electrical properties of chalcogenide glasses could not be controlled so widely before the sensational appearance of 'chemical modification' proposed by Ovshinsky." While I appreciate the statement, I wish to reiterate that my paper on modification covered elements and alloys from columns III-VI in the Periodic Chart and was not limited to chalcogenides.[8,84] I was very pleased that Davis and Mytilineou corroborated chemical modification in amorphous arsenic with nickel as the modifier.[88]

Figure 16 shows typical materials that were modified, and it can be seen that the various active modifiers are either d-orbital or multi-orbital elements. The d-orbitals act as "pin cushions" when co-deposited so that they interact with the primary elements being deposited in a manner so as to create new TIEs. These TIEs would not exist if the modify-

ing elements were deposited conventionally.[8,42] I believe the achievement of modification proves my point about multi-equilibria, since the normal structural bonds need not be affected at all by the modifying element (although they can be, if desired), i.e., the optical gap remains the same while the electrical conductivity can increase by over 10 orders of magnitude. It should be quite clear that the three-dimensional freedom of the amorphous state permits unusual and stable orbital interactions of a highly nonequilibrium nature. As can be seen from Fig. 16, various multi-orbital elements can be used and new nonequilibrium TIEs can be generated even without cosputtering, since the very fact that they are multi-orbital permits several different configurations. We have also utilized excitation as a means of having an atom or molecule enter into and interact with the matrix in such a manner as to effect modification. It is of interest that we have accomplished modification through dual nozzle melt spinning as well.[89] It should be kept in mind that the quenching process itself is a method of achieving nonequilibrium configurations.

In our technique, substantial concentrations of an appropriately chosen modifier are introduced into the amorphous network in a nonequilibrium manner so that it need not enter in its "optimal" chemical configuration. The modifier in small amounts can _decrease_ electrical conductivity, but in larger amounts ordinarily increases it. When the concentration of the modifier exceeds that of the intrinsic defect centers, the Fermi level begins to move. In other words, in small concentrations, the modifiers can compensate and convert positively charged DECs to negatively charged ones, or vice versa; however, in larger concentrations, the modifiers yield many more DECs than would have been present in an equilibrium material. Therefore, the chemical modifier alters the localized states in the gap that control transport, while alloying alters the optical gap without changing the transport properties significantly. We therefore can independently separate the electrical activation energy from the optical gap and control them individually.[90] In a sense, an alloying element is also a modifier since it modifies the overall band gap, but I have used the term chemical modifier to describe situations in which transport or active chemical sites are the properties of interest, for in such cases the purpose of modification is to

Host Material	Active Modifier
Ge Te Se As	Ni, Fe, Mo
As	Ni, W
$B_4 C$	W
Si C	W
Si	Ni, B, C
$Si_3 N_4$	W
BN	W
Te O_2	Ni
Ge	Ni
Si O_2	W
Ge Se$_2$	Ni
Se$_{95}$ As$_5$	Ni

Fig. 16. Chemical modification of amorphous semiconductors. (refs. 8 and 43.)

alter the localized states <u>within</u> rather than the positions of the mobility edges.

I utilized surface chemical modification during the 1950's when I was mostly working with oxides, particularly those of the transition metals; I used amphoteric atoms and ions to change the conductivity by over 14 orders of magnitude,[70] utilizing such interactions to design switches and memories, both digital and adaptive.[3,72]

During the early 1960's, I investigated many amorphous and disordered phases, combining primary atoms with many types of alloying elements, and was the first to make amorphous gallium arsenide films. Our laboratory also made the first amorphous silicon carbide films.[91]

Another method of modification in amorphous materials is doping. Following Chittick et al.,[85] Spear and LeComber [86] reported doping experiments on what they considered to be amorphous silicon.[92] As we have pointed out, elemental amorphous silicon is not useful as an electronic material because free-energy considerations lead to an immense density of defects, including dangling bonds and voids. How is one then to utilize the silicon atom in amorphous materials for worthwhile electronic purposes? A means must be found to allow silicon atoms to be connected so that a completed tetrahedral structure ensues. The atoms that achieve such connections must play two roles. First, they must saturate the dangling bonds, i.e., they must be chemical compensators. Equally as important, they must also fulfill the role of structural links which act to provide <u>flexibility</u> to the matrix, relieving the stresses and strains of the pure silicon matrix and compensating it <u>structurally</u> so that the local order retains the electronic properties of the completed silicon configuration. I was therefore dubious about the usefulness of "amorphous silicon" since I thought that in its elemental form it held little electronic interest. When Fritzsche and his colleagues [93] showed that the dopable "amorphous silicon" really contained a large percentage of hydrogen and was therefore an <u>alloy</u>, I was pleased since it meant that my point of view and understanding were justified and correct, and, since alloys were where our talents lay, that we could make superior alloys based upon our chemical and structural concepts. This led me to suggest fluorine as a more suitable element since elements such as hydrogen and fluorine both terminate dangling bonds and at the same time enter into the structural network. I postulated that fluorine, due to its superhalogen qualities, i.e., its extreme electronegativity, small size, specificity and reactivity, not only terminates dangling bonds and can become a bridge, but also <u>induces new local</u> order and affects the TIE by several means, including controlling the way hydrogen bonds in the material since it can bond with silicon in several different ways, some of which produce defects.[60] Fluorinated materials are intrinsically different and fluorine is responsible for new TIEs. Lee, deNeufville and I [94] showed that fluorine does induce a new configuration when combined with silicon and hydrogen. Therefore, in amorphous silicon alloys, the addition of fluorine [41,56,95-97] minimizes defects, including dangling bonds, by generating new beneficial short-range order and TIEs.

I have utilized this concept for other materials such as germanium [97] and was able to solve the problem of "anomalous" density of states of amorphous germanium alloys, which most physicists consider to be tetrahedral materials. They, indeed, are tetrahedral in terms of their NSB; however, they are <u>not</u> in their DECs. There are various divalent and other configurations due to the "inert" lone pairs found in crystalline germanium compounds as well as in those of other elements in Group IV such as tin and lead.[98] Applying our chemical approach, I was able to show that this tendency away from tetrahedralness is even more prevalent in amorphous materials. The lack of tetrahedralness leads to increased DECs and unless compensated for can make germanium-containing alloys inferior as low density-of-states electronic materials. I consider that a very important attribute of fluorine is its tendency to expand the valence of many atoms by making use of the orbitals that are within its strong chemical attraction, and it is therefore particularly valuable where defects are involved with undercoordination.

The above is particularly relevant since in previous work I had considered that germanium could be two-fold coordinated,[39] and that silicon under certain conditions in the amorphous state could also have more than one orbital available that could result in additional defects, and therefore that fluorine could terminate and compensate as well the defect states that were not available to hydrogen. Adler,[99] using thermodynamic considerations, has proposed that two-fold coordination plays a role in the defect centers of amorphous silicon-hydrogen alloys. I felt that in germanium-containing materials, fluorine would interact with the "Sedgwick" lone pairs to force germanium into a more tetrahedral structure, thereby making an intrinsic material with an inherently low number of DECs. Fluorine also introduces an ionic character to the bonding, helpful in relieving strains. It decreases the fluctuations in potential on an atomic scale caused by the disorder of the amorphous state. The results are that we now make silicon-based alloys with a concentration of localized states in the low $10^{15} cm^{-3}$ range, and achieving this quality with our germanium-based materials. By lowering the DEC noise level, we can more effectively substitutionally dope these materials, and through the use of Raman spectroscopy we have been able to show that they also have more intermediate range order. The use of fluorine assures far more stable amorphous materials and is crucial in making these and similar materials into superior microcrystalline films.[100] Free radical chemistry, the leitmotif of my work since the very beginning in the 1950's, is involved in these processes. It is a subject that cannot be covered comprehensively here. Suffice it to say that it plays a very important role in the plasma decomposition processes which lead to many of the condensed materials that are discussed here.[56] We have performed experiments which clearly show the important role that free radicals play in producing better tetrahedral materials.[101]

While amorphous silicon-hydrogen alloys can be substitutionally doped n-type, just as crystalline materials, and boron doping yields p-type material, the boron doping is not very efficient. Following the chemical arguments of this paper, one can see why. Instead of being constrained to enter the

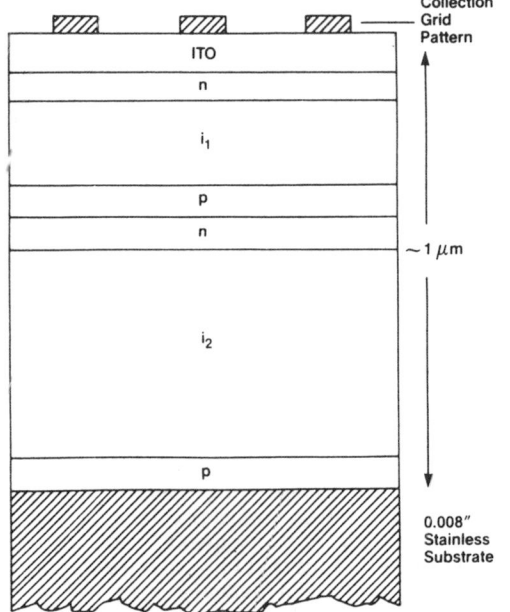

Fig. 17. Schematic cross section of 1.8eV/1.8eV ECD-Sharp Ovonic Tandem Solar Cell. (Refs. 106 and 107.)

matrix in a tetrahedral position, as in crystalline silicon, boron can form three-center bonds especially with bridging hydrogen atoms, and therefore can generate nontetrahedral structure.[8,9] Having discovered the unusual properties of boron many years ago upon reading Lipscomb's brilliant and profound work,[102] I was prepared to apply my understanding of it to amorphous materials, not only to explain boron's difficulty to adequately become a substitutional dopant in materials made from silane but also to utilize its various configurations as one of the elements to achieve chemical modification. Using this approach, we can see why it can be not only a chemical modifier even without co-deposition, because of the many configurations it can assume (this is one of the reasons it is a good glass former), but can also generate unneeded DECs, particularly in an alloy containing hydrogen. In fluorinated materials, due to the increase of intermediate-range order, both n and p substitutional doping is greatly improved. For tetrahedral materials, boron's empty orbital can also be used for coordinate bonding, making use of ordinarily non-bonded lone pairs with very interesting results. As I have pointed out, coordinate bonding can play an important role in amorphous materials.[2,8,39] Boron's natural glass-forming tendencies can be used to good avail since in proper amounts it is an excellent structural element and acts in its own way to affect coordination in a manner that can be as important as hydrogen and fluorine. These attributes give a structural stability to, for example, boron-containing tetrahedral amorphous materials. Boron and fluorine, therefore, are important elements in generating a superior photovoltaic material.

IX. CONCLUSION

We might well ask, where are we now? We have come a long way in 30 years. Photovoltaic devices based upon the superb electronic properties of amorphous silicon alloys are now in production. Figure 17 shows a commercial tandem cell which has the highest energy-to-weight ratio of any amorphous photovoltaic device. New ultra-light devices have been developed.[103] Our new generation multi-cell devices are very stable and have solar conversion efficiencies similar to those of our single band gap cells [104] which are over 10%,[105] and, because of the principles which I have outlined above, we are now making small band gap materials which are approaching the low defect concentrations of our amorphous silicon alloys. This means that we can expect efficiencies as high as 30% when we optimize the different alloys in a three-layer cell.

The information industry, the semiconductor industry, and increasingly the telecommunication industry, all are presently tied to the crystalline structure of silicon. The battle of the electronic giants is taking place on wafers that are now close to their maximum size of about six inches. More and more investment is being made to achieve higher chip densities by photolithographic means. Gordon Moore of Intel, whose expertise in crystalline materials is well known, facetiously proposed that what is needed for the circuits of the mid-1980's and 1990's is an impossible chip the size of the cardboard "wafer" shown in Fig. 18. Since wafers larger than about six inches are not to be expected from the melt-pulling techniques used to grow crystalline silicon, it seemed obvious that this could never be realized, and the industry would be limited to the approximate size of the hand-held chip in the inset. However, the discussion of amorphous silicon alloys in this paper shows that we can make materials that are not only analogous in their circuit functions to crystalline materials but have unique attributes as well. The 1000-foot long, 16-inch wide tandem cell that is shown on the right is a complete photovoltaic cell (electrical contacts are later printed on). We are making an amorphous analog of an infinitely long "crystal" of any desired width. (The width is a matter only of machine design.) We are, therefore,

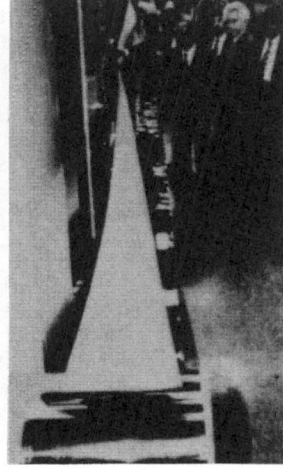

Fig. 18. Cardboard model of an imaginary "crystal wafer" on the left; a typical real wafer is shown in the inset; Ovonic thin-film semiconducting material, an "infinite crystal," on the right.

taking up the challenge and are designing large-area, totally integrated thin-film circuits which we believe will soon transform the information and telecommunication industries.

There are many other areas where I have utilized our synthetic materials approach. Our laboratory has worked successfully for years in fields as diverse as coatings, batteries, catalysts, electrochemistry, hydrogen storage, thermoelectrics and superconductors. We are already in production in a number of these areas. These fields are very important not only commercially and technically but have significant theoretical implications. Obviously, these cannot all be covered here and will be the subject of another paper; some of these areas have been described elsewhere.[108]

My intention here has been to discuss the fundamental concepts involved in amorphous materials and to show how our basic understanding can be directly related to the devices that have been and are being developed. Ours is a synthetic materials approach. When one frees oneself from the restrictions of crystalline symmetry, then not only excellent crystalline analogs such as transistors can be made,[109] but many new nonequilibrium phenomena and materials can be developed. In fact, the whole dogma of bulk homogeneity can be re-evaluated. I described exceedingly thin multi-layer and compositionally modulated devices in our patent literature years ago. Based on these concepts, we have been successfully developing devices, some of which are being utilized commercially.[110] We have reported in the scientific literature on the unusual effects seen in such materials.[48,111-114] This subject has now been rediscovered,[115] and most likely will grow into a new field.

As can be seen, even though we have arbitrarily divided this paper into electronic, chemical, and mechanical sections, one cannot really speak of one parameter divorced from another. The electronic density of states can be the equivalent of active chemical sites, etc. Amorphous materials science is a synthesis of many different disciplines, and therefore has through this process been transmuted into a discipline of its own. Amorphous materials indeed are characterized by the total interactive environment.

The voyage into the amorphous field has been one of discovery and delight, and I take great pleasure in being represented in this institute and in this volume.

ACKNOWLEDGEMENTS

Since it has been a personal odyssey and began at a time when there were few in the field, I have written this paper reflecting my own travels and travails. However, this work could not have been accomplished without my collaborators and colleagues of many years. As always, this work could not have been done without the partnership of my wife, Iris.

It is impossible to give adequate thanks to all the people with whom I have worked in the past 30 years. In the neurophysiological area where I first started, I wish to express my appreciation for their encouragement to both the late Ernest Gardner, Dean of Wayne State University Medical School, and the late Fernando Morin, Chairman of the Department of Anatomy there. I wish to also express my thanks to I.I. Rabi, Nevill Mott, and Kenichi Fukui for their encouragement through the years; also to Boris Kolomiets, an early and major figure in Russian amorphous chalcogenide work, for his statement at the IV Symposium on Vitreous Chalcogenide Semiconductors in Leningrad in 1967, when I first scientifically discussed switching. He said that this work would transform the amorphous field whose progress to that date he denoted as a very slowly-rising, essentially horizontal line to an almost vertical one, a prediction that fortunately came true. It was at that same meeting that I met Radu Grigorovici, a friend and important contributor to our field.

From the 1960's on, I benefitted immensely from the colleagual collaboration of Hellmut Fritzsche, Morrel Cohen, and David Adler. I owe David a special debt of gratitude. Our close association has been of great help to me. I acknowledge with appreciation the collaboration of Arthur Bienenstock, John deNeufville, Heinz Henisch, Marc Kastner, and Krishna Sapru, among others. The work described in this paper was made possible through the years by my colleagues, especially Wally Czubatyj, Steve Hudgens, Masat Izu, and the rest of the superb group of people that make up ECD.

REFERENCES

1. D. Adler, ed., Disordered Materials: Science and Technology, Selected Papers by S.R. Ovshinsky, (Amorphous Institute Press, Bloomfield Hills, 1-296 (1982).

2. S.R. Ovshinsky and H. Fritzsche, "Amorphous Semiconductors for Switching, Memory, and Imaging Applications," IEEE Trans. on Electron Devices ED-20, 91 (1973).

3. S.R. Ovshinsky and I.M. Ovshinsky, "Analog Models for Information Storage and Transmission in Physiological Systems," Mat. Res. Bull. 5, 681 (1970). (Mott Festschrift.)

4. S.R. Ovshinsky, "The Ovshinsky Switch," Proc. 5th Annual National Conference on Industrial Research, Chicago, 86-90 (1969). The Ovographic work was done in collaboration with P. Klose.

5. S.R. Ovshinsky, "An Introduction to Ovonic Research," J. Noncryst. Solids 2, 99-106 (1970).

6. F. Yonezawa and T. Ninomiya, eds., "Topological Disorder in Condensed Matter," Proc. 5th Taniguchi Int. Symp., Japan, 2 (1982).

7. For a discussion of short-range order, see: A.F. Ioffe and A.R. Regel, "Non-Crystalline, Amorphous, and Liquid Electronic Semiconductors" in Progress in Semiconductors, vol. 4, John Wiley, New York, 237-291 (1960).

8. S.R. Ovshinsky, "Chemical Modification of Amorphous Chalcogenides," Proc. of the 7th

<u>Int. Conf. on Amorphous and Liquid Semiconductors</u>, Edinburgh, Scotland, 519-523 (1977).

9. S.R. Ovshinsky and D. Adler, "Local Structure, Bonding, and Electronic Properties of Covalent Amorphous Semiconductors," Contemp. Phys. <u>19</u>, 109 (1978).

10. S.R. Ovshinsky, "The Chemical Basis of Amorphicity: Structure and Function," Rev. Roum. Phys. <u>26</u>, 893-903 (1981). (Grigorovici Festschrift.)

11. J.M. Ziman, <u>Principles of the Theory of Solids</u>, Cambridge University Press, 1 (1965).

12. J.M. Ziman, "How Is It Possible To Have An Amorphous Semiconductor?" J. Noncryst. Solids <u>4</u>, 426-427 (1970).

13. M.H. Cohen, H. Fritzsche and S.R. Ovshinsky, "Simple Band Model for Amorphous Semiconducting Alloys," Phys. Rev. Lett. <u>22</u>, 1065-1068 (1969). See also N.F. Mott, Adv. Phys. <u>16</u>, 49 (1967).

14. J.M. Ziman, <u>Models of Disorder</u>, Cambridge University Press, ix (1979).

15. I thank Dennis Weaire for bringing these contradictory remarks to my attention.

16. Thomas S. Kuhn, <u>The Structure of Scientific Revolutions</u>, University of Chicago Press, 6 (1962).

17. S.R. Ovshinsky, "Reversible Electrical Switching Phenomena in Disordered Structures," Phys. Rev. Lett. <u>21</u>, 1450-1453 (1968).

18. H.K. Henisch, S.R. Ovshinsky and R.W. Pryor, "Switching Effects in Amorphous Semiconductor Thin Films," in <u>Proc. of the Intl. Congress on Thin Films</u>, Cannes, October 5-10, 1970 (published by Societe Francaise des Ingenieurs et Techniciens du Vide).

19. R.W. Pryor and H.K. Henisch, "Mechanism of Threshold Switching," Appl. Phys. Lett. <u>18</u>, 324 (1971).

20. H.K. Henisch and R.W. Pryor, "Mechanism of Ovonic Threshold Switching," Solid State Elec. <u>14</u>, 765 (1971).

21. H.K. Henisch, R.W. Pryor and G.J. Vendura, "Characteristics and Mechanisms of Threshold Switching," J. Noncryst. Solids <u>8-10</u>, 415 (1972).

22. R.W. Pryor and H.K. Henisch, "Nature of the On-State in Chalcogenide Glass Threshold Switches," J. Noncryst. Solids <u>7</u>, 181 (1972).

23. W. Smith and H.K. Henisch, "Threshold Switching in the Presence of Photo-Excited Charge Carriers," Phys. Stat. Sol. A <u>17</u>, K81 (1973).

24. M.P. Shaw, S.H. Holmberg and S.A. Kostylev, "Reversible Switching in Thin Amorphous Chalcogenide Films—Electronic Effects," Phys. Rev. Lett. <u>31</u>, 542 (1973).

25. H.K. Henisch, W.R. Smith and M. Wihl, "Field-Dependent Photo-response of Threshold Switching Systems," <u>Proc. of the 5th Intl. Conf. on Amorphous and Liquid Semiconductors</u>, Garmisch-Partenkirchen, Germany, September 1973, J. Stuke and W. Brenig, eds., Taylor and Francis, London, 567 (1974).

26. W.D. Buckley and S.H. Holmberg, "Nanosecond Pulse Study of Memory Material of Different Thicknesses," Sol. State Elec. <u>18</u>, 127 (1975).

27. K.E. Petersen and D. Adler, "Probe of the Properties of the On-State Filament," J. Appl. Phys. <u>47</u>, 256 (1976).

28. K.E. Petersen, D. Adler and M.P. Shaw, "Amorphous-Crystalline Heterojunction Transistor," IEEE Trans. <u>23</u>, 471 (1976).

29. D.K. Reinhard, "Response of the OTS to Pulse Burst Waveforms (Critical Power Density)," Appl. Phys. Lett. <u>31</u>, 527 (1977).

30. D. Adler, H.K. Henisch and N. Mott, "The Mechanism of Threshold Switching in Amorphous Alloys," Rev. Mod. Phys. <u>50</u>, 209 (1978).

31. D. Adler, M.S. Shur, M. Silver and S.R. Ovshinsky, "Threshold Switching in Chalcogenide-glass Thin Films," J. Appl. Phys. <u>51</u>, 3289 (1980).

32. M.P. Shaw and N. Yildirim, "Thermal and Electrothermal Instabilities in Semiconductors," Adv. in Elec. and Electron Phys. <u>60</u>, 307-385 (1983).

33. J. Kotz and M.P. Shaw, "Thermophonic Investigation of Switching and Memory Phenomena in Thick Amorphous Chalcogenide Films," Appl. Phys. Lett. <u>42</u>, 199 (1983).

34. S.R. Ovshinsky and D. Adler, to be published.

35. S.R. Ovshinsky, "Localized States in the Gap of Amorphous Semiconductors," Phys. Rev. Lett. <u>36</u>, 1469-1472 (1976).

36. S.R. Ovshinsky, "Amorphous Materials as Interactive Systems," in <u>Proc. of the Sixth Int. Conf. on Amorphous and Liquid Semiconductors</u>, Leningrad, USSR (1975) 426-436: <u>Structure and Properties of Non-Crystalline Semiconductors</u>, B.T. Kolomiets, ed., Nauka, Leningrad 426-436 (1976).

37. H. Fritzsche, "Summary Remarks," in <u>Proc. of the Sixth Int. Conf. on Amorphous and Liquid Semiconductors</u>, Leningrad, USSR, 1975: <u>Electronic Phenomena in Non-Crystalline Semiconductors</u>, B.T. Kolomiets, ed., Nauka, Leningrad, 65-68 (1976).

38. S.R. Ovshinsky, "Lone-Pair Relationships and the Origin of Excited States in Amorphous Chalcogenides," AIP Conf. Proc. <u>31</u>, 31-36 (1976).

39. S.R. Ovshinsky and K. Sapru, "Three Dimensional Model of Structure and Electronic Properties of Chalcogenide Glasses," in Proc. of the Fifth Int. Conf. on Amorphous and Liquid Semiconductors, Garmisch-Partenkirchen, Germany, 447-452 (1974).

40. A. Bienenstock, F. Betts and S.R. Ovshinsky, "Structural Studies of Amorphous Semiconductors," J. Noncryst. Solids 2, 347 (1970).

41. S.R. Ovshinsky and A. Madan, "A New Amorphous Silicon-Based Alloy for Electronic Applications," Nature 276, 482-484 (1978).

42. R.A. Flasck, M. Izu, K. Sapru, T. Anderson, S.R. Ovshinsky and H. Fritzsche, "Optical and Electronic Properties of Modified Amorphous Materials," in Proc. 7th Intl. Conf. on Amorphous and Liquid Semiconductors, Edinburgh, Scotland 524-528 (1977).

43. S.R. Ovshinsky, "The Chemistry of Glassy Materials and Their Relevance to Energy Conversion," Proc. Intl. Conf. on Frontiers of Glass Science, Los Angeles, California; J. Noncryst. Solids 42, 335-344 (1980).

44. For references to his work see Revue Roumaine de Physique 26, No. 809 (1981). (Grigorovici Festschrift.)

45. S.R. Ovshinsky, "Electronic and Structural Changes in Amorphous Materials as a Means of Information Storage and Imaging," in Proc. of the Fourth Int. Congress for Reprography and Information, Hanover, Germany 109-114 (1975).

46. S.R. Ovshinsky, "Amorphous Materials as Optical Information Media," J. Applied Photographic Engineering 3, 35-39 (1977).

47. S.R. Ovshinsky, unpublished data, 1975; S.R. Ovshinsky and K. Sapru, 1977.

48. The Francis Bitter National Magnet Lab. Annual Report for July 1982 to June 1983, 118.

49. S.R. Ovshinsky and H. Fritzsche, "Reversible Structural Transformations in Amorphous Semiconductors for Memory and Logic," Met. Trans. 2, 641 (1971).

50. S.R. Ovshinsky, "Optical Information Encoding in Amorphous Semiconductors," Topical Meeting on Optical Storage of Digital Data, Aspen, Colorado, MB5-1-MB5-4, 1973.

51. S.R. Ovshinsky and P.H. Klose, "Reversible High-Speed High-Resolution Imaging in Amorphous Semiconductors," Proc. SID 13, 188 (1972).

52. The reason for the darkening of some of the elements in this figure will be given when we discuss chemical modification.

53. S.R. Ovshinsky and P.H. Klose, "Imaging in Amorphous Materials by Structural Alteration," J. Noncryst. Solids 8-10, 892-898 (1972).

54. D.L. Staebler and C.R. Wronski, "Reversible Conductivity Changes in Discharge Produced Amorphous Si," Appl. Phys. Lett. 31, 292 (1977).

55. S.R. Ovshinsky and P.H. Klose, "Imaging by Photostructural Changes," Proc. Symp. on Nonsilver Photographic Processes, New College, Oxford, 1973; R.J. Cox, ed., Academic Press, London, 61-70 (1975).

56. S.R. Ovshinsky, "The Role of Free Radicals in the Formation of Amorphous Thin Films," in Proc. Int. Ion Engineering Congress (ISIAT '83 & IPAT '83), Kyoto, Japan, 817-828 (1983).

57. D. Adler, "Origin of the Photo-Induced Changes in Hydrogenated Amorphous Silicon," Solar Cells 9, 133 (1983).

58. S. Guha, J. Yang, W. Czubatyj, S.J. Hudgens and M. Hack, "On the Mechanism of Light-Induced Effects in Hydrogenated Amorphous Silicon Alloys," Appl. Phys. Lett. 42, 588 (1983).

59. H. Fritzsche, "Optical and Electrical Energy Gaps in Amorphous Semiconductors," J. Noncryst. Solids 6, 49 (1971).

60. S.R. Ovshinsky, The Shape of Disorder," J. Noncryst. Solids 32, 17 (1979). (Mott Festschrift.)

61. S.R. Ovshinsky, "Principles and Applications of Amorphicity, Structural Change, and Optical Information Encoding," in: Proc. 8th Intl. Conf. on Amorphous and Liquid Semiconductors, Grenoble, France (1981): J. de Physique, Colloque C4, supplement au no. 10, 42, C4-1095-1104 (1981).

62. M. Kastner, D. Adler and H. Fritzsche, "Valence-Alternation Model for Localized Gap States in Lone-Pair Semiconductors," Phys. Rev. Lett. 37, 1504 (1976).

63. M. Kastner, "Bonding Bands, Lone-Pair Bands, and Impurity States in Chalcogenide Semiconductors," Phys. Rev. Lett. 28, 355 (1972).

64. S.R. Ovshinsky, "Electronic-Structural Transformations in Amorphous Materials--A Conceptual Model," July 13, 1972, unpublished.

65. S.C. Agarwal, "Nature of Localized States in Amorphous Semiconductors--A Study by Electron Spin Resonance," Phys. Rev. B 7, 685 (1973).

66. R.A. Street and N.F. Mott, "States in the Gap in Glassy Semiconductors," Phys. Rev. Lett. 35, 1293 (1975).

67. P.W. Anderson, "Model for the Electronic Structure of Amorphous Semiconductors," Phys. Rev. Lett. 34, 953 (1973).

68. M.P. Southworth, "The Threshold Switch: New Component for Ac Control," Control Engineering 11, 69 (1964).

69. J.R. Bosnell, "Amorphous Semiconducting Films," in <u>Active and Passive Thin Film Devices</u>, T.J. Coutts, ed., Academic Press, 288 (1978).

70. J.D. Cooney, "A Remarkable New Switching Form," Control Engineering 6, 121 (1959).

71. "How Liquid State Switch Controls A-C," Electronics 32, 76 (1959).

72. S.R. Ovshinsky, "The Physical Base of Intelligence-Model Studies," presented at the Detroit Physiological Society, 1959.

73. E.J. Evans, J.H. Helbers and S.R. Ovshinsky, "Reversible Conductivity Transformations in Chalcogenide Alloy Films," J. Noncryst. Solids 2, 339 (1970).

74. A.H. Guth, "Inflationary Universe: A Possible Solution to the Horizon and Flatness Problems," Phys. Rev. D 23, 347-356 (1981).

75. A.H. Guth and P.J. Steinhardt, "The Inflationary Universe," Sci. Am. 250, 116-128 (1984).

76. R.J. von Gutfeld and P. Chaudhari, "Laser Writing and Erasing on Chalcogenide Films," J. Appl. Phys. 43, 4688-4693 (1972).

77. A.W. Smith, "Injection Laser Writing on Chalcogenide Films," Appl. Optics. 13, 795 (1974).

78. J. Feinleib and S.R. Ovshinsky, "Reflectivity Studies of the Te(GeAs)-Based Amorphous Semiconductor in the Conducting and Insulating States," J. Noncryst. Solids 4, 564 (1970).

79. S.R. Ovshinsky, presented at the Gordon Conf. on Chemistry and Metallurgy of Semiconductors, Andover, N.H., 1969.

80. S.R. Ovshinsky, "Method and Apparatus for Storing and Retrieving Information," U.S. Patent No. 3,530,441.

81. J. Feinleib, J.P. deNeufville, S.C. Moss and S.R. Ovshinsky, "Rapid Reversible Light-Induced Crystallization of Amorphous Semiconductors," Appl. Phys. Lett. 18, 254 (1971). Earlier, Laurence Pellier and Peter Klose worked with me in this area.

82. J.P. deNeufville, "Optical Information Storage," <u>Proc. 5th Intl. Conf. on Amorphous and Liquid Semiconductors</u>, Garmisch-Partenkirchen, Germany 1973; J. Stuke and W. Brenig, eds., Taylor and Francis, London, 1351-1360 (1974).

83. Y.C. Chang and S.R. Ovshinsky, "Organo-Tellurium Imaging Materials," U.S. Patent No. 4,142,896.

84. "Amorphous Materials Modified to Form Photovoltaics," New Scientist 76, 491 (1977).

85. R.C. Chittick, J.H. Alexander and H.F. Sterling, "Preparation and Properties of Amorphous Silicon," J. Electrochem. Soc. 116, 77-81 (1969).

86. W.E. Spear and P.G. LeComber, "Electronic Properties of Substitutionally Doped Amorphous Si and Ge," Phil. Mag. 33, 935 (1976).

87. H. Okamoto and Y. Hamakawa, "Statistical Considerations on Electronic Behavior of the Gap States in Amorphous Semiconductors," J. Noncryst. Solids 33, 230 (1979).

88. E.A. Davis and E. Mytilineou, "Chemical Modification of Amorphous Arsenic," Solar Energy Mats. 8, 341-348 (1982). (Ovshinsky Festschrift.)

89. S.R. Ovshinsky and R.A. Flasck, "Method and Apparatus for Making a Modified Amorphous Glass Material," U.S. Patent No. 4,339,255.

90. We know that there is controversy as to what constitutes an optical gap in amorphous materials, but we feel that our chemical examples are quite clear.

91. E.A. Fagen, "Optical Properties of Amorphous Silicon Carbide Films," <u>Silicon Carbide-1973, Proc. 3rd Intl. Conf. on Silicon Carbide,</u> Miami Beach, Florida, R.C. Marshall, J.W. Faust, Jr. and C.E. Ryan, eds., University of Southern Carolina Press, 542-549 (1973).

92. For the work of others in the amorphous area, see, for example, <u>Science and Technology of Noncrystalline Semiconductors</u>, H. Fritzsche and D. Adler, eds., Solar Energy Materials 8, Nos. 1-3, 1-348 (1982).

93. H. Fritzsche, M. Tanielian, C.C. Tsai and P.J. Gaczi, "Hydrogen Content and Density of Plasma-Deposited Amorphous Hydrogen," J. Appl. Phys. 50, 3366 (1979).

94. H.U. Lee, J.P. deNeufville and S.R. Ovshinsky, "Laser-Induced Fluorescence Detection of Reactive Intermediates in Diffusion Flames and in Glow-Discharge Deposition Reactors," J. Noncryst. Solids 59-60, 671 (1983).

95. S.R. Ovshinsky and A. Madan, "Properties of Amorphous Si:F:H Alloys," <u>Proc. 1978 Meeting of the American Section of the Intl. Solar Energy Soc.</u>, K.W. Boer and A.F. Jenkins, eds., AS of ISES, University of Delaware, 69 (1978).

96. A. Madan and S.R. Ovshinsky, "Properties of Amorphous Si:F:H," <u>Proc. 8th Intl. Conf. on Amorphous and Liquid Semiconductors,</u> Cambridge, Massachusetts 1979; J. Noncryst. Solids 35-36, 171-181 (1980).

97. S.R. Ovshinsky and M. Izu, "Amorphous Semiconductors Equivalent to Crystalline Semiconductors," U.S. Patent No. 4,217,374; S.R. Ovshinsky and A. Madan, "Amorphous Semiconductors Equivalent to Crystalline Semiconductors Produced by Glow-Discharge Process," U.S. Patent No. 4,226,898; S.R. Ovshinsky and M. Izu, "Method for Optimizing Photoresponsive Amorphous Alloys and Devices,"

U.S. Patent No. 4,342,044; S.R. Ovshinsky and A. Madan, "Amorphous Semiconductors Equivalent to Crystalline Semiconductors," U.S. Patent No. 4,409,605; S.R. Ovshinsky and M. Izu, "Amorphous Semiconductors Equivalent to Crystalline Semiconductors," U.S. Patent No. 4,485,389.

98. E. Cartmell and G.W.A. Fowles, Valency and Molecular Structure, Van Nostrand Reinhold, New York, 1970.

99. D. Adler, "Density of States in the Gap of Tetrahedrally Bonded Amorphous Semiconductors," Phys. Rev. Lett. 41, 1755 (1978).

100. R. Tsu, S.S. Chao, M. Izu, S.R. Ovshinsky, G.J. Jan and F.H. Pollak, "The Nature of Intermediate Range Order in Si:F:H:(P) Alloy Systems," Proc. 9th Intl. Conf. on Amorphous and Liquid Semiconductors, Grenoble, France, 1981; J. de Physique, Colloque C4, supplement au no. 010, 42, C4-269 (1981).

101. R. Tsu, D. Martin, J. Gonzales-Hernandez and S.R. Ovshinsky, to be published.

102. W.N. Lipscomb, Boron Hydrides, W.A. Benjamin, New York 1963. We are honored to be working with him on some of these important problems today.

103. J. Hanak, to be published.

104. J. Yang, R. Mohr and R. Ross, "High Efficiency Amorphous Silicon and Amorphous Silicon-Germanium Tandem Solar Cells," to be presented at the First International Photovoltaic Science and Engineering conference, Kobe, Japan, November 13-16, 1984. Our devices are the only ones that have both high efficiency and great stability.

105. W. Czubatyj, M. Hack and M.S. Shur, to be published.

106. S.R. Ovshinsky, "Commercial Development of Ovonic Thin Film Solar Cells," Proc. of SPIE Symposium on Photovoltaics for Solar Energy

Applications II, Arlington, Virginia, vol. 407, 5-8 (1983).

107. M. Izu and S.R. Ovshinsky, "Production of Tandem Amorphous Silicon Alloy Solar Cells in a Continuous Roll-to-Roll Process," Proc. of SPIE Symposium on Photovoltaics for Solar Energy Applications II, Arlington, Virginia, vol. 407, 42-46 (1983).

108. S.R. Ovshinsky, Problems and Prospects for 2004, Symp. Glass Science and Technology, Vienna, 1984. To be published in J. Noncryst. Solids. (Kreidl Festschrift.)

109. Z. Yaniv, G. Hansell, M. Vijan and V. Cannella, "A Novel One-Micrometer Channel Length a-Si TFT," 1984 Materials Research Society Symposium, Albuquerque, New Mexico, 1984.

110. Ovonyx^{TM} multi-layer x-ray dispersive mirrors.

111. S.R. Ovshinsky, unpublished.

112. L. Contardi, S.S. Chao, J. Keem and J. Tyler, "Detection of Nitrogen with a Layered Structure Analyzer in a Wavelength Dispersive X-ray Microanalyzer," Scann. Electron Microscopy II, 577 (1984).

113. J. Kakalios, H. Fritzsche, N. Ibaraki and S.R. Ovshinsky, "Properties of Amorphous Semiconducting Multi-layer Films," Proc. Intl. Topical Conf. on Transport and Defects in Amorphous Semiconductors, Institute for Amorphous Studies, Bloomfield Hills, Michigan; J. Noncryst. Solids 66, 339-344, H. Fritzsche and M.A. Kastner, eds., 1984.

114. S.R. Ovshinsky and M. Izu, "Method of Optimizing Photoresponsive Amorphous Alloys and Devices," U.S. Patent No. 4,342,044.

115. Proc. Intl. Topical Conf. on Transport and Defects in Amorphous Semiconductors, Institute for Amorphous Studies, Bloomfield Hills, Michigan; J. Noncryst. Solids 66, 1-392, H. Fritzsche and M.A. Kastner, eds., 1984.

AMORPHOUS MATERIALS—PAST, PRESENT, AND FUTURE

Stanford R. Ovshinsky

Energy Conversion Devices, Inc., Troy, Michigan 48084

It is fitting that we honor Norbert Kreidl by considering the future of glass for there is a dramatic "phase" transformation presently taking place in the area of glass science and technology. The glass meetings of old were centered on the much debated subject, "What is glass?" In a sense, this reflected not only scientific uneasiness but a search for identity, for the very basis of glassy materials, their inherent disorder, was the cause of much insecurity. The crystalline field had as its bedrock the crystal lattice, the order, if not the boredom, of repetitive atoms which looking in all directions saw sharply definable and identifiable neighbors over relatively long distances. It is no wonder that glassy materials from time immemorial through the 1950's appeared to be associated only with wide band gap materials whose optical and mechanical properties were of paramount importance. Here disorder could be considered as a positive attribute yielding, for example, isotropy, and short-range order could provide some feeling of comfort as far as reproducibility was concerned. Defects, if they could be identified at all, were identified when they could be localized in large band gap materials such as oxide glasses. The study of glass, therefore, had a deceptive transparency which tended to limit the potential applications. However, after much recent work, the question "Whither glass?" can now be answered with a great deal of assurance.

We can pay no better tribute to Norbert Kreidl than by showing that today we know what glass is and, more importantly, that this knowledge provides us firm ideas about the future of its science and technology. Kreidl's ever-inquiring creative mind, his intelligence, commitment and dedication to our field are examples of what science is about at its best and how it can be utilized to serve our world society.

My tribute to Norbert Kreidl will be to show how our present understanding has led to products undreamed of during the past when glass was only "glass." Of course, there are still many important uses for glass as a purely passive material, but it is its applications as active elements which will lead us into the 21st century.

Glass, which was previously considered primarily an optical, dielectric or passivating material, can now be used to create active devices--switches, memories, solar cells, catalysts, etc. We are not interested so much in its "old" properties, but rather want to exploit new properties, its exciting electronic phenomena, chemical reactivity or inertness, unusual phonon properties, exceptional superconductivity characteristics, important magnetic properties, etc. Instead of considering only wide band gap materials as the natural area of glassy materials, we are now able to design glassy metals, an immense range of glassy semiconductors from degenerate to wide band gap materials, and new types of information encoding devices where unusual structural changes play the decisive role.

Based upon my experience I make the bold prophecy that by the year 2004, when we celebrate Norbert's 100th birthday, most of the alternative energy sources involving materials which convert light, heat, or chemical energy into electricity will be made of amorphous or disordered materials. It is generally accepted that the two largest industries in the 1990's will be energy conversion and information processing, the latter involving computers, telecommunications, etc. As I will show here, I believe that in addition to the energy conversion devices, almost all of the informational devices will also be made of amorphous materials. The third leg of the stool representing the basic industries of the world is, of course, the materials from which everything is made. Here I believe that the powerful ability to engineer at the molecular level, that is, to synthesize countless new materials, is a unique feature of the amorphous state that will make amorphous materials technology the answer for those who wish to have products that can withstand high temperatures, corrosion, abrasion, etc. Furthermore, amorphous materials will also be utilized as catalysts, membranes for water desalination, and other basic chemical processing activities.

As I have pointed out,(1,2) the ages of humankind have always been defined by the predominant materials: the Stone Age, Iron Age, Bronze Age, etc. In this paper I wish to support

---

Reprinted by permission from *Journal of Non-Crystalline Solids*, Vol. 73, pp. 395–408 (1985).

with hard evidence my contention that the next century will be the Glass Age. In fact, we are already in the transition period between the age of order and the age of disorder. But unlike the emotive meaning of disorder and the literary meaning of amorphicity, both of which presently have negative connotations, I wish to address in this paper the meaning and value of disorder or, put in another way, the value of freedom associated with lifting the restrictions of crystalline symmetry. This freedom is reflected in the many new local bonding and nonbonding configurations that give noncrystalline solids unique attributes.(2-13)

The term "glass," commonly described as a supercooled liquid is too narrow since it excludes materials that are not formed from the melt. Such a definition creates an unnatural division and diversion that confuses the important underlying concept of non-periodic materials that atoms can be placed in three-dimensional configurations with the only restrictions being those that are imposed by the chemical and electrical force fields of the constituent atoms and their local environment.(2-13) This means that atoms can be placed into the solid by many means such as sputtering, vacuum deposition, ion implantation, electrochemical deposition, plasma decomposition, chemical vapor deposition, etc.

With the new synthetic materials that can be prepared using these methods, we can make devices in which the familiar benchmarks of equilibrium, stoichiometry and homogeneity do not have the same basic importance as they do in crystalline or in old fashioned glassy materials. In fact, the only rules that atoms deposited by our various techniques necessarily follow are those imposed by chemical reactivity, steric hindrances and local environmental forces. There are in fact many local equilibrium conditions available to atoms depending upon the dynamic conditions of their assembly (see, for example, refs. 1, 12 and 13). Therefore, the structure and properties of these materials are process dependent and are affected by temperature, excitation, sticking coefficients, etc. Each of this multitude of phases is metastable rather than globally stable, but can maintain its properties for millenia at room temperature, and, of course, be stable at actual operating conditions. There are chemical and topological barriers designed into the material which prevent structural changes. I remind you that diamond is a metastable material. Moreover, concepts which apply to the ordered world have different meaning when applied to noncrystalline solids. For example, I am in the process of redefining equilibria (14) where I show that the conventional metallurgical wisdom concerned is not relevant to many of the materials that we develop. I believe that in 2004 there will be many non-equilibrium, non-homogeneous materials designed for specific uses. Our science will be stimulated and benefited by the theories involved.

Crystallinity with its very limited specific choices of configuration pales when compared to the richness of possibilities of structures allowed in amorphous materials. Instead of building up materials where atoms have to fit in and match the lattice constraints inherent in a particular crystal structure, we can now design amorphous materials where we literally place the atoms one at a time in three-dimensional space to achieve properties undreamed of in the past. We develop layered and modulated structures where we develop and utilize new solid state phenomena and chemistry.(15)

Before we describe the devices which will be the basis for many future generations in the coming age of energy, information and synthetic materials, I will discuss what has been the Rosetta Stone for deciphering the enigma of disorder.

The Rosetta Stone for understanding amorphous materials is the concept that it is the deviations from normal structural bonding (NSB) that control the transport properties of amorphous materials just as perturbations from perfect long-range order do in crystalline materials, and that in order to control the electronic properties of amorphous materials, one must understand how to remove or add states in the gap which arise from the deviant electronic configurations (DECs).(7,8,10,11,16) The ability to tailor the band gap itself is directly related to the strength of the chemical bonds between the constituent elements which represent the NSB. By suitably choosing elements that make up the amorphous alloy, we design crosslinking structures which can retard crystallization. One can design these structures to be either unstable or bistable.(4,17-19) There is a vast range of materials that can be designed and therefore the relaxation processes which are involved can differ, e.g., elements ordinarily bonding with low coordinations can tend to remove strains in fully three-dimensional networks.(4,5,7) As I have said, "Physicists have taken for granted that there is a thermodynamic drive toward crystallinity. We emphasize that there is an equally important energetic process that leads to amorphicity, that is, the preferred chemical bonding of atoms and the charge field produced by the nonbonding electrons can alter a molecular structure so that it has an anticrystalline state. Crystals by definition have geometries that allow for repetition of the basic cell structure. The shapes that I am discussing are not rigid spherical balls but complex distorted shapes formed by localized pressures, repulsions and attractions of surrounding forces, compressed here, elongated there, twisted along another axis, the very antithesis of a crystal cell model. These tangled networks are further inhibited from crystallinity by crosslinks and bridging atoms."(2) The spectrum of configurations ranging from this description to more rigid tetrahedral type amorphous materials that can have intermediate order show the power of our field for the development of new phenomena.(20-22)

Not only can amorphous materials be designed to be very stable, but they can also be tailored to exhibit structural changes ranging from subtle bond switching to reversible crystallization. The latter include materials which are the basis of information encoding devices such as electrical or optical memories that will be described subsequently. Their design depends upon an understanding of how the energy barriers to crystallization can be overcome by optical, electrical or thermal excitation.(17,18,23,24)

I will now describe three areas where amorphous, disordered and glassy materials have been made into products which are the basis for developing a

Fig. 1   Mass production Ovonic Photovoltaic Processor.

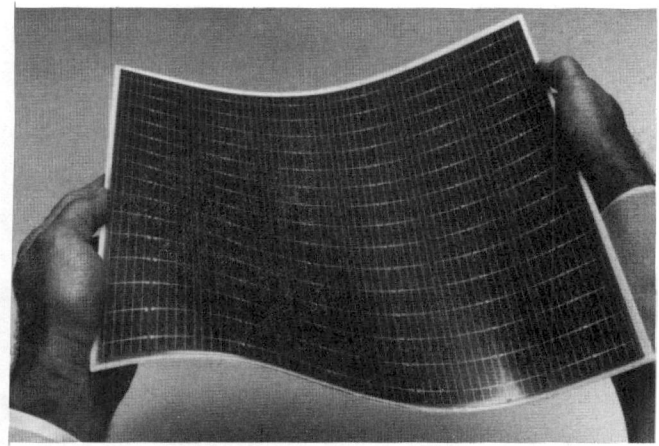

Fig. 2   Flexible one-square-foot Ovonic solar cell.

technology that will alter the way we generate electricity, store energy, and encode, transfer and distribute information. Since all of these depend upon the ability to synthesize new materials, I will also describe how synthetic coatings can advance industrial and commercial development.

In terms of energy, I will start with photovoltaics which has long been a great favorite of Kreidl in his activities to help solve the problems of providing electrical energy to meet the needs of developing countries. The ability to prepare amorphous solids inexpensively and in thin-film form over arbitrarily large areas is obviously of great value in the field of photovoltaics.(25,26)

At Energy Conversion Devices, Inc. (ECD), we have designed and built a mass production Ovonic roll-to-roll photovoltaic processor which has clearly demonstrated that the basic barrier to low-cost solar-cell production has already been broken and that one can now speak realistically of delivering power directly from the sun for under a dollar per peak watt merely by making larger versions of this continuous web, large-area thin-film machine (see Fig. 1).

We have made one-square-foot amorphous silicon alloy PIN devices (see Fig. 2) with conversion efficiencies in the 7% range and have reported smaller area laboratory PIN devices in the 10% conversion efficiency range. In addition, much higher efficiencies can be obtained with the same process by using multi-cell layered or tandem thin-film solar cell structures. These devices exhibit enhanced efficiency by utilizing a wider range of the solar spectrum.(25,26) Since the theoretical maximum efficiency for multi-cell structures is over 60%, more than twice that of a single-cell device, one can certainly realistically anticipate the production of thin-film amorphous photovoltaic devices with efficiencies as high as 30%. To accomplish this we will be utilizing narrow band gap materials which have the same low density of states as do our present silicon alloys. We already have firm experimental evidence based upon sound theoretical understanding of the material

needs that make this achievable. Our production device is already a two-cell tandem, as we have solved not only the problems of interfacing the individual cell components but also the difficulties associated with a one-foot-square format deposited on a continuous web.

Figure 3 shows a continuous roll of Ovonic solar cells. Realistic calculations for a multi-layered tandem thin-film device using our amorphous semiconductor alloys with varying optical band gaps indicate that solar energy conversion efficiencies of 20-30% can be achieved.(27) Based upon production experience and laboratory results, I can predict with a great deal of certainty that we will be in the 30% range in the year 2004. Certainly, solar power will be cheaper at that time than electricity derived from coal, gas, oil or uranium.

Figure 4 shows an interior view of a portion of the Ovonic amorphous cell production plant in

Fig. 3   Continuous roll of Ovonic photovoltaic cells.

Fig. 4 Interior view of a portion of Sharp-ECD Solar, Inc.'s Ovonic Amorphous Solar Cell Production Plant in Japan.

Japan. Note the absence of operators. This plant is highly automated, another indication of how factories producing amorphous devices will look in the next century.

Another important source of alternative energy is the utilization of the age-old dream of turning waste heat directly into electricity without moving parts, that is, the heat analog of the photovoltaic device. Again I predict that by the end of this century, thermoelectric devices will be as ubiquitous as photovoltaic devices. The problems that held back this important form of energy conversion were entirely material-based and have been solved by ongoing work at ECD. Figure 5 shows an Ovonic Thermoelectric Generator which utilizes disordered materials as the active elements. In addition, materials of the same nature have been used for Peltier Effect devices for refrigeration.

Devices such as these are already in production in the United States as well as in Japan and represent the first realistic mass production approach to an area which combines the attributes of cogeneration, conservation and pollution control. Tapping waste heat, a necessary by-product of our industrial civilization, is in a sense tapping a

nondepletable form of energy, for just as there will be photovoltaics as long as there is sunlight, there will be waste heat so long as we have civilization.

The need for batteries to store energy grows apace with every new development involving either portability or variable energy sources such as sunlight. Even without this dynamic, it has been a sad fact that batteries have not basically changed for over 100 years. Utilizing our synthetic materials approach, we have designed and built Ovonic batteries to have at least twice the energy density of conventional nickel-cadmium rechargeable batteries, as well as other improved features, while maintaining the same size and weight as conventional batteries. Initial production is being planned for 1984.

Figure 6 shows a typical battery. It is important to keep in mind that we can build these from the size of a pinhead to the size of a room, the latter for use in utility load levelling. My present prediction is that based upon laboratory studies, we will be seeing electric automobiles utilizing batteries based on these principles in widespread use by 2004. They also will fill the need for energy storage as the use of photovoltaic and thermoelectric devices grows.

Other interesting energy applications of amorphous materials include superconducting wires and magnets and the storage of hydrogen so as to make practical its use as a fuel. I predict that by 2004 we will see the widespread use of superconductivity for the storage and transmission of electricity and many vehicles will be on the road fueled by hydrogen stored in amorphous hydrides.

We have also developed a basic new approach to information technology which we believe can transform this field which until now has been dominated by the crystalline silicon industry. Our approach, utilizing proprietary amorphous materials and technology, enables us to manufacture large-area, single substrate, high density information systems which were not considered possible several years ago. For example, Ovonic technology lends itself to the development of large, flat-screen displays to replace bulky conventional cathode-ray-tube information displays. The display shown below can be made large enough to be a wall-size television set (see Fig. 7) This technology can also be applied to the fabrication of small, single substrate, completely integrated, thin-film main-frame computers which would replace room-sized ones or give small personal computers the processing capability of large mainframe units.

Fig. 5 Ovonic Thermoelectric Generator.

Fig. 6 Rechargeable Ovonic batteries.

Fig. 7   Large-area Ovonic flat screen display.

Crystalline silicon is limited to wafers no larger than six inches in diameter. By utilizing thin-film technology we can make devices many feet wide and of arbitrary length with characteristics analogous to crystalline materials at a fraction of the cost. This is the unexpected quantum leap that will transform the information and telecommunications industry by the year 2004. Instead of a circuit on a chip, we can have high density circuits vertically integrated in three-dimensional stacks of thin films on a single substrate. The various information functions we have already built are exemplified by the Ovonic write-once and reversible memory switches, PROMs, optical memories and thin-film transistors. In the 1960's I described optical memories that we were developing. In fact, we were the first to utilize lasers for what is now called "laser annealing."(28,29) The use of such optical memories is growing rapidly. Figure 8 shows an optical memory operating on the crystalline-amorphous phase-change principle. We have developed all of the other basic ingredients for an optical computer, and I believe that by the end of the century we will be seeing such new computers. I predict, based upon actual amorphous devices--both electrical and optical--already being developed and produced, a revolution sweeping the information fields as fundamental as the transition from the tube to the transistor.

Other information uses for amorphous materials include new types of photographic media that have amplification, continuous tone and very high resolution.(17,18)

Iris and I started in amorphous materials in 1955 through our interest in what is now called artificial intelligence.(30,31) Since the many products we are describing here have all developed from this approach, I would like to make another prediction. The adaptive materials (32,33) which I have used as micron-sized learning machines and the three-dimensional circuits which we are developing related to our thin-film work will, I believe, be the basis for much artificial intelligence work by the year 2004.

Another area of great interest to us is the design of passive synthetic materials. We specifically engineer coatings for a wide spectrum of other technologies, including the areas of turbines, data processing equipment, household fixtures, automotive and aviation equipment, etc.

One of the most unique coatings that we have developed is for X-ray optics for which exceedingly thin multiple layers of amorphous materials have been made into X-ray mirrors focussing elements and optical devices (15) (see Fig. 9).

There is increasing recognition that a new materials revolution is occurring. The need to overcome the limitations of existing materials technology can only be met by the ability to design, that is, synthesize, new materials with characteristics not found in either naturally occurring materials or plastics.(2)

Amorphous solids can be made by many different procedures--from powders to slabs. Of immediate technological importance is our unique ability to design and tailor-make films, coatings, wires and ribbons which can be extremely hard, resist corrosion, have increased lubricity, and/or have exceptional catalytic properties. The makers and users of industrial tools have been among the first to benefit from this new materials technology.(34)

I have permitted myself to predict what will be happening in our field in the year 2004 because the predictions I made in 1955 when there were virtually no amorphous products have now materialized into the

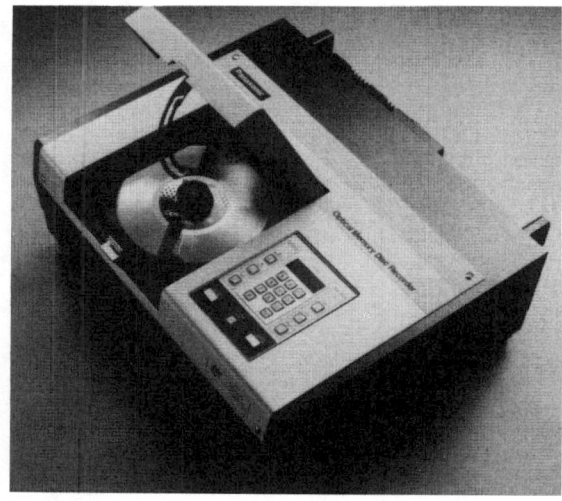

Fig. 8   Optical memory disc recorder.

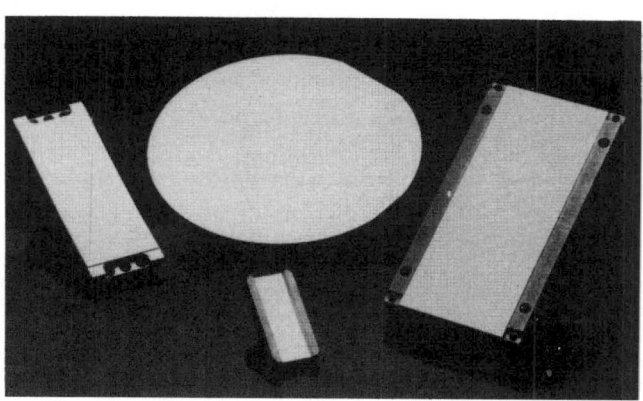

Fig. 9   Ovonyx X-ray mirrors.

many and varied applications I have described here. In 30 years we have transformed the amorphous field scientifically and technologically. I fully expect the fruition of this work to be seen in full in 20 years from now.

There are many other applications that cannot be covered in this paper. What I have sought to do is show that non-crystalline solids will be the vehicle to carry our industrial society into the new century. We all wish to see Professor Kreidl riding that vehicle urging us on.

References

1. S.R. Ovshinsky, in: Proc. 33rd Nat. Conf. on the Advancement of Research, Pennsylvania State University, State College, PA, Oct. 7-10, 1979 (Denver Res. Inst., Denver, USA, 1980) p. 15.
2. S.R. Ovshinsky, presented at 9th Int. Conf. on Amorphous and Liquid Semiconductors, Grenoble, France, July 2-8, 1981: also published in J. de Phys., Coll. C4, suppl. au no. 10, Tome 42 (October 1981) p. C4-1095.
3. S.R. Ovshinsky and H. Fritzsche, IEEE Trans. Electron Dev. ED-20 (1973) 91.
4. S.R. Ovshinsky and K. Sapru, in: Proc. 5th Int. Conf. on Amorphous and Liquid Semiconductors, Garmisch-Partenkirchen, Germany, September 1973, eds. J. Stuke and W. Brenig (Taylor and Francis, London 1974) p. 447.
5. S.R. Ovshinsky, Phys. Rev. Lett. 36 (1976) 1469.
6. S.R. Ovshinsky, in: Proc. 6th Int. Conf. on Amorphous and Liquid Semiconductors, Leningrad, USSR, Nov. 18-24, 1975 (Nauka, Leningrad,, 1979) p. 426.
7. S.R. Ovshinsky, in: AIP Conf. Proc. No. 31, Williamsburg, Virginia, March 25-27, 1976 (AIP, New York, 1976) p. 31.
8. S.R. Ovshinsky, in: Proc. 7th Int. conf. on Amorphous and Liquid Semiconductors, Edinburgh, Scotland, (June 27-July 1, 1977) p. 519.
9. R.A. Flasck, M. Izu, K. Sapru, T. Anderson, S.R. Ovshinsky and H. Fritzsche, in: Proc. 7th Int. Conf. on Amorphous and Liquid Semiconductors, Edinburgh, Scotland, (June 27-July 1, 1977).
10. S.R. Ovshinsky and D. Adler, Contemp. Phys. 19 (1978) 109; also presented at the APS March Meeting, March 27-30, 1978, Washington, D.C.
11. S.R. Ovshinsky, in: Proc. Int. Conf. on Frontiers of Glass Science, Los Angeles, California, 16-18 July 1980, J. Non-Crystalline solid 42 (1980) 3335.
12. S.R. Ovshinsky, Rev. Roum. Phys. 26 (1981) 893 (Grigorovici Festschrift).
13. D. Adler, ed., Disordered Materials: Science and Technology, Selected Papers by S.R. Ovshinsky (Amorphous Institute Press, Bloomfield Hills, Michigan, 1982).
14. S.R. Ovshinsky, to be published.
15. J. Kakalios, H. Fritzsche, N. Ibaraki and S.R. Ovshinsky, in: Proc. Int. Topical Conf. on Transport and Defects in Amorphous Semiconductors, Bloomfield Hills, Michigan, March 22-24, 1984; J. Non-Crystalline Solids 66 (1984) 339.
16. S.R. Ovshinsky, J. Non-Crystalline Solids 32 (1979) 17 (Mott Festschrift).
17. S.R. Ovshinsky and P.H. Klose, in: Proc. Soc. Inf. Display Int. Symp., May 4-6, 1971, Philadelphia, PA (Winner, New York, 1971) p. 58.
18. S.R. Ovshinsky, in: SPIE/SPSE Technical Symposium East, March 22-25, 1976, Reston, Virginia; J. Appl. Photo. Eng. 3 (1977) 35.
19. S.R. Ovshinsky, in: Proc. 4th Int. Congress for Reprography and Information, April 13-17, 1975, Hannover, Germany (Aussch. Wirtschaft, Frankfurt, 1975) p. 109.
20. S.R. Ovshinsky and A. Madan, Nature 276 (1978) 482.
21. R. Tsu, M. Izu, V. Cannella and S.R. Ovshinsky, Solid State Comm. 36 (1980) 817.
22. R. Tsu, S.S. Chao, M. Izu, S.R. Ovshinsky, G.J. Jan and F.H. Pollak, in: Proc. 9th Int. Conf. on Amorphous and Liquid Semiconductors, Grenoble, France (July 2-8, 1981); also published in J. de Phys. Coll. C4, suppl. au no. 10, Tome 42 (October 1981) C4-269.
23. S.R. Ovshinsky and H. Fritzsche, Metal. Trans. 2 (1971) 641.
24. S.R. Ovshinsky, Phys. Rev. Lett. 21 (1968) 1450.
25. S.R. Ovshinsky, presented at SPIE Symp. on Photovoltaics for Solar Energy Applications II, April 1983, Arlington, Virginia; published in Proc. SPIE 407 (1983) 5.
26. M. Izu and S.R. Ovshinsky, presented at SPIE symp. on Photovoltaics for Solar Energy Applications II, April 1983, Arlington, Virginia; published in Proc. SPIE 407 (1984) 42.
27. S.R. Ovshinsky, Invited presentation at the Int. Ion Engineering Congress ISIAT '83 & IPAT '83, September 12-16, 1983, Kyoto, Japan (Int. Ion Eng. Congr., Kyoto, 1983) p. 817.
28. S.R. Ovshinsky, in: Proc. 5th Annual National Conf. on Industrial Research, Chicago, September 19, 1969 (Ind. Res. Inc., Beverley Shores, Indiana, 1970) p. 86.
29. S.R. Ovshinsky, U.S. Patent No. 3,530,441. Applied August 22, 1968. Issued September 22, 1970.
30. S.R. Ovshinsky, The Physical Base of Intelligence—Model Studies, presented at Detroit Physiological Society (December 17, 1959) (not published).
31. S.R. Ovshinsky and I.M. Ovshinsky, Mat. Res. Bull. 5 (1970) 681.
32. S.R. Ovshinsky, J. Non-Crystalline Solids 2 (1970) 99.
33. E.J. Evans, J.H. Helbers and S.R. Ovshinsky, J. Non-Crystalline Solids 2 (1970) 334.
34. Glass Coated Drills, Ford Worldwide Productivity Idea Exchange, Issue 4 (1st quarter, 1983) p. 3.

CHEMISTRY AND STRUCTURE IN AMORPHOUS MATERIALS
The Shape of Things to Come

S.R. Ovshinsky

Energy Conversion Devices, Inc., Troy, Michigan 48084

On Sir Nevill's 80th birthday, I wish to continue the discussion which I outlined in my contributions to his 65th and 75th birthday festschrifts [1,2], and I look forward to further exploring these concepts in the festschrift for his 85th.

I viewed the term "disorder" as being divertive since it evokes so many general and emotive ideas, and have sought instead to develop ways to specify the type of atoms, their orbital arrangements and interactions when placed in three-dimensional space with "n" degrees of freedom. How the atoms utilize this freedom and how it is restricted by the topological configurations that ensue from the chemical bonding forms the basis for understanding the amorphous field [3-13].

It is difficult to conceive of and picture the resulting complex structures. In one sense, it is helpful to expand on the concept of fractals [14] and say that in the same way that the number of degrees of freedom of a coastline is expanded to greater than one because of its jagged nature, and the surface of a solid is expanded to greater than two for the same reason, we can think of the effective dimensionality of an amorphous solid as being greater than three because the surrounding atoms and their orbitals can relax rather than being necessarily tied to fixed positions.

The geometries and dimensionality of a crystal as it grows are due to its lattice structure. It takes cooperative effort to transform one crystal phase into another. In contrast, an amorphous material can have local transformations both in conformation and configuration, thus does not have one energy barrier but multiple barriers, and therefore is a multiequilibrium material. This in no way implies that it is unstable any more than diamond is because it has higher free energy than graphite under ordinary conditions.

The key to demystifying non-periodic solids is to understand that an amorphous material can have different phonon modes, electronic transport mechanisms, and excitational, nonequilibrium, electronic, and chemical properties than a crystal. The band gap of a crystal can be viewed as a "flatland," two-dimensional entity. There is no escaping the fact that the band gap, mobility gap, and density of states of an amorphous material represent a much more complex description of nature. Difficult as this may be mathematically, it can be defined, simplified, and understood if the different chemistry and geometries are taken into account. It is the great allure of the amorphous field that we are not just repeating crystalline physics and chemistry but opening up a whole world of new concepts and phenomena based on the differences from crystalline solids.

The terms "disorder" and "amorphous" imply a formlessness which is detrimental to the understanding of our field and if they ever had scientific relevance, that time is past. I have sought through the years [3] and seek here to make the point that disordered and amorphous do not mean shapeless, structureless and "empty." Instead, amorphous materials are the purest expression of the statement that chemistry cannot be separated from structure. The point is that unlike a crystal they have various local structures. Why all this is so and its consequences can be understood from first principles. In this paper, we offer a set of rules that not only can give us a qualitative understanding but can be the basis of creating synthetic materials with special properties.

The first fundamental rule distinguishing a noncrystalline solid is that its constituent atoms have bonding options [9,10,12,13]. This is the sine qua non of noncrystalline solids. Crystalline symmetry prescribes the lattice which restricts the chemical bonding choices. This constraint becomes obvious when one seeks to lay down one periodic material on top of another, for example, in superlattices. The amount of mismatching permitted is small and its effect on the transport properties great. It is not trivial to ask why and how layered materials can be made in amorphous form without such problems. To say there is no lattice in an amorphous material is to beg the question by stating the obvious. Instead we should ask why is there no need for a lattice in amorphous materials?

All the properties of an amorphous solid, its cohesive energy, its resistance to crystallization,

---

Reprinted by permission from D. Adler, H. Fritzsche, and S.R. Ovshinsky, eds., *Physics of Disordered Materials*, Plenum Press, New York, 1985, pp. 37–54.

333

its optical band gap, its mobility gap, its density of electronic states, etc., depend upon three factors: its short-range chemical "order,"* the varied topological configurations, and the total interactive environment (TIE) [8] between the configurations and the orbitals that inhabit them. All of these can be specified, the first two very succinctly, the last in broad but basic terms. These factors have little relevance to the crystalline lattice. For example, even its short-range order possibilities are limited, since crystalline symmetry can only accommodate one or at most a few deviations that affect the environment of a constituent atom. Therefore, its short-range order is determined by its long-range order. In a real sense, its short-range order is its long-range order.

An amorphous material, however, can be a nonstoichiometric alloy in a nonequilibrium configuration, consisting of many different types of atoms, providing a variety of local order and environments. It is not unusual to make a material with five or six different atoms [13,15]. Even the much-treasured concept of homogeneity is not necessary in an amorphous solid and in fact heterogeneity is still another freedom which can be a powerful design tool, allowing new dimensions of synthesis. I conceived of ultra-thin layering and compositional modulation utilizing amorphous materials years ago and have shown that even sequential layering of repeating materials need not be a limitation and that a whole new class of heterogenous material can be designed [16].

The compositions and configurations of the first few coordination shells of the constituents of a heterogeneous amorphous material can be controlled to distribute a variety of local environments throughout the material. These synthetic materials are free from the constraints of crystalline symmetry and therefore can yield new types of nonequilibrium "disordered" structures of varying complexity. This enables us to produce improved materials which have a wide range of applications such as in solar cells, x-ray optics, superconductivity, catalysis, thermoelectricity and magnetism, as well as in the development of entirely new materials having properties which make possible many new uses.

If we are to set up a taxonomy of the permitted chemical and structural configurations in amorphous materials, we must understand the basis of their nonequilibrium configurations for there are many more shapes available to us in an amorphous solid than are allowed in crystalline structures.

The varied shapes are generated by using our first rule of amorphicity, optional bonding, and are structured by the second rule which is that steric hindrances act in three-dimensional space to anchor the desired atomic and molecular configurations. These hindrances to motion are crosslinks and bridges and are important in preventing crystallization [13,15]. The ability to utilize space is

restricted in crystalline materials and open to us in amorphous materials. This additional spatial component results in a range of relaxations not permitted in crystalline materials. I suggest that the term "frustration," which is now so popular and which was used originally to describe magnetic phenomena in amorphous materials, should be relegated to the same bin that disorder will eventually be. Such emotive terms must be replaced by those which elucidate how three-dimensional space is utilized by chemical bonding. Once we describe this in a specific manner, we can show how external energy is absorbed in and affects an amorphous material fundamentally differently from a crystal. A simple way of looking at an amorphous solid is that like a crystal it utilizes a given space to connect atoms with each other with specific chemical bonding; however, unlike a crystal, the space that is left over after local bonding is optimized still has potential energy in the sense that until the entire three-dimensional space is utilized, there is no one free energy determined. In other words, until all the three-dimensional space is utilized, relaxations can take place so that it is possible to reduce the free energy further, creating configurations separated in space by various energy barriers. The steric hindrances become in effect steric facilitators as three-dimensionality is generated. Therefore, the distinctions differentiating amorphous from crystalline materials are: bonding options, crosslinks and steric hindrances to atomic movement (as opposed to the lattice constraints), and differing relaxation modes.

Let us examine how structures are created and transformed in amorphous materials, starting on an atomic level and then exploring their steric components. In order to understand the basic stereochemical rules of amorphous solids, we initially consider an empty container. First, we ask, what sort of atoms can and should we put into the container and for what purposes? To begin with, we deal mostly with covalent materials. If we were considering only single elemental materials, then we would note that for atoms in column IV of the Periodic Table, all solids but lead ordinarily bond via strong sp^3 interactions. These hybrids give the maximum possible number of bonds per atom (without using d or f electrons), four, and usually form rigid three-dimensional structures. The column V atoms have an s lone pair and bond p^3, primarily right-angle bonding (with some sp angular widening). This leads to either puckered layers or ribbon-like structures, without the three-dimensional rigidity of the column IV solids. The column VI elements are unique in possessing an outer-electron p lone pair, yet being able to bond p^2 and form solids. Since this type of bonding usually yields only two bonds per atom, these solids consist of either chain-like or ring-like structures, giving a maximum of one-dimensional rigidity. Column VII elements each have two p lone pairs, but usually can form only one bond, yielding diatomic molecules; column VIII atoms ordinarily cannot bond at all.

It would be simple enough, therefore, to draw the conclusion that Group VI makes for helixes and chains, Group V for layered structures, Group IV for three-dimensional structures. To fill our container and synthesize the most interesting materials (for

---

* Although I utilize the term because of its common usage, I would prefer to use "short-range bonding relationships."

that is the beauty of eliminating the lattice), the preferred materials are those that are not made of single elements but have atoms from other columns added. Therefore, if we add to Group VI atoms crosslinks such as germanium or arsenic from Groups IV or V, we transform the material into a fully three-dimensional solid. If we bridge the layers of Group V with elements from either IV or VI, we can transform the sheet-like structures into a three-dimensional solid. Since a Group IV element such as silicon cannot fully complete its four orbitals without introducing large strains and dangling bonds, we must add bridging, crosslinking, and strain-relieving elements from Group VII, like fluorine, or Group I, like hydrogen, to allow a more complete three-dimensional structure to be generated [2].

The use of the term "usually" in the previous bonding descriptions can now be understood since it is the deviations (DECs) from the normal structural bonding (NSB) that not only affect the bridging, crosslinking, and strain relieving mechanisms but most importantly are responsible for the states in the gap of amorphous materials which control the transport properties [4,6,17]. It is essential to understand this rule to understand amorphous materials.

Layered and chain structures not only use covalent bonds in the form of crosslinks, but also, just as importantly because of how the rings, chains, folds, and corrugated layers are formed, require that there be a number of such crosslinks. The structures are irregular and the "struts" (steric hindrances) which control their shapes do not repeat neatly. This means that the whole idea of stoichiometry which is dogma in crystalline materials must be laid aside. The rule is that we must not only define the crosslinks by specifying their chemical structures, but also know how many there are, and what their bond strengths are. Internal geometries are formed by the above factors and do not resemble a lattice in the least. As I have stated elsewhere: "The shapes that I am discussing are not rigid spherical balls, but complex distorted shapes formed by localized pressures, repulsions, and attractions of surrounding forces, compressed here, elongated there, twisted along another axis, the very antithesis of a crystal cell model" [11]. The steric elements add variety to the shapes being generated. These shapes span the spectrum from rather simple structures to very convoluted ones, and can even coexist. They affect the electronic properties of the material, for example, the sharpness of the mobility edge, and the density of states and type of traps in the gap.

The most inherently rigid matrices such as our amorphous silicon and germanium alloys made with fluorine have exceptional electronic properties, more resemble a crystal [10,12], and their rigidity carries with it a lack of ability to generate more complex forms. These materials preserve the basic tetrahedralness, both short range and long range. As one goes down column IV through germanium, tin, and lead, the Sedgewick pairs (the s lone pairs) become more chemically "inert,"* more divalence occurs, and more structural complexity (and instabilities such as the Staebler-Wronski type) can result. I have described [4,19,20] how this tendency

away from tetrahedralness in germanium, etc. is responsible not only for instabilities but for unwanted states in the gap. In tandem and multilayer photovoltaic devices this would constitute an impossible barrier to increased efficiency. I have shown how one can minimize these effects through the use of fluorine [4,19-23]. Guha [24] from our laboratory has reported that silicon-germanium alloys made with fluorine have not only exceptional stability but have electronic properties equal to the best fluorinated silicon materials. We have shown how a chemical approach has solved a major and apparently intractable problem involving the density of states as well as the Staebler-Wronski effect. Yang from our laboratory has demonstrated for the first time that the use of a narrow band gap material can actually increase solar-cell efficiency in a tandem cell (11.2%) [25], and has made a single-layer, narrow band gap device with the highest efficiency to date (8.5%) [26].

In Groups III and IV, the opportunity for optional bonding without the use of lone pairs adds new possibilities, since boron and carbon can resemble tellurium in a very important chemical way, namely, they inherently have several bonding options. These examples support our contention that bonding options and the new orbital relationships resulting from them are the basic rule of amorphous solids, both in generating noncrystalline configurations and as a fundamental building block for the synthesis of many new materials. It is interesting that boron, carbon, and tellurium are allotropic materials, and can exist quite easily either in the crystalline or amorphous form. In other words, they are natural glass-formers. The term "glass-formers" has been used for many years without a great deal of rigor. I hope we can make this term synonymous with materials that can be made with atoms that have optional bonding choices. An understanding of structural chemistry can be a powerful tool for designing amorphous materials.

Since they can enrich a material with many new configurational possibilities, I have even used boron and carbon for what I have called chemical modification [6,17,27], though most of the chemical modifiers I originally used were transition metals. How is it possible that there is a commonality between such different elements? It is because their three-dimensional orbital interactions can dramatically affect the transport properties of the material. Transition elements primarily through their d-orbitals and boron and carbon through their multi-orbital configurations permit unusual orbital intersections and interactions with nearby atomic configurations. The resulting nonequilibrium materials have unique properties such as over 10 orders of magnitude changes of conductivity without changing the optical band gap. One can generalize and state that any element that provides optional and/or additional orbital interactions can act as a modifier. I have, of course, shown that, if desired, one can have the modifier enter in an

---

* Such lone pairs can resemble those of the chalcogenides in their interactions including their lack of detection by ESR since they are compensated by being spin paired [5,15].

335

alloying mode as well. Compositional modulation and layering also can be utilized as forms of chemical modification [6,16,17,27,28].

Since we have said that the electronic parameters depend upon chemistry and structure, we discuss more fully some of the geometric concepts, and ask ourselves what geometries can describe the complex spatial arrangements and internal surfaces as the degrees of freedom of atoms are used up by their bonding and orbital interactions in three-dimensional (and possibly greater) space. Three dimensions are probably not enough to describe the relationships. Consequently, as a start we apply and extend fractal concepts as well as the possibility of new geometries that include and expand the three-dimensional space available for our varied configurations. I consider that the extra partial degree of freedom of the jagged coastline model of fractals suggests that in amorphous solids we are seeking to go beyond three-dimensionality for our understanding of the geometries generated by the interaction of the various orbitals coming in from all directions. This interaction creates not only creases and crevices, folds, chains, rings, voids, hills and valleys, but also an interaction with the "sky above" and the "earth and water" below as well as the areas abutting the coastline.

These complex interactions of orbitals in the kind of space we have described above make for difficulty in applying quantum chemistry to the problem, yet it must be done. It is encouraging that the great inventor of frontier molecular orbital theory, Fukui, and his colleagues have begun applying this theory to amorphous materials [29,30]. I believe that by adopting Fukui's theory [31], we can develop a basis for simplifying the chemical reactions in an amorphous solid [32]. Bonding considerations are dependent not upon chance but rather on the relative energies of the frontier orbitals. The discrimination between the possible reactions could be based upon the HOMO and LUMO of the potential configurations. In comparing the various choices for nearest neighbors, one preferred bonding path with a specific orientation could be chosen. When we design a material (for that is what amorphous materials are--engineered materials), we must specify the nature of the constituent atoms, the temperature of preparation, and the state of the atoms upon their arrival at the growing surface, e.g., excited or unexcited, and other preparation-dependent features. The relaxation time would then be dependent upon the dynamics of the situation as well as on the chemical system and the availability of an appropriate HOMO-LUMO combination.

In this way, the power of a computer could eventually be used to compile tables that would specify the species available under certain preparation conditions. Hopefully, we would be able to calculate what reactions were most probable, allowing for the actual degrees of freedom available. For example, the steric template feature of preparation dependence would be one restrictive and stabilizing force. The nature of the atoms and the orientation of the orbitals already in place exert a strong influence on film making because incoming atoms or molecular configurations make preferred choices. Other factors that help rearrange atoms as they develop into amorphous solids are the presence of chemical donors and acceptors, lone pairs, lone

pairs, and the electronegativity of the elements, etc. Obviously, chemical insight must guide us, for it is far from being a near-term number-crunching possibility.

We emphasize that there are inherently anticrystalline structures that can be designed in amorphous materials rather than thinking of all amorphous materials as being distorted crystalline structures [2,4,8,11]. It should be clear that there are profound differences between a material in which a lattice ensues from the chemical bonds of like atoms and one in which a lattice cannot be propagated under prescribed chemical and topological conditions. In the latter material, we utilize the concepts of steric, quantum, polymer and free-radical chemistry and by so doing we alter the physics involved. The science of amorphous materials is truly the synthesis of many different disciplines. Chemistry and geometry determine the physics.

In previous papers [8,11] I have explained why materials made from the same atoms but deposited by different preparation methods can have different electronic characteristics. This is especially true where plasmas are employed to make the materials. Following the arguments of this paper, it can be understood that new molecular configurations are formed once the extra dimension of excited and free-radical states is provided to atoms and molecules [28,33]. Their combinations and configurations differ depending on the substrate temperature, the excitation of the gases, the recombination and lifetime of the species, the nature of the catalytic activity, the recombination properties of the surface, and the sticking coefficients and diffusion lengths of the atoms and molecular fragments as they hit the surface. This understanding is particularly applicable to tetrahedral materials. Most amorphous chalcogenides, no matter how they form, have a tendency through their bonding and relaxations to compensate all dangling bonds in their structure through the use of their lone pairs. Having more available space in which to operate, the compensational processes via quenching in chalcogenides are more tolerant. However, of importance in the making of useful tetrahedral materials are the nature of free radicals, the lifetimes of the excited states in the plasma, and the rapid relaxation times of the molecules that ensue.

I have always opposed [2,17] the simplistic view that the dangling bonds of tetrahedral materials are just capped by the chemical properties of a single element, whether it is hydrogen or fluorine. Instead, I believe that a molecular structure is formed in the gas, initially on the surface and eventually in the solid, whose overall configuration as a result of the use of hydrogen and/or fluorine in the plasma and on or in the solid inherently and intrinsically does not have dangling bonds. The configuration is an entity in itself which, joined together with other such units, has a low density of localized states. To support this contention, it is clear from experiments that once a rigid structure that contains dangling bonds has been formed, the later use of the compensating elements such as atomic hydrogen and fluorine is not as effective as their incorporation while the film is being formed. I have devoted a great deal of attention to free radical and precursor chemistry [33], for they are responsible for the construction of the most desired configurations.

It should be clear that nonequilibrium processes are of great importance, both electronically and structurally, in amorphous materials, again a basic difference from crystalline materials. In a chalcogenide, nonequilibrium configurations are preferably instituted by chemical modification through which orbitals are induced to undergo unique nonequilibrium interactions by the insertion of multi-orbital elements into the matrix as they are being grown. I propose that excited or free-radical modification is a technique in which atoms or clusters are excited so as to offer new bonding conditions while a molecular configuration which would not ordinarily occur is being formed. This is an analog to chemical modification where new elements such as transition elements are introduced to create nonequilibrium materials. Using excited modification we need not introduce extra orbitals into the material, but rather get the unusual nonequilibrium extra orbital reactions by virtue of the excited states of the existing atoms and molecules. In both chemical modification and excited or free radical modification, we are designing additional orbital choice interactions. While chemical modification can eliminate states in the gap in small amounts, it has been used primarily to add states in the gap. Excited state modification offers an important new controllable parameter in the processing of amorphous materials, either to eliminate states or to add them, and the basic reason that it can do so is that unique nonequilibrium processes are possible in amorphous materials, but are rejected by the lattice.

Thus far we have been using the term "nonequilibrium" mostly to describe the preparation of amorphous solids. A better term would be "multiequilibria"! This is because there is not one distinct equilibrium state except under certain conditions such as molecule formation; in fact, there are ordinarily many equilibrium configurations available in amorphous solids. I believe that the use of the term "nonequilibrium" in this sense is only a vestigial remain of the classification of crystalline solids. From the physics point of view, the exciting part of amorphous materials is that the electron-phonon interactions and the density of states are basically different from those of crystals, the traps and recombination centers have different origins, and therefore the materials have differing responses to light, heat, and electric fields. Most important, the nonequilibrium electronic processes can be very different and exciting, as, for example, in the Ovonic threshold switch [13,15]. In the early 1960's, I saw multinonequilibrium states in some materials.

Amorphous solids can be visualized as a balance of chemical and electronic forces which, when excited, seek rearrangement of the electrons and atomic configurations. Because of their internal chemical and structural stability, excitation processes can induce relaxations, electronic and/or structural, depending upon the design of the material. The difference between an Ovonic threshold switch and an Ovonic memory switch is precisely that the former depends upon electronic

mechanisms throughout and in the latter the material is designed to permit structural changes to be initiated by electronic mechanisms.

Since in many amorphous solids all of the three-dimensional space is not completely used up and there is thus extra space or an effective additional dimensionality, the relaxation processes can yield subtle configurational changes associated with excitational processes. Where the material has been designed to be incapable of containing the excited state, larger scale structural relaxations take place that involve configurational changes and, depending upon the nearest neighbor choices, either a new amorphous phase or a crystalline phase may result (as in Ovonic memory material [13,15]). In such materials, a change of conformation can result in changes of configuration unlike, for example, in the Ovonic threshold switch or in properly designed photovoltaic materials where configurational changes are minimized.

This means that the responses to absorption of energy in an amorphous solid can be far different from those in a crystalline material, and that the "band gap," that is, both the optical gap and the electrical activation energy, can be a "breathing," flexible one. The flexibility can be localized and the mobility gap [34] must be viewed three-dimensionally rather than through the typical crystalline physicist's flatland approach. As I have said [9]: "A change of charge and occupancy of the localized states acts upon the matrix just as the matrix helps position the localized states originally." This interactive environment means that excited carriers may find that, under certain conditions, their home is not in the same place it was when they left it. This is why I have called the Staebler-Wronski effect a photostructural change [22]. This means that silicon-hydrogen alloys still have space for relaxations to occur and that their full three-dimensionality is not utilized. While there was much disagreement with this characterization at the time, there is little now.

The above bears on an earlier basic rule that energy absorption differs in amorphous and crystalline materials. Energy input is not shared throughout the amorphous material as it is in a crystalline solid, but excitational recombination processes tend to be localized. This can be an advantage since long-range phonons have difficulty in propagating and short-range phonons are the rule in amorphous solids; the relaxation processes that are available in the effective extra-dimensional space can change geometries without the breaking of bonds, that is, they can induce conformational changes. It bears repeating that in materials that are designed for structural transitions, bonds can be broken by the changes of conformation and new local configurational geometries result. The geometries, i.e., the topological units, can affect or actually be responsible for charged states, traps, recombination centers, photoluminescent centers, that is, the whole gamut of defects that control the electronic properties of a material.

We have divided amorphous materials into two basic types: unistable, in which the overall configurational change is insignificant and where conformational changes initiated by excitation are contained and shaped by the surrounding configurational structures; and bistable in which conformational changes <u>cannot</u> be contained and important configurational alterations occur in response to excitation. We have used such changes in local order for many applications in the information field, electrical and optical memories, photography, etc [11,35-39]. Conformations and configurations can change and interchange themselves under the influence of external forces such as electric field, current, light, pressure, diffusion, etc.

Amorphous materials are very different from crystalline materials for the breaking of bonds in a crystal usually results in severely distorted or dangling bonds; in amorphous materials, depending upon the chemical dynamics and the space available, new bonding configurations can take place. The spatial aspects are very important and can be illustrated by the limited rearrangements that can and do take place on the surface of the crystal which has sites which resemble the interior of an amorphous solid where extra dimensionality is available for bonding and configurational reconstructions. A surface reconstruction can be looked at as a two-dimensional geometry changing into a three-dimensional one. It is therefore instructive to consider the backbonding and other reconstructions that occur at the surface of a crystal. Such rearrangements take place because there is spatial freedom allowed for them and there are no counteracting chemical forces in the space above the surface, such as oxygen, nitrogen or other elements, for if they were there, they could attach to a dangling bond and in a sense create a micro-surface alloy, that is, a very localized new chemical composition. Without such competition, as in a vacuum, the space above and the nearby repulsive and attractive forces of the orbital configurations on the surface are sufficient to affect the local configuration (for example, via backbonding).

When atoms are in a constrained space, that is, a region with a surface potential barrier, we can associate a total energy with the contained atoms. This energy is the sum of the kinetic energy of the motion of the ion cores and the electrons and the potential energy that describes their potential motion. In an amorphous material, the ability to use <u>all</u> of the available three-dimensional space enables an overall minimization of the local energy in that spatial continuum. Space and energy in an amorphous solid are thus directly related, and therefore in a material that is designed by the rules that we have mentioned, the free energy of the atoms in the available space can be optimized. This manner of looking at energy and spatial relationships in amorphous materials is not only a theoretical construct but has far-reaching implications, for example, as atoms are deposited to grow a film, the ability of an atom to move also means that that movement can <u>stimulate</u> a response from other atoms causing them to move. Such relaxations can be considered to add an extra dimensionality to the space being utilized.

On the surface of a crystal, atoms and orbitals can pop in and out and rotate in various orientations depending upon the spatial freedom and the chemical energy of nearby atoms. In an amorphous material, internal freedoms (as contrasted to the limited external freedoms at the surface of crystals) lead not only to new types of bonding arrangements but by the very nature of the chemical and spatial interactions, new topological units are generated. I believe that this freedom of material design is the motivating force that will shape much of our science and technology in the coming years.

There can be very subtle, completely reversible changes effected by relaxations from an excitational process, changing the character of the states in the gap and moving them in space and energy, or there can be larger structural changes such as cooperative phase transitions that completely alter the nature of the material without changing its chemical constituents; in other words, relaxation processes can be of such a nature as to cause changes of local order and structural rearrangements.

I find it fascinating that cosmologists are discovering and utilizing phase changes [40,41] as a means of explaining the origin of our universe or universes, and I have suggested that, rather than the liquid-to-crystal transformation they are presently considering, the theories that we have utilized to explain the liquid/plasma-to-amorphous, amorphous-to-amorphous and amorphous-to-crystalline transformations are more pertinent [4]. In this way, one can freeze in various effects and take into account clustering phenomena.

I feel that the above concept can explain how cosmological defects can be created and frozen in. It has important ramifications since it predicts that such defects can be locally altered and/or annealed by thermal or charged spikes or other effects. This implies that there still can be local events including phase changes with great exotherms resembling to a certain extent the initial expansionary phenomena. There occurs in the universe great unexplained energetic activity and the absorption of energy in nearby spatial environments could have alterations and relaxations similar to those which I have outlined in this paper. It is exciting to think that there even could be some local "reversibility."

The "disorder" of the universe, that is, its lack of periodicity and its short-range order relationships, are to me a more accurate description of the physical reality with more far reaching and profound implications than the proposed crystalline models. In this sense, we can assume that as the universe runs down, energy barriers, defects and relaxations would be affected and new phenomena could occur. Since there are different zones of time in the universe associated with its expansion of space, it is possible that we can observe and classify some of these quenching activities and therefore get a snapshot of state of entropy.

To come back to terrestrial amorphous solid state physics and chemistry, when considering subtle changes, for example, when conformational shapes are altered by excitation, we must again leave the "flatland" physics of the crystalline world with its rigid band gap and particularly rigid density of states. The effective density of states in an amorphous solid is the solution of a complex many-body problem, the basis of which is its three-dimensionality. In an amorphous material, the conformational changes can make states appear, interact, move and disappear depending upon the material. For example, if one has a chalcogenide material, and therefore a normal structural bonding (NSB) that is primarily divalent, but with some bonded as well as non-bonded lone pairs, then application of a dc electric field across such a material would move the lone-pair electrons. As they move, they alter their local conformational geometry, and indeed can even transform the lone-pair bonds. The interconversion of conformational geometries is due to the aforementioned "breathing" ability of the flexible polymeric structure of a primarily Group VI material. In an amorphous solid, the electric field and photochemical responses have different pathways from those of thermal activation alone. The unusual dc effect in some Ovonic threshold materials is caused by the "polarization" implied above. The changes of resistance that can be seen in chalcogenide materials in response to electric field are related to this effect.

How can we classify amorphous materials? It is understood that they can be metals, have a wide range of semiconducting band gaps from near degenerate to dielectric, that they can be ferromagnetic, diamagnetic, or paramagnetic, that they have no structural change or can change their structure both subtly and cooperatively. As discussed here, the basis for all of these characteristics lies in the spatial relationships and orientation of orbitals where rigidity or internal structure is determined by the manner in which three-dimensional space is utilized. The chemical forces which bond atoms and distribute orbitals can create underline inherent anticrystalline configurations [2,11], which generate intertwining chains and rings whose three-dimensionality is assured by strongly bonded crosslinks. They take one-dimensional chains or rings or two-dimensional sheets or layered structures and so intertwine, separate and anchor them that crystallization can be either difficult or virtually impossible. Stable configurations that could only happen by chance can now be deliberately "frozen in," that is, designed by understanding how to control relaxations either chemically or by quenching methods. We contrast this approach to the still current thinking in our field which persists in considering that all amorphous materials are just distorted crystal structures. We emphasize that negative correlation energy is associated with the ability to accommodate relaxations through the utilization of spatial-energy geometric considerations discussed in this paper. Therefore, the three-dimensional rigidity of the system is the important factor rather than the Anderson [42] approach in which disorder is the criterion.

Due to our ability to use the powerful tool of synthesis inherent in amorphous materials, we have been able to make many materials which have little or no relationship to crystal structures and I believe by Nevill's next festschrift such materials will become more and more important and that current "conventional wisdom" will have shifted to this point of view. I recall when I used the term "steric hindrances" in the past in discussing amorphous materials, the approach was met with little understanding. Nevertheless, I see the concept and term increasingly used now, especially in the Japanese literature. The same can be said for the terms "amorphize" and "amorphization." I believe that by

using a more precise vocabulary, such terms reflect our more profound understanding of the amorphous state. Young scientists working in the field who do not have the burdensome heritage of crystalline physics will think more and more in purely amorphous terms.

It is considered that the difference between glasses and amorphous materials is that glasses are made from the liquid state. It is understood that amorphous structures can be generated in other ways than from the melt. Since quenching rates are central to both glasses and amorphous materials, when does a quenching rate become a fast relaxation? It is my position that in various plasma depositions, quenching plays an important role in the freezing in of free radicals, reactive fragments, and other varied molecular configurations. These exceedingly fast quenching relaxations provide the most interesting species. This process is related to recombination, radical lifetime, surface reactions, etc. Looking at plasmas in this manner solves the problem of how they produce films. Unlike plasma deposition, film deposition by vacuum and sputtering is straightforward. Utilizing the plasma quenching concept, we can show that film-making processes are part of the same continuum with quenching mechanisms playing the central role in all cases. The lifetime of excited species and their recombination rates are important parameters in extending the concept of amorphicity.

I have argued that the term "melting" is inadequate for many of the processes that it has been used to describe in amorphous materials [7,14]. Diffusional energy, impact energy, and excitational energy can all be utilized to create amorphous and even crystalline structures because they can alter the positional local chemical bonding relationships in a selective manner.

As already pointed out, elemental materials such as silicon and certainly boron have several possible local structural configurations in the amorphous state, even without any alloying [6,17]. The rigidity of the mostly tetrahedral structures such as elemental amorphous silicon results in large concentrations of dangling bonds and voids [9,10,12,18]. They obviously are materials with high densities of localized states. In order to minimize the density of gap states, one must understand their origin and use the chemical rules outlined previously for alloying and bridging where, through the addition of flexibility, new configurations are generated and the proper coordination established, resulting in a minimum of dangling bonds [2,6,8,10-13,15,17,34]. The establishment of the proper coordination is the needed factor to minimize density of states.

The nonequilibrium transport properties are greatly affected by the density of traps and recombination centers. There are three ways to make an amorphous semiconductor with a low density of states in the gap: (1) chemically, by removing the states, optimizing the required coordination. This results in allowing excitation processes to propagate throughout the material with a minimum of loss; (2) by effectively making a material with a high density of gap states think it is a material with a low density of gap states by transiently swamping the existing states with carriers (thus filling carrier traps and increasing the Fermi energy), and allowing new incoming carriers generated, for example, by injection, to have large mean free paths so that they do not "see" any encumbrances to their flow; and (3) the subtlest method, by transiently shifting the recombination centers in space and therefore in energy through the slight relaxation processes discussed here, so that the recombination centers are effectively changed into traps and the carriers therefore remain alive longer. Of course, doping (except for the proximity type) and most especially chemical modification are methods for increasing the states in the gap.

There are not only switching, memory and photoconducting uses for amorphous materials based upon these concepts, but also such diverse applications as superconductors, magnetic materials and catalysts. Most amorphous materials are the antithesis of crystalline materials in that they do not require stoichiometry, homogeneity, "equilibrium," extended phonons, lattice matching, and all of the other sacred precepts of periodicity. They are indeed a new area of science and technology. One does not understand amorphous materials because one understands crystallinity; rather, one can understand crystallinity as a special case when one understands the basic rules of amorphicity.

I have stated [4,6,17] that the Rosetta Stone of amorphous materials is the understanding of the relationship between the normal structural bonding (NSB) which characterizes the great majority of atoms and is responsible for the cohesiveness of the amorphous solid and the deviant electronic configurations (DECs) that control the transport properties and provide the active chemical sites of the material. Tying these two together is the concept of the total interactive environment (TIE) [8] which takes into account the special nature of various local, chemical, topological, and electronic interactions in amorphous solids. We can write a new language of materials if we make use of this new alphabet. We need not be limited to the old dogmas of homogeneity and equilibrium chemistries. We have a new world of nonequilibrium chemistry and varying topological structures. We can deliberately create combinations and geometries in which even the local short-range order can vary subtly or drastically from one part of the material to the other. Such new structural chemistry again produces new electronic phenomena and new sites for chemical activity. Amorphous materials with unique cluster configurations and those containing crystalline inclusions and layers are also part of the spectrum of engineered materials discussed here. Such designed materials have far-reaching applications. We can synthesize and engineer materials where we mismatch and compensate atoms without the problems of mismatching lattices. We can carry this further through the use of layering and compositional modulation [16]. In fact, heterogeneity then becomes a welcome tool rather than a scare word.

In summary, an amorphous solid has extra dimensionality in all available dimensional space through unusual orbital relationships freed from the tyranny of the crystalline lattice. Exciting new mechanisms and phenomena are available and a new science with its own theory and understanding, classification and language is developing. The formless and structureless world of amorphous mate-

rials is no more formless and structureless than that of the nerve cell and its interactions, a field that I entered in 1955. I based my approach in neurophysiology on attempting to understand the disorder of the surfaces of nerve cells (where much of the action occurs) [43]. This led me to the study of amorphicity generally [1,3,15,43-45]. In building switching and memory models of amorphous materials, I sought to understand and utilize the various energy transformations that occur in nature and how energy is absorbed in a material and converted into other forms of energy and information. In neurophysiology, I can recall a meeting as late as 1960 [46] when McCullough, a pioneer in the field, said that he was just then beginning to think that the neuron was not "just an empty bag." This statement was a great shock to me who had been operating not only on the opposite assumption but on the basis that short-range relationships and structure were playing an important role. Since that time, the world of neurophysiology has been greatly expanded by understanding the surface properties, the chemistry, and the structure of the neuron.

In terms of amorphous materials, we are still living in historical times. I feel that our approach is in keeping with the career and example of Mott [47]. Nevill's work has spanned such areas as nuclear physics, theory of metals, theory of photography, theory of crystalline semiconductors, theory of disordered and amorphous materials, the metal-nonmetal transition, and many, many others. I believe that devices, whether a universe, a nerve cell, a memory, a switch, or a solar cell , should result from new concepts. Just as I have tried to show in this paper that amorphous materials because of their optional bonding and orbital relationships can have unusual anisotropic and directional effects that have not been fully recognized, I believe that the pioneering contributions of Mott have given a directionality to our field that will be an example to others for generations to come.

ACKNOWLEDGEMENTS

As always, this work was accomplished with the other half of our lone pair, Iris.

I owe a great debt of gratitude to David Adler with whom I have discussed and clarified many of the ideas expressed here.

REFERENCES

1. Ovshinsky, S.R. and Ovshinsky I.M., "Analog Models for Information Storage and Transmission in Physiological Systems," Mat. Res. Bull. 5, 681-690 (1970). (Mott Festschrift.)

2. Ovshinsky, S.R., "The Shape of Disorder," J. Non-Cryst. Solids 32, 17-28 (1979). (Mott Festschrift.)

3. Adler,D., ed., Disordered Materials: Science and Technology, Selected Papers by S.R. Ovshinsky, Bloomfield Hills, Michigan, Amorphous Institute Press, (1982) 1-296. (See for more complete references.)

4. Ovshinsky, S.R., "Fundamentals of Amorphous Materials," in <u>Physical Properties of Amorphous Materials</u>, Institute for Amorphous Studies Series, vol. 1, D. Adler, B.B. Schwartz and M.C. Steele, eds., Plenum Press, New York, (1985) 105-155.

5. Ovshinsky, S.R. and Sapru, K., "Three Dimensional Model of Structure and Electronic Properties of Chalcogenide Glasses," in <u>Proc. of the Fifth Int. Conf. on Amorphous and Liquid Semiconductors</u>, Garmisch-Partenkirchen, Germany, (1974) 447-452.

6. Ovshinsky, S.R. and Adler, D., "Local Structure, Bonding, and Electronic Properties of Covalent Amorphous Semiconductors," Contemp. Phys. <u>19</u>, 109-126 (1978).

7. Ovshinsky, S.R. and Fritzsche, H., "Amorphous Semiconductors for Switching, Memory and Imaging Applications," IEEE Trans. Electron Devices <u>ED-20</u>, 91-105 (1973).

8. Ovshinsky, S.R., "The Chemical Basis of Amorphicity: Structure and Function," Rev. Roum. Phys. <u>26</u>, 893-903 (1981). (Grigorovici Festschrift.)

9. Ovshinsky, S.R., "Localized States in the Gap of Amorphous Semiconductors," Phys. Rev. Lett. <u>36</u>, 1469-1472 (1976).

10. Ovshinsky, S.R., "Amorphous Materials as Interactive Systems," in <u>Proc. of the Sixth Int. Conf. on Amorphous and Liquid Semiconductors</u>, Leningrad, USSR (1975) 426-436: <u>Structure and Properties of Non-Crystalline Semiconductors</u>, B.T. Kolomiets, ed., Nauka, Leningrad (1976) 426-436.

11. Ovshinsky, S.R., "Principles and Applications of Amorphicity, Structural Change, and Optical Information Encoding," J. de Physique <u>42</u>, C4-1095-1104 (1981).

12. Ovshinsky, S.R., "Lone-Pair Relationships and the Origin of Excited States in Amorphous Chalcogenides," in <u>Proc. of the Int. Conf. on Structure and Excitation of Amorphous Solids</u>, Williamsburg, Virginia (1976) 31-36.

13. Ovshinsky, S.R., "An Introduction to Ovonic Research," J. Non-Cryst. Solids <u>2</u>, 99-106 (1970).

14. Mandelbrot, B.B., <u>The Fractal Geometry of Nature</u>, W.H. Freeman, New York, (1983).

15. Ovshinsky, S.R., "Reversible Electrical Switching Phenomena in Disordered Structures," Phys. Rev. Lett. <u>21</u>, 1450-1453 (1968).

16. Ovshinsky, S.R., "Compositionally Varied Materials and Method for Synthesizing the Materials," U.S. Patent Application Serial No. 422,155, filed 9/23/82.

17. Ovshinsky, S.R., "Chemical Modification of Amorphous Chalcogenides," in <u>Proc. of the Seventh Int. Conf. on Amorphous and Liquid Semiconductors</u>, Edinburgh, Scotland (1977) 519-523.

18. Fritzsche, H., "Summary Remarks," in <u>Proc. of the Sixth Int. Conf. on Amorphous and Liquid Semiconductors</u>, Leningrad, USSR, 1975: <u>Electronic Phenomena in Non-Crystalline Semiconductors</u>, B.T. Kolomiets, ed., Nauka, Leningrad, (1976) 65-68.

19. Internal ECD Report, January 1, 1983.

20. Ovshinsky, S.R., "Roll-to-Roll Mass Production Process for a-Si Solar Cell Fabrication," in Technical Digest of the International Photovoltaic Science and Engineering Conference, November 13-16, 1984, Kobe, Japan (in oral presentation).

21. Ovshinsky, S.R. and Madan, A., "Properties of Amorphous Si:F:H Alloys," in <u>Proc. of 1978 Meeting of the American Section of the International Solar Energy Society</u>, K.W. Boer and A.F. Jenkins, eds., AS of ISES, University of Delaware, (1978) 69-73.

22. Ovshinsky, S.R. and Madan, A., "A New Amorphous Silicon-Based Alloy for Electronic Applications," Nature <u>276</u>, 482-484 (1978).

23. Ovshinsky, S.R. and Izu, M., "Amorphous Semiconductors Equivalent to Crystalline Semiconductors," U.S. Patent No. 4, 217, 374, filed 3/8/78; Ovshinsky, S.R. and Madan, A., "Amorphous Semiconductors Equivalent to Crystalline Semiconductors Produced by Glow-Discharge Process," U.S. Patent No. 4,226,898, filed 3/16/78; Ovshinsky, S.R. and Izu, M., "Method for Optimizing Photoresponsive Amorphous Alloys and Devices," U.S. Patent No. 4,342,044, filed 9/9/80.

24. Guha, S., "Light-Induced Effects in Amorphous Silicon Alloys - Design of Solar Cells with Improved Stability," to be presented at the Eleventh Int. Conf. on Amorphous and Liquid Semiconductors, Rome, Italy, September 2-6, 1985.

25. Yang, J., Mohr, R., Ross, R. and Fournier, J., to be published.

26. Yang, J., Mohr, R., Ross, R. and Fournier, J., to be published.

27. Ovshinsky, S.R., "The Chemistry of Glassy Materials and Their Relevance to Energy Conversion," J. Non-Cryst. Solids <u>42</u>, 335-344 (1980).

28. Ovshinsky, S.R. and Flasck, R.A., "Method and Apparatus for Making a Modified Amorphous Glass Material," U.S. Patent No. 4,339,255, filed 9/9/80.

29. Tachibana, A., Yamabe, T., Miyake, M., Tanaka, K., Kato, H. and Fukui, K., "Electronic Behavior of Amorphous Chalcogenide Models," J. Phys. Chem. <u>82</u>, 272-277 (1978).

30. Tanaka, K., Yamabe, T., and Fukui, K., "A Role of the Lowest Unoccupied Molecular Orbital of

the Local Structure of Amorphous Materials," Solar Energy Mats. $\underline{8}$, 9-13 (1982).

31. Fukui, K., "Role of Frontier Orbitals of Chemical Reactions, Science $\underline{218}$, 747-754 (1982).

32. Ovshinsky, S.R., "Intuition and Quantum Chemistry," to be published by D. Reidel in Proc. of the Nobel Laureate Symposium of Applied Quantum Chemistry, 1984 Int. Chemical Congress of Pacific Basin Societies, Honolulu, Hawaii, December 18, 1984.

33. See, for example, Ovshinsky, S.R., "The Role of Free Radicals in the Formation of Amorphous Thin Films," in Proc. Int. Ion Engineering Congress (ISIAT '83 & IPAT '83), Kyoto, Japan, (1983) 817-828. Ovshinsky, S.R. and Izu, M., "Amorphous Semiconductors Equivalent to Crystalline Semiconductors," U.S. Patent No. 4,217,374, filed 3/8/78; Ovshinsky, S.R., Allred, D., Walter, L. and Hudgens, S., "Method of Making Amorphous Semiconductor Alloys and Devices Using Microwave Energy," U.S. Patent No. 4,504,518, filed 4/30/84.

34. Cohen, M.H., Fritzsche, H. and Ovshinsky, S.R., "Simple Band Model for Amorphous Semiconducting Alloys," Phys. Rev. Lett. $\underline{22}$, 1065-1068 (1969).

35. Ovshinsky, S.R., "The Ovshinsky Switch," in Proc. Fifth Annual National Conf. on Industrial Research, Chicago (1969) 86-90.

36. Ovshinsky, S.R. and Klose, P.H., "Imaging in Amorphous Materials by Structural Alteration," J. Non-Cryst. Solids $\underline{8-10}$, 892-898 (1972).

37. Ovshinsky, S.R., "Electronic and Structural Changes in Amorphous Materials as a Means of Information Storage and Imaging," in Proc. of the Fourth Int. Congress for Reprography and Information, Hanover, Germany (1975) 109-114.

38. Ovshinsky, S.R., "Amorphous Materials as Optical Information Media," J. Applied Photographic Engineering $\underline{3}$, 35-39 (1977).

39. Ovshinsky, S.R., "Amorphous Materials--Past, Present and Future," Problems and Prospects for 2004, Symposium on Glass Science and Technology, Vienna, Austria, July 3, 1984; J. Non-Cryst. Solids (in press). (Kreidl Festschrift.)

40. Guth, A.H., "Inflationary Universe: A Possible Solution to the Horizon and Flatness Problems," Phys. Rev. D $\underline{23}$, 347-356 (1981).

41. Guth, A.H. and Steinhardt, P.J., "The Inflationary Universe," Sci. Am. $\underline{250}$, 116-128 (1984).

42. Anderson, P.W., "Model for the Electronic Structure of Amorphous Semiconductors," Phys. Rev. Lett. $\underline{34}$, 953-955 (1973).

43. Ovshinsky, S.R., "The Physical Base of Intelligence-Model Studies," presented at the Detroit Physiological Society (1959).

44. Southworth, M.P., "The Threshold Switch: New Component for Ac Control," Control Engineering $\underline{11}$, 69-72 (1964).

45. Ovshinsky, S.R., "Nerve Impulse," (1955), unpublished.

46. Bionics Meeting, Dayton, Ohio, Spring 1960.

47. Mott, N.F., "Electrons in Glass," 1977 Nobel Prize Lecture, Science $\underline{201}$, 871-875 (1978).

INTUITION AND QUANTUM CHEMISTRY

S.R. Ovshinsky

Energy Conversion Devices, Inc., Troy, Michigan 48084

The field of science to which Professors Fukui, Herzberg, Hoffmann, Lipscomb and Mulliken have contributed so much is an excellent example of how intuition is the source of quantum leaps in science, especially in quantum chemistry. These Nobelists have been much honored for their scientific achievements. My purpose here is to discuss their extraordinary intuition without which there would be no quantum chemistry, for quantum chemistry is nothing more than quantum mechanics plus intuition.

Many people consider that science is the body of existing knowledge and scientists add to this knowledge in a straightforward, logical manner. This commonly accepted viewpoint is at variance with what another Nobelist, Szent-Gyorgyi, said, "A discovery must be, by definition, at variance with existing knowledge."[1] The fact that well-meaning people and good scientists can have such opposing views shows that C.P. Snow's division of our society into two cultures of arts and science is wrong; there are two cultures in science itself. However, there is truly but one culture in which art, literature, music, and science are one, for all the basic attributes of the arts--of beauty, aesthetics, simplicity and the wonderment of the human condition-- can be expressed in many ways, but are an essential part of our civilization.

I would like to illustrate the two cultures in science by citing Yukawa, the Nobelist whose creativity was so instrumental in advancing the field of high-energy particle physics with his invention of the meson. He described his concern about declining creativity by contrasting those who have "an excessive conservatism in the realm of ideas" against "those who have a spirit of adventure." He expressed it thus: "The number of research workers has increased, and among them are men of great ability. The fault surely lies in an excessive conservatism in the realm of ideas, retaining the same preconceptions and pursuing the same lines of development--an unfortunate and inefficient process. Even while keeping to one particular line of development, the objective will rarely be attained without some radical leap forward along the way."[2]

The presence of Professors Fukui, Lipscomb and Hoffmann at this meeting allows us to especially honor them. They are among the most adventuresome men who have made those "great leaps forward," and in so doing give us examples of the importance of intuition and illustrate the courage of the "frontiersman," if I can borrow Professor Fukui's terminology.

Why do I use the term "courage"? Physicists applying quantum mechanics were only able to describe the hydrogen atom and the harmonic oscillator, yet these brave men have utilized quantum mechanics in complicated chemical systems. Their work has advanced the field of chemistry in a profound manner. There is only one explanation of their ability to accomplish these path-breaking advances--that is, intuition, for in order to utilize quantum mechanics in chemistry one has to simplify complexity, to make the right choices out of many options. This could not be done by a computer; this could only have been accomplished by the artistry of their approach, their genius in "knowing" what paths to take. Their approaches to science, therefore, have a commonality that, while differently expressed, not only explains phenomena but allows for prediction.

In these days of computers and their overwhelming use by scientists, we should hearken back to what Wigner and Seitz said long ago, [3] "if one had a great calculating machine, one might apply it to the problem of solving the Schrodinger equation for each metal and obtain thereby the interesting physical quantities, such as the cohesive energy, the lattice constant, and similar parameters. It is not clear, however, that a great deal would be gained by this. Presumably the results would agree with the experimentally determined quantities and nothing vastly new would be learned from the calculation. It would be preferable instead to have a vivid picture of the behavior of the wave functions, a simple description of the essence of the factors which determine cohesion and an understanding of the origins of variation in properties from metal to metal."

Reprinted by permission from R.S. Mulliken, K. Fukui, W. Lipscomb, and R. Hoffmann, eds., *Applied Quantum Chemistry* (Proceeding of the Nobel Laureate Symposium on Applied Quantum Chemistry in Honor of G. Herzberg), D. Reidel, Dordrecht, 1986, pp. 27–31.

The theories, the simplifications, and the new ways of looking at the problems of quantum chemistry have permitted these men that we honor here to utilize quantum mechanics not only to desribe but to explain. Structure and function are combined, thus I would add to my previous definition of quantum chemistry as quantum mechanics plua intuition the additional characteristic of structure, that is, the ability to conceive in three-dimensional space configurations that are responsible for reactivity.

The style in which these men do science therefore is of great interest. Notice that I have used terms such as adventure, bravery, style. These are terms occasionally employed in literature and the expression of the artist, but are not ordinarily applied to scientists. To me, the fact that they can be accurately used here emphasizes the one culture of art and science.

Professor Fukui has written a book of autobiographical essays unfortunately not yet translated into English. In it his approach to science can be seen. One discerns his unaffected curiosity, his early excitement and fascination with nature, the stimulation of his finding Fabre's Book of Insects, the joy he felt in chasing butterflies. He was both fascinated and humble when confronting nature. His feelings about creativity can be illustrated by his admiration of the Japanese sculptor, Unkei, who carved buddhas out of wood and felt that one does not create a buddha but looks very carefully at the wood and then digs the buddha out.

Fukui Sensei is in his own way a sculptor--his predictions of the properties of a molecule are related to the shape that their outer frontier electrons assume. In fact, the _insight_ as to how structure and function are related is a great commonality among all three of these men and supports my view of the importance of structure in quantum chemistry.

When we speak of structures, we can appreciate Professor Lipscomb, for God figured out the complexities of carbon; it took Lipscomb to do it for boron! In Bill's beautiful essay on the aesthetics of science, [4] one gets a sense of the importance of the intuitive process as well as an appreciation of how truly science is an enobling part of our culture. He is a person who shows great delight in discovery and great delight in music. I think there is a connection, for like his science, his love of music is deeply felt and unaffected. He feels it and he does it. Certainly he is extremely well organized, but his creative spontaneity can be likened to the riffs in jazz that he so much appreciates. (I sometimes wonder if his pride in being a Kentucky colonel makes that title as important as the Nobel Prize!)

Just as we have been discussing the commonality of intuition, the subject of aesthetics also links our honorees. Professor Hoffmann in his Nobel Prize lecture [5] spoke about the importance of aesthetics in science and the construction of conceptual bridges between inorganic and organic chemistry. By the way, humans made disciplines, Nature did not; and Professor Hoffmann's thinking shows how physics and chemistry are truly opposite sides of the same coin. His approach to density of states and Fermi levels discussed at this meeting is still building conceptual bridges which are of great importance to understanding not only the chemical pathways but the electronic mechanisms that are associated with the active sites in a solid. Einstein said that intuition is being "sympathetically in touch with experience." One would be hard put to find a better description of Professor Hoffmann and his work. Nature took many millenia to find its pathways for chemical expression--Hoffmann is doing it in one lifetime.

One has heard the criticism that the quantum chemistry of our honorees is involved with approximations. How silly, for it is exactly their intuition which has allowed them to simplify the most complex problems and find the correct insights and expressions. Number crunching does not do that, and number crunching does not give one that extra dimension that is expressed by Hoffmann and our other honorees. In the beginning there was chemistry and with it came structure. It is structure in chemistry that must be given increasing attention.

I discussed the problems of disorder and amorphicity with Professor Fukui who, with his colleagues, then addressed his frontier orbital theory to amorphous materials; [6,7] I work with Professor Lipscomb on various chemical problems related to amorphous materials; I am very pleased to know that Professor Hoffmann is now bringing the power of his intellect to the consideration of the role of disorder.

In a way, I dislike using these terms since disorder and amorphicity are both emotive. Literary people use amorphicity pejoratively for formlessness, that is, structurelessness, and disorder connotes riots in the streets. As a long-time worker in this area, let me assure you that the field of amorphous materials is based on quantum chemistry with short-range order and varied structures. It has been my position that the sine qua non of amorphous solids is the ability of atoms to have optional bonding. You as chemists should not be frightened of materials that do not have periodicity for here intuition is again our guide. Those of us who can think about optional atomic relationships and chemical bonding in three-dimensional space are rewarded with the excitement and the pleasure of discovery and the opening up of a new field of science which combines chemistry and physics.

As I have written elsewhere,[8] "The universal intuitive process, whether expressed in art, literature, music or science, is the ability to see connections between facts or concepts which to others are unrelated. Creativity links insights in such a way that a meaningful pattern leaps out of interlocking steps and becomes a bridge or pathway."

"Many people who are merely imaginative are not insightful or creative, for the path has to lead somewhere and the bridge must be a means of fording a stream. Intuition, the basis of science, is therefore not an exotic tool but the most utilitarian of arts and its practitioners are the craftsmen of imagination."

I am honored to be in the position of honoring the three men present here for they represent the best of our culture and therefore of our civilization. By making their "radical leap forward," they became exemplary figures. We can all benefit by understanding that they represent a very critical part of science and that this requires daring, bravery, adventure and dedication. Their lives are lights that illuminate the beauty of science.

References

1. Szent-Gyorgyi, A., "Dionysians and Apollonians," Science <u>176</u> (1972) 966.

2. Yukawa, H., <u>Creativity and Intuition: A Physicist Looks at East and West</u>, (Kodansha International Ltd., Tokyo, 1973).

3. Wigner, E.P. and Seitz, F., "Qualitative Analysis of the Cohesion in Metals," in <u>Solid State Physics, Advances in Research and Applications</u>, Frederick Seitz and David Turnbull, eds., vol. 1, (Academic Press, New York, 1955) 97.

4. Lipscomb, W.N., "Aesthetic Aspects of Science," in <u>The Aesthetic Dimension of Science</u>, 1980 Nobel Conference, Curtin, D.W., ed., (Philosophical Library, New York, 1982) 1-24.

5. Hoffmann, R., "Building Bridges Between Inorganic and Organic Chemistry," (Nobel lecture), Angewandte Chemie <u>21</u>, 711-724 (1982).

6. Tachibana, A., Yamabe, T., Miyake, M., Tanaka, K., Kato, H. and Fukui, K., "Electronic Behavior of Amorphous Chalcogenide Models," J. Phys. Chem. <u>82</u>, 272-277 (1978).

7. Tanaka, K., Yamabe, T. and Fukui, K., "A role of the Lowest Unoccupied Molecular Orbital of the Local Structure of Amorphous Materials," Solar Energy Mats. <u>8</u>, 9-13 (1982).

8. D. Adler, ed., <u>Disordered Materials: Science and Technology</u>, Selected Papers by S.R. Ovshinsky, (Amorphous Institute Press, Bloomfield Hills, 1982) 295-296.

THE QUANTUM NATURE OF AMORPHOUS SOLIDS

Stanford R. Ovshinsky

Energy Conversion Devices, Inc., 1675 West Maple Road, Troy, Michigan 48084

I. INTRODUCTION

It is fitting to discuss in this festschrift for Hellmut Fritzsche the phenomena which first interested him in the amorphous field.

The elimination of the crystalline lattice provides unique opportunities to understand and investigate quantum phenomena in amorphous materials. When we consider the basic physics, it becomes clear that the early prejudices and antiquated notions that were and still are held concerning the nature of amorphous solids represent the last gasps of a conventional mindset that is inherent in the crystalline ways of thinking that dominated condensed matter physics for so many profitable years. This mindset must be replaced by a new paradigm, one which I have been advocating for many years.[1-3] This paradigm can be illustrated by examining some of the more important arguments that have been used to describe the physics of amorphous materials. My approach is to turn the arguments on their heads, so to speak. In so doing, I shall try to illustrate how viewing a problem from a different physical perspective is the preferred way of stimulating innovative concepts in this field.

In this paper I show how amorphous materials should be viewed basically as materials which can be designed to express unusual quantum activity. In Section II, I analyze the electronic structure of these materials, using chalcogenide alloys as an illustrative example. Threshold switching is discussed in Section III.

II. ELECTRONIC STRUCTURE OF AMORPHOUS MATERIALS

Let us start with the statement that has been so often repeated that it is believed to be a truism: electrons and holes in amorphous materials have low mobilities and therefore cannot possibly lead to high-speed devices. After all, in a semiconductor with many localized states in the gap that act as traps and recombination centers, the mean free path of an electron or a hole must be very small.

If we now rephrase this apparently objective fact and examine it in a more fundamental manner, we can indeed reach the opposite conclusion, namely, that an amorphous material can be designed to be a medium in which electrons and holes can propagate unhindered at optimum speeds. How is this so? How can these seemingly diametrically opposite statements be reconciled? In fact, it is the synthesis of these contradictions which makes for exciting physics, unusual chemistry, interesting molecular geometries, and unique high field and optical phenomena.

Because of its periodic structure, all of the electronic states of a crystalline solid are Bloch states, with equal probability of being found in the vicinity of every equivalent atom. Electrons in Bloch states are scattered by phonons, which represent deviations from perfect periodicity, and it is this scattering that controls the mobility in high-quality crystals at ordinary temperatures. In contrast, an amorphous solid has no long-range order, and the electronic wave functions can be initially considered to be localized in the vicinity of a particular atom. Generally, however, the density of states, $g(E)$, is sufficiently large at most energies that electrons can propagate through the material by quantum-mechanical tunneling. Above a critical value of $g(E)$, the effective mobility sharply increases,[4] resulting in the existence of mobility edges.[5] Beyond the mobility edge, electrons propagate in a band-like manner, just like in a crystal; however, because of the absence of a lattice in amorphous semiconductors, deviations from periodicity, such as long-range phonons, should not scatter propagating electrons, and thus do not limit the mobility.[1,2]

The existence of localized states in the gap of amorphous materials makes for an array of new phenomena not possible in periodic crystals. For example, as discussed above, in a pure crystalline semiconductor, the mobility is limited by phonons. Since at any finite temperature an equilibrium concentration of phonons is always present, there is no way to increase the mobility by any means. In contrast, in an amorphous material, as I shall discuss in Section III, the effective mobility can be increased by effects such as screening, charged-trap filling, and other techniques.

Reprinted by permission from M.A. Kastner, G.A. Thomas, and S.R. Ovshinsky, eds., *Disordered Semiconductors*, Plenum Press, New York, 1987, pp. 195–204.

III. THRESHOLD SWITCHING

Even though we had been working on amorphous switching since the 1950's, [1] we did not publish in physics journals until 1968.[6] According to the prevailing mindset of the time (a) disordered and amorphous materials were considered to be ununderstandable, (b) there was no need to understand them since there could be no interesting electronic phenomena involved, and (c) certainly the use of quantum concepts to elucidate amorphous materials would be a contradiction since it did not appear to be reasonable to assume that electronic activity on a quantum basis in a nonperiodic system would be imaginable. We first used the name Quantrol [7,8] for what are now called Ovonic Threshold Switches (OTS) on my assumption that I was observing quantum phenomena in these switches.

We describe in this paper the physical basis for a unique room-temperature switching phenomenon in amorphous materials in the picosecond range, speeds that have been associated only with Josephson-junction devices, which must be operated at liquid helium temperatures and were developed much later.

The OTS has been described many times. It is sufficient here to summarize its characteristics. A typical composition is $Te_{40}As_{35}Si_{18}Ge_7$, and it is ordinarily prepared in thin-film form with a thickness of the order of $1\mu m$. It has a mobility gap of about 1.1 eV and has of the order of $10^{18}cm^{-3}$ charged defect centers, which yield states in the gap that effectively pin the Fermi energy. The drift mobility of holes is of the order of $0.01-0.1\ cm^2/V-s$ and it has no detectable concentration of unpaired spins. Fritzsche personally made many of the early measurements, and was involved in the characterization of the switches.[9-12] His plenary presentation at the March 1969 APS meeting in Philadelphia received great attention.[13]

The material behaves as an intrinsic semiconductor in the sense that its resistance decreases exponentially with temperature with an activation energy of about half the mobility gap, transport taking place via band-like propagation.[14] This material is thermally stable and has a high glass transition temperature. Recent measurements have confirmed that it remains amorphous even if taken to $400^{\circ}C$ for a half hour.[15] It would therefore seem that the possibility of being able to switch the material at room temperature on a picosecond time scale to a conducting state is unlikely, if not impossible.

The I-V characteristics of a 1 μm thick film of $Te_{40}As_{35}Si_{18}Ge_7$ is shown in Fig. 1.[6,16] It is clear that rapid switching does take place. The reason that it occurs has to do with the fact that these materials have mobility gaps as described above as well as positively and negatively charged states in the gap that act as traps. The solid can then sustain high-field injection, and both electrons and holes can propagate in a band-like manner, yielding a plasma. Once the charged traps are filled, the conductivity is essentially metallic, with about $10^{19}cm^{-3}$ electrons moving with

mobility of the order of $10cm^2/V-s$.[14] It is important to realize that the switching time, i.e., the time it takes for the current to increase from its OFF-state to its ON-state values, is not limited by the transit time and is much faster. The switching time is limited by the carrier trapping time which is typically of the order of 1 picosecond. The transit time is also in the picosecond range until a thickness of 1 micron or over is reached. For a 1 μm-thick device with an applied voltage of 1 V, the transit time is $L^2/\mu V \approx 1$ nanosecond.

Within the mobility gap, states with energies within kT of each other are at least 10Å apart, and conduction occurs only by hopping. When electric fields greater than a critical value of about $2\times10^5 V/cm$, are applied, injected electrons and holes fill the charged traps. The mean free path of injected electrons increases sharply to values greater than the thickness of the film, and conduction occurs with mobilities of the order of those in good metals like copper. The return to the nonconducting state occurs when the current is sufficiently small that recombination occurs faster than free carriers are replenished. The critical field for turning the switch ON defines a threshold voltage, while the minimum current necessary to sustain the ON state defines a holding current.

Such a reversible transition is far from evident within conventional crystalline semiconducting theory. The ON state does not represent the increase of the Fermi energy (as in doping), or the quasi-Fermi energy (as in photoconductivity). Instead, it represents an increase in nonequilibrium free carriers and their drift mobility, with the Fermi energy remaining the same.

Let us now look at a different regime, equally mysterious from the crystalline point of view. Figure 2 shows the I-V characteristics of exactly the same material, except the film thickness is under 150Å. Note that the threshold "wings" have disappeared, so that only the conducting state remains; the relationship between the wingless device and the TONC experiments of Pryor and Henisch [17], and Petersen and Adler [18] will be discussed in a subsequent paper. The device resembles a symmetric Zener diode (recall that the Zener diode turned out not to be a tunneling phenomenon as was first thought).

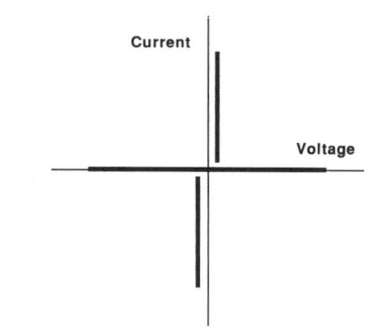

Fig. 1. The I-V characteristics of a 1 μm-thick film of $Te_{40}As_{35}Si_{18}Ge_7$.

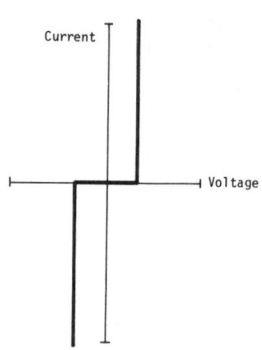

Fig. 2. The I-V characteristics of 125Å thick film of $Te_{40}As_{35}Si_{18}Ge_7$.

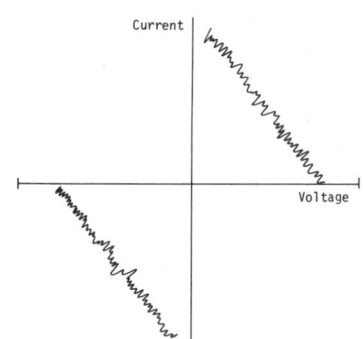

Fig. 3. The Ovonic Threshold Switch in the oscillatory mode.

The disappearance of the wings can be understood from a simple analysis. A critical field of about 2×10^5 V/cm is necessary to achieve switching. When a voltage is applied to the material, about 0.4 eV falls across the interface region near the contacts. The remaining 0.3 V necessary to achieve the minimum holding voltage of 0.7 V appears across the bulk of the chalcogenide. When the chalcogenide is approximately 150Å thick, the average field is then 2×10^5 V/cm, just the critical field for switching. Thus, in this case, the threshold voltage equals the holding voltage, and the wings disappear. Note that for a thickness, L, of 150Å and an applied voltage, V, of 0.3 V, the transit time, $L^2/\mu V$, is less than 1 picosecond.

Figure 3 shows a threshold switch in the oscillatory mode. This regime is achieved when the load is sufficiently high that neither state is completely stable, because the critical holding current is not attained in the ON state. The ON-state current is then insufficient to keep the traps filled on a steady-state basis. As in the other regimes discussed, switching here is clearly electronic in nature and occurs at speeds on the order of a picosecond.

Note that conduction in the material occurs by either phonon-assisted hopping or band-like propagation (i.e., tunneling) in both the OFF and ON states. The only difference is the greater mean free path of the ON state, due to trap filling from the high electric field throughout the material. This field provides a substantial increase in the carrier drift mobility as well as providing an increase in the effective number of carriers.

If the threshold switch is made very thick, of the order of 10 μm, there is a widespread belief that the switching mechanism changes from electronic to thermal. However, this is not the case. The switching phenomenon is, in fact, the same in thicker films. The only modification is the existence of what I have called trimming.[12] For example, in thick films, the low thermal conductivity of amorphous chalcogenides can lead to a small temperature increase in the center of the film, far from both contacts. Since chalcogenides are intrinsic semiconductors, even a small increase in temperature can induce an exponential rise in current. Because of this current rise, the field is enhanced

over the remainder of the material, leading to an overall reduction in the threshold voltage. However, the critical field that induces the switching is exactly the same in thick as in thin films, and the ON state is identical in both cases. Instead of a temperature increase, field-induced generation or photogeneration can accomplish similar types of trimming. It is only necessary to provide enough carriers so that a field redistribution takes place. As I have pointed out,[19-22] the lone-pair nature of chalcogenides shown by Kastner [23] is the basis for unique electronic activity since the lone pairs are not only a source of charged states [22,24] but, most importantly, they are a source of nonbonded electrons whose excitation in an Ovonic Threshold Switch material does not affect the structural bond integrity. Of course, the excitation process is exceedingly fast. The bonding electrons are much deeper in energy and are not affected by the electric field.[1] In contrast, Ovonic memory materials have been deliberately designed so that structural changes can occur, initiated by threshold switching, and the bonding is not stable against the excitation process.

The variations of the Ovonic Threshold Switch shown in Figs. 1-3 show that quantum phenomena operate just as well in amorphous as in crystalline materials. The recent need for switches with pico-second response times has triggered a great deal of work in Josephson junctions and ballistic transistors, much of it frustrating and uneconomical. It is ironic that well over twenty years ago we discussed picosecond switching in amorphous materials, without the need for devices at liquid helium temperatures or with submicron photolithographic processes.

In this paper I have described quantum effects and electronic activity in amorphous solids which appeared to be revolutionary when I first discussed them, and which are necessary to understand the true significance of amorphous materials now that they have become so popular. It was the new physics that attracted Hellmut Fritzsche. His decision that this was an important area has been amply justified over the years.

ACKNOWLEDGEMENTS

I wish to thank Rosa Young for the use of her measurements and Napo Formigoni for recreating my original wingless switches. I would like to extend my gratitude to David Adler with whom I am writing a more detailed paper on this subject.

REFERENCES

1. S.R. Ovshinsky, _Disordered Materials: Science and Technology_, ed. D. Adler, Amorphous Institute Press, 1982.

2. S.R. Ovshinsky, "Fundamentals of Amorphous Materials," in _Physical Properties of Amorphous Materials_, Institute for Amorphous Studies Series, vol. 1, eds. D. Adler, B.B. Schwartz, and M.C. Steele, Plenum Press, New York, 1985.

3. S.R. Ovshinsky, "Chemistry and Structure in Amorphous Materials: The Shape of Things to Come," in _Physics of Disordered Materials_, Institute for Amorphous Studies Series, eds. D. Adler, H. Fritzsche, and S.R. Ovshinsky, Plenum Press, New York, 1985. (Mott Festschrift.)

4. N.F. Mott, "Electrons in Disordered Structures," Adv. Phys. 16, 49 (1967).

5. M.H. Cohen, H. Fritzsche and S.R. Ovshinsky, "Simple Band Model for Amorphous Semiconducting Alloys," Phys. Rev. Lett. 22, 1065 (1969).

6. S.R. Ovshinsky, "Reversible Electrical Switching Phenomena in Disordered Structures," Phys. Rev. Lett. 21, 1450 (1968).

7. M.P. Southworth, "The Threshold Switch: New Component for Ac Control," Control Engineering 11, 69 (1964)

8 J.R. Bosnell, "Amorphous Semiconducting Films," in _Active and Passive Thin Film Devices_, ed. T.J. Coutts, Academic Press, 1978, p. 288.

9. H. Fritzsche and S.R. Ovshinsky, "Electronic Conduction in Amorphous Semiconductors and the Physics of the Switching Phenomena," J. Non-Cryst. Solids 2, 393 (1970).

10. E.A. Fagen and H. Fritzsche, "Electrical Conductivity of Amorphous Chalcogenide Alloy Films," J. Non-Cryst. Solids 2, 170 (1970).

11. H. Fritzsche, "Physics of Instabilities in Amorphous Semiconductors," presented at the Symposium on Instabilities in Semiconductors, IBM Watson Research Lab., Yorktown Heights, N.Y., 1969: IBM J. Res. & Dev. 13, 515 (1969).

12. S.R. Ovshinsky and H. Fritzsche, "Amorphous Semiconductors for Switching, Memory, and Imaging Applications," IEEE Trans. Electron Devices ED-20, 91 (1973).

13. H. Fritzsche, "Recent Experiments on Amorphous Semiconductors," presented at the APS March Meeting, Philadelphia 1969, Bull. APS II, 14, 342 (1969).

14. D. Adler, H.K. Henisch and N.F. Mott, "The Mechanism of Threshold Switching in Amorphous Alloys," Rev. Mod. Phys. 50, 209 (1978).

15. R. Young, to be published.

16. D. Adler, M. Shur, M. Silver and S.R. Ovshinsky, "Threshold Switching in Chalcogenide-Glass Thin Films," J. Appl. Phys. 51, 3289 (1980).

17. R.W. Pryor and H.K. Henisch, "Nature of the ON-State in Chalcogenide Glass Threshold Switches," J. Non-Cryst. Solids 7, 181 (1972).

18. K.E. Petersen and D. Adler, "Probe of the Properties of the On-State Filament," J. Appl. Phys. 47, 256 (1976).

19. S.R. Ovshinsky, "Localized States in the Gap of Amorphous Semiconductors," Phys. Rev. Lett. 36, 1469 (1976).

20. S.R. Ovshinsky, "Amorphous Materials As Interactive Systems," _Proc. 6th Intl. Conf. on Amorphous and Liquid Semiconductors_, Leningrad, 1975: _Structure and Properties of Non-Crystalline Semiconductors_, ed., B.T. Kolomiets, Nauka, Leningrad, 1976, p. 426.

21. S.R. Ovshinsky, "Lone-Pair Relationships and the Origin of Excited States in Amorphous Chalcogenides," AIP Conf. Proc. 31, 31 (1976).

22. S.R. Ovshinsky and K. Sapru, "Three-Dimensional Model of Structure and Electronic Properties of Chalcogenide Glasses," _Proc. 5th Intl. Conf. on Amorphous and Liquid Semiconductors_, Garmisch-Partenkirchen, Germany 1973: eds. J. Stuke and W. Brenig, Taylor and Francis, London, 1974, p. 447.

23. M. Kastner, "Bonding Bands, Lone-Pair Bands, and Impurity States in Chalcogenide Semiconductors," Phys. Rev. Lett. 28, 355 (1972).

24. M. Kastner, D. Adler and H. Fritzsche, "Valence-Alternation Model for Localized Gap States in Lone-Pair Semiconductors," Phys. Rev. Lett. 37, 1504 (1976).

PROGRESS IN THE SCIENCE AND APPLICATION OF AMORPHOUS MATERIALS

Stanford R. Ovshinsky[1] and David Adler[2]

[1]Energy Conversion Devices, Inc., 1675 West Maple Road, Troy, Michigan 48084
[2]Department of Electrical Engineering and Computer Science, Massachusetts Institute of Technology, Cambridge, Massachusetts 02139

1. INTRODUCTION

The way we work, indeed, our entire industrial civilization, depends upon materials. We have long used noncrystalline solids in the form of glasses for their passive properties, such as their inertness, transparency, and in everyday use as containers. They are artifacts of an ancient time, not only on earth but in the heavens--on the moon glassy materials are widespread.

The transistor revolution, which started in 1947 and affected both how we work and how we utilize our intelligence, that is, how we encode and transfer information, was originally based upon crystalline materials, the physics of which started to be understood in the 1930's. In a sense, there was an inevitability to the discovery of transistor action once it was appreciated that atomic periodicity was not only a physical occurrence in many materials but also had to be controlled in order to apply the technology. When a semiconducting material such as silicon or germanium was made in crystalline form and its periodicity was not hampered by defects or unintentional impurities, then deliberate impurities, that is, dopants, could be introduced and the electronic current could be controlled. After this development, it was not the use of new principles that continued the fueling of the transistor revolution, but rather the creation of new technologies, as, for example, the giant step forward of high-density integrated circuits.

In the 1950's, noncrystalline solids, because of their lack of periodicity, seemed to have no possibilities as electronic materials except on an occasional empirical basis, such as the development of the xerox drum. Indeed, noncrystalline solids were automatically deemed not to be the materials of choice, since they did not seem to have any advantages over crystals.

We shall describe a growing momentum that is taking place as more and more companies throughout the world are rushing to join the many universities that are prospecting in this new field of amorphous materials. However, the uninitiated, who are prospecting with crude hammer and chisel, will be surprised to find that there already exists a very

sophisticated understanding based on more than 25 years of experience, as well as analytical equipment, experimental tools, and production technology. The fact is that amorphous materials are far from structureless; they have local structures and environments that can be defined precisely in three-dimensional space, and the understanding of this total interactive environment of electronic, structural, and chemical activity can provide both rich scientific discoveries and technological applications. The products that are now being made or developed from amorphous materials are of such uniqueness, diversity, and fundamental importance that they signal the shifting of one mindset to another; the characteristic of a scientific revolution is the change of paradigmic value.

The three most important areas for application of amorphous materials are energy, information, and materials. These three areas are the bedrock of the post-industrial revolution, and all three are currently in crisis. Amorphous materials can now provide the solutions.

2. ENERGY

2.1 Solar Cells

There are three key issues in any solar-cell technology: efficiency, stability, and production. We will review how we have broken through these barriers.

Single-junction PIN devices using conventional hydrogenated amorphous silicon alloys suffer severe degradation in their performance after prolonged exposure to sunlight. Laboratory-scale devices with initial efficiencies as high as 10% rapidly degrade to 5-6%, hence making claims of initial efficiencies irrelevant. The relatively low value of the stabilized efficiency coupled with the low throughput of the conventional batch manufacturing process form an insurmountable barrier to low-cost power usage.

We have recognized the problems associated with hydrogenated amorphous silicon alloys, and have developed a new fluorinated material which addresses

Reprinted by permission from *Journal of Non-Crystalline Solids,* Vol. 90, pp. 229–242 (1987).

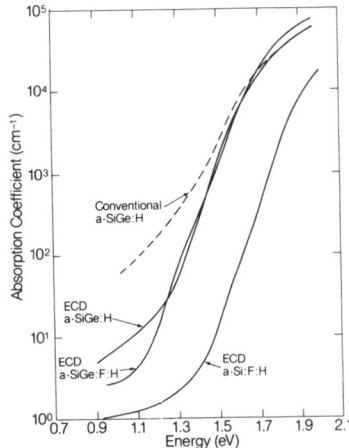

Fig. 1 Absorption coefficient as a function of photon energy for a-Si:Ge:F:H and a-Si:Ge:H alloy films compared to that of an a-Si:F:H film with a 0.3 eV larger band gap. Also shown is a conventional a-Si:Ge:H curve.

the problems described above.(1-4) We have also recognized the limitations of single-junction solar cells and put into practice, not only in research and development but in production, the tandem structure which minimizes recombination losses and reduces degradation.(5) The Sharp-ECD solar plant established in 1981 in Japan is a good example. The amorphous silicon processor employs a roll-to-roll process in which a 1000-foot long, 16-inch wide stainless steel substrate roll is coated by six layers of amorphous silicon materials in a continuous fashion and produces lightweight and flexible tandem cells. Millions of calculator cells have been manufactured.

Using fluorinated amorphous silicon and silicon-germanium alloys, we have already achieved a record 13% conversion efficiency in our laboratory.(4,6) This device has a triple-junction stacked-cell configuration, a configuration that offers exceedingly good stability. We have successfully developed new narrow band gap fluorinated materials with excellent sub-band gap absorption properties, which are an absolute requirement for spectrum splitting 20% conversion efficiency. The fluorinated alloys, the improved stability with tandem structure, and the unique manufacturing technology based on our flexible roll-to-roll process have allowed us to break the efficiency-stability-production barrier, which has held back the field of amorphous photovoltaics.

In order to achieve high efficiencies in a tandem structure, one must first develop high-quality materials from single-junction devices. For PIN configuration, this necessarily means that not only does one need high-quality intrinsic material but also high-quality n^+ and p^+ layers. We have previously reported on the development of high-quality intrinsic and n^+ materials with the incorporation of fluorine.(1,7,8) We have recently reported on the development of a fluorinated microcrystalline p^+ silicon alloy which has high dark conductivity and low optical loss.(9) Using these fluorinated materials, we have fabricated single-junction PIN devices on stainless steel substrates with light entering the p^+ layer and

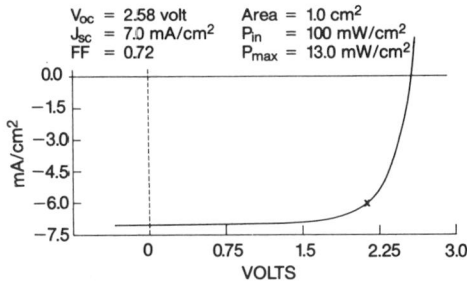

Fig. 2 Current-voltage characteristic of a triple-junction solar cell with an efficiency of 13%.

achieved an 11.3% conversion efficiency for a 1 cm^2 active-area device. Tandem and triple same-gap devices have conversion efficiencies of 11.4% and 12.0%, respectively. Although one-, two-, and three-cell same-band gap devices have similar conversion efficiencies, the stability of multi-junction devices is much superior to that of the single-junction devices.

The device efficiency using our amorphous silicon alloy with 1.7 eV band gap has approached its theoretical limit. We have developed proprietary narrow band gap materials to broaden the spectral response and increase the efficiency; for example, a 1.5 eV fluorinated amorphous silicon-germanium alloy that exhibits very low sub-band gap absorptions.(10) Figure 1 plots absorption coefficient versus energy for a-Si:F:H, a-Si:Ge:F:H, and a-Si:Ge:H made at ECD and for conventional a-Si:Ge:H. It is clear that the fluorinated materials are of higher quality. It has also been shown by Guha (10) that fluorinated material exhibits much reduced light-induced effects. Using the 1.5 eV fluorinated amorphous silicon-germanium alloy, we have achieved a 10.0% efficiency in a single-junction device. Dual-band gap two-cell tandem, three-cell triple, and four-cell quadruple devices were fabricated using the fluorinated 1.5 eV and 1.7 eV materials; efficiencies of 12.5%, 13.0% and 11.7% were achieved (6) for their respective structures. The 13% efficiency represents the highest value reported for amorphous solar cells. Figure 2 shows the J-V characteristic of this device.

In terms of stability, we have shown (6) that a dual-band gap triple device with initial efficiency of 11.2% retains 90% of its initial value after 2500 hours of continuous AM1 exposure, as shown in Fig. 3. The 10% stabilized efficiency value is the highest achieved to date. Our best stability data on quadruple devices show no loss in efficiency

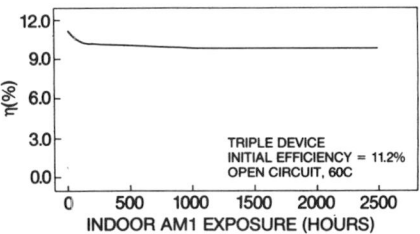

Fig. 3 Conversion efficiency vs. light-soaking time for a dual-band triple device.

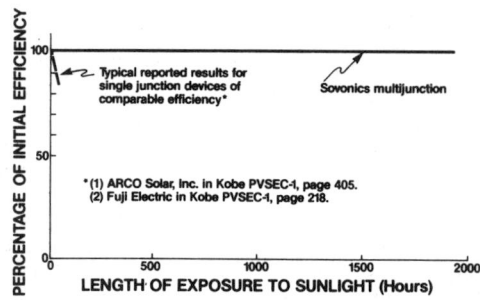

Fig. 4   Stability of a Sovonics multi-junction solar cell after 2500 hours of continuous exposure compared to recent results on typical single-junction a-Si:H cells.

after 2500 hours of continuous light exposure, as shown in Fig. 4. This should be compared with the 50% reduction reported by others.

To further increase device efficiencies beyond 13%, one must obtain high-quality materials with band gap narrower than 1.5 eV. We have successfully developed fluorinated materials that have band gaps of 1.40 eV, 1.34 eV, 1.25 eV,(11) and 1.15 eV.(12) These materials exhibit excellent sub-band gap absorption properties as shown in Fig. 5. It should be pointed out that curves 1 and 2 represent 1.7 eV and 1.5 eV materials that allowed us to obtain the 13% efficiency. These two curves have very similar slopes as well as low defect densities. It is further noted that curves 3-6 also have similar slopes and low defect densities. As these new fluorinated materials are fully incorporated into multi-junction device configurations, efficiencies approaching 20% or even beyond are made practical.

While there is no question that amorphous photovoltaics can provide a nondepletable,

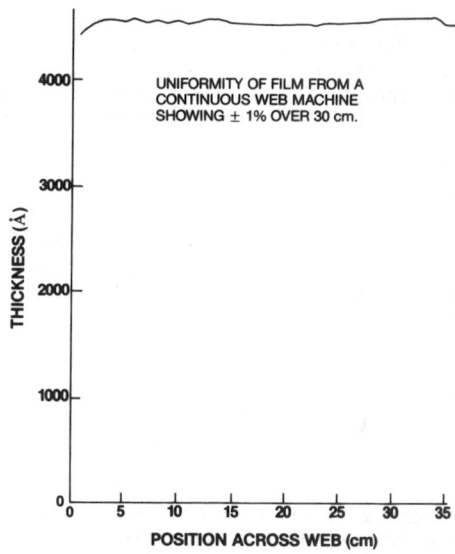

Fig. 6   Film thickness as a function of position across a strip 35-cm wide, deposited by ECD's continuous web process.

pollution-free, worldwide-attainable source of energy, a low-cost mass production technology had to be developed to serve this purpose. As early as the 1970's, ECD recognized this need and has since designed and built four generations of the roll-to-roll amorphous silicon processors. This efficient and continuous process can be compared with the printing of newspapers. Since the roll-to-roll process is a continuous deposition process, ECD designed a proprietary means to effectively minimize dopant contamination. This is best illustrated by SIMS data on the dopant concentrations in the intrinsic layer obtained from Sharp's roll-to-roll processor built by ECD, as reported by Hirobe et al.(13) They found that the boron and phosphorous concentrations were less than 5×10^{16} atoms/cc in the intrinsic materials, indicating the excellent isolation between deposition regions.

The uniformity of the deposition across the web is also a crucial parameter in ensuring the product uniformity. Figure 6 shows the uniformity of the film across the web obtained from our roll-to-roll

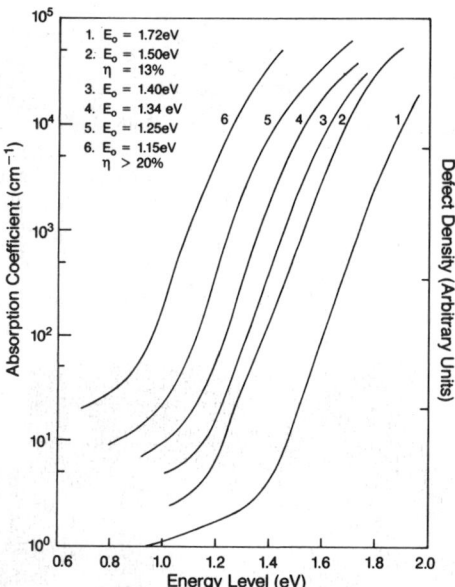

Fig. 5   Absorption coefficient as a function of photon energy for a-Si:Ge:F:H alloy films with various band gaps.

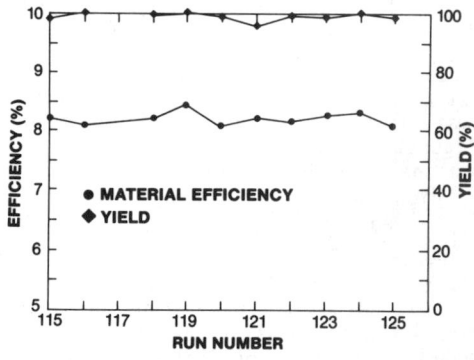

Fig. 7.   Efficiency and yield as a function of run number for a typical set of runs on a roll-to-roll production machine.

production machine. Our production machine is currently producing two-cell tandem devices using fluorinated amorphous silicon alloys. Figure 7 illustrates the excellent yield and consistency in efficiency of several consecutive production runs.

To address the issue of power delivered per unit weight, we have developed an ultralight module (14) which produces power at the impressive density of 2,418 W_p/kg, more than 10 times the previous record and nearly twice NASA's 1995 goal. As our new materials are incorporated, this figure will go up even more dramatically.

We have also studied the effect of air mass on various device structures.(15) Our outdoor performance data indicate that multi-junction devices deliver higher power output than single-junction devices.

Our tandem modules have been independently shown to be superior to single-junction modules in the market in terms of accelerated stress testing.(16) Our performance data have also been confirmed in Florida and India. The dream of low-cost solar power can now be a reality.

2.2 Thermoelectric Devices

Ovonic Thermoelectric Company is now marketing an array of devices capable of delivering from 1W to 120W of power. These devices are based upon our ability to design alloys that are chosen to maximize thermoelectric power and electrical conductivity, while minimizing thermal conductivity. Excellent results are obtained when we design disordered bismuth-tellurium based alloys that are chemically modified. For example, a solid-fuel irrigation system that weighs 55 lbs. (including the pump) and occupies about 3 ft^3 of space produces 120W at a voltage of 13V, yielding water flows in excess of 1000 gallons per hour. Thermoelectric generators are lightweight, completely portable, and have the high reliability and low maintenance costs attendant to any system with no movable parts.

2.3 Batteries

ECD has developed a rechargeable hydride battery with the same power output as conventional NiCd but with only half the size, thus twice the power density. These batteries run the gamut in size from small to room size for load-levelling applications. Finally, a practical electric automobile may be in the offing. It is almost a certainty that the markets for batteries will sharply increase with the growing solar photovoltaic technology.

Figure 8 shows several of ECD's energy products.

2.4 Hydrogen Production and Energy Storage

Despite its many advantages as a potential fuel, the problems of hydrogen storage are well known. ECD has developed lightweight, high-storage density hydrides with excellent low-temperature thermodynamic reversibility. This has been made possible by the application of principles of amorphicity.(17) With the problems of energy generation solved, we have put great emphasis on energy storage, utilizing our new batteries and hydrides.

3. INFORMATION

The sine qua non of our information-based society has been the crystalline transistor. It has replaced magnetic memories, vacuum tubes, and electromechanical relays in computer and other information systems. The integrated circuit has become ubiquitous and the basic building block of the electronic revolution. However, a crisis has been reached in the information field of the same importance as the oil crisis in the energy field. Everyone is making the same devices, and Silicon Valley is slowly turning into Death Valley. The leading electronics user, the computer industry, which depends upon crystalline technology, is, of course, undergoing the same trauma for similar reasons.

A new approach is needed--that of large-area thin-film devices based on amorphous technology.(18) This technology is not limited by the size of a wafer or constrained to two dimensions by lattice mismatches. It opens up the opportunity of three-dimensional structures in which one cubic meter of memory has the potential of replacing one square kilometer of disk drives!

The dream of a three-dimensionally integrated computer requires the development of a high-quality amorphous thin-film transistor (TFT). Although there has been a great deal of effort toward fabricating TFTs based on amorphous-silicon alloys, the source-to-drain currents have been quite low, tens of microamperes for typical source-drain and gate voltages of approximately 25V. This has resulted in relatively poor device performance. In section 3.2, we describe how a new device developed at ECD, a transistor based on entirely new physical principles, has solved the problem of an amorphous TFT.

3.1 Large-Area-Flat-Screen Displays

We have become accustomed to the bulkiness and power demands of CRTs, often without realizing the possibility of alternatives. Liquid-crystal displays (LCDs) are most promising because of their low power requirements, low cost, wide grey-scale availability, low voltage demands, and capability of extension to full color. LCDs have already achieved a dominant position in the small-area display market, such as in watches and hand calculators, but the concept of flat-screen television and large computer displays has till now been an unfulfilled promise. The major problem with existing LCD technology is the requirement of multiplexed address lines, which in high-density displays results in low

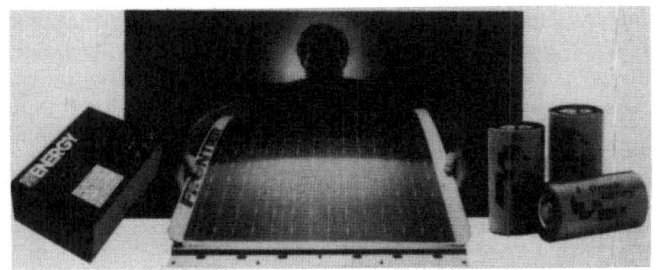

Fig. 8.   Some of ECD's energy products.

353

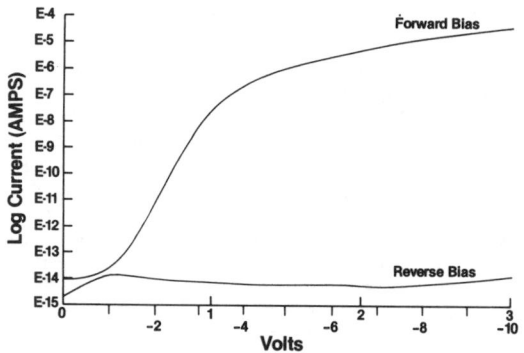

Fig. 9. Current-voltage dependence of a typical 20 μm x 20 μm ODS diode after processing. The upper curve is for forward bias to +3V, the lower curve is for reverse bias to -10V.

contract, narrow viewing angle, poor visibility, loss of grey scale, and very poor color imaging. An ideal solution to this problem is an active matrix LCD, in which we employ a proprietary amorphous semiconductor switch at each display element (pixel). This results in high speed, high contrast, wide-viewing angle, and good color capability.

ECD has already demonstrated the capability of rapidly and inexpensively depositing wide-area low-defect-density thin films of these materials for solar-cell applications, with the major fringe benefit of low temperature (< 300°C) processing. In principle, amorphous-silicon-alloy thin-film transistors (TFTs) would appear to be ideal for the active elements in a flat-panel LCD. However, TFTs require many processing steps, including high-precision photolithography over large areas, which have thus far led to threshold-voltage instabilities, frequent shorts caused by crossing bus lines, and poor electrical quality of the gate dielectric. The alternative solution presented by ECD involves replacing the TFTs by amorphous-silicon-alloy <u>diodes</u> in a special switching configuration.(19) This involves a much simpler device structure, with no gate dielectric or crossing bus lines, fewer processing steps, and lower alignment requirements. The overall result is a high-yield, low-cost manufacturing process that has already produced reproducible, uniform devices with excellent long-term stability. Figure 9 shows the I-V characteristic of a typical ECD diode, processed using VLSI techniques and having an active area of 20 μm x 20 μm. This diode has an ideality factor of n = 1.6 and a reverse saturation current density of 10^{-13} A/cm^2. At 3V, the rectification factor is ~ 10^{10}, and the reverse-bias current density remains below 10^{-8} A/cm^2 out to -15V. The resolution is 52 lines per inch, with an active area of 3.7 x 7.7 inches and a measured constant ratio of 12:1. It thus appears that economical, reliable, large-area flat-screen displays are finally on the threshold of commerciality.

3.2 <u>Thin-Film Transistors</u>

ECD's new transistor (20) is called a DIFET (double injection field effect transistor) and

incorporates the most desirable features of both bipolar transistors and MOSFETs. The amorphous DIFET exhibits the high current and high speed capabilities of bipolar devices as well as the high input impedance of the MOSFET. Accordingly, it requires very low input power.

The DIFET operates by amplifying a double-injection current of both electrons and holes by application of electric charge to the control gate of the device. The great value in modulating both kinds of charge carriers in a DIFET (as opposed to a single carrier in a MOSFET) is that since electrons and holes are charged in opposite polarities, the same amount of control charge applied to the gate electrode can amplify the flow of large numbers of electrons and holes, thereby increasing its conductance and current-carrying capabilities by a very large amount. The result of these features is the DIFET's faster switching capability, which is extremely important to present-day amorphous electronics applications, and will become even more important in the high-bit-density computers of the near future.

The new DIFET will have a very large impact on amorphous-silicon-alloy transistors because it overcomes the low current-carrying limitation of conventional amorphous-silicon-alloy field effect transistors. It already has been shown to increase the anode-cathode current by a factor of 15 to about 2000 μA at a gate voltage of 25V (see Fig. 10). In addition, the presence of both electrons and holes can result in visible light emission from the DIFET. This capability will provide modulated optical output for communications applications, requiring only very low power drive signals. An additional product may be a visible light, thin-film solid-state laser.

The profound changes that the now-possible three-dimensional amorphous integration will spur in computer technology are so revolutionary that we no longer refer to fifth-generation or sixth-generation computers. The coming quantum leap can be described only as a phase transition to the nth generation.

4. SYNTHETIC MATERIALS

While the two pillars of the future in terms of science, technology, and commercial importance are information and energy (they are becoming the two

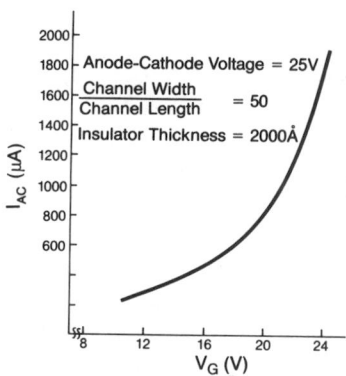

Fig. 10. Anode-cathode characteristics of amorphous silicon alloy DIFET as a function of gate voltage.

largest industries in the world), both are built on the foundation of materials technology. The ages of humankind have always been identified with materials—for example, the Stone Age, the Iron Age, and the Bronze Age. However, we keep bumping up against limitations when dealing with naturally occurring materials--for example, the widespread use of wood depleted the forests and hastened the development of other building materials and fuels. All industrialists would like materials to be able to withstand more heat, better resist abrasion and corrosion, be more lubricious, etc., but essentially all naturally occurring materials have already been investigated for these properties.

Synthetic materials until recently have primarily included organic solids, such as plastics which are passive materials used for their protective values and low cost, but which cannot withstand as much temperature, moisture, and light as we would wish.

Once we overcame the constraints of crystalline symmetry and were able to place atoms in three-dimensional space in ways which had never been achieved in nature, we attained the ability to design and synthesize, that is, engineer and tailor-make, many materials with specific attributes simply not possible previously. Of equal, if not greater, importance is the ability to fabricate active devices, that is, where energy is not only received by the material but where various transformations occur electrically or structurally that can generate electricity, store energy, encode, switch, and transmit information, etc. The present uses for active disordered materials range from catalysis to superconductivity. For example, utilizing its ability to synthesize needed materials, ECD has developed a new magnetic material that has twice the output per unit cost of even the most advanced magnets. It is known that X-rays are absorbed by almost all materials and thus cannot be controlled optically in the same way as visible light. However, ECD makes novel multilayer thin films for both X-ray diffraction and X-ray mirrors ("amorphous superlattices"). In addition to their many present spectroscopic uses throughout the world, these materials should eventually also find importance in the long-desired X-ray laser, which needs mirrors to attain the threshold for laser action.

5. CONCLUSIONS

We have briefly summarized some of the recent applications of amorphous technology to the areas of energy, information, and synthetic materials. However, space limitations preclude our detailing of other important developments, which include tool coatings, decorative coatings, optical fibers, optical memory systems, video disks, electronic memories, electron-beam-addressable memory systems, switching devices, imaging systems, photocopiers, image sensors, scanners, and electronic whiteboards. It should be clear from this paper that amorphous materials will soon form the basis

for a host of new industries that will help the world's economy by providing new science, new technology, and new jobs. The world has lived and prospered under the benevolent tyranny of the crystalline lattice. We are now at the beginning of a new age of freedom, made possible by the "disorder" of the amorphous state.

References

1. S.R. Ovshinsky and A. Madan, Nature 276 (1978) 482.
2. S.R. Ovshinsky, in Proceedings of the International Ion Engineering Congress (ISIAT '83 & IPAT '83), Kyoto, Japan (1983) p. 817.
3. S.R. Ovshinsky, in Physical Properties of Amorphous Materials, ed. D. Adler, B.B. Schwartz and M.C. Steele, New York: Plenum Press (1985) p. 105.
4. S.R. Ovshinsky, in Proceedings of the 18th IEEE PV Specialists Conference, Las Vegas (1985) p. 1365.
5. S. Guha, J. Yang, W. Czubatyj, S.J. Hudgens and M. Hack, Appl. Phys. Lett. 42 (1983) 588.
6. J. Yang, R. Ross, R. Mohr and J.P. Fournier, in Materials Research Society Symposia Proceedings, Palo Alto, CA (1986). In press.
7. R. Tsu, M. Izu, V. Cannella, S.R. Ovshinsky, G.J. Jan and F.H. Pollak, J. Phys. Soc. Jpn. 49, Suppl. A (1980) 1249.
8. R. Tsu, S.S. Chao, M. Izu, S.R. Ovshinsky, G.J. Jan and F.H. Pollak, J. de Physique 42 (1981) C4-269.
9. S. Guha, J. Yang, P. Nath and M. Hack, Appl. Phys. Lett. 49, (1986) 218.
10. S. Guha, J. Non-Cryst. Solids 77-78 (1985) 1451.
11. S. Guha, to be published.
12. R. Yang, to be published.
13. T. Hirobe, M. Katayama, Y. Shimada, T. Nagayasu, H. Oka, H. Izawa, H. Nanou, N. Shiozaki, H. Morimoto, T. Takemoto and N. Nakajima, in Proceedings of the 2nd International PVSEC, Beijing, China, August 19-22 (1986) p. 471.
14. J. Hanak, in Proceedings of the 18th IEEE PV Specialists Conference, Las Vegas (1985) p. 89.
15. J. Yang, T. Glatfelter, J. Burdick, J.P. Fournier, L. Boman, R. Ross and R. Mohr, in Proceedings of the 2nd International PVSEC, Beijing, China, August 19-22 (1986) p. 361.
16. J. Lathrop and E. Royal, in Proceedings of the 2nd International PVSEC, Beijing, China, August 19-22 (1986) p. 386.
17. S.R. Ovshinsky, "The physical understanding which is the basis for the design of unique catalysts, hydrides, and various electrodes from amorphous materials," May 20, 1980, unpublished.
18. S.R. Ovshinsky, SPIE Proceedings Vol. 617, ed. D. Adler, Bellingham, Washington (1986) p. 2.
19. Z. Yaniv, V. Cannella, A. Lien, J. McGill and W. den Boer, SPIE Proceedings Vol. 617, ed. D. Adler, Bellingham, Washington (1986) p. 16.
20. M. Hack, M. Shur and W. Czubatyj, Appl. Phys. Lett. 48 (1986) 1386.

A PERSONAL ADVENTURE IN STEREOCHEMISTRY, LOCAL ORDER, AND DEFECTS
Models for Room-Temperature Superconductivity

Stanford R. Ovshinsky

Energy Conversion Devices, Inc., 1675 West Maple Road, Troy, Michigan 48084

I. INTRODUCTION

Heinz Henisch's contributions to semiconductor physics have had a great effect on me. Many years before I met him 20 years ago, I often used his well-known book "Rectifying Semiconductor Contacts."[1] His important contributions to the amorphous field are many; we cite just a few [2-6] which did much to establish the electronic nature of Ovonic threshold switching.[7,8]

Heinz's work is intertwined with ours both in the past and currently with his suggestion that the Ovonic Threshold Switch (OTS) could be a superconducting device.[9] In this paper, I discuss some of our work in superconductivity and propose two models, one of which speculates on such a possibility for the OTS and the other which, as is my wont, does not emphasize the crystalline nature of the new high-temperature superconducting materials but rather structural chemical concepts that I have developed for amorphicity, such as normal structural bonding (NSB), deviant electronic configurations (DECs), and total interactive environment (TIE).[10-13] I describe how the unusual electronic properties of multi-elemental materials can be related to their steric chemistry in such a way that high-temperature superconductors ensue.

The lack of theoretical understanding of the fundamental mechanism of the new high-temperature superconductors has led to many theories and models. It is an exhilarating time because there is a freshness of thinking that has swept the physics world and new ideas are stimulating much work. I believe that there is not just one but several mechanisms for achieving high-temperature superconductors.

In this paper I shall discuss two quite different systems in which superconductivity can occur. The first may be termed equilibrium superconductors. These are the conventional superconductors as well as the new high-T_c superconducting materials. In these, the superconducting state below T_c is a state of lowest free energy. The second system may be called nonequilibrium superconductors. I propose that in these, the conducting charge plasma is created by external means and may become superconducting below T_c. This could be accomplished by a large gate voltage producing a metallic state in the inversion layer of a semiconductor, or by a strong light excitation producing, above the Mott insulator-metal criterion, a metallic electron-hole plasma and electron-hole liquid droplets in crystalline semiconductors, or by strong double injection which yields a conducting filament in an amorphous semiconductor OTS. Even though the nonmetal-metal transition has been established in these nonequilibrium systems, the transition to the superconducting state has not yet been observed with certainty. This theme will therefore be highly speculative.

I will begin by discussing the arguments for and against the possibility that the conducting On-state of an OTS is carried by a superconducting filament. I then proceed by explaining the stereochemical relationships which characterize the unique properties of amorphous materials and which make chemical modification of the conductivity of amorphous semiconductors possible. I close by describing a stereochemical model for the high-T_c superconductivity observed in the yttrium-barium-copper-oxide ceramic materials, which I believe has general applicability to other mixed valence systems. Finally, the evidence for raising the critical temperature T_c to 155K and higher by fluorination will be presented.

II. CONDUCTING STATE OF THE OVONIC THRESHOLD SWITCH

Figure 1 shows the current-voltage characteristic of an Ovonic Threshold Switch (OTS). This device is typically made of a 0.5μm thick multi-component non-crystalline chalcogenide semiconductor contacted on both sides by non-alloying contacts having a diameter of a few microns.[7] Without an applied voltage, the OTS is in the high resistance or OFF-state. At the threshold voltage, V_T, the OTS switches to the ON-state which will be examined in this section. The switching occurs so fast that the speed was never accurately established because of limitations due to external inductances and capacitances. The switching time is less than a picosecond which makes this the fastest room-

Reprinted by permission from R. Pryor, B.B. Schwartz, and S.R. Ovshinsky, eds., *Disorder and Order in the Solid State,* Plenum Press, New York, 1988, pp. 143-178.

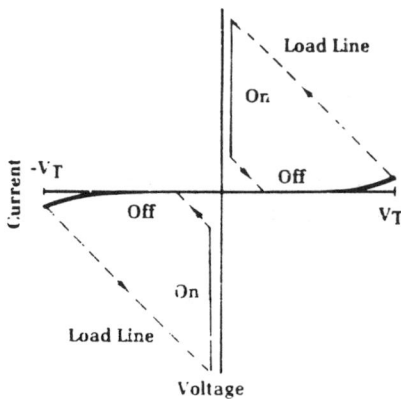

Fig. 1. The current-voltage characteristic of an Ovonic Threshold Switch (OTS).

temperature switch. The Josephson junction later reached this speed at liquid helium temperatures.

There are several factors suggesting superconductivity as the origin of the ON-state. Henisch [9] reviews "In the ON-state, the resistance is independent of electrode area, signifying that the current flows in a filament that is ordinarily much smaller, e.g. 10^{-8} cm^2, than conventional electrode areas. Accordingly, the current densities are enormous.* In the ON-state, the resistance is almost independent of system thickness, signifying that the potential drop is close to the electrodes, not in the bulk of the semiconducting material. The electric field free bulk is the region of special interest here."

The conducting ON-state is stable as long as the current exceeds a minimum holding current as indicated in Fig. 1. This conducting state is believed to be established by double injection of electrons and holes from the electrodes. The injected carriers first fill the localized gap states and in so doing smooth out existing potential fluctuations and enhance the mobility of free carriers. Depending on the exciton binding energy, a fraction of the free electrons and holes may form bound excitons. Since such excitons are neutral, they do not promote the conducting state. It is more likely that the density of electron-hole pairs exceeds the Mott criterion for the nonmetal-metal transition. We estimate that this concentration is about 10^{19}/cc with carrier effective masses of $0.5m_e$ and a dielectric constant of 12.

As previously mentioned, essentially all of the voltage drop across a threshold switch in the ON-state occurs at the contacts. Such would be the case if the electron-hole plasma in the conducting filament were actually superconducting at room temperature or above. Ovonic threshold switches contain chalcogenides in which the highest unoccupied

---

* In thin-film devices, we at ECD measure a minimum of 2×10^4 amps/cm^2. Our early data suggest that shaped electrodes which focus the electric field in very small areas can produce higher current densities. Threshold values are of the order of 10^5 V/cm.[14]

states form a lone-pair band. In the ON-State, the Fermi level splits into two quasi-Fermi levels with the hole level deep enough for there to be a large hole density in the lone-pair band and similarly for electrons in the antibonding conduction band. If one now invokes a negative effective correlation energy in these materials showing that there is a large, attractive Hubbard interaction, U, between two holes or between two electrons, then such an interaction could lead to superconductivity by Bose condensation of electron pairs, hole pairs, or both, the pairs being bipolaron-like, [15] or of the Bardeen-Cooper-Schrieffer (BCS) type [16] where the unique very strong electron-phonon interaction of these materials could be invoked to provide pairing.

I have pointed out in many papers the physical basis of the negative effective correlation energy in the chalcogenides in relation to the lone pairs and their relevance to the excitation process of the OTS [17-22] and superconductivity. Lone pairs are not only unbonded but can have various bonding configurations such as one- and three-center bonds as well as dative/coordinative bonds [21] and the elegant valence alternation pairs of the Kastner-Adler-Fritzsche (KAF) model.[22] The point that I want to emphasize here is that in the unbonded state the lone-pair configuration with its spin up and spin down can be viewed as a localized Bose particle.

The Bose particles formed by excitation then follow Bose-Einstein statistics, leading to condensation to the ground state at a particular temperature and density of excited particles. In fact, the large number of particles in such an excited state cannot exceed a certain critical value without reaching a saturation condition that would result in Bose-Einstein condensation and superconductivity. The assumed velocity and effective mass of the particles support the model.

In the OTS one can see excitation processes prior to the formation of the filament which involve the generation of carriers from the bulk of the material (see Fig. 2). Only when the filament is formed and the self-regulating plasma confinement results in a high constant current density with a fixed volume and excitation temperature is it possible that the Bose-Einstein conditions can be met.

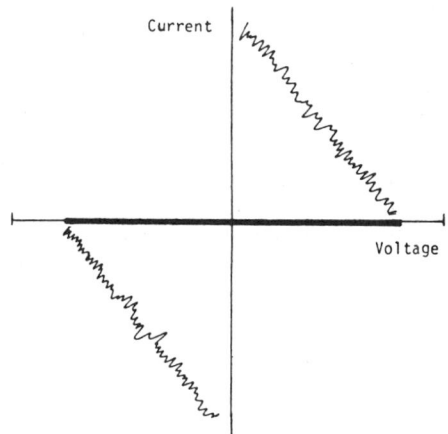

Fig. 2. The Ovonic Threshold Switch in the oscillatory mode.[23]

It is interesting that the actual OTS "ON-state" plasma is not affected by an external magnetic field and that these amorphous chalcogenide materials are normally diamagnetic due to their lone-pair character with compensated spin up, spin down configurations. Of interest to our stereochemical model of the copper oxide superconductors to be discussed later is the commonality of chain structures, crosslinking/bridging atoms and mixed valency and coordination states which I have used to describe the physics and chemistry of amorphicity.[8] Of passing interest, the OTS would like to undergo a structural phase transition but is prevented from doing so by strongly bonded crosslinking atoms.

The fact that, as a result of their interaction with the local environment, lone pairs can move without breaking [24] has led me to a thought problem which can be outlined as follows: While the energetics do not favor it, considering the Cohen-Fritzsche-Ovshinsky (CFO) model [25] and allowing for a spectrum of lone pair-states, some strongly localized, some weakly so, could it be possible that under a very high electric field in a very thin amorphous lone-pair device with a high density of states, a sufficient number of weakly localized Bose pairs could tunnel as pairs, providing another device possibility. It is interesting to consider that the "wingless" OTS [23] can represent the tunneling mechanism while a thicker version would favor excitation.

I have considered Little's [26], Ginzburg's [27] and Bardeen's [28] excitonic models and have the following thoughts about them. Obviously, the many materials and configurations tried through the years to achieve their type of excitonic superconductivity have not been fruitful. However, if one considers a conducting plasma as a "surface," replacing the layer postulated by Ginzburg, and then interfaces and encloses that plasma with an ionizable medium such as a high-resistance chalcogenide or other dielectric, then it would seem to me that the requirements for pairing of their hypothesized excitonic mechanism might be met.

Some historical reflections before I discuss our latest work on high T_C superconductors. I was never able to go so far myself as to publicly consider, as Henisch now boldly does, that the OTS is a room-temperature superconductor. However, I did consider in the 1960's that the conducting state was so unusual, for the reasons that I have discussed above, that at some temperature probably lower than room-temperature the ON state might go superconducting. There were several attempts made by others to find superconducting correlations with our devices. One mistake made was that a researcher confused the memory crystalline ON-state with the threshold plasma ON-state and did not find any high-temperature superconductivity, as indeed there was no reason to assume that he would.[29] On the other hand, Fritzsche and Sakai addressed the problem by taking arsenic telluride and subjecting it to high pressure so as to move the chains together and to reach a superconducting state.[30] The results were interesting in that they showed that an amorphous chalcogenide alloy could have a superconducting transition although it was only at 3K. In my view, what was missing were the excitation processes necessary to provide the conditions for high-temperature superconductivity.

The motivating factor for our superconducting work with amorphous materials in the 1960's and 1970's was that I believed that without lattice restrictions there were possible new orbital relationships and interactions necessary to develop high-temperature superconductors. This was contrary to the dogma of that time that superconductivity depended upon specific crystal structure.

I showed that amorphous materials could have orders of magnitude higher critical magnetic field capability than had been reported till then and we later were able to get a very respectable zero resistance superconductivity of 9K.[31] J.T. Chen, who spent a sabbatical at ECD, developed with us the technique of the inverse Josephson junction effect that has been so helpful in indirectly measuring high-temperature superconductivity.[32] Strongin at Brookhaven had developed a precursor of the technique.[33]

Why would I have considered the chalcogenides as suitable candidates for superconductors? My chemical reasoning goes back to my work begun in the mid-1950's on designing conductivity changes in rare earth oxides and particularly in transition metal oxides including copper oxides.[8,34] My interest in f- and especially d-orbitals was connected with my interest in disorder where, depending upon the local environment, one could have a variety of electronic configurations that are not available in crystalline materials. Of special interest in this history is my Ovitron device (see Fig. 3) of 1957 where I utilized transition metal ("valve") oxides such as tantalum to achieve an over 14 orders of magnitude drop of resistance in response to a small signal through "...the interaction of the metallic ion with the amorphous film..."[35] The metallic ion was divalent zinc (which can be amphoteric). This device was based upon a thin film of electro-chemically formed amorphous oxide of tantalum containing small amounts of other elements such as a halogen. I was uncertain of the oxidation state of the sub-oxide, and the effect of the minor constituents in the electrolyte upon the oxidation state

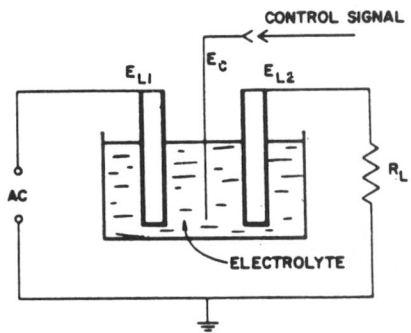

Fig. 3. Schematic diagram of amorphous dielectric film switch and modulator. The load resistor R_L and the amorphous film on the anodized tantalum electrodes E_{L1} and E_{L2} submersed in electrolyte form the load circuit. Current flows through the load circuit if a positive signal is applied to the control electrode E_C. Gain and, under certain conditions, memory are observed when metallic ions influence the blocking properties of the amorphous dielectric films which are shown for AC operation back to back.

of the tantalum was never fully determined. I felt that the interaction of the zinc orbitals with the tantalum oxide altered the coordination, transformed the valency and therefore the band structure so that the oxide went from a dielectric to a conductor which carried large current densities. This dynamic interaction can be considered to have similarities to chemical modification which I developed in the 1970's [36] and may be related to the mixed valency, charge balance, superconducting mechanism to be discussed.

In the Ovitron device, there had to be a balance of charge as the metal ion interaction altered the oxidation state with a simultaneous reduction action at another electrode. I originally used a liquid electrolyte. It was of great interest to me that if the metal ions were removed from the electrolyte, the unique features were lost. In the late 1950's, I adapted these principles to the solid state.[35]

Relevant to our model for high-temperature super-conductivity to be discussed, there was a reservoir of carriers, a chemical "pump" and a simultaneous electron transfer and exchange resulting in a valence transformation utilized for these unique conductivity changes.[8,35]

III. CONDUCTANCE CHANGES BY CHEMICAL MODIFICATION

Unusual conductivity changes intrigued me (I even investigated the conductivity changes of metal liquid ammonia systems), and I set out to show that one could independently control the conductivity in amorphous and disordered semiconductors and in wide band gap dielectrics without altering other important parameters unless desired. I achieved conductivity changes of over 10 orders of magnitude through chemical modification in materials that spanned the Periodic Chart (Fig. 4). In the fully modified state, metallic-like conduction was achieved (Fig. 5).[13,36-38] This chemical modification was reproduced by many prominent scientists such as Kolomiets et al., Davis and Mytilineou, etc.[39-41]

The important factor was that I showed that one could transform (chemically modify) what would normally be considered large band gap dielectrics consisting of elements and alloys utilized in

CHEMICAL MODIFICATION
OF AMORPHOUS SEMICONDUCTORS

Host Material	Active Modifier
Ge Te Se As	Ni, Fe, Mo
As	Ni, W
$B_4 C$	W
Si C	W
Si	Ni, B, C
$Si_3 N_4$	W
BN	W
Te O_2	Ni
Ge	Ni
Si O_2	W
Ge Se$_2$	Ni
Se$_{95}$ As$_5$	Ni

Fig. 4. A listing of elements that chemically modify amorphous semiconductor hosts.

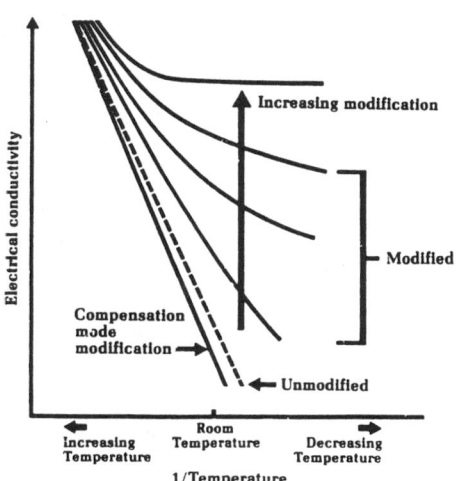

Fig. 5. The effect of chemical modification on the electrical conductivity of amorphous films.

ceramics such as silicon carbide and silicon nitride into surprisingly good conductors while still keeping their large band gap intact. Chemical modification was not only counter-intuitive but demonstrated that the conductivity of amorphous chalcogenides could be controlled, contrary to the generally-held view that large changes of conductivity such as those obtained by doping of crystalline materials were not possible. I showed that new orbital relationships could be established in a material in a stable, nonequilibrium manner so as to control its electronic properties without resorting to substitutional dopants and dramatically emphasized the importance that I attributed to stereochemistry in the solid state. I feel that chemical modification is relevant to the new high-temperature ceramic superconductors. The d-orbital interactions of yttrium and like elements with copper oxide can also affect pairing and provide increased conductivity.

What was of interest to me was not only the unusual transformation of conductivity achieved by chemical modification but that, despite the transition elements used, the materials could be diamagnetic. When one considers that deposited films have a high density of states, such as dangling bonds and configurations that would normally be paramagnetic, one can assume that the transition metal d orbitals are being utilized for pairing. In terms of the chemistry of transition metals, maximum use is made of all available parti-cles, that is, holes and electrons, interacting so that pairing becomes the preferred mode as in low spin crystal field splitting where various states are close in energy to one another.

I decided to work with amorphous chalcogenides in 1960 because of my dissatisfaction with the rigidity of structures involved in the mixed valence oxides. Because of my biological motivation, I looked particularly toward helical and chain struc-tures. I chose tellurium because of its helical/ chain configuration. In order to stabilize these one-dimensional structures, I utilized the princi-ples of bridging and crosslinking from polymer chemistry. In 1977, I wrote for a polymeric chemical lecture series "Charge transfer in semi-conductors is almost always forbidden by the Coulomb repulsion between the excess electron and the other electrons already on that site, often called the correlation energy. Because of the lowering in energy resulting from the bond-breaking described above, in chalcogenide glasses the effective correlation energy is negative. (The resulting effective attraction between electrons is also analogous to superconductivity.) This leads to many of the unique properties of chalcogenide glasses, including the OTS switching phenomena."[42,43] I was particularly interested in the anisotropic qualities of the chain-like structures and how to utilize them electronically while structurally transforming them into configurations of higher dimensionality.

The valency and coordination transformations in the Ovitron and my other mixed valency oxide work are a bridge between my early work and our model for high-temperature superconductors which follows. Much of my past and present work depends upon the structural chemistry of mixed valence materials and on the orbital interactions in one-, two- and three-dimensional space of elements that under ordinary conditions would not "see," that is, interact with, each other in space and energy. With its dependence on a rigid monolithic lattice, conventional crystalline thinking seems to me to be of little use to the new high-temperature superconductors.[24,44]

As I wrote in Mott's Festschrift in 1985, "I have stated ... that the Rosetta Stone of amorphous materials is the understanding of the relationship between the normal structural bonding (NSB) which characterizes the great majority of atoms and is responsible for the cohesiveness of the amorphous solid and the deviant electronic configurations (DECs) that control the transport properties and provide the active chemical sites of the material. Tying these two together is the concept of the total interactive environment (TIE) ... which takes into account the special nature of various local, chemical, topological, and electronic interactions in amorphous solids. We can write a new language of materials if we make use of this new alphabet. We need not be limited to the old dogmas of homogeneity and equilibrium chemistries. We have a new world of nonequilibrium chemistry and varying topological structures. We can deliberately create combinations and geometries in which even the local short-range order can vary subtly or drastically from one part of the material to the other. Such new structural chemistry again produces new electronic phenomena and new sites for chemical activity. Amorphous materials with unique cluster configurations and those containing crystalline inclusions and layers are also part of the spectrum of engineered materials discussed here. Such designed materials have far-reaching applications. We can synthesize and engineer materials where we mismatch and compensate atoms without the problems of mismatching lattices. We can carry this further through the use of layering and compositional modulation.... In fact, heterogeneity then becomes a welcome tool rather than a scare word."[24] (Emphasis in original.)

IV. SUPERCONDUCTIVITY AT HIGH TEMPERATURES

Starting with my basic structural chemical concept explaining high-temperature superconducting

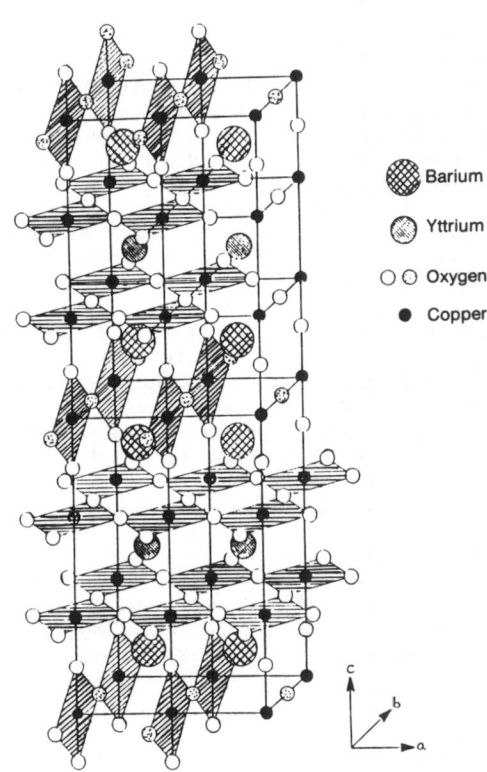

Fig. 6. The crystal structure of $YBa_2Cu_3O_7$. In this figure cross-hatched planes are used to clearly delineate the "sheet" and "chain" copper sites. Although the "sheets" are shown as planar, the Cu-O-Cu bonds are actually puckered with a bond angle of 165°. Bridging oxygen atoms in the "chains" are shaded in the figure for ease in identification.

materials, my collaborators and I developed a model which describes the fundamental mechanism that produces the high-temperature superconducting state. It is very specific, testable, self consistent and has predictive properties. We have published a paper presenting it in greater depth and detail [45] and portions of it appear here. In the present paper I shall advance some new thoughts which seek to dispel the mysteries of sheets and chains and show how as long as one has interacting and transformable oxidation states of +2's and +3's in desired geometric relationships, whether a material is lanthanum-yttrium- or bismuth-based copper oxide, [46] it can show high-temperature superconductivity. The physics follows directly from the stereochemically-determined structures, defects and local and total interactive environments just as it does in amorphous materials.

As seen in Fig. 6, the crystal structure of orthorhombic $YBa_2Cu_3O_7$ has copper atoms occupying two distinct sites. Copper is present both in dimpled two dimensional copper-oxygen sheets located between the yttrium and barium layers, and in one dimensional copper-oxygen-copper chains, formed by the ordered oxygen vacancies located between the two barium layers. It is exceptional that not only are the chains created by the vacancies but that the barium, or a like element such as strontium, does not only provide proper overall charge compensation, specific valency and size but is in place to stabilize what might otherwise be an unstable chain structure. I suggest that barium, which has such an affinity to oxygen that it is classically used as a getter, interacts with the lone pairs of oxygen assuring the unique chain structure.

I feel that the ferroelectric characteristics of these superconducting materials are also relevant. As Von Hippel pointed out years ago in connection with conventional ferro-electric materials, "Furthermore, the role of the barium ions is more important than that of a simple charge compensator and glue... These cations add their own dipole system to the overall balance, alter the size, shape, and deformability of the oxygen octahedra, and may tilt their mutual orientation."[47]

Of the three copper atoms in the unit cell, two are present in sheet sites and one is in a chain site. The chemical bonding and, particularly, the valence in these two sites are different. It is this difference of valency that permits the transfer of electrons and holes as well as setting up the conditions for antiferromagnetic coupling which is necessary for high T_c superconductivity. Such mixed valency can exist in sheet materials (e.g., lanthanum or bismuth) and in mixed sheet and chain materials (e.g., yttrium). Elements such as yttrium (or other rare earths of the same valency) and barium (or like elements such as strontium) provide the proper charge balance to assure the mixed valence system and are structural elements as well. The pairing mechanism and resulting Bose particle condensation are fundamental to the stereochemistry detailed in this paper.

Anisotropic superconduction in the sheets has been considered a mystery. It is caused by insufficient transfer sites between the sheets and limits the attainment of the optimum superconductivity by inhibiting the required Bose particle density. Sheet conductivity is basically dependent on mixed valency with additional contributions from d-orbital interactions when present. In the lanthanum sheet materials, excess oxygen, the presence of defects, or the substitution of some of the lanthanum by strontium or barium can generate mixed valency and balance charges.

To achieve the density requirements of the Bose-Einstein formula, there must be short-range connecting sites joining the sheets/chains, promoting the carrier exchange between them. They provide the needed change from anisotropic to isotropic geometries and conductivities. Lone pairs, because of their spatial flexibility, are helpful in this respect.[8] Materials with chain structures, with their dimensionally critical orbital overlaps, are formed by the ordered oxygen vacancies. Without these vacancies, the ~ 90K zero resistance superconductivity could not be achieved in the YBaCuO systems. Twinning and other structural distortions can generate new localized chemical configurations and connections which can represent new phases, some of which may be the hard-to-achieve, very high-temperature ones (over 125K).

The interactions described below are essential for high-temperature superconductivity in the YBaCuO systems. The copper in the dimpled two-dimensional copper-oxygen sheets has a nominal valency of two and in the chain sites a nominal valency of three. It is important to note that the copper oxygen bond angle for the layers is 165° and for the chain it is exactly 180°. Copper III is an unusual state and the 180° angle exceptional. The majority of copper II spins in the sheets is coupled antiferromagnetically, resulting in no local magnetic moment. Antiferromagnetic coupling comes about through an oxygen intermediary by the superexchange process. The antiparallel spin alignment utilizes the oxygen lone-pair p orbitals to form the bond and is favored by the large copper-oxygen-copper bond angle. Some of my favorite configurations, lone pairs and coordinate bonds, play an important role through the oxygen-mediated superexchange process. Pairing strength depends upon bond angles and bond lengths. Undesirable spatial separations weaken the antiferromagnetic coupling and disrupt the continuity of the valence transformation process and affects the density of Bose particles.

The copper atoms sit at the center of a planar array of oxygen atoms, forming copper-oxygen-copper chains between the "tunnels" created by the ordered oxygen vacancies which are present in the ortho-rhombic structure. Our identification of the valence state of these atoms as spinless copper III with a d^8 configuration is consistent with observed coordination and bond angles, and is dictated by our previous assignment of copper II to the copper atoms in the sheets. The four nearest neighbor oxygen atoms lie roughly in the sheet with the fifth, the apex oxygen atom, showing a Jahn-Teller distortion [48] and located at a larger bond length below the copper along the crystallographic c axis. In my view, this apex oxygen transfer configuration is critical.

The basis for high-temperature superconductivity is the establishment of simultaneous valence transformation processes involving two atoms at a time throughout the unit cell. This is accomplished by the interactions between the sheets and the chains through the aforementioned "pump"-like electron transfer mechanism of the pyramidal linking structure. As electrons are transferred up and down from the adjacent sheets to the copper-oxygen-copper chains, a mixed copper II, copper III valence state is formed in the sheets and the chains. Each electron transferred from the sheet leaves behind a copper III and converts one of the copper III atoms on the chain to copper II. This dynamic process is the heart of the new high-temperature superconducting materials.

This process does not produce localized copper II atoms on the chains, nor does it leave copper II atoms with unpaired spins on the sheets since this would produce local magnetic moments. Rather the d orbital holes present in this mixed valence state are delocalized, giving rise to the observed Pauli-Landau temperature independent paramagnetic susceptibility. These delocalized holes, however, still interact with one another via the oxygen atom intermediaries and, at a particular temperature, two spins on alternate sides of a bridging oxygen atom can be in a favorable position, because of the large bond angle, to interact through the superexchange process to produce an anti-parallel spin pair. The spin pairs which are thus formed can also _migrate_, now in a _bound state_, in what can be considered to be a _simultaneous valence transformation_ process involving two atoms at a time.

The superexchange coupled spin pairs on the chains and sheets are mobile, strongly bound, spinless composite particles which obey Bose statistics. At any given concentration of these spin pairs, there will exist a Bose condensation temperature at which a transition to a superconducting ground state will occur. There are, therefore, three important temperatures to be considered. The first two are the spin pairing temperatures for the d orbital holes in the sheets and in the chains respectively and the third is the Bose condensation temperature. The superconducting state is achieved at the lowest of these temperatures. It is possible, of course, that the lowest temperature could be the Bose condensation temperature. This novel situation would have dramatic consequences for the electrical properties of the normal state, to say the least, resulting in the occurrence of charge transport through uncondensed spin pairs! It is possible that under certain conditions such uncondensed charge carrier pairs can participate in "normal" conduction.

We can obtain an estimate of the Bose condensation temperature for the model system in the following way. The volume occupied by the wave function of a Bose particle can be approximated by a cube with dimensions equal to the particle's de Broglie wavelength. The de Broglie wavelength, in turn, is determined by the particle's momentum and, therefore, its thermal energy. This temperature dependent Bose particle interaction volume therefore increases with decreasing temperature. When the interaction volume grows to become equal to the volume available per Bose particle in the system, the Bose particles interact so as to bring about condensation. Using this approach, calculations show that the greatest density of Bose particles will occur when two-thirds of the chain copper atoms have copper II valence. This will, of course, also require that each sheet is one-third empty. One obtains, for this occupancy, a density of carrier pairs, $n=2.9 \times 10^{21} cm^{-3}$, or roughly one pair per 6 copper atoms. This carrier density is in close agreement with the free carrier density which is measured [49] above T_c by Hall Effect in $YBa_2Cu_3O_7$. Bose condensation temperatures exceeding room temperature are possible.

Although it is difficult to obtain a realistic estimate of the spin pairing temperatures in terms of this simple model, it is clear that, because of the increased orbital overlap between copper d electrons and oxygen p electrons which occurs in the 180° bond angles along the chains, we would expect spin pairs to be more strongly bound through superexchange in these structures than in the sheets. Support for this is obtained from recent nuclear spin lattice relaxation experiments on the yttrium-barium-copper-oxygen system reported by Warren et al. [50] which clearly indicate the presence of two distinct pairing energies, with substantially larger energies for quasiparticle formation (pair breaking) in the chains than in the sheets.

There is unnecessary confusion as to the role of sheets and chains. The same antiferromagnetic spin pairing mechanism is operative in both. The sheets can contain a large number of +3 configurations interacting with +2's. The mixed valence copper system is established in the sheets either by doping with strontium or barium or through oxygen excess. The chains are unusual stereochemical means of achieving a +3 valency to add to the overall valence transformation possible throughout the sheet-chain apex molecular system.

The above concepts show the importance of local dimensionality and total interactive environment. The spatial-energetic relationships, reflected in bond angles, bond lengths and antiferromagnetic pairing strengths, are important to the mechanisms of both pair formation and Bose particle condensation. I emphasize here my personal view that one must establish the necessary density for Bose condensation, and that this requires not sheets alone but a mechanism transforming two dimensionality into three dimensionality—that is, intersheet connections which establish the necessary containment for the Bose particles.

Mixed valency can be provided by various elements, for example, the amount of oxygen in bismuth materials maintains the +2/+3 ratio of valence states proposed in our model. However, the bismuth materials lack sufficient valence transformation linkages _between_ the sheets. This explains why even with more sheets than the 1,2,3 materials, they do not have optimal zero resistance superconductivity. However, the bond angles in the bismuth sheets appear to be closer to the 180° ideal, and therefore the pair breaking energy would be higher.

The superexchange interaction energy, which produces pairing, becomes the dominant factor as the temperature-dependent lattice vibrations, which destroy pairing, grow smaller with decreasing temperature. At T_c, the spin pairs form and condense into the superconducting Bose state. Large spacings between the sheets and difficult communication between them are antagonistic to high-temperature superconductivity. The high conductivity of anisotropic sheets is misleading since the sheets may contain the proper number of particles for high-temperature superconductivity but lack the proper density to fit the basically isotropic formula.

In our work in high-temperature superconducting materials, I chose fluorine to make a new alloy in which it could be a factor in the control of charge and valency, increase carriers available for superconductivity, make for stronger bonding and be involved not only in the chain configurations but also, under certain conditions, in the bridging apex and sheets. I felt that it would promote stability in a material notorious for its weak oxygen bonding. The results were dramatic. We reported 155-168K zero resistance superconductivity in fluorinated copper oxide ceramics.[51] These results have been confirmed by several groups, for example, in China, [52,53] Taiwan [54,55] and Sweden.[56] Despite our reports [51,57-59] and their confirmation, one can still find statements that there are no confirmed reports of superconductivity above 125K. Figure 7

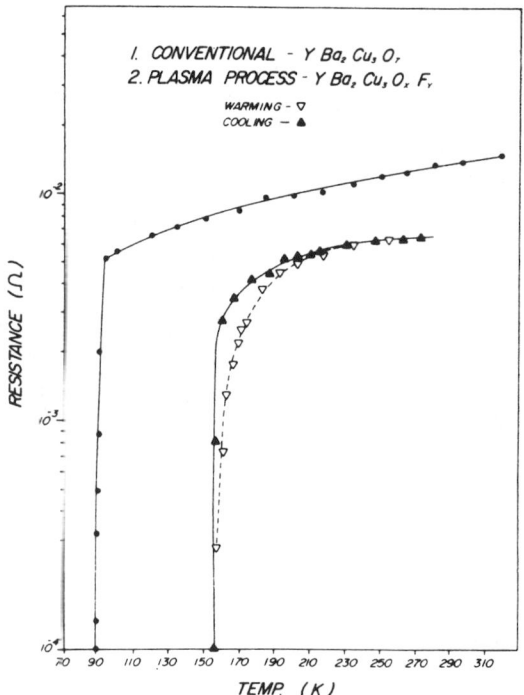

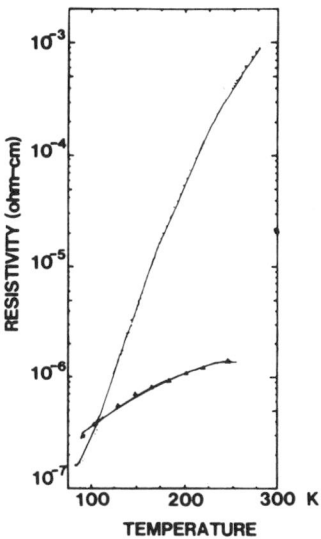

Fig. 9. The logarithm of the average of YBaCuOF sample. Average resistivity was calculated on the assumption of uniform current density. Resistivity was found to follow T^n, where n = 8.3. The ideal resistivity of pure copper is also plotted (triangles).

Fig. 7. Resistance vs. temperature plot showing 154K zero resistance transition of a microwave-treated YBaCuOF sample.

shows a fluorinated sample (T_c=154K) made by a plasma process [59] rather than the solid phase reaction [51] that we had previously reported.

What is equally exciting is our observation of diamagnetic signals and flux trapping at temperatures as high as 305K,[60] indicating that there are higher than room-temperature phases in our fluorinated materials (see Fig. 8). Supporting this magnetic data is Fig. 9 where we were able to show evidence by conductivity measurements of the existence of a phase exhibiting superconductivity onset above room temperature. It is important to point out that the resistivity of this sample <u>above</u> T_c is four times lower than that of copper![57]

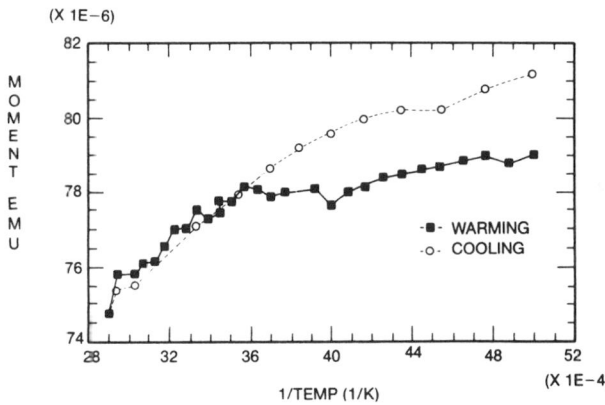

Fig. 8. Magnetic moment vs. 1/T for a YBaCuOF sample. Measurements made with use of 40G field. Data for warming after zero-field cooling are indicated by ■ and data from cooling with field applied are indicated by o.

In the YBaCuO material, the role of the chains is that they uniquely provide +3's, add three dimensionality to the system and are in communication with the sheets through the apical structure in such a manner that valence transformation can take place throughout the three configurations making up the molecular structure. They demonstrate the unique 180° bridging bond angle between the copper and the oxygen, and the oxygen-vacancy-generated and barium-stabilized quasi-one-dimensional chain structure which had been considered not to be possible.(61) The chain-sheet dichotomy is a false one since it is the mixed valence molecular structure of which the chain is a component which becomes superconducting.

Crystal structures are relevant to the extent that they geometrically provide maximum particle availability and pair interaction. Whether materials are tetragonal or orthorhombic is not basic but what is important is how the atoms are related to each other in three-dimensional space. They must have not only a mixed valency but also must be able to communicate and interact three dimensionally in space and energy to optimally meet the Bose condensation criteria. Local configurational spacing is important since it determines pairing energy. Overall spacing involved in the molecular structure is important to density. There is much room for improving high-temperature superconductors.

It is important to understand that the short coherence length is a fundamental clue to the described mechanism of high-temperature superconductivity. That the new superconducting materials are not ordinary metals is apparent from consideration of the weak temperature dependence of conductivity in the normal state. We do not see either the electronic delocalization and long mean-free path that one expects in a metal in the normal state or the long interaction length that one sees in normal BCS-type metallic superconductors. I suggest that this is due to the fact that the carriers responsible for superconductivity do not originate from the same source as the metal, i.e., they are very much more localized coming from the mixed valency transfer mechanism. These materials emphasize the small dimensional, localized, tight binding antiferromagnetic interactions that make for the strongest Bose formation pairing energy. There are two spatially, and therefore energetically, controlled, volumetric configurations---the first reflecting the local antiferromagnetic couplings; the second, a larger but still constricted container of Bose particles whose density, and therefore condensation to the superconducting state, is dependent upon their small mean-free path and low mobility. As in semiconductors, unwanted defects can act as recombination centers and even scatter the Bose particles.*

The high-temperature ceramic superconductors with their reliance upon connectivity, varied short and intermediate range order, their relatively low mobility and their process dependency are reminiscent of amorphous and disordered materials more than they are of the conventional crystalline materials. Subtle structural relaxations can play a significant role in determining T_c through coupling strength and volumetric changes.

One should learn from miracles. The attempt to continue making superconducting ceramic materials in the conventional way is basically flawed from a materials and mechanism viewpoint. God did not just point his finger at the yttrium-barium-copper oxide crystal structure and say "I have done my work." It is up to us to synthetically design new materials--laying them down almost atomically, layer by layer, as we have done in our x-ray mirrors and even our amorphous "superlattices."[8]

It is easy to be misled by Fig. 6, for it does not take into account unwanted defects, sheet and chain breaks, vacancies, the effect of subtle structural relaxations, bond reconstructions, twinning, interfaces, and the changes of charge that are involved with these factors, all of which are process related and in my view can be controlled or eliminated by the methods of synthesis described above.

Superconductivity exists at and above room temperature. It is difficult to produce such superconducting materials in substantial amounts by present methods. The first transistor illustrates the problems and opportunities of the historical process of creating new devices (Fig. 10). The surface, chemical and structural problems involved in this primitive device are now all but forgotten. I believe that the superconductivity concepts and model outlined above are a guide for our technological and scientific advances. There is a naivete in thinking that the happenstance of the technique of making a bulk T_c = 95K material is relevant to the task of

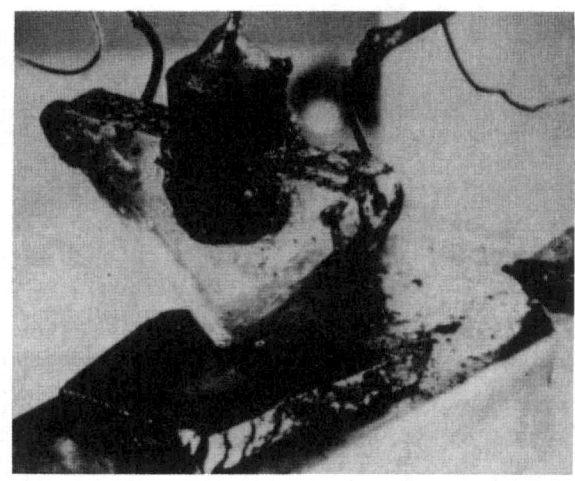

Fig. 10. Photograph of the first transistor.

achieving higher T_c materials needed for the next great step forward.

We have made four important contributions through the use of fluorine--the first, utilizing fluorine to make a new high-temperature superconducting material results in the highest confirmed <u>zero resistance</u> superconducting temperature,[51,57] albeit the material is multiphasic with a small volume fraction of the very high-temperature phases (such phases have fortuitously transposed and/or juxtaposed atoms in favorable configurations so that the sheets have optimal spacings and interconnections); the second, conductivity and magnetic measurements of our fluorinated materials show superconductivity above room temperature. The third solves the deleterious oxygen diffusion problem by replacing some oxygen atoms with fluorine and thereby assure thermal stability,[57,58] incidentally demonstrating thereby that the oxygen mobility is not connected with the superconducting mechanism. The fourth, the use of dopant amounts of fluorine to get over 90% oriented crystals as against the conventional non-fluorinated random 1,2,3 crystallites. This in principle solves the important problem of achieving high critical current densities. This has been confirmed by Mankiewich et al.[63]

V.   CONCLUSION

In conclusion, for Heinz's and my sake, I hope that the OTS can be found to be the superconductor that he suggests. Perhaps we have been "speaking prose," as Molière said, all the time without knowing it.[64] Certainly, my early transition metal oxide and chemical modification work need revisiting.

It is well known that metallic materials can go superconducting at low temperatures. I can see no reason why nonequilibrium metallic states produced by excitation, including double injection, photoexcitation and space charge accumulation, cannot do the same. Therefore, I suggest that these are interesting systems for exploring high-temperature superconductivity as well as new device structures.

---

* Based upon this concept, I will discuss in another paper the fundamental differences in mechanism between BCS superconductors and the new high-temperature ceramic superconductors.[62]

In any case, there is no unique material that is accidentally found in crystal form with the right chemical configurations and spatial relationships that will give us the best high-temperature superconductors. What is needed is a basic understanding of the mechanism and the "freezing in" of metastable configurations. What I have offered here is an expansion of my approach to amorphous materials which I feel can extend the Rosetta Stone of understanding local order to deciphering high-temperature superconductivity. Its universal stereochemical principles should be useful in synthesizing new materials which help make room-temperature and above superconductors possible and practical. Valency and local coordination control are key to the future as they have been to our past. The freedom of lone pairs to adjust optimally and interactively to their local environment, the importance of separate but interactive local order, defects, controlled carrier density, the design advantages of utilizing multi-elemental synthetic materials, and the stereochemistry involved in all of the above, are not only the basis for our work in amorphous materials, but can be viewed as means for the understanding and advancement of high-temperature superconductors which after all have varied configurations and positional, translational, compositional and interfacial disorder.

Acknowledgements

I acknowledge with gratitude the collaboration through the years of David Adler. Working with him was such a pleasure. I miss him deeply. I thank Stephen Hudgens for his collaboration and contributions to the previously published high temperature superconductor model (also for discussions on the Ovonic threshold switch) and Richard Lintvedt and David Rorabacher for their contributions to it, particularly in assuring its chemical soundness. The fluorinated high-temperature superconductor experimental work owes much to Rosa Young's collaboration. My appreciation to Hellmut Fritzsche not only for his discussions, suggestions, advice and encouragement on the work discussed here but also for our collaboration over the past 25 years. My thanks to Iris for her loving support and continuous help.

Postscript

Up until the very last, I.I. Rabi was excited about physics. His encouragement and support has meant so much to me. Our last conversations were about this new high-temperature superconductivity work.

References

1. H.K. Henisch, "Rectifying Semiconductor Contacts," Clarendon Press, Oxford (1957).

2. H.K. Henisch and R.W. Pryor, "Mechanism of Ovonic Threshold Switching," Solid State Elec. 14:765 (1971).

3. R. W. Pryor and H.K. Henisch, "Nature of the On-State in Chalcogenide Glass Threshold Switches," J. Non-Cryst. Solids 7:181 (1972).

4. H.K. Henisch, R.W. Pryor and G.J. Vendura, "Characteristics and Mechanisms of Threshold Switching,: J. Non-Cryst. Solids 8-10:415 (1972).

5. H.K. Henisch, W.R. Smith and M. Wihl, "Field-Dependent Photo-Response of Threshold Switching Systems," in: Proc. the 5th Intl. Conf. on Amorphous and Liquid Semiconductors, Garmisch-Partenkirchen, Germany, J. Stuke and W. Brenig, eds., Taylor and Francis, London, 567 (1974).

6. D. Adler, H.K. Henisch and N. Mott, "The Mechanism of Threshold Switching in Amorphous Alloys," Rev. Mod. Phys. 50:209 (1978).

7. S.R. Ovshinsky, "Reversible Electrical Switching Phenomena in Disordered Structures," Phys. Rev. Lett. 21:1450 (1968).

8. See S.R. Ovshinsky, "Fundamentals of Amorphous Materials," in: Physical Properties of Amorphous Materials, D. Adler, B.B. Schwartz and M.C. Steele, eds., Institute for Amorphous Studies Series, Plenum Publishing Corporation, New York (1985) for early references.

9. H.K. Henisch, "Threshold Switching––A Form of Superconductivity?", unpublished (1987).

10. S.R. Ovshinsky, "Principles and Applications of Amorphicity, Structural Change, and Optical Information Encoding," in: Proc. 8th Intl. Conf. on Amorphous and Liquid Semiconductors, Grenoble, France (1981): J. de Physique, Colloque C4, supplement au no. 10, 42:C4-1095 (1981).

11. S.R. Ovshinsky, "The Chemical Basis of Amorphicity: Structure and Function," Revue Roumaine de Physique 26:893 (1981); also in: Disordered Materials: Science and Technology, Selected Papers by S.R. Ovshinsky, D. Adler, ed., Amorphous Institute Press, Bloomfield Hills, MI (1982). (Grigorovici Festschrift.)

12. S.R. Ovshinsky, "The Shape of Disorder," J. Non-Cryst. Solids 32:17 (1979). (Mott Festschrift.)

13. S.R. Ovshinsky and D. Adler, "Local Structure, Bonding, and Electronic Properties of Covalent Amorphous Semiconductors," Contemp. Phys. 19:109 (1978).

14. S.R. Ovshinsky and H. Fritzsche, "Amorphous Semiconductors for Switching, Memory, and Imaging Applications," IEEE Trans. on Electron Devices ED-20:91 (1973).

15. I wish to thank Morrel Cohen for his clarifying comments on the negative correlation argument and discussion of bipolarons.

16. J. Bardeen, L.N. Cooper and J.R. Schrieffer, "Theory of Superconductivity," Phys. Rev. 108:1175 (1957).

17. S.R. Ovshinsky and K. Sapru, "Three-Dimensional Model of Structure and Electronic Properties of Chalcogenide Glasses," in Proc. 5th Intl. Conf. on Amorphous & Liquid Semiconductors, Garmisch-

Partenkirchen, Germany 1973; J. Stuke and W. Brenig, eds., Taylor and Francis, London (1974).

18. I hope that Heinz forgives me my emphasis on lone pairs. Certainly other carriers can initiate and make up a highly dense plasma, but the presence of lone pairs also in nonchalcogenide materials such as in group V, albeit not as pronounced or available, still must be taken into account. In any case, we have seen some forms of threshold switching in nonchalcogenide materials early at ECD and later at Penn State (K. Homma, H.K. Henisch and S.R. Ovshinsky, J. Non-Cryst. Solids 35&36:1105 (1980)) which indicates to me that there is a possibility of the critical on-state plasma being present in a spectrum of materials. However, there is no question that the switching mechanism is best seen and most stable in the lone-pair chalcogenides since the excitation process occurs in the nonbonded lone pairs rather than in the structural bonds as it does in other materials. It is in these materials that the effective negative correlation energy reigns supreme and that the volumetric control of the constant current density of the filament is best expressed.

19. S.R. Ovshinsky, "Amorphous Materials As Interactive Systems," Proc. 6th Intl. Conf. on Amorphous & Liquid Semiconductors, Leningrad, 1975: Structure and Properties of Non-Crystalline Semiconductors, B.T. Kolomiets, ed., Nauka, Leningrad (1976); and oral presentation (see H. Fritzsche, Proc. 6th Intl. Conf. on Amorphous & Liquid Semiconductors, Leningrad, 1975: Electronic Phenomena in Non-Crystalline Semiconductors, B.T. Kolomiets, ed., Nauka, Leningrad (1976)).

20. S.R. Ovshinsky, "Lone-Pair Relationships and the Origin of Excited States in Amorphous Chalcogenides," Proc. of the Intl. Topical Conference on Structure and Excitation of Amorphous Solids, Williamsburg, Virginia (1976).

21. S.R. Ovshinsky, "Localized States in the Gap of Amorphous' Semiconductors," Phys. Rev. Lett. 36:1469 (1976).

22. Kastner, Adler and Fritzsche took up this theme of the one- and three-electron pairs and made an elegant and important model based upon valence alternation pairs: M. Kastner, D. Adler and H. Fritzsche, "Valence-Alternation Model for Localized Gap States in Lone-Pair Semiconductors," Phys. Rev. Lett. 37:1504 (1976).

23. S.R. Ovshinsky, "The Quantum Nature of Amorphous Solids," in: Disordered Semiconductors, M.A Kastner, G.A. Thomas and S.R. Ovshinsky, eds., Institute for Amorphous Studies Series, Plenum Publishing Corporation (1987). (Fritzsche Festschrift.)

24. S.R. Ovshinsky, "Chemistry and Structure in Amorphous Materials: The Shapes of Things to Come," in Physics of Disordered Materials, D. Adler, H. Fritzsche and S.R. Ovshinsky, eds., Institute for Amorphous Studies Series, Plenum Publishing Corporation (1985). (Mott Festschrift.)

25. M.H. Cohen, H. Fritzsche and S.R. Ovshinsky, "Simple Band Model for Amorphous Semiconducting Alloys," Phys. Rev. Lett. 22:1065 (1969).

26. W.A. Little, "The Possibility of Synthesizing an Organic Superconductor," Phys. Rev. A 134:1416 (1964).

27. V.L. Ginzburg, "On Surface Superconductivity," Phys. Lett. 13:101 (1964).

28. D. Allender, J. Bray and J. Bardeen, "Model for an Exciton Mechanism of Superconductivity," Phys. Rev. B 7:1020 (1973).

29. We cannot locate the reference. Dave, how we miss your encyclopedic memory!

30. N. Sakai and H. Fritzsche, "Semiconductor-Metal and Superconducting Transitions Induced by Pressure in Amorphous As_2Te_3," Phys. Rev. B 15:973 (1977).

31. Internal ECD reports, 1975 and 1977. Samples remeasured at the Francis Bitter National Magnet Lab, Report dated July 1982-June 1983, p. 118.

32. H. Sadate-Akhavi̇, J.T. Chen, A.M. Kadin, J. E. Keem and S.R. Ovshinsky, "Observation of RF-Induced Voltages in Sputtered Binary Superconducting Films," Solid State Commun. 50:975 (1984).

33. A.M. Saxena, J.E. Crow and M. Strongin, "Coherent Properties of a Macroscopic Weakly Linked Superconductor," Solid State Commun. 14:799 (1974).

34. S.R. Ovshinsky, "Resistance Switches and the Like," U.S. Patent No. 3,271,719 (original filed June 21, 1961), issued September 6, 1966.

35. See S.R. Ovshinsky and I.M. Ovshinsky, "Analog Models for Information Storage and Transmission in Physiological Systems," Mat. Res. Bull. 5:681 (1970) for early references. (Mott Festschrift.)

36. S.R. Ovshinsky, "Chemical Modification of Amorphous Chalcogenides," in: Proc. of 7th Intl. Conf. on Amorphous and Liquid Semiconductors, Edinburgh, Scotland (1977).

37. R.A. Flasck, M. Izu, K. Sapru, T. Anderson, S.R. Ovshinsky and H. Fritzsche, "Optical and Electronic Properties of Modified Amorphous Materials," in Proc. 7th Intl. Conf. on Amorphous and Liquid Semiconductors, Edinburgh, Scotland (1977).

38. S.R. Ovshinsky, "The Chemistry of Glassy Materials and their Relevance to Energy Conversion," in: Proc. Intl. Conf. on Frontiers of Glass Science, Los Angeles, California; J. Non-Cryst. Solids 42:335 (1980).

39. B.T. Kolomiets, V.L. Averyanov, V.M. Lyubin and O.Ju. Prikhodko, "Modification of Vitreous As_2Se_3," Solar Energy Mats. 8:1 (1982). (Ovshinsky Festschrift.)

40. E.A. Davis and E. Mytilineou, "Chemical Modification of Amorphous Arsenic," _Solar Energy Mats._ 8:341 (1982). (Ovshinsky Festschrift.)

41. Hamakawa called chemical modification "sensational." H. Okamoto and Y. Hamakawa, "Gap States in Amorphous Semiconductors," _J. Non-Cryst. Solids_ 33:225 (1979).

42. S.R. Ovshinsky, "Polymeric Semiconductors," Lecture Notes For "Recent Advances in Polymeric Materials (March 1977).

43. In the early 1960's, we called the OTS the Quantrol. J.R. Bosnell, "Amorphous Semiconducting Films," in Active and Passive Thin Film Devices, T.J. Coutts, ed., Academic Press (1978).

44. How orbitals interact differently, perhaps fractally, in amorphous materials is discussed, for example, here and in reference 22. S.R. Ovshinsky, "Basic Anticrystalline Chemical Configurations and Their Structural and Physical Implications," _J. Non-Cryst. Solids_ 75:161 (1985).

45. S.R. Ovshinsky, S.J. Hudgens, R.L. Lindvedt and D.B. Rorabacher, "A Structural Chemical Model for High T_c Ceramic Superconductors," _Modern Physics Letters B_, Vol. 1, Issue 7/8 (October/November 1987).

46. (a) J.G. Bednorz and K.A. Muller, "Possible High T_c Superconductivity in the Ba-La-Cu-O System," _Z. Phys. B - Condensed Matter_ 64:189 (1986); (b) M.K. Wu, J.R. Ashburn, C.J. Tong, P.H. Hor, R.L. Wong, L. Gao, Z.J. Huang, Y.Q. Wang and C.W. Chu, "Superconductivity at 93K in a New Mixed-Phase Y-Ba-Cu-O Compound System at Ambient Pressure," _Phys. Rev. Lett._ 58:908 (1987) and P.H. Hor, L. Gao, R.L. Meng, Z.J. Huang, Y.O. Wang, K. Forster, J. Vassiliow and C.W. Chu, "High-Pressure Study of the New Y-Ba-Cu-O Superconducting Compound System," _Phys. Rev. Lett._ 58:911 (1987); (c) News reports on bismuth materials.

47. A.R. Von Hippel, "Molecular Science and Molecular Engineering," The Technology Press of M.I.T and John Wiley & Sons, Inc., New York (1959), p. 259.

48. H.A. Jahn and E. Teller, "Stability of Polyatomic Molecules in Degenerate Electron States," _Proc. Roy. Soc._ A161:220 (1937).

49. A.I. Braginski, "Carrier Density Measurement Using Hall Effect," in: _Proc. Intl. Workshop on Novel Mechanisms of Superconductivity_, V. Kresin and S.A. Wolf, eds., Plenum Press, New York (1987).

50. W.W. Warren, Jr., R.E. Walstedt, G.F. Brennert, G.P. Espinosa and J.P. Remeika, "Evidence for Two Pairing Energies from Nuclear Spin-Lattice Relaxation in Superconducting $Ba_2YCu_3O_{7-\delta}$," _Phys. Rev. Lett._ 59:1860 (1987).

51. S.R. Ovshinsky, R.T. Young, D.D. Allred, G. DeMaggio and G.A. Van der Leeden, "Superconductivity at 155K," _Phys. Rev. Lett._ 58:2579 (1987).

52. X.R. Meng, Y.R. Ren, M.Z. Lin, Q.Y. Tu, Z.J. Lin, L.H. Sang, W.Q. Ding, M.H. Fu, Q.Y. Meng, C.J. Li, X.H. Li, G.L, Qiu and M.Y. Chen, "Zero Resistance at 148.5K in Fluorine Implanted Y-Ba-Cu-O Compound," _Solid State Commun._ 64:325 (1987).

53. Z.X. Zhao, Academia Sinica, Beijing, personal communication.

54. J.H. Kung, in: _Proc. 1987 Symposium on Low-Temperature Physics_, September 7-8, 1987, Hsin-Chu, Taiwan.

55. P.T. Wu, R.S. Liu, S.M. Suhng, Y.C. Chen and J.H. Kung, "Possibility of High T_c Copper Fluoride Oxide Superconductors," presented at the 1987 Materials Research Society Meeting, November 30-December 5, 1987, Boston, MA.

56. C. Krontiras, personal communication.

57. S.R. Ovshinsky, R.T. Young, B.S. Chao, G. Fournier and D.A. Pawlik, "Superconductivity in Fluorinated Copper Oxide Ceramics," presented at the Intl. Conf. on High-Temperature Superconductivity, July 29-30, 1987, Drexel University, Philadelphia, PA; in: _Proc. of the Drexel Intl. Conf. on High-Temperature Superconductivity_, S. Bose and S. Tyagi, eds., World Scientific Publishing Co., Singapore (January 1988).

58. S.R. Ovshinsky, "Superconductivity at 155K and Room Temperature," presented at Superconductors in Electronics Commercialization Workshop, San Francisco, California, September, 1987.

59. R.T. Young, S.R. Ovshinsky, B.S. Chao, G. Fournier and D.A. Pawlik, "Superconductivity in the Fluorinated YBaCuO," presented at the Materials Research Society Meeting, November 30-December 5, 1987, Boston, MA.

60. We have one 370K measurement.

61. V.L. Ginzburg, "High-Temperature Superconductivity: Some Remarks," November 1987, to be published in _Progress in Low-Temperature Physics_.

62. S.R. Ovshinsky, in Collection of papers on amorphous materials in honor of Professor David Adler, China; to be published.

63. P.M. Mankiewich, J.H. Scofield, W.J. Skocpol, R.E. Howard, A.H. Dayem and E. Good, "Reproducible Technique for Fabrication of Thin Films of High Transition Temperature Superconductors," _Appl. Phys. Lett._ 51:1753 (1987).

64. Molière, Le Bourgeois Gentilhomme (1670).

65. Happy 65th Birthday, Heinz!

UNUSUAL FLUORINATION EFFECTS ON SUPERCONDUCTING FILMS

Stanford R. Ovshinsky and Rosa T. Young

Energy Conversion Devices, Inc., 1675 West Maple Road, Troy, Michigan 48084

In this paper we extend our previous fluorination work presenting an unusual fluorination effect in growing device quality superconducting film directly on sapphire. The superiority of the film quality is attributed to the fact that fluorine plays a significant role in the control of nucleation and in the enhancement of the growth rate of the superconducting film in the basal plane, therefore, an "epitaxial" film is obtained. Furthermore, the high quality fluorinated film can be grown at a lower temperature. As a result, the grain boundary weak link effect and the interface diffusion between the superconducting film and the substrate are minimized. The superconducting film with a high critical current density and a very smooth surface is achieved. With our technique, we believe that device quality superconducting film could be grown not only on sapphire but on other flexible inexpensive continuous substrates for high field applications which could lead to a major technological advancement.

1. INTRODUCTION

The ability for high T_c superconductors to carry enough critical current in a commercially useful form is not only of great scientific importance but has immense economic significance.

The extremely anisotropic nature of the high temperature superconductors, where the current flows preferentially along the Cu-O plane, and the strong chemical reactivity of the material have been the major stumbling blocks in the commercial development of high T_c superconductors. It is clear that randomly oriented polycrystalline film, tape or wire could never provide the current density required for most active device applications. Up to now, the high current carrying capability of the high T_c material has only been demonstrated with tiny single crystals or on films epitaxially grown on perovskite substrates with lattice mismatch of less than 2% such as $SrTiO_3$, $LaAlO_3$, $LaGaO_3$, etc. These materials, however, have no practical use. They are very expensive, available only in small wafers and they also have high dielectric constants and high dielectric losses.

As is well known, it is impossible to grow long flexible single crystal wires or tapes and it is not feasible to deposit thin film devices and interconnects where epitaxy is required. Therefore a new approach to grow high quality film on a commercially useful substrate is urgently needed. Of these, sapphire (Al_2O_3) has extraordinary dielectric properties and has been a commonly used substrate in many microelectronic applications and has attracted considerable attention. Because the YBaCuO film interacts chemically with sapphire and because of the large lattice mismatch, the quality of the superconducting films directly deposited on sapphire was not yet satisfactory; the $T_c(R=0)$ is about 80 K and the transport critical current at liquid nitrogen temperature is usually low.[1,2]

To make high quality non-epitaxially grown superconducting film, which can carry high critical current, it is crucial 1) to reduce the chemical reaction between the substrate and the film by reducing the deposition temperature 2) to grow film with a high degree of grain alignment so that current can flow not only along the preferred orientation but with minimal hindrance from the high angle boundaries. In this paper, we report our approach by utilizing fluorine to solve these basic problems. We demonstrated that by using a fluorinated YBCOF target during laser ablation, high quality superconducting films not only with the c-axis completely oriented perpendicular to the substrate but having a high degree of in-plane grain alignment can be grown on bare Al_2O_3 with no barrier

Reprinted by permission from the *Proceedings of the SPIE Symposium on Modeling of Optical Thin Films II*, San Diego, California, 12–13 July 1990, Vol. 1324, pp. 32–41.

layer needed. The film exhibits excellent normal state metallic properties and very sharp superconducting transition with T_c (R=O) of 86-88 K. The critical current density is ~ 10^5 A/cm^2 at liquid nitrogen temperature. The magnetic field dependence of Jc indicates the film is similar to the high quality epitaxial film on $SrTiO_3$.

Previously (3-5) through the incorporation of fluorine in the copper oxide superconductors, we have observed and reported superconductivity zero resistance transition at 155-168 K. Electrical and magnetic measurements also show indications of existence of some phases with even higher temperatures. We also reported that a small amount of fluorine in the YBaCuO compound can promote the crystalline alignment and improve the critical current density of sintered pellets.(6) These findings are subsequently confirmed by others.(7-11) This is the first time the aforementioned results have been demonstrated in an in-situ grown film.

The superiority of the film quality is attributed to the fact that fluorine plays a significant role in the control of nucleation and in the enhancement of the growth rate of the superconducting film on the a-b plane, therefore, an epitaxial-like film growth is obtained. Furthermore, the high quality fluorinated film can be grown at a lower temperature. As a result, the grain boundary weak link effect and the interface diffusion between the superconducting film and the substrate are minimized and a high critical current density is achieved. We believe that our technique can also be applied to grow thick epitaxial-like films on flexible substrate for large scale high field applications. In particular, this process can be readily adapted to our advanced manufacturing continuous roll-to-roll process for large volume low-cost production.

2. EXPERIMENTAL

The superconducting films were prepared with the laser ablation technique. A schematic diagram of the laser ablation system is shown in Figure 1. The rotating fluorinated target is ablated with pulses

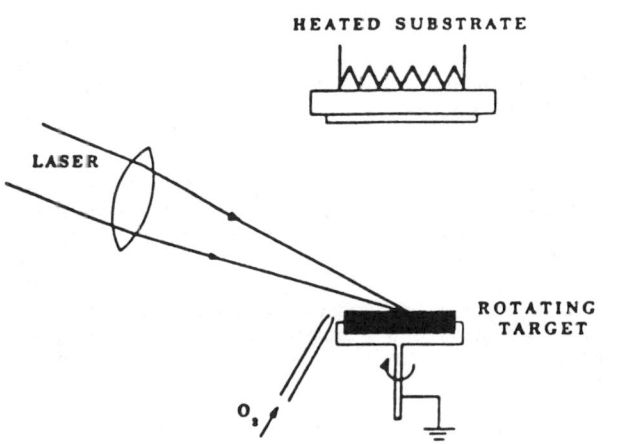

Fig. 1. Schematic diagram of the laser ablation system.

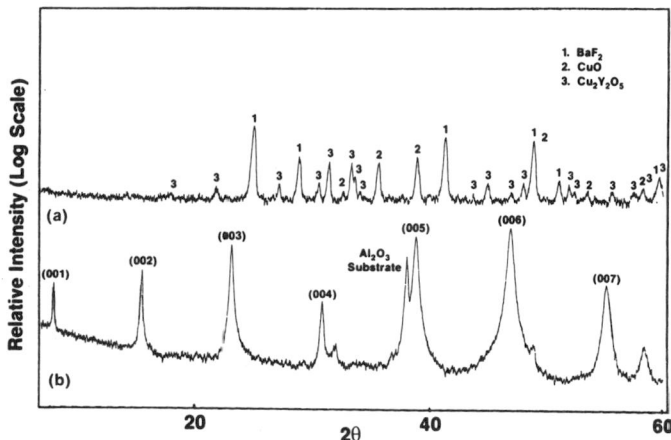

Fig. 2. X-ray diffraction pattern of (a) A multiphase fluorinated target; (b) A "single phase" oriented YBaCuO(F) film on sapphire. The film is deposited by laser ablation from the multiphase fluorinated target.

from a XeCl excimer laser (λ=308nm, τ=30ns) fired at 0.3 Hz. Typically the laser energy density is ~ 3J/cm^2. During the deposition, the chamber was maintained at a constant oxygen partial pressure with the substrate holder held at 680°C. The substrate was mechanically clamped to the holder. The surface temperature was measured to be 70-100°C lower than the substrate holder temperature. After deposition, the sample was slowly cooled to room temperature in approximately 40 minutes.

The electrical transport properties of the films were measured by a standard four-probe dc measurement. The critical current density was measured on a 0.2 x 2 mm line defined by laser scribing technique. The film quality and interface reactions were studied using a JEOL 2000FX analytical electron microscope equipped with a Kevex Quantum light element x-ray detection system. The cross-sectional specimen was prepared by bonding two slices of the substrate together with epoxy with the film sides face to face, followed by mechanical polishing, dimpling and argon ion milling with liquid nitrogen cooling. Raman spectroscopy was used to determine the in-plane alignment and axis orientation.

3. EXPERIMENTAL RESULTS AND DISCUSSION

Figure 2(a) shows the x-ray diffraction pattern of a fluorinated target. The multiphasic nature of the target which consists of BaF_2, CuO, and $Y_2Cu_2O_5$ is clearly seen from the diffraction data. The X-ray diffraction pattern of a laser ablated film from the fluorinated target is shown in Figure 2(b). It is interesting to note that the film generated from the multiphasic target which contains no superconducting phase is virtually "single phase." Only the (00ℓ) diffraction peaks along with an Al_2O_3 peak are detected. The two most intense diffraction peaks of a completely random diffraction such as (103) and (013)/(110) are hardly seen, indicating the c-axis of the superconducting film is predominately normal to the substrate.

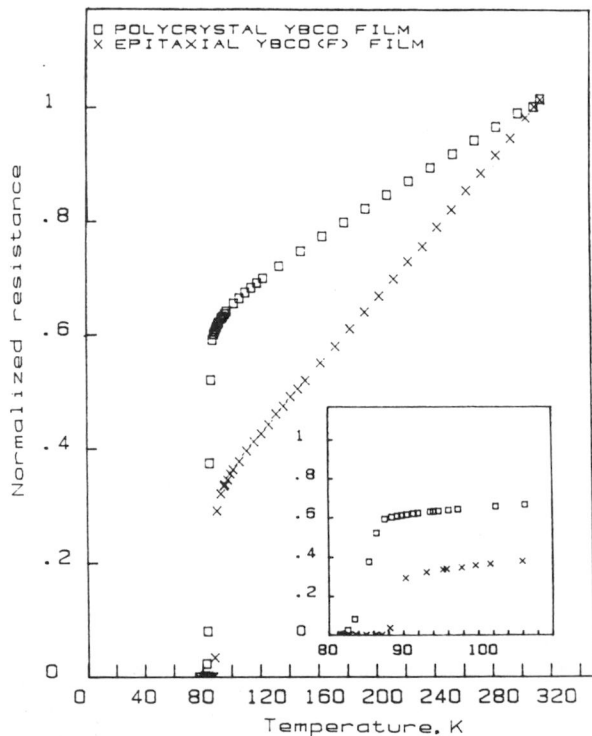

□ POLYCRYSTAL YBCO FILM
X EPITAXIAL YBCO(F) FILM

Fig. 3. Resistivity vs. temperature from in-situ ablated films on sapphire. (x) The YBCOF film from a multiphase fluorinated target. (□) The YBCO film from a single phase 123 target.

Figure 3 shows a comparison of the transition to the superconducting states of two laser ablated films. The YBaCuO(F) film was deposited from the multiphase fluorinated target and the YBaCuO film was from a single phase 123 target without fluorine. The deposition conditions were kept the same. The films are ~ 0.3μm thick. The YBaCuO film has an onset temperature of 88K and $T_c(R=0)$ at 82K, whereas the YBaCuO(F) film shows the onset at 94K and $T_c(R=0)$ at 88K. Furthermore, the much steeper normal state resistivity-temperature slope of the YBaCuOF film indicates a superior metallic behavior. We shall point out that these results represent data from a group of films which were deposited over a period of time from several different targets. The $T_c(R=0)$ of the YBaCuO film on sapphire is in good agreement with the reported data.(1,2) The critical current density of the YBaCuO and the YBaCuO(F) film, measured at 77K on a laser patterned stripe of 0.2 x 2mm, are 5 x 10^3 and 1.5 x 10^5 A/cm^2 respectively.

The superior electrical transport property of the fluorinated YBCO film is attributed to the fact that the fluorine promotes the epitaxial type of growth, whereas the columnar growth is dominated in the non-fluorinated films. The bright field TEM images from the cross-sections of the YBaCuO and the YBaCuO(F) film on sapphire are shown in Figure 4a and 4b. The nature of the columnar grains and the sharp vertical boundaries across the entire film thickness are clearly seen in Figure 4a. However, such grain boundaries are not observable in Figure

3b, but a few impurity phases such as $Y_2Cu_2O_5$ and Y_2BaCuO_5 are present. It is interesting to note that these foreign inclusions did not present any obstacles to the epitaxial growth. The surface morphology of the fluorinated epitaxial-like film and non-fluorinated texturized film is shown in Figure 5. A much smoother surface of the fluorinated film is clearly seen from this figure.

The information on the crystal structure and in-plane axes orientation is further studied with Raman spectroscopy. It is known that if the sample has good crystal properties, the polarization of the scattered radiation can be analyzed subject to the

Fig. 4a. Bright field TEM micrograph of an as-deposited polycrystalline YBaCuO film on sapphire. The top surface of the film was thinned by ion milling process.

Fig. 4b. Bright field TEM micrograph of an as-deposited epitaxial YBaCuO(F) film on sapphire.

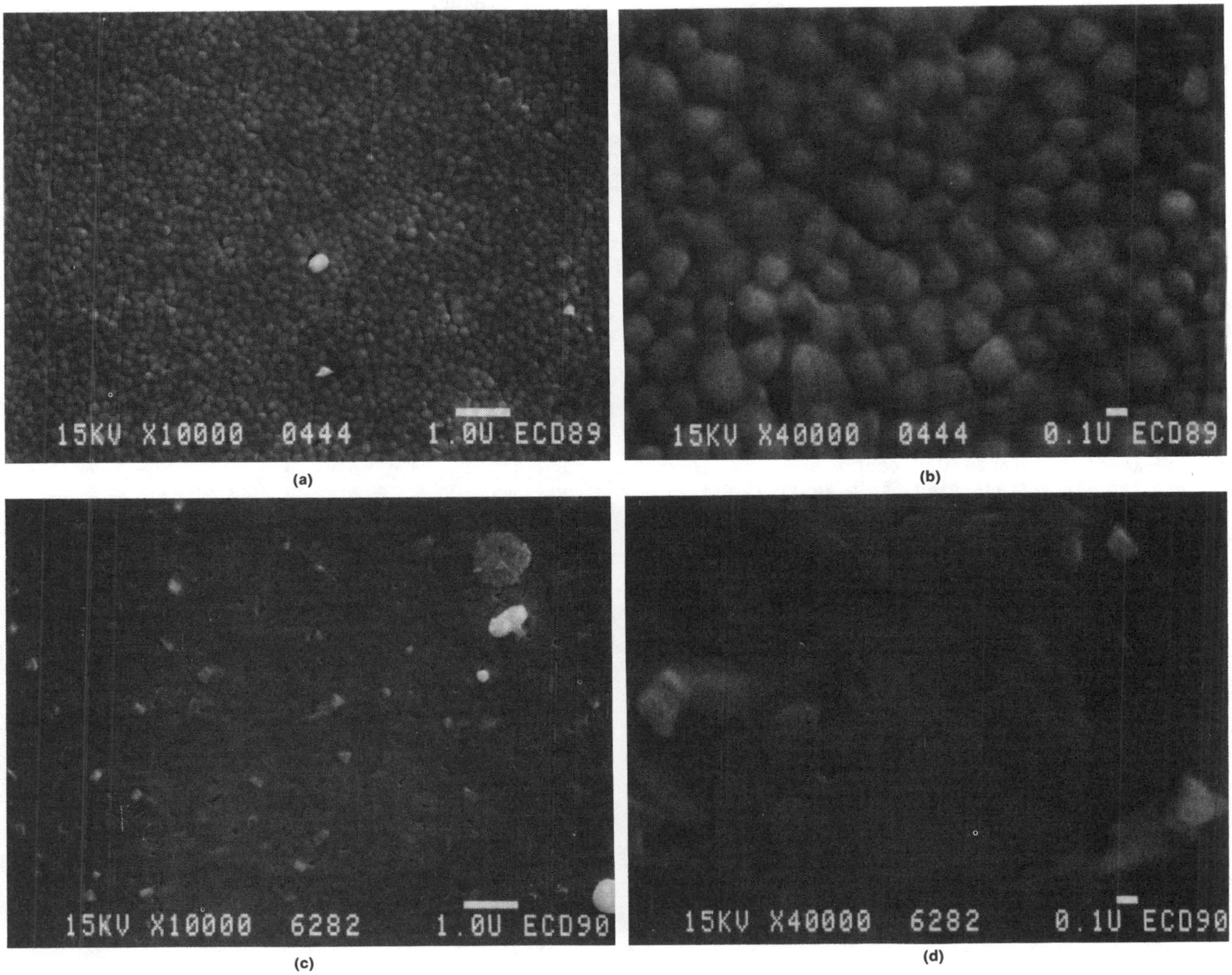

15KV X10000 0444 1.0U ECD89
(a)
15KV X40000 0444 0.1U ECD89
(b)
15KV X10000 6282 1.0U ECD90
(c)
15KV X40000 6282 0.1U ECD90
(d)

Fig. 5. Surface morphology of the polycrystalline film (a & b) vs. the epitaxial film (c & d) on sapphire.

selection rules which depend on the crystal structure.(12) It has been shown by Farrow et al.(13) that the polarization characteristics of the 500 cm^{-1} and 335 cm^{-1} Raman lines of the $Y_1Ba_2Cu_3O_7$ crystal can be used to determine whether or not the crystal axes are oriented and if so, in giving the direction of that orientation. Figure 6 shows a set of Raman spectra for a fluorinated YBCO film on sapphire. The electric vector of the incident (E_i) and scattered (E_s) radiation are always in the plane of the film. The film is about 2000A thick, 4mm wide and 12mm long. The c-axis of the film is perpendicular to the substrate as shown in Figure 1, however, the orientation of the a-b axes is unknown. As can be seen from Figure 6, the 335 cm^{-1} mode is completely absent in (a) and (d) whereas the 500 cm^{-1} mode is absent in (b) and (d). These spectra exhibit the same polarization behavior as single crystal as shown by Farrow et al. It therefore further confirms that the film is epitaxial-like. The a-b axes of the film is found to be nearly 45° to the side of the substrate.

RAMAN INTENSITY (ARB. UNIT)
300
400
500
RAMAN FREQUENCY (cm-1)

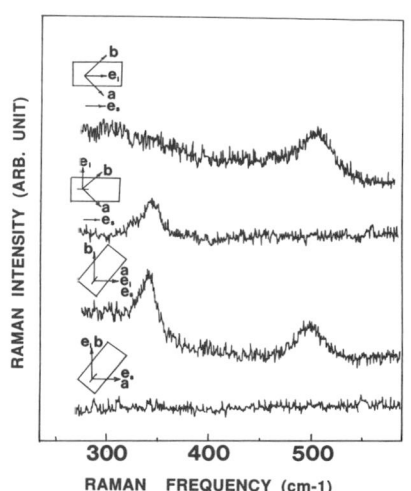

Fig. 6. Raman spectra for an "epitaxial" film on sapphire E_i and E_s are in the film plane directed as shown.

We also studied the magnetic field (H) dependence of Jc at 77K with H perpendicular to the film. Our preliminary data (14) indicated that the dependence of Jc(H)/Jc(O) on H of the fluorinated film on sapphire is identical to that of high quality epitaxial film on SrTiO$_3$.

In summary, we have demonstrated in this work that a high quality "epitaxial" superconducting film can be grown reproducibly on sapphire by using a multiphase fluorinated YBaCuOF target. Evidently, fluorine plays an important role in promoting the epitaxy. We suggest that the role of fluorine is to control the nucleation and to enhance the growth in the a-b axes. Technologically, this finding is particularly important because we believe that with our approach the epitaxial-like film will grow not only on sapphire but on other flexible metallic substrates which can readily be incorporated with our advanced manufacturing continuous roll-to-roll process for large scale high field applications.

4. ACKNOWLEDGMENTS

We acknowledge B. Chao, D. Pawlik, J. Hernandez, K. More, D. K. Christen and J. Budai for the research collaboration. We also thank H. Fritzsche for valuable discussion.

5. REFERENCES

1. C.C.R. Chang, X.D. Wu, A. Inan, D.M. Hwang, T. Venkatesan, P. Barbous and J.M. Tarascon, Appl. Phys. Lett. 53, 517 (1988).

2. S. Witanachchi, S. Patel, D.T. Shaw and H.S. Kwok, Appl. Phys. Lett. 55, 295 (1989).

3. S.R. Ovshinsky, R.T. Young, D.D. Allred, G. DeMaggio and D.A. Van der Leeden, Phys. Rev. Lett. 58, 2579 (1987).

4. S.R. Ovshinsky, Disorder and Order in the Solid State, Concepts and Devices, eds. R. Pryor, B.B. Schwartz and S.R. Ovshinsky, p. 143, Plenum Press (Institute for Amorphous Studies Series), (1988).

5. S.R. Ovshinsky, "An Approach to the Puzzle of High Temperature Superconductivity—A Letter to David Adler," to be published in Disordered Materials: Science and Technology, Selected Papers by S. R. Ovshinsky, Second Edition, eds. M. Silver and B.B. Schwartz, Plenum Press, New York (Institute for Amorphous Studies Series), (1990).

6. R.T. Young, S.R. Ovshinsky, B.S. Chao, G. Fournier, and D. Pawlik, Mater. Res. Soc. Symp. Proc. 99, 549 (1988); S.R. Ovshinsky, R.T. Young, B.S. Chao, G. Fournier, D. Pawlik, Rev. Sol. St. Sc. 1, #2, 207 (1987).

7. J.R. LaGratt, E.C. Behrman, J.A.T. Taylor, F.J. Rotella and J.D. Jorgensen, L.Q. Wang and P.G. Mattocks, Phys. Rev. B 39, 347 (1989).

8. Y. Hakuraku, F. Sumiyoshi and T. Ogushi, Appl. Phys. Lett. 52, 1528 (1988).

9. R. Sugise et al., Japanese J. Appl. Phys. 27, L1254 (1988).

10. P.K. Davies et al., Solid State Comm. 6, 1441 (1987).

11. F. Sumiyoshi, T. Hamada and S. Kawabata, Cryogenics 28, 3 (1988).

12. D.L. Rousseau, R.P. Bauman and S.P.S. Porto, J. Raman Spectrosc. 10, 253 (1981).

13. L.A. Farrow, Siu-Wai Chan, L.H. Greene, W.L. Geldmann, T. Venkatesan, W.A. Bonner, R.R. Krchnavek and S.J. Allen, SPIE Proc. 1187, 282 (1989).

14. Private communication. D. Christen, Oak Ridge National Laboratory.

EPILOGUE

AN APPROACH TO THE PUZZLE OF HIGH TEMPERATURE SUPERCONDUCTIVITY
A Letter to David Adler

Stanford R. Ovshinsky

Energy Conversion Devices, Inc., 1675 West Maple Road, Troy, Michigan 48084

A working model is offered to explain the mechanism of high temperature superconductivity in ceramic materials. In this model there are several interdependent, subtly balanced processes which combine to create new and unique phenomena due to the stereochemical nature of these materials.

We describe how a superexchange antiferromagnetic configuration such as alternating copper and oxygen atoms when constrained to a plane can be viewed as a a virtual composite boson where the pairing energy, while weaker than covalent bonds, can be quite strong. When this already paired configuration is given movement by hopping, the dynamic magnetic pairing energy is even weaker. However, the basic spin orientation is retained by virtue of the valence transformations that occur in the copper-oxygen planes. Such a mobile antiferromagnetic configuration becomes an actual composite boson.

We describe the steric chemistry and electronic configurations that are vital in achieving and confining these composite Bose particles so that under the proper conditions of temperature and density, they fulfill the Bose-Einstein criteria and achieve their lowest free energy by condensing into the superconducting state.

This model can be used as a tool to create new materials which can have higher temperature properties, better current carrying capability and new device possibilities.

David Adler spent his early years studying transition metal oxides and was working with me on their relation to the new high temperature superconductors when he died. There is no question that as always he would have made important contributions.

Working with David was a pleasure. He was not only one of the finest scientists who contributed greatly to the understanding and use of amorphous and disordered materials but was a perfect colleague and collaborator. I could speak to him about any area of our work engaging his open and inquiring mind to explore new ideas. He did this without fear of the reaction of his "peers" nor did he require sanction for his work. He loved the excitement and creativity of exploring new science.

I met Dave over 20 years ago. I had looked forward to meeting him because of my work dating back to the mid-1950's in achieving many orders of magnitude changes in conductivity of transition metal oxides. As I had read several preprints of his important papers in the area of transition-metal oxides,(e.g., 1,2) I arranged to meet him at the March 1968 meeting of the American Physical Society in San Francisco.

I was particularly interested in such materials because I had developed devices from them including those made of copper oxide which became unusual conductors and had unique switching properties reversibly going from a highly resistive semiconducting state to a unique highly conducting state.(3-6) In this regard, I speculated that some of the oxides might be interesting superconductors. I saw a great similarity between the transition metal oxides and the chalcogenides and other amorphous materials with which I had also been working.

Through the years, Dave and I discussed disorder, amorphicity, defects, metastability, electronic and structural phase changes, the role and interactions of f- and particularly d-orbitals, conducting mixed valence oxides, p-orbital and lone pair relationships, switching and superconductivity. We continued these discussions for many years in great friendship and stimulating and rewarding collaboration.(e.g., 7,8)

This paper will be in the form of a letter to Dave describing my intuitive approach* to the very important subject of room temperature and higher superconductivity which we were discussing until his untimely death.

Dear Dave,

The understanding of the mechanism of high temperature superconductivity is currently a most exciting activity in solid state physics. With a multitude of theorists attacking the problem from so many contradictory positions, it is beneficial to remember what we have often spoken about. Despite the formal elegance of the BCS theory, it has not been predictive of new superconductors or useful as a means of improving existing materials. The present excitement in high temperature superconductivity occurred as a result of inspired materials work without the least benefit of any theory.

As a synthetic materials person, I have always been comfortable dealing with systems containing many elements and with disorder and various types of local order. I have been working hard to extend what you have called my "physical-chemical insight" to high temperature superconductors to evolve a set of chemical design rules which will permit us to do experiments and in the process develop in a reproducible manner the much-sought after room temperature superconductor.

As we have often discussed, a model to be useful need not be entirely theoretically exact for, after all, it is not yet a formal theory. It should, however, accelerate the discovery process by suggesting new materials based upon it. Despite its simplifications, if its general principles are correct, it can have predictive power.

I have developed a unifying stereochemical concept which addresses the fundamental mechanism. This has been a work in progress.(6,9,10) It is the general consensus of the scientific community that pairing energy is still not understood and that without it one cannot have a theory. I seek to show here that pairing can be understood, and that it is only a necessary and not sufficient basis for high temperature superconductivity. Unlike conventional superconductors, condensation to the superconducting state does not automatically follow.

My model explains in structural, chemical and electronic terms how the pairs form, how they achieve mobility and how they reach the critical density needed for high temperature superconductivity. It seeks to explain the mechanism in understandable physical terms and use it for the design of materials.

I therefore propose to answer the questions: Why do these materials exhibit such a short coherence length? Why do they show anisotropic conduction? Why and how can sheets be involved in superconductiv-

ity? Why are all the materials high T_C ceramics rather than conventional metals? What is the role of antiferromagnetism? Can there be room temperature and higher superconductivity? Is the copper-oxygen configuration unique? What are the roles of yttrium, barium, thallium, bismuth, etc.? All these questions have been raised by the recent work in this area and they should be answerable with one unifying concept.

In the high T_C materials, the archetypical parent nonsuperconducting ceramic compounds are inherently antiferromagnetic, that is, their free spins are paired by virtue of the superexchange process between copper d- and oxygen p-orbitals. This is an attractive interaction which provides the initial energy holding the carriers in a static, spin-paired, non-conducting condition. We consider the static configuration to be, in effect, a virtual composite boson which, unlike the spin pairs in low temperature superconducting materials, exists above T_C.

When the materials are chemically modified so as to promote mixed valency, enabling them to become superconductors, the virtual boson becomes an actual boson. This concept was boldly set forth in our 1987 paper.(10) Two new possibilities were suggested. The first "would give rise to a situation in which, above T_C, normal conduction would occur through both ordinary Fermi particles and uncondensed spin pairs." A second possibility is "that the lowest temperature is the Bose condensation temperature. This novel situation would have dramatic consequences for electrical properties of the normal state, to say the least, resulting in the occurrence of charge transport exclusively through uncondensed spin pairs!"(11)

In an earlier draft of this letter, I stated: "We start by asking ourselves 'Does high temperature superconductivity differ from conventional low temperature superconductivity?' Setting aside the obvious difference that the conventional phonon pairing interaction does not apply to the high temperature materials, there is an even more basic assumption that must be examined--that is whether the Bose particles in the high T_C materials are just like conventional Cooper pairs in that once they are formed, they automatically condense. It is my thesis that the mechanism of high temperature superconductivity is contrary to the conventional understanding that 'there is no analogue [in a superconducting metal] of the He-I phase, where the bosons exist but are not condensed--... the Cooper pairs either do not exist at all or are automatically condensed.'(12) I propose that in the copper oxide materials, uncondensed composite Bose particles, which are charged analogs of the helium atom, exist above T_C and can undergo Bose condensation at T_C similar to the superfluid transition in He."(13)

The seemingly insoluble problem of pairing becomes understandable; it does not imply that all antiferromagnetic materials can become superconductors or, for that matter, that any paired groups of carriers can be virtual bosons. Nothing is further from the truth. The pairs must have a mechanism to make them actual, that is mobile. They then must meet the boson density requirements necessary for

---

* Hopefully, I am following Poincaré's dictum: "It is by logic that we prove, but by intuition that we discover."

condensation. I hope that my paper can end the speculation that these materials should be considered spin glasses.

I suggest that the mechanism that makes for high temperature superconductivity is composed of several intertwined processes made possible by structural chemistry. The material must be made up of alternating elements such as copper and oxygen which are strongly chemically bonded in a plane. There is a superexchange energy which is responsible for the antiferromagnetic configuration that is generated by the copper d- and the oxygen p-orbitals (lone pairs). This magnetic bonding energy is weaker than that of the covalent/ionic chemical bonds of these elements which provide the structural integrity of the material. Other elements in such a multielemental material assure the mixed valent and planar nature of the copper-oxygen configuration, separating the planes and affecting their charge, that is, their valency. They are a source of or receptor of carriers and are also a means of controlling the formation of oxygen vacancies.

The copper oxide materials have similarities to amorphous chalcogenides in that the normal structural bonds are deeper in energy than the nonbonded or more weakly bonded lone pair configurations that are responsible for the conduction process of such materials. Compositional, positional and translational disorder make for a spectrum of bonding energies with the creation of a mobility gap. The CFO model (14) can be adapted to take into account the various spin compensated antiferromagnetic and lone pair configurations just as it accommodates the lone pair configurations in amorphous chalcogenides. I believe that if one understands this mobility gap concept with its pinning of the Fermi level, it will relieve the confusion as to where the Fermi level lies in high temperature superconducting materials. It would be more productive to call it a boson level.

The superexchange magnetic configurations of the copper and oxygen atoms form virtual composite bosons. In order for them to be transformed into high temperature superconductors via the creation of actual (mobile) composite bosons, they must be modified/doped so that mixed valency is created making for spin-paired hopping to acceptor sites. In order for this "ripple" effect to occur, there must be a sufficient number of carriers relative to empty orbital acceptors.

There are various types of high temperature superconducting materials with varied superconducting transitions. The model that I propose applies to all of them, whether they are simply layered or a combination of layers and chains. In all of these materials, two copper atoms can interact antiferromagnetically through the intermediary of an oxygen atom as well as having the ability to exist reversibly in two different valencies. When the copper-oxygen atoms are placed repetitively in a plane, the conduction process is through hopping. The accessible equivalent sites for the hopping carriers to move to are related to a valence transformation which also assures charge balance. What is important in these materials is that the hopping is not one carrier at a time, nor even two carriers at a time, but two carriers at a time that are spin-correlated.

In order for this unusual circumstance to occur, there are several requirements. The first is that there must be a plane constraining and directing the movement of the carriers, there must be repetitive atoms within hopping distance of each other, and the acceptor sites have to retain and control the spin orientation of the carriers as they hop. The latter is assured by the basic antiferromagnetic configuration of the atomic structure from which the carriers originate, that is, that the copper-oxygen-copper planar structure presents a template for the carriers which is energetically more favorable for the motion of two spin-correlated electrons than any other configuration.* As the carriers move from atom to atom, they retain their spin relationships so that at any one time, if a snapshot is taken, even though carriers have been exchanged, the superexchange antiferromagnetic configuration appears to be unchanged, i.e., charge and magnetic spin configurations are conserved, and the criteria of an actual composite boson are met.

Conduction takes place by the hopping of two spin-correlated carriers. This pair hopping is from one valence state to another transforming the valencies in the process. The mobile carriers with their spin up-spin down configuration are actual composite Bose particles. Such bosons under the requisite temperature and density conditions condense into their lowest free energy state--the superconducting state. We will show how the density requirements are also dictated by the stereochemistry of the materials since the crossover from two-dimensionality to three-dimensionality assures the required boson density.

A mixed valence system is necessary but not sufficient for one can have mixed valency in an isotropic material. In order for two carriers to hop in a spin-correlated manner, the valence transformations must occur in anisotropic structures where, for example, as the carriers hop between stationary copper and oxygen atoms which are laid out in an antiferromagnetic plane, Cu^{II} is transformed into Cu^{III} (and vice versa). This means that as the holes leave the Cu^{II} and the configuration becomes Cu^{III}, the orbital template action of the underlying atoms that lay down the spin pattern assures that the magnetic pairing energy is maintained. (Cu^{II} in a copper oxide system is common; Cu^{III} is unusual and a result of stereochemical factors that I have discussed.)(6,10) The bond angle between copper and oxygen which is reflected in Cu^{III} is 180°. That is why the flatness of the plane is an important parameter.

What is basic in any of the materials utilized is a plane structure, whether a two-dimensional sheet or a one-dimensional chain. In all cases, there is communication between copper atoms through an oxygen intermediary and the copper has valence transformations so that resulting mobile antiferromagnetic configurations become composite bosons. Intervening vertical structures or other

_____

* There are other factors that have to be taken into account that will be discussed such as the need for a minimum carrier density.

non-copper atomic layers are there to assure the mechanism of valence transformation.

The coupling in the 1,2,3 materials is from the layered Cu^{II} sheets through the apical Cu^{II} configurations to the Cu^{III} chains. In the lanthanum type materials which are primarily composed of sheets, the communication is between Cu^{II}s and Cu^{III}s in and between the sheets. The highest superconducting temperatures are connected with the pairing energy which is strongest in the Cu^{III} configuration due to the $180°$ bond angle associated with it. The amount and ratio of Cu^{II}s and Cu^{III}s affect the volume and therefore the density required for Bose condensation.

In order for the dynamic superexchange antiferromagnetic configurations to propagate and become bosons, a plane is required where the alternating atoms such as oxygen and copper are in a row with equivalent sites for the carriers to move to. Such movement of two carriers at a time is correlated with changes of valency of two atoms at a time. Randomly distributed antiferromagnetic configurations cannot be tolerated; planar configurations which do not have accessible sites or in which sites for transfer are further apart than their hopping probability do not meet the criteria. It is the basic antiferromagnetic structure underlying the movement of the carriers which controls and assures spin orientation of the hopping pairs.

Anisotropic structures are a necessity for other reasons as well since the superexchange energy is related to the distance between the two carriers making up the composite Bose particle. The two carrier movement would be dissipated and disrupted by any other structure and spins could be readily flipped. The short coherence length is dictated by the local order retention of the superexchange energy as the carriers hop.

Other elements cannot fit into the sheet structure, for the antiferromagnetic pair movement would not only be disrupted, the spacing would be changed and the foreign elements would act as defects, spin frustrators and sinks, that is, recombination centers, and the bucket brigade would be broken. The plane with acceptor sites in a mixed valence system embedded in an antiferrogmagnetic background is the basis of spin-correlated pair hopping and the basic building block for high temperature superconductivity.

The mobile antiferromagnetic configuration is in reality a new kind of bond, of course not a strong stationary covalent bond, but a mobile magnetic one, and depends upon d- and p-orbital relationships that are more common in transition metal oxides and, in the case of f-orbitals, rare earth oxides. I suggest that similar, weaker, "nonbonding" pairing mechanisms are the common link between the copper oxide ceramic superconductors and the heavy fermion superconductors.[15]

Summarizing, the importance of the copper-oxygen configuration is its ability by chemical modification to create a mixed valence state which is the basis for conduction by the spin-pair hopping mechanism. We have proposed that the copper be required to be in two different oxidation states,

for example Cu^{II}, which is quite usual, and Cu^{III}, which is quite unusual.*[10] The difference in valence states allows the mobile carriers to transfer between copper orbitals via the oxygen intermediary, through its lone pair orbitals so that a mobile dynamic composite boson is the result. Such valence transformation and alternation in a plane and/or between planes encourages the movement of two carriers rather than one and the local structural chemical environment described supports the assumption that it costs less energy for hopping to occur in pairs and for the antiferromagnetic spin alignment to be preserved.

It should be obvious that we are not discussing a metal with a sea of electrons where parentage is of no importance to the carriers, but, rather, a special mixed valence material poised between an antiferromagnetic insulator and an unusual antiferromagnetic metal. As Mott pointed out long ago in his discussion of the Mott metal-insulator transition,[16] there is a narrow region where antiferromagnetism can continue to exist even in the metallic state in conventional antiferromagnetic metals without superexchange. I extrapolate that this also happens in materials which are antiferromagnetic because of the superexchange process. The extra carriers which "tip the balance" are those that define the difference between localization and delocalization and are provided by the elements between the sheets of the copper oxide. They can either contribute electrons or take them away creating holes, making it possible for the material to become either conducting or superconducting.** These holes become mobile by hopping to equivalent but unoccupied sites. There is therefore a critical concentration of carriers required for a transition between the nonconducting and the conducting state assuring the integrity of the mobile composite boson. This transition can be controlled by utilizing the principles of chemical modification.[17-19]

These modifying elements serve the functions previously described: to provide the copper-oxygen mixed valency through their charge interaction; to structurally separate the sheets; to be a reservoir or receiver of carriers and to set up the three-dimensional nature of the material necessary to meet the boson density requirements.

The critical carrier concentration can be understood by our concept of screening which is that if there are too many carriers made available by modification, then they will swamp and destroy the antiferromagnetic configurations which are the source of the composite bosons. Too few carriers

---

* I believe that if copper III were usual, we would be living in a world where high temperature superconductivity would be common.

** In materials in which there is both hole and electron conduction, whether the normal conductivity is p-type or n-type is related to whether the hole or electron is more mobile which depends upon the elements used for modification/doping and the defects and stoichiometry introduced by these elements.

inhibit composite boson formation and therefore limit or destroy superconductivity. What makes these materials different from normal metals is that there must be just sufficient carriers to feed and complete the mobile composite boson configuration. This is a source of the short coherence length; in fact, it should be axiomatic that short coherence length based on local chemical and magnetic order is the hallmark of high temperature superconductivity.

The coherence length must of necessity be short due to the relatively localized nature of the valency transformations, the hopping distances available and, particularly, the antiferromagnetic configurations involved. Pairing energy is related to distance between orbitals involved in superexchange. A long coherence length such as occurs in conventional superconductors implies a much larger volume and lower density of Bose particles and is much more conducive to the phonon interactions necessary for the expression of the BCS theory. It certainly is contrary to antiferromagnetic pairing which does not require phonon action, but does require intermediary atoms to express the mechanism. The superexchange energy which pairs the hopping carriers is strong enough to withstand other localized effects such as coulombic repulsion.

A clear indication of the reasonableness of the electron coupling mechanism proposed here is apparent from the following simple observation. Within the Cu-O plane in all high T_C superconductors, electrons in the copper d-orbitals are antiferromagnetically coupled through a neighboring oxygen atom by the superexchange mechanism. The effective spin-spin distance for this interaction is, therefore, about twice the Cu-O bond length. One of the distinctive features of the copper oxide ceramic superconductors besides their high transition temperature is their very small superconducting correlation lengths. Measurements which allow one to infer this length yield a value which is approximately equal to the effective antiferromagnetic spin-spin interaction length. This strongly supports the antiferromagnetic origin of high T_C superconductivity and the mechanism which we suggest.

The thesis of this paper is that high temperature superconductivity depends upon structural chemistry, that is, stereochemistry. It is useless to try to find links between the normal metallic superconductors and the high temperature superconductors. However, my model can be relevant to heavy fermion superconductors.(15)

The new high temperature materials which, by nature of the mechanism which I have described, have a much more constricted effective volume and therefore higher densities make for the most efficient use of the number of available Bose particles.

The stereochemical geometries not only set up the short range relationships but the volumetric considerations necessary for condensation. It is interesting that since density and coherence length are inversely related, the high T_C superconductors, unlike the conventional ones, rely upon steric considerations to aid them in achieving their highest temperatures. Therefore, one can distinguish between the relatively simple metallic materials to which the BCS model applies, and the multielemental, geometrically complex, high-temperature ceramic materials where structural considerations and short-range antiferromagnetic electronic interactions help in constructing and constraining the composite Bose particles and aid in establishing the necessary densities.

For the most part, we have been describing electronic configurations and structures which permit superconducting fluctuations. What is needed to complete the picture of mobile antiferromagnetically formed composite Bose particles is to establish the conditions under which these particles can fulfill the Bose-Einstein criteria and achieve their lowest free energy by condensing to the superconducting state.

The antiferromagnetic superexchange and, indeed, the directionality of the movement of the majority carriers, are constrained and restricted by the very nature of the sheet configuration and the separation between the sheets. These anisotropic planar configurations are important in that while the superconductive fluctuations originate in them, a degree of three-dimensionality is needed to stabilize them and establish the density requirements for condensation to take place.

It is well known that a perfectly two-dimensional assembly of non-interacting Bose particles will exhibit condensation to the ground state only at zero temperature. As one begins to introduce three-dimensionality, the Bose transition temperature becomes finite. Of course, no physical system is actually perfectly two-dimensional but it can be demonstrated that in a multi-plane system, for example, as the system becomes more three-dimensional by virtue of increasing coupling between the planes, the temperature for Bose condensation will rise. This can also be regarded theoretically as a straightforward consequence of the reduction of effective mass in the direction perpendicular to the planes which occurs with increased interplane coupling. Such coupling is emphasized by chemical linkages.

The spacing between the sheets is critical in terms of wave function overlap. Since the de Broglie wavelength is determined by the momentum of the particle and therefore its thermal energy, the volumetric considerations become paramount. The sheets help determine the containment of the composite Bose particles so at the proper temperature and density, Bose condensation can occur.

What can we learn from the anisotropic sheet configuration that would be helpful for the design of new and optimal materials? As discussed, the anisotropic sheet configuration energetically favors the movement of two carriers along the sheets and minimizes vertical hopping. There should be structural linkages between the sheets which not only assure structural integrity but also are composed of copper-oxygen configurations which participate in valence transformation and pump carriers into the sheets. Therefore, what is also necessary to achieve increased three-dimensionality is to augment the connections between the sheets by bridging configurations such as the Jahn-Teller apical Cu^{II} configuration and/or to make the perpendicular correlation length become comparable with the spacing between the sheets. In this way, all sheets can act in a correlated manner resulting in the proper degree of three-dimensionality for the Bose particles to meet the density needs.

This can be facilitated structurally by interaction among the sheets if the sheets are close enough. As more and more planes are brought into proximity, interaction among the composite Bose particles on adjacent planes has the effect of increasing the superconducting transition temperature. This effect, which results from increasing three-dimensionality of the system, can explain the experimental trend to higher T_C which is observed in the multi-plane thallium high T_C superconductor system (20) and it is a consequence of a very general phenomenon which has analogs in magnetically ordered and other systems which exhibit phase transitions. Because of their short coherence length, we emphasize the need to effectively constrain the composite Bose particles to reach optimal density.

The dynamic, mobile composite Bose particles have either a charge of 2e or 2h or possibly both, since once they are established as composite Bose particles, one can have a mixed system. I feel that the new dynamic pairing mechanism for the composite Bose particles, because it differs fundamentally from the traditional Cooper pairing via phonon interaction in BCS theory, opens up the potential for important new device developments.(6) With this new mechanism, one should be better able to control through chemistry or external fields both the formation and breaking of the composite Bose particles as well as affect the Bose particle density necessary for condensation.

This thought brings us to the crux of our working model--the design of new materials. Having a very formal and mathematical understanding of the BCS theory never once led to the development of new materials. This was a favorite complaint of Bernd Matthias. I am afraid that the new particles that are being invented and the mathematical descriptions of the various models will not be much aid to us either. While these efforts by theoreticians should be encouraged, they are not likely to be useful to develop room temperature materials.

The original complex materials were discovered by accident. What we must do is to create new structures which can express a mechanism which will explain not only the present materials but point the way for new ones. Indeed, since there is new physics here resulting from the composition of the Bose particles, short coherence lengths and mobile antiferromagnetic mechanisms, we should not be trying to just reproduce, for example, Josephson junctions based on the old superconducting physics, but create new switching, control and storage devices.(6)

We call this a working model since we can use it to make new materials. For example, the material parameters that we can affect are the replacement of oxygen, either partially or wholly, with fluorine which plays a vital role through its small size and extreme electronegativity, providing for stronger polarization and antiferromagnetic coupling between the p- and d-orbitals which can make for a stronger pairing interaction and therefore higher temperature

materials.* Fluorine is also able to provide more carriers and permits a larger selection of compensating modifying elements which can lead to higher temperature superconductors. This ability to affect modifying elements is important as the modifiers yttrium, barium, strontium, lanthanum, and lately bismuth and thallium, serve several important purposes. Their charges affect the valency of the copper atoms so that the necessary balance and ratio of Cu^{II} atoms to Cu^{III} atoms are achieved and they establish three-dimensional spatial/structural relationships which permit carrier pairing interaction to take place. In other words, the elements not only structurally fill up space and serve as a valency control, but they also help determine the geometries that control volume and provide the needed reservoir of carriers to the sheets so that the conduction process is assured and the density requirements are met so that superconductivity can take place.

Structurally, fluorine not only improves crystal growth and crystallite alignment necessary to meet the need for high current,** but as we report, is the basis for an entirely new and unique way of growing epitaxial-like crystals nonepitaxially, an extremely important development.(21) Electronically, its partial replacement of oxygen at the vacancy sites adds to hole conduction and its partial replacement of oxygen at other sites can increase electron conduction. It is a powerful new tool in the development of new alloys.(6,21-24)

I consider that fluorine can increase the superexchange pairing energy so that the composite Bose particles can exist at much higher temperatures than the conventional high temperature superconducting materials. At the same time, it can, when charge compensated by another element, open up new possibilities to increase the carriers available for pairing which can positively affect the density parameter. It can be a controlling factor in establishing the optimal ratio of the two valencies. It can be useful in the local structures such as the apical one which transfer carriers and make for the containment of the composite bosons. It can be looked upon as part of a more powerful template that can orient the spins as they move to make up the composite Bose particles.

Crystal structure is important to the extent that it places the interacting atoms and their orbitals in preferred steric relationships, setting up the conditions for the creation of composite Bose particles, enhancing their mobility through planar configurations and geometrically providing for Bose condensation to occur. In this basic sense, it is not relevant whether the material is orthorhombic or tetragonal. What is important is the total interactive environment (25) which means that one requires not so much a crystal as an engineered

---

* Fluorine in the plasma, on the surface and in the bulk is a valuable tool in making high temperature superconductors.

** There is no basic reason to consider that these materials are limited to low currents and low magnetic field interactions.

structure with various types of local order. We are certainly not discussing a conventional crystal or a multielemental disordered material, but one with various types of local order interacting with each other.

Other important manipulatable parameters for materials design are the number of layers, controlling the distance between layers and linkages between the layers. There is a temperature dependence as well that affects the mobility of the composite Bose particles relating to their wave function which is intimately tied to their ability to interact, to their density and therefore to the containment vessel in which they are placed.

If we analyze the materials that are being utilized, we see that there is a problem with all of those investigated so far in that their copper-oxygen sheets are not uniformly spaced. Bismuth materials are particular examples of such irregularity. Conventional lanthanum superconductors have more regularity but not enough sheets in communication to establish the optimal confinement nor do they have the necessary ratio of $Cu^{II}s$ to $Cu^{III}s$. A material may have the necessary prerequisites for antiferromagnetism but either too few or too many carriers.

From a structural viewpoint, there is still room for considering other elements since multielemental complex materials are a necessity. They provide the positional, translational and compositional freedoms which give the materials the flexibility to juxtapose the various structural and electronic configurations necessary for high temperature superconductivity.

I believe that the manipulation of these structures and the creation of new ones have just begun and the optimum material will not be achieved by throwing darts at the Periodic Table. Most solid state physicists are not happy when they leave the security of silicon or gallium arsenide. Six or seven elements with many different local orders leave them without any prior experience or guidelines to direct them.

Since this is the area of our knowledge, we welcome the opportunity to make the materials even more complex by the use of nonequilibrium methods which give us new local and structural order.

We are working to achieve superconducting action in periodically layered and compositionally modulated synthetic materials in which local order plays a critical role.(6) By putting down atomic layers, a layer at a time, one designs the stereochemically preferred nature of the materials. (This is a continuation of my earlier work in developing multilayered superconductors. See, for example, ref. 26.)

New high current room temperature superconductors can be synthesized if they are deliberately made in a nonequilibrium manner. It is my contention that nonequilbrium methods of material preparation add an important weapon to our armamentarium, expanding the field of superconductivity. Metastable fluorinated compounds have shown indications of very high temperature superconducting

phases (21-24) and, as we have pointed out, fluorine plays several other very important roles. It makes for a better material for it can eliminate oxygen diffusion and it is extremely helpful in aligning crystallites and removing interfacial barriers. It is helpful in contributing and controlling carriers which aid the normal conducting and superconducting properties.(21-24) Of great importance is our ability to make new single crystals by the use of new chemistry and physics.(15,21)

I have been emphasizing structural chemistry since structure and function in these materials are literally coupled. Therefore, if structures do not naturally exist that are needed to meet all the requirements that we postulate, they have to be invented and we believe this paper provides the mechanism and guidelines for doing so.

For some time before the discovery of high temperature superconductivity, various materials existed which were found to be high temperature superconductors once they were measured for this effect. Since we cannot depend upon accidents, we must rely upon design principles based upon a relevant model. We must rely not on the miracle of nature but on our creativity. The model that I have outlined has evolved by looking at the complex materials and, so to speak, asking them what they are trying to teach us. They speak in a chorus of atomic voices, each with its own distinctive sound and timbre. I find the music beautiful.

We are at the beginning of the revolution in superconductivity. The long-term consequences of current efforts will be different than many of the prognostications.

There are different types of superconductors, semiconductors and metallic states with different phenomena available to them. If we apply the principles discussed here, we should be able to design superconductors even out of organic materials where anisotropic high conductivity phenomena are known to exist. I believe that mixed valent, "weakly" paired configurations that utilize d-, f- and p-orbitals can be developed which do not necessarily utilize copper and oxygen to establish the mixed valent mobile antiferromagnetic state.

There is much other interesting physics and chemistry in these oxides that is not connected to conventional crystalline materials but very much related to the work which we have done in amorphous and disordered materials and the glassy state generally. For example, nonequilibrium local order, intermediate range order, lone-pair relationships and polarizations, d-and f-orbital interactions, defects, hopping and tunneling, chemical modification (which also shows pairing phenomena) and the various kinds of instabilities associated with cross-linked and phase-change materials.

I repeat, new physics must bring new kinds of devices, not just extensions of the old. The ability that we have developed in the amorphous and disordered field to understand d-orbitals, p-bonding and lone-pair interactions and their excitations and to synthesize entirely new multielemental materials can, I believe, illuminate the way.

We at ECD, especially my collaborator, Rosa Young, with her small dedicated group, are following the concepts set forth here and the results are very encouraging.

My collaborators in our original paper (10) or, for that matter, those whom I gratefully acknowledge here, should not be held responsible for "radical leaps forward" (27) that I have made here. My discussions with Hellmut Fritzsche were, as always, most helpful and very much appreciated. Discussions with Steve Hudgens have aided me in clarifying my position. I acknowledge the critical comments of Morrel Cohen on an early draft. As always, I owe very much to Iris.

Dave, it is an incredible loss to all of us that you are not with us in making the final charge to reach the dream of room temperature superconductivity that we often spoke about through the years. I so miss the benefit of your encyclopedic knowledge, your important contributions, your incisive, insightful comments and enthusiastic collaboration.

Iris and all of your friends at ECD join me in honoring you not only in remembrance but in our continuing work.

REFERENCES

1. D. Adler, "Insulating and Metallic States in Transition Metal Oxides," Solid State Physics 21, 1 (1968).

2. D. Adler, "Mechanisms for Metal-Nonmetal Transitions in Transition-Metal Oxides and Sulfides," Rev. Mod. Phys. 40, 714 (1968).

3. My first transition metal oxide switches were achieved in 1957. For early references see S. R. Ovshinsky and I. M. Ovshinsky, "Analog Models for Information Storage and Transmission in Physiological Systems," Mat. Res. Bull. 5, 681 (1970). (Mott Festschrift).

4. S.R. Ovshinsky, U.S. Patent No. 3,271,719, "Resistance Switches And The Like." Filed June 1961.

5. For references see S.R. Ovshinsky, "Fundamentals of Amorphous Materials," in Physical Properties of Amorphous Materials, ed. D. Adler, B. B. Schwartz and M. S. Steele, Institute for Amorphous Studies Series, Plenum Publishing Corporation, New York, 1985.

6. S.R. Ovshinsky, "A Personal Adventure in Stereochemistry, Local Order, and Defects: Models for Room-Temperature Superconductivity," in Disorder and Order in the Solid State: Concepts and Devices, ed. R.W. Pryor, B.B. Schwartz and S.R. Ovshinsky, Institute for Amorphous Studies Series, Plenum Publishing Corporation, New York, 1988. (Henisch Festschrift.)

7. S.R. Ovshinsky and D. Adler, "Local Structure, Bonding, and Electronic Properties of Covalent Amorphous Semiconductors," Contemp. Phys. 19, 109 (1978).

8. S.R. Ovshinsky and D. Adler, "Present Status of the Science and Technology of Amorphous Solids", published in Nikkei Science (Japanese Scientific American) August 1983.

9. A shorter version of this paper appears in Topics in Non-Crystalline Semiconductors; In Memory of David Adler 1937-87, ed. H. Fritzsche and Ailien Jung, Beijing, China, BUAA, 1988.

10. S.R. Ovshinsky, S.J. Hudgens, R.L. Lintvedt and D.B. Rorabacher, "A Structural Chemical Model for High T_c Ceramic Superconductors," Mod. Phys. Lett. B 1, 275 (1987).

11. Ibid, p. 282.

12. A. Leggett, "Low Temperature Physics, Superconductivity, and Superfluidity," in: The New Physics, ed. P. Davies, Cambridge, Univ. Press, N. Y., 1989, p. 283.

13. S.R. Ovshinsky, unpublished, May 15, 1989.

14. M.H. Cohen, H. Fritzsche and S.R. Ovshinsky, "Simple Band Model for Amorphous Semiconducting Alloys," Phys. Rev. Lett. 22, 6 (1969).

15. S.R. Ovshinsky, to be published.

16. N.F. Mott, Metal-Insulator Transitions, Taylor and Francis, London, 1974.

17. S.R. Ovshinsky, "Chemical Modification of Amorphous Chalcogenides," in Proc. of the 7th Int. Conf. on Amorph. and Liq. Semiconductors, Edinburgh, Scotland, 519-523, 1977.

18. R.A. Flasck, M. Izu, K. Sapru, T. Anderson, S.R. Ovshinsky and H. Fritzsche, "Optical and Electronic Properties of Modified Amorphous Materials," in Proc. 7th Intl. Conf. on Amorph. and Liq. Semiconductors, Edinburgh, Scotland, 524-528, 1977.

19. S.R. Ovshinsky, "The Chemistry of Glassy Materials and Their Relevance to Energy Conversion," in Proc. Int. Conf. on Frontiers of Glass Science, Los Angeles, California; J. Noncryst. Solids 42, 335 (1980).

20. Z.Z. Sheng and A.M. Hermann, "Bulk Superconductivity at 120 K in the Tl-Ca/Ba-Cu-O System," Nature 332, 138 (1988).

21. S.R. Ovshinsky and R.T. Young, "Unusual Fluorinated Effects on Superconducting Films," SPIE, 1324, Modeling of Optical Thin Films II (1990).

22. S.R. Ovshinsky, R.T. Young, D.D. Allred, G. DeMaggio and G.A. Van der Leeden, "Superconductivity at 155 K," Phys. Rev. Lett. 58, 2579 (1987).

23. S.R. Ovshinsky, R.T. Young, B.S. Chao, G. Fournier and D.A. Pawlik, "Superconductivity in Fluorinated Copper Oxide Ceramics," presented at the Int. Conf. on High Temperature Superconductivity, July 29-30, 1987, Drexel University,

Philadelphia, PA; Reviews of Solid State Science 1, 207 (1987).

24. R.T. Young, S.R. Ovshinsky, B.S. Chao, G. Fournier and D.A. Pawlik, "Superconductivity in the Fluorinated YBaCuO," presented at the Materials Research Society Meeting, November 30-December 5, 1987, Boston, Massachusetts.

25. S.R. Ovshinsky, "The Chemical Basis of Amorphicity: Structure and Function," Revue Roumaine de Physique 26, 893 (1981); also in: Disordered Materials: Science and Technology, Selected Papers by S. R. Ovshinsky, ed. D. Adler, Amorphous Institute Press, Bloomfield Hills, MI, 1982. (Grigorovici Festschrift.)

26. A.M. Kadin, R.W. Burkhardt, J.T. Chen, J. E. Keem and S. R. Ovshinsky, "Superconducting Properties of Amorphous Multilayer Metal-Semiconductor Composites," in Layered Structures Epitaxy and Interfaces, ed. J.M. Gibson and L.R. Dawson, (Vol. 37 of Materials Research Society Symposia Proceedings), Pittsburgh, PA, 1985, p. 503.

27. H. Yukawa, "Creativity and Intuition: A Physicist Looks at East and West." Kodansha, Tokyo, dist. by Harper and Row, New York, 1973, p. 184. "...the objective will rarely be attained without some radical leap forward along the way."

CREATIVITY AND INTUITION
A Physicist Looks at East and West
By Hideki Yukawa

Translated by John Bester

206 pp., published by Kodansha International Ltd.,
Tokyo, New York, and San Francisco;
distributed by Harper and Row, New York, 1973

Reviewed by Sanford R. Ovshinsky

Hideki Yukawa, the Japanese theoretical physicist and Nobelist, has written a provocative book of essays including personal memoirs and thoughts on peace. He unselfconsciously discusses his inspiration for doing science with all of the beauty and contradictions inherent in the subject. Straight away let us understand that Yukawa states that he achieved his inspiration and insight from reading the early Chinese philosophers from Confucius to Lao Tzu and Chuangtse. It is "well known" that these philosophers are in fact anti-scientific. He speaks of romance, imagination, the world of fancy, the value of defeat, and the utilization of fables. Do these terms apply to science? That they do is an experimental fact as witness Yukawa's ability to utilize them in conceiving his pioneering work in elementary particles--his prediction of the meson. All new ideas must be proven and Yukawa had to wait 12 years for final confirmation. To understand the value of the process of elucidating a new idea, in Yukawa's case, several other particles were discovered in the investigation of his concept. It is in this rigorous process that many scientists can participate and which greatly adds to existing knowledge.

The precepts implicit in Yukawa's thinking show intuition to be dependent upon the ability to draw analogies. How or whence one draws them depends upon the individual. Yukawa can use the famous Japanese novel, "The Tale of the Genji," and the parables of Taoism of Chuangtse because they inspire his imagination. Great philosophies are vague enough so that the individual brings to them his own interpretations. Chinese philosophers see Taoism as mysticism, but Needham, the famous historian, can read into it exemplary early class-consciousness with proto-anarchistic aims. Such ambiguities can be understood when one considers that Darwin's inspiration was Malthus.

This is not another book about two cultures. It is about two cultures in science, and inferentially in society, about people who have "an excessive conservatism in the realm of ideas" as against those who have "a spirit of adventure."

Intuition has been a little-understood but powerful means of advancing science in quantum steps, yet somehow not considered legitimate. There are a few honorable exceptions to this thinking. Holton, explaining Einstein, emphasizes that major advances in science owe much to intuition. Szent-Gyorgyi's, Watson's, Medawar's, Tribus's and Ermenc's ideas can be at least indicated by Szent-Gyorgyi's statement, "A discovery must be, by definition, at variance with existing knowledge." Yukawa would heartily agree. This is confusing to many scientists since they consider that science _is_ existing knowledge. Imagine the confusion to the rest of the world. Most scientists are uncomfortable with the use of such terms as creativity and intuition, and if they practice them, it is done in Marrano-like secrecy.

Is there then some utopia where creativity can find a home? Nowhere, for its home is in the mind of the individual. It is the "seed beneath the snow," most often beneath the concrete. Newton could express it, isolated on a farm by bubonic plague; Einstein, unknown and alienated in a Swiss patent office, without friends or scientific dialogue, could articulate it. Yukawa writes of his own early isolation. Each was alone, not part of the mass. Each was self-motivated and persisted in the face of apparent contradictions in his work.

At most a few people can be convinced of the soundness of a new idea. Japan, which has also produced Yukawa, has become sensitive to its lack of innovators. The answer is its dependence on authority in all strata of society and its reliance on consensus which, although an effective means of mobilizing mass effort, puts peer pressure on individual thought.

Under the best of conditions, it will always be difficult for new ideas to be introduced, for skepticism is a necessary and first reaction to them. History is replete with great minds, such as Einstein's, resisting the intuition of others, e.g., his refusal to accept quantum theory.

The United States in the past, because of its multiple and conflicting groups in government, industry and the university, allowed some options so that in the search for institutional support for new concepts, there was a possibility for innovation. This was the crack in the concrete, but is rapidly disappearing. American society is becoming more corporate and conformist, resembling the homogeneous and monolithic societies that do not have readily available institutional means for the expression of innovation. While Japan has, through its culture, achieved psychological conformity, China and the USSR are attempting to create such a consensus but must enforce it. Channeling free ideas into science and technology without overlap into the more sensitive areas of art, culture, and politics is virtually impossible.

Mankind normally shares its inhumanity. A humanist such as Yukawa transcends the specialities and national barriers which he deplores. The universal intuitive process, whether expressed in art, literature, music or science, is the ability to "see" connections between facts or concepts which to others are unrelated. Creativity links insight in such a way that a meaningful pattern leaps out of interlocking steps and becomes a bridge or pathway.

Many people who are merely imaginative are not insightful or creative, for the path has to lead somewhere and the bridge must be a means of fording a stream. Intuition, the basis of science, is therefore not an exotic tool but the most utilitarian of arts and its practitioners are the craftsmen of imagination.

Stanford R. Ovshinsky

385